Cytochrome P450

Structure, Mechanism, and Biochemistry

Second Edition

Cytochrome P450

Structure, Mechanism, and Biochemistry

Second Edition

Edited by

Paul R. Ortiz de Montellano

University of California, San Francisco
San Francisco, California

Plenum Press • New York and London

Library of Congress Cataloging-in-Publication Data

```
Cytochrome P450 : structure, mechanism, and biochemistry / edited by
  Paul R. Ortiz de Montellano. -- 2nd ed.
      p.   cm.
  Includes bibliographical references and index.
  ISBN 0-306-45141-7
  1. Cytochrome P-450.   I. Ortiz de Montellano, Paul R.
QP671.C83C98  1995
574.19'25--dc20
                                                     95-25126
                                                        CIP
```

Cover Photograph: Solvent accessible surface of a 10Å-thick slice through the substrate access channel and the heme site of cytochrome P450BM-3. The proposed redox partner docking region is at the bottom of the figure. The C-alpha backbone is indicated by the white lines and the heme by the green lines. The surface of the uncharged residues is shown in white, positively charged residues in blue, and negatively charged residues in red.

ISBN 0-306-45141-7

© 1995, 1986 Plenum Press, New York
A Division of Plenum Publishing Corporation
233 Spring Street, New York, N. Y. 10013

10 9 8 7 6 5 4 3 2 1

Printed in the United States of America

Contributors

JORGE H. CAPDEVILA • Departments of Medicine and Biochemistry, Vanderbilt University Medical Center, Nashville, Tennessee 37232

THOMAS K. H. CHANG • Division of Pharmacology and Toxicology, Faculty of Pharmaceutical Sciences, The University of British Columbia, Vancouver, British Columbia, V6T 1Z3, Canada

MARIA ALMIRA CORREIA • Department of Cellular and Molecular Pharmacology, University of California, San Francisco, San Francisco, California 94143

JILL CUPP-VICKERY • Department of Molecular Biology and Biochemistry, University of California, Irvine, Irvine, California 92717

MICHAEL S. DENISON • Department of Environmental Toxicology, University of California, Davis, Davis, California 95616

JOHN R. FALCK • Department of Molecular Genetics, Southwestern Medical Center, Dallas, Texas 75235

SANDRA E. GRAHAM-LORENCE • Department of Biochemistry, The University of Texas Southwestern Medical Center at Dallas, Dallas, Texas 75235

JOHN T. GROVES • Department of Chemistry, Princeton University, Princeton, New Jersey 08544

F. PETER GUENGERICH • Department of Biochemistry and Center in Molecular Toxicology, Vanderbilt University Medical Center, Nashville, Tennessee 37232

YUAN-ZHANG HAN • Department of Chemistry, Princeton University, Princeton, New Jersey 08544. *Present address*: ARCO Chemical Company, Newtown Square, Pennsylvania 19073

ANNE V. HODGSON • Department of Biochemistry and Molecular Biology, The University of Texas Medical School at Houston, Houston, Texas 77225

ERIC F. JOHNSON • Division of Biochemistry, Department of Molecular and Experimental Medicine, The Scripps Research Institute, La Jolla, California 92037

NORIO KAGAWA • Department of Biochemistry, Vanderbilt University School of Medicine, Nashville, Tennessee 37232

ARMANDO KARARA • Department of Medicine, Vanderbilt University Medical School, Nashville, Tennessee 37232

TODD A. KENNEDY • A.B. Hancock Jr. Memorial Laboratory for Cancer Research, Center in Molecular Toxicology, Departments of Biochemistry and Chemistry, Vanderbilt University School of Medicine, Nashville, Tennessee 37232

HUIYING LI • Department of Physiology and Biophysics, University of California, Irvine, Irvine, California 92717

PAUL J. LOIDA • Department of Biochemistry and Beckman Institute for Advanced Science and Technology, University of Illinois, Urbana, Illinois 61801

KEIKO MAKITA • Department of Medicine, Vanderbilt University Medical School, Nashville, Tennessee 37232

DANIEL MANSUY • Laboratoire de Chimie et Biochimie Pharmacologiques et Toxicologiques, URA 400 CNRS, Université Paris V, F-75270 Paris Cedex 06, France

LAWRENCE J. MARNETT • A.B. Hancock Jr. Memorial Laboratory for Cancer Research, Center in Molecular Toxicology, Departments of Biochemistry and Chemistry, Vanderbilt University School of Medicine, Nashville, Tennessee 37232

ERNEST J. MUELLER • Department of Biochemistry and Beckman Institute for Advanced Science and Technology, University of Illinois, Urbana, Illinois 61801

DAVID R. NELSON • Department of Biochemistry, The University of Tennessee, Memphis, Memphis, Tennessee 38163

PAUL R. ORTIZ de MONTELLANO • Department of Pharmaceutical Chemistry, School of Pharmacy, University of California, San Francisco, San Francisco, California 94143

JULIAN A. PETERSON • Department of Biochemistry, The University of Texas Southwestern Medical Center at Dallas, Dallas, Texas 75235

THOMAS L. POULOS • Department of Physiology and Biophysics, University of California, Irvine, Irvine, California 92717

JEAN-PAUL RENAUD • Laboratoire de Chimie et Biochimie Pharmacologiques et Toxicologiques, URA 400 CNRS, Université Paris V, F-75270 Paris Cedex 06, France. *Present address*: UPR 9004 CNRS, Institut de Génétique et Biologie Moléculaire et Cellulaire, F-67400 Illkirch-Graffenstaden, France

SIJIU SHEN • Department of Biochemistry and Molecular Biology, The University of Texas Medical School at Houston, Houston, Texas 77225

STEPHEN G. SLIGAR • Department of Biochemistry and Beckman Institute for Advanced Science and Technology, University of Illinois, Urbana, Illinois 61801

HENRY W. STROBEL • Department of Biochemistry and Molecular Biology, The University of Texas Medical School at Houston, Houston, Texas 77225

CLAES VON WACHENFELDT • Division of Biochemistry, Department of Molecular and Experimental Medicine, The Scripps Research Institute, La Jolla, California 92037

MICHAEL R. WATERMAN • Department of Biochemistry, Vanderbilt University School of Medicine, Nashville, Tennessee 37232

DAVID J. WAXMAN • Division of Cell and Molecular Biology, Department of Biology, Boston University, Boston, Massachusetts 02215

JAMES P. WHITLOCK, JR. • Department of Molecular Pharmacology, Stanford University School of Medicine, Stanford, California 94305

DARRYL ZELDIN • Department of Medicine, Vanderbilt University Medical School, Nashville, Tennessee 37232

Preface

In the ten years that have elapsed since the first edition of this book went to press, the cytochrome P450 field has completed the transition to a discipline in which structure and mechanism, even regulation and biological function, are dealt with in molecular terms. The twin forces that have propelled this remarkable progress have been the widespread adoption of molecular biological approaches and the successful application of modern structural techniques. Only a few P450 primary sequences were available in 1985, whereas hundreds of P450 sequences are now available. Site-specific mutagenesis was then mostly a proverbial gleam in the eye of the P450 community, but it is now a standard technique in the research repertoire. The first crystal structure of a cytochrome P450 enzyme had just been solved in 1985 and appeared on the cover of the first edition. Today, the high-resolution crystal structures of four soluble bacterial P450 enzymes are available and the race is on to develop approaches that will permit us to determine the structures of the membrane-bound forms of the enzyme. The past ten years has seen phenomenal progress—let us hope that the next ten will prove equally exciting.

The book is informally divided into four sections. In order to hold the book close to the advancing front of research, some of the chapter topics from the first edition have been dropped to make room for new or expanded topics. The first section reviews recent work with model systems (Chapter 1) and peroxidases (Chapter 2) that, even today, provides much of the foundation for our understanding of P450 structure and mechanism. The second section deals with the structurally defined bacterial enzymes. It includes a chapter on the role of protein residues in oxygen activation and catalysis (Chapter 3) and two chapters that analyze different aspects of the four available P450 crystal structures (Chapters 4 and 5). The third section summarizes our knowledge of the structures and mechanisms of membrane-bound P450 enzymes. This section includes chapters on the structures of the proteins and their active sites (Chapter 6), the structures of the electron transfer proteins and their interactions with cytochrome P450 (Chapter 7), and the mechanisms of substrate oxidation (Chapter 8) and enzyme inhibition (Chapter 9). In the final section of the book, the regulatory mechanisms and physiological roles of cytochrome P450 enzymes are discussed. The first three chapters of this section cover receptor-mediated induction (Chapter 10), hormonal regulation (Chapter 11), and the specific regulation of P450 enzymes in steroidogenic tissues (Chapter 12). This is followed by a chapter on the P450-catalyzed oxidation of arachidonic acid to physiologically active substances (Chapter 13), a discussion of human P450 enzymes and their critical role in drug metabolism (Chapter 14), and a chapter on hemoproteins with a thiolate–iron ligand that differ in fundamental ways from conventional P450 enzymes. The book closes with two appendices that offer practical information on cytochrome P450 nomenclature, sequences, and specificities. It is my hope that the information and insights gathered in this book will serve the P450 community as well as those in the first edition.

Cytochrome P450 has occupied a large part of my waking hours, fostered my career, put bread on my table, and introduced me to an exciting company of colleagues. None of this would have been possible without the hard work and creativity of the many graduate students and postdoctoral fellows who have labored in my laboratory over the past two decades. To all of them, from those who soared in response to the challenge to those who struggled but eventually managed to fly, I dedicate this book.

Paul R. Ortiz de Montellano

San Francisco

Contents

Part I

Model Systems

Models and Mechanisms of Cytochrome P450 Action

JOHN T. GROVES and YUAN-ZHANG HAN

1. Introduction

The reactions catalyzed by the cytochrome P450 family of enzymes have challenged and intrigued chemists for more than two decades. Alkane hydroxylation and olefin epoxidation, particularly, have attracted a sustained worldwide effort, the allure deriving from both a desire to understand the details of biological oxygen activation and transfer and, as well, the sense that catalysts for the practical application of these principles to organic synthesis and to large-scale process chemistry could be of considerable economic value. Cytochrome P450 is able to incorporate one of the two oxygen atoms of an O_2 molecule into a broad variety of substrates with concomitant reduction of the other oxygen atom by two electrons to H_2O.[1] Cytochrome P450 enzymes have been isolated from numerous mammalian tissues (e.g., liver, kidney, lung, intestine, adrenal cortex), insects, plants, yeasts, and bacteria.[2] Cytochrome P450 is known to catalyze hydroxylations, epoxidations, N-, S-, and O-dealkylations, N-oxidations, sulfoxidations, dehalogenations, and other reactions.[1] The reactive site of all of these enzymes is extraordinarily simple, containing only an iron protoporphyrin IX (1) (Fig. 1) with cysteinate as the fifth ligand, leaving the sixth coordination site to bind and activate molecular oxygen. The local environment of oxygen binding and activation is also very simple, with mostly hydrophobic protein residues and a single threonine hydroxyl which is essential for catalysis for some but not all P450s. The principal catalytic cycle of cytochrome P450 has been much discussed and often reviewed, but the essential features have been agreed upon now for some time. The essential steps involve (Scheme I): (1) binding of the substrate, (2) reduction of the ferric, resting

---

JOHN T. GROVES and YUAN-ZHANG HAN • Department of Chemistry, Princeton University, Princeton, New Jersey 08544. *Present address of Y.-Z. H.*: ARCO Chemical Company, Newtown Square, Pennsylvania 19073.

Cytochrome P450: Structure, Mechanism, and Biochemistry (Second Edition), edited by Paul R. Ortiz de Montellano. Plenum Press, New York, 1995.

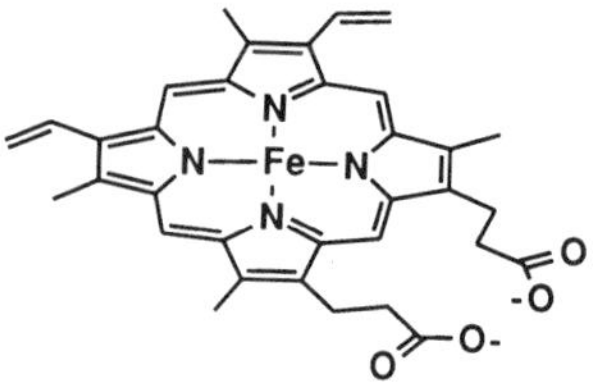

FIGURE 1. Iron protoporphyrin IX.

cytochrome P450 to the ferrous state, (3) binding of molecular oxygen to give a ferrous cytochrome P450–dioxygen complex, (4) transfer of the second electron to this complex to give a peroxoiron(III) complex, (5) protonation and cleavage of the O–O bond with the concurrent incorporation of the distal oxygen atom into a molecule of water and the formation of a reactive iron–oxo species, (6) oxygen atom transfer from this oxo complex to the bound substrate, and (7) dissociation of the product. While the first three steps of the enzymatic process have been monitored spectroscopically, the transfer of the second electron, the O–O bond cleavage, and the oxidation of the substrate occur too rapidly and have yet to be observed.[3] The identity of the active oxidant is believed to be an oxoiron(IV) porphyrin cation radical **2** (Fig. 2), based on both model studies and biological analogy with horseradish peroxidase Compound I, although no physical characterization of this species is presently available for cytochrome P450 itself. In a few cases (*vide infra*) the *peroxo*iron(III) complex formed in step 4 is considered to react with substrates (**S**) to form

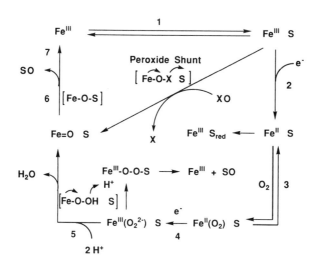

SCHEME I. Catalytic cycle of cytochrome P450.

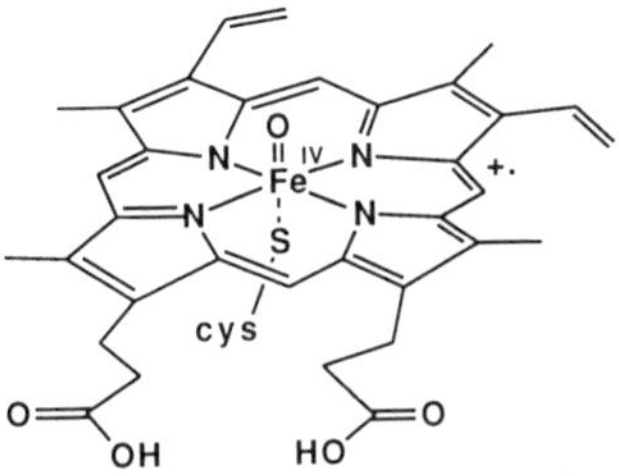

FIGURE 2. Proposed reactive species of cytochrome P450.

intermediates of the type Fe-O-O-S. Likewise, some substrates such as alkyl halides are *reduced* by the iron(II) form of P450.

The purpose of this review is to survey the *oxidative* reactions mediated by cytochrome P450 and those of model metalloporphyrin systems. Particular attention is to be afforded the mechanisms of O–O bond cleavage (step 5), the nature of reactive oxometalloporphyrin intermediates, and the interactions of these reactive intermediates with typical substrates (step 6). The mechanisms of alkane hydroxylation and olefin epoxidation are treated in detail. Together, the enzymatic and model studies have afforded important insights into the nature of biological oxygen activation and transfer.

Terminal iron–oxo complexes such as that depicted in Fig. 2 are rare. In fact, the only bona fide examples in all of chemistry, aside from the trivial case of the ferrates, are those of heme proteins and their metalloporphyrin models. Of these the celebrated Compound I of horseradish peroxidase is certainly the most famous and of the most significance. Compound I was the first intermediate observed by the spectrographic technique, and later by stop-flow spectrophotometry, on addition of hydrogen peroxide to HRP.[4e] While the nature of Compound I was not known for many years, a relationship between oxygen activation by cytochrome P450 and the peroxidases was suggested by the demonstration that other oxygen donors such as alkyl hydroperoxides, hydrogen peroxide, peroxyacids, $NaIO_4$, $NaClO_2$, and iodosylbenzene could substitute for NADPH and molecular oxygen.[5] We have called this the "peroxide shunt"[5f] mechanism as depicted in Scheme I.

It is instructive to look carefully at the properties of HRP Compound I and those of the few related model compounds for indications of what the reactive P450 intermediate might be like. Electronic spectra[6] and magnetic susceptibility studies[7] of Compound I of HRP indicate an $S = 3/2$ spin state which is consistent with an oxoiron(IV) center ferromagnetically coupled to a porphyrin cation radical. EXAFS spectra have revealed the presence of a short, terminal iron–oxygen bond (1.6 Å).[16a] By contrast, however, *mangano*-HRP has been oxidized to a Compound I analogue which, while it retains the analogous manganyl ($Mn^{IV}=O$) moiety, displays the EPR signature of a protein radical rather than an oxidized porphyrin ring.[4d] The electronic distribution in the five highest filled (or half filled) and the lowest empty molecular orbitals have been calculated for HRP Compound I and for a cytochrome P450 active oxidant $[FeO]^{3+}$ with an axial sulfur ligand and are shown in Fig. 3A and 3B.[8] For the P450 model in Fig. 3B, one unpaired electron is in an A_{2u}-type orbital centered at pyrrole nitrogens, oxygen and sulfur ligands, and *meso*

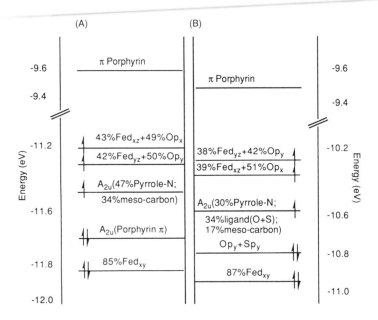

FIGURE 3. Calculated eigenvalues and electron distribution in highest occupied and lowest unoccupied orbitals of (A) HRP Compound I and (B) a cytochrome P450 $[FeO]^{3+}$ analogue with axial sulfur.

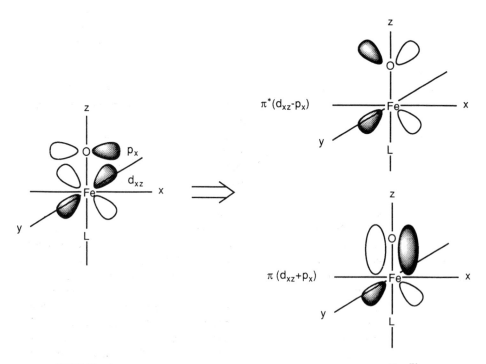

FIGURE 4. The orbital diagram of the active oxidant of cytochrome P450, $(Por^{+\bullet})Fe^{IV}(O)$.

carbons. The other two unpaired electrons are in higher-energy, nearly degenerate and highly delocalized (Fe-d_π + O-p_π) orbitals. Loew *et al.* have proposed that the electrophilicity of such a ferryl group is "covalently controlled" since the calculations indicated that oxygen carries a net negative charge (–0.31 *e*). Low-lying orbitals which are partially occupied and yet contain considerable electron density on the oxygen atom can serve as primary electron acceptors in covalent interactions leading to the transfer of oxygen to a nucleophilic substrate.[8,9] The orbital diagram[10,13d] of the active oxidant of cytochrome P450, $(Por^{+\bullet})Fe^{IV}(O)$, is shown in Fig. 4.

2. Synthetic Metalloporphyrins and Oxometalloporphyrins as Models for Cytochrome P450

Synthetic metalloporphyrins, especially *meso*-tetraaryl porphyrins (Fig. 5), have been extensively studied as models of the active site of cytochrome P450.[2] Since the first report of the hydroxylation and epoxidation reactions catalyzed by FeTPPCl with iodosylbenzene as the oxygen donor from our laboratories in 1979,[11a] there have been numerous studies of metalloporphyrin-catalyzed oxidation reactions. Three different metalloporphyrin-catalyzed oxidation systems have been developed. The first system contains metalloporphyrins such as iron, manganese, and chromium porphyrins together with oxygen donors such as iodosylbenzene, NaClO, amine *N*-oxides, and periodate, similar to the cytochrome P450 "peroxide shunt" process.[11–13] The second genre contains metalloporphyrin, molecular oxygen, and coreductants such as $NaBH_4$, ascorbic acid, Zn(Hg), colloidal platinum-H_2, and others, which mimic the natural P450 oxidation process because it involves the activation of molecular oxygen through the use of reducing agents.[14] The third type of system uses a *trans*-dioxoruthenium(VI) porphyrin complex as the catalyst together with molecular oxygen without the need of a reducing agent to catalytically oxidize olefins into epoxides.[15f–h]

The reactive, high-valent oxometalloporphyrin complexes in these model systems have been characterized spectroscopically. A green oxoiron(IV) porphyrin cation radical species $Fe^{IV}(O)TMP^{+\bullet}(X)$ (**3a**)[16a–d] and the 2,6-dichlorophenyl derivative[16f] have been characterized by visible spectroscopy, NMR, EPR, Mössbauer, and EXAFS data. The iron–oxygen distance in **3a** was determined from EXAFS data to be 1.64 Å, characteristic of a bond order greater than one, and consistent with the electronic structure described above. Significantly, complex **3a** was shown to be highly reactive toward oxygen atom transfer, reacting as a powerful electrophile with olefins to afford epoxides and with alkanes to give alcohols. By contrast, the HRP Compound II analogue, **3b**,[12f–h] has been shown recently to have only moderate reactivity toward olefins and it reacts with the selectivity of an oxygen radical.[11e] Other iron porphyrin complexes to be generated for the first time

3a   **3b**

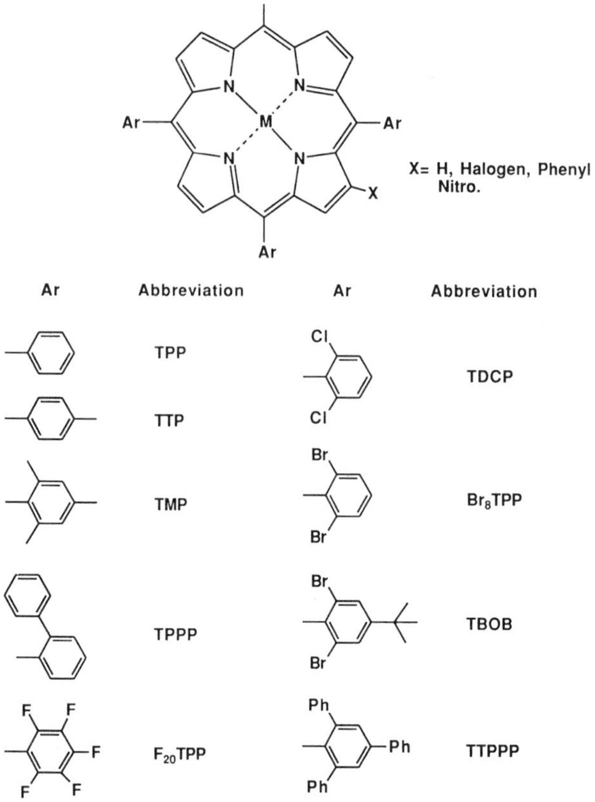

X= H, Halogen, Phenyl
Nitro.

Ar	Abbreviation	Ar	Abbreviation
	TPP		TDCP
	TTP		Br_8TPP
	TMP		TBOB
	TPPP		TTPPP
	$F_{20}TPP$		

FIGURE 5. Abbreviations of *meso*-tetraarylporphyrin with X=H.

recently are an oxoiron(IV) chlorin cation radical[16h] analogous to **3a**, a peroxoiron(III) chlorin,[16i] and an unusual *N*-alkyl-oxoiron(IV) porphyrin.[12i] High-valent oxomanganese porphyrin species,[13] oxochromium(V) porphyrin complexes,[15a–e] and *trans*-dioxoruthenium(VI) porphyrin complexes[15f–h] have also been isolated, characterized, and shown to be the reactive species of oxygen transfer reactions present in these metalloporphyrin-catalyzed systems.[16e]

3. The O–O Bond Cleavage Step

The crucial step in the formation of reactive iron–oxo intermediates is the O–O bond cleavage of an iron–peroxo precursor (step 5 in Scheme I). This process has received extensive attention in a large number of laboratories and, while there are still controversial aspects, the fundamental processes are now reasonably well understood. The discussion here will be short since we have reviewed this reaction in great detail elsewhere.[3a]

The reduction of a model iron(II)dioxygen adduct with ascorbate to afford a putative *hydroperoxy* iron(III) complex has recently been reported.[17a] This species is of particular interest both because it was formed by reduction in analogy to step 4 in Scheme I of the consensus cytochrome P450 mechanism and because it differs from several η^2-peroxometalloporphyrin complexes.[16i,17f,g] Evidence for the hydroperoxo complex was based on a unique EPR spectrum characteristic of a low-spin, six-coordinate iron(III) species (g = 2.286, 2.171, 1.953) and the formation of a similar complex on adding alkaline hydrogen peroxide to chloroiron(III)OEP.

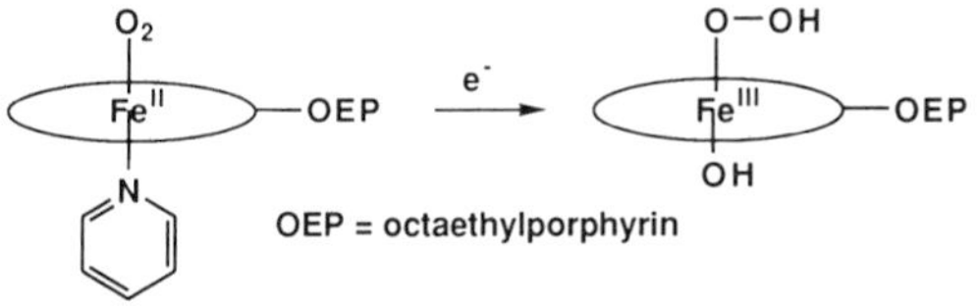

OEP = octaethylporphyrin

A stable acylperoxoiron(III) porphyrin has been prepared in our laboratories and the kinetics of its transformation to an oxoiron(IV) porphyrin cation radical have been examined in detail.[11d]

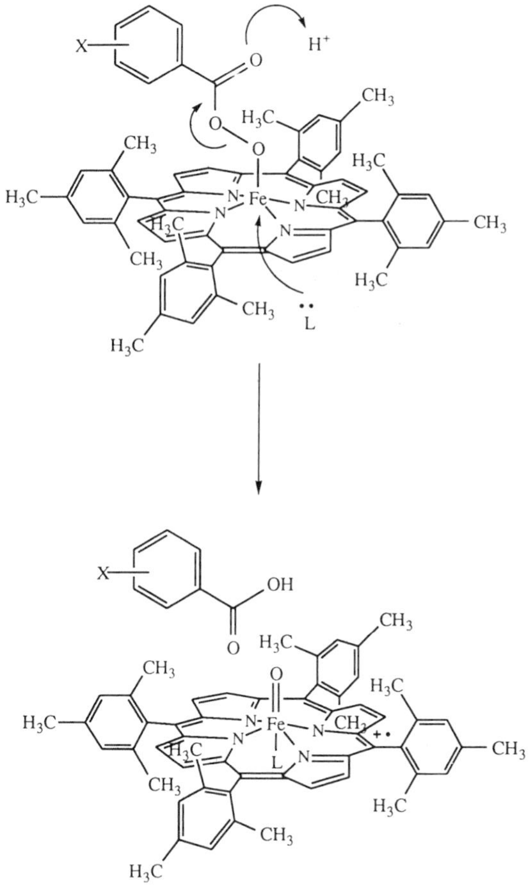

O-O bond scission

By direct observation of this O–O bond scission event a number of insights could be obtained about the process uncomplicated by preequilibria and catalytic turnover. First, the temperature dependence of the rate revealed an extraordinarily low activation enthalpy ($\Delta H^{\ddagger} = 3.6$ kcal/mol) and a large, negative entropy term ($\Delta S^{\ddagger} > -25$ eu). Electron-withdrawing substituents on the aryl portion of the peroxyacyl ligand facilitated O–O bond cleavage ($\rho = +0.5$) consistent with a heterolytic bond cleavage relating to the nature of the leaving group. The unfavorable entropy for O–O bond scission supports the view that association of the axial ligand **L** and positioning of the proton on the leaving group are important in controlling the reaction rate. Acid catalysis of the O–O bond heterolysis was also evident. Taken together, the picture that emerges for O–O bond heterolysis is as depicted in the idealized orbital interaction diagram below. Sixth-ligand coordination would be expected to favor a low-spin electronic configuration for iron(III). The maximum assistance for heterolysis of the O–O bond would result from a doubly occupied $d_{xz(yz)}$ orbital interacting favorably with the σ^{*} of the peroxide as shown. Further, antibonding lone-pair interactions of the sixth ligand **L** would be expected to increase the d_{xy}–σ^{*} interaction. This role for the thiolate ligand in cytochrome P450 was suggested some time ago by Dawson *et al.*[17h] and this "push–pull" effect has been examined more recently by Yamaguchi *et al.*[17b]

Significantly, the reaction of a coordinated peroxyacid was observed to take an entirely different course in dry toluene and in the absence of protons. Substituent effects in the peroxyacid were opposite those observed for the heterolysis above and the product obtained was an unusual iron(III) porphyrin *N*-oxide.[11d] Thus, under these conditions a homolytic O–O bond scission was proposed.

The role of a coordinated thiolate has been probed directly by Hirobe and colleagues.[17c,d] In this work, phenylperoxyacetic acid was used as a probe of the nature of O–O bond cleavage. For this peroxyacid O–O bond homolysis will lead to an unstable carboalkoxy radical which, on decarboxylation, will produce products derived from benzyl radicals. By contrast, O–O bond heterolysis will produce phenylacetic acid. Consistent with the discussion above, added acid favored heterolytic O–O bond scission over homolysis for simple porphyrins such as iron(III)TPP. However, for a model porphyrin with a coordinated thiolate tail, exclusive O–O bond heterolysis was observed even in nonpolar, aprotic environments. Thus, it is clear that thiolate ligation in cytochrome P450 will facilitate the formation of an oxoiron(IV) porphyrin cation radical even in the hydrophobic active site environment of these enzymes.

Confirmation of this conclusion for a heme protein environment comes from axial ligand mutagenesis studies with human myoglobin.[17e] Thus, replacement of the axial histidine-93 with cysteine was shown to enhance the heterolytic O–O bond cleavage in added alkyl hydroperoxides while the phenolate ligand had a negligible effect compared to the native histidine imidazole. Likewise, for a model study with water-soluble iron porphyrins, the axial nitrogen base was shown to be influential in the catalysis of O–O bond scission in alkyl hydroperoxides.[12e]

An interesting, and elaborate model system has been described recently by Patzelt and Woggon[17i] in which an iron(III) porphyrin thiophenolate complex binds and activates dioxygen. A methylene strap on the oxygen binding face of this molecule eventually

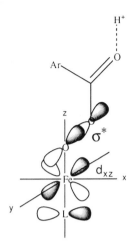

became hydroxylated. The result seems to confirm that a properly designed synthetic metalloporphyrin catalyst can mimic the entire reaction sequence outlined in Scheme I.

3.1. Mechanism of Hydroxylation by Cytochrome P450

Cytochrome P450-catalyzed hydroxylation of hydrocarbons is considered to occur by a mechanism involving hydrogen atom abstraction from the substrate (R–H) followed by rapid transfer of the metal-bound hydroxy radical to an intermediate alkyl radical (R·) (Scheme II).[3] This so-called oxygen rebound mechanism[18a,b] is consistent with the stereochemical, regiochemical, and allylic scrambling results observed in the oxidation of norbornane, camphor, and cyclohexene by cytochrome P450. The hydroxylation of a saturated methylene (CH_2) in norbornane was accompanied by a significant amount of epimerization at the carbon center. Thus, the hydroxylation of exo-exo-exo-exo-tetradeuterionorbornane by $P450_{LM}$ and the hydroxylation of camphor by $P450_{cam}$ gave exo-alcohol with retention of the exo-deuterium label (Scheme III). The hydroxylation of selectively deuterated cyclohexene proceeded with substantial allylic scrambling (Scheme III).[18b] The intrinsic isotope effects for the oxygen insertion into a C–H bond are very large: (1) k_H/k_D = 11.5 ± 1.0 for the hydroxylation of tetra-exo-deuterated norbornane,[18a] (2) k_H/k_D = 11 for the benzylic hydroxylation of [1,1-D]-1,3-diphenylpropane,[18c] (3) k_H/k_D = 10 for the O-demethylation of p-trideuteriomethoxyanisole,[18d] and (4) k_H/k_D = 13.5 for the demethylation of 7-methoxycoumarin.[18e] These large isotope effects are inconsistent with an insertion process and indicate that the C–H bond is essentially half-broken in a linear [O· · ·H· · ·C] transition state, thus providing strong evidence for a nonconcerted mechanism. Importantly, model iron porphyrin systems displayed the same behavior for both the

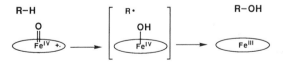

SCHEME II. Rebound mechanism for the hydroxylation by cytochrome P450.

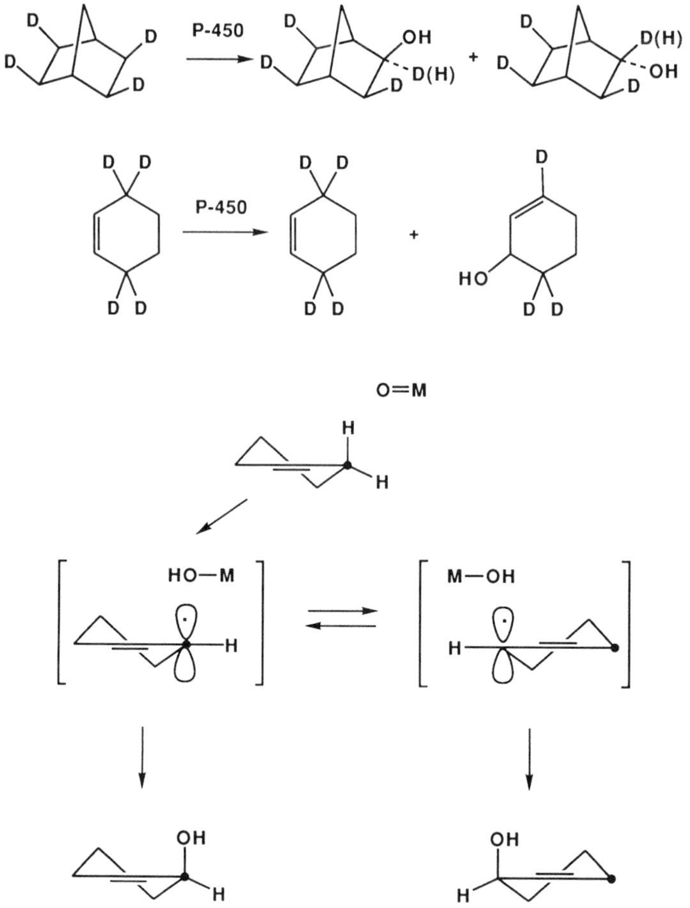

SCHEME III. Epimerization and allylic scrambling observed for cytochrome P450-catalyzed hydroxylation.

norbornane[18f] and cyclohexene[18b] substrates. Thus, it can be concluded that the epimerization and allylic scrambling processes are intrinsic properties of the oxygen transfer event from an oxoiron complex.

Another revealing probe of the nature of P450-mediated hydroxylation is a study of the rearrangement of alkyl hydroperoxides.[18g] 1-Hydroperoxyhexane has been shown to afford 1,2-dihydroxyhexane on exposure to P450$_{LM2}$. This is an unusual reaction since the oxidizing equivalents of the hydroperoxide have been used in this case to hydroxylate the *neighboring* methylene group. A mixed, double oxygen label experiment established that the rearrangement was intramolecular. Thus, the terminal hydroperoxide oxygen was incorporated into the adjacent C–H bond. Analysis of the diols derived from chirally labeled 2-deuterio-1-hydroperoxyhexane showed that there was a loss of stereochemistry at the hydroxylated carbon center. Accordingly, the results support a mechanism involving initial peroxide heterolysis, hydrogen abstraction at the adjacent methylene, and radical recombination to afford the product diol. The result is revealing since O–O bond homolysis to form a hexyloxy radical should lead to γ-hydrogen abstraction and products derived therefrom. Recently, the oxidation of benzylic alcohols by cytochrome P450 has been described by Vaz and Coon.[18h] The results were consistent with a stepwise process here as well, involving an initial hydrogen atom abstraction to form an α-hydroxy radical and subsequent oxygen rebound to form a *gem*-diol which dehydrates to produce the observed carbonyl product.

3.2. The Rate of the Recombination Step

In 1987, Ortiz de Montellano and Stearns applied the "radical-clock" method[19a] to alkane hydroxylation using phenobarbital-pretreated rat liver microsomal P450 to measure the rate of the rebound process.[19b] The oxidations of nortricyclane and methylcyclopropane proceed to nortricyclanol and cyclopropylmethanol without giving detectable rearranged products, indicating that the radical recombination step is much faster than the ring-opening rate of the cyclopropylmethyl radical to 3-butenyl radical (which has a rate constant of 1.2×10^8 sec^{-1} at 37 °C). However, bicyclo[2.1.0]pentane was oxidized by rat liver microsomes to a 7:1 mixture of *endo*-2-hydroxybicyclo[2.1.0]pentane and 3-cyclopenten-1-ol (Scheme IV). The rate constant for the rearrangement of bicyclo[2.1.0]pent-2-yl radical to 3-cyclopenten-1-yl radical was recently determined to be 2.4×10^9 sec^{-1} at room temperature by using laser flash photolysis techniques.[19c] Thus, a rate constant of $k_{OH} = 1.7 \times 10^{10}$ M^{-1} sec^{-1} was estimated for the rebound process, more rapid than many typical conformational and configurational rearrangements and also, perhaps, more rapid than the gross molecular motion of many enzyme-bound substrates.[19c] The numerous reported cases of retention of configuration and stereochemistry in cytochrome P450-catalyzed hydroxylations, and in

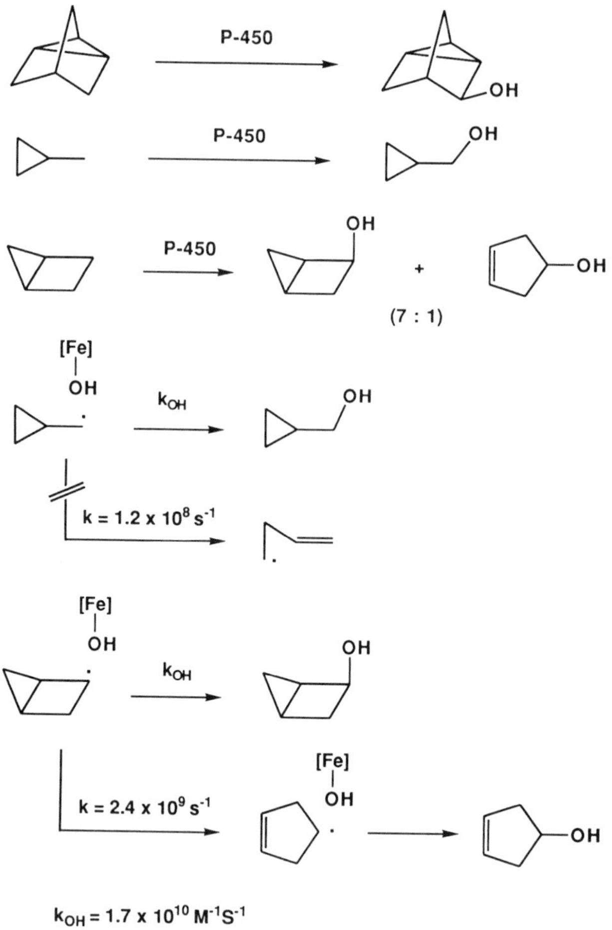

SCHEME IV. Determination of the rebound rate in the hydroxylation by cytochrome P450.

particular, the remarkable regioselectivity observed in the hydroxylation of bicyclo[2.1.0]pentane (which undergoes abstraction and rebound only at the *endo* face), therefore can be explained in these terms.[19c]

3.3. Hydrogen Atom Transfer versus Electron Transfer in Hydroxylation Reactions

Hydrogen atom transfer is a satisfying and consistent initial step for the hydroxylation of unactivated alkanes by cytochrome P450 and metalloporphyrin model catalysts.[1,3] On the other hand, oxidation of a hydrocarbon by initial removal of an electron rather than a hydrogen atom has recently been demonstrated in the special case of a highly oxidizable

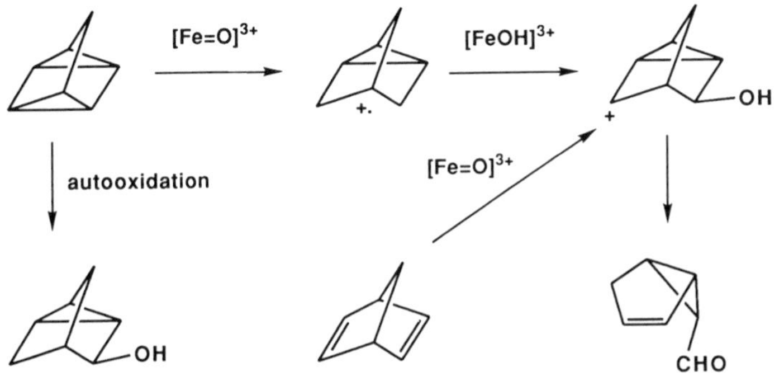

SCHEME V. Electron transfer in the cytochrome P450-catalyzed oxidation of quadricyclane.

hydrocarbon.[20] The NADPH-dependent microsomal oxidation of quadricyclane, a strained hydrocarbon with a very low oxidation potential ($E_{1/2}$ = 0.92 V),[21a] yields a rearranged aldehyde which is also obtained as the principal product in the enzymatic oxidation of norbornadiene. Autoxidation of quadricyclane in the presence of unreduced microsomes or trace metals gives nortricyclanol (Scheme V). Control experiments have shown that norbornadiene is not an intermediate in the quadricyclane oxidation, but rather that a common intermediate is probably generated from both substrates.[20] Oxidation of quadricyclane to the radical cation, followed by capture of the carbon radical by the enzymatically activated oxygen, rationalizes these results because the resulting cationic intermediate is known to rearrange to the observed aldehyde.[21b]

The scope of hydrogen atom transfer and electron transfer mechanisms has been studied for the benzylic oxidation of alkylaromatic compounds and benzyltrimethylsilanes induced by iron(III) porphyrin complexes.[22] Product compositions for side-chain oxidations of 4-substituted 1,2-dimethylbenzenes compared with those of the reactions of the same substrates with cerium(IV) ammonium nitrate (CAN), a bona fide one-electron transfer oxidant, and with Br·(NBS), a species that certainly reacts by a HAT mechanism, showed that the intramolecular selectivity of the FeTFPPCl-catalyzed reactions is close to that observed in the free radical bromination, but significantly different from that found in the ET reactions with CAN (Scheme VI). Thus, most probably these reactions take place by a HAT mechanism. On the other hand, although the oxidation of the more electron-rich silane **4** in methylene chloride gave only 4-methoxybenzaldehyde (HAT mechanism), in CH_2Cl_2–MeOH–H_2O substantial formation of benzylic derivatives (4-methoxybenzyl alcohol and 4-methoxybenzyl methyl ether) was observed, which can be rationalized by an ET mechanism to give the 4-methoxybenzylsilane radical cation which forms the 4-methoxybenzyl radical by a fast desilylation reaction (Scheme VII). It is evident that an ET mechanism can be turned on in the oxidation of electron-rich hydrocarbons in polar solvents.

Lindsay Smith and colleagues have studied substituent effects on the α-hydroxylation of *para*-substituted toluenes and cumenes by FeTPPCl and PhIO.[23a] The results suggest that the active oxidant is reacting as an electrophilic radical and in the transition state both

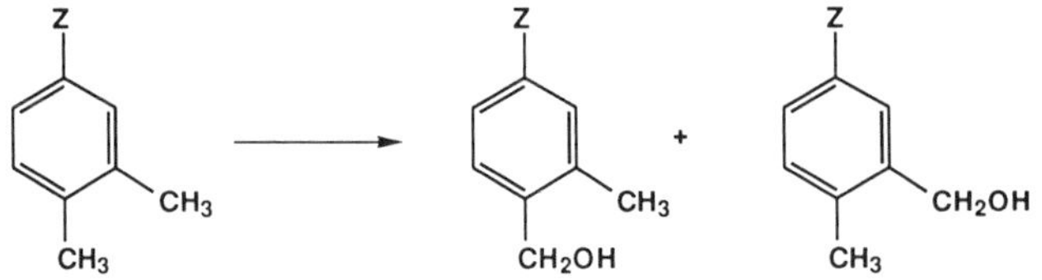

	Z = -C(CH₃)₃,	-Cl
FeTFPPCl/PhIO/CH₂Cl₂	1.9 : 1	2.2 : 1
FeTFPPCl/PhIO/CH₂Cl₂-CH₃OH-H₂O	2.0 : 1	2.2 : 1
NBS/CCl₄	1.6 : 1	3.0 : 1
CAN/AcOH	6.0 : 1	13.0 : 1

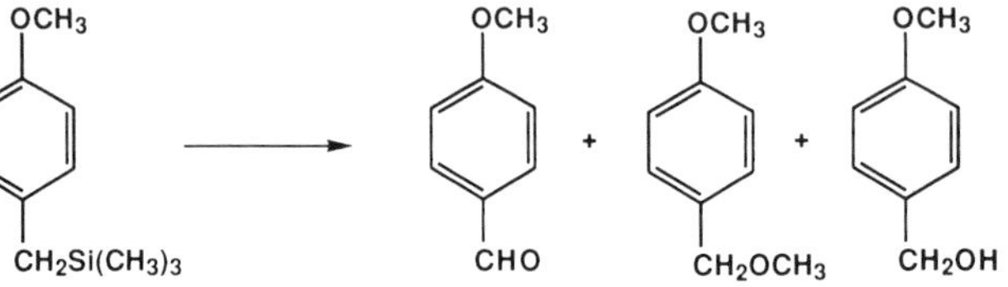

4

FeTFPPCl/PhIO/CH₂Cl₂	1 : 0 : 0
FeTFPPCl/PhIO/CH₂Cl₂-CH₃OH-H₂O	4 : 1 : 3

SCHEME VI. The benzylic oxidation of alkylaromatic compounds and benzyltrimethylsilanes induced by Fe(III)TFPPCl.

positive charge and radical character develop on the benzylic carbon (Fig. 6), typical of an electrophilic radical.

To quantitatively assess the significance of charge transfer in the transition states of hydroxylation reactions, Khanna et al.[23b] have studied the regioselectivity of the hydroxylation of the chemical probe 5-nitroacenaphthene (**5**). The photochlorination of **5** with t-BuOCl yielded 1-chloro-5-nitroacenaphthene and 2-chloro-5-nitroacenaphthene in a

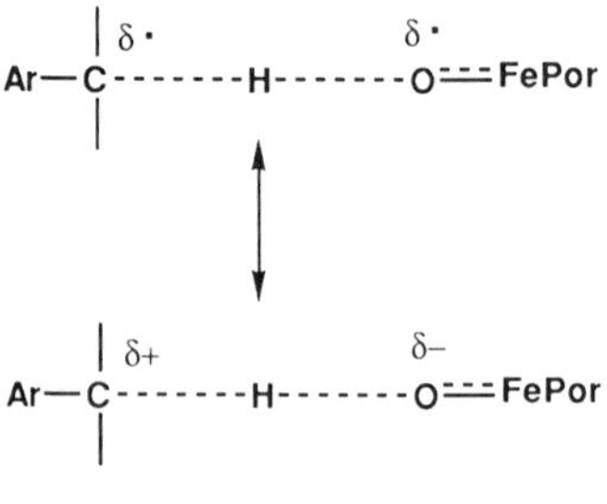

FIGURE 6. Transition state for hydrogen abstraction from saturated C–H bond by active oxidant in Fe(Por)Cl/PhIO system.

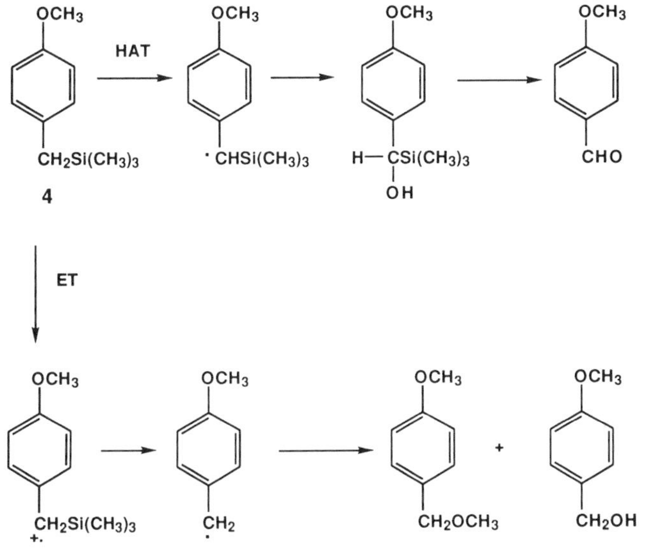

SCHEME VII. Electron transfer (ET) and hydrogen atom transfer (HAT) in the oxidation of **4** by iron porphyrin.

ratio of 1.6:1. Hydrogen atom abstraction by the electrophilic t-BuO· intermediate was preferred from carbon 1 of the substrate possibly because of a greater stabilization of the partial positive charge at this position in the polar transition state. By contrast, one-electron photooxidation of **5** with ceric ammonium nitrate (CAN) yielded a mixture of 1-RONO$_2$ and 2-RONO$_2$, and subsequently, 1-hydroxy-5-nitroacenaphthene and 2-hydroxy-5-nitroacenaphthene in a ratio of 1:5.8. This result suggests that rate-determining deprotonation from the initially generated radical cation of **5** occurs preferably from carbon 2 (Scheme VIII). Product yields and regioselectivities for chromium, iron, and manganese porphyrin-catalyzed oxidations of **5**, as compared with reactions with t-BuOCl and CAN, indicated a similarity with the reactivity of t-butoxy radical and suggested that a hydrogen atom abstraction pathway is preferred in the oxidation of **5** for all of the metalloporphyrins. The observed halogenated products were attributed to radicals that had escaped the solvent cage.[13e]

4. Epoxidation of Olefins by Cytochrome P450 and Metalloporphyrins

There have been extensive studies regarding mechanisms for olefin epoxidation catalyzed by cytochrome P450 and metalloporphyrins.[3] *Cis*-olefins are generally more reactive toward metalloporphyrin catalysts than the corresponding *trans*-olefins.[10a] Typical competitive selectivities for iron and ruthenium porphyrins are compared with that of mCPBA epoxidation in Table I. A twofold preference for *cis*-olefins by mCPBA has been attributed to methyl-eclipsing strain relief in the transition for the *cis*-olefin. As can be seen, the oxidations of *cis*-olefins catalyzed by iron and ruthenium porphyrins are 14 to 40

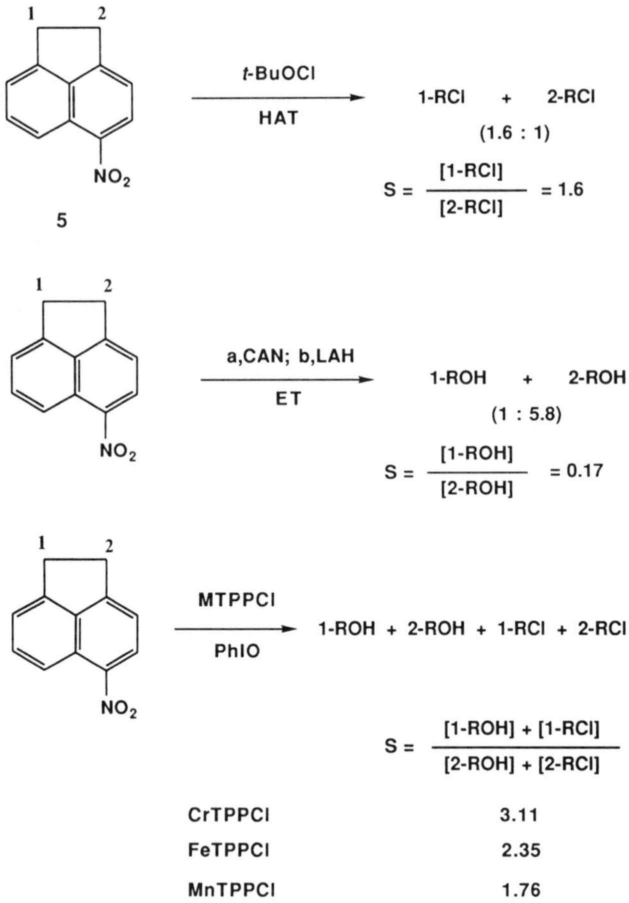
1
2
NO2
5
t-BuOCl
HAT
1-RCl + 2-RCl
(1.6 : 1)
S = [1-RCl] / [2-RCl] = 1.6
a,CAN; b,LAH
ET
1-ROH + 2-ROH
(1 : 5.8)
S = [1-ROH] / [2-ROH] = 0.17
MTPPCl
PhIO
1-ROH + 2-ROH + 1-RCl + 2-RCl
S = [1-ROH] + [1-RCl] / [2-ROH] + [2-RCl]
CrTPPCl 3.11
FeTPPCl 2.35
MnTPPCl 1.76

SCHEME VIII. Regioselectivities in the reactions of 5-nitroacenaphthene.

times faster than those of the corresponding *trans*-olefins. By contrast, the olefin reactivity reported for $[(bpy)_2pyRu^{IV}{=}O]^{2+}$ is styrene > *trans*-stilbene > *cis*-stilbene.[24a] Thus, there must be some unique features of oxometalloporphyrins. The consensus view is that oxygen transfer is concerted for most systems with the initial, rate-determining step involving some degree of charge transfer. Three possible geometries for the oxygen transfer transition state are depicted in Fig. 7a. Approach of the olefin to the (Por)M=O moiety could occur from the top of the bound oxygen with the olefin plane parallel to the porphyrin plane, "parallel" to the M=O bond, or "side-on" with the olefin π-bond perpendicular to the metal–oxo bond. That 1,1-disubstituted olefins react more slowly than *cis*-olefins argues against the parallel approach geometry and the four-membered ring transition state this mode of approach suggests. Some time ago we proposed a "side-on" transition state in which the

TABLE I. *Cis–Trans* Selectivities of Epoxidations by *m*CPBA and Metalloporphyrin Systems

Oxidant	Reactivities		
*m*CPBA		2:1	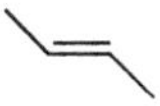
Ru^{VI}TMP(O)$_2$/O$_2$		14:1	
	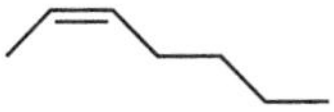	32:1	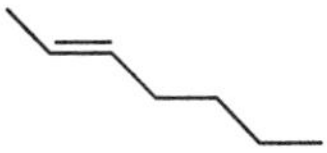
	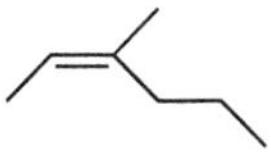	40:1	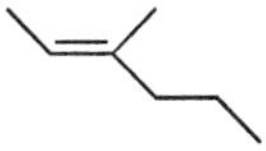
Fe^{III}TPPCl/PhIO		14:1	
		4.5:1	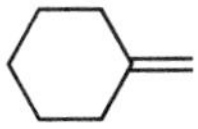

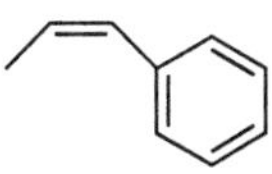

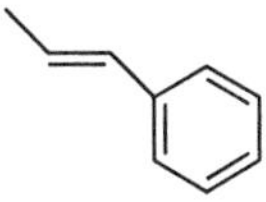

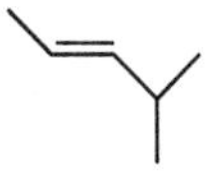

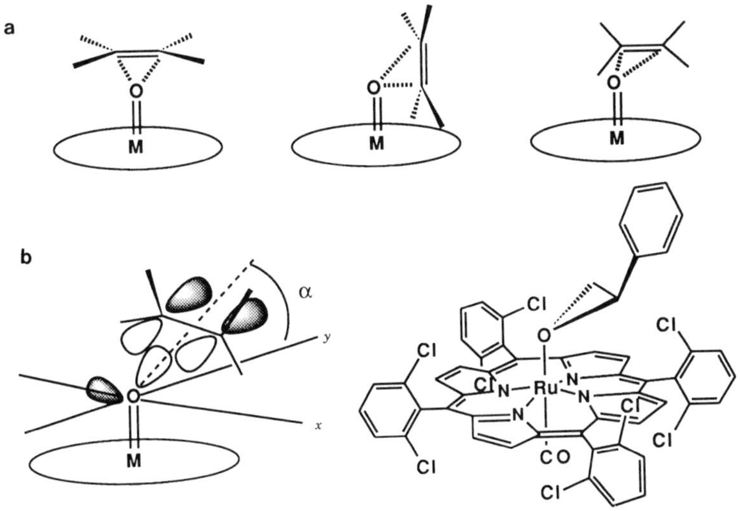

FIGURE 7. (a) Three possible transition state structures proposed for the epoxidation of olefins by oxometalloporphyrins; left, "top-on" approach; middle, parallel approach; right, "side-on" approach. (b) The "side-on" transition state for olefin epoxidation by oxometalloporphyrins compared to ruthenium(II)–epoxide coordination complex.

olefin π-bond approaches the bound oxygen of (Por)M=O in a bent fashion, because it best explained the *cis*-olefin selectivity. Thus, for a *cis*-olefin, the bulky substituents of the olefin can point away from the porphyrin ring, while for a *trans*-olefin, one of the substituents must be directed toward the porphyrin ring. Interestingly, however, one of the largest selectivities (40:1) has been observed for the two configurational isomers of a *trisubstituted* olefin. The "side-on" approach to the metal–oxo group is also appealing from a stereoelectronic point of view since this geometry will allow favorable interactions between the filled π-orbital of the approaching olefin and the metal–oxygen π-antibonding orbitals (Fig. 7b).[10a] This geometry is suggested as well by the X-ray crystal structure of a styrene-oxide-ruthenium(II) porphyrin coordination complex[24b] and it has been useful in predicting the stereoselectivities of chiral epoxidation catalysts (*vide infra*).

A number of electronic structures have been considered as possible intermediates during olefin epoxidation, as shown in Fig. 8.[25] Although the direct insertion of oxygen as illustrated in **6a** is consistent with the high stereospecificity of the reaction (retention of the configuration of olefins) as in the oxidation of *trans*-[1-D]-1-octene by cytochrome P450,[26] it cannot account for all of the events in cytochrome P450-catalyzed oxidation of all olefins, such as the small but usually detectable amounts of formation of the *trans*-epoxides from *cis*-olefins and the formation of other side products. For example, the liver microsomal oxidation of trichloroethylene gives trichloroacetaldehyde as well as trichloroethylene oxide. Synthetic trichloroethylene oxide does not rearrange to trichloroacetaldehyde under physiological conditions. Accordingly, the epoxide metabolite is not responsible for the formation of trichloroacetaldehyde.[27] Similarly, the microsomal oxidation of *trans*-1-phenylbutene yields 1-phenyl-1-butanone and 1-phenyl-2-butanone as minor products,[27b] and that of styrene a trace of 2-phenylacetaldehyde (Scheme IX).[28a] Nevertheless, **6a** is still an attractive intermediate (or transition state) possibly involved in the reaction pathway of cytochrome P450-catalyzed oxidation of olefins.

Bruice has presented results of a study with the cyclopropyl olefin **7** which seem to rule out a long-lived radical species **6b** as a discrete intermediate in iron porphyrin-catalyzed epoxidation at least for that substrate.[29] This compound should undergo a cyclopropylcarbinyl-to-homoallyl radical rearrangement with a rate constant $> 2 \times 10^7$ sec^{-1} on formation of a radical species, such as **6b** or **6e**. However, none of the oxygen-containing rearrangement products expected from intermediate **6b** were detected when olefin **7** was subjected to epoxidation by iron porphyrins.

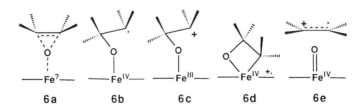

FIGURE 8. Proposed intermediates in epoxidations by cytochrome P450 and metalloporphyrins.

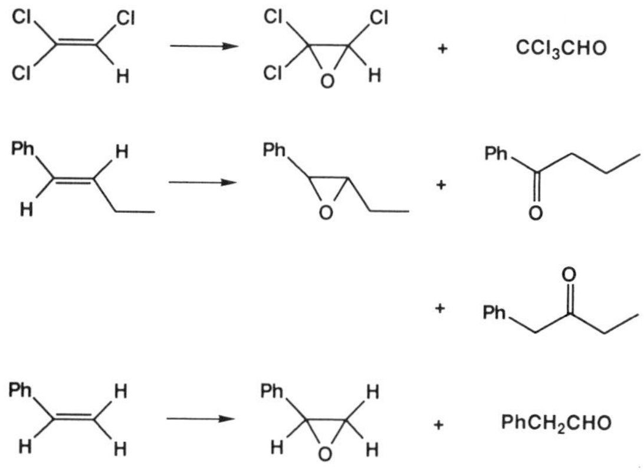

SCHEME IX. Rearrangement observed during the oxidation of olefins.

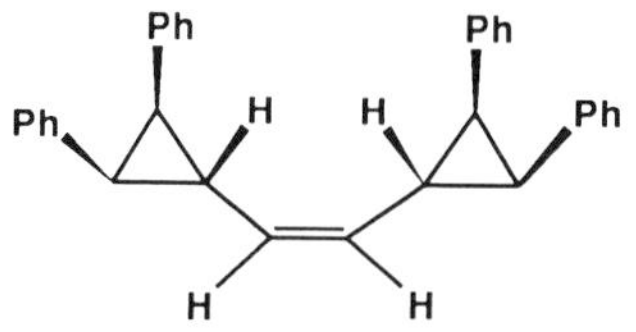

7

The possibility of the formation of cation intermediate **6c** by the direct attack is ruled out largely by the observation that the ρ^+ values for epoxidation of a series of *para*-substituted styrenes with metalloporphyrins are too low for an electrophilic addition mechanism.[30,31] The ρ^+ values for epoxidation of substituted styrenes by metalloporphyrins are determined to vary from –0.83 to –1.9. The ρ^+ values for some known electrophilic reactions that are believed not to involve carbon cationic intermediates are: –0.62 (:CCl$_2$ insertion into the double bond of substituted styrenes),[32] –1.61 (:CH$_2$ insertion into the double bond of substituted styrenes using C$_2$H$_5$ZnCH$_2$I),[33] and –1.2 (epoxidation of substituted stilbenes with peroxybenzoic acid).[34] On the other hand, the reported ρ^+ values for known electrophilic addition to substituted styrenes that involve the formation of carbocation intermediates as their rate-determining steps are much greater (–3.58 for hydration,[35] and –4.8 for bromination[36]). This indicates that although an electrophilic mechanism obtains in the epoxidation of olefins by high-valent oxometalloporphyrins, an early transition state with only fractional charge transfer is necessary to accommodate the observed electronic effects. Traylor and co-workers have found that both *exo* and *endo* epoxides are produced by the reaction of a range of iron tetraarylporphyrins with norbornene.[37] Direct electrophilic, radical, or molecular attack on norbornene is known to proceed exclusively from the *exo* side.[38] Thus, this result has been advanced as evidence against intermediates **6a**, **6b**, as well as **6c**. In a comparison of several chromium, manganese, and iron porphyrins, Traylor found an increase in the amount of *endo* epoxide on proceeding from chromium to iron systems. It was proposed that as the electron density of the oxometalloporphyrin oxidant decreased from chromium to manganese to iron, the extent of the bond formation in the transition state decreased (i.e., $r^{\neq}$ increased),[37] so lower selectivity between the *exo* and *endo* faces was observed.

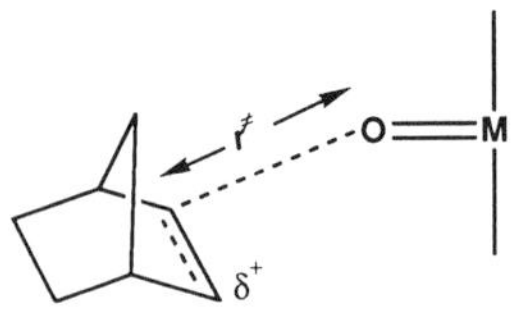

The possible intermediacy of an oxometallacycle intermediate **6d** in oxometal chemistry has been both alluring and elusive.[39] Thus, saturation kinetics observed in a hypochlorite–manganese porphyrin system was initially used to suggest that an oxometallo–olefin complex was formed reversibly and that its breakdown to epoxide was the rate-limiting step.[39b] However, subsequent analysis has indicated that this system is extremely complex and that the kinetic results were misleading.[39d]

We have reported spectral evidence for the formation of an intermediate between high-valent oxoiron porphyrin and olefins; however, this intermediate appears to be some sort of charge-transfer complex.[31a,40] Metallacycles have been proposed in the reaction of oxochromium complexes with olefin,[41] and a few such intermediates have been characterized in olefin metathesis[42] and other transition-metal-catalyzed reactions.[43] However, it has been shown that extremely sterically hindered olefins such as adamantylideneadamantane and tetramethyl ethylene can be epoxidized even by sterically hindered metalloporphyrins in which the formation of oxametallacycle intermediates are sterically not possible.[10a, 37] Thus, there remains no definitive evidence supporting a metallacycle intermediate in model oxometalloporphyrin chemistry.

We have reported that the oxidation of small terminal olefins (propene and 1-butene) by a cytochrome P450 LM$_2$ reconstituted system was accompanied by a stereoselective proton exchange of the E-proton.[44] Propene was oxidized in D$_2$O to *trans*-[1-D]-propene oxide (80:20) and *trans*-[1-D]-propene was oxidized in H$_2$O to propene oxide (98:2). Very high precision in the deuterium inventory was obtained in these cases by microwave spectroscopy of the volatile products. The epoxidation of 1-butene in D$_2$O generated a mixture of 86% 1-butene oxide and 14% *trans*-[1-D]-1-butene oxide (Scheme X). Larger

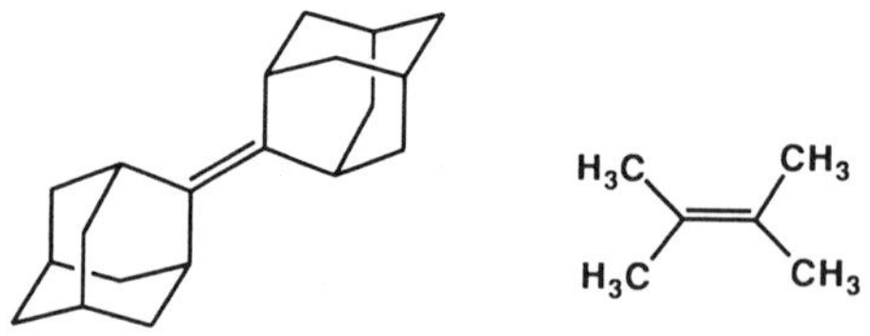

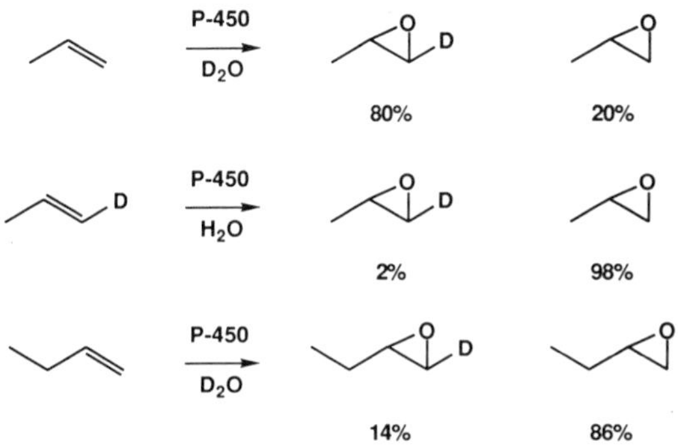

SCHEME X. Stereospecific D–H exchange during the epoxidation of propene and 1-butene by cytochrome P450 LM$_2$ reconstituted system.

terminal alkenes in the homologous series did not undergo such an exchange. It is difficult to explain this unusual result unless a mechanism involving the formation of *organoiron* porphyrin intermediates is invoked. An alkyl iron intermediate (**9**) could give an iron carbene (**10**) on deprotonation, which could in turn be deuterated to account for the observed proton exchange (Scheme XI). Such an alkyl iron intermediate could reasonably result from initial formation of either a metallacycle (**8**) or an *N*-alkyl heme precursor (**11**).

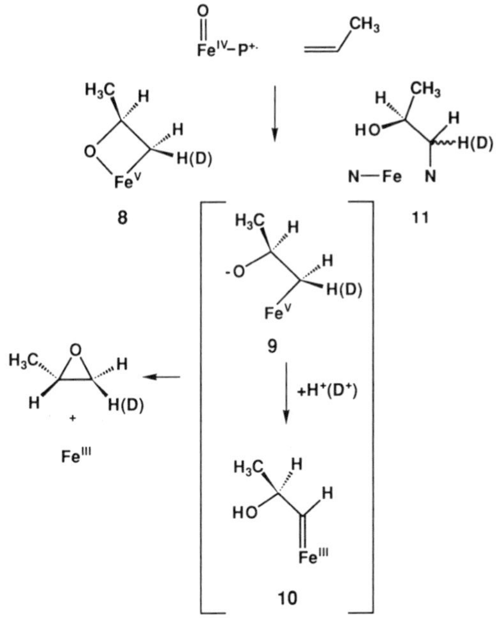

SCHEME XI. Proposed mechanism for D–H exchange in the oxidation of propene by cytochrome P450.

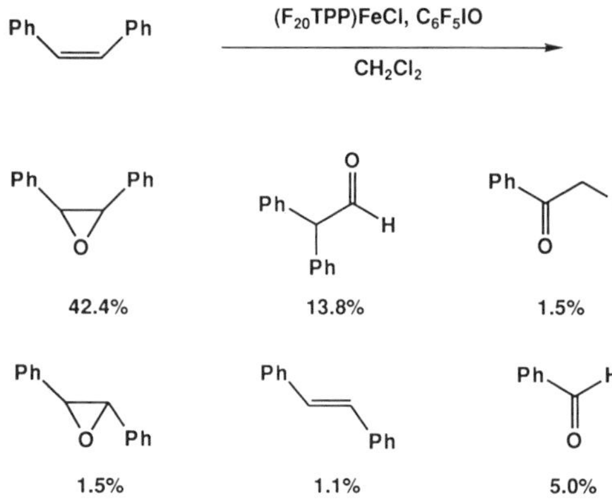

SCHEME XII. The oxidation of *cis*-stilbene by iron porphyrin and pentafluoroiodosylbenzene.

A charge transfer mechanism for olefin epoxidation has been supported by several studies.[25] Both Traylor and Bruice have independently described the linear relationships between the free energy of activation and one-electron oxidation potential of olefins.[30,31d] In addition, Bruice has argued that the rate-determining formation of alkene π-cation-radical intermediate is not likely because the linear relationship extends even to substrates for which the ionization potential is greater than the activation energy for the epoxidation. Accordingly, it was argued that a full charge transfer complex could not be the transition state for the epoxidation of these olefins. The fact that there is no break in the plot for log(rate) versus ionization potential, however, argues that the same mechanism applies to all substrates. Bruice proposed a partial charge transfer in the transition states of these epoxidation reactions.[30] Traylor found a continuous increase of the formation of the *endo*-norbornene epoxide as the electron density of the reactive oxometalloporphyrin decreases from chromium to manganese to iron.[37] There may be a continuous change in mechanism from direct electrophilic attack to full electron transfer, or, in other words, a continuous change in the degree of charge transfer in the transition state, depending on the one-electron oxidation potentials of substrates and the oxidation potential of active metalloporphyrin oxidants. The oxidation of *cis*-stilbene with iron porphyrins and pentafluoroiodosylbenzene gave *trans*-stilbene oxide, diphenylacetaldehyde, deoxybenzoin, *trans*-stilbene, and benzaldehyde in addition to *cis*-stilbene oxide (Scheme XII).[45]

Benzaldehyde is formed from the reaction of carbocation radical escaped from the solvent cage with molecular oxygen. *Trans*-stilbene derives from stereoisomerization and electron capture of the olefin cation radical intermediate. These results indicate that the formation of a carbocation radical intermediate by full electron transfer is involved in the epoxidation of *cis*-stilbene by this very electron-deficient iron porphyrin. However, Bauld and colleagues have shown that an intramolecular Diels–Alder reaction is a sensitive probe of cation radical intermediates. The application of this probe to iron and manganese porphyrins gave only epoxides as products. Accordingly, it appears unlikely that *free* olefin cation radicals are formed during the epoxidation process.[45b]

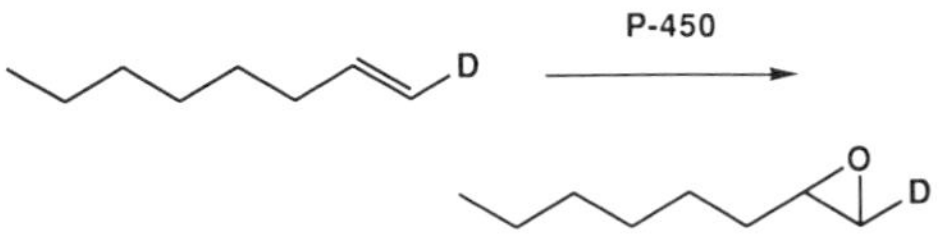

5. Suicide Inactivation of Cytochrome P450 by Terminal Olefins

Epoxidation of terminal olefins by cytochrome P450 has been found to be accompanied by N-alkylation of the prosthetic heme group of the enzyme.[46] The epoxide metabolite was shown not to be responsible for heme alkylation (Scheme XIII). The results thus require epoxidation and heme alkylation to diverge at a point prior to epoxide formation.[1,26] Linear olefins (ethylene, propene, octene) alkylate only pyrrole ring D of the prosthetic group of the phenobarbital-inducible isozymes from rat liver, but heme alkylation by two globular olefins (2-isopropyl-4-pentenamide and 2,2-dimethyl-4-pentenamide) is less regiospecific.[46b] A detailed study of the regiochemistry and stereochemistry of heme alkylation by $trans$-[l-D]-1-octene showed that (1) the olefin stereochemistry was preserved during the alkylation reaction, i.e., the oxygen and the pyrrole-N add to the same face of the olefin, and (2) oxygen was delivered only to the re face of the double bond in the heme alkylation reaction even though the stereochemical analysis of the epoxide metabolite indicated that the oxygen had been delivered almost equally to both faces of the π bond during epoxidation.[46b] Heme alkylation is thus a highly regio- and stereospecific process while the cytochrome-P450-catalyzed epoxidation reaction shows very low enantioselectivity.[28] A mechanism consistent with all of these results is shown in Scheme XIII, similar to the mechanism proposed by Ortiz de Montellano.[28c] The initial formation of a charge transfer

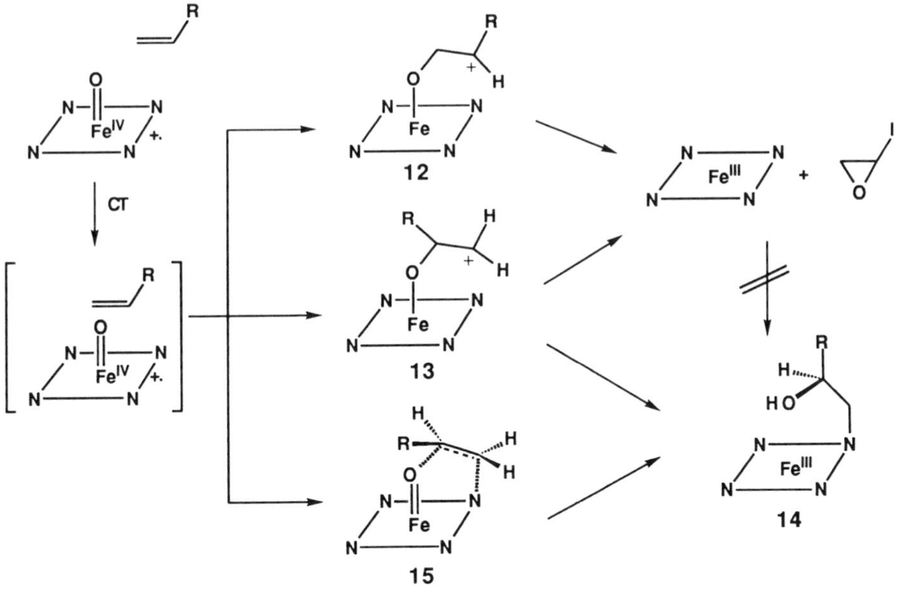

SCHEME XIII. Heme alkylation during the oxidation of terminal olefins.

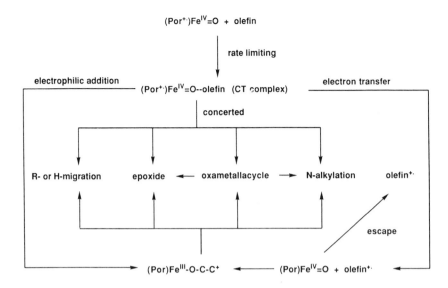

SCHEME XIV. Mechanism of the oxidation of olefins by cytochrome P450 or metalloporphyrins.

complex by partial charge transfer could lead to intermediates **12** or **13**. An N-alkylated complex, **14**, can be formed from intermediate **13** or directly from a charge transfer complex in a concerted manner. The high stereospecificity of the epoxidation requires a slow C–C bond rotation in **12** and **13**. The enantioselectivity of the epoxidation reaction indicates the very low selectivity of the enzyme pocket toward the two faces of the olefin along the epoxidation pathway. On the other hand, the extremely high enantioselectivity and regioselectivity (only D ring for linear terminal olefins) indicate the strict requirements for the steric interaction between the substrate and the protein pocket in order for the alkylation to occur. Significantly, heme-N-alkylation has also been observed to accompany epoxidation for a number of olefin substrates.

A unified mechanism for the epoxidation of olefins by cytochrome P450 and metalloporphyrins can be summarized in Scheme XIV. The partial charge transfer is the rate-limiting step for all of the reactions (epoxide formation, side product formation, N-alkylation, and iron–carbon bond formation).[25] Subsequent to charge transfer, either the pathways diverge in concerted manners or an electrophilic addition gives a carbocation intermediate which leads to products. The relative importance of each pathway depends on the olefin. The corresponding stereoelectronic diagram[24b,45,47,48] of these processes is shown in Fig. 9. It should be pointed out that the steric interactions between the substrate and the protein structure surrounding the active site of cytochrome P450 play an important role in controlling the fate of the charge transfer complex, such as N-alkylation of the heme group and the unusual D-H exchange in the oxidation of propylene. It should be noted that almost all aspects of the cytochrome-P450-catalyzed oxidation of olefins such as retention of the stereochemistry of olefins and N-alkylations have been modeled by using synthetic iron and other metalloporphyrins[49] except the observed D-H exchange.[44]

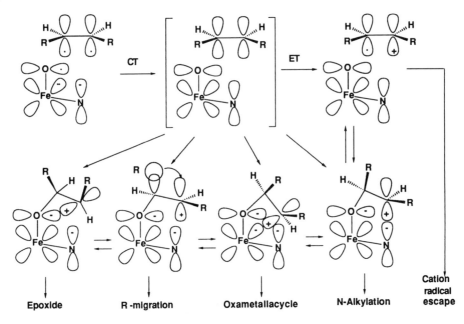

FIGURE 9. Orbital diagram of oxidation of olefins by cytochrome P450 and metalloporphyrins.

6. Synthetic Metalloporphyrins as Regioselective and Stereoselective Catalysts

The elaboration of the cytochrome P450 mechanism with model systems has afforded a significant opportunity to develop useful regioselective and stereoselective catalysts. A very recent review has surveyed a wide range of such systems.[50] The stereoselectivity of metalloporphyrin-catalyzed oxidations is sensitive to the metal, the porphyrin ligand, its superstructure, the sixth axial ligand, and the host matrix.[37,51] For example, the reactivity of synthetic manganese porphyrins is very sensitive to the sixth axial ligand used.[51d,e,52,53]

Since (Por)M=O is the reactive center for the oxidation of hydrocarbons, metalloporphyrins with special steric features have been constructed to achieve stereoselectivity of the oxidation such as regioselectivity and enantioselectivity based on the steric interactions between the substrate and porphyrin catalysts. Thus, high regioselectivity (C-4 and C-5) in the oxidation of the octyl chain of **16** by iodosylbenzene, in which the substrate was covalently linked to heme propionates, was the first indication that an oxoiron intermediate had been prepared.[11a] Likewise, Chang and Kuo[54] subsequently reported the selective oxidation of the middle carbon C–H bonds in the strapped porphyrin **17a**. More recently, Patzelt and Woggon[17i] have reported the hydroxylation of the alkyl strap in the thiophenolate ligated porphyrin **17b** and Grieco and Stuk have reported the regioselective hydroxylation of steroidal substrates attached to a manganese porphyrin.[55] Breslow *et al.* reported that the remote prebinding of substrates through copper ions to the porphyrin catalyst **18** increased the reactivity by 40-fold compared to those substrates that could not bind.[56]

16   17a

17b

Suslick and colleagues [57] have reported that a bulky "bis-pocket" porphyrin catalyst **19** exhibits marked selectivity for terminal versus internal double bonds. For example, using 1,4-octadiene as a substrate, FeTPPCl produces 20-fold more internal than terminal epoxide, whereas **19** actually produces a slight excess of terminal epoxide. We have shown that catalyst **20** epoxidized 1-octene about 6-fold faster than *cis*-cyclooctene while FeTPPCl and FeTMPCl showed higher reactivity toward *cis*-cyclooctene. [47] Collman *et al.* [58a] reported a series of "picnic basket" porphyrins **21a–f** which show extremely high shape selectivity (>1000 for *cis*-2-octene versus *cis*-cyclooctene) and, more recently, the enantioselective epoxidation of simple olefins has been reported with ee's in the 70–85% range. [58b]

We have described a membrane-spanning iron porphyrin **22** that epoxidizes diolefinic sterols exclusively at the side chain. [59] A similar manganese porphyrin selectively hydroxylated cholesterol at the C-25 tertiary carbon.

Several chiral porphyrin catalysts have been synthesized. In 1983 we reported the use of chiral porphyrin **23** to carry out asymmetric epoxidation, giving 50% ee (enantiomeric excess) with *p*-chlorostyrene. [60] Mansuy and colleagues reported 50% ee for the oxidation of *p*-chlorostyrene by a chiral "basket-handle" porphyrin catalyst. [61] O'Malley and Kodadek [62] recently reported a "chiral wall" porphyrin **24**, which gives 40% ee for the epoxidation of *cis*-β-methylstyrene. Naruta *et al.* [63] reported a chiral "twin cornet" porphyrin **21g** which epoxidizes 2-nitrostyrene with 80% ee.

We have reported a very robust chiral vaulted binaphthyl porphyrin **25** which gave an ~70% ee for epoxidation of *cis*-β-methylstyrene. More importantly, this catalyst afforded

18

19

20

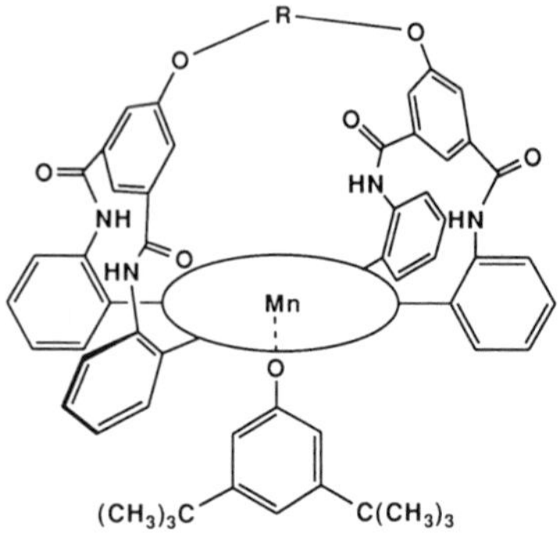

-(CH$_2$)$_2$-   21a

-(CH$_2$)$_4$-   21b

-(CH$_2$)$_6$-   21c

-(CH$_2$)$_8$-   21d

-(CH$_2$)$_{10}$-   21e

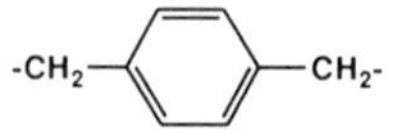

-CH$_2$-⟨benzene⟩-CH$_2$-   21f

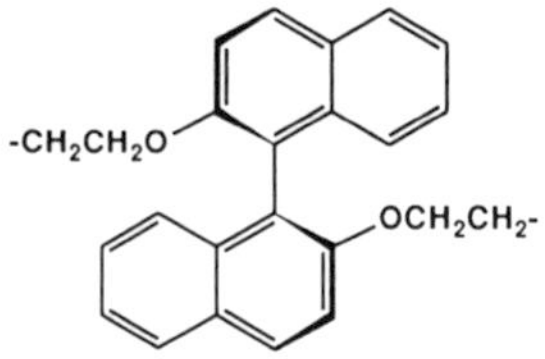

-CH$_2$CH$_2$O⟨binaphthyl⟩OCH$_2$CH$_2$-   21g

the first case of catalytic asymmetric *hydroxylation* by a model system, giving a 70% ee for hydroxylation of ethylbenzene and related hydrocarbons.[64] A detailed analysis of the stereoselectivity and the deuterium inventories for *R*- and *S*-deuterioethyl benzene revealed that the initial hydrogen atom abstraction was *less* stereoselective than the subsequent oxygen atom transfer step. This result shows how a nonselective hydrogen abstraction can still lead to stereoselective hydroxylation by a high selectivity for the capture of the intermediate carbon radical in the oxygen rebound mechanism.

Some nonporphyrin metal complexes[65–69] have also been examined as catalysts for epoxidation with iodosylbenzene, alkyl hydroperoxides, and hypochlorite oxidants. Jacobsen and colleagues[69] have recently achieved very high ee's (80–97%) for the epoxidation of simple, unfunctionalized olefins by chiral Mn(salen) catalysts 26. The ready availability of the chiral salen ligand, and the good isolated yields of the product epoxides make this an attractive approach to practical asymmetric synthesis. The reported formation of *trans*-epoxides in high enantioselectivity from *cis*-olefins in one of these systems is mechanistically significant.[69b] Manganese porphyrin catalysts have been reported by us to afford epoxides via either a stereoretentive pathway or with loss of stereochemistry depending on the reaction conditions.[13c,d] Thus, oxo-manganese(V) intermediates were

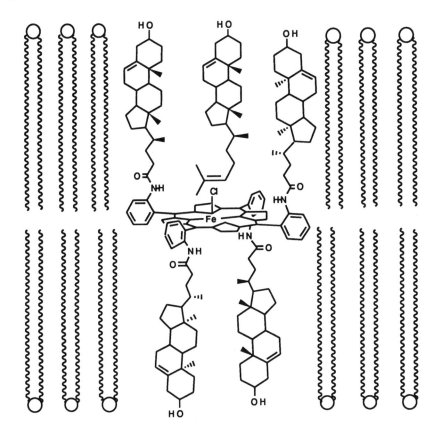

22

R =

23

24

25

26a

26b

invoked in a concerted epoxidation with retention of the configuration of the double bond. However, oxo-manganese(IV) intermediates were proposed to produce epoxides via a stepwise process involving carbon radical intermediates. While the nature of the change in mechanism for the manganese-salen catalysts on addition of chiral quaternary ammonium salts is not yet known, it is tempting to invoke a rationale similar to the porphyrin case.

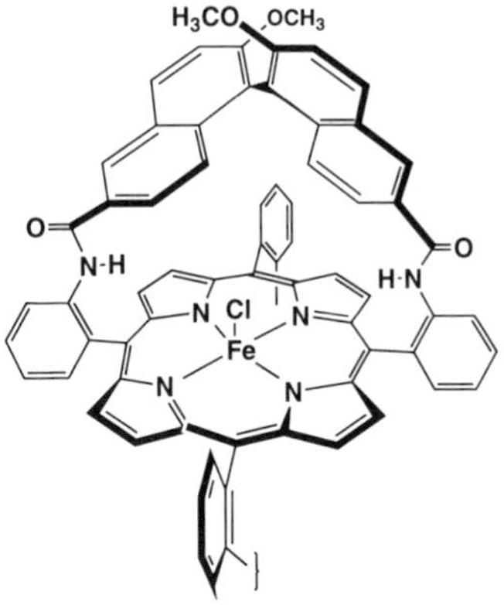

25-Fe(III)Cl

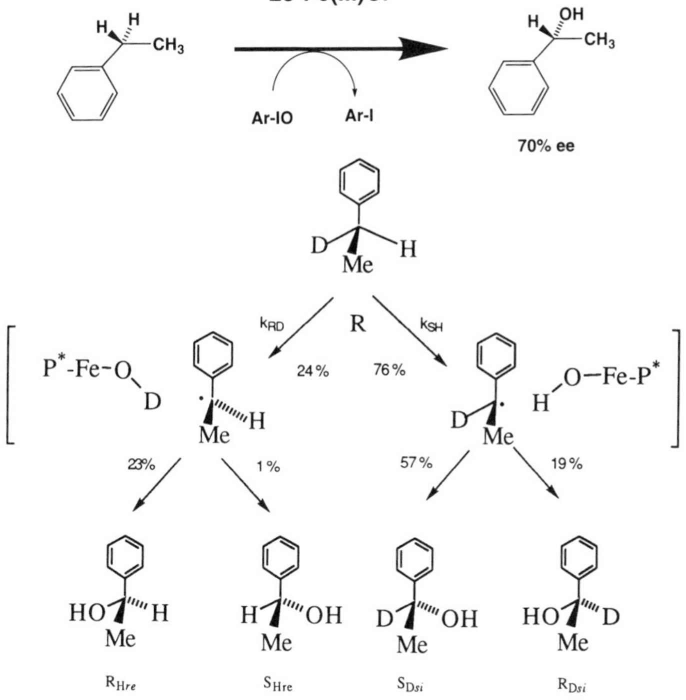

Ar-IO   Ar-I

70% ee

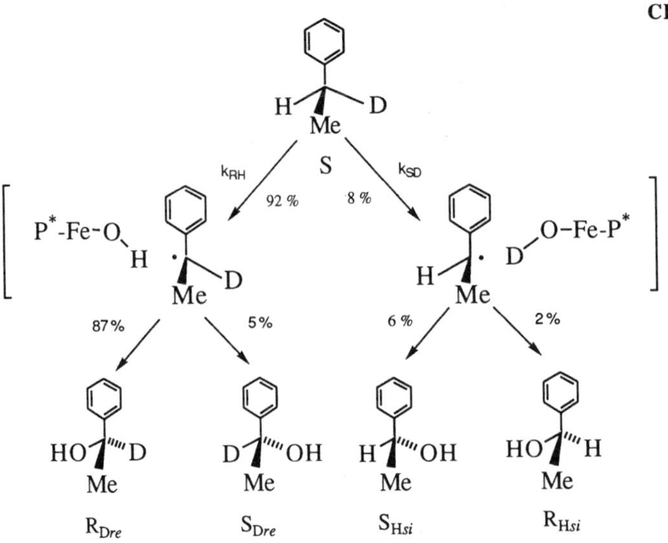

7. Mechanisms of Nitric Oxide Synthase (NOS)

The intense current interest in ·NO originates from its recent identification as a messenger molecule that is involved in a variety of critical physiological processes such as vasodilation, macrophage immune response, and some aspects of neurotransmission. Because of the obvious impact of these functions and their regulation on human health, there is an intense worldwide effort to understand the biochemistry of ·NO and to devise protocols for the *selective* pharmacological intervention in these processes. A detailed understanding of the mechanisms of NOS action is a prerequisite for advances in this area. It is instructive to review what is currently known about the function of NOS and to reflect those facts onto the generally accepted mechanisms for cytochrome P450 action.

The biosynthesis of ·NO has been shown to involve unusual oxidation chemistry and has led to a new repertoire in eukaryotic enzymology. Native NOS isoforms are dimeric (monomer M.W. = 130,000 to 160,000). The N-terminal domain of each NOS monomer binds iron protoporphyrin. The C-terminus is homologous to P450 reductase and binds to one equivalent each of FAD and FMN.[70] NOS catalyzes the formation of ·NO and citrulline from L-arginine at the expense of NADPH and dioxygen via N^G-hydroxy-L-arginine(NHA) as an intermediate (Scheme XV).[71] Convincing spectral evidence has been presented that NOS is a cytochrome P450-type hemeprotein and that the sole heme group

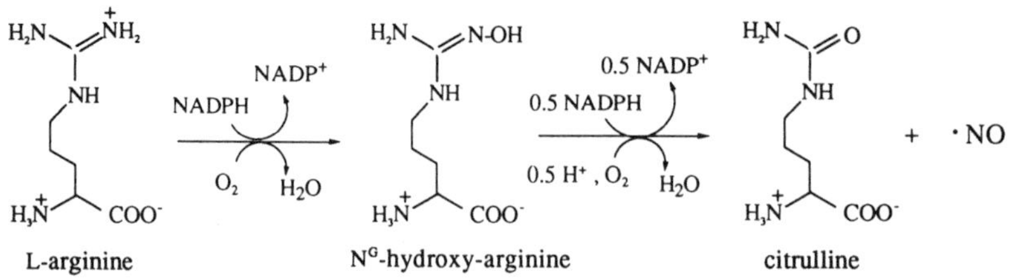

SCHEME XV.

in each NOS monomer carries out both steps of the NOS reaction.[70,72] The enzyme also contains a pteridine which assists in homodimer formation but while it is tempting to invoke a role for this cofactor in substrate oxidation, no strong evidence has emerged as yet to support that view.

8. The Three-Electron Problem

While the first step of the NOS reaction, the N-hydroxylation of L-arginine, is typical of a number of P450-mediated transformations,[73] it is not clear how the second step, the oxidative cleavage of NHA to citrulline and ·NO, is effected by the heme group of NOS. Certainly one of the most intriguing and least understood aspects of NOS chemistry is the three-electron oxidation that occurs in the ·NO-releasing step. Stuehr and colleagues have reported that the N-hydroxylation of L-arginine consumes one equivalent of NADPH while the conversion of NHA to ·NO consumes only half an equivalent of NADPH in reconstituted NOS reactions.[71] It was also demonstrated that the ·NO produced comes exclusively from the hydroxy-bearing guanidino nitrogen of ^{15}N-labeled NHA. These results support the view that NOS first carries out a two-electron oxidation on L-arginine at the expense of one equivalent of NADPH and dioxygen. Subsequently, NOS consumes half an equivalent of NADPH and a second dioxygen to furnish a three-electron oxidation of NHA to generate ·NO and citrulline.

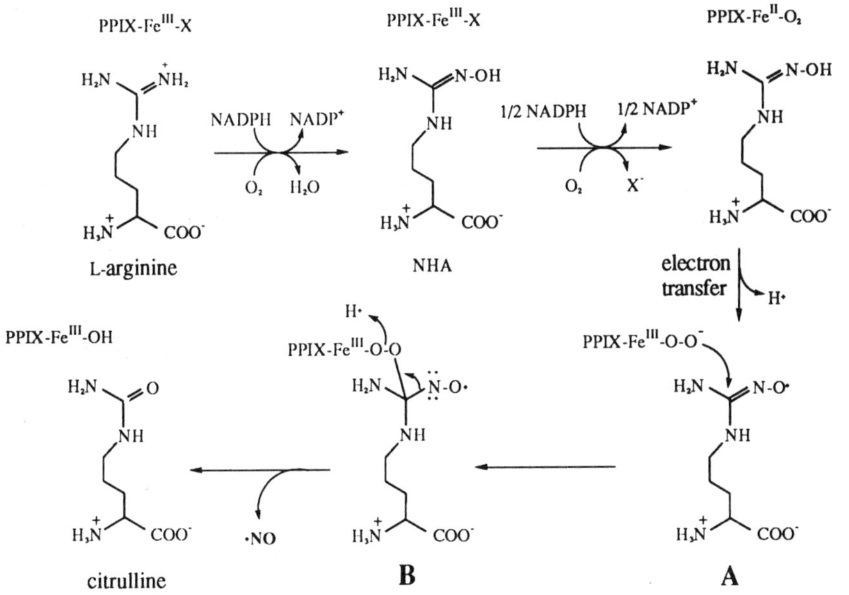

SCHEME XVI.

The implications are apparent. The substrate NOS must provide the second electron in the second step, but how this happens is unknown. Marletta[70] and Stuehr et al.[71] have independently proposed the mechanism sequence shown in Scheme XVI. The second step of this mechanism involves a nucleophilic attack by Fe^{III}-OO^- on an *oxidized* hydroxy-guanidine group followed by O–O bond cleavage. While there are several appealing aspects of this process, other aspects are unusual and need comment and experimental verification. First, the electron transfer from NHA to the oxy form of NOS is probably contrathermodynamic. Nitroxyl radicals of the type **A** are known but their formation requires stronger oxidizing agents than Enz-S-Fe(II)-O_2. Indeed, it has recently been estimated from the oxidation potential of NHA that such an electron transfer would be endothermic by 12 kcal/mol.[74] Abstraction of the *N*-hydroxy hydrogen by Enz-S-Fe(II)-O_2 ↔ Enz-S-Fe(III)-O_2· has been suggested as an alternative. Next, in the decomposition of the tetrahedral intermediate **B**, Enz-S-Fe(III) acts only as a Lewis acid. This is analogous to the mechanism proposed by Coon and colleagues for P450 aromatase,[75] which involves a nucleophilic attack of Fe^{III}-OO^- on the carbonyl group of cyclohexane carboxaldehyde to explain the very unusual formation of cyclohexene and formate. However, there are at

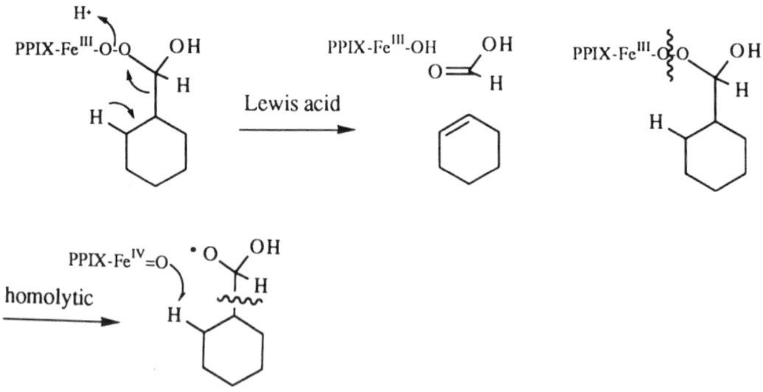

H·

PPIX-Fe^{III}-O–O   OH
                    H
H

Lewis acid
→

PPIX-Fe^{III}-OH   OH
        O=   H

PPIX-Fe^{III}-C–O   OH
                    H
H·

homolytic
→

PPIX-Fe^{IV}=O   ·O   OH
                        H
H·

least two other likely pathways that intermediate **B** could take after an initial nucleophilic attack on the substrate by Enz-S-Fe(III)-OO^-. *Homolytic* cleavage of the O–O bond in **B** would afford an oxoiron(IV) species analogous to HRP Compound II and O–O bond *heterolysis* would generate the oxenoid analogous to Compound I. Nucleophilic attack of Fe^{III}-OO^- on carbonyl substrates followed by a homolysis pathway has been suggested for the P450 aromatase reaction by Watanabe and Ishimura.[76] And the heterolysis pathway is exactly analogous to the "normal" peroxide shunt pathway for P450. As discussed above, model studies of O–O bond heterolysis of peroxoiron(III) species[3a] have shown a very low activation energy for this process. So it is not at all clear how the enzyme could prevent the formation of oxoiron intermediates if a peroxide such as **B** should ever form.

An alternative NOS mechanism with FeO^{3+} as the active oxidant has been offered by Mayer (Scheme XVII).[77] In this process, NHA provides one electron to the formation of the ferryl (FeO^{3+}) species, which hydroxylates the C–H tautomer of a nitrosyl cation radical generated from NHA to give, consequently, the products ·NO and citrulline.

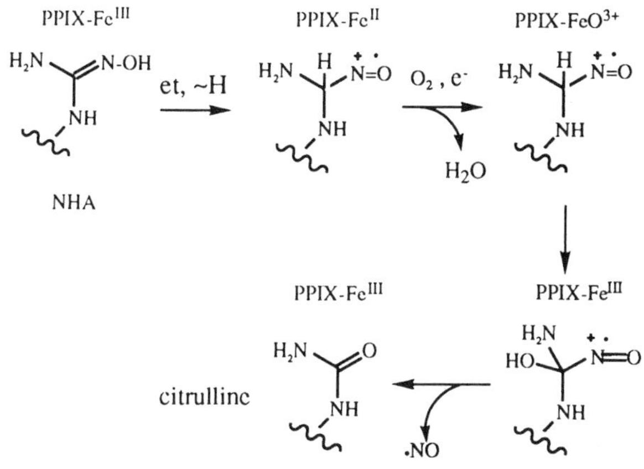

SCHEME XVII.

Lessons from Model Studies

There are very few close analogies in the literature to the kinds of oxidations described here for NHA. N-(N-hydroxyamidino)piperidine, an analogue of NHA, is known to be oxidized by peroxyacids to give HNO and piperidinecarboamide, the citrulline equivalent product.[78] Nucleophilic attack of the peroxyacid, in analogy to the putative ferric peroxo species (Fe-OO$^-$), at the guanidino carbon of NHA would reasonably give rise to the observed products. N-hydroxyguanidines afford variously ·NO [with Pb(OAc$_4$) or Fenton's reagent] and HNO (with ferric ion and PbO$_2$). However, the organic product is an N-cyanoamine rather than a urea. The catalytic oxidation of NHA by HRP and microsomal cytochrome P450 has been reported.[79] Both hemeproteins catalyze the formation of citrulline and NO$_2^-$, the oxidation product of ·NO. The iron-peroxo pathway does seem unlikely for HRP, however. Microsomal P450 also catalyzes the oxidation of an amidoxime to give ·NO and the corresponding amide.[80] These results not only confirm that the oxidative cleavage reactions of NOS are common to other hemeproteins but also indicate that the substrate of oxidative cleavage is not limited to N-hydroxyguanidine.

It was shown some time ago that simple cobalt salen complexes mediate the oxygenation of phenols as shown below. Likewise, oximes are converted to the corresponding ketones by the same complexes, likely by an analogous mechanism.[81]

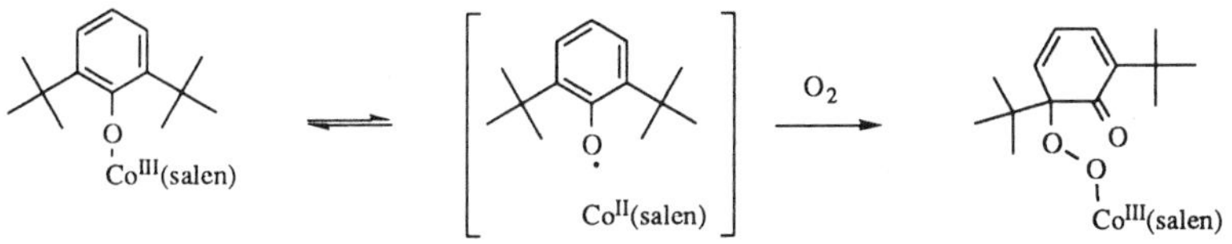

Thus, it is clear that there is a need for model studies with metalloporphyrin catalysts that can illuminate these interesting processes.

9. Summary

An oxoiron(IV) porphyrin cation radical intermediate similar to Compound I of HRP but with a thiolate axial ligand is believed to be the reactive oxidant of cytochrome P450-catalyzed oxidation of hydrocarbons. The hydroxylation of alkanes by cytochrome P450 involves an initial hydrogen atom abstraction from the alkane followed by a rapid transfer of the resulting metal-bound hydroxy radical to an intermediate alkyl radical, the oxygen rebound mechanism. Electron transfer, on the other hand, was proposed to be the initial step for the oxidation of alkanes with very low oxidation potentials. The oxidation of alkenes involves partial charge transfer as the initial rate-determining step. Subsequent to charge transfer, the reaction pathways (epoxide formation, side product formation, N-alkylation, and iron–carbon bond formation) diverge in concerted manners or electrophilic addition gives a carbocation intermediate which leads to products. The relative importance of each pathway depends on the alkene. Almost all aspects of cytochrome P450-catalyzed oxidation of alkenes have been modeled by using iron and other metalloporphyrins except the observed D–H exchange observed in the oxidation of propene and 1-butene. Metalloporphyrins with special steric features have been constructed; moderate to high stereoselectivities have been achieved in the hydroxylation of alkanes and epoxidation of olefins with these catalysts. Such stereoselective metalloporphyrin catalysts may soon find applications in synthetic organic chemistry.

Finally, it is becoming clear that the mechanisms for the manipulation of oxygen deduced for cytochrome P450 may represent a general biological strategy. Thus, there are a number of similarities between the P450 processes discussed here and the energy transduction processes of cytochrome oxidase[82] and the hydroxylating activity of the nonheme, diiron enzyme methane monooxygenase.[83]

ACKNOWLEDGMENTS. Research described herein which was carried out at Princeton University was generously supported by the National Institutes of Health and the National Science Foundation.

References

1. Ortiz de Montellano, P. R. (ed.), 1986, *Cytochrome P-450: Structure, Mechanism, and Biochemistry,* Plenum Press, New York.
2. Dawson, J. H., and Eble, K. S., 1986, Cytochrome P-450: Heme iron coordination structure and mechanisms of action, *Adv. Inorg. Bioinorg. Mech.* **4:**1–64.
3. (a) Watanabe, Y., and Groves, J. T., 1992, Molecular mechanism of oxygen activation by cytochrome P-450, in: *The Enzymes,* 3rd ed., Vol. XX, (D. Sigman, ed.), Academic Press, New York, pp. 406–453. (b) McMurry, T. J., and Groves, J. T., 1986, Metalloporphyrin models for cytochrome-P-450, in: *Cytochrome P-450: Structure, Mechanism, and Biochemistry* (P. R. Ortiz de Montellano, ed.) Plenum Press, New York, pp. 1–28.
4. (a) Roberts, J. E., Hoffman, B. M., Rutter, R., and Hager, L. P., 1981, Electron-nuclear double resonance of horseradish peroxidase compound I. Detection of the porphyrin π-cation radical, *J. Biol. Chem.* **256:**2118–2121. (b) Roberts, J. E., Hoffman, B. M., Rutter, R., and Hager, L. P., 1981, Oxygen-17 ENDOR of horseradish peroxidase compound I, *J. Am. Chem. Soc.* **103:**7654–7656. (c) Penner-Hahn, J. E., McMurry, T. J., Renner, M., Latos-Grazynsky, L., Eble, K. S., Davis, I. M., Balch, A. L., and Groves, J. T., 1983, X-ray absorption spectroscopic studies of high-valent iron porphyrins,

J. Biol. Chem. **258:**12761–12764. (d) Nick, R. J., Ray, G. B., Fish, K. M., Spiro, T. G., and Groves, J. T., 1991, Evidence for a weak Mn=O bond and a non-porphyrin radical in manganese-substituted horseradish peroxidase compound I, *J. Am. Chem. Soc.* **113:**1838–1840. (e) Mann, T., 1988, An exercise in nostalgia on the theme of David Keilin, in: *Oxidases and Related Redox Systems* (T. E. King, H. S. Mason, and M. Morrison, eds.), Liss, New York, pp. 29–49.

5. (a) Hrycay, E. G., Gustafsson, J.-Å., Ingelman-Sundberg, M., and Ernster, L., 1975, Sodium periodate, sodium chlorite, organic hydroperoxides, and hydrogen peroxide as hydroxylating agents in steroid hydroxylation reactions catalyzed by partially purified cytochrome P 450, *Biochem. Biophys. Res. Commun.* **66:**209–216. (b) Danielsson, H., and Wikvall, K., 1976, On the ability of cumene hydroperoxide and sodium periodate to support microsomal hydroxylations in biosynthesis and metabolism of bile acids, *FEBS Lett.* **66:**299–302. (c) Gustaffson, J.-Å., Hrycay, E. G., and Ernster, L., 1976, Sodium periodate, sodium chlorite, and organic hydroperoxides as hydroxylating agents in steroid hydroxylation reactions catalyzed by adrenocortical microsomal and mitochondrial cytochrome P450, *Arch. Biochem. Biophys.* **174:**438–451. (d) Gustaffson, J.-Å., Rondahl, L., and Bergman, J., 1979, TI iodosylbenzene derivatives as oxygen donors in cytochrome P-450 catalyzed steroid hydroxylations, *Biochemistry* **18:**865–870. (e) Gustaffson, J.-Å., and Bergman, J., 1976, Iodine- and chlorine-containing oxidation agents as hydroxylating catalysts in cytochrome P-450-dependent fatty acid hydroxylation reactions in rat liver microsomes, *FEBS Lett.* **70:**276–280. (f) Groves, J. T., Krishnan, S., Avaria, G. E., and Nemo, T. E., 1980, Studies of the hydroxylation and epoxidation reactions catalyzed by synthetic metalloporphyrinates. Models related to the active oxygen species of cytochrome P-450, *Adv. Chem. Ser. Series* **191:**277–289.

6. (a) Dolphin, D., Forman, A., Borg, D. C., Fayer, J., and Felton, R. H., 1971, Compounds I of catalase and horse radish peroxidase: π cation radicals, *Proc. Natl. Acad. Sci. USA* **68:**614–618. (b) Dolphin, D., and Felton, R. H., 1974, Biochemical significance of porphyrin π cation radicals, *Acc. Chem. Res.* **7:**26–32.

7. Peisach, J., Blumberg, W. E., Wittenberg, B. A., and Wittenberg, J. B., 1968, Electronic structure of protoheme proteins. III. Configuration of the heme and its ligands, *J. Biol. Chem.* **243:**1871–1880.

8. Loew, G. H., Kert, C. J., Hjelmeland, L. M., and Kirchner, R. F., 1977, Active site models of horseradish peroxidase compound I and a cytochrome P-450 analog: Electronic structure and electric field gradients, *J. Am. Chem. Soc.* **99:**3534–3536.

9. Sawyer, D. T., 1988, Formation, characterization, and reactivity of the oxene adduct of [tetrakis(2,6-dichlorophenyl)porphinato]-iron(III)perchlorate in acetonitrile. Model for the reactive intermediate of cytochrome P-450, *J. Am. Chem. Soc.* **110:**2465–2470.

10. (a) Groves, J. T., and Nemo, T. E., 1983, Epoxidation reactions catalyzed by ferric porphyrins. Oxygen transfer from iodosylbenzene, *J. Am. Chem. Soc.* **105:**5786–5791. (b) Sevin, A., and Fontecave, M., 1986, Oxygen transfer from iron oxo porphyrins to ethylene. A semiempirical MO/VB approach, *J. Am. Chem. Soc.* **108:**3266–3272. (c) Ostovic, D., and Bruice, T. C., 1988, Transition state geometry in epoxidation by iron-oxo porphyrin at the compound I oxidation level. Epoxidation of alkenes catalyzed by a sterically hindered meso-tetrakis(2,6-dibromophenyl)-porphinato iron(III) chloride, *J. Am. Chem. Soc.* **110:**6906–6908. (d) Czernuszewicz, R. S., Su, Y. O., Stern, M. K., Macor, K. A., Kim, D., Groves, J. T., and Spiro, T. G., 1988, Oxomanganese(IV) porphyrins identified by resonance Raman and infrared spectroscopy: Weak bonds and the stability of the half filled t_{2g} subshell, *J. Am. Chem. Soc.* **110:**4158–4165.

11. (a) Groves, J. T., Nemo, T. E., and Myers, R. S., 1979, Hydroxylation and epoxidation catalyzed by iron–porphyrin complexes, *J. Am. Chem. Soc.* **101:**1032. (b) Groves, J. T., Haushalter, R. C., Nakamura, M., Nemo, T. E., and Evans, B. J., 1981, High-valent iron–porphyrin complexes related to peroxidase and cytochrome P-450, *J. Am. Chem. Soc.* **103:**2884. (c) Groves, J. T., Nemo, T. E., 1983, Hydroxylation reactions catalyzed by ferric porphyrins, *J. Am. Chem. Soc.* **105:**6243–6248. (d) Groves, J. T., and Watanabe, Y., 1988, Reactive iron porphyrin derivatives related to the catalytic cycles of cytochrome P450 and peroxidase. Studies of the mechanism of oxygen activation, *J. Am.*

Chem. Soc. **110**:8443–8452. (e) Groves, J. T., Gross, Z., and Stern, M. K., 1994, Preparation and reactivity of oxoiron(IV) porphyrins, *Inorg. Chem.* **33**:5065–5072.

12. (a) Dicken, C. M., Lu, F.-L., Nee, M. W., and Bruice, T. C., 1985, Kinetics and mechanisms of oxygen transfer in the reaction of p-cyano-N,N-dimethylaniline N-oxide with metalloporphyrin salts. 2. Amine oxidation and oxygen transfer to hydrocarbon substrates accompanying the reaction of p-cyano-N,N-dimethylaniline N-oxide with meso-(tetraphenylporphinato)iron(III) chloride, *J. Am. Chem. Soc.* **107**:5776–5789. (b) Balasubramanian, P. N., Lee, R. W., and Bruice, T. C., 1989, Reaction of [meso-tetrakis(2,6-dimethyl-3-sulfonatophenyl)porphinato]iron(III) hydrate with various acyl and alkyl hydroperoxides in aqueous solution, *J. Am. Chem. Soc.* **111**:8714–8721. (c) Lindsay-Smith, J. R., Balasubramanian, P. N., Lee, R. W., and Bruice, T. C., 1988, The dynamics of reaction of a water soluble and non-μ-oxo dimer forming iron(III) porphyrin with *tert*-butyl hydroperoxide in aqueous solution. 1. Studies using a trap for immediate oxidation products, *J. Am. Chem. Soc.* **110**:7411–7418. (d) Gopinath, E., and Bruice, T. C., 1991, Dynamics of reaction of [5,10,15,20-tetrakis(2,6-dimethyl-3-sulfonatophenyl)porphinato]iron(III) hydrate with tert-butyl hydroperoxide in aqueous solution. 3. Comparison of refined kinetic parameters and D_2O solvent isotope effects to those for [5,10,15,20-tetrakis(2,6-dichloro-3-sulfonatophenyl)porphinato]iron(III) hydrate and iron(III) hydrate, *J. Am. Chem. Soc.* **113**:4657–4665. (e) Beck, M. J., Gopinath, E., and Bruice, T. C., 1993, Influence of nitrogen base ligation on the rate of reaction of [5,10,15,20-tetrakis(2,6-dimethyl-3-sulfonato-phenyl)porphinato]iron(III) hydrate with *t*-BuOOH in aqueous solution, *J. Am. Chem. Soc.* **115**:21–29. (f) Chin, D. H., Balch, A. L., and LaMar, G. N., 1980, Mechanism of autoxidation of iron(II) porphyrins. Detection of a peroxo-bridged iron(III) porphyrin dimer and the mechanism of its thermal decomposition to the oxo-bridged iron(III) porphyrin dimer, *J. Am. Chem. Soc.* **102**:4344–4350; Formation of porphyrin ferryl (FeO^{2+}) complexes through the addition of nitrogen bases to peroxo-bridged iron(III) porphyrins, *J. Am. Chem. Soc.* **102**:1446–1448. (g) Balch, A. L., Chan, Y.-W., Cheng, R.-J., LaMar, G. N., Latos-Grazynsky, L., and Renner, M. W., 1984, Oxygenation patterns for iron(II) porphyrins. Peroxo and ferryl (Fe^{IV}O) intermediates detected by proton nuclear magnetic resonance spectroscopy during the oxygenation of (tetramesityl-porphyrin)iron(II), *J. Am. Chem. Soc.* **106**:7779–7785. (h) Balch, A. L., Latos-Grazynsky, L., and Renner, M. W., 1985, Oxidation of red ferryl [(Fe^{IV}O)$^{2+}$] porphyrin complexes to green ferryl [(Fe^{IV}O)$^{2+}$] porphyrin radical complexes, *J. Am. Chem. Soc.* **107**:2983–2985. (i) Balch, A. L., Cornman, C. R., Latos-Grazynsky, L., and Renner, M. W., 1992, Highly oxidized iron complexes of *N*-methytletra-*p*-tolylporphyrin, *J. Am. Chem. Soc.* **114**:2230–2237.

13. (a) Meunier, B., 1986, Metalloporphyrin-catalyzed oxygenation of hydrocarbons, *Bull. Soc. Chim. Fr.* **1986**:578–594. (b) Meunier, B., 1983, Homogeneous-phase oxidations catalyzed by transition metals: Recent advances, *Bull. Soc. Chim. Fr.* **1983**:345–366. (c) Groves, J. T., and Stern, M. K., 1987, Olefin epoxidation by manganese(IV) porphyrins. Evidence for two reaction pathways, *J. Am. Chem. Soc.* **109**:3812–3814. (d) Groves, J. T., and Stern, M. K., 1988, Synthesis, characterization, and reactivity of oxomanganese(IV) porphyrin complexes, *J. Am. Chem. Soc.* **110**:8628–8638. (e) Brown, R. B., Jr., and Hill, C. L., 1988, Catalytic homogeneous functionalization of adamantane. Influence of electronic and structural features of the metallo-porphyrin catalyst on atom transfer selectivity (oxygenation versus acidification/halogenation), *J. Org. Chem.* **53**:5762–5768. (f) De Poorter, B., Ricci, M., and Meunier, B., 1985, Oxone as oxygen donor in the catalytic hydroxylation of saturated hydrocarbons, *Tetrahedron Lett.* **26**:4459–4462. (g) De Poorter, B., Ricci, M., and Meunier, B., 1985, Catalytic hydroxylation of saturated hydrocarbons with the sodium hypohalite/manganese porphyrin system, *J. Mol. Catal.* **31**:221–224. (h) Mansuy, D., Bartoli, J. F., and Momenteau, M., 1982, Alkane hydroxylation catalyzed by metalloporphyrins: Evidence for different active oxygen species with alkylhydroperoxides and iodosobenzene as oxidants, *Tetrahedron Lett.* **23**:2781–2784. (i) Battioni, P., Renaud, J.-P., Bartoli, J. F., and Mansuy, D., 1986, Hydroxylation of alkanes by hydrogen peroxide: An efficient system using manganese porphyrins and imidazole as catalysts, *J. Chem. Soc. Chem. Commun.* **1986**:341–343. (j) Stern, M. K., and Groves, J. T., 1992, Oxygen transfer reactions of

oxo-manganese porphyrins, in: *Manganese Redox Enzymes,* (V. Pecoraro, ed.), Verlag Chemie, Weinheim, pp. 233–259

14. (a) Tabushi, I., 1988, Reductive dioxygen activation by use of artificial P-450 systems, *Coord. Chem. Rev.* **86:**1–42. (b) Mansuy, D., Fontecave, M., and Bartoli, J. F., 1983, Monooxygenase-like dioxygen activation leading to alkane hydroxylation and olefin epoxidation by a manganese(III)(porphyrin) –ascorbate biphasic system, *J. Chem. Soc., Chem. Commun.* **1983:**253–254. (c) Battioni, P., Bartoli, J. F., Leduc, P., Fontecave, M., and Mansuy, D., 1987, A new and efficient biomimetic system for hydrocarbon oxidation by dioxygen using manganese porphyrins, imidazole, and zinc, *J. Chem. Soc. Chem. Commun.* **1987:**791–792. (d) Leduc, P., Battioni, P., Bartoli, J. F., and Mansuy, D., 1988, A biomimetic electrochemical system for the oxidation of hydrocarbons by dioxygen catalyzed by manganese porphyrins and imidazole, *Tetrahedron Lett.* **29:**205–208. (e) Creager, S. E., and Murray, R. W., 1987, Electrochemical reactivity of manganese(II) porphyrins. Effects of dioxygen, benzoic anhydride, and axial ligands, *Inorg. Chem.* **26:**2612. (f) Mansuy, D., 1993, Activation of alkanes: The biomimetic approach, *Coord. Chem. Rev.* **125:**129–141.

15. (a) Groves, J. T., and Kruper, W. J., Jr., 1979, Preparation and characterization of an oxoporphinato-chromium(V) complex, *J. Am. Chem. Soc.* **101:**7613. (b) Groves, J. T., and Haushalter, R. C., 1981, E.S.R. evidence for chromium(V) porphyrinates, *J. Chem. Soc. Chem. Commun.* **1981:**1165–1166. (c) Takahashi, T., 1985, The generation, characterization and reaction of high valent oxo-, imido-, and nitrodometalloporphyrins of chromium, manganese, and ruthenium (catalytic amination, aziridine, electrochemical oxidation, photolysis), Ph.D. dissertation University of Michigan.(d) Creager, S. E., and Murray, R. W., 1985, Electrochemical studies of oxo(meso-tetraphenylporphinato)chromium(IV). Direct evidence for epoxidation of olefins by an electrochemically generated formal chromium(V) state, *Inorg. Chem.* **24:**3824–3828. (e) Garrison, J. M., and Bruice, T. C., 1989, Intermediates in the epoxidation of alkenes by cytochrome P-450 models. 3. Mechanism of oxygen transfer from substituted oxochromium(V) porphyrins to olefinic substrates, *J. Am. Chem. Soc.* **111:**191–198. (f) Groves, J. T., and Quinn, R., 1984, Models of oxidized heme proteins. Preparation and characterization of a *trans*-dioxoruthenium(VI) porphyrin complex, *Inorg. Chem.* **23:**3844–3846. (g) Groves, J. T., and Quinn, R., 1985, Aerobic epoxidation of olefins with ruthenium porphyrin catalysts, *J. Am. Chem. Soc.* **107:**5790–5792. (h) Groves, J. T., and Ahn, K.-H., 1987, Characterization of an oxoruthenium(IV) porphyrin complex, *Inorg. Chem.* **26:**3831–3833.

16. (a) Penner-Hahn, J. E., McMurry, T. J., Renner, M., Latos-Grazynsky, L., Eble, K. S., Davis, I. M., Balch, A. L., Groves, J. T., Dawson, J. R., and Hodgson, K. O., 1983, X-ray absorption spectroscopic studies of high-valent iron porphyrins: Horseradish peroxidase (HRP) compounds I and II, *J. Biol. Chem.* **258:**12761–12764. (b) Groves, J. T., Quinn, R., McMurry, T. J., Lang, G., and Boso, B., 1984, Porphyrins from iron(III) porphyrin cation radicals, *J. Chem. Soc., Chem. Commun.* **1984:**1455–1456. (c) Boso, B., Lang, G., McMurry, T. J., and Groves, J. T., 1983, Mössbauer effect study of tight spin coupling in oxidized chloro-5,10,15,20-tetra-(mesityl)-porphyrinato-iron(III), *J. Chem. Phys.* **79:**1122–1126. (d) Penner-Hahn, J. E., Eble, K. S., McMurry, T. J., Renner, M., Balch, A. L., Groves, J. T., Dawson, J. H., and Hodgson, K. O., 1986, Structural characterization of horseradish peroxidase using EXAFS spectroscopy. Evidence for Fe=O ligation in compounds I and II, *J. Am. Chem. Soc.* **108:**7819–7825. (e) Watanabe, Y., Yamaguchi, K., Morishima, I., Takehira, K., Shimizu, M., Hayakawa, T., and Orita, H., 1991, Remarkable solvent effect on the shape-selective oxidation of olefins catalyzed by iron(III) porphyrins, *Inorg. Chem.* **30:**2581–2582. (f) Mandon, D., Weiss, R., Jayaraj, K., Gold, A., Terner, J., Bill, E., and Trautwein, A. X., 1992, Models for peroxidase compound I: Generation and spectroscopic characterization of new oxoferryl porphyrin π cation radical species, *Inorg. Chem.* **31:**4404–4409. (g) Tsuchiya, S., 1991, Stable oxo-iron(IV) Porphyrin π radical cation related to the oxidation cycles of cytochrome P-450 and peroxidase, *J. Chem. Soc. Chem. Commun.* **1991:**716–717. (h) Ozawa, S., Watanabe, Y., and Morishima, I., 1992, Preparation and characterization of a novel oxoiron(IV) chlorin π-cation radical complex. The first model for compound I of chlorin-containing heme enzymes, *Inorg. Chem.* **31:**4042–4043. (i) Ozawa, S., Watanabe, Y., and

Morishima, I., 1994, Spectroscopic characterization of peroxo-iron(III) chlorin complexes. The first model for a reaction intermediate of cytochrome d, *Inorg. Chem.* **33**:306–313.

17. (a) Tajima, K., Shigematsu, M., Jinno, J., Ishizu, K., and Ohya-Nishiguchi, H., 1990, Generation of Fe(III)OEP-hydrogen peroxide complex (OEP = octaethylporphyrinato) by reduction of Fe(II)OEP-oxygen with ascorbic acid sodium salt, *J. Chem. Soc. Chem. Commun.* **2**:144–145. (b) Yamaguchi, K., Watanabe, Y., and Morishima, I., 1992, Push effect on the heterolytic O–O bond cleavage of peroxoiron(III) porphyrin adducts, *Inorg. Chem.* **31**:156–157. (c) Higuchi, T., Uzu, S., and Hirobe, M., 1990, Synthesis of a highly stable iron porphyrin coordinated by alkylthiolate anion as a model for cytochrome P-450 and its catalytic activity in oxygen–oxygen bond cleavage, *J. Am. Chem. Soc.* **112**:7051–7053. (d) Higuchi, T., Shimada, K., Maruyama, N., and Hirobe, M., 1993, Heterolytic oxygen–oxygen bond cleavage of peroxy acid and effective alkane hydroxylation in hydrophobic solvent mediated by an iron porphyrin coordinated by thiolate anion as a model for cytochrome P-450, *J. Am. Chem. Soc.* **115**:7551–7552. (e) Adachi, S.-i., Nagano, S., Ishimori, K., Watanabe, Y., Morishima, I., Egawa, T., Kitagawa, T., and Makino, R., 1993, Roles of proximal ligand in heme proteins: Replacement of proximal histidine of human myoglobin with cysteine and tyrosine by site-directed mutagenesis as models for P-450, chloroperoxidase, and catalase, *Biochemistry* **32**:241–252. (f) McCandlish, E., Miksztal, A. R., Nappa, M., Sprenger, A. Q., Valentine, J. S., Stong, J. D., and Spiro, T. G., 1980, Reactions of superoxide with iron porphyrins in aprotic solvents. A high spin ferric porphyrin peroxo complex, *J. Am. Chem. Soc.* **102**:4268–4271. (g) Burstyn, J. N., Roe, J. A., Miksztal, A. R., Schaevitz, B. A. Lang, G., and Valentine, J. S., 1988, Magnetic and spectroscopic characterization of an iron porphyrin peroxide complex. Peroxoferrioctaethylporphyrin(l-), *J. Am. Chem. Soc.* **110**:1382–1388. (h) Dawson, J. H., Holm, R. H., Trudell, J. R., Barth, G., Linder, R. E., Bunnenberg, E., Djerassi, C., and Tang, S. C., 1976, Magnetic circular dichroism studies. 43. Oxidized cytochrome P-450. Magnetic circular dichroism evidence for thiolate ligation in the substrate-bound form. Implications for the catalytic mechanism, *J. Am. Chem. Soc.* **98**:3707–3709. (i) Patzelt, H., and Woggon, W.-D., 1992, Oxygen insertion into nonactivated carbon–hydrogen bonds: The first observation of O_2 cleavage by a P-450 enzyme model in the presence of a thiolate ligand, *Helv. Chim. Acta* **75**:523–530.

18. (a) Groves, J. T., McClusky, G. A., White, R. E., and Coon, M. J., 1978, Aliphatic hydroxylation by highly purified liver microsomal cytochrome P-450, *Biochem. Biophys. Res. Commun.* **81**:154. (b) Groves, J. T., and Subramanian, D. V., 1984, Evidence for radical intermediates in allylic hydroxylation by cytochrome P-450, *J. Am. Chem. Soc.* **106**:2177. (c) Hjelmeland, L. M., Aronow, L, and Trudell, J., 1977, Intramolecular determination of primary kinetic isotope effects in hydroxylations catalyzed by cytochrome P-450, *Biochem. Biophys. Res. Commun.* **76**:541–549. (d) Foster, A. B., Jarman, M., Stevens, J. D., Thomas, P., and Westwood, J. H., 1974, Isotope effects in O- and N-demethylations mediated by rat liver microsomes. Application of direct insertion electron impact mass spectrometry, *Chem. Biol. Interact.* **9**:327–340. (e) Miwa, G. T., Walsh, J. S., and Lu, A. Y., 1984, Kinetic isotope effects on cytochrome P-450-catalyzed oxidation reactions. The oxidative O-dealkylation of 7-ethoxycoumarin, *J. Biol. Chem.* **259**:3000–3004. (f) Traylor, T. G., Hill, K. W., Fann, W.-P., Tsuchiya, S., and Dunlap, B. E., 1992, Aliphatic hydroxylation catalyzed by iron(III) porphyrins, *J. Am. Chem. Soc.* **114**:1308–1312. (g) Fish, K. M., Avaria, G. E., and Groves, J. T., 1988, Rearrangement of alkyl hydroperoxides mediated by cytochrome P-450: Evidence for the oxygen rebound mechanism, in: *Microsomes and Drug Oxidations* (J. O. Miners, D. J. Birkett, R. Drew, B. K. May, and M.E. McManus, eds.), Taylor & Francis, London, pp.176–183. (h) Vaz, A. D. N., and Coon, M. J., 1994, On the mechanism of action of cytochrome P450: Evaluation of hydrogen abstraction in oxygen-dependent alcohol oxidation, *Biochemistry* **33**:6442–6449.

19. (a) Griller, D., and Ingold, K. U., 1980, Free-radical clocks, *Acc. Chem. Res.* **13**:317–323. (b) Ortiz de Montellano, P. R., and Stearns, R. A., 1987, Timing of the radical recombination step in cytochrome P-450 catalysis with ring-strained probes, *J. Am. Chem. Soc.* **109**:3415–3420. (c) Bowry, V. W., Lusztyk, J., and Ingold, K. U., 1989, Calibration of the bicyclo[2.1.0]pent-2-yl radical ring opening and an oxygen rebound rate constant for cytochrome P-450, *J. Am. Chem. Soc.* **111**:1927–1928. (d)

Bowry, V. W., Lusztyk, J., and Ingold, K. U., 1991, Calibration of a new horologery of fast radical clocks. Ring-opening rates for ring- and α-alkyl-substituted cyclopropylcarbinyl radicals and for the bicyclo[2.1.0]pent-2-yl radical, *J. Am. Chem. Soc.* **113**:5687–5698. (e) Bowry, V. W., and Ingold, K. U., 1991, A radical clock investigation of microsomal cytochrome P-450 hydroxylation of hydrocarbons. Rate of oxygen rebound, *J. Am. Chem. Soc.* **113**:5699–5707.

20. Stearns, R. A., and Ortiz de Montellano, P. R., 1985, Cytochrome P-450 catalyzed oxidation of quadricyclane. Evidence for a radical cation intermediate, *J. Am. Chem. Soc.* **107**:4081–4082.

21. (a) Gassman, P. G., and Yamaguchi, R., 1982, Electron transfer from highly strained polycyclic molecules, *Tetrahedron* **38**:1113–1122. (b) Meinwald, J., Labana, S. S., and Chadha, M. S., 1963, Peracid reactions. III. The oxidation of bicyclo[2.2.1]heptadiene, *J. Am. Chem. Soc.* **85**:582.

22. Baciocchi, E., Crescenzi, M., and Lanzalunga, O., 1990, Hydrogen atom transfer versus electron transfer in iron(III) porphyrin catalyzed benzylic oxidations, *J. Chem. Soc. Chem. Commun.* **1990**:687–688.

23. (a) Inchley, P., Lindsay Smith, J. R., and Lower, R. J., 1989, Model systems for cytochrome P450 dependent monooxygenases. Part 6. The hydroxylation of saturated carbon–hydrogen bonds with etraphenylporphyrinatoiron(III) chloride and iodosylbenzene, *New J. Chem.* **13**:669–676. (b) Khanna, R. K., Sutherlin, J. S., and Lindsey, D., 1990, Mechanisms in a biomimetic hydroxylation of a chemical probe: 5-Nitroacenaphthene, *J. Org. Chem.* **26**:6233–6234.

24. (a) Dobson, J. C., Seok, W. K., and Meyer, T. J., 1986, Epoxidation and catalytic oxidation of olefins based on a RU^{IV}=O/Ru^{II}=OH$_2$ couple, *Inorg. Chem.* **25**:1513. (b) Groves, J. T., Han, Y., and Van Engen, D., 1990, Co-ordination of styrene oxide to a sterically hindered ruthenium(II) porphyrin, *J. Chem. Soc. Chem. Commun.* **1990**:436–437.

25. (a) Ostovic, D., and Bruice, T. C., 1989, Intermediates in the epoxidation of alkenes by cytochrome P-450 models. 5. Epoxidation of alkenes catalyzed by a sterically hindered (meso-tetrakis(2,6-dibromophenyl)porphinato)iron(III)chloride, *J. Am. Chem. Soc.* **111**:6511–6517. (b) Bruice, T. C., 1988, The mechanisms of oxygen transfer from acyl and alkyl hydroperoxides to metal(III) porphyrins and the epoxidation of alkenes by the resultant hypervalent metal-oxo porphyrin products, *Aldrichimica Acta* **21**:87–94. (c) White, P. W., 1990, Mechanistic studies and selective catalysis with cytochrome P-450 model systems, *Bioorg. Chem.* **18**:440–456. (d) Ostovic, D., and Bruice, T. C., 1992, Mechanism of alkene epoxidation by iron, chromium, and manganese higher valent oxo-metalloporphyrins, *Acc. Chem. Res.*, **25**:314–320. (e) Arasasingham, R. D., He, G.-X., and Bruice, T. C., 1993, Mechanism of manganese porphyrin-catalyzed oxidation of alkenes. Role of manganese(IV)-oxo species, *J. Am. Chem. Soc.* **115**:7985–7991.

26. Ortiz de Montellano, P. R., Mangold, B. L. K., Wheeler, C., Kunze, K. L., and Reich, N. O., 1983, Stereochemistry of cytochrome P-450-catalyzed epoxidation and prosthetic heme alkylation, *J. Biol. Chem.* **258**:4208–4213.

27. (a) Guengerich, F. P., and Macdonald, T. L., 1984, Chemical mechanisms of catalysis by cytochromes P-450: A unified view, *Acc. Chem. Res.* **17**:9–16. (b) Liebler, D. C., and Guengerich, F. P., 1983, Olefin oxidation by cytochrome P-450: Evidence for group migration in catalytic intermediates formed with vinylidene chloride and trans-1-phenyl-1-butene, *Biochemistry* **22**:5482–5489. (c) Miller, R. E., and Guengerich, F. P., 1982, Oxidation of trichloroethylene by liver microsomal cytochrome P-450: Evidence for chlorine migration in a transition state not involving trichloroethylene oxide, *Biochemistry* **21**:1090–1097.

28. (a) Mansuy, D., Leclaire, J., Fontecave, M., and Momenteau, M., 1984, Oxidation of monosubstituted olefins by cytochromes P-450 and heme models: Evidence for the formation of aldehydes in addition to epoxides and allylic alcohols, *Biochem. Biophys. Res. Commun.* **119**:319–325. (b) Wistuba, D., Nowotny, H.-P., Trager, O., and Schurig V., 1989, Cytochrome P-450-catalyzed asymmetric epoxidation of simple prochiral and chiral aliphatic alkenes: Species dependence and effect of enzyme induction on enantioselective oxirane formation, *Chirality* **1**:127–136. (c) Ortiz de Montellano, P. R., Fruetel, J. A., Collins, J. R., Camper, D. L., and Loew, G. H., 1991, Theoretical and experimental

analysis of the absolute stereochemistry of cis-β-methylstyrene epoxidation by cytochrome P450$_{cam}$, *J. Am. Chem. Soc.* **113**:3195–3196.

29. Bruice, T. C., and Castellino, A. J., 1988, Intermediates in the epoxidation of alkenes by cytochrome P-450 models. 2. Use of the trans-2,trans-3-diphenylcyclopropyl substituent in a search for radical intermediates, *J. Am. Chem. Soc.* **110**:7512–7519.

30. Garrison, J. M., Ostovic, D., and Bruice, T. C., 1989, Is a linear relationship between the free energies of activation and one-electron oxidation potential evidence for one-electron transfer being rate determining? Intermediates in the epoxidation of alkenes by cytochrome P-450 models. 4. Epoxidation of a series of alkenes by oxo(meso-tetrakis(2,6-dibromophenyl)porphinato)chromium(V), *J. Am. Chem. Soc.* **111**:4960–4966.

31. (a) Groves, J. T., and Watanabe, Y., 1986, On the mechanism of olefin epoxidation by oxo-iron porphyrins, *J. Am. Chem. Soc.* **108**:507–508. (b) Lindsay Smith, J. R., and Sleath, P. R., 1982, Model systems for cytochrome P450 dependent mono-oxygenases. Part 1. Oxidation of alkenes and aromatic compounds by tetraphenylporphinatoiron(III) chloride and iodosylbenzene, *J. Chem. Soc. Perkin Trans. II* **1982**:1009–1015. (c) Bortolini, O., and Meunier, B., 1984, Enhanced selectivity by an 'open-well effect' in a metalloporphyrin-catalyzed oxygenation reaction, *J. Chem. Soc. Perkin Trans. II,* **1984**:1967. (d) Traylor, T. G., and Xu, F., 1988, Model reactions related to cytochrome P-450. Effects of alkene structure on the rates of epoxide formation, *J. Am. Chem. Soc.* **110**:1953–1958. (e) Samsel, E. G., Srinivasan, K., and Kochi, J. K., 1985, Mechanism of the chromium-catalyzed epoxidation of olefins. Role of oxochromium(V) cations, *J. Am. Chem. Soc.* **107**:7606–7617.

32. Seyferth, D., Mui, J. Y.-P., and Damrauer, R., 1968, Halomethyl-metal compounds. XIX. Further studies of the aryl(bromodichloromethyl)mercury-olefin reaction, *J. Am. Chem. Soc.* **90**:6182–6186.

33. Nishimura, J., Furukawa, J., Kawabata, N., and Kitayama, M., 1971, Relative reactivity of olefins in cycloaddition with zinc carbenoid, *Tetrahedron* **27**:1799–1806.

34. Ogata, Y., and Tabushl, I., 1961, Kinetics of the epoxidation of substituted α-methylstilbenes, *J. Am. Chem. Soc.* **83**:3440.

35. Schubert, W. M., and Keefe, J. R., 1972, Acid-catalyzed hydration of styrenes, *J. Am. Chem. Soc.* **94**:559–566.

36. Yates, K., McDonald, R. S., and Shapiro, S. A., 1973, Kinetics and mechanisms of electrophilic addition. I. Comparison of second- and third- order brominations, *J. Org. Chem.* **38**:2460–2464.

37. (a) Traylor, T. G., and Miksztal, A. R., 1989, Alkene epoxidations catalyzed by iron(III), manganese(III), and chromium(III) porphyrins. Effects of metal and porphyrin substituents on selectivity and regiochemistry of epoxidation, *J. Am. Chem. Soc.* **111**:7443–7448. (b) Traylor, T. G., Tsuchiya, S., Byun, Y. S., and Kim, C. 1993, High-yield epoxidations with hydrogen peroxide and *tert*-butyl hydroperoxide catalyzed by iron(III) porphyrins: Heterolytic cleavage of hydroperoxides, *J. Am. Chem. Soc.* **115**:2775–2781.

38. Traylor, T. G., Nakano, T., Dunlap, B. E., and Traylor, P. S., 1986, Mechanisms of hemin-catalyzed alkene epoxidation. The effect of catalyst on the regiochemistry of epoxidation, *J. Am. Chem. Soc.* **108**:2782–2784.

39. (a) Collman, J. P., Kodadek, T., Raybuck, S. A., and Meunier, B., 1983, Oxygenation of hydrocarbons by cytochrome P-450 model compounds: Modification of reactivity by axial ligands, *Proc. Natl. Acad. Sci. USA* **80**:7039–7044. (b) Collman, J. P., Brauman, J. I., Meunier, B., Hayashi, T., Kodadek, T., and Raybuck, S. A., 1985, Epoxidation of olefins by cytochrome P-450 model compounds: Kinetics and stereochemistry of oxygen atom transfer and origin of shape selectivity, *J. Am. Chem. Soc.* **107**:2000–2005. (c) Collman, J. P., Kodadek, T., and Brauman, J. I., 1986, Oxygenation of styrene by cytochrome P-450 model systems. A mechanistic study, *J. Am. Chem. Soc.* **108**:2588–2594. (d) Collman, J. P., Brauman, J. I., Hampton, P. D., Tanaka, H., Bohle, D. S., and Hembre, R. T., 1990, Mechanistic studies of olefin epoxidation by a manganese porphyrin and hypochlorite: An alternative explanation of saturation kinetics, *J. Am. Chem. Soc.* **112**:7980–7984, and references therein.

40. Watanabe,Y., and Groves, J. T., 1988, Oxygen activation by metalloporphyrins, heterolytic and homolytic O–O bond cleavage reactions of (acylperoxo)manganese(III) porphyrins, in: *Studies in Organic Chemistry. The Role of Oxygen in Chemistry and Biochemistry*, Vol. 33 (W. Ando and Y. Moro-Oka, eds.) Elsevier, Amsterdam, pp. 471–476.

41. Sharpless, B., Teranishi, A. Y., and Bäckvall, J. E., 1977, Chromyl chloride oxidations of olefins. Possible role of organometallic intermediates in the oxidations of olefins by oxo transition metal species, *J. Am. Chem. Soc.* **99**:3120–3128.

42. (a) Grubbs, R. H., 1978, The olefin metathesis reaction, *Prog. Inorg. Chem.* **24**:1–50. (b) Grubbs, R. H., and Tumas, W., 1989, Polymer synthesis and organotransition metal chemistry, *Science* **243**:907–915. (c) Schrock, R. R., 1990, Living ring-opening metathesis polymerization catalyzed by well-characterized transition-metal alkylidene complexes, *Acc. Chem. Res.* **23**:158–165.

43. (a) Tjaden, E. B., and Stryker, J. M., 1990, Nucleophilic addition of enolates to the central carbon of transition-metal η 3-allyl complexes. Metallacyclobutane formation, reversibility of nucleophilic addition, and synthesis of α cyclopropyl ketones, *J. Am. Chem. Soc.* **112**:6420–6422. (b) Ivin, K. J., Rooney, J. I., Stewart, C. D., Green, M. L. H., and Mahtab, R., 1978, Mechanism for the stereospecific polymerization of olefins by Ziegler–Natta catalysts, *J. Chem. Soc., Chem. Commun.* **1978**:604–606. (c) Brookhart, M. H., Timmers, D., Tucker, J. R., Williams, G. D., Husk, G. R., Brunner, H., and Hammer, B., 1983, Enantioselective cyclopropane synthesis using the chiral carbene complexes (SFeSC)- and (RFeSC)-(C5H5)(CO)(Ph2R*P)Fe:CHCH3+ (R* = (S)-2-methyl-butyl). Role of metal vs. ligand chirality in the optical induction, *J. Am. Chem. Soc.* **105**:6721–6723. (d) Yang, G. K., and Bergman, R. G., 1983, Characterization and evidence for alkylation of hydridodicarbonylcyclopentadienylrhenate(I) ion [CpRe(CO)2H]⁻ in the conversion of dihydrodicarbonylcyclopentadienylrhenium [CpRe(CO)2H2] to CpRe(CO)2R2. Synthesis of a rhenacyclopentane and its thermolysis to methylcyclopropane, *J. Am. Chem. Soc.* **105**:6500–6501. (e) Klein, D. P., Hayes, J. C., and Bergman, R. G., 1988, Insertion of (η⁵-C5Me5)(PMe5)Ir into the carbon–hydrogen bonds of functionalized organic molecules: A C-H activation route to 2-oxa- and 2-azametallacyclobutanes, potential models for olefin oxidation intermediates, *J. Am. Chem. Soc.* **110**:3704–3706. (f) Hayasi, Y., and Schwartz, J., 1981, Reaction between epoxides and β-diketonate complexes of low-valent vanadium and molybdenum, *Inorg. Chem.* **20**:3473. (g) Lenarda, M., Pahor, N. B., Calligaris, M., Graziani, M., and Radaccio, L., 1978, Synthesis and crystal structure of 3,3,4-tricyano-2,2-bis(triphenyl-phosphine)-1-oxa-2-platinacyclobutane, *J. Chem. Soc. Dalton Trans.* **1978**:279–282. (h) Osborne, R. B., and Ibers, J. A., 1982, The reactions of platinum(0) and palladium(0) tertiary phosphine complexes with phenyl dicyanooxiranes, *J. Organomet. Chem.* **232**:371–385. (i) Schlodder, R., Ibers, J. A., Lenarda, M., and Graziani, M., 1974, Structure and mechanism of formation of the metallooxacyclobutane complex bis(triphenylarsine)tetracyanooxiraneplatinum, the product of the reaction between tetracyanooxirane and tetrakis (triphenylarsine)platinum, *J. Am. Chem. Soc.* **96**:6893–6900. (j) Su, F.-M., Cooper, C., Geib, S. J., Rheingold, A. L., and Mayer, J. M., 1986, Synthesis and characterization of high-valent oxo olefin and oxo carbonyl complexes. Crystal and molecular structure of W(O)Cl2(CH2:CH2)(PMePh2)2, *J. Am. Chem. Soc.* **108**:3545–3547. (k) Bryan, J. C., Geib, S. J., Rheingold, A. L., Mayer, J. A., 1987, Oxidative addition of carbon dioxide, epoxides, and related molecules to WCl2(PMePh2)⁴ yielding tungsten(IV) oxo, imido, and 8 sulfido complexes. Crystal and molecular structure of W(O)Cl2(CO)(PMePh2)², *J. Am. Chem. Soc.* **109**:2826–2828. (l) Atagi, L. M., Over, D. E., McAlister, D. R., and Mayer, J. A., 1991, On the mechanism of oxygen-atom or nitrene-group transfer in reactions of epoxides and aziridines with tungsten(II) compounds, *J. Am. Chem. Soc.* **113**:870–874.

44. (a) Groves, J. T., Avaria-Neisser, G. E., Fish, K. M., Imachi, M., and Kuczkowski, R. L., 1986, Hydrogen–deuterium exchange during propylene epoxidation by cytochome P-450, *J. Am. Chem. Soc.* **108**:3837–3838. (b) Groves, J. T., Fish, K. M., Avaria-Neisser, G. E., Imachi, M., and Kuczkowski, R. L., 1988, A unique deuterium/proton exchange during cytochrome P-450 mediated epoxidation of propene and butene, *Prog. Clin. Biol. Res.* **274**:509–524.

45. (a) Castellino, A. J., and Bruice, T. C., 1988, Intermediates in the epoxidation of alkenes by cytochrome P-450 models. 1. cis-Stilbene as a mechanistic probe, *J. Am. Chem. Soc.* **110**:158–162. (b) Mirafzal, G. A., Kim, T., Liu, J., and Bauld, N. L., 1992, Cation radical probes. Development and application to metalloporphyrin-catalyzed epoxidation, *J. Am. Chem. Soc.* **114**:10968–10969.

46. (a) Komives, E. A., and Ortiz de Montellano, P. R., 1987, Mechanism of oxidation of π bonds by cytochrome P-450. Electronic requirements of the transition state in the turnover of phenylacetylenes, *J. Biol. Chem.* **262**:9793–9802. (b) Kunze, K. L., Mangold, B. L. K., Beilan, H. S., Ortiz de

Montellano, P. R., 1983, The cytochrome P-450 active site. Regiospecificity of prosthetic heme alkylation by olefins and acetylenes, *J. Biol. Chem.* **258**:4202–4207. (c) Ortiz de Montellano, P. R., Beilan, H. S., Kunze, K. L., and Mico, B. A., 1981, Destruction of cytochrome P-450 by ethylene. Structure of the resulting prosthetic heme adduct, *J. Biol. Chem.* **1981**:4395–4399. (d) Luke, B. T., Collins, J. R., Loew, G. H., and Mclean, A. D., 1990, Theoretical investigations of terminal alkenes as putative suicide substrates of cytochrome P-450, *J. Am. Chem. Soc.* **112**:8686–8691. (e) Mashiko, T., Dolphin, D., Nakano, T., and Miksztal, A. R., 1985, N-Alkyl-porphyrin formation during the reactions of cytochrome P-450 model systems, *J. Am. Chem. Soc.* **107**:3735–3736.

47. Ahn, K.-H., and Groves, J. T., 1994, Shape-selective oxygen transfer to olefins catalyzed by sterically hindered iron porphyrins, *Korean J. Chem.* **15**(11):957–961.

48. Groves, J. T., Ahn, K.-H., and Quinn, R., 1988, Cis-trans isomerization of epoxides catalyzed by ruthenium(II) porphyrins, *J. Am. Chem. Soc.* **110**:4217–4220.

49. (a) Collman, J. P., Hampton, P. D., and Brauman, J. I., 1990, Suicide inactivation of cytochrome P-450 model compounds by terminal olefins. 1. A mechanistic study of heme N-alkylation and epoxidation, *J. Am. Chem. Soc.* **112**:2977–2986. (b) Collman, J. P., Hampton, P. D., and Brauman, J. I., 1990, Suicide inactivation of cytochrome P-450 model compounds by terminal olefins. 2. Steric and electronic effects in heme N-alkylation and epoxidation, *J. Am. Chem. Soc.* **112**:2986–2998. (c) Meunier, B., 1988, Are intermediates with a metal–carbon bond involved in oxygenation reactions catalyzed by metalloporphyrins? *Gazz. Chim. Ital.* **118**:485–493. (d) Dolphin, D., Matsumoto, A., and Shortman, C., 1989, β-Hydroxy-alkyl σ-metallophyrins. Models for epoxide and alkene generation from cytochrome P-450, *J. Am. Chem. Soc.* **111**:411–413. (e) Nakano, T., Traylor, T. G., and Dolphin, D., 1990, The formation of N-alkyl-porphyrins during epoxidation of ethylene catalyzed by iron(III) meso-tetrakis(2,6-dichlorophenyl)porphyrin, *Can. J. Chem.* **68**:1504–1506. (f) Mashiko, T., Dolphin, D., Nakano, T., and Traylor, T. G., 1985, N-Alkylporphyrin formation during the reactions of cytochrome P-450 model systems, *J. Am. Chem. Soc.* **107**:3735–3736.

50. Collman, J. P., Zhang, X., Lee, V. J., Uffelman, E. S., and Brauman, J. I., 1993, Regioselective and enantioselective epoxidation catalyzed by metalloporphyrins, *Science* **261**:1404–1411.

51. (a) Meunier, B., Carvalho, M. E., Bortolini, O., and Momenteau, M., 1988, Proximal effect of the nitrogen ligands in the catalytic epoxidation of olefins by the sodium hypochlorite/manganese(III) porphyrin system, *Inorg. Chem.* **27**:161–164. (b) Creager, S. E., and Murray, R. W., 1987, Electrochemical reactivity of manganese(II) porphyrins. Effects of dioxygen, benzoic anhydride, and axial ligands, *Inorg. Chem.* **26**:2612–2618. (c) Battioni, J. P., Renaud, J. F., Bartoli, J. F., Reina-Artiles, M., Fort, M., and Mansuy, D., 1988, Monooxygenase-like oxidation of hydrocarbons by hydrogen peroxide catalyzed by manganese porphyrins and imidazole: Selection of the best catalytic system and nature of the active oxygen species, *J. Am. Chem. Soc.* **110**:8462–8470. (d) Nappa, M. J., and McKinney, R. J., 1988, Selectivity control by axial ligand modification in manganese porphyrin-catalyzed oxidations, *Inorg. Chem.* **27**:3740–3745. (e) Nappa, M. J., and Tolman, C. A., 1985, Steric and electronic control of iron porphyrin catalyzed hydrocarbon oxidations, *Inorg. Chem.* **24**:4711–4719.

52. Champion, P. M., 1989, Elementary electronic excitations and the mechanism of cytochrome P450, *J. Am. Chem. Soc.* **111**:3433–3434.

53. (a) Dawson, J., and Sono, M., 1987, Cytochrome P-450 and chloroperoxidase: Thiolate-ligated heme enzymes. Spectroscopic determination of their active-site structures and mechanistic implications of thiolate ligation, *Chem. Rev.* **87**:1255–1276. (b) Dawson, J., 1988, Probing structure–function relations in heme-containing oxygenases and peroxidases, *Science* **240**:433–439.

54. Chang, C. K., and Kuo, M.-S., 1979, Reaction of iron(III) porphyrins and iodosoxylene. The active oxene complex of cytochrome P-450, *J. Am. Chem. Soc.* **101**:3413–3415.

55. Grieco, P. A., and Stuk, T. L., 1990, Remote oxidation of unactivated carbon–hydrogen bonds in steroids via oxometalloporphinates, *J. Am. Chem. Soc.* **112**:7799–7801.

56. Breslow, R., Brown, A. B., McCullough, R. D., and White, P. W., 1989, Substrate selectivity in epoxidation by metalloporphyrin and metallosalen catalysts carrying binding groups, *J. Am. Chem. Soc.* **111**:4517–4518.

57. (a) Cook, B. R., Reinert, T. J., and Suslick, K. S., 1986, Shape-selective alkane hydroxylation by metalloporphyrin catalysts, *J. Am. Chem. Soc.* **108**:7281–7286. (b) Suslick, K. S., and Cook, B. R.,

1987, Regioselective epoxidations of dienes with manganese(III) porphyrin catalysts, *J. Chem. Soc. Chem. Commun.* **1987**:200–202.

58. (a) Collman, J. P., Zhang, X., Hembre, R. T., and Brauman, J. I., 1990, Shape-selective olefin epoxidation catalyzed by manganese picnic basket porphyrins, *J. Am. Chem. Soc.* **112**:5356–5357. (b) Collman, J. P., Lee, V. J., Zhang, X., Ibers, J. A., and Brauman, J. I., 1993, Enantioselective epoxidation of unfunctionalized olefins catalyzed by threitol-strapped manganese porphyrins, *J. Am. Chem. Soc.* **115**:3834–3835.

59. (a) Groves, J. T., and Neumann, R., 1989, Regioselective oxidation catalysis in synthetic phospholipid vesicles. Membrane spanning steroidal metallo-porphyrins, *J. Am. Chem. Soc.* **111**:2900–2909. (b) Groves, J. T., and Neumann, R., 1988, Enzymic regioselectivity in the hydroxylation of cholesterol catalyzed by a membrane spanning metalloporphyrin, *J. Org. Chem.* **53**:3891–3893. (c) Groves, J. T., and Neumann, R., 1987, Membrane-spanning steroidal metalloporphyrins as site-selective catalysts in synthetic vesicles, *J. Am. Chem. Soc.* **109**:5045–5047.

60. Groves, J. T., and Myers, R. S., 1983, Catalytic asymmetric epoxidation with chiral iron porphyrins, *J. Am. Chem. Soc.* **105**:5791–5796.

61. Mansuy, D., Battioni, P., Renaud, J.-P., and Guerin, P., 1985, Asymmetric epoxidation of alkenes catalyzed by a basket-handle iron-porphyrin bearing amino acids, *J. Chem. Soc. Chem. Commun.* **1985**:155–156.

62. O'Malley, S., and Kodadek, T., 1989, Synthesis and characterization of the "chiral wall" porphyrin: A chemically robust ligand for metal-catalyzed asymmetric epoxidations, *J. Am. Chem. Soc.* **111**:9116–9117.

63. Naruta, Y., Tani, F., and Maruyama, K., 1989, Synthesis of chiral "twin coronet" porphyrins and catalytic and asymmetric epoxidation of olefins, *Chem. Lett.* **1989**:1269–1272.

64. (a) Groves, J. T., and Viski, P., 1990, Asymmetric hydroxylation, epoxidation, and sulfoxidation catalyzed by vaulted binaphthyl metalloporphyrins, *J. Org. Chem.* **55**:3628–3634. (b) Groves, J. T., and Viski, P., 1989, Asymmetric hydroxylation by a chiral iron porphyrin, *J. Am. Chem. Soc.* **111**:8537–8538.

65. Sheldon, R. A., and Kochi, J. K., 1981, *Metal Catalyzed Oxidations of Organic Compounds,* Academic Press, New York.

66. Koola, J. D., and Kochi, J. K., 1987, Nickel catalysis of olefin epoxidation, *Inorg. Chem.* **26**:908–916. (b) Samsel, E. G., Srinivasan, K., and Kochi, J. K., 1985, Mechanism of the chromium-catalyzed epoxidation of olefins. Role of oxochromium(V) cations, *J. Am. Chem. Soc.* **107**:7606–7617. (c) Srinivasan, K., and Kochi, J. K., 1985, Synthesis and molecular structure of oxochromium(V) cations. Coordination with donor ligands, *Inorg. Chem.* **24**:4671–4679.

67. (a) Jorgensen, K. A., 1989, Transition-metal-catalyzed epoxidations, *Chem. Rev.* **89**:431. (b) Holm, R., 1987, Metal-centered oxygen atom transfer reactions, *Chem. Rev.* **87**:1401–1449.

68. (a) Collins, T. J., and Gorden-Wylie, S. W., 1990, Enantioselective epoxidation of unfunctionalized olefins catalyzed by salen manganese complexes. *J. Am. Chem. Soc.* **112**:2801–2803. (b) Ozaki, S., Mimura, H., Yasuhara, N., Masui, M., Yamagata, Y., Yomita, K., and Collins, T. J., 1990, Synthesis of chiral square planar cobalt(III) complexes and catalytic asymmetric epoxidations with these complexes, *J. Chem. Soc. Perkin Trans. II* **1990**:353–360. (c) Kinnery, J. F., Albert, J. S., and Burrows, C. J., 1988, Mechanistic studies of alkene epoxidation catalyzed by nickel(II) cyclam complexes. Oxygen-18 labeling and substituent effects, *J. Am. Chem. Soc.* **110**:6124–6129. (d) Leung. W.-H., and Che, C.-M., 1989, Oxidation chemistry of ruthenium–salen complexes, *Inorg. Chem.* **28**:4619–4622.

69. (a) Zhang, W., Loebach, J. L., Wilson, S. R., and Jacobsen, E. N., 1989, A manganese(V)-oxo complex, *J. Am. Chem. Soc.* **111**:4511–4513. (b) Chang, S., Galvin, J. M., and Jacobsen, E. N., 1994, Effect of chiral quaternary ammonium salts on (salen)Mn-catalyzed epoxidation of cis-olefins. A highly enantioselective, catalytic route to trans-epoxides, *J. Am. Chem. Soc.* **116**:6937–6938. (c) Brandes, B. D., and Jacobsen, E. N., 1994, Highly ennantioselective, catalytic epoxidation of trisubstituted olefins, *J. Org. Chem.* **59**: 4378–4380. (d) Deng, L., and Jacobsen, E. N., 1992, A practical, highly enantioselective synthesis of the taxol side chain via asymmetric catalysis, *J. Org. Chem.* **57**:4320–4323. (e) Zhang, W., and Jacobsen, E. N., 1991, Asymmetric olefin epoxidation with sodium hypochlorite catalyzed by easily prepared chiral Mn(III) salen complexes, *J. Org. Chem.* **56**:2296–2298.

70. Marletta, M. A., 1993, Nitric oxide synthase structure and mechanism, *J Biol. Chem.* **268**:12231.
71. Stuehr, D. J., Kwon, N. S., Nathan, C. F., and Griffith, O.W., 1991, N^{ω}-hydroxy-L-arginine is an intermediate in the biosynthesis of nitric oxide from L-arginine, *J. Biol. Chem.* **266**:6259.
72. (a) White, K. A., and Marletta, M. A., 1992, Nitric oxide synthase is a cytochrome P-450 type hemoprotein, *Biochemistry* **31**:6627. (b) McMillan, K., Bredt, D. S., Hirsch, D. J., Snyder, S. H., Clark, J. E., and Masters, B. S. S., 1992, Cloned, expressed rat cerebellar nitric oxide synthase contains stoichiometric amounts of heme, which binds carbon monoxide, *Proc. Natl. Acad. Sci. USA* **89**:1141–1145.
73. Parli, C. J., Wang, N., and McMahon, R. E., 1971, The enzymatic N-hydroxylation of an imine, *J. Biol. Chem.* **246**:6953.
74. Korth, H.-G., Sustmann, R., Thater, C., Butler, A. R., and Ingold, K. U., 1994, On the mechanism of the nitric oxide synthase-catalyzed conversion of N^{ω}-hydroxy-L-arginine to citrulline and nitric oxide, *J. Biol. Chem.* **269**:17776–17779.
75. Vaz, A. D. N., Roberts, E. S., and Coon, M. J., 1991, Olefin formation in the oxidative deformylation of aldehydes by cytochrome P-450. Mechanistic implication for catalysis by oxygen-derived peroxide, *J. Am. Chem. Soc.* **113**:5886.
76. Watanabe, Y., and Ishimura, Y., 1989, A model study on aromatase cytochrome P-450 reaction: Transformation of androstene-3,17,19-trione to 10b-hydroxyester-4-ene-3,17-dione, *J. Am. Chem. Soc.* **111**:8047.
77. Klatt, P., Schmidt, G. U., and Mayer, B., 1993, Multiple catalytic function of brain nitric oxide synthase, *J. Biol. Chem.* **268**:14781.
78. Fukuto, J. M., Stuehr, D. J., Feldman, P. L., Bova, M. P., and Wong, P., 1993, Peracid oxidation of an N-hydroxyguanidine compound: A chemical model for the oxidation of N^{ω}-hydroxy-L-arginine by nitric oxide synthase, *J. Med. Chem.* **36**:2666.
79. Boucher, J. L., Genet, A., Vadon, S., Delaforge, M., and Mansuy, D., 1992, Formation of nitric oxides and citrulline upon oxidation of N^{ω}-hydroxyl-L-arginine by hemeprotein, *Biochem. Biophys. Res. Commun.* **184**:1158.
80. Andronik-Lion, V., Boucher, J. L., Delaforge, M., and Mansuy, D., 1992, Formation of nitric oxide by cytochrome P450-catalyzed oxidation of aromatic amidoximes, *Biochem. Biophys. Res. Commun.* **184**:452.
81. Nishinaga, A., Yamazaki, S., Miwa, T., and Matsuura, T., 1991, Co(salen) catalyzed oxidation of oximes with t-butyl hydroperoxide, *React. Kinet. Catal. Lett.* **43**:273.
82. (a) Babcock, G. T., and Varotsis, C., 1993, Discrete steps in dioxygen activation—The cytochrome oxidase/O2 reaction, *J. Bioenerg. Biomembr.* **25(2)**:71–80.(b) Han, S., Ching, Y.-c., and Rousseau, D. L., 1990, Ferryl and hydrox intermediates in the reaction of oxygen with reduced cytochrome c oxidase, *Nature* **348**:89–90. (c) Varotsis, C., Zhang, Y., Appelman, E. H., and Babcock, G. T., 1993, Resolution of the reaction sequence during the reduction of O2 by cytochrome oxidase, *Proc. Natl. Acad. Sci. USA* **90**:237–241.
83. (a) DeRose, V. J., Liu, K. E., Kurtz, D. M., Jr., Hoffman, B. M., and Lippard, S. J., 1993, Proton ENDOR identification of bridging hydroxide ligands in mixed-valent diiron centers of proteins: Methane monooxygenase and semimet azidohemerythrin, *J. Am. Chem. Soc.* **115**:6440. (b) Lee, S.-K., Fox, B. G., Froland, W. A., Lipscomb, J. D., and Münck, E., 1993, A transient intermediate of the methane monooxygenase catalytic cycle containing an $Fe^{IV}Fe^{IV}$ cluster, *J. Am. Chem. Soc.* **115**:6450. (C) Fox, B. G., Hendrich, M. P., Surerus, K. K., Andersson, K. K., Froland, W. A., Lipscomb, J. D., and Münck, E.,1993, Mössbauer, EPR, and ENDOR studies of the hydroxylase and reductase components of methane monooxygenase from *Methylosinus trichosporium* OB3b , *J. Am. Chem. Soc.* **115**:3688.

Comparison of the Peroxidase Activity of Hemoproteins and Cytochrome P450

LAWRENCE J. MARNETT and TODD A. KENNEDY

> *Osric:* You are not ignorant of what excellence Laertes is—
> *Hamlet:* I dare not confess that, lest I should compare with him in excellence; but to know a man well were to know himself.
> WILLIAM SHAKESPEARE,
> *The Tragedy of Hamlet, Prince of Denmark*

1. Introduction

Shakespeare recognized centuries ago the value of a foil for highlighting similarities and differences between characters. Comparative analysis is also an effective mechanism for studying related proteins to better understand the structural basis of their functions. Indeed, there are many similarities between peroxidases and cytochromes P450 (P450s) in terms of prosthetic group and catalytic mechanism. However, there are also important differences in three-dimensional structure, reactions catalyzed, and interactions with other proteins. As a result, these two classes of oxidizing enzymes evolved to fulfill very different functions. Peroxidases and P450s are hemeproteins; both react with peroxides to generate higher oxidation states; and both catalyze oxidation and oxygenation reactions that involve electron transfer from the substrate to the higher oxidation states. The major differences between peroxidases and P450s include the ability of P450s to accept electrons from a reductase; the redox potentials of the higher oxidation states; and the ability of P450s to generate the equivalent of a metal-bound peroxide by reduction of O_2 at the heme center during catalytic turnover.

LAWRENCE J. MARNETT and TODD A. KENNEDY • A.B. Hancock Jr. Memorial Laboratory for Cancer Research, Center in Molecular Toxicology, Departments of Biochemistry and Chemistry, Vanderbilt University School of Medicine, Nashville, Tennessee 37232.

Cytochrome P450: Structure, Mechanism, and Biochemistry (Second Edition), edited by Paul R. Ortiz de Montellano. Plenum Press, New York, 1995.

$$ROOH \; + \; XH_2 \longrightarrow ROH \; + \; X/XH_2O \tag{1}$$

The basis for the functional similarity of peroxidases and P450s is their ability to catalyze hydroperoxide-dependent substrate oxidations [reaction (1)].[1-5] Similarities in the regiochemistry and stereochemistry of NADPH-dependent and hydroperoxide-dependent oxidations catalyzed by P450s suggest that the oxidizing agent generated in both reactions is the same.[3] Since the hydroperoxide-dependent oxidations are mechanistically less complex than the NADPH-dependent oxidations because of the absence of electron transfers from P450 reductase, they have been intensively studied to shed light on the oxidation steps of the normal P450 catalytic cycle.

Hydroperoxide-dependent oxidations by peroxidases were investigated long before the discovery of P450s and have provided spectroscopic, mechanistic, and structural information that was useful in formulating hypotheses for the action of the latter. In fact, the ability of proteins to react with H_2O_2 to generate a stronger oxidant is one of the oldest known catalytic activities, having been first indirectly described in 1810 by Planche who reported the blueing of an alcoholic tincture of guaiac resin by various plant materials including an extract of horseradish root.[6] Peroxidases have been the subject of numerous reviews and an excellent two-volume set was published recently that provides a comprehensive picture of their structure and mechanism.[4,7-10] The current review will highlight advances that update the comparison of peroxidases and P450s published nine years ago.[5] Emphasis is placed on structural information that has become available in the intervening period.

Four peroxidase structures have been solved recently of which two are of mammalian proteins.[11-15] This brings the total available peroxidase structures to six (counting the peroxide-cleaving enzyme catalase).[16,17] The two mammalian peroxidases are myeloperoxidase (MPO), which plays an important role in the killing of invading pathogens by

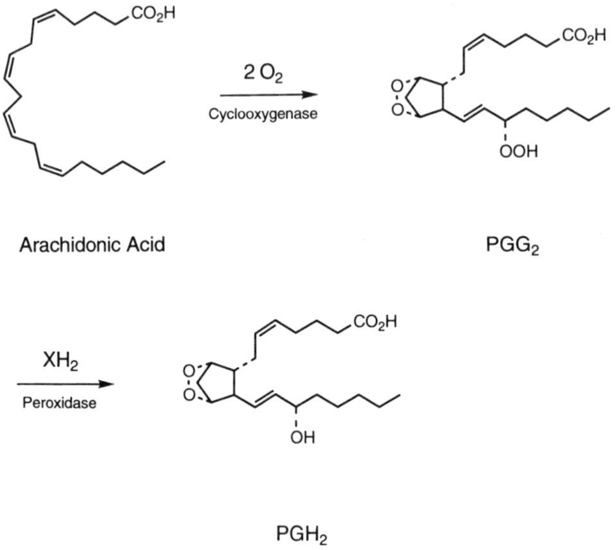

Arachidonic Acid PGG_2

$$\tag{2}$$

PGH$_2$

neutrophils,[18] and prostaglandin endoperoxide synthase (PGH synthase), which catalyzes the first two steps in the biosynthesis of prostaglandins, thromboxanes, and prostacyclin [reaction (2)].[19,20] The first step catalyzed by PGH synthase is the oxygenation of arachidonic acid to the hydroperoxy endoperoxide, PGG_2 (cyclooxygenase activity), and the second is reduction of PGG_2 to the hydroxy endoperoxide, PGH_2 (peroxidase activity).[19,20] PGH synthase is a membrane protein and demonstrates a novel mechanism of membrane insertion that appears to differ from the mechanism employed by microsomal or mitochondrial P450s.[14]

2. Primary Structure

A number of peroxidase sequences have been reported including several from mammalian sources. These include the human forms of MPO,[21,22] eosinophil peroxidase (EPO),[23] thyroid peroxidase (TPO),[24] and PGH synthase.[25–27] A partial sequence for human lactoperoxidase (LPO) is also available [28]; for sequence comparison, we have used the full-length sequence of bovine LPO.[29] MPO and EPO are derived from a precursor that undergoes proteolytic cleavage whereas TPO, LPO, and PGH synthase are composed of a single polypeptide. Excluding PGH synthase, the mammalian heme peroxidases exhibit ~45–70% sequence identity with each other.[11] According to the rules proposed for P450s, this degree of homology would place MPO, EPO, TPO, and LPO in the same family. MPO and EPO are similar enough (~70%) to be placed in the same subfamily.[23] The regions in the heme-binding sites and a calcium-binding loop are well conserved among MPO, EPO, TPO, and LPO (Tables I and II).[30] All peroxidases contain His as the proximal ligand to the heme iron and a His residue distal to the heme that is not ligated to the iron. The proximal ligand to the heme iron is a cysteine thiolate in P450s and there are no His residues in the distal region.

PGH synthase exhibits low homology to the other mammalian heme peroxidases (<20 %).[25–27] Although it contains His residues proximal and distal to the heme group, the sequence around the proximal His is not conserved between PGH synthase and the other mammalian peroxidases (Table I).[14] It is important to note that the peroxidase activity of PGHS is not merely related to the fact that it is a hemeprotein. Its peroxidase activity has a turnover number similar to those of highly efficient plant peroxidases and several orders of magnitude higher than those of hemeproteins that are not efficient peroxidases (such as myoglobin and P450).[31–36] Furthermore, the peroxidase activity is required to activate its

TABLE I
Sequences Containing Proximal Histidine Residues of Peroxidases

MPO	Ala	Phe	Arg	Tyr	Gly	His[a]	Thr	Leu	Ile
EPO	Ala	Phe	Arg	Phe	Gly	His	Thr	Met	Leu
LPO	Ala	Phe	Arg	Phe	Gly	His	Met	Glu	Val
TPO	Ala	Phe	Arg	Phe	Gly	His	Ala	Thr	Ile
PGHS	His	Leu	Tyr	His	Trp	His	Pro	Leu	Met

[a]Proximal His.

TABLE II
Sequences Containing Distal Histidine Residues of Peroxidases

MPO	Gly	Gln	Leu	Leu	Asp	His	Asp	Leu	Asp
EPO	Gly	Gln	Phe	Ile	Asp	His	Asp	Leu	Asp
LPO	Gly	Gln	Ile	Val	Asp	His	Asp	Leu	Asp
TPO	Gly	Gln	Tyr	Ile	Asp	His	Asp	Ile	Ala
PGHS	Ala	Gln	His	Phe	Thr	His	Gln	Phe	Phe
CCP	Val	Arg	Leu	Ala	Trp	His	Ile	Ser	Gly
LigPO	Leu	Arg	Met	Val	Phe	His	Asp	Ser	Ile

cyclooxygenase activity for maximal catalytic turnover and to protect the protein from hydroperoxide-induced decomposition.[31,37]

The only obvious sequence homology between PGH synthase and other heme peroxidases is a highly conserved decapeptide (Table III), which misled investigators as to its role in protein structure and function.[38] Because of the high degree of sequence conservation in this peptide and the presence of a His residue, it was proposed to participate in heme binding and the His was postulated to be the proximal ligand.[38] This hypothesis was strengthened by the finding that site-directed mutagenesis of the histidine in PGH synthase (His[309]) abolished heme binding.[39] However, solution of the crystal structures of canine MPO [11] and ovine PGH synthase [14] revealed that this histidine is not the proximal ligand to the heme iron. In fact, His[309] of PGH synthase and His[257] of MPO are located over 25 Å from the heme iron in each of the two proteins.

Although this region of mammalian peroxidases does not appear to have a direct role in enzyme function, it is very important for protein structure. As mentioned above, when His[309] of ovine PGH synthase is changed to Gln or Ala, the enzyme has no detectable cyclooxygenase or peroxidase activity and does not appear to bind heme.[39] In addition, when Cys[313] is mutated to Ser, the enzyme loses 90% of its cyclooxygenase activity but retains its sensitivity to cyclooxygenase inhibitors and its ability to dimerize.[40] Thus, changing a single sulfur atom to an oxygen drastically reduces catalytic activity presumably as the result of a long-range conformational effect. The highly conserved decapeptides of PGH synthase and MPO are at the top of sharp bends between two long α-helices that are the dominant structural features of both proteins (see below). Small changes in structure in this region are apparently translated to remote regions of the protein and affect heme binding and catalysis.

TABLE III
Highly Conserved Decapeptide Sequence of Mammalian Peroxidases

MPO	Thr	Leu	Leu	Leu	Arg	Glu	His	Asn	Arg	Leu
EPO	Thr	Leu	Phe	Met	Arg	Glu	His	Asn	Arg	Leu
LPO	Thr	Leu	Leu	Leu	Arg	Glu	His	Asn	Arg	Leu
TPO	Thr	Leu	Trp	Leu	Arg	Glu	His	Asn	Arg	Leu
PGS	Thr	Leu	Trp	Leu	Arg	Glu	His	Asn	Arg	Val

A detailed comparison of prokaryotic, fungal, and plant peroxidase sequences has been published.[41] CCP from yeast and LigPO from *Phanerochaete chrysosporium* are prototypical of the peroxidases and crystal structures of both are available. At the sequence level, they exhibit 18% identity to each other, but no homology to their mammalian counterparts aside from a small region around the distal His (Table II). Peroxidases have no sequence homology to bacterial or mammalian P450s.

3. Tertiary Structure

3.1. Peroxidases

The crystal structures of five heme peroxidases—CCP,[16] LigPO,[12,13] *Arthromyces ramosus* peroxidase (ARP),[15] MPO,[11] and PGH synthase[14] have been solved. In addition, the structure of another peroxide-metabolizing enzyme, catalase, is available[17] and the folding pattern of an isoenzyme of horseradish peroxidase (HRP) has been established.[42] As discussed previously, there is a remarkable conservation of tertiary structure among the peroxidases despite the limited homology of the primary sequences. These findings illustrate the principle that tertiary structure is more conserved than primary structure. The tertiary structures of MPO and PGH synthase appear to be slight variations on the fold of CCP[11,14,30] (Fig. 1). The overall structures are irregular packings of α-helices to form ellipsoids; there is no organized β-sheet structure. A core packing of five α-helices is conserved among MPO, PGH synthase, CCP, LigPO, ARP, and HRP. Two of these helices form an "A-frame" over the heme pocket. These two helices (designated H5 and H6 in PGH synthase and MPO) are much longer in the mammalian peroxidases than in the yeast, plant, and fungal peroxidases and are the dominant features in the structures. The highly conserved decapeptide discussed above is found at the top of H6. The angle between helices H5 and H6 is ~ 30°. At the open end of the "A-frame" are three helices that form the heme pocket. H2 lies across the distal face of the heme and H8 and H12 lie parallel to each other on the proximal face of the heme.

The overall similarity between MPO and PGH synthase might give the impression that the identity of individual amino acid residues is not particularly important to structure. However, as noted above, subtle changes in critical regions of the protein, such as substitution of an oxygen for a sulfur as far as 25 Å from the active site, can dramatically alter catalytic efficiency.[40] This may not seem too surprising because this substitution is in a highly conserved region of the protein. However, a similar 90% decrease in catalytic activity is observed when Cys^{540} is changed to Ser.[40] Cys^{540} is located at the end of helix H17 where the aspirin acetylation site, Ser^{530}, resides.[14] This helix runs perpendicular to the proposed arachidonic acid-binding channel and intersects the channel at Ser^{530}. Cys^{540} is 27 Å from the heme iron and 13 Å from Ser^{530}. Examination of the crystal structure indicates that Cys^{540} is close to helix H5 and may contact some of its side chains. Thus, both mutations of Cys residues that lead to loss of cyclooxygenase and peroxidase activities are at positions close to a major structural element in the protein.

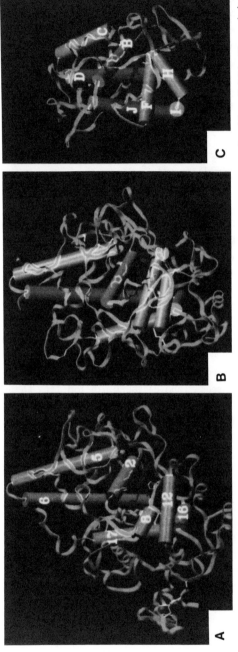

FIGURE 1. Comparison of the three-dimensional structures of (A) PGH synthase, (B) MPO, and (C) CCP. Reproduced from Ref. 14 with permission. The color version of this figure appears facing page 68.

3.2. P450s

The three-dimensional structures of P450s appear quite different from those of peroxidases. Three structures of bacterial P450s have been determined [P450$_{cam}$ (CYP101); P450$_{terp}$ (CYP108); and P450$_{BM-3}$ (CYP102)].[43–45] Although the sequence identity for these three proteins is less than 10%, the overall folding of the proteins is quite similar (Fig. 2).[46] The proteins exhibit a two domain structure composed of a β-sheet-containing domain and an α-helix-rich domain. The dominant element is an α-helix (the I helix) that traverses the entire protein and participates with three other helices (D, E, and L) in defining the heme binding region and the substrate oxidation site. In addition, there is a depression in the protein surface that allows close approach (8–10 Å) of a redox partner (a flavoprotein or an iron-sulfur protein) to the heme group. The path of the electrons from the reductase to the heme is not certain.

Several P450s have been discovered that metabolize peroxides but do not accept electrons from a proteinaceous reducing partner.[47–50] Alignment of the amino acid sequences of these P450s (e.g., thromboxane synthase, CYP5A1; allene oxide synthase, CYP74) with bacterial P450s reveals insertions or other mutations in the protein-binding cavity of the P450 which are anticipated to prevent binding with the redox partner.[46]

CCP is the only peroxidase that interacts with a protein redox partner. Cytochrome c provides the electrons that reduce the higher oxidation states of CCP. Pelletier and Kraut have solved the structure of a cytochrome c•CCP cocrystal and defined the protein contact points.[51] Electron transfer is between the two heme groups which are located ~ 24 Å from each other. The path that the electron takes from cytochrome c to CCP has not been defined but is the subject of intense investigation.

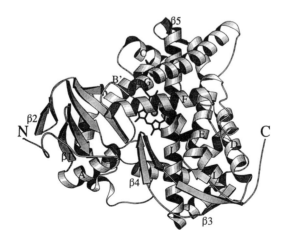

FIGURE 2. Three-dimensional structure of P450$_{BM-3}$(CYP102). Reproduced from Ref. 46 with permission.

4. Active Sites

4.1. Peroxidases

There is a high degree of similarity in the active site structures of CCP, LigPO, and ARP (Figure 3).[12] His[175] of CCP and His[176] of LigPO on helix F serve as the proximal ligands to the heme and are hydrogen-bonded to the carboxylate oxygens of Asp[235] (CCP) and Asp[238] (LigPO). The distal side of the heme contains conserved His and Arg residues provided by helix B. His[52] of CCP is 5.5 Å from the iron and His[47] of LigPO is 5.3 Å from the iron.[13,16] Arg[48] of CCP and Arg[43] of LigPO are situated immediately above the plane of the heme. The major difference in active site structure between the two enzymes is the substitution of Phe for Trp above and below the plane of the heme in LigPO (Phe[46] and Phe[193] in LigPO and Trp[51] and Trp[191] in CCP). The overall folding pattern of the HRP isoenzyme that has been solved is very similar to that of CCP.[42] Based on sequence homology, it is highly likely that the proximal ligand to the heme iron of the widely studied HRP isoenzyme C is His[170] and the distal residues are His[42], Arg[38], and Phe[41].[41]

The three-dimensional structures of MPO and PGH synthase (Figs. 4 and 5) reveal that His is the proximal ligand to the heme, thus confirming the results of previous biophysical studies.[52,53] In MPO, His[336] in helix H8 is the proximal ligand to the iron and this residue is hydrogen-bonded to the side-chain amide oxygen of Asn[421] in helix H12. This Asn residue is conserved in EPO, TPO, and LPO, which suggests that this hydrogen bond on the proximal side is also present in these three enzymes.[11] In PGH synthase, the proximal ligand is donated by His[388], which is three residues to the carboxy-terminus of helix H8.[14] His[388] is hydrogen-bonded to the amide oxygen of the peptide backbone of His[386], 3 Å away. This difference in hydrogen bonding of the proximal His between MPO and PGH synthase suggests that the hydrogen bond between Asn and His is not required for peroxidase catalysis *per se* (see below).

His[95] and Gln[91] of MPO lie just above the distal face of the heme, 5.7 and 4.5 Å from the iron, respectively. A third residue, Arg[239], lies 7 Å from the iron. Overlay of the distal

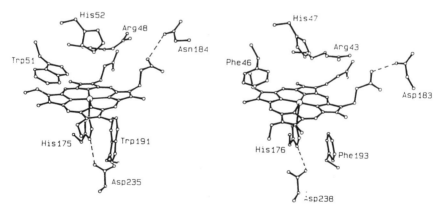

FIGURE 3. Comparison of the active site structures of CCP and LigPO. Reproduced from Ref. 12 with permission.

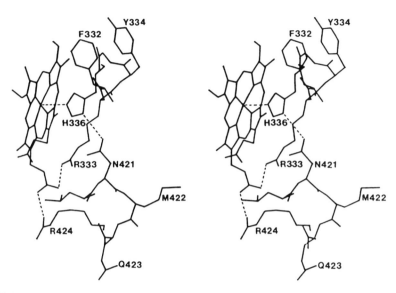

FIGURE 4. Proximal side of the active site structure of MPO. Reproduced from Ref. 11 with permission.

pocket structures of MPO and CCP clearly shows that the Gln in MPO is analogous to the Arg in CCP (Fig. 6). The distal Gln (Gln^{203}) in PGH synthase also superimposes over the Gln in MPO (Fig. 7). Arg^{239} of MPO does not superimpose well with the distal Arg of CCP or the distal Gln of PGH synthase. In PGH synthase, His^{207} lies 4.5 Å above the heme iron and Gln^{203} is 4.7 Å above the iron. No other residues in the distal pocket of PGH synthase appear to be in positions to assist in peroxidase catalysis. Presumably the Gln's in MPO and PGH synthase play the role of the catalytic Arg in CCP, but there is no direct evidence to support this.

Zeng and Fenna have recently compared the amino acid residues in MPO that are 4.5 Å from the heme iron with the corresponding residues in EPO, LPO, and TPO.[11] A remarkable conservation was observed; only 2 out of 24 residues displayed nonconservative substitutions. This result indicates that these peroxidases probably share similar heme environments. In MPO, the heme is located in a 15- to 20-Å crevice. A channel, 10 Å in diameter, provides solvent access to the distal face of the heme. This is the only conduit to the active site and probably restricts the size and shape of peroxides that can access the heme. This channel also probably limits the type of substrates that may be oxidized directly by Compound I in MPO (see below).

In contrast, the heme in PGH synthase is found in a shallow (~14 Å) crevice, which leaves an entire heme edge exposed to solvent. This open geometry is consistent with the variety of hydroperoxide substrates and reducing substrates that react with PGH synthase.[31,54,55] The exposed heme edge may also provide a binding site for the physiological substrate, PGG_2. When the heme environment of PGH synthase is compared with that of other mammalian peroxidases, there is limited similarity except for the abundance of hydrophobic residues. Residues from helix H2, which comprise most of the distal heme

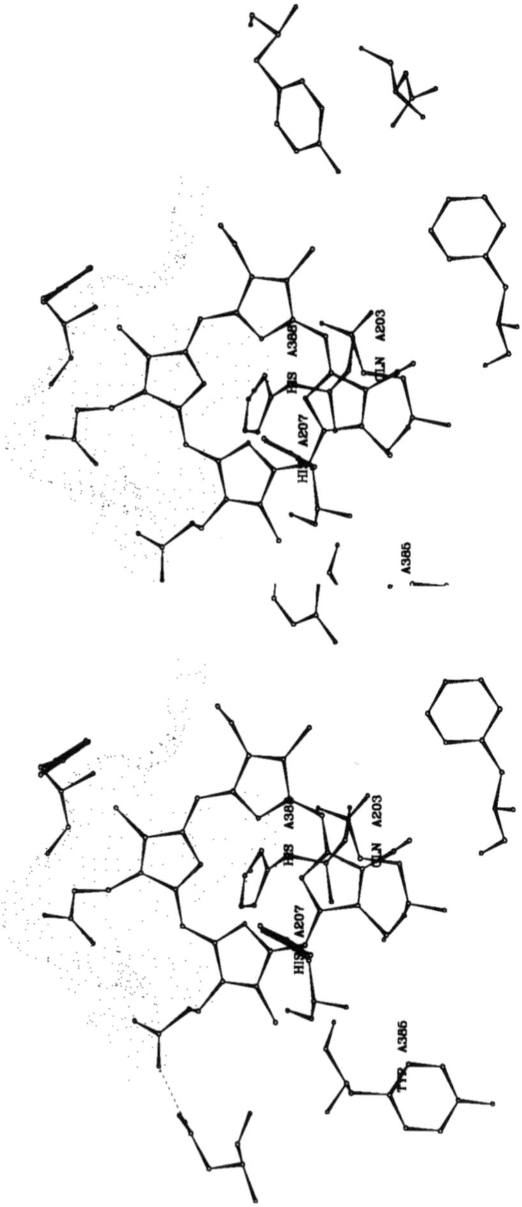

FIGURE 5. Active site structure of PGH synthase. The shaded area is the solvent-accessible surface.

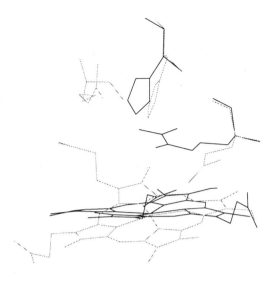

FIGURE 6. Comparison of the orientation of the distal residues of CCP and MPO. Solid lines, CCP.

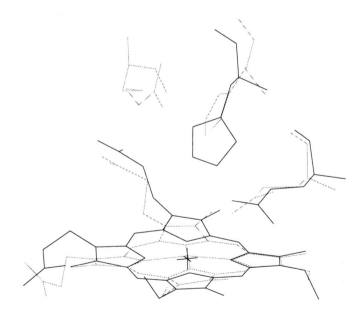

FIGURE 7. Comparison of the distal residues of MPO and PGH synthase. Solid lines, PGH synthase.

pocket, show the most conservation between PGH synthase and the other enzymes. The least similar region is that provided by helix H8, where the proximal His is located. In PGHS, this helix is just 7 residues long, much shorter than the 13-residue H8 helix in MPO. This helix appears to provide the interface between the peroxidase and cyclooxygenase activities in PGHS. At the end of H8 is Tyr^{385}, the proposed catalytic residue of the cyclooxygenase.[56] The geometry required for cyclooxygenase catalysis may preclude the extension of H8 to the length found in MPO.

4.2. P450s

Comparison of the heme binding sites and the distal heme pockets in P450s with those of the peroxidases offers insight as to the structural features that may lead to differences in catalytic activity. One obvious difference between P450s and peroxidases is the proximal ligand to the heme iron. The cysteine ligand in the P450s cannot hydrogen bond to other residues as the proximal histidines of peroxidases do. The importance of the nature of the proximal ligand will be discussed later. The most dramatic difference between peroxidases and P450s is the lack of polar amino acids near the distal side of the heme in the P450s. In CYP102 the only polar residue within 5 Å of the heme on the distal side is Thr^{268}.[45] In CYP101 and CYP108 there is also a striking absence of hydrophilic residues except for the conserved Thr.[43,44] Peroxidases have both a distal His and a distal Arg or Gln that are critical for efficient hydroperoxide heterolysis. The absence of these critical residues may preclude P450s from the rapid heterolysis of peroxides (see below).

P450s also appear to have a more restricted access to the heme than peroxidases. Indeed, CYP101 does not have an obvious opening that connects the heme to the surface of the protein. Two possibilities have been proposed,[57] both of which require movement of the protein to allow a continuous channel to the external solvent. In CYP102 substrate access to the heme is more apparent.[45] A long, hydrophobic channel ~ 8–10 Å in diameter leads to the distal face of the heme. CYP108 has a deep hydrophobic pocket that leads to the distal heme pocket.[44] A shallow hydrophobic pocket is located at the mouth of the deeper pocket. A 16-residue loop between helices F and G is disordered in the crystal and has been suggested to serve as a flexible lid over the substrate channel. In all P450s, the heme is otherwise insulated from the bulk solvent by protein. Such a structure suggests the substrates available for oxidation by the higher oxidation state of P450 are limited by access to the heme.

5. Quaternary Structure

Both MPO and PGH synthase crystallize as multimeric proteins.[11,14] PGH synthase is a dimer of identical subunits. MPO is synthesized as a single polypeptide that is proteolytically cleaved into heavy and light chains. The mature protein consists of a tetramer containing two heavy chains and two light chains. Each heavy chain is linked to its own light chain by a single disulfide bond. PGH synthase is purified as a dimer in detergent solution. The monomer is difficult to prepare and is inactive. MPO is purified as a tetramer, yet may be reductively cleaved to identical halves that are catalytically active.[58] Recombinant MPO expressed in Chinese hamster ovary cells does not form dimers and is also catalytically active.[59] Thus, it appears that quaternary structure is required for PGH synthase, but not for MPO. This difference may result from the relative hydrophobicity of PGH synthase, a membrane protein. Dimerization may be required to keep the protein in the membrane and therefore in a soluble state.

6. Structural Basis of Catalysis

The overall catalytic cycle of peroxidases is depicted in Fig. 8. The hydroperoxide substrate is reduced by the heme prosthetic group to an alcohol with concomitant generation of a higher oxidation state. This reaction transfers the two oxidizing equivalents of the hydroperoxide to the enzyme producing a metal–oxo derivative of the enzyme called Compound I.[60,61] One of the oxidizing equivalents in Compound I resides in the iron which is oxidized from the ferric state in the resting enzyme to the ferryl state in the iron–oxo complex.[4,30] The other oxidizing equivalent resides either in the porphyrin ring as a cation radical or in an amino acid residue as a protein radical.[62,63] The formation of a protein radical on hydroperoxide reduction requires the presence of an easily oxidizable residue in proximity to the heme. In CCP, Trp[191] is the residue oxidized in the formation of Compound I.[64,65] As stated above, the major difference in active site structure between CCP and LigPO is substitution of Phe residues for Trp residues. Phe[193] of LigPO is not oxidizable so the second redox equivalent of Compound I of LigPO is a porphyrin radical cation.[12] PGH synthase forms a Compound I that contains a porphyrin cation radical.[52] However, PGH synthase contains several Tyr residues in the vicinity of the heme and in the absence of reducing substrates one of these Tyr residues intramolecularly reduces the porphyrin cation radical and forms a spectroscopically detectable tyrosyl radical.[14,66] The identity of the Tyr residue oxidized has not been established.[67] The intramolecular electron transfer observed with PGH synthase has no significance for peroxidase catalysis other than perhaps leading to irreversible inactivation of the protein.[68] The spectroscopically detectable tyrosyl radical has been proposed to play a key role in cyclooxygenase catalysis but this hypothesis remains controversial.[66,68–71]

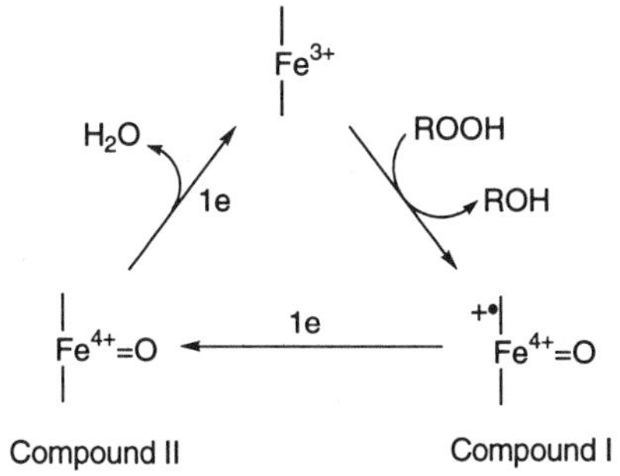

FIGURE 8. Peroxidase catalytic cycle.

The second stage of peroxidase turnover is regeneration of the resting enzyme by reduction of the oxidized enzyme. This requires input of two electrons from a reducing substrate.[4] Reduction usually occurs in a sequential fashion (i.e., two one-electron transfers), but in some cases, two-electron reduction from a single donor takes place.[72] One-electron reduction of Compound I produces Compound II, which contains a ferryl–oxo complex but a fully covalent porphyrin or protein. Reduction of Compound II by a single electron regenerates resting enzyme. This step is approximately one to two orders of magnitude slower than reduction of Compound I and usually represents the rate-limiting-step in the overall peroxidase catalytic cycle.[4]

Direct two-electron reduction of Compound I is normally associated with transfer of the oxo ligand from the peroxidase higher oxidation state to the substrate. This hydroperoxide-dependent oxygenation represents a model for substrate oxygenation by the metal–oxo intermediate of P450 generated either by reaction with a hydroperoxide or with NADPH and O_2. Two-electron reduction of MPO Compound I by chloride ion generates hypochlorite, a major antibacterial metabolite of neutrophils.[73]

6.1. Mechanisms of Hydroperoxide Cleavage by Metal Complexes and Metalloproteins

All of the steps in the peroxidase catalytic cycle can be recapitulated by reaction of hydroperoxides with model heme complexes in organic or aqueous solution.[74,75] In many ways, the peroxidase chemistry of model porphyrins is more complex than that of peroxidases. Depending on the nature of the heme, hydroperoxide or solvent, the hydroperoxide oxygen–oxygen bond is cleaved homolytically [reaction (3)] or heterolytically [reaction (4)].[76–80] For example, iron(III)tetraphenylporphyrin reduces fatty acid hydro-

$$\overset{|}{\underset{|}{Fe^{3+}}} + ROOH \longrightarrow \overset{+\bullet|}{\underset{|}{Fe^{4+}}}=O + ROH \tag{3}$$

$$\overset{|}{\underset{|}{Fe^{3+}}} + ROOH \longrightarrow \overset{|}{\underset{|}{Fe^{4+}}}=O + RO\bullet \tag{4}$$

peroxides homolytically in the absence of a strongly electron-donating proximal ligand but heterolytically in the presence of an excess of imidazole.[80] The ability of imidazole to stimulate heterolysis is related to its ligation of iron and its function as a general base–acid catalyst. The latter function facilitates proton transfers required for peroxide heterolysis. In contrast to the multiple reaction pathways exhibited by simple porphyrins, the peroxidase activity of PGH synthase quantitatively reduces fatty acid hydroperoxides by two electrons.[31,81] This implies that amino acid residues at the active site of peroxidases maximize the heme's ability to reduce hydroperoxides by two electrons.

Some hemeproteins that are not peroxidases reduce hydroperoxides but the mechanism is complex. For example, myoglobin and hemoglobin reduce hydroperoxides and generate a higher oxidation state.[82] The rate of hydroperoxide reduction is several orders of magnitude slower than those exhibited by typical peroxidases and the higher oxidation state contains only one of the oxidizing equivalents of the hydroperoxide.[83,84] This suggests hydroperoxide reduction occurs mainly by homolytic scission. A tyrosyl radical is generated on myoglobin during its reaction with hydroperoxides but whether it is formed by electron transfer to an extremely short-lived Compound I produced by heterolysis or to an alkoxyl radical produced by homolysis is uncertain.[85-87] The products of hydroperoxide reduction by myoglobin and hemoglobin suggest that homolytic and heterolytic peroxide bond scission occur simultaneously although homolysis predominates.[88] Thus, even though the coordination chemistry of myoglobin is similar to that of peroxidases (i.e., His as proximal heme ligand and His distal to the heme), the mechanism of reaction with hydroperoxides is kinetically and mechanistically different. This highlights the importance of other protein–heme interactions in determining the rate and mechanism of hydroperoxide cleavage.

Similar results to those described for myoglobin have been obtained with mammalian P450s. The rate of reaction with hydroperoxides is low and the products of hydroperoxide reduction suggest the simultaneous operation of one-electron and two-electron reduction.[3,83,89-92] Overall, hemoproteins such as myoglobin and P450s resemble model porphyrin complexes more closely than peroxidases in their reactions with hydroperoxides.

As mentioned earlier, several P450s have evolved to metabolize peroxides without input of electrons from a reductase. In all cases, the peroxide substrates are products of polyunsaturated fatty acid oxygenases. Thromboxane synthase and prostacyclin synthase convert the PGH synthase product PGH_2 to the rearranged products thromboxane and prostacyclin [reaction (5) and (6)]. Allene oxide synthase converts the lipoxygenase product 15-hydroperoxy-8-hydroxy-eicosapentanoic acid to an allene oxide [reaction (7)].

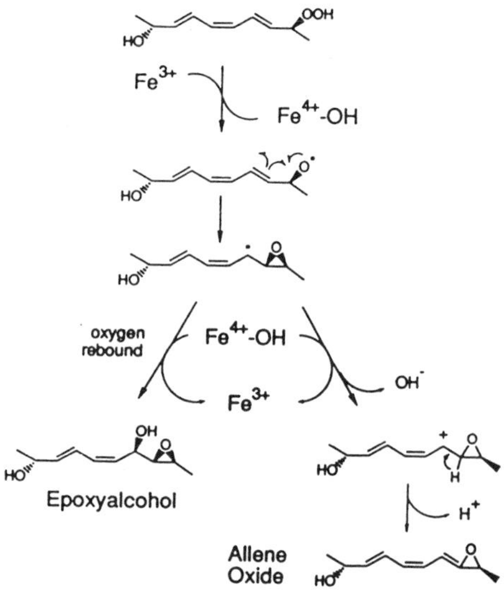

FIGURE 9. Mechanism of action of allene oxide synthase. Reproduced from Ref. 93 with permission.

An elegant investigation of the mechanism of hydroperoxide reduction by allene oxide synthase has been published by Brash and colleagues.[93] They find that allene oxide synthase converts 15(S)-hydroperoxy-8(S)-hydroxy-eicosapentaenoic acid to an allene oxide, but 15(S)-hydroperoxy-8(R)-hydroxy-eicosapentaenoic acid to an epoxyalcohol (Fig. 9). Epoxyalcohol formation proceeds via an intramolecular oxygen rebound with an epoxyallylic radical as an intermediate. This requires that the peroxide bond be cleaved homolytically to form an alkoxyl radical. The formation of allene oxides is also likely to occur via hydroperoxide homolysis because the rate of reaction of 15(S)-hydroperoxy-8(R)-hydroxy-eicosapentaenoic acid is similar to that of 15(S)-hydroperoxy-8(S)-hydroxy-eicosapentaenoic acid. Presumably, the inversion of configuration of the 8-hydroxyl group alters the conformation of the epoxyallylic radical in the enzyme active site which facilitates oxygen rebound from the heme–oxo complex rather than oxidation to a carbonium ion that can proceed to an allene oxide. Rearrangement of polyunsaturated fatty acid hydroperoxides to epoxyalcohols via homolytic scission and oxygen rebound is catalyzed by simple heme complexes in organic and aqueous solutions. This suggests that incorporation of the heme into allene oxide synthase does not change the mechanism of its interaction with peroxides. Thus, although the rate of this reaction is higher than the rate of reaction of hydroperoxides with other P450s, the mechanism of peroxide cleavage is similar, i.e., homolytic scission. Similar conclusions have been drawn by Ullrich and colleagues for the mechanism of peroxide cleavage by thromboxane synthase and prostacyclin synthase.[48]

It is clear from the foregoing discussion that a major difference between peroxidases and P450s is the mechanism of their reaction with hydroperoxides. Specifically, peroxidases catalyze heterolytic hydroperoxide scission at extremely high rates. Since P450s catalyze predominantly homolytic scission and at reduced rates, it is clear that the protein components of peroxidases have evolved to increase the rate of heterolytic hydroperoxide scission. The identities of the amino acid residues responsible for this property are of considerable interest.

Role of Active Site Residues

The proximal ligand to the heme iron is thought to control many of the biochemical and biophysical properties of hemeproteins. Recent work by Adachi and co-workers has shown that changing the proximal His ligand in myoglobin to either Cys or Tyr alters the spectroscopic, catalytic,and redox properties of the protein.[94] All mutants are high-spin, pentacoordinate, and display the spectral characteristics of the proteins that have the same axial ligation. Both the Cys and Tyr mutations dramatically decrease the reduction potential of the protein (from +50 mV to –200 mV). This decrease is thought to result from the greater anionic character of the proximal thiolate or phenolate ligand versus the imidazole ligand in wild-type myoglobin. The Cys mutation increases the amount of heterolysis (21-fold) and homolysis (9.5-fold) of cumene hydroperoxide when compared to wild-type myoglobin. The Tyr mutation does not significantly change the catalytic properties of myoglobin. These results suggest that the proximal ligand controls redox and spectral properties in hemeproteins, but makes only a minor contribution to catalytic activity.

Hemeproteins that use His as the proximal ligand differ dramatically in their redox potential. For example, the redox potential of HRP is –250 mV, much more negative than that of myoglobin (+50 mV). The broad range in redox potential is thought to be modulated by hydrogen bonding of other residues to the proximal His. CCP, LigPO, ARP, and MPO all contain a conserved hydrogen bond from either an Asp or Asn residue to the proximal His. Based on sequence homologies, this hydrogen bond appears to be conserved in TPO, LPO, and EPO. Mutation of Asp^{235} in CCP to an Asn results in a protein with dramatic structural changes around the heme pocket.[95] These structural changes include movement of the iron toward the heme plane and placement of water in the sixth coordination site. In addition, Trp^{191} rotates 180°. This mutant is inactive in the oxidation of horse heart cytochrome c, suggesting the native conformation of Trp^{191} is required for transfer of electrons from cytochrome c.[96] Indeed, the major effect of Asp^{235} in CCP may be to maintain electron transfer pathways from the heme through Trp^{191} to cytochrome c.[95,97]

Recently, more detailed studies have shown that the Asp^{235}Asn mutant does form Compound I with hydrogen peroxide at a rate 4-fold lower than wild-type CCP.[97] In these experiments the formation of Compound I was directly monitored by stopped-flow techniques. The authors suggest that the decreased rate of Compound I formation is related to the hexacoordination of the heme iron in the mutant protein as a result of the structural perturbations caused by the mutation. Thus, one role of the Asp^{235} may be to keep the heme in the pentacoordinate state in the native enzyme.[97] Asp^{235} may not significantly influence peroxide bond heterolysis and stabilization of Compound I, because the rate of peroxide reduction in the mutant enzyme is still significantly greater (10,000-fold) than the corresponding rate in myoglobin. In myoglobin, the axial His is hydrogen-bonded only to a backbone carbonyl oxygen.

Despite the evidence that Asp^{235} of CCP is not involved in Compound I formation or stabilization, the presence of a hydrogen bond to the proximal His alters the properties of the iron center. In CCP mutants where the hydrogen bond to His has been disrupted, the midpoint reduction potential is increased to ~ −80 mV from −182 mV in the wild-type enzyme.[96] Changing the Asp^{235} to Glu decreases the angle of the hydrogen bond to the proximal His, although the length of the bond is unaffected.[96] The decreased strength of the hydrogen bond is thought to decrease the imidazolate character of the proximal ligand. This slight change in geometry results in a redox potential of −113 mV. Thus, the redox potential of His-ligated hemeproteins appears to be exquisitely sensitive to the hydrogen bonding of the axial ligand. However, this modulation does not appear to be a major factor in determining the rate or mechanism of hydroperoxide reduction.

Perhaps more surprising is the finding that proximal His ligation is not required for rapid peroxide heterolysis. In CCP, the proximal His has been changed to Gln with no reduction in the rate of reaction with hydrogen peroxide.[98,99] Apparently the His ligand is not required for peroxide bond scission. A spectral intermediate is detected that is characteristic of a Compound I. The electronic spectra of both the resting enzyme and Compound I of this mutant are very similar to those of wild-type CCP.[99] However, the stability of the Compound I is drastically decreased. Gln^{175} appears to coordinate the iron via the side chain oxygen. These investigators suggest that the decreased stability of Compound I is caused by the faster migration of oxidation equivalents throughout the protein. This hypothesis is based on the observation that, in contrast to wild-type CCP, no isobestic point is observed spectrally when the mutant Compound I decays.

Recently, the amino acids proposed to play critical roles in the peroxidase reaction of CCP, the distal His and Arg, have been mutated. The mutant proteins were purified, crystallized, and analyzed biochemically and structurally. The distal His had been proposed to act as a general acid/base catalyst that aids deprotonation of the substrate peroxide and protonation of the product water after heterolytic cleavage of the hydroperoxide (Fig. 10).[100] As predicted, changing this His^{52} to a Leu decreases the rate of Compound I formation by 5 orders of magnitude.[101] Indeed, the rate of catalysis in the mutant is similar to that for the reaction of soluble Fe(III)tetraphenylporphyrins with hydrogen peroxide[74] or metmyoglobin and hydrogen peroxide.[102–105] X-ray analysis of the mutant demonstrates that structural changes are limited to the vicinity of the mutated His.[101] As discussed by Erman and colleagues, this finding suggests that the distal His alone accounts for the rapid rate of peroxide scission. However, the role of the distal Arg and/or the proximal Asp

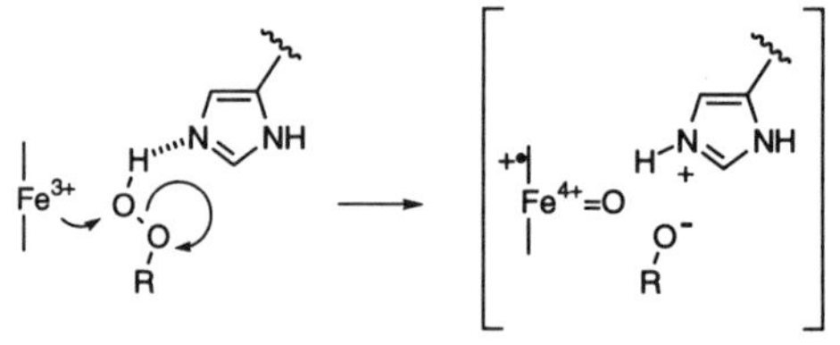

FIGURE 10. Proposed role of distal residues of CCP in facilitating heterolysis of peroxide O–O bond.

in accelerating the reaction cannot be dismissed. These residues may help to increase the rate of reaction relative to that of myoglobin. Myoglobin has a His in the distal peroxide binding pocket but exhibits a very low peroxidase activity. The poor peroxidase activity of metmyoglobin may be related to the lack of particular residues (e.g., residues analogous to Arg^{48} or Asp^{235} of CCP) or to steric factors. Double mutants of CCP will allow evaluation of the role of other residues in the His^{52} mutant.

The distal Arg has been hypothesized to assist Compound I formation by stabilizing the developing negative charge on the leaving group oxygen during heterolytic cleavage.[100] In CCP, Arg^{48} has been changed to either a Lys or Leu residue.[106] Structural changes in the protein are confined to the immediate vicinity of the mutations. The Lys^{48} mutant reduces hydrogen peroxide at a 2-fold lower rate than the wild-type enzyme. The Leu^{48} mutant reduces hydrogen peroxide at a rate 55-fold lower than wild-type CCP. The results are consistent with the proposed role of the distal Arg, despite the relatively small effect on catalytic rate when compared to the distal His mutation. However, under the stopped-flow conditions used for analysis of the mutants, the rate of complex formation between H_2O_2 and the heme iron is the rate-limiting step of Compound I formation.[106] With wild-type CCP, the rate-limiting step is hydroperoxide reduction. Because the distal Arg is proposed to participate during the hydroperoxide reduction step, mutations of this residue would be expected to have little effect on the overall rate of catalysis if the rate-limiting step is changed to substrate binding. Indeed, a relatively minor effect is observed. The analogous Arg-to-Lys mutation has been made in HRP.[107] In contrast to CCP, this conservative substitution causes a 1000-fold decrease in the rate of the peroxidase reaction. However, in these experiments activity was measured under steady-state conditions in the presence of a reducing substrate. Thus, besides a slower rate of hydroperoxide reduction, the decrease in activity may be the result of differences in the interaction of the reducing substrate with the enzyme or in the stability of Compound I. Consequently, the precise role of the distal Arg in CCP or HRP is unresolved.

All of the mutants described above generate a Compound I intermediate with reduced stability. Compound I of the $His^{52}Leu$ mutant of CCP decays 66 to 133 times faster than native Compound I, depending on the buffer constituents.[101] The stability of Compound I in both Arg mutants of CCP is decreased about two to three orders of magnitude relative to the wild-type enzyme.[106] These observations suggest that the structure of the distal pocket influences the formation and stability of Compound I.

6.2. Reduction of Peroxidase Higher Oxidation States

The second half of the peroxidase catalytic cycle is oxidation of the reducing substrate with concomitant regeneration of the resting enzyme. As stated earlier, reduction of the higher oxidation state occurs by electron transfer from the reducing substrate to the ferryl–oxo complex or oxygen transfer from the ferryl–oxo complex to the reducing substrate. The difference between substrates that are oxidized by electron transfer and by oxygen transfer may be related to substrate access to the oxo ligand of the higher oxidation state.

6.2.1. Electron Transfer

Peroxidases are notoriously accommodating with respect to the range of substrates they oxidize.[9] This is particularly true for substrates oxidized by electron transfer. An astonishing range of large aromatic molecules are oxidized that seem unlikely to be binding to a well-defined site on the inside of the protein immediately adjacent to the heme–oxo ligand. The ability of a molecule to serve as an electron transfer substrate appears to be dependent mainly on its redox potential. Smooth Hammett plots have been measured for a range of aromatic amines and phenols for oxidation by both Compound I and Compound II.[4] The ρ values for oxidation of amines and phenols by Compound I are -6.9 and -7.0, respectively, which indicates a high degree of sensitivity to the electron-releasing ability of the *para* substituent.[108] The high correlation coefficients for these Hammett plots suggest that the electronic properties of the substrate are the major determinant of the extent of reaction and, conversely, that steric effects are relatively minor.

The minimal dependence of electron transfer on steric effects suggests that substrates capable of reducing Compound I by this mechanism may bind to a site on the surface of the protein or at least a site remote from the oxo ligand of the higher oxidation states. As mentioned above, precedent for a surface interaction of reducing substrate and peroxidase is provided by the interaction of CCP and cytochrome c.[51] The site on CCP where small molecular substrates bind is at the heme 8–12 Å from the iron whereas cytochrome c binds on the surface of the protein.[109]

Experiments by Ortiz de Montellano and colleagues suggest that electron transfer substrates are oxidized by an exposed edge of the porphyrin of CCP rather than by the heme–oxo complex.[110] Hydrazines and several other peroxidase substrates inactivate CCP during oxidation. Among the products of inactivation are δ-*meso*-porphyrin derivatives which arise by addition of an oxidation intermediate to the porphyrin (Fig. 11).[110] Similar results are obtained with several other peroxidases including HRP and LigPO. In contrast, inactivation of P450s with hydrazines leads to the formation of iron–alkyl derivatives by coupling of intermediate alkyl radicals to the heme iron.[111–113] The difference in inactivation products between peroxidases and P450s suggests there are considerable differences in accessibility to the heme–oxo complex in these two classes of enzymes. The distal regions of P450s are large enough to allow oxidizable substrates direct access to the oxo ligand, thereby accounting for the wide range of monooxygenation reactions catalyzed by these enzymes. On the other hand, peroxidases restrict access of substrates to the heme–oxo ligand so that electron transfer must occur to the heme edge. These results are complemented by nuclear Overhauser studies that estimate aromatic substrates to be 8.4–11.0 Å from the iron of HRP and in the vicinity of the heme 8-methyl group.[114,115]

Reconstitution of HRP with δ-*meso*-ethylheme produces an enzyme that reacts with H_2O_2 to generate higher oxidation states but does not oxidize substrates.[72] In contrast, reconstitution with δ-*meso*-methylheme produces a catalytically active enzyme. This suggests that the inability of the δ-*meso*-ethylheme to oxidize substrates is related to steric hindrance rather than an electronic effect and supports the involvement of the δ-*meso*-heme edge in the oxidation of electron transfer substrates.

Mutation of Trp[51] of CCP to Phe or Ala produces an enzyme that demonstrates increased reactivity toward cytochrome c and classical peroxidase substrates such as

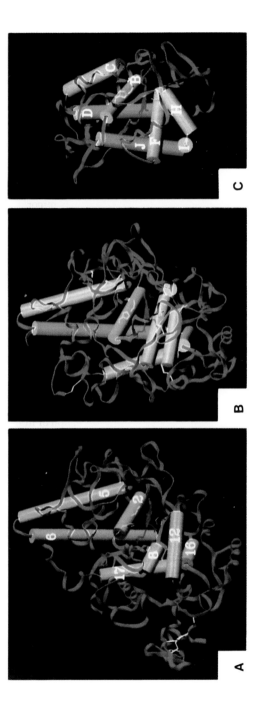

FIGURE 2.1. Comparison of the three-dimensional structures of (A) PGH synthase, (B) MPO, and (C) CCP. Reproduced from Ref. 14 with permission.

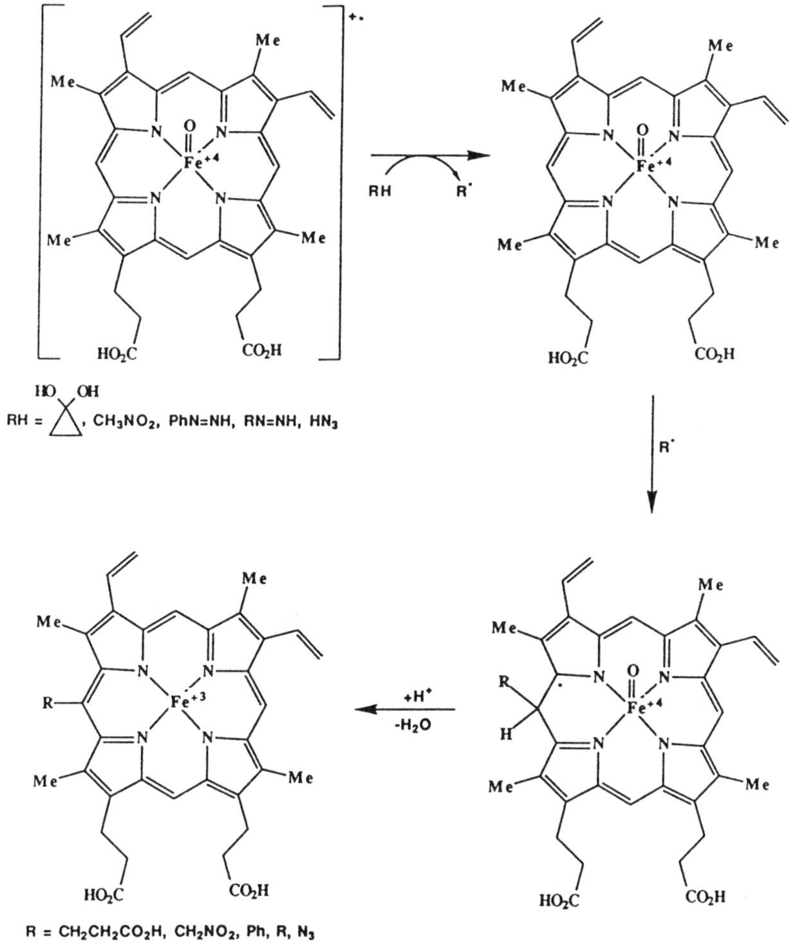

FIGURE 11. Alkylation of δ-*meso* position of porphyrin during inactivation of peroxidases by hydrazines. Reproduced from Ref. 110 with permission.

guaiacol and aromatic amines.[106,116] This is not the result of an increase in the rate of formation of Compound I but rather an increase in reactivity of both Compound I and Compound II toward reducing substrates. Since Trp[51] sits immediately over the distal side of the heme near an edge, one interpretation of these findings is that reducing the steric bulk in this region increases reaction with the substrate. Roe and Goodin point out that Trp[51] is hydrogen-bonded to the oxo ligand of Compound I and that removing this interaction may increase its reactivity independent of steric factors.[106] However, Miller *et al.* have shown that the Trp[51]Ala mutant is able to catalyze the hydroperoxide-dependent epoxidation of styrene and that the epoxide oxygen is hydroperoxide-derived.[117] This establishes that CCP can be converted from an enzyme that catalyzes only electron transfer to one that also catalyzes oxygen transfer by reducing the volume of a residue near the substrate access site. The Trp[51]Ala mutant is also able to form iron–alkyl complexes on reaction with alkylhydrazines.[117] This further substantiates the role of Trp[51] in limiting access of substrates to the heme–oxo ligand of CCP higher oxidation states.

As discussed earlier, Phe[41] of HRP is likely to be in a similar location to Trp[51] of CCP. A mutant HRP, Phe[41]Val, was found to have the opposite substrate specificity as native and recombinant enzyme with respect to 2,2′-azido-bis(3-ethylbenzthiazoline-6-sulfonate) and p-aminobenzoic acid.[107,118] Moreover, the Phe[41]Val mutant exhibits twice the affinity for the substrate benzhydroxamic acid than either the native or recombinant enzyme. This observation is consistent with a substrate binding site for electron transfer near the heme.

6.2.2. Oxygen Transfer

Oxidation of substrates by oxygen transfer from Compound I is analogous to substrate oxidation by P450. Oxygen transfer by peroxidases is limited to substrates bearing heteroatoms α to aromatic rings whereas oxygen transfer by P450s occurs to aliphatic hydrocarbons, aliphatic olefins, and aromatic rings in addition to heteroatoms. This suggests that the metal–oxo complexes of P450s are much stronger oxidants than those of peroxidases. In fact, estimates of the redox potential of the Compound I equivalent of P450 are in the range of 1.5–2 V whereas the redox potential of peroxidase Compounds I are around 1 V.[119,120] For reference, the redox potential of the strongly oxidizing hydroxyl radical is 2.3 V.[121]

Substrates oxidized by oxygen transfer must have direct access to the oxo ligand of Compound I because isotopic labeling studies indicate that the oxygen atom incorporated into the substrate originates in the hydroperoxide group [reaction (8)]. This is true for

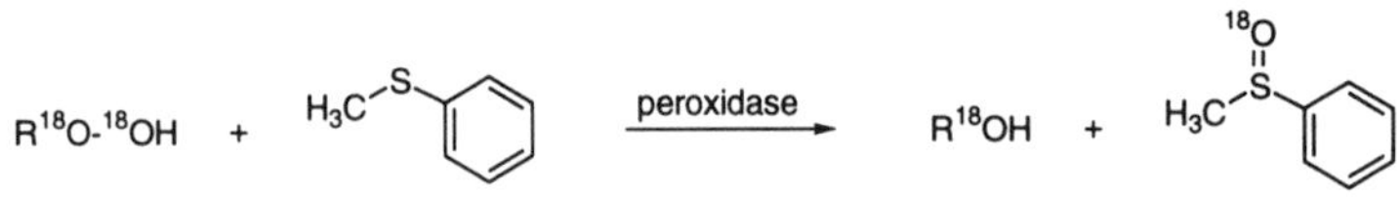

$$R^{18}O\text{-}^{18}OH \quad + \quad H_3C{-}S{-}\bigcirc \quad \xrightarrow{\text{peroxidase}} \quad R^{18}OH \quad + \quad H_3C{-}\overset{\overset{^{18}O}{\|}}{S}{-}\bigcirc \quad (8)$$

oxidation of sulfides to sulfoxides and aromatic amines to hydroxylamines.[122–124] Amine oxidation is complicated by the fact that the hydroxylamine products are also excellent peroxidase substrates and are further oxidized to nitroso compounds which are themselves oxidized to nitro compounds.[124,125] For oxygen transfer from the heme–oxo complex, one might anticipate sizable steric effects on oxygenation. For example, PGH synthase catalyzes the hydroperoxide-dependent oxygenation of methylphenylsulfide and ethylphenylsulfide at comparable rates but the oxygenation of isopropylphenylsulfide at a tenfold reduced rate.[126] In addition, the enzyme displays up to a tenfold higher rate of oxygenation of one enantiomer over another when chiral sulfides are tested as substrates.[126] Incorporation of carboxylic acids into alkylarylsulfides significantly reduces their oxidation by PGH synthase, suggesting the enzyme also discriminates against polar substrates.[126] Examination of the crystal structure of PGH synthase reveals a high concentration of aromatic residues at the mouth of the channel to the peroxidase active site which may favor the approach of hydrophobic substrates to the heme–oxo group.[14]

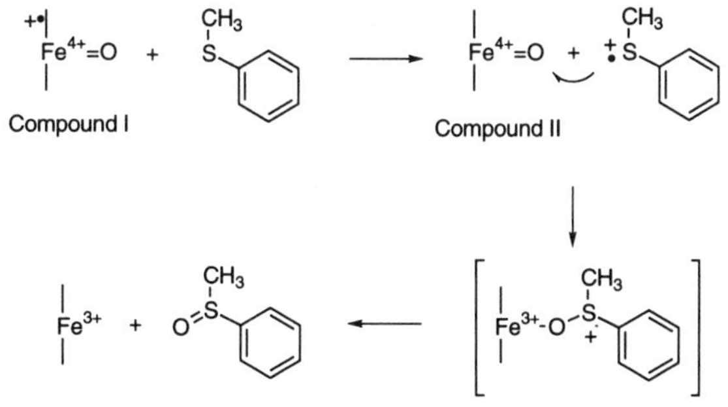

FIGURE 12. Mechanism of sulfide oxygenation by peroxidase higher oxidation states.

Oxygen transfer to aromatic sulfides appears to occur by electron transfer followed by oxygen delivery to a substrate radical cation (Fig. 12). Hammett plots relating the rate of oxygenation of p-substituted thioanisoles to the electron-releasing ability of the p-substituent are linear with ρ values that range from -0.35 (HRP) to -0.8 (PGH synthase).[122,126] This is consistent with development of a significant amount of positive charge on the sulfide during oxidation. The ρ value for oxidation of thioanisoles by P450 is -0.16 but cannot be compared to those for peroxidases because it was measured for the NADPH-dependent reaction instead of the hydroperoxide-dependent reaction.[127]

Perez and Dunford and Casella $et\ al.$ have detected Compound II spectroscopically during the oxygenation of thioanisole by HRP.[128,129] This is consistent with initial electron transfer from the sulfide to Compound I to form Compound II and a sulfide radical cation. The radical cation then couples to the oxo ligand of Compound II to form the sulfoxide, which diffuses from the enzyme. However, Ortiz de Montellano and Doerge $et\ al.$ have pointed out that the oxygenation of sulfides by HRP is complex so that detection of Compound II during the reaction does not necessarily mean it is a catalytic intermediate.[110,130]

Studies described above implicate Phe[41] of HRP in controlling substrate binding to the enzyme. Mutation of this residue to Leu causes a remarkable increase in the enantioselectivity of sulfoxidation of alkylarylsulfides and increased rates of oxidation.[131] The magnitude of these changes varies with substrate. The greatest effect is observed with cyclopropylmethylphenylsulfide, which is oxidized 10 times faster and displays an increase in enantioselectivity from 7% to 94% by the Phe[41]Val mutant. Hammett analysis indicates that the catalytic mechanism is not altered by the mutation. An HRP mutant in which Phe[41] is changed to Thr also displays altered sulfoxidation of these substrates.[131] This mutant oxidizes thioanisole and p-chlorothioanisole at higher rates than the native enzyme, yet oxidation is less enantioselective. These results provide strong evidence that Phe[41] is positioned near the substrate binding site. Indeed, based on these data, the authors postulate that alkylarylsulfides can bind in either of two pockets near the ferryl oxygen. Changing Phe[41] to Val enlarges one pocket and retains hydrophobicity, thus favoring increased binding at that site. Substitution of Thr in the same site enlarges it, but decreases its hydrophobicity which may decrease substrate binding.

HRP and PGH synthase represent extremes in the accessibility of substrates to their active sites and this differential appears related to the size of the distal pocket to the heme. Both enzymes appear to have much smaller distal pockets than P450s which partially accounts for the differences in reactions catalyzed by the two classes of enzymes. Peroxidases do exist that appear to have quite large distal pockets. Pea seed peroxygenase is an enzyme that catalyzes polyunsaturated fatty acid hydroperoxide-dependent oxygenation of a range of substrates including indole, alkylarylsulfides, and polyunsaturated fatty acids.[132,133] In the absence of an oxidizable substrate, the enzyme catalyzes intramolecular oxygen transfer to isolated *cis* double bonds in the hydroperoxide molecule [reaction (9)].[134] Olefin epoxidation proceeds with retention of configuration. Isotopic labeling

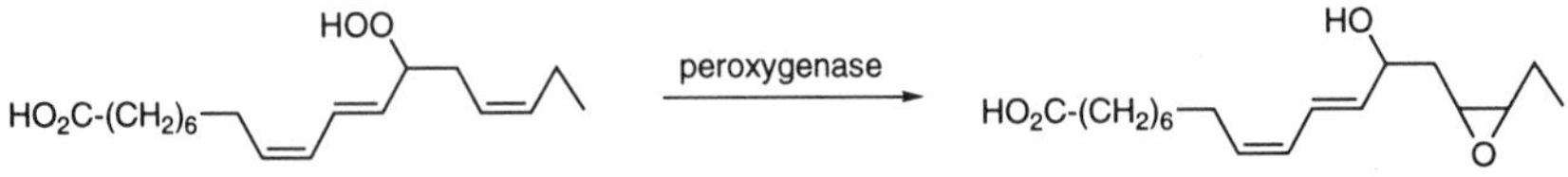

$$\text{(9)}$$

studies indicate that the oxygen introduced originates in the hydroperoxide molecule, suggesting that the oxidizing agent is a heme–oxo complex. Pea seed peroxygenase displays the spectral characteristics of a typical b-type cytochrome, suggesting histidine is the fifth ligand to the heme.

Chloroperoxidase catalyzes the H_2O_2-dependent oxidation of chloride to hypochlorite.[135,136] In the absence of chloride, it catalyzes substrate oxygenation such as olefin epoxidation. Studies with alkylhydrazines indicate it is capable of forming iron–alkyl complexes analogous to those formed by P450.[137] Both of these observations suggest that chloroperoxidase has a large pocket distal to the heme. The proximal ligand to the heme iron of chloroperoxidase is a cysteine thiolate making it analogous to the P450s that do not have redox partners. Interestingly, the rate of reaction of chloroperoxidase with hydroperoxides is significantly slower than typical peroxidases and is in a range exhibited by P450s. It will be interesting to define the coordination chemistry of chloroperoxidase and explore the role of specific amino acid residues in its catalytic cycle.

7. Conclusions

The nine years since the publication of the first volume of this book has witnessed a major increase in our knowledge of the structure, function, and catalytic mechanisms of peroxidases and P450s. With that knowledge has come a measure of understanding. The proximal ligands to the heme iron are a distinguishing feature of these two classes of enzymes and appear important for determining the oxidation potential of the metal–oxo derivatives generated during catalytic turnover. The oxidation potential establishes the types of oxidations that can be catalyzed, e.g., P450s hydroxylate aliphatic hydrocarbons and epoxidize aromatic hydrocarbons whereas peroxidases do not. However, the residues on the distal side of the heme appear equally important for the catalytic mechanism, at least with regard to hydroperoxide-dependent oxidations. The presence of His and Arg or His and Gln residues gives peroxidases the ability to effectively reduce hydroperoxides by two electrons and produce the oxidizing agent Compound I. P450s lack these residues and reduce hydroperoxides mainly by one electron, thereby producing Compound II and an alkoxyl radical. The substrate binding regions of peroxidases appear to be smaller than the substrate binding sites of P450s, which restricts the range of compounds that peroxidases oxidize by direct oxygen transfer. The ability of peroxidases to oxidize substrates by electron transfer to the heme edge appears to account for the broad range of compounds that are termed classical peroxidase substrates.

Nine years ago the ability of P450s to react with hydroperoxides was an experimental curiosity that provided useful mechanistic insights into the "normal" P450 reactions driven by NADPH. However, it is now clear that homolytic cleavage of hydroperoxides and cyclic peroxides is an essential reaction in several metabolic pathways that convert polyunsaturated fatty acid derivatives to signaling and defense molecules. One anticipates that the number of such reactions will grow. In addition, comparative analysis of the structures of myeloperoxidase and PGH synthase has revealed how divergent evolution created a protein that retains peroxidase activity and acquired dioxygenase activity (PGH synthase). The subtle structural alterations that wrought this functional change also created opportunities for modulating enzyme action that have been exploited by nature and the pharmaceutical industry to alleviate human pain and suffering.[138] The parallel paths in chemistry and physiology that led to these insights are intersecting and offer the promise of novel chemicals to treat a variety of human diseases as well as engineered proteins to improve chemical processing.

ACKNOWLEDGMENTS. I am grateful to J.A. Peterson and M. Garavito for helpful discussions, to C. Smith for assistance with molecular modeling, to L. Landino for a critical reading, and to C. Riley for editorial assistance. Work in my laboratory has been supported by a research grant from the National Institutes of Health (CA47479).

References

1. Hrycay, E., and O'Brien, P. J., 1971, Cytochrome P-450 as a microsomal peroxidase utilizing a lipid peroxide substrate, *Arch. Biochem. Biophys.* **147**:14–27.
2. Hrycay, E. G., and O'Brien, P. J., 1972, Cytochrome P450 as a microsomal peroxidase in steroid hydroperoxide reduction, *Arch. Biochem. Biophys.* **153**:480–494.
3. White, R. E., and Coon, M. J., 1980, Oxygen activation by cytochrome P-450, *Annu. Rev. Biochem.* **49**:315–356.
4. Dunford, H. B., and Stillman, J. S., 1976, On the function and mechanism of action of peroxidases, *Coord. Chem. Rev.* **19**:187–251.
5. Marnett, L. J., Weller, P. A., and Battista, J. R., 1986, Comparison of the peroxidase activity of hemeproteins and cytochrome P 450, in: *Cytochrome P-450* (P. R. Ortiz de Montellano, ed.), Plenum Press, New York, pp. 29–76.
6. Fruton, J. S., 1972, *Molecules and Life: Historical Essays on the Interplay of Chemistry and Biology*, Wiley–Interscience, New York.
7. Saunders, B. C., Holmes-Siedel, A. G., and Stark, B. P., 1964, *Peroxidases*, Butterworths, London.
8. Yamazaki, I., 1974, Peroxidase, in: *Molecular Mechanisms of Oxygen Activation* (O. Hayaishi, ed.), Academic Press, New York, pp. 535–558.
9. Saunders, B. C., 1975, Peroxidases and catalases, in: *Inorganic Biochemistry* (G. L. Eichhorn, ed.), Elsevier, Amsterdam, pp. 988–1021.
10. Everse, J., Everse, K. E., and Grisham, M. B. (eds.), 1991, *Peroxidases in Chemistry and Biology*, CRC Press, Boca Raton, FL.

11. Zeng, J., and Fenna, R. E., 1992, X-ray crystal structure of canine myeloperoxidase at 3 Å resolution, *J. Mol. Biol.* **226:**185–207.

12. Edwards, S. L., Raag, R., Wariishi, H., Gold, M. H., and Poulos, T. L., 1993, Crystal structure of lignin peroxidase, *Proc. Natl. Acad. Sci. USA* **90:**750–754.

13. Poulos, T. L., Edwards, S. L., Wariishi, H., and Gold, M. H., 1993, Crystallographic refinement of lignin peroxidase at 2 Å, *J. Biol. Chem.* **268:**4429–4440.

14. Picot, D., Loll, P. J., and Garavito, R. M., 1994, The X-ray crystal structure of the membrane protein prostaglandin H_2 synthase-1, *Nature* **367:**243–249.

15. Kunishima, N., Fukuyama, K., Matsubara, H., Hatanaka, H., Shibano, Y., and Amachi, T., 1994, Crystal structure of the fungal peroxidase from *Arthromyces ramosus* at 1.9Å resolution. Structural comparisons with the lignin and cytochrome c peroxidases, *J. Mol. Biol.* **235:**331–344.

16. Poulos, T. L., Freer, S. T., Alden, R. A., Edwards, S. L., Skogland, U., Takio, K., Eriksson, B., Xuong, N.-H., Yonetani, T., and Kraut, J., 1980, The crystal structure of cytochrome c peroxidase, *J. Biol. Chem.* **255:**575–580.

17. Murthy, M. R. N., Reid, T. J., III, Sicignano, A., Tanaka, N., and Rossmann, M.G., 1981, Structure of beef liver catalase, *J. Mol. Biol.* **152:**465–499.

18. Klebanoff, S. J., 1991, Myeloperoxidase: Occurrence and biological function, in: *Peroxidases in Chemistry and Biology* (J. Everse, K. E. Everse, and M. B. Grisham, eds.), CRC Press, Boca Raton, FL, Vol. 1, pp. 1–35.

19. Hamberg, M., Svensson, J., Wakabayashi, T., and Samuelsson, B., 1974, Isolation and structure of two prostaglandin endoperoxides that cause platelet aggregation, *Proc. Natl. Acad. Sci. USA* **71:**345–349.

20. Nugteren, D. H., and Hazelhof, E., 1973, Isolation and properties of intermediates in prostaglandins biosynthesis, *Biochim. Biophys. Acta* **326:**448–461.

21. Johnson, K. R., Nauseef, W. M., Care, A., Wheelock, M. J., Shane, S., Hudson, S., Koeffler, H. P., Selsted, M., Miller, C., and Rovera, G., 1987, Characterization of cDNA clones for human myeloperoxidase: Predicted amino acid sequence and evidence for multiple mRNA species, *Nucleic Acids Res.* **15:**2013–2028.

22. Morishita, K., Kubota, N., Asano, S., Kaziro, Y., and Nagata, S., 1987, Molecular cloning and characterization of cDNA for human myeloperoxidase, *J. Biol. Chem.* **262:**3844–3851.

23. Sakamaki, K., Tomonaga, M., Tsukui, K., and Nagata, S., 1989, Molecular cloning and characterization of a chromosomal gene for human eosinophil peroxidase, *J. Biol. Chem.* **264:**16828–16836.

24. Kimura, S., Kotani, T., McBride, O. W., Umeki, K., Hirai, K., Nakayama, T., and Ohtaki, S., 1987, Human thyroid peroxidase: Complete cDNA and protein sequence, chromosome mapping, and identification of two alternately spliced mRNAs, *Proc. Natl. Acad. Sci. USA* **84:**5555–5559.

25. Merlie, J. P., Fagan, D., Mudd, J., and Needleman, P., 1988, Isolation and characterization of the complementary DNA for sheep seminal vesicle prostaglandin endoperoxide synthase (cyclooxygenase), *J. Biol. Chem.* **263:**3550–3553.

26. DeWitt, D. L., and Smith, W. L., 1988, Primary structure of prostaglandin G/H synthase from sheep vesicular gland determined from the complementary DNA sequence, *Proc. Natl. Acad. Sci. USA* **85:**1412–1416.

27. Yokoyama, C., Takai, T., and Tanabe, T., 1988, Primary structure of sheep prostaglandin endoperoxide synthase deduced from cDNA sequence, *FEBS Lett.:* **231:**347–351.

28. Dull, T. J., Uyeda, C., Strosberg, A. D., Nedwin, G., and Seilhamer, J. J., 1990, Molecular cloning of cDNAs encoding bovine and human lactoperoxidase, *DNA Cell Biol.* **9:**499–509.

29. Cals, M.-M., Mailliart, P., Brignon, G., Anglade, P., and Dumas, B. R., 1991, Primary structure of bovine lactoperoxidase, a fourth member of a mammalian heme peroxidase family, *Eur. J. Biochem.* **198:**733–739.

30. Poulos, T. L., and Fenna, R. E., 1994, Peroxidases: Structure, function, and engineering, in: *Metal Ions in Biological Systems: Metalloenzymes Involving Amino Acid-Residue and Related Radicals* (H. Sigel and A. Sigel, eds.), Dekker, New York, pp. 25–75.

31. Markey, C. M., Alward, A., Weller, P. E., and Marnett, L. J., 1987, Quantitative studies of hydroperoxide reduction by prostaglandin H synthase, *J. Biol. Chem.* **262**:6266–6279.

32. Kulmacz, R. J., 1986, Prostaglandin H synthase and hydroperoxides: Peroxidase reaction and inactivation kinetics, *Arch. Biochem. Biophys.* **249**:273–285.

33. Hsuanyu, Y., and Dunford, H. B., 1992, Prostaglandin H synthase kinetics. The effect of substituted phenols on cyclooxygenase activity and the substituent effect on phenolic peroxidatic activity, *J. Biol. Chem.* **267**:17649–17657.

34. Hsuanyu, Y., and Dunford, H. B., 1990, Kinetics of the reaction of prostaglandin H synthase compound II with ascorbic acid, *Arch. Biochem. Biophys.* **281**:282–286.

35. Bakovic, M., and Dunford, H. B., 1993, Kinetics of the oxidation of *p*-coumaric acid by prostaglandin H synthase and hydrogen peroxide, *Biochemistry* **32**:833–840.

36. Sun, W., and Dunford, H. B., 1993, Kinetics and mechanism of the peroxidase-catalyzed iodination of tyrosine, *Biochemistry* **32**:1324–1331.

37. Smith, W. L., Eling, T. E., Kulmacz, R. J., Marnett, L. J., and Tsai, A., 1992, Tyrosyl radicals and their role in hydroperoxide-dependent activation and inactivation of prostaglandin endoperoxide synthase, *Biochemistry* **31**:3–7.

38. DeWitt, D. L., El-Harith, E. A., Kraemer, S. A., Andrews, M. J., Yao, E. F., Armstrong, R. L., and Smith, W. L., 1990, The aspirin and heme-binding sites of ovine and murine prostaglandin endoperoxide synthases, *J. Biol. Chem.* **265**:5192–5198.

39. Shimokawa, T., Kulmacz, R. J., DeWitt, D. L., and Smith, W. L., 1990, Tyrosine 385 of prostaglandin endoperoxide synthase is required for cyclooxygenase catalysis, *J. Biol. Chem.* **265**:20073–20076.

40. Kennedy, T. A., Smith, C. J., and Marnett, L. J., 1994, Investigation of the role of cysteines in catalysis by prostaglandin endoperoxide synthase, *J. Biol. Chem.* **269**:27357–27364.

41. Welinder, K. G., 1992, Superfamily of plant, fungal and bacterial peroxidases, *Curr. Opin. Struct. Biol.* **2**:388–393.

42. Morita, Y., Mikami, B., Yamashita, H., Lee, J. Y., Aibara, S., Sato, M., Katsube, Y., and Tanaka, N., 1991, Primary and crystal structures of horseradish peroxidase isozyme E5, in: *Biochemical, Molecular and Physiological Aspects of Plant Peroxidases* (J. Lobarzewski, H. Greppin, C. Penel, and T. Gaspar, eds.), University of Geneva, Geneva, pp. 81–88.

43. Poulos, T. L., Finzel, B. C., Gunsalus, I. C., Wagner, G. C., and Kraut, J., 1985, The 2.6Å crystal structure of *Pseudomonas putida* cytochrome P 450, *J. Biol. Chem.* **260**:16122–16130.

44. Hasemann, C. A., Ravichandran, K. G., Peterson, J. A., and Diesenhofer, J., 1994, Crystal structure and refinement of cytochrome P450$_{terp}$ at 2.3Å resolution, *J. Mol. Biol.* **236**:1169–1185.

45. Ravichandran, K. G., Boddupalli, S. S., Hasemann, C. A., Peterson, J. A., and Deisenhofer, J., 1993, Crystal structure of hemoprotein domain of P450$_{BM-3}$, a prototype for microsomal P450's, *Science* **261**:731–736.

46. Hasemann, C. A., Ravichandran, K. G., Boddupalli, S. S., Peterson, J. A., and Deisenhofer, J., 1995, Structure and function of cytochromes P450: A comparative analysis of the three-dimensional structures of P450$_{terp}$, P450$_{cam}$, and the hemoprotein domain of P450$_{BM-3}$, *Structure*, in press.

47. Haurand, M., and Ullrich, V., 1985, Isolation and characterization of thromboxane synthase from human platelets as a cytochrome P-450 enzyme, *J. Biol. Chem.* **260**:15059–15067.

48. Hecker, M., and Ullrich, V., 1989, On the mechanism of prostacyclin and thromboxane A$_2$ biosynthesis, *J. Biol. Chem.* **264**:141–150.

49. Ullrich, V., Castle, L., and Weller, P., 1981, Spectral evidence for the cytochrome P450 nature of prostacyclin synthase, *Biochem. Pharmacol.* **30**:2033–2036.

50. Song, W.-C., Funk, C. D., and Brash, A. R., 1993, Molecular cloning of an allene oxide synthase: A cytochrome P450 specialized for the metabolism of fatty acid hydroperoxides, *Proc. Natl. Acad. Sci. USA* **90**:8519–8523.

51. Pelletier, H., and Kraut, J., 1992, Crystal structure of a complex between electron transfer partners, cytochrome c peroxidase and cytochrome c, *Science* **258**:1748–1755.

52. Lambeir, A. M., Markey, C. M., Dunford, H. B., and Marnett, L. J., 1985, Spectral properties of the higher oxidation states of prostaglandin H synthase, *J. Biol. Chem.* **260:**14894–14896.
53. Kulmacz, R. J., Tsai, A.-L., and Palmer, G., 1987, Heme spin states and peroxide-induced radical species in prostaglandin H synthase, *J. Biol. Chem.* **262:**10524–10531.
54. Marnett, L. J., and Reed, G. A., 1979, Peroxidatic oxidation of benzo[a]pyrene and prostaglandin biosynthesis, *Biochemistry* **18:**2923–2929.
55. Ohki, S., Ogino, N., Yamamoto, S., and Hayaishi, O., 1979, Prostaglandin hydroperoxidase, an integral part of prostaglandin endoperoxide synthetase from bovine vesicular gland microsomes, *J. Biol. Chem.* **254:**829–836.
56. Shimokawa, T., Kulmacz, R. J., DeWitt, D. L., and Smith, W. L., 1990, Tyrosine 385 of prostaglandin endoperoxide synthase is required for cyclooxygenase catalysis, *J. Biol. Chem.* **265:**20073–20076.
57. Poulos, T. L., 1986, The crystal structure of cytochrome P-450$_{cam}$, in: *Cytochrome P-450: Structure, Mechanism, and Biochemistry* (P. R. Ortiz de Montellano, ed.), Plenum Press, New York, pp. 505–539.
58. Andrews, P. C., and Krinsky, N. I., 1981, The reductive cleavage of myeloperoxidase in half, producing enzymically active hemi-myeloperoxidase, *J. Biol. Chem.* **256:**4211–4218.
59. Moguilevsky, N., Garcia-Quintana, L., Jacquet, A., Tournay, C., Fabry, L., Pierard, L., and Bollen, A., 1991, Structure and biological properties of human recombinant myeloperoxidase produced by Chinese hamster ovary cell lines, *Eur. J. Biochem.* **197:**605–614.
60. Chance, B., 1943, The kinetics of the enzyme–substrate compound of peroxidase, *J. Biol. Chem.* **151:**553–577.
61. George, P., 1953, Intermediate compound formation with peroxidase and strong oxidizing agents, *J. Biol. Chem.* **201:**413–426.
62. Dolphin, D., and Felton, R. H., 1974, The biochemical significance of porphyrin π cation radicals, *Acc. Chem. Res.* **7:**26–32.
63. Yonetani, T., 1976, Cytochrome c peroxidase, in: *The Enzymes* Vol.13 (P. D. Boyer, ed.), Academic Press, New York, pp. 345–362.
64. Sivaraja, M., Goodin, D. B., Smith, M., and Hoffman, B. M., 1989, Identification by ENDOR of Trp[191] as the free-radical site in cytochrome c peroxidase compound ES, *Science* **245:**738–740.
65. Houseman, A. L. P., Doan, P. E., Goodin, D. B., and Hoffman, B. M., 1993, Comprehensive explanation of the anomalous EPR spectra of wild-type and mutant cytochrome *c* peroxidase compound ES, *Biochemistry* **32:**4430–4443.
66. Karthein, R., Dietz, R., Nastainczyk, W., and Ruf, H. H., 1988, Higher oxidation states of prostaglandin H synthase. EPR study of a transient tyrosyl radical in the enzyme during the peroxidase reaction, *Eur. J. Biochem.* **171:**313–320.
67. Tsai, A.-L., Hsi, L. C., Kulmacz, R. J., Palmer, G., and Smith, W. L., 1994, Characterization of the tyrosyl radicals in ovine prostaglandin H synthase-1 by isotope replacement and site-directed mutagenesis, *J. Biol. Chem.* **269:**5085–5091.
68. Lassmann, G., Odenwaller, R., Curtis, J. F., DeGray, J. A., Mason, R. P., Marnett, L. J., and Eling, T. E., 1991, Electron spin resonance investigation of tyrosyl radicals of prostaglandin H synthase. Relation to enzyme catalysis, *J. Biol. Chem.* **266:**20045–20055.
69. Tsai, A.-L., Palmer, G., and Kulmacz, R. J., 1992, Prostaglandin H synthase. Kinetics of tyrosyl radical formation and of cyclooxygenase catalysis, *J. Biol. Chem.* **267:**17753–17759.
70. DeGray, J. A., Lassmann, G., Curtis, J. F., Kennedy, T. A., Marnett, L. J., Eling, T. E., and Mason, R. P., 1992, Spectral analysis of the protein-derived tyrosyl radicals from prostaglandin H synthase, *J. Biol. Chem.* **267:**23583–23588.
71. Kulmacz, R. J., Ren, Y., Tsai, A.-L., and Palmer, G., 1990, Prostaglandin H synthase: Spectroscopic studies of the interaction with hydroperoxides and with indomethacin, *Biochemistry* **29:**8760–8771.
72. Harris, R. Z., Newmyer, S. L., and Ortiz de Montellano, P. R., 1993, Horseradish peroxidase-catalyzed two-electron oxidations. Oxidation of iodide, thioanisoles, and phenols at distinct sites, *J. Biol. Chem.* **268:**1637–1645.

73. Thomas, E. L., and Learn, D. B., 1991, Myeloperoxidase-catalyzed oxidation of chloride and other halides: The role of chloramines, in: *Peroxidases in Chemistry and Biology* (J. Everse, K. E. Everse, and M. B. Grisham, eds.), CRC Press, Boca Raton, FL, Vol. 1, pp. 83–103.

74. Bruice, T. C., 1991, Reactions of hydroperoxides with metallotetraphenylporphyrins in aqueous solutions, *Acc. Chem. Res.* **24:**243–249.

75. Ostovic, D., and Bruice, T. C., 1992, Mechanism of alkene epoxidation by iron, chromium, and manganese higher valent oxo-metalloporphyrins, *Acc. Chem. Res.* **25:**314–320.

76. Traylor, T. G., and Xu, F., 1988, Model reactions related to cytochrome P-450. Effects of alkene structure on the rates of epoxide formation, *J. Am. Chem. Soc.* **110:**1953–1958.

77. Traylor, T. G., and Xu, F., 1990, Mechanisms of reactions of iron(III)porphyrins with hydrogen peroxide and hydroperoxides: Solvent and solvent isotope effects, *J. Am. Chem. Soc.* **112:**178–186.

78. Traylor, T. G., Lee, W. A., and Stynes, D. V., 1984, Model compound studies related to peroxidases. Mechanisms of reactions of hemins with peracids, *J. Am. Chem. Soc.* **106:**755–764.

79. Traylor, T. G., Tsuchiya, S., Byun, Y.-S., and Kim, C., 1993, High-yield epoxidations with hydrogen peroxide and *tert*-butyl hydroperoxide catalyzed by iron(III) porphyrins: Heterolytic cleavage of hydroperoxides, *J. Am. Chem. Soc.* **115:**2775–2781.

80. Labeque, R., and Marnett, L. J., 1989, Homolytic and heterolytic scission of organic hydroperoxides by meso-tetraphenylporphinato-iron(III) and its relation to olefin epoxidation, *J. Am. Chem. Soc.* **111:**6621–6627.

81. Marnett, L. J., Chen, Y.-N. P., Maddipati, K. R., Plé, P., and Labeque, R., 1988, Functional differentiation of cyclooxygenase and peroxidase activities of prostaglandin synthase by trypsin treatment: Possible location of a prosthetic heme binding site, *J. Biol. Chem.* **263:**16532–16535.

82. George, P., and Irvine, D. H., 1954, Reaction of metmyoglobin with strong oxidizing agents, *Biochem. J.* **58:**188–195.

83. McCarthy, M. B., and White, R. E., 1983, Functional differences between peroxidase compound I and the cytochrome P-450 reactive oxygen intermediate, *J. Biol. Chem.* **258:**9153–9158.

84. George, P., and Irvine, D. H., 1955, A possible structure for the higher oxidation state of metmyoglobin, *Biochem. J.* **60:**596–604.

85. Catalano, C. E., Choe, Y. S., and Ortiz de Montellano, P. R., 1989, Reactions of the protein radical in peroxide-treated myoglobin. Formation of a heme-protein cross-link, *J. Biol. Chem.* **264:**10534–10541.

86. Wilks, A., and Ortiz de Montellano, P. R., 1992, Intramolecular translocation of the protein radical formed in the reaction of recombinant sperm whale myoglobin with H_2O_2, *J. Biol. Chem.* **267:**8827–8833.

87. Rao, S. I., Wilks, A., and Ortiz de Montellano, P. R., 1993, The roles of His-64, Tyr-103, Tyr-146, and Tyr-151 in the epoxidation of styrene and β-methylstyrene by recombinant sperm whale myoglobin, *J. Biol. Chem.* **268:**803–809.

88. Allentoff, A. J., Bolton, J. L., Wilks, A., Thompson, J. A., and Ortiz de Montellano, P. R., 1992, Heterolytic versus homolytic peroxide bond cleavage by sperm whale myoglobin and myoglobin mutants, *J. Am. Chem. Soc.* **114:**9744–9749.

89. Weiss, R. H., and Estabrook, R. W., 1986, The mechanism of cumene hydroperoxide-dependent lipid peroxidation: The function of cytochrome P-450, *Arch. Biochem. Biophys.* **251:**348–360.

90. Weiss, R. H., and Estabrook, R. W., 1986, The mechanism of cumene hydroperoxide-dependent lipid peroxidation: The significance of oxygen uptake, *Arch. Biochem. Biophys.* **251:**336–347.

91. Vaz, A. D. N., and Coon, M. J., 1987, Hydrocarbon formation in the reductive cleavage of hydroperoxides by cytochrome P-450, *Proc. Natl. Acad. Sci. USA* **84:**1172–1176.

92. Vaz, A. D. N., Roberts, E. S., and Coon, M. J., 1990, Reductive β-scission of the hydroperoxides of fatty acids and xenobiotics: Role of alcohol-inducible cytochrome P-450, *Proc. Natl. Acad. Sci. USA* **87:**5499–5503.

93. Song, W.-C., Baertschi, S. W., Boeglin, W. E., Harris, T. M., and Brash, A. R., 1993, Formation of epoxyalcohols by a purified allene oxide synthase. Implications for the mechanism of allene oxide synthesis, *J. Biol. Chem.* **268:**6293–6298.

94. Adachi, S., Nagano, S., Ishimori, K., Watanabe, Y., Morishima, I., Egawa, T., Kitagawa, T., and Makino, R., 1993, Roles of proximal ligand in heme proteins: Replacement of proximal histidine of human myoglobin with cysteine and tyrosine by site-directed mutagenesis as models for P-450, chloroperoxidase, and catalase, *Biochemistry* **32:**241–252.

95. Wang, J. M., Mauro, M., Edwards, S. L., Oatley, S. J., Fishel, L. A., Ashford, V. A., Xuong, N. H., and Kraut, J., 1990, X-ray structures of recombinant yeast cytochrome c peroxidase and three heme-cleft mutants prepared by site-directed mutagenesis, *Biochemistry* **29:**7160–7173.

96. Goodin, D. B., and McRee, D. E., 1993, The Asp-His-Fe triad of cytochrome *c* peroxidase controls the reduction potential, electronic structure, and coupling of the tryptophan free radical to the heme, *Biochemistry* **32:**3313–3324.

97. Vitello, L. B., Erman, J. E., Miller, M. A., Mauro, J. M., and Kraut, J., 1992, Effect of Asp-235→Asn substitution on the absorption spectrum and hydrogen peroxide reactivity of cytochrome c peroxidase, *Biochemistry* **31:**11524–11535.

98. Sundaramoorthy, M., Choudhury, K., Edwards, S. L., and Poulos, T. L., 1991, Crystal structure and preliminary functional analysis of the cytochrome c peroxidase His[175]Gln proximal ligand mutant, *J. Am. Chem. Soc.* **113:**7755–7757.

99. Choudhury, K., Sundaramoorthy, M., Mauro, J. M., and Poulos, T. L., 1992, Conversion of the proximal histidine ligand to glutamine restores activity to an inactive mutant of cytochrome c peroxidase, *J. Biol. Chem.* **267:**25656–25659.

100. Poulos, T. J., and Kraut, J., 1980, The stereochemistry of peroxidase catalysis, *J. Biol. Chem.* **255:**8199–8205.

101. Erman, J. E., Vitello, L. B., Miller, M. A., Shaw, A., Brown, K. A., and Kraut, J., 1993, Histidine 52 is a critical residue for rapid formation of cytochrome *c* peroxidase compound I, *Biochemistry* **32:**9798–9806.

102. Dalziel, K., and O'Brien, J. R. P., 1954, Spectrophotometric studies of the reaction of methaemoglobin with hydrogen peroxide, 1. The formation of methaemoglobin-hydrogen peroxide, *Biochem. J.* **56:**648–659.

103. George, P., and Irvine, D. H., 1956, A kinetic study of the reaction between ferrimyoglobin and hydrogen peroxide, *J. Colloid Sci.* **11:**329–339.

104. Yonetani, T., and Schleyer, H., 1967, Studies on cytochrome c peroxidase. IX. The reaction of ferrimyoglobin with hydroperoxides and a comparison of peroxide-induced compounds of ferrimyoglobin and cytochrome c peroxidase, *J. Biol. Chem.* **242:**1974–1979.

105. Fox, J. B., Jr., Nicholas, R. A., Ackerman, S. A., and Swift, C. E., 1974, A multiple wavelength analysis of the reaction between hydrogen peroxide and metmyoglobin, *Biochemistry* **13:**5178–5186.

106. Roe, J. A., and Goodin, D. B., 1993, Enhanced oxidation of aniline derivatives by two mutants of cytochrome *c* peroxidase at tryptophan 51, *J. Biol. Chem.* **268:**20037–20045.

107. Smith, A. T., Sanders, S. A., Greschik, H., Thorneley, R. N. F., Burke, J. F., and Bray, R. C., 1992, Probing the mechanism of horseradish peroxidase by site-directed mutagenesis, *Biochem. Soc. Trans.* **20:**340–345.

108. Job, D., and Dunford, H. B., 1976, Substituent effect on the oxidation of phenols and aromatic amines by horseradish peroxidase compound I, *Eur. J. Biochem.* **66:**607–614.

109. DePillis, G. D., Sishta, B. P., Mauk, A. G., and Ortiz de Montellano, P. R., 1991, Small substrates and cytochrome c are oxidized at different sites of cytochrome c peroxidase, *J. Biol. Chem.* **266:**19334–19341.

110. Ortiz de Montellano, P. R., 1992, Catalytic sites of hemoprotein peroxidases, *Annu. Rev. Pharmacol. Toxicol.* **32:**89–107.

111. Raag, R., Swanson, B. A., Poulos, T. L., and Ortiz de Montellano, P. R., 1990, Formation, crystal structure, and rearrangement of a cytochrome P-450$_{cam}$ iron–phenyl complex, *Biochemistry* **29**:8119–8126.

112. Swanson, B. A., Dutton, D. R., Lunetta, J. M., Yang, C. S., and Ortiz de Montellano, P. R., 1991, The active sites of cytochromes P450 IA1, IIB1, IIB2, and IIE1. Topological analysis by *in situ* rearrangement of phenyl–iron complexes, *J. Biol. Chem.* **266**:19258–19264.

113. Ortiz de Montellano, P. R., 1987, Control of the catalytic activity of prosthetic heme by the structure of hemoproteins, *Acc. Chem. Res.* **20**:289–294.

114. Sakurada, J., Takahashi, S., and Hosoya, T., 1986, Nuclear magnetic resonance studies on the spatial relationship of aromatic donor molecules to the heme iron of horseradish peroxidase, *J. Biol. Chem.* **261**:9657–9662.

115. Modi, S., Behere, D. V., and Mitra, S., 1989, Interaction of thiocyanate with horseradish peroxidase. ^{1}H and ^{15}N nuclear magnetic resonance studies, *J. Biol. Chem.* **264**:19677–19684.

116. Goodin, D. B., Davidson, M. G., Roe, J. A., Mauk, A. G., and Smith, M., 1991, Amino acid substitutions at tryptophan-51 of cytochrome c peroxidase: Effects on coordination, species preference for cytochrome *c*, and electron transfer, *Biochemistry* **30**:4953–4962.

117. Miller, V. P., DePillis, G. D., Ferrer, J. C., Mauk, A. G., and Ortiz de Montellano, P. R., 1992, Monooxygenase activity of cytochrome *c* peroxidase, *J. Biol. Chem.* **267**:8936–8942.

118. Smith, A. T., Sanders, S. A., Thorneley, R. N. F., Burke, J. F., and Bray, R. R. C., 1992, Characterisation of a haem active-site mutant of horseradish peroxidase, Phe41→Val, with altered reactivity towards hydrogen peroxide and reducing substrates, *Eur. J. Biochem.* **207**:507–519.

119. MacDonald, T. L., Gutheim, W. G., Martin, R. B., and Guengerich, F. P., 1989, Oxidation of substituted N,N-dimethylanilines by cytochrome P-450: Estimation of the effective oxidation potential of cytochrome P-450, *Biochemistry* **28**:2071–2077.

120. Hayashi, Y., and Yamazaki, I., 1979, The oxidation–reduction potentials of compound I/II and II/ferric couples of horseradish peroxidases A$_2$ and C, *J. Biol. Chem.* **254**:9101–9106.

121. Fee, J. A., and Valentine, J. S., 1977, Chemical and physical properties of superoxide, in: *Superoxide and Superoxide Dismutases* (A. M. Michelson, J. M. McCord, and I. Fridovich, eds.), Academic Press, New York, pp. 19–60.

122. Kobayashi, S., Nakano, M., Kimura, T., and Schaap, P. A., 1987, On the mechanism of the peroxidase catalyzed oxygen transfer reaction, *Biochemistry* **26**:5019–5022.

123. Egan, R. W., Gale, P. H., Vandenheuvel, W. J., Baptista, E. M., and Kuehl, F. A., 1980, Mechanism of oxygen transfer by prostaglandin hydroperoxidase, *J. Biol. Chem.* **255**:323–326.

124. Doerge, D. R., and Corbett, M. D., 1991, Peroxygenation mechanism of chloroperoxidase-catalyzed N-oxidation of arylamines, *Chem. Res. Toxicol.* **4**:556–560.

125. Hughes, M. F., Smith, B. J., and Eling, T. E., 1992, The oxidation of 4-aminobiphenyl by horseradish peroxidase, *Chem. Res. Toxicol.* **5**:340–345.

126. Ple, P., and Marnett, L. J., 1989, Alkylaryl sulfides as peroxidase reducing substrates for prostaglandin H synthase: Probes for the reactivity and environment of the ferryl–oxo complex, *J. Biol. Chem.* **264**:13983–13993.

127. Watanabe, Y., Iyanagi, T., and Oae, S., 1980, Kinetic study on enzymatic S-oxygenation promoted by a reconstituted system with purified cytochrome P-450, *Tetrahedron Lett.* **21**:3685–3688.

128. Perez, U., and Dunford, H. B., 1990, Transient-state kinetics of the reactions of 1-methoxy-4-(methylthio)benzene with horesradish peroxidase compounds I and II, *Biochemistry* **29**:2757–2763.

129. Casella, L., Gullotti, M., Ghezzi, R., Poli, S., Beringhelli, T., Colonna, S., and Carrea, G., 1992, Mechanism of enantioselective oxygenation of sulfides catalyzed by chloroperoxidase and horseradish peroxidase. Spectral studies and characterization of enzyme–substrate complexes, *Biochemistry* **31**:9451–9459.

130. Doerge, D. R., Cooray, N. M., and Brewster, M. E., 1991, Peroxidase-catalyzed S-oxygenation: Mechanism of oxygen transfer for lactoperoxidase, *Biochemistry* **30**:8960–8964.

131. Ozaki, S.-I., and Ortiz de Montellano, P. R., 1994, Molecular engineering of horseradish peroxidase. Highly enantioselective sulfoxidation of aryl alkyl sulfides by the Phe-41→Leu mutant, *J. Am. Chem. Soc.* **116:**4487–4488.

132. Blee, E., and Schuber, F., 1989, Mechanism of S-oxidation reactions catalyzed by a soybean hydroperoxide-dependent oxygenase, *Biochemistry* **28:**4962–4967.

133. Blee, E., and Schuber, F., 1990, Efficient epoxidation of unsaturated fatty acids by a hydroperoxide-dependent oxygenase, *J. Biol. Chem.* **265:**12887–12894.

134. Blee, E., Wilcox, A. L., Marnett, L. J., and Schuber, F., 1993, Mechanism of reaction of fatty acid hydroperoxides with soybean peroxygenase, *J. Biol. Chem.* **268:**1708–1715.

135. Hewson, W. D., and Hager, L. P., 1979, Peroxidases, catalases, and chloroperoxidase, in: *The Porphyrins. Part B, Vol. 7* (D. Dolphin, ed.), Academic Press, New York.

136. Griffin, B. W., 1991, Chloroperoxidase: A review, in: *Peroxidases in Chemistry and Biology* (J. Everse, K. E. Everse, and M. B. Grisham, eds.), CRC Press, Boca Raton, FL, pp. 85–137.

137. Samokyszyn, V. M., and Ortiz de Montellano, P. R., 1991, Topology of the chloroperoxidase active site: Regiospecificity of heme modification by phenylhydrazine and sodium azide, *Biochemistry* **30:**11646–11653.

138. Vane, J., and Botting, R., 1988, Inflammation and the mechanism of action of anti-inflammatory drugs, *FASEB J.* **2:**89–96.

Part II

Structurally Defined Bacterial Enzymes

CHAPTER 3

Twenty-five Years of P450cam Research

Mechanistic Insights into Oxygenase Catalysis

ERNEST J. MUELLER, PAUL J. LOIDA, and
STEPHEN G. SLIGAR

1. Introduction

It has been nearly 25 years since the discovery of the first microbial cytochrome P450 system by the Gunsalus laboratory. Since that initial discovery, seminal breakthroughs have been made in mammalian P450s through work with microsomal fractions[1,2] and the astounding tour de force accomplishments in the purification of membrane-bound P450 systems originally initiated by Coon and his colleagues.[3,4] However, it was generally realized by the early 1980s that the availability of a purified cytochrome P450 and its redox transfer partners offered an unparalleled opportunity for detailed structure–function studies of this important class of hemoproteins. Through the research efforts of numerous laboratories over the last 25 years we have reached a point where essentially every spectroscopy currently known has been applied to the first of these bacterial P450s, cytochrome P450cam.

It has been interesting to watch the development and explosive growth of the cytochrome P450 field. Because of the obvious importance of the P450s in the metabolism of xenobiotics and in the critical steps of steroid hormone biosynthesis, study of this enzyme from the standpoint of regulation and mechanism has occupied a central place in the interests of pharmacologists and toxicologists. From the earliest studies of Klingenberg[5] and Garfinkle[6] and the perception of Mason,[7] and Sator,[8] it was clear that this hemoprotein had fundamental inorganic chemical properties that were quite distinct from the normally encountered globins. Many of these unique spectral properties were ulti-

ERNEST J. MUELLER, PAUL J. LOIDA, and STEPHEN G. SLIGAR • Department of Biochemistry and Beckman Institute for Advanced Science and Technology, University of Illinois, Urbana, Illinios 61801.

Cytochrome P450: Structure, Mechanism, and Biochemistry (Second Edition), edited by Paul R. Ortiz de Montellano. Plenum Press, New York, 1995.

mately traced to the presence of a thiolate group as the sixth axial ligand as first suggested by Mason[9] to explain the unique rhombic distortion observed in the ferric high-spin EPR signal. The identity of this sixth ligand was proved by the observation of a perturbation in the Fe–S stretch frequency after direct sulfur and iron isotopic substitution by Champion *et al.*,[10] and finally by the solution of the three-dimensional crystal structure of $P450_{cam}$ by Poulos.[11] Elegant studies using magnetic circular dichroism,[12] X-ray absorption spectroscopy,[13] solution of two additional crystal structures,[14,15] and numerous other direct and indirect techniques have unambiguously identified cysteine as the axial ligand of the P450s. Despite this certain assignment of the axial ligand, the role of cysteine in catalysis remains poorly understood. Considering the enormous amount of information known about the mechanisms of P450, it is humbling how little is actually understood at the level that would satisfy the inorganic or physical chemist.

Looking back at 25 years of work with a single P450 isozyme it is interesting to note how the field has grown and the focus on individual aspects changed. In many instances the force driving a change in focus has been the emergence of new technology. Because of its importance in medicine, it is not surprising that when recombinant DNA technology emerged, P450s were one of the first enzymes cloned and expressed[16–21] and mutated.[22–28] When molecular biological tools were developed to provide quantitative information on gene regulation and transcription control, the various P450 systems provided clear insight into general mechanisms. When the quest for evolutionary relationships became fashionable, it was the prior work in assimilating the great diversity of P450 phylogeny that provided a key experimental input.[29] When the general problems of membrane–protein interactions and the mechanisms of multiprotein complex assembly in bilayers became fashionable, it was the P450s with their associated electron transfer cofactors that provided the opportunity for elucidation of the fundamental mechanistic information on the key steps of recognition and function in mesoscale assemblies. And now, when sophisticated theoretical advances in the general problem of protein folding demand the largest data base on similar structures, it is again the P450 systems that provide an optimal test case.

A simple examination of the scope and attendance of P450 meetings shows the explosive growth of the field. In the beginning, which for this discussion we define as from discovery through the early to mid-1970s, scientific gatherings were defined solely by the protein "P450" and not by the discipline or subdiscipline of investigation. By today's standards, meeting attendance was relatively modest. Special topics sessions at the Federation meetings were attended by 100 or so people. Remarkably, most of the talks were given by students and beginning scientists. Chemists intermingled with pharmacologists, biochemists, physicists, and toxicologists. As knowledge about this wonderful enzyme expanded, it became clear that the P450 system posed an ideal system to study a wide variety of fundamental problems in molecular biology, chemistry, physics, evolution, and the like. Meetings proliferated and became more focused. This evolution brought the decided advantage of collateral input from related disciplines that would serve to be critical in the advancement of the field, but the advantage for the advancement of understanding through what the information scientist calls the "browsing factor" has been pushed aside. Currently, P450 occupies a central place in the schedules of inorganic, organic, physical, medicinal, and biochemists; physicists, molecular biologists, oncologists, toxicologists, and the subdisciplines of life sciences: entomology, bacteriology, cell biology, etc.—the list extends to the horizon. It now takes a focused effort to bring the disparate disciplines together to assemble an overall picture that is absolutely required in order to advance the understanding of the P450 field. Yet, this must be done to elucidate the molecular details of enzyme action. By understanding the details of P450 substrate recognition and reaction mechanism, we hope to push back the frontiers of molecular design and harness P450 catalysis for application in bioremediation and synthesis of pharmacological substances for the direct benefit of mankind. Understanding the interrelationships of P450s in human function will open the frontiers of clinical relevance where disease states can be understood and alleviated.

This chapter is but a small part of the second edition of a volume that has become the most referenced review of the P450 literature. The purpose of such a work should not be to provide a myopic view of one's own research interests, but rather to provide some easy starting place for exciting the interests of the next generation of scientists in pursuing what has certainly become one of the most interesting enzymatic systems in biochemistry. For one of us (S.G.S.), who has been a continuous part of the 25-year history of P450 research, this enzyme has not only "paid the rent" but provided intellectual stimulation and scientific friendships. Hopefully this chapter, and the entire volume, will serve to entice the young investigator into the most exciting field of P450 mechanisms.

2. The Cytochrome P450 Reaction Cycle

For 25 years the reaction cycle of P450 has remained nearly invariant.[30] Mimetic inorganic model systems and research efforts in related enzyme systems that activate molecular dioxygen or process various reduced states of oxygen such as hydrogen peroxide have provided great insight into the mechanism that nature employs to catalyze oxygen–oxygen bond scission, carbon–hydrogen bond activation, and ultimately produce a functionalized substrate. Twenty-five years! Most of the interesting catalytic mechanisms of cofactor enzymology, such as pyridoxal, biotin, or thymine pyrophosphate, were essentially solved in far less time. Over coffee one morning many years ago, a good friend and colleague pointed out that the problem with P450 catalysis is that the substrates include electrons, protons and dioxygen, while one of the products is water—indistinguishable from the reaction solvent. These are not normal bioorganic substrates. The single classical organic participant in the entire reaction cycle of P450 is a low-molecular-weight hydrocarbon. It should come as no surprise that much of the bioorganic input into the mechanism has come from studies that have focused on this one reactant.

It has become commonplace to begin reviews of the cytochromes P450 by discussing the currently envisioned reaction cycle step by step, as shown in Fig. 1. Although our review follows this established structure, we encourage the reader to approach P450cam by focusing on the inorganic chemistry of the iron-porphyrin oxygen system and the organic chemistry of hydrogen atom abstraction and radical recombination. In doing this, it is important to mentally highlight the features of the mechanism that are unknown, unproven, or just wishful thinking.

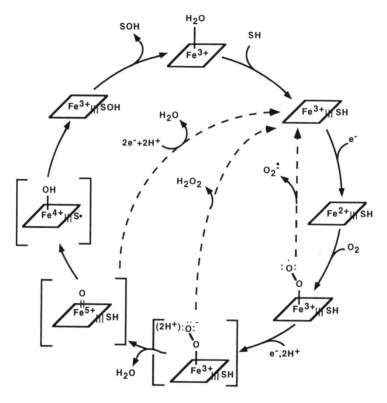

FIGURE 1. P450$_{cam}$ reaction cycle with uncoupling side reactions shown. SH = substrate, and the heme is shown as a trapezoid. Intermediates depicted in brackets are proposed structures and have not been observed. Charges shown in these bracketed structures represent one of several possible electronic states, and are not meant to indicate the actual charge on the metal center.

   Keeping this latter viewpoint in mind, it is important to emphasize that the proposed reaction cycle contains hypothetical states that may have *no basis in reality*. The sole basis for their existence in a book devoted to P450 mechanism is one of the following: Following good freshman-level chemistry, the proposed states balance electrons and nuclei throughout the reaction cycle. Or, there have been identified similar charge states in other heme systems (e.g., peroxidases) or inorganic model systems that *do not have a thiolate ligand or active site structure that in any way resemble* that of the P450 monooxygenases. Having now made their way into the textbooks, these assumptions allow us to propose a mechanism that is chemically plausible but still unproven. This situation must again be interpreted as a call for young investigators into the field of P450 research. Recent documentation of a nucleophilic peroxyanion reaction mechanism for the P450s reinforces the notion that we are only beginning to understand detailed mechanisms.[31]

   Thus, we continue the typical format for a P450 review by presenting an overview of the P450 reaction cycle and discuss each sequential step from the standpoint of current knowledge. Figure 1 focuses on the hemoprotein component of the P450 monooxygenase system, augmenting the well-defined and isolated intermediates with those suggested from a wide variety of indirect studies of other enzyme and model systems. Each of these hypothetical states maintains faith in the balance of electron and nuclei equivalents, but should be taken with the largest grain of salt and serve only to focus experimental design toward the ultimate isolation and characterization of the undefined intermediates in the P450 reaction cycle. A mechanism can never be proven beyond a doubt, it can only be disproven by experiment.

Discussion of P450$_{cam}$ catalysis typically begins with the substrate-free, oxidized (ferric iron) state. Easily isolated, this protein has a redox potential of about –300 mV with the heme iron in the low-spin six-coordinated form.[32] Solution of the X-ray structure of substrate-free cytochrome unambiguously documented water or OH⁻ as the sixth axial ligand *trans* to the thiolate proximal ligand provided by cysteine coordination. Because of the exhaustive review and discussion of ligand identification in the P450s,[4,33–36] no summary of this extensive literature will be presented in this chapter. The first key step in the reaction cycle is the binding of substrate. This fundamental aspect of molecular recognition not only defines completely the regio- and stereochemistry of oxygenation, but also provides the key "hand in glove" fit in the buried active site that excludes bulk water and ensures the efficient coupling of reducing equivalents from the pyridine nucleotide-coupled flavin–iron–sulfur electron transfer proteins. Only through recent extensive mutagenesis studies has the central nature of the protein–small molecule recognition event become appreciated. The first and last sections of this review will examine our current knowledge of the importance of bound substrate–protein coupled dynamic recognition.

The binding of camphor to the P450 active site results in dehydration of the heme iron environment, and a concomitant shift of the enzyme redox potential from –300 mV to about –170 mV. The free energy relationships between the spin state equilibrium, substrate binding, and redox potential shift are now well documented.[37] Even at first glance, a teleological mechanism can be appreciated when it is realized that the potential of the iron–sulfur donor in the subsequent first electron transfer operates at about –200 mV. Hence, a substrate-induced alteration in oxidation/reduction potential of the heme component could serve as a physiologically relevant "gate" for electron transfer. Through detailed analysis of kinetic processes, we now know that the spin state of the P450 heme iron is regulated by the accessibility of water to the six coordination position. Elegant theoretical investigation by Harris and Lowe[38] has suggested that nature has utilized a sophisticated application of electrostatic stabilization to poise the ferric $S = 5/2 \rightarrow S = 1/2$ spin transition such that simple water ligation can alter the dynamic spin state equilibrium significantly. Certainly the bacterial P450$_{cam}$ has provided the best studied system of spin-redox-substrate linkages. Since camphor metabolism in *Pseudomonas* utilizes P450$_{cam}$ as the first step in an advantageous growth pathway, perhaps this "switch" in terms of reducing equivalent flow offers a selective evolutionary advantage. Other P450 systems, which perhaps do not need to control so tightly the catalytic throughput of the oxygenase in response to the presence of a given substrate, have not evolved such an elegant regulatory scheme.

Oxygen binding forms a ferric-superoxide species, the so-called "oxy-P450," which has been characterized.[30,39] The next step in the P450 reaction cycle depicted in Fig. 1 involves the transfer of an electron from the iron–sulfur protein putidaredoxin to the P450 hemoprotein fraction. Implicit in this event is the formation of a specific complex between these two macromolecules. Intensive investigations in other multicomponent electron transfer systems have suggested that the dynamics and structure of the donor–acceptor complex play key roles in the overall efficiency and regulation of the redox transfer process.

The events occurring after the delivery of the second electron have been the subject of debate in the P450 community for many years. The mechanism detailed in Fig. 1 depicts one possible reaction mechanism, which we believe is most consistent with the available data. None of the bracketed species have been directly observed, although many different spectroscopic methods have been applied to this end. The mechanism involves reduction to a ferric-peroxy species, which may exist alternatively as a ferrous-superoxide species. From this state two protons are delivered to the active site to facilitate the loss of one atom of oxygen as water and generate a high-valent iron–oxo species. Although depicted as an $Fe^{5+}=O$ moiety, this species can be stabilized through a number of resonance structures involving the heme ring and oxygen atom; the most stable of these is the $Fe^{4+}=O$-porphyrin.$^{+\bullet}$.[40] Another equivalent depiction of this species is $[FeO]^{3+}$, which avoids any implication of a specific iron–oxo bonding. The abstraction of a hydrogen atom by this compound from the substrate generates a substrate radical species, which collapses via an oxygen rebound,[41] to generate the ferric enzyme–product complex. The ferric-superoxy, ferric-peroxy, and high-valent iron species are each thought to be unstable, able to decompose by their own pathways. It is not known if any of these bracketed, hypothetical species exist as a discrete intermediate, and it is possible that reaction from the ferric-superoxide intermediate is concerted up to the point of producing a ferric–alcohol complex.

We now examine the current mechanistic information for each of these first steps in the catalytic reaction cycle of P450$_{cam}$. We cover recent information that has allowed the partial elucidation of the events leading to substrate binding. As a major question of molecular recognition, we address the structure and dynamics of the putidaredoxin–P450$_{cam}$ electron transfer complex in some detail in this review. We will concentrate on work accomplished since the last publication of this series which demonstrates the link between oxygen binding and proton transfer in P450$_{cam}$. Finally, this review will present the evidence for the existence of a high-valent iron intermediate as well as the existence of a discrete substrate radical within the P450$_{cam}$ reaction cycle.

2.1. Substrate Association

2.1.1. Substrate Access: Control by Protein Dynamics

The solution of the crystal structure for substrate-bound p450$_{cam}$[11] revealed that camphor binds to an active site with no obvious channel to the aqueous milieu. It was apparent that substrate binding could only occur in conjunction with a fairly large reorganization of the protein structure, and the protein dynamics necessary to allow substrate binding have been a subject of conjecture. The first postulate as to how P450$_{cam}$ binds substrate was published with the substrate-free X-ray structure determination, in which Poulos and co-workers[42] identified three regions in the substrate-free protein which showed increased thermal mobility relative to the structure of the substrate-bound enzyme. The three sections include the B' helix, containing residues 87–96; the loop joining helices F and G, involving residues 185–192; and amino acid 251, an aspartate residue that forms a salt bridge with Arg^{186}. This hypothesis was strengthened after the discovery of an enzyme inhibitor for $P450_{cam}$ containing a long aliphatic tail, and the subsequent solution of the inhibitor-bound enzyme complex.[43] This inhibitor–enzyme structure differed from previous structures in that the orientation of the side chains of residues Phe^{193}, Tyr^{96}, and Phe^{87} were all substantially altered. The interpretation of these differences was that these aromatic residues line the substrate access channel.

The postulate that substrate binding is possible only on protein reorganization has recently been confirmed using photoacoustic calorimetry. This work builds on the fact that expulsion of camphor is coupled to the photolysis of CO from $P450_{cam}$.[44] This phenomenon allows the induction of a nearly simultaneous egress of substrate from the active site of a population of substrate-bound CO $P450_{cam}$. Di Primo et al.[45] used this technique to demonstrate that loss of substrate from $P450_{cam}$ results in overall enthalpy and volume changes of -15.9 kcal/mole and 10.3 ml/mole respectively. Moreover, an intermediate was found to form during this process, with a lifetime of 170 nsec at 17°C. These experiments again point to substantial protein reorganization during substrate release, and by microscopic reversibility a similar process must occur during substrate binding. This work hinges on the assumption of camphor expulsion on CO dissociation, and further work is necessary to show that this linked dissociation is operant.

The role of salt bridges in regulating camphor access was explored by Deprez et al.[46] using a cosolvent system and varying ionic strength to perturb the dielectric of the protein environment. By careful choice of experimental conditions, the authors were able to measure the contribution of camphor binding separate from the microscopic spin state equilibrium. It was determined that the microscopic equilibrium constant for substrate dissociation decreases slightly as ionic strength is increased. When $P450_{cam}$ was placed in a cosolvent system with a lower dielectric, the microscopic dissociation equilibrium constant was found to substantially increase. Such behavior is consistent with a model in which one or more salt bridges must be broken, or at least weakened, for substrate to occupy the active site. Although the experiments of Deprez et al.[46] cannot assign the role of specific salt bridges to substrate binding, these findings are in accord with the proposal of Poulos et al.,[42] which implicated the Asp^{251}–Lys^{178} salt bridge as important in substrate binding.

Site-directed mutagenesis carried out in our laboratory has also implicated the above residues in substrate binding.[47] The on and off rates of substrate binding were determined for a number of mutants using stop-flow techniques first described by Peterson.[48] These rates are shown in Table I. The Tyr^{96} Phe mutant can be thought of as a control in these experiments, as the effective steric bulk changes very little in this mutation. Most mutations resulted in a large increase in k_{off}, and are therefore thought to be important in partially limiting the rate of substrate egress. The Phe^{87}Trp mutant, which results in a residue sizably larger than wild type, slows the rate of substrate binding, consistent with a model where substrate enters the active site after passing by Phe^{87}.

A further set of mutagenesis experiments carried out by Gerber[47] were designed to engineer an intramolecular tether in $P450_{cam}$, in an attempt to at least partially block

TABLE I
Effect of Specific Mutations on Substrate Access[a]

	k_{on} (M^{-1} sec^{-1})	k_{off} (sec^{-1})
Wild type	2.5×10^7	21
Phe^{87}Trp	0.5×10^7	94
Tyr^{96}Phe	3.1×10^{7b}	47
Thr^{101}Met	3.0×10^{7c}	192^c
Phe^{193}Cys	3.6×10^{7c}	104^c

[a]Determined at 20°C in 50 mM KPi pH 7.2 with either 20 or 100 μM camphor, except as noted.
[b]Determined at 10°C. Under these conditions $k_{on} = 1.0 \times 10^7$ M^{-1} sec^{-1}.
[c]Determined as in footnote a, but KCl was added to give [K$^+$] = 0.5 M. Under these conditions $k_{on} = 3.2 \times 10^7$ M^{-1} sec^{-1}, $k_{off} = 4.6$ sec^{-1}.

substrate access. The double mutant Ala^{194}Cys–Ala^{95}Cys was generated and shown by both selective radiolabeling and fluorescence techniques to form a viable disulfide linkage. This tethered protein, with linked B′ and G helices, resulted in an enzyme with increased substrate binding kinetics. The k_{on} for the disulfide-linked protein was 2.5 times faster, the k_{off} was 4.4 times faster than for the wild-type enzyme. Since this modification links two regions that are thought to be required to be mobile to allow substrate access, it was proposed that formation of the disulfide bond permanently opens, at least partially, the substrate's route to the active site. A similar set of mutants was made to link Tyr^{29} to Ala^{194}, this time using the bifunctional mercaptan bis-maleimidohexane (BMI). The generation of a Tyr^{29}Cys–Ala^{194}Cys double mutant allows, with the use of BMI, the tethering of the disordered N-terminal region of P450$_{cam}$ to the G helix. This mutant, once reacted with bifunctional mercaptan, results in a 2-fold reduction of k_{off}. This finding was interpreted to signify that the linkage of residues 29 and 196 through a six-carbon chain results in partial immobilization of protein residues required for substrate binding and a partial constriction of the substrate access channel.

The results of the mutagenesis and tethering experiments provide support for the location of the access channel originally proposed by Poulos et al.[42] In addition, these experiments specifically implicate Tyr^{29} and Phe^{193}, as important in modulating substrate access, with Phe^{87} playing a lesser role in this process. Phe^{96}, Phe^{87}, and Thr^{101} are all implicated as being involved in active site access, but their rolls are difficult to assess, as all three are involved in enzyme–substrate contacts (see Fig. 2).

The recent solution of X-ray structures for two other P450s also supports the idea that substrate access in P450$_{cam}$ occurs as suggested by Poulos. The heme domain of P450$_{BM-3}$ shares 17% identity with P450$_{cam}$, and the X-ray structure reveals that the active site heme is accessible through a long hydrophobic channel formed primarily by the β domain and the B′ and F helices of the α domain.[15] Furthermore, an analogous mode of substrate access is thought to operate for P450$_{terp}$, which shares 24% identity with P450$_{cam}$.[14]

The current evidence as to how camphor enters the active site of P450$_{cam}$ thus agrees, in general, with the model proposed by Poulos and co-workers.[42] The picture that emerges is one in which camphor squeezes into the active site via a hydrophobic corridor defined by the B′ helix and the F–G loop. Both Phe^{192} and Tyr^{29} are implicated to lie along the path of access. Once inside the active site, camphor is poised for hydroxylation with the required regio- and stereospecificity. The functional consequence of this complicated binding strategy is that water is excluded from the active site while chemistry occurs, an advantage that will be discussed further below. Future work will hopefully elucidate the dynamics necessary for P450cam to bind substrate while excluding water from the enzyme active site.

2.1.2. Substrate Binding

Substrate binding to P450cam elicits a spectral change related to a spin state conversion of the heme iron.[37,49] This convenient handle has spurred prolific studies of the events that occur on camphor binding, reviews of which are published.[50] Briefly, binding of substrate displaces water from the active site, changing the heme iron from a six-coordinate, low-spin configuration with water as an axial ligand to a five-coordinate, high-spin state. In the resulting complex, substrate sterically blocks water access to the active site. Camphor is not the only molecule able to produce this series of events, and a series of alternate substrates and competitive inhibitors have been discovered, the structures of which are depicted in Fig. 2.

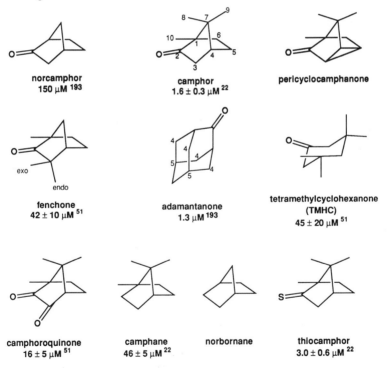

FIGURE 2. Structures for a number of P450cam substrates and competitive inhibitors. Compounds with reported K_d values for P450cam are labeled, along with the original reference.

The kinetics of this conversion has been examined by Fisher and Sligar.[51] In this work the spin equilibrium of $P450_{cam}$ substrate complex was perturbed using a < 20 μsec temperature jump of 3°C. The rate of spin relaxation was monitored by visible spectroscopy over a 5-msec period for several different substrate analogue–enzyme complexes. The researchers discovered a correlation between the percentage of high-spin species present at equilibrium and the transition rates of high- to low-spin iron, suggesting this transition is governed by a kinetically restricted access of water to the axial position.

The ability of camphor to induce a shift to high spin had been postulated to be indicative of either an in-plane heme iron, or the presence of a hydroxyl ligand at the active site.[52] The crystal structure shows that both substrate-free and substrate-bound forms contain iron out-of-plane.[11,42,53] In addition, spin-state conversion is ordinarily studied at ~pH 7, and thus the existence of a hydroxyl ligand seemed unlikely. Harris and Loew have shown, using Hartree–Fock calculations and molecular dynamics, that the protein itself creates an active site environment necessary and sufficient to allow spin-state conversion to occur.[38] This intriguing study implies that the protein can control the redox potential of the active species, a previously unidentified determinant of P450 specificity.

Besides the six crystallographic water molecules identified in the substrate-free crystal structure, a number of other waters have been implicated as important in the enzymatic reaction. Theoretical analysis suggests that mobile water may also be present in the active site, stabilizing the correct orientation of camphor in $P450_{cam}$.[54] Further work, perhaps with substrate analogues, will be necessary to confirm this hypothesis. In addition, preliminary studies using osmotic pressure experiments imply that a total of approximately 28 waters are involved in the $P450_{cam}$ catalytic reaction.[55] Some of these waters may be involved in the $P450_{cam}$–putidaredoxin interface.

$P450_{cam}$ also contains a cation binding site, noted over 20 years ago.[48,49] Studies have shown that potassium has the greatest affinity for the enzyme.[56] In solving the crystal structure, Poulos noted an unusual octahedral coordination site about water 515, and suggested that this "water" was rather the cation binding site.[53] The proposed potassium coordination sphere consists of the carbonyl moieties of Tyr^{96}, Gly^{93}, Glu^{94}, and Glu^{98} as well as two ordered waters, and suggests that the effect of potassium on substrate binding[48] is mediated via Tyr^{96} which forms a hydrogen bond with the C-2 carbonyl of camphor. Atkins and Sligar[22] have generated the $Tyr96^{Phe}$ mutant, which was found to have an altered potassium affinity.[56,57] The loss of binding free energy for potassium in the mutant was found to be the same as the loss of binding free energy for camphor,[22] and thus it was concluded that Tyr^{96} did indeed couple potassium and substrate binding.[57] However, work by these same authors in determining the K_d of camphor in the wild type and $Tyr^{96}Phe$ mutants as a function of varying potassium concentrations yielded conflicting results. The reported K_d of camphor in the mutant enzyme is lower than for the wild type under certain potassium concentrations. Further work remains to be done to unquestionably identify the means by which cation and substrate binding are linked.

The means by which $P450_{cam}$ binds heme is poorly understood. Although the crystal structure shows the protein–heme contacts, how the protein folds about the heme has not been studied. Recent work by Yoshikawa et al.[58] suggests that the residues on the proximal side of the heme, and particularly Glu^{286}, form a hydrogen-bonding network that stabilizes the proper heme positioning. More work is needed to determine how heme is incorporated in P450$_{cam}$.

The substrate-bound crystal structure of P450$_{cam}$ provides the best look at how camphor binds enzyme.[11] Structures of P450$_{cam}$ with several alternate substrates have also been solved.[59] The binding pocket is composed of mostly hydrophobic residues, as one might expect given the nature of the natural substrate (see Table II). None of these residues, except for the hydroxyl moiety of Tyr[96], have direct contacts with the substrate. However, this series of residues can be thought of, as defining a fence or barrier about the substrate, which allows 1–2 Å of movement in any direction.

2.2. Electron Transfer

The reducing equivalents that drive the hydroxylation of camphor are, in *P. putida,* ultimately derived from NADH. The flow of electrons proceeds through what is now the well-known redox proteins putidaredoxin reductase and putidaredoxin to P450$_{cam}$.[60,61] A similar system operates in eukaryotes, but requires NADPH as the electron source.[62] Perhaps the best studied, nonbacterial electron transfer enzymes in P450s are the adrenal mitochondrial enzymes adrenodoxin reductase and adrenodoxin, which are chemically competent electron carriers for P450 11A1 (P450scc) and P450 11B1 (P450$_{11\beta}$).

Because of adrenodoxin's similarity with the putidaredoxin redox chain, a brief review of this system may help the reader put the putidaredoxin system in perspective. The interactions of adrenodoxin with adrenodoxin reductase and the P450 redox partner have been suggested to be electrostatic in nature.[63,64] Modification studies of adrenodoxin[65,66] and P450 11A1[67,68] as well as mutagenesis of adrenodoxin[69] and P450 11A1[70] have attempted to map out the adrenodoxin reductase–adrenodoxin and adrenodoxin–P450

TABLE II
Nearest P450$_{cam}$ Residues to Camphor, by Atom

Camphor carbon	P450$_{cam}$ residue	Distance (Å)
C-2 carbonyl	Y[96]-OH	1.6
	F[87] Cε2	3.8
	T[101] Oγ	5.0
C-3	T[101] Cγ	4.1
C-5	L[244] Cδ	4.0
	Fe	4.5
C-6	G[248] Cα	4.0
C-8	V[295] Cγ1	3.5
	I[395] Cδ	4.1
	D[297] Oγ	4.1
	F[87] Cε2	4.6
C-9	V[396] Cγ2	4.0
	T[252] Cγ	4.1
	V[295] Cγ2	4.2
C-10	T[185] Cγ	3.8
	V[247] Cγ	3.8
	V[396] Cγ2	3.9
	F[87] Cζ	4.0

binding sites. Many of the modifications affect either binding or electron transport only, and thus the mechanism for each of these phenomena are just starting to be understood. Preliminary crystallographic data have been reported for bovine adrenodoxin[71] and adrenodoxin reductase,[72] and the fairly recent cloning of the human adrenodoxin reductase gene[73] gives hope that structural information will soon be forthcoming for this redox isozyme.

Bovine adrenodoxin shares 37% sequence homology with putidaredoxin, yet is not chemically competent to receive reducing equivalents from putidaredoxin reductase.[74] Bovine ferredoxin, however, functions as the intermediate electron transfer reagent in either the adrenodoxin[75] or putidaredoxin[76] systems. The adrenodoxin-mediated transfer of electrons from adrenodoxin reductase to P450 has been proposed to proceed via both a ternary complex[77,78] and a ferredoxin shuttle.[63,64,79,80]

In the $P450_{cam}$ electron transport system, putidaredoxin reductase, the first electron transport protein, is an FAD-containing protein with a molecular weight of 48,000, as determined by SDS-PAGE and gel filtration. NADH delivers two electron equivalents to the protein which mediates two, one-equivalent transfers to putidaredoxin, although a stable semiquinone species has not been detected.[81] The reaction of putidaredoxin with putidaredoxin reductase has been proposed to proceed through an unusual two-site, modified Ping-Pong mechanism.[82] The relatively recent engineered overexpression of putidaredoxin reductase[83] will hopefully spur future work on this redox protein as a starting point for site-directed mutagenesis and a source for the quantities of protein necessary for structure determination.

In contrast to the redox reaction between putidaredoxin reductase and putidaredoxin, the kinetics of electron transfer from putidaredoxin to $P450_{cam}$ is one of the best studied in the field of biological electron transfer. Gunsalus and co-workers elucidated the fundamental aspects of redox kinetics via this Fe_2S_2 protein,[30,61,84,85] and showed that the steps involving electron transfer to $P450_{cam}$ were the two slowest steps in the hydroxylation of camphor, with the second electron transfer limiting the overall reaction.[86] While a number of reducing agents are competent to reduce substrate-bound, ferric $P450_{cam}$, only putidaredoxin has been shown to reduce the ferric-superoxide enzyme species; thus, it was concluded that putidaredoxin acts not only as a redox shuttle, but also as an effector in camphor hydroxylation.[50,76] It was subsequently discovered that substrate binding regulated the redox potential of $P450_{cam}$, with camphor binding shifting the potential of the enzyme from -300 to -170 mV.[32,37]

Careful study of the first[87,88] and second[89,90] electron transfer events by Peterson and co-workers has shown that putidaredoxin binding is not rate determining for either process; rather some step subsequent to binding governs the rate of these reactions. Although the exact rates of these steps are kinetically masked, bounds have been determined for these steps, and a simple model distinguishing the binding from the electron transfer steps has been proposed. The Michaelis constants of $P450_{cam}$ for putidaredoxin were found to differ significantly; the K_m of ~1 μM for the first redox reaction at 0.16 M buffer, 10°C, is sevenfold lower than that of the second electron transfer, when measured in 0.1 M KCl at 4°C. This finding suggests that the two electron transfer reactions may differ fundamentally in their affinity, sites, or reaction mechanism, as suggested previously.[91] The first electron transfer reaction was shown to proceed via a bimolecular putidaredoxin–P450$_{cam}$complex, with a k_{obs} of 10–15 sec^{-1} at 4°C when saturated with putidaredoxin. Ionic strength was found to alter the binding of these two proteins, with a tenfold increase in ionic strength resulting in a sixfold decrease in the K_m between the redox partners,[81,87] suggesting an electrostatic association. The second electron transfer was found to have a maximal rate of 75–80 sec^{-1} at 4°C. Thus, at higher putidaredoxin concentrations the first redox reaction is rate limiting, but at physiological concentrations the second electron transfer becomes the overall rate-limiting step.[90]

Chemical modification of putidaredoxin with a water-soluble carbodiimide abolishes the ability to receive reducing equivalents from putidaredoxin reductase. However, modification does not affect the UV/visible spectrum of the enzyme, which led Geren et al.[74] to conclude that the Fe$_2$S$_2$ center is unperturbed by this modification. Trypsin digests of the modified protein were isolated by HPLC and sequenced. The most heavily modified residues, Asp^{58} and Glu^{65}, Glu^{67}, Glu^{72}, and Glu^{77,74}, are located in the analogous region important in adrenodoxin reductase–adrenodoxin binding.[65]

The Gunsalus laboratory discovered that a tryptophan at the carboxy-terminus of putidaredoxin was required for efficient electron transfer,[92] an observation also followed up later by Davies et al.[93] Using site-directed mutagenesis, these researchers were able to demonstrate that Trp^{106} was required for putidaredoxin to mediate electron transfer between putidaredoxin reductase and P450$_{cam}$. Although this mutation did not affect the redox potential of this protein, it was discovered that mutation of this residue primarily affects the K_d of putidaredoxin for P450$_{cam}$.[94] The fact that Trp^{106} is the sole tryptophan residue in putidaredoxin allows the use of fluorescence to monitor the environment of this residue. Both steady-state and time-resolved fluorescence experiments were performed on the oxidized and reduced forms of the Fe$_2$S$_2$ protein. Multiple species were detected for both forms of putidaredoxin, suggesting that oxidized and reduced proteins exist in a variety of states.[95] This conformational microheterogeneity is consistent with a mechanism of electron transfer regulation termed conformational gating. In this process protein motions allow several possible conformations, with one more favorable for electron transfer, and in this way the rate of electron transfer is limited. However, the multiple tryptophans in P450$_{cam}$preclude the extension of this technique to sample the environment of Trp^{106} in the putidaredoxin–P450$_{cam}$ complex. Thus, an alternate mechanism of regulation by different docking conformations and electron transfer via covalent[96] or electrostatic bonds cannot be ruled out by these experiments.

Until recently, elucidation of the docked structure of the redox active complex between putidaredoxin and P450$_{cam}$ was stymied by the lack of structural data for putidaredoxin. Initial studies relied on the bimolecular complex models proposed for the structurally characterized proteins cytochrome c and cytochrome b_5,[97] and capitalized on the fact that rat liver b_5 is a chemically competent redox partner for P450$_{cam}$.[76] Stayton et al.[98] were able to study the association of b_5 and P450$_{cam}$ by fluorescence, after engineering surface cysteines on b_5 which were subsequently labeled with the environmentally sensitive fluor acrylodan. Using b_5 labeled at two different locations, these researchers were able to measure a dissociation constant for P450$_{cam}$ at pH 7.15 of 0.65 ± 0.05 μM. A structure for the P450$_{cam}$–b_5 complex was proposed[99] relying on the assumption that P450$_{cam}$ binds b_5

in the same fashion as in the electrostatic model proposed by Salemme for the c–b_5 complex. This model suggested that the basic $P450_{cam}$ residues Arg^{112}, Arg^{364}, Arg^{72}, and Lys^{344} formed electrostatic interactions with b_5. Electrostatic modeling of the $P450_{cam}$ surface shows a striking area of cationic charge directly above the proximal cysteine ligand, with a diameter of ~20 Å, which includes the above-mentioned residues.[100]

The finding that putidaredoxin competitively inhibits b_5–$P450_{cam}$ association was interpreted by Stayton *et al.* to suggest that $P450_{cam}$ binds b_5 and putidaredoxin at the same, or overlapping sites.[99] Subsequent site-directed mutagenesis of $P450_{cam}$ residues Arg^{72} and Lys^{344} to either glutamate or glutamine have all resulted in a modest lowering (20–60%) of binding affinities for putidaredoxin to $P450_{cam}$.[100] However, recent work by Koga *et al.*[101] provides the best evidence for the involvement of one of these basic residues in forming the $P450_{cam}$–putidaredoxin complex. These researchers found that the $P450_{cam}$ mutant $Arg^{112}Cys$ has a K_m for putidaredoxin of 106 µM, six times higher than the wild-type enzyme. The V_{max} for this mutant is within a factor of two of that of the wild type. The spectral characteristics of the $Arg^{112}Cys$ reaction intermediates, the K_d for camphor, and the ability to bind CO are comparable to those found for the wild type, but the autoxidation rate of the oxy-P450 is roughly four times faster than in the native enzyme. If significant homodimer formation has not skewed these results, it appears that at least Arg^{112} is involved in forming the $P450_{cam}$–putidaredoxin complex. Further work is required before the protein contacts are unambiguously established.

The effort to determine the structure of putidaredoxin has begun to bear some fruit within the past few years, but solution of the complete structure has been slowed by our inability to crystallize the protein and the paramagnetic broadening effected by the Fe_2S_2 cluster. Site-directed mutagenesis of the cysteine residues[102] has confirmed the proposals of Tuls *et al.*[103] and Cupp and Vickery[104] that the putidaredoxin cysteine residues 39, 45, 48, and 86 are liganded to iron. Pochapsky and co-workers have applied NMR methodology to obtain a partial structure of putidaredoxin,[105] and their further work has allowed the derivation of a complete structural model.[106] The structural changes elicited on reduction of the iron center have also recently been elucidated.[107,108] These recent advances should usher in future work on electron transfer in the putidaredoxin system, with the ultimate goal of determining whether the mechanism of electron transfer regulation in the protein occurs by conformational gating,[109,110] covalent switching,[96] or a yet undescribed process.

Recently, Pochapsky proposed a structure for the putidaredoxin:P450cam complex.[194] This model was generated from a starting complex in which the metal centers were as close together as possible, and the Trp^{106} of putidaredoxin was able to interact with an aromatic residue on P450cam. This initial complex was subjected to an energy minimization routine. The final structure resulted in a 12Å Fe–Fe distance, with Trp^{106} of putidaredoxin interacting with Tyr^{78} of P450cam. Three residues on putidaredoxin, Asp^{34}, Asp^{38}, and the Trp^{106} carboxylate, were proposed to form intermolecular salt bridges with the P450cam residues Arg^{109}, Arg^{112}, and Arg^{79} respectively. Five intermolecular hydrogen bonds were also proposed. Further investigations will be needed to test the validity of this proposed electron transfer structure, and to elucidate the path of electron transfer.

2.3. Diatomic Ligand Binding

2.3.1. Ligation of Carbon Monoxide

The binding of oxygen to P450$_{cam}$ induces a well-defined UV/visible spectroscopic shift, which has allowed characterization of the substrate-bound ferric-superoxide (or oxy-) and carbon monoxide complexes (see Fig. 1).[111] Historically, CO has been an extremely useful tool in the study of hemoproteins, acting as a dead-end inhibitor in the place of oxygen. Unlike the oxy-complex, CO complexes are extremely stable, and the reaction of CO with P450 enzymes is well known to give a red-brown species for which this class of enzymes was named. The crystal structure for the P450$_{cam}$ substrate-bound CO complex has been solved,[112] and has served as a model for the functional ferric-superoxide species. In this structure, CO is bound such that the Fe–C–O angle is 166°, much like in other CO-globin structures. CO makes nonbonding contacts with camphor, which appears to be more loosely bound than in the substrate-bound ferric structure.[11] In addition, the oxygen of the diatomic substrate analogue contacts I-helix residues G248 and T252, a region containing an unusual midhelix bend. This kink is stabilized by an apparent hydrogen bond between the T252 alcohol moiety and the amide of G248, and for reasons discussed below has been identified as the putative O$_2$ binding site.

2.3.2. Dioxygen Association

Direct observation of O$_2$ binding in P450$_{cam}$ has now been observed by Raman spectroscopy.[113] These authors generated the ferric-superoxide complex using dithionite, and recorded the resonance Raman spectrum of the unstable species ($t_{1/2}$ ~90 sec at ambient temperature) in a –60°C cell. Excitation of the sample at 420 nm produced a 1140 cm^{-1} peak which shifted to 1074 cm^{-1} when ^{18}O$_2$ was used to generate the intermediate, and this isotope shift allowed this signal to be unambiguously assigned as the v_{o-o} stretch. This finding, and Mössbauer spectroscopy accomplished earlier,[114] both suggest that the P450 "oxy" intermediate is more like a ferric-superoxide species than the alternate ferrous oxy species (Fig. 1). More recently, the complex has been detected using time-resolved resonance Raman on a sample at steady state.[115] The authors found that the addition of NADH, putidaredoxin, and putidaredoxin reductase to the sample did not significantly alter the resonance Raman spectrum, and interpreted this to signify that binding of putidaredoxin does not measurably affect the environment of oxygen in the ferric-superoxide intermediate. However, they were unable to detect the buildup of any other isotope-sensitive intermediate along the reaction pathway.

In 1989 the laboratories of Ishimura[25] and Sligar[28] reported that mutation of the Thr^{252} residue in P450$_{cam}$ resulted in an uncoupling of NADH consumption from 5-*exo*-hydroxy-camphor production. Imai *et al.*[25] measured O$_2$ consumption, 5-*exo*-hydroxycamphor production, and peroxide generation; their results are summarized in Table III. The Thr^{252}Ala mutant has been the most completely characterized of these mutants, and for this species it was discovered that only 5 ± 2% of the reducing equivalents from NADH were utilized to form 5-*exo*-hydroxycamphor. Of the remaining electrons, the majority were found to be used to generate peroxide, and less than 10% were found to be consumed in the four-electron reduction of O$_2$. From the above data, it is clear that Thr^{252} is important

TABLE III
Effect of Site-Directed Mutants at Position 252 on Catalysis

Mutant	O_2 consumption min^{-1}	% H_2O_2 produced[a]	% ROH produced[a]
WT (Thr^{252})	1330	none	100
Thr^{252}Thr-OMe	410	none	100
Thr^{252}Ala	1100	83	6
Thr^{252}Val	420	45	22
Thr^{252}Gly	1090	88	3
Thr^{252}Ser	1100	15	81

in stabilizing the ferric-peroxide species in the P450$_{cam}$ reaction cycle. The analogous substitution Thr^{268}Ala has been made in CYP102 (BM-3), and was found to uncouple NADH oxidation to H_2O_2 production for the substrates laurate and palmitate.[195] Peterson, however, has reported tight coupling for this mutant with arachidonic acid.[196] Recent work by Shimada et al.[197] with P450cam has extended the above work by substituting a number of both natural and unnatural amino acid substitutions at position 252 (Table III). In general, mutation to residues with polar groups allows efficient coupling of NADH derived electrons to alcohol production. The above data is consistent with a mechanism in which O–O bond scission is mediated by a water molecule which is hydrogen bonded to the Thr^{252} residue.

The crystal structure of Thr^{252}Ala has been accomplished by Poulos and co-workers[116] and shows that the area identified as the putative O_2 binding site is indeed distorted from that found in the wild-type enzyme. The I-helix "kink" seen in the wild-type mutant is still apparent in the Thr^{252}Ala mutant, but the center of this feature is shifted one residue toward the N-terminus in the mutant enzyme. A new solvent molecule is located in this bent helix region of the mutant, in direct contact with camphor. Furthermore, a water channel previously buried near Glu^{366} in the wild-type structure is now accessible to the active site. From this finding, Raag et al.[116] postulated that the solvent in the active site could sterically interfere with and destabilize dioxygen binding. Additionally, the increased exposure of solvent channel to active site residues of the mutant was postulated to enable easier delivery of protons to the active site, and facilitate H_2O_2 production during the catalytic cycle. Work continues to further define the possible contributions of other residues to oxygen binding in P450$_{cam}$.

CO binding in P450$_{cam}$ has proven to be a sensitive probe of the overall environment of the active site, and the contribution that substrate plays in stabilizing the substrate-bound ferric-CO structure. In 1978 O'Keefe et al. showed that substrate binding to this intermediate effected a change in the CO FTIR stretching modes.[117] Further work in this area has shown that the active site of the reduced P450$_{lin}$–CO species can also form several different environments, as judged by CO stretching.[118] This view is also supported by the analysis of several P450$_{cam}$ substrate-bound ferric structures with alternate substrates.[59,119] Resonance Raman studies of substrate-bound ferric-CO samples have shown that both the Fe–C(O) stretching[44] and C–O stretching modes[120] are extremely sensitive to the substrate, indicating a direct interaction between CO and substrate in the intermediate. Jung et al.[120]

have suggested that the active site environment is different depending on the type of substrate bound in the active site. Substrate-free enzyme is the most malleable, displaying four different active-site environments (CO stretching modes) contributing to the overall structure. Purely aliphatic substrates such as adamantane or camphane and substrates such as camphoroquinone or 3-*endo*-bromocamphor (see Fig. 2) all have multiple allowed binding orientations, but in general bind in a preferred conformation with one or two minor alternate conformations. Substrates such as norcamphor, fenchone, adamantanone, tetramethylcyclohexanone, and the natural substrate camphor bind in one preferred orientation, and show the lowest mobility within the active site.

2.4. Cytochrome P450$_{cam}$ Catalysis

The catalytic mechanism of the P450s has been extensively reviewed since the last publication of this series.[33,34,121–124] The basic catalytic scheme, shown in Fig. 1, has been supported by a wealth of spectral analysis as the substrate-bound ferric,[125,126] substrate-bound ferrous,[85,126] and substrate-bound ferric-superoxide[125–128] species have characteristic UV/visible spectroscopy. Pederson *et al.* demonstrated that this latter complex is the last observable intermediate in the cycle, with oxidation of Pd^r and regeneration of the substrate-bound ferric state occurring at the same rate (within 5 msec).[86] Since the introduction of the second electron is kinetically rate-limiting to the overall reaction, the bracketed species in Fig. 1 do not build up, and thus none have been directly characterized. A large number of investigators have unsuccessfully applied various techniques to identify a chemical intermediate occurring after ferric-superoxide formation in P450$_{cam}$.[76,115,129] In the past 10 years, indirect evidence has been gathered for the existence of several of these intermediates, and will be presented below.

For the reader interested in P450 enzymes other than P450$_{cam}$, it should be noted that the liver microsomal enzyme system varies significantly from the adrenal and bacterial systems with respect to substrate selectivity and stereochemical control.[33] The former class generally accepts a wide variety of substrates and exhibits overlapping selectivities.[130] The latter classes are apparently "designed" for a specific reaction, one that requires the stereo- and regiospecific oxidation of the physiological substrate.[131,132] These observations led to the hypothesis that the active sites of the microsomal enzymes are relatively open and therefore present few steric restrictions on the substrate as it approaches the oxidizing species.[133] In contrast, in the adrenal and bacterial systems, intermolecular constraints should limit the orientational freedom of the substrate in the binary complex.[132,134] Thus, the experiments described below and mechanistic conclusions drawn may not correspond precisely to other systems, although all available evidence suggests that P450 enzymes share an essentially common mechanism.[122]

We will focus on recent advances in the elucidation of structural features of the enzyme–substrate complex which dictate functional activity for the system. This is natural in light of the rigorous experimental and theoretical analysis that has been developed for P450$_{cam}$ with respect to specific substrate active site contacts, regio, stereo, and reaction specificities, and enzyme active site engineering.

2.4.1. Dioxygen Bond Scission

Historically, the hottest debate in P450 mechanism has concerned dioxygen bond scission and the steps that follow to give hydroxylated product. Heterolytic cleavage of the dioxygen bond would yield a formal $Fe^{5+}=O$ (perferryl) species that would react to give oxidized product. Conversely, homolytic cleavage would proceed via a formal $Fe^{IV}=O$ (ferryl) species. The two possibilities are outlined in Fig. 3. P450 model compounds have been synthesized by a number of laboratories,[135,136] but depending on the oxidant used, both heterolytic and homolytic mechanisms are operant.[137] As these studies demonstrate that either mechanism is chemically feasible for P450s, it is unlikely that this line of investigation will resolve the issue as to whether the enzyme catalyzes homo- or heterolytic chemistry, since the models will never be able to completely mimic the environment of the enzyme active site. The apoprotein, which is too complicated to be

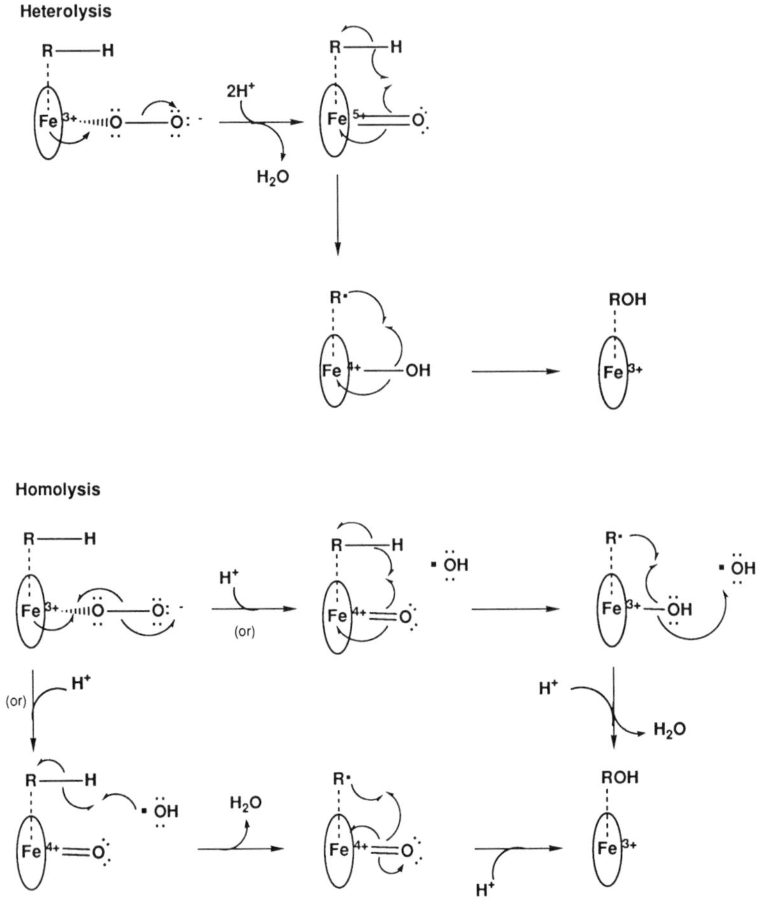

FIGURE 3. Alternative mechanisms for O–O bond scission in cytochromes P450.

modeled in a simple system, is obviously important in determining the environment of the heme,[52] as well as determining substrate specificity and reaction regiospecificity.[2] As such, the proposed role of the cysteine ligand in catalysis remains unresolved.[34]

The issue of whether a high-valent iron–oxo intermediate actually exists has been addressed in a series of experiments by Atkins and Sligar.[138,139] This work utilized the camphor analogue norcamphor (see Fig. 2), which is hydroxylated at the 3-, 5-, and 6-*exo* positions in the proportions shown in Table IV. The V_{max} for this substrate is one-third that determined for camphor, but the efficiency of the enzyme in utilizing reducing equivalents is greatly reduced.[138] Stoichiometry measurements of NADH and O_2 consumption showed that ~1.5 NADH molecules were utilized for every O_2, indicating that oxidase activity was occurring. Careful measurements showed that peroxide is also generated when norcamphor is used as a substrate. The reaction of norcamphor deuterated at the 3, 5-*exo*, and 6-*exo* positions (see Fig. 2) alters the amount of hydroxylated product formed in favor of an increased oxidase activity. In viewing Table IV, it is obvious that there is an isotope effect in the hydroxylation of norcamphor. In fact, the isotope effects on oxygen uptake and peroxide formation were found to be 1.22 and 1.16, respectively.[139] Close examination shows that there is an increase in oxidase activity when deuterated substrates are offered to the enzyme. Thus, P450$_{cam}$ proceeds through a branch point in product formation for deuterated substrates, which implies that a common intermediate must exist. This intermediate must be able to either oxidize substrate or accept two electrons to form water. An apparent inverse isotope effect for NADH consumption of 0.77 reinforces this interpretation.[139] This analysis assumes that deuteration of norcamphor does not alter the mechanism, and that camphor and norcamphor react with P450$_{cam}$ by the same mechanism. If one accepts these caveats, the existence of a single oxygen atom-containing intermediate, two electrons oxidized with respect to water, is indicated. Recent work by Dawson and coworkers with 9,9,9-trideutero-5,5-difluorocamphor has yielded similar results to those for norcamphor, and has led these investigators to the same conclusions.[199]

The phenomenon of uncoupling, that is, production of peroxide or water by cytochromes P450, is a well-known side reaction that has been documented in most, if not all, enzymes of this class. The means by which these reactions are thought to occur are shown in Fig. 1. Autoxidation, the decomposition[39,140] of the ferric-superoxide species to give ferric enzyme and superoxide, is the "first" branch point from the normal catalytic

TABLE IV
P450$_{cam}$ ^{2}H Isotope Effects for Camphor and Norcamphor

Substrate	Consumed				Produced			
	NADH nmole	O_2 nmole	H_2O_2 nmole	ROH nmole	% 3- *exo*	% 5- *exo*	% 6- *exo*	H_2O nmole[a]
Camphor	313 ± 8	280 ± 8	20 ± 12	290 ± 2	0	100	0	0
Norcamphor	577 ± 8	354 ± 5	88 ± 22	88 ± 12	8	45	47	189
5,6-*exo,exo-d₂*-norcamphor	583 ± 1	338 ± 5	84 ± 22	44 ± 2	52	27	21	210
3,3-*d₂*-norcamphor	562 ± 4	ND	ND	78 ± 9	2	49	49	ND
3,3,5,6-*exo,exo-d₄*-norcamphor	564 ± 15	319 ± 4	74 ± 28	27 ± 4	10	44	46	218

[a]Calculated from the difference between NADH consumption and product (ROH, H_2O_2) formation.

mechanism. Peroxide is formed by decomposition of the ferric-peroxide species at the second branch point,[140,141] while water is produced from the $[FeO]^{3+}$ species at the third branch point.[142] Production of superoxide, which dismutates rapidly, peroxide, or water indicates a reduction in the efficiency of the enzyme to couple reductive equivalents from NADH to the production of alcohol product. Although uncoupling is seen in many P450 systems, P450$_{cam}$ uses NADH to produce 5-*exo*-hydroxycamphor at >95% efficiency. Uncoupling is seen in this system when the enzyme turns over substrates other than camphor (as seen above) when mutations are introduced that affect the stability of one of the relevant intermediates (as discussed for Thr^{252} mutants in Section 2.3.2) or when mutations allow the introduction of water into the active site, as will be discussed (Section 2.4.2).

This work has given new hope for the isolation and characterization of a kinetically competent P450 high-valent iron species . If one assumes that all P450s proceed through a common mechanism, and that different enzymes will have different rate-limiting steps,[2] then given the myriad of P450 enzymes there is a chance that one at least will allow buildup of the oxidizing intermediate. Perhaps one of the intermediates reported by Blake and Coon,[143] Larroque *et al.*[144] or from some yet unreported system will finally allow characterization and resolution of the nature of the P450 high-valent iron intermediate.

Another mechanistic question concerns the delivery of protons necessary for substrate turnover. It had been reported that P450$_{cam}$ takes up a proton on substrate binding,[145] but whether this proton was involved in catalysis is still unknown. In searching for the mechanism of active site proton delivery, one of the first clues came with the alignment of multiple P450 sequences and the discovery of several highly conserved regions.[146] Some of these regions correspond in P450$_{cam}$ to the distal ligand Cys^{357} as well as Thr^{252} and Asp^{251}. Mutagenesis of Thr^{252}, discussed above, suggested that a hydroxyl moiety was necessary at this position to allow efficient O–O bond scission.[25]

Generation of the P450$_{cam}$ mutant Asp^{251}Asn was reported by Gerber and Sligar.[147] The major effect of the mutation was to decrease the overall rate of the reaction by an order of magnitude. Intriguingly, the measurable kinetic parameters up to and including introduction of the second electron were the same, within a factor of two (see Table V). This finding which agreed with reports of Shimada *et al.*,[148] allowed Gerber and Sligar to propose that Asp^{251}Asn was part of a distal charge relay, creating an H-bonding network for the transfer of protons from the bulk solvent to the active site.[147] These authors further proposed that Asp^{182}, Lys^{178} and Arg^{186} were likely candidate residues connecting the active site to the aqueous milieu.

Further work has shown that at neutral pH, wild-type P450$_{cam}$ and the mutants Thr^{252}Ala and Lys^{178}Gln all support NADH oxidation rates of ~ 600–800 min^{-1}, indicating that the results with Asp^{251}Asn are not related to some gross structural change induced by the mutant.[149] In contrast to the wild-type enzyme, the rate of NADH consumption for Asp^{251}Asn increases as the pH is lowered from 8 to 5 by a factor of >20. This observation is consistent with the proposed function. Finally, the P450$_{cam}$ Asp^{251}Asn mutant displays a steady-state intermediate with UV/visible spectral characteristics differing from the wild-type ferric-superoxide species which usually accumulates. At this time, it is unknown

TABLE V
Rate Constants for Wild-Type and Asp^{251} Asn Mutant Enzyme

Assay		Wild type	Asp^{251} Asn
Camphor binding	k_{on}, $\mu M^{-1} sec^{-1}$	24.5	22.4
	k_{off}, sec^{-1}	21.1	15.0
	K_d, μM	0.86	0.76
1st electron transfer	k_{obsd}, sec^{-1}	60.7	145.2
Oxygen binding	k_{on}, $\mu M^{-1} sec^{-1}$	81.0	56.0
	k_{off}, sec^{-1}	2.7	2.9
Autoxidation	k_{obsd}, sec^{-1}	4.1×10^{-3}	2.7×10^{-3}
Catalysis[a]	k_{obsds} sec^{-1}	34	0.15
NADH oxidation rate, $\mu M/min/\mu M$ heme		847	8
5-OH/NADH		0.95	0.90^{b}
$NADH/O_2$		1.03	1.40

[a]Includes reaction from the ferric superoxide intermediate to the substrate-bound ferric state.

whether this difference reflects an altered species in the mutant, a buildup of a steady-state intermediate previously uncharacterized, or Pd binding.

Results similar to those for $P450_{cam}Asp^{251}Asn$ have also been published for CYP1A2 $(P450_d)^{150}$ and aromatase.[151] In addition, mutations at Arg^{184} in $P450_{cam}$ also show a reduced apparent V_{max}.[152] Mutations at Lys^{178} seem to have no readily apparent effect on NADH consumption.[149,152] Thus, it appears that the proton relay mechanism may extend to these other P450 systems, and extends via Arg^{184} in $P450_{cam}$ to the bulk solvent.

The mechanism for proton transfer in wild-type $P450_{cam}$ has been further probed using kinetic solvent isotope effect (KSIE) methodology.[153] Camphor hydroxylation was found to display a KSIE of 1.8 ± 0.05. This isotope effect was determined, by using stopped-flow single turnover experiments, to be linked to second electron transfer and/or subsequent catalytic steps (see Fig. 1). As there was no KSIE found for the first electron transfer step (KSIE = 1.03 ± 0.04), it was argued that $Pd/P450_{cam}$ binding does not contribute to the observed effect. A proton inventory was constructed by varying the mole fraction of D_2O in the reaction solvent, implicating at least two protons in the proton transfer transition state. From this finding the authors proposed that proton transfer must occur from the ferric-peroxide intermediate via both Thr^{252} and Asp^{251}, but could not rule out a possible role for mobile, active-site water in the transfer.

2.4.2. Substrate Hydroxylation ($P450_{cam}$ Reactivity: The Substrate Side)

An effective approach to investigating the $P450_{cam}$ mechanism is from the view of the hydrocarbon substrate undergoing oxidation. The following discussion begins with a brief summary of the evidence for a substrate radical intermediate and its implications for a nonconcerted P450 mechanism. In addition, the considerable data that have been accumulated on how $P450_{cam}$ discriminates between potential substrates, and the efforts at reengineering new substrate specificity into this enzyme will also be discussed. The focus here is on recent investigations of the noncovalent interactions within the substrate–$P450_{cam}$ complex. Site-directed mutagenesis, a host of substrates, and extensive theoretical analysis have been utilized in the last 10 years to characterize the mechanism of small molecule recognition in this system.

Within the P450 literature at large, the evidence for a substrate radical intermediate is extensive.[122] Several elegant demonstrations of the nonconcerted nature of the reaction have been reported.[154,155] For $P450_{cam}$, the relevant example was the analysis of intramolecular isotope effects in the hydroxylation of 5-*exo*-deuterocamphor and 5-*endo*-deuterocamphor by the reconstituted $P450_{cam}$ system.[156] Significant amounts of deuterium were found in the product alcohol regardless of the substrate. While *exo*-hydrogen abstraction is preferred, the selectivity of substrate activation permits this step to occur with either *exo* or *endo* stereochemistry. The recombination event is strictly enforced to the *exo* face of the bicyclic substrate. The observed loss of stereochemistry rules out the possibility of a concerted reaction mechanism such as that expected for the insertion of singlet oxygen in a C–H bond. The observed isotopic ratios were similar with exogenous oxygen donors such as hydrogen peroxide, iodosobenzene, or *m*-chloroperbenzoic acid, suggesting a reactive oxygen species similar to that of the reconstituted enzyme in the presence of dioxygen and its redox partner proteins. This mechanism of substrate activation followed by ligand association with loss of stereochemistry is similar to that proposed for porphyrin model systems.[41,157]

The prospect of a tetrahedral carbanion intermediate was addressed by utilizing the camphor analogue pericyclocamphanone[158] (see Fig. 2). With this substrate the hydroxylation reaction occurs primarily at the 6-*exo* and 6-*endo* positions. The observed regiospecificity suggests that a planar intermediate at the 5-carbon is formed easily at the secondary 5-carbon of *d*-camphor. A trigonal species would be highly disfavored at the 5-carbon of pericyclocamphanone because of ring strain resulting from the additional covalent bond between the 3- and 5-carbons. It must be noted that the lack of 5-hydroxylation may also reflect the higher bond strength of a cyclopropyl C–H bond. Additional support for a planar intermediate is provided by the observation that the analogue in which the 6-carbon is blocked, 6-keto-pericyclocamphanone, also does not result in 5-carbon hydroxylation. In this system the consumption of reducing equivalents is uncoupled from the hydroxylation reaction, resulting in the side product hydrogen peroxide .

One approach to differentiating between possible radical and carbonium intermediates is to consider the selective *exo* capture of norbornane carbonium ions[159] (see Fig. 2). Studies on the liver microsomal $P450_{LM2}$ system revealed 2-*endo*-norborneol as a product, implicative of a substrate radical. While the stereospecificity of enzymatic oxidation may be complicated by noncovalent interactions between the substrate and active site, the previous example is a case of reduced stereospecificity on the part of the enzyme relative to the reaction free in solution. The mechanism of hydrocarbon oxidation by iron porphyrin model systems is known to proceed through a substrate radical intermediate, and the investigators in the field have been quick to point out the chemical and structural similarities to P450.[136] In contrast, recent radical-clock experiments by Newcomb and coworkers have called into question whether a radical intermediate is preferred over a carbocation mechanism.[198]

The means by which enzymes recognize their substrates have traditionally been probed using substrate analogues. In the P450 field, the large number of well-studied enzymes makes this tool instrumental as a means of comparing the different active sites of enzyme within the class. The high regio- and stereospecific reaction of P450$_{cam}$ with camphor, and comparison of the product profiles of camphor, adamantanone, and adamantane hydroxylation by P450$_{cam}$ and P450 2B4 indicate that substrates are generally more constrained by the former isozyme and limited to a unique orientation with restricted access of alternate carbons to the iron.[160] P450$_{cam}$ was found to be regiospecific with the substrates adamantanone and adamantane producing 5-hydroxyadamantanone and 1-hydroxyadamantane, respectively. This is in contrast to the multiple products obtained with P450 2B4 which include 3-hydroxycamphor, 5-*exo*-hydroxycamphor, and 5-*endo*-hydroxycamphor; 4-*anti*-hydroxyadamantanone; 5-hydroxyadamantanone; 1-hydroxyadamantane and 2-hydroxyadamantane, respectively. The authors pointed out that the rank order of products formed with the microsomal P450 is roughly that predicted on the basis of relative chemical reactivities free in solution estimated from radical stability and steric hindrance. Hence, prior to the X-ray structure, there was evidence from substrate selectivity studies to suggest that topological features of the P450$_{cam}$ active site were responsible for orienting camphor for stereo- and regiospecific hydroxylation.

A combination of substrate analogues and site-directed mutants have been used to quantitate the specific steric and hydrogen bonding interactions with respect to the reaction rate, regiospecificity, and efficiency.[22,23] The X-ray structure of P450$_{cam}$ implicates a hydrogen bond between the ketone of *d*-camphor and the side chain hydroxyl of Tyr[96].[53] The substrate analogues thiocamphor and camphane (Fig. 2) are incapable of forming hydrogen bonds and were utilized as structural probes of the putative interaction.[22] Also, the P450$_{cam}$ mutant Tyr[96]Phe was constructed to create a conservative, yet nonpolar amino acid substitution which would not be expected to alter the structure of the protein. Wild-type enzyme reaction with thiocamphor and camphane, and Tyr[96]Phe camphor hydroxylation occurred at multiple positions on the camphor skeleton, indicating a less specific interaction and increased mobility of the substrate. In addition to the 5-*exo* alcohol, the 6-*exo* and 3-hydroxy products accounted for up to 30% of the hydroxylation products. NADH consumption was tightly coupled to hydroxylation in all cases except for camphane, in which 10% of the reducing equivalents were presumably utilized to form hydrogen peroxide or water via the oxidase pathway. This is the only enzyme–substrate complex that lacks the potential for both hydrogen bonding and steric contacts with the ketone. The observation that camphane resulted in the largest decrease in specificity for all measured reaction parameters suggests a significant steric contribution of the camphor ketone to substrate recognition. An estimation of the relative efficiency for the series of substrates and enzymes can be derived in the form of k_{cat}/K_m by utilizing the steady-state rate of NADH oxidation, as an approximate measure of k_{cat}, divided by the dissociation constant, which was assumed to scale roughly with the K_m. To correct for the decrease in regiospecificity, the value is multiplied by the fraction of 5-*exo* alcohol produced in the reaction as shown in Table VI. When all aspects of substrate binding and hydroxylation are considered, the Tyr[96] hydrogen bond contributes at least a factor of five to the efficiency of the reaction.

Camphor analogues 1-methylnorcamphor and norcamphor were used to assess the role of hydrophobic contacts in maintaining the correct juxtaposition of substrate relative to the iron-bound oxidizing species.[23] 1-Methylnorcamphor lacks the 8,9-*gem*-dimethyl

TABLE VI
Effect of the Tyr[96]-Substrate Hydrogen Bond on Hydroxylation

System	k_2, NADH consumption (nmole NADH/min per nmole P450)	Relative yield of 5-*exo*-isomer	Corrected k_2 (nmole/min per nmole P450)	k_2/Kd [a] relative specificity $(min^{-1}M^{-1} \times 10^{-6})$
Wild type/camphor	64 ± 3	1.0	64	40
Y[96]F/camphor	29 ± 23	0.92	27	8.2
Wild type/thiocamphor	36 ± 4	0.64	23	6.5

[a] k_2 is the relative rate of 5-*exo* product formation, not the rate of hydrogen abstraction by the oxidizing intermediate. Also, the optically determined K_d is not a true K_m.

groups which protrude from the 2.2.1 bicyclic skeleton of camphor. With norcamphor all three methyl groups are absent, thus the analogues represent a series of sterically altered substrates, each of which successively deviates from the native camphor structure to a greater extent (see Fig. 2). In the X-ray structure of P450$_{cam}$ the *gem*-dimethyls appear to make contacts with Val[295] (see Table II). Detailed analysis via rigid docking experiments suggests that the methyl groups of the valine residue interdigitate with the 8,9-methyls of the substrate. This motif could be rationalized by invoking the finely tuned stereo- and regiospecificity of the recombination reaction, a process that is very likely dependent on the rotational freedom of the substrate and/or its potential for multiple binding orientations. As a measure of substrate recognition, the regiospecificity of the reaction with camphor analogues is expected to produce the 5-hydroxy alcohol. The fraction of alternate products represents the degree of substrate mobility and the loss of specificity within the enzyme–substrate complex. In this sense, the product profile is utilized as a probe of specific active site–substrate contacts. A correlation is observed between the loss of substrate–active site complementarity in the camphor, 1-methylnorcamphor, norcamphor series, and the percentage of 3-hydroxy and 6-hydroxy alcohols produced for the series of camphor analogues (Table VII). This relationship demonstrates the dependence of the wild-type specificity on the complete set of substrate–enzyme contacts. These investigations were complemented by studies utilizing other camphor analogues.[161] Of particular interest is a study that showed that P450$_{cam}$ hydroxylates 5,5-difluorocamphor at the 9-carbon.[162] This was the first example of methyl group hydroxylation by the bacterial isozyme. The results suggest that camphor itself is unrestrained in the active site to the extent that multiple substrate carbons approach the iron, and, if the normal site of oxidation is blocked, are competent to undergo the hydroxylation reaction.

TABLE VII
Effect of Substrate–Active Site Steric Contacts on Hydroxylation

	Norcamphor	1-Methylnorcamphor	*d*-Camphor
% high spin	45%	48%	95%
Total yield hydroxy products	0.12	0.45	1.0
% 5-*exo*- of total products	45%	82%	100%
Yield 5-*exo*-/NADH	0.05	0.37	1.0

As a complement to the substrate analogues, active site residues Val^{295} and Val^{247} were mutated into isoleucine and alanine, respectively. The larger, hydrophobic side chain at position Val^{295} would be expected to restore steric contacts eliminated on removal of the 8,9-*gem* dimethyl groups. The smaller amino acid at Val^{247} serves as a probe of the steric contributions of this residue, as contacts with the 6- and 10-carbon of camphor are evident from the X-ray structure (see Table II). As seen in Table VIII, these mutations have a negligible effect on camphor hydroxylation, but result in predictable changes in the product profile of the 1-methylnorcamphor. 5-*exo*-hydrox-1-methylnorcamphor accounts for 82% of the oxidized products with wild-type enzyme. A 10% increase in 5-*exo*-hydroxy product is observed with Val^{295}Ile and a 10% decrease with Val^{247}Ala. The larger effects on the substrate analogue relative to camphor underscore the importance of the additive contributions of multiple active site–substrate contacts in maintaining the correct orientation of substrate with respect to the $[FeO]^{3+}$ species. The uncoupling of NADH to hydrogen peroxide and water is also increased as intermolecular contacts are removed. The recognition of methyl groups is essential for maintaining the efficient hydroxylation of substrate. This seminal work established proof of principle for further investigations that focused on (1) the elucidation of important enzyme–substrate contacts, (2) establishing a base of substrate specificity data for a broad range of compounds, and (3) approaches to the engineering of P450 active sites for novel substrate specificities.

As a further experiment in the design of $P450_{cam}$ enzymes with altered specificities, a series of mutations were made to affect the coupling of NADH to hydroxylated product for the substrate ethylbenzene.[163] While little was known previously about the activity of $P450_{cam}$ with substrates structurally unlike *d*-camphor, it was found that ethylbenzene is hydroxylated almost exclusively at the benzylic carbon to afford 1-phenylethanol in 5% yield. *O*-ethylphenol is detected in trace amounts. The remaining reducing equivalents are recovered as hydrogen peroxide, 80%, and water via the oxidase activity, 15%. The first step to altering the reaction coupling was to recognize an important structural feature of the active site, that is, the ordering of amino acid residues in parallel planes relative to the

TABLE VIII
Active Site Reconstruction of P450cam

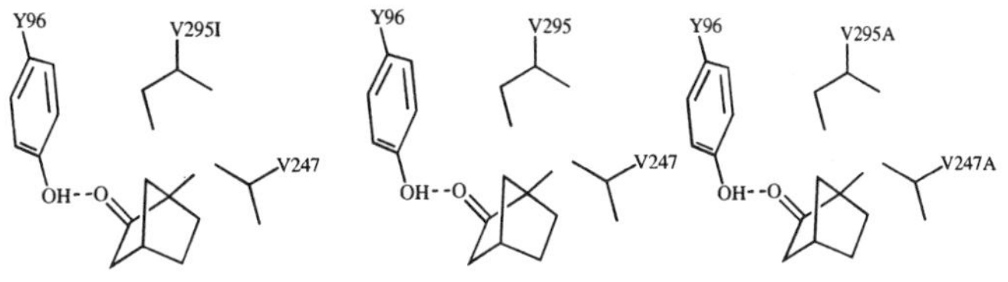

	V295I	V295	V295A
% high spin	49%	48%	37%
Total yield hydroxy products/NADH	0.45	0.45	0.40
% 5-*exo*- of total products	0.90	0.82	0.73
Yield 5-*exo*-/NADH	0.41	0.37	0.29

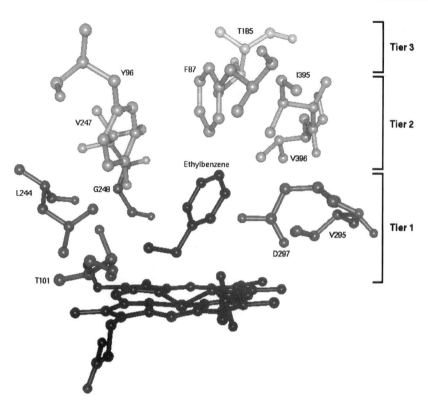

FIGURE 4. Side-on view of the P450$_{cam}$ active site with a model of ethylbenzene oriented for benzylic hydroxylation. Residues in Tier 1, shown in dark gray, are positioned near the heme and include Thr[101], Leu[244], Val[295], Gly[248], and Asp[297]. Tiers 2 and 3, shaded light gray, form the upper region of the binding pocket and are comprised of residues Phe[87], Tyr[96], Thr[185], Val[247], Ile[395], and Val[396].

heme (Fig. 4). As such, "Tier 1" refers to those residues that form the inside surface of the active site and are positioned in the first layer of residues above the heme. "Tier 2" includes the amino acids located in the plane parallel to and above Tier 1. "Tier 3" consists only of T185 which serves as the "cap" at the top of the binding cleft. Based on the position of Tier 1 relative to Tiers 2 and 3 in substrate docking experiments, it was reasoned that a mutation that increased the size of an amino acid side chain should block substrate access to the putative $[FeO]^{3+}$ species if introduced in a residue belonging to Tier 1. The opposite would be expected for similar substitutions in Tiers 2 and 3. Thr[101] and Val[295] from Tier 1 and Val[247] and Thr[185] from Tiers 2 and 3 were substituted with the larger hydrophobic side chains leucine, isoleucine, methionine, and phenylalanine. In the reconstituted system, the observed coupling of NADH oxidation to benzylic alcohol correlates well with the position of the amino acid substitution.[164] Tier 1 mutations result in diminished recovery of 1-phenylethanol while similar changes in Tiers 2 and 3 increase the yield of hydroxylated product by almost threefold. Given the tight correlation observed between the reaction efficiency and the position of the mutation, a detailed analysis of the partitioning of reducing equivalents at the individual reaction branch points was carried out.[165]

As described previously, the pathway for utilization of pyridine nucleotide-derived reducing equivalents in the P450 monooxygenase systems has three major branch points[139–142,166] (See Fig. 1). The first is a partitioning between autoxidation of a ferrous, oxygenated heme adduct and input of the second reducing equivalent required for monooxygenase stoichiometry. The second is between dioxygen bond scission and release of two-electron reduced O_2 as hydrogen peroxide. The third is between substrate hydrogen abstraction initiated by a putative higher-valent iron–oxo species and reduction of this intermediate by two additional electrons to produce water in an overall oxidase stoichiometry. The stoichiometry of NADH oxidation to hydrogen peroxide production and four-electron reduction of oxygen to form water via the oxidase activity was determined for each mutant. For all mutants investigated, the direct release of superoxide at the first branch point never competes with second electron input. The reaction specificity at the second and third branch points is effected by site-directed mutations that alter the topology of the binding pocket.

The percent partitioning to produce peroxide at branch point two is shown in Fig. 5A. For nearly all mutant enzymes the fraction of uncoupling to peroxide is decreased with the largest effect observed for the triple substitution, Thr^{101}Met–Thr^{185}Phe–Val^{247}Met. The results were taken as evidence that the key determinant controlling uncoupling at the ferric-superoxide species is water distribution in the active site and solvent access to the iron. The calculated volume changes on substituting a single amino acid side chain are 30 Å^3 for Val to Ile or Met, 40 Å^3 for Thr to Leu, Ile, or Met, and 60 Å^3 for Thr to Phe. These correspond to a decrease of about 10, 15, and 25% in the size of the P450$_{cam}$ binding site, respectively. The triple mutant reduces the pocket size by 50%. The extra bulk of the leucine, isoleucine, methionine, and phenylalanine side chains makes the active site more hydrophobic by filling volume otherwise occupied by water.[42] With many substrates water is present in the active site as shown by X-ray structural analysis of the enzyme with camphor analogues bound. Furthermore, equilibrium and kinetic analysis of the ferric spin-state equilibrium for a range of camphor analogues demonstrates that water access to the heme iron is dependent on the structure of the particular substrate.[51] The increased polarity of the water-filled pocket would favor charge separation at the iron promoting the release of superoxide or hydrogen peroxide anion. Similar mechanisms have been proposed for myoglobin based on kinetic and thermodynamic studies of the oxygen-bound form of the wild-type protein and several active site mutants.[167,168]

The effect of active site mutants on reaction specificity and uncoupling to form water at branch point three is depicted in Fig. 5B. Engineering steric bulk around the perimeter of the active site causes a switch in chemistry from hydroxylation to oxidase activity, which is observed in a 400-fold change of the ratio of hydroxylated substrate to water production. This result is strong evidence for the concept that substrate positioning is the key factor for tight coupling at the level of the putative iron–oxo species. Adding large nonpolar groups in Tier 1 reduces the hydroxylation of ethylbenzene by "pushing" the substrate away from the heme. Under these conditions the longer lifetime of the [FeO]$^{3+}$ allows the input of two additional electrons from the redox partner protein putidaredoxin and

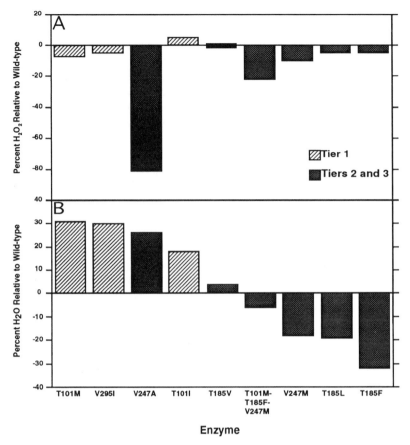

FIGURE 5. Reaction specificity differences between wild-type P450$_{cam}$ and a series of active site mutants in the hydroxylation of ethylbenzene. The large hydrophobic side chains of Ile, Leu, Met, and Phe were substituted for Val and Thr at four positions in the binding pocket. Residues in Tier 1 are located near the heme and are shown by hatched bars, and residues in Tiers 2 and 3 are distant from the heme and are indicated by solid bars. (A) Each bar represents the difference in the partitioning of oxygen to peroxide at the second branch point for the mutants minus wild-type P450$_{cam}$. The hydrogen peroxide is expressed as a percentage of the total oxygen consumed in the reaction. (B) Each bar represents the difference in water formed at the oxidase pathway relative to wild-type P450$_{cam}$. The water is expressed as a percentage of the total oxygen equivalents passing through the third reaction branch point. For active site residues in Tiers 2 and 3 there is a negative correlation between side chain packing volume and the production of water. This relationship is reversed for mutations in Tier 1.

subsequent release of water. With Thr[101]Met and Val[295]Ile, the hydroxylation of ethylbenzene by P450$_{cam}$ is essentially eliminated. Extended side chains are more disruptive than β-branched amino acids as shown by the additional increase in water production in going from the isoleucine to methionine substitution at Thr[101]. Substituting larger residues at positions in the pocket that are distant from the heme pushes the substrate toward the oxidizing intermediate. The greater residence time of the ethylbenzene benzylic carbon at the $[FeO]^{3+}$ species is reflected in a faster rate of hydroxylation relative to electron transfer and an increase in the production of 1-phenylethanol. With $Thr^{185}Phe$ the hydroxylation of ethylbenzene becomes the predominant reaction pathway accounting for two-thirds of the flux through the third branch point. The effects of individual substitutions are approximately additive when combined in the triple mutant $Thr^{101}Met–Thr^{185}Phe–Val^{247}Met$. This suggests that there are no significant secondary effects on the protein structure as a result of combining the $Thr^{101}Met$, $Thr^{185}Phe$, and $Val^{247}Met$ mutations and that these residues, despite their proximity, behave independently. Introducing $Thr^{185}Phe$ and $Val^{247}Met$ into the background of $Thr^{101}Met$ reverses the effect of the initial substitution. $Thr^{101}Met$ virtually eliminates ethylbenzene oxidation by denying substrate access to the heme, but steric bulk introduced in Tiers 2 and 3 forces the substrate toward the iron and perhaps causes the flexible methionine side chain to assume an alternate conformation and the hydroxylation activity to be restored. The positioning of ethylbenzene in the active site is primarily determined by the location and steric packing volume of the altered amino acid and also depends on the side chain's shape and flexibility. The results suggest that substrate access to the iron can be altered in a predictable fashion.

The results for the "Tier" mutations were generalized to other substrates and reaction chemistries by utilizing the reductive dehalogenation of chlorinated ethanes and methanes. The capacity of $P450_{cam}$ for reductive and oxidative dehalogenation reactions has been known for quite some time.[169–174] Initial studies on the substrate selectivity of the purified enzyme system with respect to a wide range of halogenated methanes and ethanes were conducted only recently.[175–177] These investigations revealed that certain members of this group were substrates for $P450_{cam}$ and others were not, with no obvious relationship distinguishing the two sets of compounds. The rate of dehalogenation was determined for a subset of the $P450_{cam}$ mutants representing Tiers 1, 2, and 3 and several of the chlorinated ethanes and methanes. Initial rates were derived from measurements of product formation determined by head space analysis.[178] The reduction of chlorinated ethanes by $P450_{cam}$ generally results in elimination products, i.e., trichloroethylene (TCE) from pentachloroethane and dichloroethylene (DCE) from 1,1,1,2-tetrachloroethane. As expected, the Tier 1 substitution, $Thr^{101}Met$, decreases the rate of dechlorination with both substrates. Tier 2 and 3 mutations consistently increase the reaction rate with the tetrachlorinated compound, with the difference in rates ranging over an order of magnitude. The correlation with pentachloroethane is not absolute, and the discrepancy is explained by its larger molecular volume. As in the case of d-camphor, this substrate appears to be too large to respond to changes in the topology of the active site.[23] Thus, similar efficiency for substrates with different reaction chemistries lends further support to the conclusion that the effect of the Tier mutations on the observed changes in enzyme efficiency is related to alterations in the access of the substrate to the heme iron.

The above work has led to the development of a set of parameters that have been applied to predict the stereochemistry of ethylbenzene hydroxylation in different $P450_{cam}$ mutants. Ethylbenzene hydroxylation by wild-type $P450_{cam}$ is stereospecific with 73% (R)-1-phenylethanol formed.[163] Since the pro-R and pro-S hydrogens appear to be equally accessible to the heme iron, the asymmetric hydroxylation cannot be attributed to any obvious structural feature of the wild-type $P450_{cam}$ active site. All of the Tier mutations increase the excess (R)-1-phenylethanol produced, with the maximum change from 46% to 80% for the $Thr^{101}Met$ and $Thr^{101}Ile$ substitutions. A firm structural basis for the observed changes in stereospecificity is also not detected by means of a static representation of the active site region. Molecular dynamics trajectories were successful in reproducing the experimentally determined stereospecificities with the wild-type enzyme.[179] The theoretically determined ratio of 76:24 for the (R)- and (S)-1-phenylethanol is in excellent agreement with experiment, suggesting that the dynamics of substrate–active site complex can be represented by a molecular mechanics-level calculation.[180–182] These simulations suggested that smaller substrates such as ethylbenzene, which are not contoured to bind in the active site with complementary steric contacts, sample a large fraction of the available rotational and conformational space. The results are consistent with a conceptual model of enzyme–small molecule recognition in which substrate mobility and/or accessible binding orientations increase as the structure of the substrate diverges from that of d-camphor.[23]

A collection of similar investigations are consistent with the interpretations drawn in the previous study and establish computational analysis as a valid approach to the characterization and prediction of small-molecule binding and metabolism with $P450_{cam}$.[182–186] Ortiz de Montellano, Loew, and their co-workers have shown that $P450_{cam}$ epoxidizes styrene, $trans$-(β-methylstyrene, and cis-(β)-methylstyrene stereospecifically with the $(1S,2R)$ enantiomer favored in all cases; 83, 75 and 89%, respectively. The predicted ratios of 65:35, 75:25, and 84:16 were derived from molecular dynamics simulations and are in good agreement with experiment. The overall coupling of NADH to epoxide is very low in these systems with overall yields of 2, 0.5, and 1%, respectively, and hydrogen peroxide accounting for most of the reducing equivalents. More recently, these techniques have been used to explain the observation that pentachloroethane undergoes reductive dechlorination by $P450_{cam}$ and 1,1,1-trichloroethane does not.[175] In the simulated system, 1,1,1-trichloroethane rotates such that the 1-carbon and bonded chlorine atoms are positioned as distant from the iron as possible.[187] This reorientation occurs regardless of the position of the substrate at the start of the simulation. Additional simulations to determine the specific amino acids responsible for the inactivity of 1,1,1-trichloroethane and the nature of the steric and electrostatic interactions will provide additional information as to the mechanism of recognition with this class of substrates. In addition, molecular dynamics simulations and free energy calculations have been used to predict the relative binding free energy and regiospecificity of (S)- versus (R)-nicotine hydroxylation.[184] The calculated $\Delta\Delta G$ of 0.4–0.6 kcal is good agreement with the difference in spectrally derived dissociation constants, $\Delta\Delta G=0.3$ kcal. Preferential hydroxylation of the 5′ methylene relative to the N-methyl carbon was also predicted from the calculations. Analogous theoretical investigations utilizing the $P450_{cam}$ binding site and valproic acid suggested that the fundamental difference in substrate recognition mediated by mammalian P450 isozymes versus $P450_{cam}$ is the alternate set of steric contacts resulting from variations in active site structure.[188] Thus, the $P450_{cam}$ active site has served as a "model" for investigations aimed at determining the metabolites of the P450-mediated oxidation of physiologically relevant small molecules.

One example of molecular simulation builds from the work of Collins and Loew[183] to yield the *a priori* prediction of the stereoselective hydroxylation of $1R$- and $1S$-norcamphor by Bass and Ornstein.[185,189] 5-Hydroxy- and 6-hydroxynorcamphor are produced in equal amounts utilizing wild-type $P450_{cam}$ with a racemic mixture of norcamphor. Molecular dynamics simulations of the individual enantiomers predicted selective hydroxylation of $1R$-norcamphor at the 5-carbon and $1S$-norcamphor at the 6-carbon. As the initial reports were published 2 years before the preparation of the enantiomers, the validity of these predictions is undisputed. The chiral resolution of $1R$- and $1S$-norcamphor and subsequent reconstitution with the $P450_{cam}$ system confirms the predicted stereoselectivity of hydroxylation[164] (Table IX). Theoretically determined ratios for the 5-, 6-, and 3-hydroxy products not only are consistent with experiment in a qualitative sense, but are also in reasonable agreement quantitatively.[189]

An important objective of the extensive experimental and theoretical characterizations of the substrate specificity with $P450_{cam}$ is to achieve a level of predictability that allows active sites to be engineered for novel specificities. The unique aspects of biological catalysts such as P450 might be harnessed for practical applications such as bioremediation[190,191] or organic synthesis. In an attempt to extend our understanding of $P450_{cam}$–substrate recognition, active site mutations were designed to alter the regiospecificity of $1R$-norcamphor hydroxylation in a predictable fashion by replacing the hydrogen bond donor group at Tyr[96]Phe with a new hydrogen bond donor at residue Phe[87]Tyr. Based on ligand docking experiments, as shown in Fig. 6, this putative "hydrogen bond switch" is expected to increase the fraction of 6-hydroxynorcamphor by stabilizing a substrate orientation in which the 6-carbon is adjacent to the iron. The product profile of the single mutants, Tyr[96]Phe and Phe[87]Tyr, are shown along with the double mutant in Table IX. The results suggest that the engineered hydrogen bond donor competes for the substrate ketone and favors the 6-hydroxy product.[192] This interpretation is complicated by the fact that the single mutation, Tyr[96]Phe, results in a product profile similar to that of Tyr[96]Phe–Phe[87]Tyr. Thus, the isolated steric interactions with $1R$-norcamphor serve to orient the substrate in favor of 6-hydroxylation. The contribution of the native hydrogen bond to the ratio of 5- and 6-carbon hydroxylation is characterized by a difference in Gibbs free energy of over 1 kcal/mole. This is in contrast to the case of *d*-camphor, in which the 5-*exo*-alcohol

TABLE IX

Effect of Tyr[96]Phe on the Regiospecificity of $P450_{cam}$ Camphor and Norcamphor Hydroxylations

$P450_{cam}$	1S-Norcamphor		1R-Norcamphor		1R-Camphor	
	% coupling NADH to alcohol	5 : 6 : 3	% coupling NADH to alcohol	5 : 6 : 3	% coupling NADH to alcohol	5 : 6 : 3 : 9
Wild type	12	28:62:10	14	67:31:2	97	100:0:0:0
Wt—theory[186,189]		27:73:0		83:13:0[a]	100	100:0:0:0
Tyr[96]Phe	13	22:56:22	5	36:53:11	98	95:3:1:1
Phe[87]Tyr	12	40:51:9	11	38:58:4	100	96:1:0:3
Tyr[96]Phe–Phe[87]Tyr	10	40:49:11	12	37:60:3	93	100:0:0:0

[a]These values are averages of several runs and consequently do not all equal 100%.

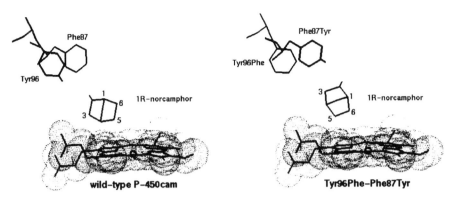

FIGURE 6. Representation of the substrate binding in the H-bond switch mutants in $P450_{cam}$. The left side depicts wild-type $P450_{cam}$ with the heme edge-on, and the camphor 5-carbon lying closest to the heme iron. The right side depicts a modeled substrate–enzyme complex for the H-bonding mutant $Tyr^{96}Phe–Phe^{87}Tyr$, with the 6-carbon now closest to the heme iron.

accounted for greater than 90% of the hydroxylated products with the $Tyr^{96}Phe$ mutant.[22] Clearly, the hydrogen bond with Tyr^{96} plays a major role in the recognition of small camphor analogues by the $P450_{cam}$ active site. Further investigations, including additional hydrogen bond donors, will be necessary to substantiate the role of specifically engineered interactions in the enzyme–substrate complex.

3. Conclusions

In this review we have attempted to document the vast information available on the structure and functioning of the P450 monooxygenase system that has been gleaned through study of a single bacterial P450 isozyme over the past 25 years. While there is obviously a wealth of data concerning the detailed mechanisms of action for the $P450_{cam}$ system, we are clearly far short in describing the detailed picture of electron transfer and dioxygen activation that a physicist or chemist might desire.

The first steps of the catalytic cycle involving substrate binding, electron transfer, and dioxygen association have been fairly well characterized for the simple reason that these intermediates are relatively stable and have a UV/visible spectrum that makes them amenable to investigation. Less is known about the proposed species shown in Fig. 1, although it now seems clear that the mechanism is not concerted and some of these species must therefore exist as discrete intermediates. It is hoped that work with other P450 isozymes, mutant species of those isoforms now known, or further experiments with exogenous oxidants will allow the direct observation of these intermediates. The dogma that has been generated through tenuous comparisons with peroxidases and inorganic model systems may place the field on thin ice when a detailed picture of the catalytic mechanism of dioxygen activation finally emerges.

Far too little is known about the role of protein and substrate dynamics in the formation of productive protein–small molecule or protein–protein recognition complexes. $P450_{cam}$

and its redox partners seem a perfect system for the study of these interactions. Hidden in these processes are the fundamental workings of substrate access to the buried active sites of the P450 isozymes, the role of substrate motions in determining stereo- and regiospecificity, and the functioning of dynamic motions across the energy landscapes of multiprotein complex formation. Future investigations must use advances in technology to directly observe the subsequent protein states in the proposed catalytic cycle.

Clearly the availability of an X-ray structure has helped immeasurably in forming a road map of the active site and in providing suggestions as to mechanisms. New structures from the Poulos and Deisenhofer laboratories provide key insights into the universality of mechanistic bioinorganic chemical details such as the possible role of a hydrogen bond "pull" in aiding heterolytic cleavage of the bound atmospheric dioxygen. In the future, Laue diffraction promises to yield detailed information on subsequent states of the P450 cycle.

The next 5 years will probably shed as much light on the functioning of the P450 enzymes as has the last 25 years—provided that room is made for young investigators to carry the field into new dimensions and that this new talent is given the opportunity of presenting their work to receptive audiences. The 1990s should indeed be a golden decade for P450 research.

ACKNOWLEDGMENTS. The authors wish to thank those researchers who made available manuscripts not yet published. Work by the Sligar lab discussed in this review was supported in part by NIH Grants GM-33775 and GM-31756.

References

1. Capdevila, J. H., Falck, J. R., and Estabrook, R. W., 1992, Cytochrome P450 and the arachidonate cascade, *FASEB J.* **6**:731–736.
2. Guengerich, F. P., 1991, Reactions and significance of cytochrome P-450 enzymes, *J. Biol. Chem.* **266**:10019–10022.
3. Coon, M. J., Ding, X., Pernecky, S. J., and Vaz, A. D. N., 1992, Cytochrome P450: Progress and predictions, *FASEB J.* **6**:669–673.
4. Porter, T. D., and Coon, M. J., 1991, Cytochrome P-450: Multiplicity of isoforms, substrates, and catalytic and regulatory mechanisms, *J. Biol. Chem.* **266**:13469–13472.
5. Klingenberg, M., 1958, Pigments of rat liver microsomes, *Arch. Biochem. Biophys.* **75**:376–386.
6. Garfinkel, D., 1957, Isolation and properties of cytochrome b5 from pig liver, *Arch. Biochem. Biophys.* **71**:111–120.
7. Hashimoto, Y., Yamano, T., and Mason, H. S., 1962, An electron spin resonance study of microsomal electron transport, *J. Biol. Chem.* **237**:3843–3844.
8. Omura, T., and Sato, R., 1962, A new cytochrome in liver microsomes, *J. Biol. Chem.* **237**:1375–1376.
9. Muramaki, K., and Mason, H. S., 1967, An electron spin resonance study of microsomal Fex, *J. Biol. Chem.* **242**:1102–1110.
10. Bangcharoenpaurpong, O., Champion, P. M., Hall, K. S., and Hager, L. P., 1986, Resonance Raman studies of isotopically labelled chloroperoxidase, *Biochemistry* **25**:2374–2378.
11. Poulos, T. L., Finzel, B. C., Gunsalus, I. C., Wagner, G. C., and Kraut, J., 1985, The 2.6 Å crystal structure of *Pseudomonas putida* cytochrome P-450, *J. Biol. Chem.* **260**:16122–16130.
12. Vickery, L., Salmon, A., and Sauer, K., 1975, Magnetic circular dichroism studies on microsomal aryl hydrocarbon hydroxylase: Comparison with cytochrome b5 and cytochrome P-450cam, *Biochim. Biophys. Acta* **386**:87–98.

13. Cramer, S. P., Dawson, J. H., Hodgson, K. O., and Hager, L. P., 1978, Studies on the ferric forms of cytochrome P-450 and chloroperoxidase by extended X-ray absorption fine structure. Characterization of the Fe–N and Fe–S distances, *J. Am. Chem. Soc.* **100:**7282–7290.

14. Hasemann, C. A., Ravichandran, K. G., Peterson, J. A., and Diesenhofer, J., 1994, Crystal structure and refinement of cytochrome P450terp at 2.3Å resolution, *J. Mol. Biol.* **236:**1169–1185.

15. Ravichandran, K. G., Boddupalli, S. S., and Hasemann, C. A., 1993, Crystal structure of hemoprotein domain of P450BM-3, a prototype for microsomal P450's, *Science* **261:**731–736.

16. Koga, H., Rauchfuss, B., and Gunsalus, I. C., 1985, P-450cam gene: Cloning and expression in *Pseudomonas putida* and *Escherichia coli, Biochem. Biophys. Res. Commun.* **130:**412–417.

17. Ropp, J. D., Gunsalus, I. C., and Sligar, S. G., 1992, Cloning and expression of a member of a new cytochrome P-450 family: P-450lin from *Pseudomonas putida (incognita), J. Bacteriol.* **175:**6028–6037.

18. Shimizu, T., Sogawa, K., Fujii-Kuriyama, Y., Takahashi, M., Ogoma, Y., and Hatano, M., 1986, Expression of cytochrome P-450d by *Saccharomyces cerevisiae, FEBS Lett.* **207:**217–221.

19. Trower, M. K., Lenstra, R., Omer, C., Buchholz, S. E., and Sariaslani, F. S., 1992, Cloning, nucleotide sequence, determination and expression of the genes encoding cytochrome P-450soy (*soyC*) and ferredoxin-soy (*soyB*) from *Streptomyces griseus, Mol. Microbiol.* **6:**2125.

20. Unger, B. P., 1988, Sequence, expression and mutagenesis of the *Pseudomonas putida* cytochrome P-450cam gene in *Escherichia coli,* Ph.D. thesis, University of Illinois.

21. Zhou, D., Pompon, D., and Chen, S., 1991, Structure–function studies of human aromatase by site-directed mutagenesis: Kinetic properties of mutants Pro-308→Phe, Tyr-361→Phe, Tyr-361→Leu, and Phe-406→Arg, *Proc. Natl. Acad. Sci. USA* **88:**410–414.

22. Atkins, W. M., and Sligar, S. G., 1988, The roles of active site hydrogen bonding in cytochrome P-450cam as revealed by site-directed mutagenesis, *J. Biol. Chem.* **263:**18842–18849.

23. Atkins, W. M., and Sligar, S. G., 1989, Molecular recognition in cytochrome P-450: Alteration of regioselective alkane hydroxylation via protein engineering, *J. Am. Chem. Soc.* **111:**2715–2717.

24. Imai, Y., and Nakamura, M., 1988, The importance of threonine-301 from cytochromes P-450 (laurate (ω-1)-hydroxylase and testosterone 16-alphahydroxylase) in substrate binding as demonstrated by site-directed mutagenesis, *FEBS Lett.* **234:**313–315.

25. Imai, M., Shimada, H., Watanabe, Y., Matsushima-Hibiya, Y., Makino, R., Koga, H., Horiuchi, T., and Ishimura, Y., 1989, Uncoupling of the cytochrome P-450cam monooxygenase reaction by a single mutation, threonine-252 to alanine or valine: A possible role of the hydroxy amino acid in oxygen activation, *Proc. Natl. Acad. Sci. USA* **86:**7823–7827.

26. Ishida, N., Aoyama, Y., Hatanaka, R., Oyama, Y., Imajo, S., Ishiguro, M. Oshima, T., Nakazato, H., Naguchi, T., Maitra, U. S., Mohan, V. P., Sprinson, D. B., and Yoshida, Y., 1988, A single amino acid substitution converts cytochrome P-450cam to an inactive form, cytochrome P-450SG1: Complete primary structures deduced from cloned DNAs, *Biochem. Biophys. Res. Commun.* **155:**317–323.

27. Shimizu, T., Hirano, K., Takahashi, M., Hatana, M., and Fuji-Kuriyama, Y., 1988, Site-directed mutagenesis of rat liver cytochrome P-450d: Axial ligand and heme incorporation, *Biochemistry* **27:**4138–4141.

28. Martinis, S. A., Atkins, W. M., Stayton, P. S., and Sligar, S. G., 1989, A conserved residue of cytochrome P-450 is involved in heme–oxygen stability and activation, *J. Am. Chem. Soc.* **111:**9252–9253.

29. Nebert, D. W., and Gonzales, F. J., 1987, P450 genes: Structure, evolution, and regulation, *Annu. Rev. Biochem.* **56:**945–993.

30. Tyson, C. A., Lipscomb, J. D., and Gunsalus, I. C., 1972, The roles of putidaredoxin and P-450cam in methylene hydroxylation, *J. Biol. Chem.* **247:**5777–5784.

31. Roberts, E. S., Vaz, A. D. N., and Coon, M. J., 1991, Catalysis by cytochrome P-450 of an oxidative reaction in xenobiotic aldehyde metabolism: Deformylation with olefin formation, *Proc. Natl. Acad. Sci. USA* **88:** 8963–8966.

32. Sligar, S. G., and Gunsalus, I. C., 1976, A thermodynamic model of regulation: Modulation of redox equilibria in camphor monooxygenase, *Proc. Natl. Acad. Sci. USA* **73**:1078–1082.

33. Black, S. D., and Coon, M. J., 1987, P-450 cytochromes: Structure and function, *Adv. Enzymol. Relat. Areas Mol. Biol.* **60**:35–87.

34. Dawson, J. H., and Sono, M., 1987, Cytochrome P-450 and chloroperoxidase: Thiolate-ligated heme enzymes. Spectroscopic determination of their active site structures and mechanistic implications of thiolate ligation, *Chem. Rev.* **87**:1255–1276.

35. Poulos, T. L., and Finzel, B. C, 1984, in: *Peptide and Protein Reviews* (M. T. W. Hearn, ed.), Dekker, New York, Vol. IV, pp.115–171.

36. Paine, A. J., 1991, The cytochrome-P450 gene superfamily, *Int. J. Exp. Pathol.* **72**:349–363.

37. Sligar, S. G., 1976, Coupling of spin, substrate and redox equilibria in cytochrome P-450, *Biochemistry* **15**: 5399–5406.

38. Harris, D., and Loew, G., 1993, Determinants of the spin state of the resting state of cytochrome P450cam, *J. Am. Chem. Soc.* **115**:8775–8779.

39. Sligar, S. G., Lipscomb, J. D., Debrunner, P. G., and Gunsalus, I. C., 1974 Superoxide anion production by the autoxidation of cytochrome P-450cam, *Biochem. Biophys. Res. Commun.* **61**:290–296.

40. White, R. E., and Coon, M. J., 1980, Oxygen activation by cytochrome P-450, *Annu. Rev. Biochem.* **49**:315–356.

41. Groves, J. T., and McCluskey, G. A., 1976, Aliphatic hydroxylation via oxygen rebound. Oxygen transfer catalyzed by iron, *J. Am. Chem. Soc.* **98**:859–861.

42. Poulos, T. L., Finzel, B. C., and Howard, A. J., 1986, Crystal structure of substrate-free *Pseudomonas putida* cytochrome P-450, *Biochemistry* **25**:5314–5322.

43. Raag, R., Li, H., Jones, B. C., and Poulos, T. L., 1993, Inhibitor induced conformational change in cytochrome P-450cam, *Biochemistry* **32**:4571–4578.

44. Wells, A. V., Li, P., Champion, P. M., Martinis, S. A., and Sligar, S. G., 1992, Resonance Raman investigations of *Escherichia coli* cytochrome P450 and P420, *Biochemistry* **31**:4384–4393.

45. DiPrimo, C., Hui Bon Hoa, G., Deprez, E., Douzou, P., and Sligar, S. G., 1993, Conformational dynamics of cytochrome P-450cam as monitored by photo-acoustic calorimetry, *Biochemistry* **32**:3671–3676.

46. Deprez, E., Gerber, N. C., Di Primo, C., Douzou, P., Sligar, S. G., and Hui Bon Hoa, G., 1994, Electrostatic control of the substrate access channel in cytochrome P-450cam, *Biochemistry* **33**:14464–14468.

47. Gerber, N. C., 1994, Bioorganic activation of cytochrome P-450cam, Ph.D. thesis, University of Illinois, Urbana.

48. Peterson, J. A., 1971, Camphor binding by *Pseudomonas putida* cytochrome P-450, *Arch. Biochem. Biophys.* **144**:678–693.

49. Tsai, R., Yu, C.-A., Gunsalus, I. C., Peisach, J. Blumberg, W. E., Orme-Johnson, W. H., and Beinert, H., 1970, Spin-state changes in cytochrome P-450cam on binding specific substrates, *Proc. Natl. Acad. Sci. USA* **66**:1157–1163.

50. Sligar, S. G., and Murray, R. I., 1986, in: *Cytochrome P-450: Structure, Mechanism, and Biochemistry* (P. R. Ortiz de Montellano, ed.), Plenum Press, New York, pp. 429–503.

51. Fisher, M. T., and Sligar, S. G., 1987, Temperature jump relaxation kinetics of the P-450cam spin equilibrium, *Biochemistry* **26**:4797–4803.

52. Loew, G. H., Collins, J., Luke, B., Waleh, A., and Pudzianowski, A., 1986, Theoretical studies of cytochrome P450: Characterizations of stable and transient active states, reaction mechanisms and substrate-enzyme interactions, *Enzyme* **36**:54–78.

53. Poulos, T. L., Finzel, B. C., and Howard, A. J., 1987, High-resolution crystal structure of cytochrome P-450cam, *J. Mol. Biol.* **195**:687–700.

54. Wade, R. C., 1990, Solvation of the active site of cytochrome P450-cam, *J. Comp. Aided Mol. Des.* **4**:199–204.

55. Di Primo, C., Sligar, S. G., Hui Bon Hoa, G., and Douzou, P., 1992, A critical role of protein-bound water in the catalytic cycle of cytochrome P-450 camphor, *FEBS Lett.* **312:**252–254.

56. Deprez, E., Di Primo, C., Hui Bon Hoa, G., Sligar, S. G., and Douzou, P., 1995, Effects of monovalent cations on cytochrome P-450 camphor: Evidence for preferential binding of potassium, manuscript in preparation.

57. Di Primo, C., Hui Bon Hoa, G., Douzou, P., and Sligar, S., 1990, Mutagenesis of a single hydrogen bond in cytochrome P-450 alters cation binding and heme solvation, *J. Biol. Chem.* **265:**5361–5363.

58. Yoshikawa, K., Noguti, T., Tsujimura, M., Koga, H., Yasukochi, T., Horiuchi, T., and Go, M., 1992, Hydrogen bond network of cytochrome P-450cam: A network connecting the heme group with helix K, *Biochim. Biophys Acta* **1122:**41–44.

59. Raag, R., and Poulos, T. L., 1991, Crystal structures of cytochrome P-450cam complexed with camphane, thiocamphor, and adamantane: Factors controlling P-450 substrate hydroxylation, *Biochemistry* **30:**2674–2684.

60. Gunsalus, I. C., Pederson, T. C., and Sligar, S. G., 1975, Oxygenase-catalyzed biological hydroxylations, *Annu. Rev. Biochem.* **44:**377.

61. Gunsalus, I. C., Meeks, J. R., Lipscomb, J. D., Debrunner, P., and Munck, E., 1974, in: *Molecular Mechanisms of Oxygen Activation* (O. Hayaishi, ed.), Academic Press, New York.

62. Duppel, W., Poensgen, J., Ullrich, V., and Dahl, G., 1978, in *Microsomes and Drug Oxidations* (V. Ullrich, I. Roots, A. Hildebrandt, R. W. Estabrook, and A. H. Conney, eds.), Pergamon Press, Elmsford, NY, pp. 31–38.

63. Lambeth, J. D., and Kamin, H., 1979, Adrenodoxin reductase–adrenodoxin complex: Flavin to iron–sulfur electron transfer as the rate-limiting step in the NADPH-cytochrome c reductase reaction, *J. Biol. Chem.* **254:**2766–2774.

64. Lambeth, J. D., and Kriengsiri, S., 1985, Cytochrome P-450scc–adrenodoxin interactions: Ionic effects on binding, and regulation of cytochrome reduction by bound steroid substrates, *J. Biol. Chem.* **260:**8810–8816.

65. Geren, L. M., O'Brien, P., Stonehuerner, J., and Millett, F., 1984, Identification of specific carboxylate groups on adrenodoxin that are involved in the interaction with adrenodoxin reductase, *J. Biol. Chem.* **259:**2155–2160.

66. Miura, S., Tomita, S., and Ichikawa, Y., 1991, Modification of histidine 56 in adrenodoxin with diethyl pyrocarbonate inhibited the interaction with cytochrome P-450scc and adrenodoxin reductase, *J. Biol. Chem.* **266:**19212–19216.

67. Tsubaki, M., Iwamoto, Y., Hiwatashi, A., and Ichikawa, Y., 1989, Inhibition of electron transfer from adrenodoxin to cytochrome P-450scc by chemical modification with pyridoxal 5′-phosphate: Identification of adrenodoxin-binding site of cytochrome P-450scc, *Biochemistry* **28:**6899–6907.

68. Tuls, J., Geren, L., and Millett, F., 1989, Fluorescein isothiocyanate specifically modifies lysine 338 of cytochrome P-450scc and inhibits adrenodoxin binding, *J. Biol. Chem.* **264:**16421–16428.

69. Beckert, V., Dettmer, R., and Bernhardt, R., 1994, Mutations of tyrosine 82 in bovine adrenodoxin that affect binding to cytochromes P45011A1 and P45011B1 but not electron transfer, *J. Biol. Chem.* **269:**2568–2573.

70. Wada, A., and Waterman, M. R., 1992, Identification by site-directed mutagenesis of two lysine residues in cholesterol side chain cleavage cytochrome P450 that are essential for adrenodoxin binding, *J. Biol. Chem.* **267:**22877–22882.

71. Marg, A. Kuban, R. J., and Behlke, J., 1992, Crystallization and x-ray examination of bovine adrenodoxin, *J. Mol. Biol.* **227:**945–947.

72. Kuban, R. J., Marg, A., and Resch, M., 1993, Crystallization of bovine adrenodoxin reductase in a new unit cell and its crystallographic characterization, *J. Mol. Biol.* **234:**245–248.

73. Lin, D., Shi, Y., and Miller, W. L., 1990, Cloning and sequence of the human adrenodoxin reductase gene, *Proc. Natl. Acad. Sci. USA* **87:**8516–8520.

74. Geren, L., Tuls, J., O'Brien, P., Millett, F., and Peterson, J. A., 1986, The involvement of carboxylate groups of putidaredoxin in the reaction with putidaredoxin reductase, *J. Biol. Chem.* **261:**15491–15495.

75. Coghan, V. M., Cupp, J. R., and Vickery, L. G., 1988, Purification and characterization of human placental ferredoxin, *Arch. Bioch. Biophys.* **264:**376–382.

76. Lipscomb, J. D., Sligar, S. G., Namtvedt, M. J., and Gunsalus, I. C., 1976, Autoxidation and hydroxylation reactions of oxygenated cytochrome P-450cam, *J. Biol. Chem.* **251:**1116–1124.

77. Turko, I. B., Adamovich, T. B., Krilliova, N. M., Usanov, S. A., and Chashchin, V. L., 1989, Cross-linking studies of the cholesterol hydroxylation systems from bovine adrenocortical mitochondria, *Biochim. Biophys. Acta* **996:**37–42.

78. Kido, T., and Kimura, T., 1979, The formation of binary and ternary complexes of cytochrome P-450scc with adrenodoxin and adrenodoxin reductase–adrenodoxin complex, *J. Biol. Chem.* **254:**11806–11815.

79. Lambeth, J. D., and Pember, S. O., 1983, Cytochrome P-450scc–adrenodoxin complex. Reduction properties of the substrate-associated cytochrome and relation of the reduction states of heme and iron–sulfur centers to association of the proteins, *J. Biol. Chem.* **258:**5596–5602.

80. Hanukoglu, I., and Gutfinger, T., 1989, cDNA sequence of adrenodoxin reductase: Identification of NADP-binding sites in oxidoreductases, *Eur. J. Biochem.* **180:**479–484.

81. Roome, P. W., Philley, J. C., and Peterson, J. A., 1983, Purification and properties of putidaredoxin reductase, *J. Biol. Chem.* **258:**2593–2598.

82. Roome, P. W., and Peterson, J. A., 1988, The oxidation of reduced putidaredoxin reductase by oxidized putidaredoxin, *Arch. Biochem. Biophys.* **266:**41–50.

83. Peterson, J. A., Lorence, M. C., and Amarneh, B., 1990, Putidaredoxin reductase and putidaredoxin: Cloning, sequence determination, and heterologous expression of the proteins, *J. Biol. Chem.* **265:**6066–6073.

84. Gunsalus, I. C., Tyson, C. A., and Lipscomb, J. D., 1973, in: *Oxidases and Related Redox Systems* (T. E. King, H. S. Mason, and M. Morrison, ed.), University Park Press, Baltimore, Vol. 2, pp. 583–603.

85. Tyson, C. A., Tsai, R., and Gunsalus, I. C., 1970, Fast reaction studies on the camphor P-450 hydroxylase system, *J. Am. Oil Chem. Soc.* **74:**343A–344A.

86. Pederson, T. C., Austin, R. H., and Gunsalus, I. C., 1977 in: *Microsomes and Drug Oxidations* (V. Ullrich, ed.), Pergamon Press, Elmsford, NY, pp. 275–283.

87. Hintz, M. J., and Peterson, J. A., 1981, The kinetics of reduction of cytochrome P-450cam by reduced putidaredoxin, *J. Biol. Chem.* **256:**6721–6728.

88. Hintz, M. J., Mock, D. M., Peterson, L. L., Tuttle, K., and Peterson, J. A., 1982, Equilibrium and kinetic studies of the interaction of cytochrome P-450cam and putidaredoxin, *J. Biol. Chem.* **257:**14324–14332.

89. Brewer, C. B., and Peterson, J. A., 1986, Single turnover studies with oxycytochrome P-450cam, *Arch. Biochem. Biophys* **249:**515–521.

90. Brewer, C. B., and Peterson, J. A., 1988, Single turnover kinetics of the reaction between oxycytochrome P-450cam and reduced putidaredoxin, *J. Biol. Chem.* **263:**791–798.

91. Sligar, S. G., 1975, A kinetic and equilibrium description of camphor hydroxylation by the P-450cam monooxygenase system, Ph.D. thesis, University of Illinois, Urbana.

92. Sligar, S. G., Debrunner, P. G., Lipscomb, J. D., Namtvedt, M. J., and Gunsalus, I. C., 1974, A role for putidaredoxin COOH-terminus in P-450cam (cytochrome m) hydroxylations, *Proc. Natl. Acad. Sci. USA* **71:**10.

93. Davies, M. D., Qin, L., Beck, J. L., Suslick, K. S., Koga, H., Horiuchi, T., and Sligar, S. G., 1990, Putidaredoxin reduction of cytochrome P-450cam: Dependence of electron transfer on the identity of putidaredoxin's C-terminal amino acid, *J. Am. Chem. Soc.* **112:**7396–7398.

94. Davies, M. D., and Sligar, S. G., 1992, Genetic variants in the putidaredoxin–cytochrome P-450cam electron transfer complex: Identification of the residue responsible for redox-state-dependent conformers, *Biochemistry* **31:**11383–11389.

95. Stayton, P. S., and Sligar, S. G., 1991, Structural microheterogeneity of a tryptophan residue required for efficient biological electron transfer between putidaredoxin and cytochrome P-450cam, *Biochemistry* **30**:1845–1851.

96. Baldwin, J. E., Morris, G. M., and Richards, W. G., 1991, Electron transport in cytochromes-P-450 by covalent switching, *Proc. R. Soc. Lond. Ser. B* **245**:43–51.

97. Salemme, F. R., 1976, An hypothetical structure for an intermolecular electron transfer complex of cytochromes c and b5, *J. Mol. Biol.* **102**:563–568.

98. Stayton, P. S., Fisher, M. T., and Sligar, S. G., 1988, Determination of cytochrome b5 association reactions: Characterization of metmyoglobin and cytochrome P450cam binding to genetically engineered cytochrome b5, *J. Biol. Chem.* **263**:13544–13548.

99. Stayton, P. S., Poulos, T. L., and Sligar, S. G., 1989, Putidaredoxin competitively inhibits cytochrome b5-cytochrome P-450cam association: A proposed molecular model for a cytochrome P-450cam electron-transfer complex, *Biochemistry* **28**: 8201–8205.

100. Stayton, P. S., and Sligar, S. G., 1990, The cytochrome P-450cam binding surface as defined by site-directed mutagenesis and electrostatic modeling, *Biochemistry* **29**:7381–7386.

101. Koga, H., Sagara, Y., Yaoi, T., Tsujimura, M., Nakamura, K., Sekimizu, K., Makino, R., Shimada, H., Ishimura, Y., Yura, K., Go, M., Ikeguchi, M., and Horiuchi, T., 1993, Essential role of the Arg112 residue of cytochrome P450cam for electron transfer from reduced putidaredoxin, *FEBS Lett.* **331**:109–113.

102. Gerber, N. C., Horiuchi, T., Koga, H., and Sligar, S. G., 1990, Identification of 2Fe2S cysteine ligands in putidaredoxin, *Biochem. Biophys. Res. Commun.* **169**:1016–1020.

103. Tuls, J., Geren, L., Lambeth, J. D., and Miliett, F., 1987, The use of a specific fluorescence probe to study the interaction of adrenodoxin with adrenodoxin reductase and cytochrome P-450scc, *J. Biol. Chem.* **262**:10020–10025.

104. Cupp, J. R., and Vickery, L. E., 1988, Identification of free and [Fe2S2]-bound cysteine residues of adrenodoxin, *J. Biol. Chem.* **263**:17418–17421.

105. Pochapsky, T. C., and Ye, X. M., 1991, 1H NMR identification of a β-sheet structure and description of folding topology in putidaredoxin, *Biochemistry* **30**:3850–3856.

106. Pochapsky, T. C., Ye, X. M., Ratnaswamy, G., and Lyons, T. A., 1994, An NMR-derived model for the solution structure of oxidized putidaredoxin, a 2-Fe 2-S ferredoxin from *Pseudomonas, Biochemistry* **33**:6424–6432.

107. Ratnaswamy, G., and Pochapsky, T. C., 1993, Characterization of hyperfine-shifted 1H resonances in oxidized and reduced putidaredoxin, an Fe2S2 ferredoxin from *P. putida, Magn. Reson. Chem.* **31**:S73–S77.

108. Pochapsky, T. C., Ratnaswamy, G., and Patera, A., 1994, Redox dependent 1H NMR spectral features and tertiary structural constraints on the C-terminal region of putidaredoxin, *Biochemistry* **33**:6433–6441.

109. McLendon, G., Pardue, K., and Bak, P., 1987, Electron transfer in the cytochrome c/cytochrome b2 complex: Evidence for "conformational gating," *J. Am. Chem. Soc.* **109**:7540–7541.

110. Liang, N., Mauk, A. G., Pielak, G. J., Johnson, J. A., Smith, M., and Hoffman, B. M., 1988, Regulation of interprotein electron transfer by residue 82 of yeast cytochrome c, *Science* **240**:311–313.

111. Gunsalus, E. C., Meeks, J., Lipscomb, J., Debrunner, P., and Munck, E., 1979, in: *Molecular Mechanisms of Oxygen Activation* (O. Hayaishi, ed.), Academic Press, New York, pp. 559–613.

112. Raag, R., and Poulos, T. L., 1989, Crystal structure of the carbon monoxide–substrate–cytochrome P-450cam ternary complex, *Biochemistry* **28**:7586–7592.

113. Bangcharoenpaurpong, O., Rizos, A., Champion, P., Jollie, D., and Sligar, S., 1986, Resonance Raman detection of bound dioxygen in cytochrome P-450cam, *J. Biol. Chem.* **261**:8089–8090.

114. Sharrock, M., Munck, E., Debrunner, P. G., Marshall, V., Lipscomb, J. D., and Gunsalus, I. C., 1973, Mossbauer studies of cytochrome P-450cam, *Biochemistry* **12**:258–265.

115. Egawa, T., Ogura, T., Makino, R., Ishimura, Y., and Kitagawa, T., 1991, Observation of the O-O stretching Raman band for cytochrome-P-450cam under catalytic conditions, *J. Biol. Chem.* **266**:10246–10248.

116. Raag, R., Martinis, S. A., Sligar, S. G., and Poulos, T. L., 1991, Crystal structure of the cytochrome P-450cam active site mutant Thr252Ala, *Biochemistry* **30**: 11420–11429.

117. O'Keefe, D. H., Ebel, R. E., Peterson, J. A., Maxwell, J. C., and Caughey, W. S., 1978, An infrared spectroscopic study of CO bonding to ferrous cytochrome P-450, *Biochemistry* **17**:5845–5852.

118. Jung, C., and Marlow, F., 1987, Dynamic behavior of the active site structure in bacterial cytochrome P-450, *Stud. Biophys.* **120**:241–251.

119. Raag, R., and Poulos, T. L., 1989, The structural basis for substrate-induced changes in redox potential and spin equilibrium in cytochrome P-450cam, *Biochemistry* **28**:917–922.

120. Jung, C., Hui Bon Hoa, G., Schroeder, K.-L., Simon, M., and Doucet, J. P., 1992, Substrate analogue induced changes of the CO-stretching mode in the cytochrome P450cam-carbon monoxide complex, *Biochemistry* **31**:12855–12862.

121. Guengerich, F. P., and MacDonald, T. M., 1990, Mechanisms of cytochrome P-450 catalysis (review), *FASEB J.* **4**:2453–2459.

122. White, R. E., 1991, The involvement of free radicals in the mechanisms of monooxygenases, *Pharmacol. Ther.* **49**:21–42.

123. Sariaslani, F. S., 1991, Microbial cytochromes P-450 and xenobiotic metabolism, *Adv. App. Microbiol.* **36**:133–178.

124. Guengerich, F. P., 1990, Enzymatic oxidation of xenobiotic chemicals, *CRC Crit. Rev. Biochem.* **25**: 97–153.

125. Gunsalus, I. C., 1968, A soluble methylene hydroxylase system: Structure and role of cytochrome P-450 and iron–sulfur protein components, *Hoppe-Seylers. Physiol. Chem.* **349**:1610–1613.

126. Gunsalus, I. C., and Lipscomb, J. D., 1973, in: *Iron–Sulfur Proteins* (W. Lovenberg, ed.), Academic Press, New York, Vol. 1, pp. 151–171.

127. Ishimura, Y., Ullrich, V., and Peterson, J. A., 1971, Oxygenated cytochrome P-450 as reaction intermediate in enzymatic hydroxylation of d-camphor, *Fed. Proc.* **30**:1092.

128. Peterson, J. A., Ishimura, Y., and Griffin, B. W., 1972, *Pseudomonas putida* cytochrome P-450: Characterization of an oxygenated form of the hemoprotein, *Arch. Biochem. Biophys.* **149**:197–208.

129. Hui Bon Hoa, G., Begard, E., Debey, P., and Gunsalus, I. C., 1978, Two univalent electron transfers from putidaredoxin to bacterial cytochrome P-450cam at subzero temperature, *Biochemistry* **17**:2835–2839.

130. Guengerich, F. P., 1992, Human cytochrome P-450 enzymes, *Life Sci.* **50**:1471–1478.

131. Coon, M. J., and Koop, D. R., 1983, in: *The Enzymes* (P. D. Boyer, ed.), Academic Press, New York, Vol. 14, pp. 645–677.

132. Jefcoate, C. R., 1986, *Cytochrome P-450 Enzymes in Sterol Biosynthesis and Metabolism*, Plenum Press, New York, pp. 387–428.

133. Miwa, G. T., and Lu, A. Y. H., 1986, in: *Cytochrome P-450: Structure, Mechanism, and Biochemistry* (P. R. Ortiz de Montellano, ed.), Plenum Press, New York, pp. 77–88.

134. Poulos, T. L., 1986 in: *Cytochrome P-450: Structure, Mechanism, and Biochemistry* (P. R. Ortiz de Montellano, ed.), Plenum Press, New York, pp. 505–523.

135. Gunter, M. J., and Turner, P., 1991, Metalloporphyrins as models for the cytochromes P-450, *Coord. Chem. Rev.* **108**:115–161.

136. White, P. W., 1990, Mechanistic studies and selective catalysis with cytochrome P-450 model systems, *Bioorg. Chem.* **18**:440–456.

137. Lee, W. A., and Bruice, T., 1985, Homolytic and heterolytic oxygen–oxygen bond scissions accompanying oxygen transfer to iron (III) porphyrins by percarboxylic acids and hydroperoxides. A mechanistic criterion for peroxidase and cytochrome P-450, *J. Am. Chem Soc.* **107**:513–-514.

138. Atkins, W. M., and Sligar, S. G., 1987, Metabolic switching in cytochrome P-450cam: Deuterium isotope effects on regiospecificity and the monooxygenase/oxygenase ratio, *J. Am. Chem. Soc.* **109:**3754–3760.

139. Atkins, W. M., and Sligar, S. G., 1988, Deuterium isotope effects in norcamphor metabolism by cytochrome P-450cam: Kinetic evidence for the two-electron reduction of a high-valent iron-oxo intermediate, *Biochemistry* **27:**1610–1616.

140. Zhukov, A. A., and Arachov, A. I., 1982, Complete stoichiometry of free NADPH oxidation in liver microsomes, *Biochem. Biophys. Res. Commun.* **109:**813–818.

141. Kuthan, H., and Ullrich, V., 1982, Oxidase and oxygenase function of the microsomal cytochrome P-450 monooxygenase system, *Eur. J. Biochem.* **126:**583–588.

142. Gorsky, L. D., Koop, D. R., and Coon, M. J., 1984, On the stoichiometry of the oxidase and monooxygenase reactions catalyzed by liver microsomal cytochrome P-450: Products of oxygen reduction, *J. Biol. Chem.* **259:**6812–6817.

143. Blake, R. C., and Coon, M. J., 1989, On the mechanism of action of cytochrome P-450: Spectral intermediates in the reaction with iodosobenzene and its derivatives, *J. Biol. Chem.* **264:**3694–3701.

144. Larroque, C., Lange, R., Maurin, L., Bienvenue, A., and van Lier, J. E., 1990, On the nature of the cytochrome P450scc "ultimate oxidant": Characterization of a productive radical intermediate, *Arch. Biochem. Biophys.* **282:**198–201.

145. Sligar, S. G., and Gunsalus, I. C., 1979, Proton coupling in the cytochrome P-450 spin and redox equilibria, *Biochemistry* **18:**2290–2295.

146. Nelson, D. R., and Strobel, H. W., 1987, Evolution of cytochrome P-450 proteins, *Mol. Biol. Evol.* **4:**572–593.

147. Gerber, N. C., and Sligar, S. G., 1992, Catalytic mechanism of cytochrome P-450: Evidence for a distal charge relay, *J. Am. Chem. Soc.* **114:**8742–8743.

148. Shimada, H., Makino, R., Imai, M., Horiuchi, T., and Ishimura, Y., 1990, in: *International Symposium on Oxygenases and Oxygen Activation*, Yamada Science Foundation, pp. 133–136.

149. Gerber, N. C., and Sligar, S. G., 1994, A role for Asp-251 in cytochrome P-450cam oxygen activation, *J. Biol. Chem.* **269:**4260–4266.

150. Ishigooka, M., Shimizu, T., Hiroya, K., and Hatano, M., 1992, Role of Glu318 at the putative distal site in the catalytic function of cytochrome P450d, *Biochemistry* **31:**1528–1531.

151. Zhou, D., Korzekwa, K. R., Poulos, T., and Chen, S., 1992, A site-directed mutagenesis study of human placental aromatase, *J. Biol. Chem.* **267:**762–768.

152. Shimada, H., Makino, R., Horiuchi, T., and Ishimura, Y., 1993, in: *Second International Symposium on Cytochrome P450 of Microorganisms and Plants*, Hachioji, Tokyo.

153. Aikens, J., and Sligar, S. G., 1994, Kinetic solvent isotope effects during oxygen activation by cytochrome P-450cam, *J. Am. Chem. Soc.* **116:**1143–1144.

154. Groves, J. T., McCluskey, G. A., White, R. E., and Coon, M. J., 1978, Aliphatic hydroxylation by highly purified liver microsomal cytochrome P-450: Evidence for a carbon radical intermediate, *Biochem. Biophys. Res. Commun.* **81:**154–160.

155. White, R. E., Miller, J. P., Faureau, C. V., and Bhattacharyya, A. J., 1986, Stereochemical dynamics of aliphatic hydroxylation by cytochrome P-450, *J. Am. Chem. Soc.* **108:**6024–6029.

156. Gelb, M. H., Heimbrook, D. C., Mälkönen, P., and Sligar, S. G., 1982, Stereochemistry and deuterium isotope effects in camphor hydroxylation by the cytochrome P-450cam monooxygenase system, *Biochemistry* **21:**370–377.

157. Groves, J. T., 1985, Key elements of the chemistry of cytochrome P-450: The oxygen rebound mechanism, *J. Chem. Educ.* **62:**928–931.

158. Sligar, S. G., Gelb, M. H., and Heimbrook, D. C., 1982, in: *Microsomes, Drug Oxidations and Drug Toxicity* (R. Sato and R. Kato, eds.), Wiley–Interscience, New York, pp. 155–161.

159. White, R. E., Groves, J. T., and McClusky, G. A., 1979, Electronic and steric factors in regioselective hydroxylation catalyzed by purified cytochrome P-450, *Acta Biol. Med. Germ.* **38:**475–482.

160. White, R. E., McCarthy, M. B., Egeberg, K. D., and Sligar, S. G., 1984, Regioselectivity in the cytochromes P-450: Control by protein constraints and by chemical reactivities, *Arch. Bioch. Biophys.* **228:**493–502.

161. Maryniak, D. M., Kadkhodayan, S., Crull, G. B., Bryson, T. A., and Dawson, J. H., 1993, The synthesis of 1R- and 1S-5-methylenylcamphor and their epoxidation by cytochrome P-450cam, *Tetrahedron* **49:**9373–9384.

162. Eble, K. S., and Dawson, J. H., 1984, Novel reactivity of cytochrome P450cam: Methyl hydroxylation 4of 5,5-difluorocamphor, *J. Biol. Chem.* **259:**14389–14393.

163. Loida, P. J., and Sligar, S. G., 1993, Engineering cytochrome P-450cam to increase the stereospecificity and coupling of aliphatic hydroxylation, *Prot. Eng.* **2:**207–212.

164. Loida, P. J., Paulsen, M. D., Arnold, G. E., Ornstein, R. L., and Sligar, S G., 1994, Stereoselective hydroxylation of norcamphor by cytochrome P-450cam: Experimental verification of molecular dynamics simulations, *J. Biol. Chem.* **270:**5326–5330.

165. Loida, P. J., and Sligar, S. G., 1993, Molecular recognition in cytochrome P-450cam: Mechanism for the control of uncoupling reactions, *Biochemistry* **32:**11530–11538.

166. Staudt, H., Lichtenberger, F., and Ullrich, V., 1974, The role of NADH in uncoupled microsomal monooxygenations, *Eur. J. Biochem.* **46:**99–106.

167. Springer, B. A., Egeberg, K. S., Sligar, S. G., Rohlfs, R. J., Mathews, A. J., and Olson, J. S., 1989, Discrimination between oxygen and carbon monoxide inhibition of autooxidation by myoglobin, *J. Biol. Chem.* **264:**3057–3060.

168. Carver, T. E., Brontley, R. E., Singleton, E. W., Arduini, R. M., Quillin, M. L., Phillios, G. N., and Olson, J. S., 1992, A novel site-directed mutant of myoglobin with an unusually high oxygen affinity and low auto-oxidation rate, *J. Biol. Chem.* **267:**14443–14450.

169. Gould, P., Gelb, M. H., and Sligar, S. G., 1981, Interaction of 5-bromocamphor with cytochrome P-450cam, *J. Biol. Chem.* **256:**6686–6691.

170. Castro, C. E., Wade, R. S., and Belser, N. O., 1983, Biodehalogenation: The metabolism of chloropicron by *Pseudomonas sp., J. Agric. Food Chem.* **31:**1184–1187.

171. Castro, C. E., Wade, R. S., and Belser, N. O., 1985, Biodehalogenation: Reactions of cytochrome P-450 with polyhalomethanes, *Biochemistry* **24:**204–210.

172. Castro, C. E., Wade, R. S., and Belser, N. O., 1988, Dehalogenation: Reductive reactivities of microbial and mammalian cytochromes P-450 compared with heme and whole-cell models, *J. Agric. Food Chem.* **36:**915–919.

173. Logan, M. S. P, Newman, L. M., Schanke, C. A., and Wackett, L. P., 1993, Cosubstrate effects in reductive dehalogenation by *Pseudomonas putida G786* expressing cytochrome P450cam, *Biodegradation* **4:**39–50.

174. Wackett, L. P., Logan, M. S. P., Blocki, F. A., and Bao-li, C., 1992, A mechanistic perspective on bacterial metabolism of chlorinated methanes, *Biodegradation* **3:**19–36.

175. Li, S., and Wackett, L. P., 1993, Reductive dehalogenation by cytochrome P450cam: Substrate binding and catalysis, *Biochemistry* **32:**9355–9361.

176. Lefever, M. R., and Wackett, L. P., 1994, Oxidation of low molecular weight chloroalkanes by cytochrome P-450cam, *Biochem. Biophys. Res. Commun.* **201:**373–378.

177. Koe, G. S., and Vilker, V. L., 1993, Dehalogenation by cytochrome P-450cam: Effect of oxygen levels on the decomposition of 1,2-dibromo-3-chloropropane, *Biotech. Prog.* **9:**608–614.

178. Thompson, J. A., Ho, B., and Mastovich, S. L., 1985, Dynamic headspace analysis of volatile metabolites from the reductive dehalogenation of trichloro- and tetrachloroethanes by hepatic microsomes, *Anal. Biochem.* **145:**376–384.

179. Filipovic, D., Paulsen, M. D., Loida, P. L., Sligar, S. G., and Ornstein, R. L., 1992, Ethylbenzene hydroxylation by cytochrome P-450cam, *Biochem. Biophys. Res. Commun.* **189:**488–495.

180. Paulsen, M. D., and Ornstein, R. L., 1991, A 175 psec molecular dynamics simulation of camphorbound cytochrome P-450cam, *Proteins Struct. Funct. Genet.* **11:**184–204.

181. Paulsen, M. D., Bass, M. B., and Ornstein, R. L., 1991, Analysis of active site motions from a 175 picosecond molecular dynamics simulation of cytochrome P-450cam, *J. Biomol. Struct. Dyn.* **9**:187–203.

182. Fruetel, J. A., Collins, J. R., Camper, D. L., Loew, G. H., and Ortiz de Montellano, P. R., 1992, Calculated and experimental absolute stereochemistry of the styrene and beta-methylstyrene epoxides formed by cytochrome P450 cam, *J. Am. Chem. Soc.* **114**:6987–6993.

183. Collins, J. R., and Loew, G. H., 1988, Theoretical study of the product specificity in the hydroxylation of camphor, norcamphor, 5,5-difluorocamphor, and pericyclocamphor by cytochrome P-450cam, *J. Biol. Chem.* **263**:3164–3170.

184. Jones, J. P., Trager, W. F., and Carlson, T., 1993, The binding and regiospecificity of reaction of (R)- and (S)-nicotine with cytochrome P-450cam: Parallel experimental and theoretical studies, *J. Am. Chem. Soc.* **115**:381–387.

185. Bass, M. B., Paulsen, M. D., and Ornstein, R. L., 1992, Substrate mobility in a deeply buried active site: Analysis of norcamphor bound to cytochrome P-450cam as determined by a 201 psec molecular dynamics simulation, *Proteins Struct. Funct. Gene.* **13**:26–37.

186. Paulsen, M. D., and Ornstein, R. L., 1992, Predicting the product specificity and coupling of cytochrome P450cam, *J. Comput. Aided Mol. Des.* **6**:449–460.

187. Paulsen, M. D., and Ornstein, R. L., 1994, Active site mobility inhibits reductive dehalogenation of 1,1,1-trichloroethane by cytochrome P450cam, *J. Comput. Aided Mol. Des.* in press.

188. Collins, J. R., Camper, D. L., and Loew, G. H., 1991, Valproic acid metabolism by, cytochrome-P450: A theoretical study of stereoelectronic modulators of product distribution, *J. Am. Chem. Soc.* **113**:2736–2743.

189. Bass, M. B., and Ornstein, R. L., 1993, Substrate specificity of cytochrome P450cam for l- d-norcamphor as studied by molecular dynamics simulations, *J. Comp. Chem.* **14**:541–548.

190. Wackett, L. P., Sadowsky, M. J., Newman, L. M., Hur, H., and Shuying, L., 1994, Metabolism of polyhalogenated compounds by a genetically engineered bacterium, *Nature* **368**:627–629.

191. Kulisch, G. P., and Vilker, V. L., 1991, Application of *Pseudomonas putida* pPG 786 containing P-450 cytochrome monooxygenase for removal of trace naphthalene concentrations, *Biotechnol. Prog.* **7**:93–98.

192. Loida, P. L., and Sligar, S. G., 1995, Molecular recognition in cytochrome P-450cam: Engineering a hydrogen bond switch, in preparation.

193. DiPrimo, C., Hui Bon Hoa, G., Douzou, P., and Sligar, S. G., 1992, Heme-pocket-hydration change during the inactivation of cytochrome P-450camphor by hydrostatic pressure, *Eur. J. Biochem.* **209**:583–588.

194. Pochapsky, T. C., Lyons, T., Ratnaswamy, G., Kazanis, S., Ye, X.-M., and Arakaki, T., 1995, Structure-activity relationships in putidaredoxin reduction of cytochrome P-450cam. *9th International Conference on Cytochrome 450*, Zurich, Switzerland, July 23–27, 1995, p. 161.

195. Yeom, H., Sligar, S. G., Li, H., Poulos, T., and Fulco, A., 1995, The role of Thr268 in oxygen activation of cytochrome P450BM-3, *Biochemistry*, in press.

196. Peterson, J. A., 1995, personal communication.

197. Kimata, Y., Shimada, H., Hirose, T., and Ishimura, Y., 1995, Role of Thr-252 in cytochrome P450cam: A study with unnatural amino acid mutagenesis, *Biochem. Biophys. Res. Commun.* **208**:96–102.

198. Newcomb, M., Le Tadic, M-H., Pitt, D. A., and Hollenberg, P. F., 1995, An incredibly fast apparent oxygen rebound rate constant for hydrocarbon hydroxylation by cytochrome P-450 enzymes, *J. Am. Chem. Soc.* **117**:3312–3313.

199. Dawson, J. H., Coulter, E. D., Maryniak, D., Kadhodayam, S., and Bryson, T. A., 1995, Two modes of uncoupling electron transfer and oxygen transfer in the metabolism of "slow" substrates by cytochrome P450-cam, *9th International Conference on Cytochrome 450*, Zurich, Switzerland, July 23–27, 1995, p. 195.

CHAPTER 4

Structural Studies on Prokaryotic Cytochromes P450

THOMAS L. POULOS, JILL CUPP-VICKERY, and HUIYING LI

1. Introduction

The camphor monooxygenase from *Pseudomonas putida*, $P450_{cam}$, has been the single best paradigm for P450 structure and function studies for over two decades.[1] Following a wealth of biochemical and biophysical studies on $P450_{cam}$, the high-resolution crystal structure became available in 1987.[2] This was followed by a series of structures on various inhibitor/substrate complexes which revealed some key structure–function relationships in P450s. In addition, with the development of recombinant expression systems for $P450_{cam}$, it has been possible to use site-directed mutagenesis[3,4] with reference to the crystal structure to probe questions of how structure relates to function.

Fortunately, $P450_{cam}$ is no longer the only P450 crystal structure available. The crystal structure of the heme domain of the *Bacillus megaterium* fatty acid hydroxylase, $P450_{BM-3}$,[5,6] and $P450_{terp}$,[7] a *Pseudomonad* α-terpineol monooxygenase, have been determined. Very recently, a fourth P450 structure, $P450_{eryF}$, has been solved.[8,9] Although the primary focus of this chapter is on $P450_{cam}$, it is instructive to include comparisons with the other two P450s under active investigation in our lab, $P450_{BM-3}$ and $P450_{eryF}$.

2. Background on P450eryF

Since $P450_{eryF}$ has not been as thoroughly studied as either $P450_{cam}$ or $P450_{BM-3}$, it is useful to provide some additional background information. $P450_{eryF}$ is produced by the actinomycete, *Saccharopolyspora erythaea*. This organism produces the antibiotic erythromycin A (ErA) and $P450_{eryF}$ catalyzes the first of two P450-dependent hydroxylation

THOMAS L. POULOS, JILL CUPP-VICKERY, and HUIYING LI • Departments of Molecular Biology & Biochemistry and Physiology & Biophysics, University of California, Irvine, Irvine, California 92717.

Cytochrome P450: Structure, Mechanism, and Biochemistry (Second Edition), edited by Paul R. Ortiz de Montellano. Plenum Press, New York, 1995.

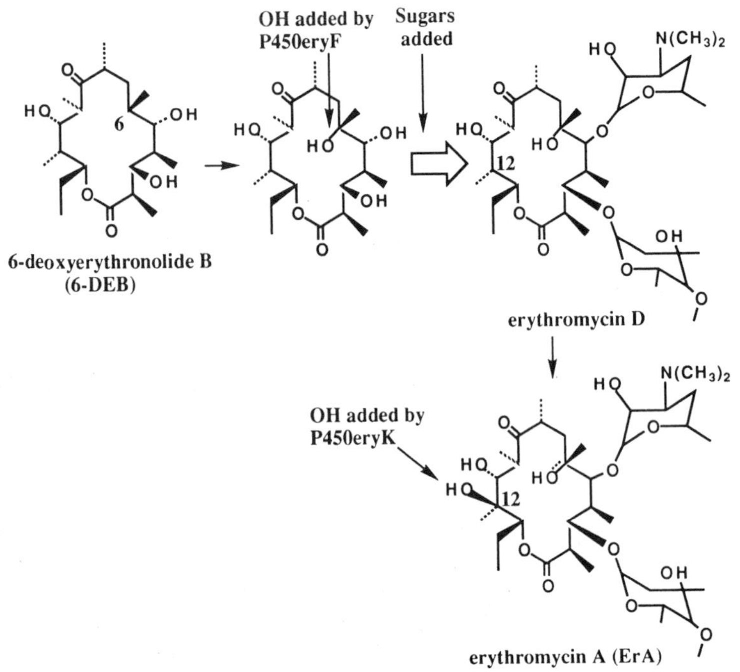

OH added by   Sugars
P450eryF   added

6-deoxyerythronolide B
(6-DEB)

erythromycin D

OH added by
P450eryK

erythromycin A (ErA)

FIGURE 1. Biosynthetic scheme for the production of erythromycin. The hydroxylation steps involving P450s are indicated.

steps in ErA biosynthesis. The overall biosynthetic scheme is outlined in Fig. 1. Utilizing two electron transfer proteins, $P450_{eryF}$ catalyzes the stereospecific insertion of an OH group on C-6 of 6-deoxyerythronolide B (6DEB).[10] The specificity of $P450_{eryF}$ has been investigated[11] and, as might be expected for an important biosynthetic step, $P450_{eryF}$ resembles $P450_{cam}$ in exhibiting a high level of specificity. The supporting electron transport proteins only recently have been identified,[12] although the fully reconstituted system has not been well characterized. What makes $P450_{eryF}$ of particular interest is the lack of a Thr residue near the proposed O_2 binding site found in the vast majority of P450s sequenced to date and the large size and shape of the substrate, 6DEB, compared to the $P450_{cam}$ or $P450_{BM-3}$ substrates (Fig. 2). Because of these differences, one might expect the active site and regions near the heme group to be considerably different from other known P450 structures. We have determined the crystal structure of $P450_{eryF}$ to 2.1 Å

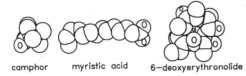

camphor   myristic acid   6-deoxyerythronolide

FIGURE 2. Space-filled models of three P450 substrates illustrating the differences in size and shape. Oxygen atoms are labeled.

resolution with the substrate bound,[9] which allows for direct comparisons with the P450$_{cam}$-substrate complex.

3. Overall Fold

The structures of the three P450s are shown in Fig. 3. Despite having sequence identities less than 20%, the overall fold and topography are very much the same in all three. Therefore, it is very likely that the same fold is maintained in all P450s including membrane-bound P450s. P450s thus conform to a classic "rule" stating that protein tertiary structure is far more conserved than is primary structure.

The overall fold does not consist of the usual domains found in many other globular proteins. Very often enzymes on the order of the size of a P450 will fold into N- and C-terminal domains. P450s do not follow this pattern. Instead, the N-terminus begins on the "left" side (Fig. 3) with the polypeptide crossing to the right once, back to the left, and one more time to the right where the polypeptide terminates. Nevertheless, P450 is divided into a helical-rich region located primarily on the right side while the left side contains much of the beta structure. Most of the long helices and sheets lie in a plane parallel to the heme plane. This pattern generates a flat triangular-shaped molecule.

P450 exhibits an interesting folding pattern with respect to how the C- and N-terminal sequences are arranged around the heme. As shown in Fig. 4, the C-terminal half of the molecule forms the inner core and especially the two key helices, I and L, that bracket either side of the heme. The C-terminal β structure, however, is at the surface, which is characteristic for most antiparallel β pairs and sheets. The N-terminal half of P450 consists primarily of helices surrounding the C-terminal sequences. This folding pattern is quite different from that found in peroxidases where the heme is embedded in a cleft between

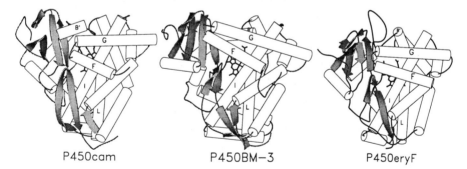

P450cam   P450BM-3   P450eryF

FIGURE 3. Schematic diagrams of P450$_{cam}$, the heme domain of P450$_{BM-3}$, and P450$_{eryF}$. Some of the key helices are labeled according to Ref. 2. The P450$_{BM-3}$ model shown was taken from the structure solved in our laboratory and consists of residues 1–455 and all 455 residues have been located in the electron density maps. The structure reported by Ravichandran *et al.*[5] consists of residues 1–471 but since the C-terminus is disordered in this structure and residues 458–471 were not visible, this 16-residue difference in the two structures is probably insignificant. The overall topography is very much the same in all three P450s. Note especially the similarity in the division of helical segments (cylinders) and β sheets (shaded arrows). This and other figures were produced with the programs MOLSCRIPT[45] and SETOR.[46]

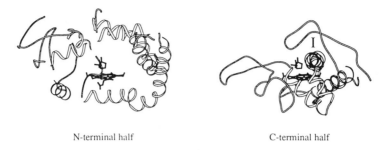

N-terminal half   C-terminal half

FIGURE 4. The left panel shows P450$_{cam}$ with residues in the first half of the sequence and the right panel in the second half of the sequence. The main heme contact regions consist of the C-terminal I and L helices. These also are the regions that exhibit the closest structural homology between P450s for which crystal structures are available.

two clearly defined N- and C-terminal domains. It thus becomes much easier to envision how the heme might insert into apoperoxidase as opposed to apo-P450. This might be one reason why little is known about apo-P450s or P450s reconstituted with various metalloporphyrins. The P450 field is fortunate in that, for whatever reason, heme insertion has not proven a problem in at least a small handful of recombinant P450 expression systems.

4. Structural Comparisons with Peroxidases

There is an obvious relationship between peroxidases and P450s since both are heme enzymes that are thought to utilize similar iron–oxo intermediates for oxidative reactions. Indeed, peroxidases are often the leaping off point for understanding P450 reactions since the various peroxidase intermediates like Compounds I and II thought to be paralleled by P450s are readily trapped and subject to detailed studies. For nearly a decade the only basis for comparison between peroxidases and P450s at the atomic level has been cytochrome c peroxidase (CCP) and P450$_{cam}$. Now, however, not only are additional P450 structures available, but the lignin peroxidase (LiP)[13] and one other fungal peroxidase structure[14] also are available. Like CCP, LiP is a traditional heme peroxidase in that the enzyme reacts with H_2O_2 to generate Compound I which then is reduced by substrates. The substrate for LiP is lignin, a complex aromatic polymer, that provides the protective coating around cellulose in woody plants. LiP also is able to oxidize small aromatic compounds[15] and, unlike CCP, is expected to have an aromatic binding site.

There are some interesting parallels in comparing CCP versus LiP and P450$_{cam}$ versus P450$_{BM-3}$. The degree of sequence identity between the two peroxidases (15% based on the X-ray structures) is about the same as the two P450s (17%[5]). Moreover, LiP is 49 residues larger than CCP while the heme domain of P450$_{BM-3}$ (residues 1–455) is 41 residues larger than P450$_{cam}$. A priori one might anticipate even greater differences between the two peroxidases since LiP contains four disulfide bonds, two structural calcium ions, and at least two sites of glycosylation while CCP has none of these features. These parallel differences provide a good beginning for addressing the question on the

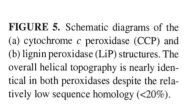

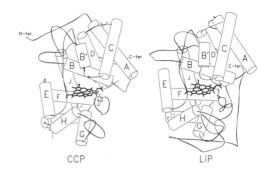

FIGURE 5. Schematic diagrams of the (a) cytochrome c peroxidase (CCP) and (b) lignin peroxidase (LiP) structures. The overall helical topography is nearly identical in both peroxidases despite the relatively low sequence homology (<20%).

degree of structural variability within the respective heme enzyme families and how these differences might relate to functional differences.

The LiP and CCP structures shown in Fig. 5 are to be compared with the P450$_{cam}$ and P450$_{BM-3}$ structures in Fig. 3. Both the peroxidases and P450s are composed of approximately 50% helix. At first glance the helical topography within each enzyme group looks very similar. However, a closer examination shows that this is not the case (Fig. 6). Figure 6 was generated by first carrying out a least squares fit between the hemes and then applying the resulting orientation matrix to the rest of the protein atoms. In both the peroxidases and P450s, those helices (distal and proximal helices) bracketing either side of the heme are positioned very much the same although the similarity between the two peroxidases is greater than in the two P450s. It is interesting to note that in the peroxidases and globins the distal and proximal helices are oriented approximately antiparallel to one another while in P450, these helices are approximately parallel. Moreover, in P450, the proximal Cys ligand is at the N-terminal end of the proximal helix while in the peroxidases and globins the proximal His ligand is at the C-terminal end.

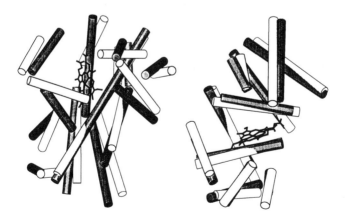

FIGURE 6. Comparison of helical topographies in CCP versus LiP and P450$_{cam}$ versus P450$_{BM-3}$. These figures were generated by carrying out a least squares fit of the hemes and then applying the resulting orientation and translation matrix to the remainder of the protein atoms. For clarity of comparison, only the helices are shown as cylinders.

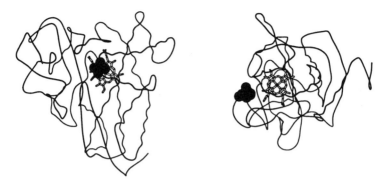

FIGURE 7. Right panel shows the predicted location of the LiP substrate 3,4-dimethoxy benzyl alcohol[15] and the left panel shows the location of camphor in P450$_{cam}$. Note that in the peroxidase, the substrate binds closer to the surface while in P450, the substrate must bind closer to the "heart" of the protein near the oxyferryl center. This functional requirement is likely to be one reason why the two P450s exhibit greater structural variability than do the two peroxidases.

As we move farther from the heme, the contrast in helical topography becomes more striking. The overall helical topography in the two peroxidases is much more similar than the helical topography in the two P450s and the helices in the two peroxidases are much closer to the same length while P450$_{BM-3}$ contains longer helical stretches than P450$_{cam}$. The overall root mean square difference in Cα carbons for helical segments in the two peroxidases is ~1.6 Å while for the two P450s the difference is ~2.8 Å. This simple visual and semiquantitative comparison of helical topographies serves to illustrate that given the same degree of sequence identity, the two peroxidases more closely resemble one another than do the two P450s. One other factor that contributes to this difference is that the 49 extra residues in LiP versus CCP occur almost entirely at the C-terminal end, giving an extra "tail" in LiP, while the extra residues in P450$_{BM-3}$ are spread throughout the molecule.

These differences might have been anticipated considering the mechanism by which each enzyme encounters and oxidizes its respective substrates. Ortiz de Montellano and co-workers[16] have shown that small-molecule peroxidase substrates have restricted access to the peroxide binding pocket and are most likely oxidized at the heme edge. In sharp contrast, P450 substrates must bind directly adjacent to the iron-linked oxo center (Fig. 7). Furthermore, the stereochemistry of P450s is more tightly controlled owing, at least in many cases, to the requirements of strict regio- and stereoselectivity. As a result, the P450 active site must be designed to provide not only an entry for substrates to the "heart" of the enzyme but also specific contact points for correct substrate orientation. The situation with peroxidases is quite different. Most heme peroxidases catalyze electron transfer from aromatic substrates at or near the heme edge. Such an electron transfer reaction has far fewer stereochemical restrictions than does a P450 reaction resulting in less stringent requirements for the peroxidase substrate binding pocket. Moreover, since the peroxidase substrate site is closer to the surface (Fig. 7), much less topological variability is required to alter substrate specificity from one peroxidase to the next.

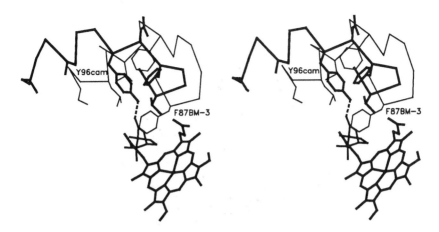

FIGURE 8. A comparison of the B′ helix in $P450_{cam}$ (thick lines) and $P450_{BM-3}$ (thin lines). This region is important in conferring substrate specificity and in $P450_{cam}$, the B′ helix provides important contact points with camphor. This region also is one of the most highly variable in P450s based on sequence alignments. Not surprisingly, the B′ helix is positioned quite differently with respect to the heme and substrate in the two P450s.

If this view is correct, it has some interesting implications for protein engineering and design. First, and most obviously, it should be easier to correctly model unknown peroxidase structures based on known structures even with relatively low sequence homology. For P450s, however, it will likely prove much more difficult to predict the correct structure. Such modeling will be especially difficult in the most interesting parts of the structure, namely those regions that control substrate specificity. Comparing the B′ helix in $P450_{cam}$ and $P450_{BM-3}$ underscores the problem. The B′ helix contains key contact points with camphor and especially the Tyr^{96}–camphor H-bond. Note (Fig. 8) that the B′ helix in $P450_{BM-3}$ is positioned quite differently from that in $P450_{cam}$. Moreover, several residues after the end of the B′ helix, $P450_{BM-3}$ contains a Phe (F87, Fig. 8) whose topologically equivalent residue in $P450_{cam}$ is Thr^{101}, a smaller residue. Phe^{87} should severely restrict the size of substrates available to $P450_{BM-3}$ and is probably important in preventing hydroxylation of the terminal methyl group in fatty acid substrates. The B′ helix in $P450_{eryF}$ also is positioned much differently than in $P450_{cam}$ and is oriented approximately perpendicular to the $P450_{cam}$ B′ helix (Fig. 3). In addition, the B′ helix in $P450_{eryF}$ is preceded by a loop extending out in solution not present in $P450_{cam}$. Such differences would have been extremely difficult if not impossible to correctly predict based on the $P450_{cam}$ structure alone. Even with four P450 structures in hand, the structural data base is probably still too small to confidently predict what structural features control substrate specificity in those P450s with low sequence identity to any of the P450s whose structures are known. At least now, however, we have a better understanding of the degree of topological variability possible for P450s with very different specificities.

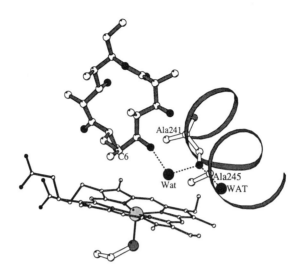

FIGURE 9. Substrate binding site of P450$_{eryF}$. Oxygen atoms are highlighted as dark spheres. Note that a water molecule forms an H-bonded bridge between the substrate C-5 OH group and the carbonyl oxygen atom of Ala[241] in the distal helix I. Also note the location of Ala[245] which is the highly conserved Thr residue found in most other P450s that is thought to play a role in oxygen binding and activation. The water labeled in uppercase (WAT) occupies a position similar to that of the Thr[252] side chain OH in P450$_{cam}$. This water could play the same role in O_2 activation as does the Thr side chain OH group in other P450s.

5. P450$_{eryF}$ Substrate Binding

Key features of the P450$_{eryF}$–substrate complex are highlighted in Fig. 9. As expected, the active site is considerably larger in P450$_{eryF}$, but like P450$_{cam}$, the pocket consists primarily of hydrophobic groups. Also like P450$_{cam}$, the carbon atom to be hydroxylated, C-6, is close to the expected position of the iron-linked oxygen ligand. A new feature in P450$_{eryF}$, however, is the participation of water molecules in substrate binding. One of the key water molecules is shown in Fig. 9. This water forms an H-bonding bridge between the OH group attached to C-5 of the substrate and a protein peptide carbonyl. A substrate where the C-5 OH is converted to a keto group turns over about 100 times slower than 6DEB,[11] clearly indicating some important role for the C-5 OH. This will be probed in more detail when considering the catalytic mechanism.

6. Cys Ligand Loop

Despite these differences between P450s in functionally important regions, those regions closest to the heme are, not surprisingly, the most highly conserved. The Cys ligand loop (Fig. 10) consisting of residues 350–357 (P450$_{cam}$ numbering) is conserved in the other P450s and owing to strong sequence homologies in this region, the loop structure is expected to be the same in all P450s. Residues 350–357 define a β bulge which is similar to an antiparallel β pair except that an H-bond is missing. A β bulge is defined where two

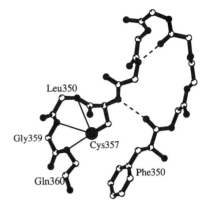

FIGURE 10. The cysteine ligand loop in P450$_{cam}$ consisting of residues 357–400. Dashed lines indicate peptide hydrogen bonds and the thin lines highlight the Cys ligand sulfur-peptide NH interactions <3.6Å. For clarity, most of the side chains have been removed.

residues in one strand lie opposite one residue in the other strand. In P450$_{cam}$, residues 351 and 352 lie opposite 356. Residues 353 to 356 form a tight turn while the carbonyl oxygen atom of Phe[351] H-bonds with the peptide NH of Cys[357]. The peptide NH groups of residues 358–360 are not H-bonded to any protein groups or ordered solvent molecules. Note, however, that the sulfur atom of the Cys ligand is 3.6 Å, 3.3 Å, and 3.4 Å from the peptide NH groups of residues 358, 359, and 360, respectively. Thus, the electronegative sulfur atom is close to electropositive centers. Despite the fact that the geometry is not especially good for sulfur–peptide NH H-bonding geometry, the sulfur should be able to "feel" the partial positive charges on the peptide NH groups. Such an arrangement is reminiscent of the oxyanion hole in serine proteases. Here, the developing negative charge on the carbonyl oxygen atom of the scissile peptide group is stabilized by H-bonding with two peptide NH groups.[17] The similar arrangement in P450s may help to stabilize the anionic form of the Cys ligand.

Another factor that could help to stabilize the anionic form of the sulfur ligand is its location at the N-terminal end of the proximal helix. This places the Cys ligand near the positive end of the helical dipole. Contrasting this is the proximal His ligand in peroxidases and globins. In these hemoproteins, the proximal His ligand is at the C-terminal end of the proximal helix, placing the His ligand at the negative end of the helical dipole, exactly the opposite of what is found in P450. Although the case for helical dipoles often is overstated, these differences are perhaps worth exploring with some theoretical calculations. Of greater surety is the role of H-bonds regarding the His ligand. In both the globins and peroxidases, protein groups accept an H-bond from the imidazole nitrogen that is not coordinated with the iron, which not only aids in stabilizing the orientation of the His ligand but also aids in controlling the charge on the imidazole group.

Stabilization of the anionic form of the sulfur ligand can explain, in part, the relationship between substrate binding and redox potential. As discussed in a later section, the binding of substrates to P450$_{cam}$ leads to an increase in redox potential and a shift in spin state from low to high.[18] P450$_{BM-3}$ may follow a similar pattern since the heme can be fully reduced by its redox partner only in the presence of substrate.[19] Crystallographic structures show that in the absence of camphor, the substrate pocket fills with solvent molecules, one of which is a heme ligand.[20] These solvent molecules, including the ligand, are displaced by the substrate. If we consider that the porphyrin core has a net charge of -2 and that the Cys ligand sulfur is effectively neutralized in the unprotonated form by the surrounding protein environment, the Fe^{3+} redox state will have a net charge of $+1$ and that of Fe^{2+} is 0. Water in the active site pocket, including the aqua ligand, should stabilize the extra $+1$ charge in the Fe^{3+} state and hence lead to a lowering of the redox potential.

7. Distal Helix I

Another important structural feature close to the heme that is conserved is the local distortion in helix I (Fig. 11). At the local widening of the helix, the side chain of Thr^{252} ($P450_{cam}$ numbering) donates an H-bond to the peptide carbonyl oxygen atom of Gly^{248}. Thus, if Thr^{252} is residue i, the H-bond is to residue i-4. Thr^{268} in $P450_{BM-3}$ is the homologue to Thr^{252} in $P450_{cam}$. In this case, however, the H-bonding pattern is both i to i-3 and i to i-4. Note, too, that the i-4 carbonyl oxygen H-bonds with the water ligand in $P450_{BM-3}$. One final difference is that $P450_{BM-3}$ also has a Thr at position 269 while $P450_{cam}$ has a Val. This Thr also forms an H-bond with the i-3 carbonyl group. This second Thr is found in many eukaryotic P450s so we might expect a similar set of H-bonding interactions in these P450s. Despite these subtle differences in H-bonds involving $Thr^{252/268}$, both of these P450s and presumably many others experience a very similar distortion in local helical topography near the heme and O_2 binding site.

$P450_{eryF}$ provides the first detailed look at a P450 where this Thr residue is not conserved. Instead, $P450_{eryF}$ has an Ala at this position (Fig. 9). This is an important

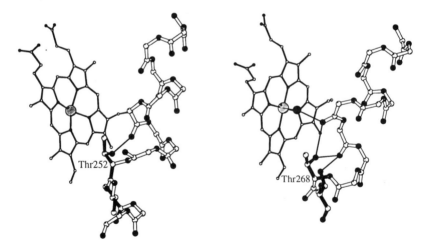

FIGURE 11. A comparison between $P450_{cam}$ (left) and $P450_{BM-3}$ (right) of the local helical distortion in helix I near the O_2 binding site. Thin lines denote possible H-bonds. In both P450s, the side chain OH of the Thr residue at position i H-bonds with the carbonyl oxygen of residue i-4. In $P450_{BM-3}$ the Thr OH group also is close enough to the carbonyl oxygen of residue i-3 for a second H-bond. In addition, the i-4 carbonyl oxygen H-bonds with the axial aqua ligand. In the substrate-free form of $P450_{cam}$, we observed no H-bond between the aqua ligand and protein groups.[20]

difference considering the role that Thr^{252} is thought to play in catalysis. $P450_{eryF}$ however, may compensate for the missing Thr side chain by using a water molecule rather than the Thr side chain OH group. We consider this possibility in the next section.

8. Catalytic Groups

It usually is assumed that once a crystal structure becomes available, it will be relatively straightforward to identify key catalytic groups involved in bond making and breaking reactions. Unfortunately, this is not the case with P450. Cleavage of the O–O bond will require protonation of the leaving oxygen atom at some point in the catalytic cycle and the identity of the group(s) that provide the required proton(s) has proven elusive. This is in sharp contrast to peroxidases where the peroxide binding pocket contains a conserved His and Arg that are critical for rapid heterolysis of the peroxide O–O bond. Mutagenesis has proven to be an invaluable tool for testing the roles of various active site groups in peroxidases. Erman et al.[21] found that converting the distal His in cytochrome c peroxidase to Leu lowers the rate of O–O bond cleavage by 5 orders of magnitude. Choudhury et al.[22,23] have shown that the proximal His ligand can be converted to Gln or Glu with little change in the rate of O–O bond cleavage, clearly demonstrating that the precise nature of the proximal ligand is not important for high rates of O–O bond fission. Therefore, in peroxidases it is the distal pocket that is most important in controlling the rate of O–O bond cleavage.

The mechanism of O–O bond fission must be quite different in P450 since the oxygen distal pocket contains no obvious acid–base catalytic groups such as the peroxidase His or Arg. As suggested by Dawson,[24] the catalytic push for breaking the O–O bond may come from the P450 thiolate ligand, which is a better electron-donating group than is His. Unfortunately, it has not been possible to replace the Cys ligand in P450 with His or the peroxidase His with Cys to directly test this idea.

Independent of the proximal thiol ligand role in catalysis, the question still remains of how dioxygen becomes protonated. The crystal structure suggests two possibilities. The first involves two conserved residues, Thr^{252} and Glu^{366}. The counterparts in $P450_{BM-3}$ are Thr^{268} and Glu^{409}. Note that Thr^{252} is the conserved (but not strictly) Thr at the local distortion in helix I and also contacts CO in the ferrous P450–CO–camphor ternary complex (see Fig. 17). As shown schematically in Fig. 12, a necklace of ordered solvent

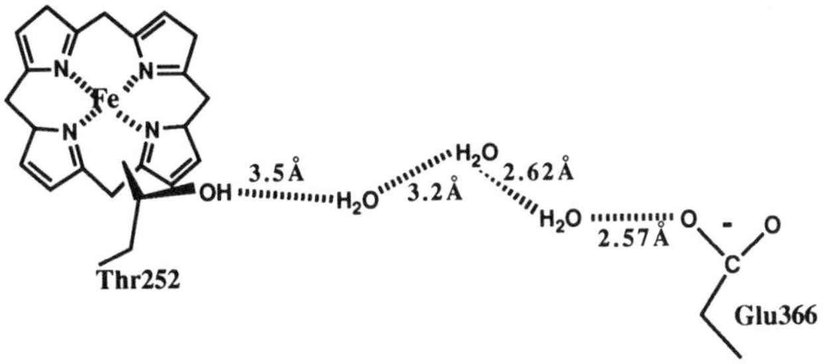

FIGURE 12. Schematic drawing showing the internal solvent channel connecting Glu^{366} to Thr^{252} in $P450_{cam}$. Glu^{366} is buried and not accessible to solvent molecules except via the internal solvent channel.

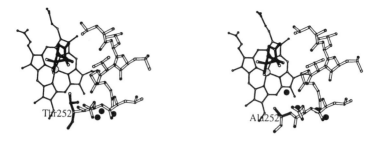

FIGURE 13. Comparison of the P450$_{cam}$ wild-type (left) and Thr^{252}Ala mutant (right) crystal structures. In the mutant, the local widening in helix I increases allowing a new solvent molecule to move closer to the O$_2$ binding site.

connects Thr^{252} to the buried side chain of Glu^{366}. The conformation in this region is similar in P450$_{BM-3}$ and P450$_{eryF}$ except that in P450$_{BM-3}$ the channel between Glu^{409} and Thr^{268} is smaller and no ordered solvent is observed between the Thr and Glu side chains. This is one possible source of protons for the leaving oxygen atom.

If the Thr^{252} side chain directly donates a proton or operates as part of a proton shuttle to molecular oxygen, movement of the Thr^{252} side chain and possibly parts of the distal helix must occur. This is not unreasonable since we know that there is considerable flexibility in helix I. In the Thr^{252}Ala mutant, the mutated side chain moves 1.4 Å away from the oxygen binding site, providing additional room for a new water molecule in the distal pocket (Fig. 13). Furthermore, a similar 1.1-Å shift has been observed in wild-type P450$_{cam}$ crystals soaked for an extensive period in adamantane, a poor substrate for P450$_{cam}$. Finally, the structures of P450$_{cam}$ complexed with metyrapone and the various isomers of phenylimidazole exhibit shifts of up to 2 Å in the I helix.[25] The most interesting of these was the 2-phenylimidazole structure (Fig. 14). Of the five inhibitor structures determined, four have a heterocyclic nitrogen atom bonded to the iron with the remainder

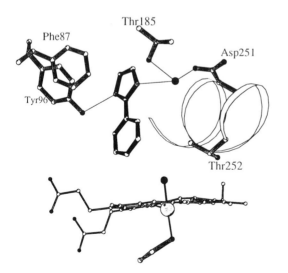

FIGURE 14. Crystal structure of the complex formed by 2-phenylimidazole and P450$_{cam}$. The thin lines denote H-bonds. Thr^{185} and Asp^{251} must adopt new conformations in order to H-bond with the new solvent molecule that enters the active site pocket to H-bond with the inhibitor. The mobility of Asp^{251} could be important since this residue has been implicated in the catalytic mechanism.[18] The Tyr^{96} side chain OH normally H-bonds with the camphor carbonyl oxygen and in the inhibited complex, the imidazole nitrogen accepts the H-bond from Tyr^{96}. Note that an aqua ligand remains coordinated to the iron.

of the inhibitor situated roughly where the camphor normally binds. 2-Phenylimidazole cannot bind in this way since formation of an N–Fe bond would result in a steric clash between the phenyl ring and the heme. Instead, the inhibitor binds such that the phenyl ring occupies the camphor pocket while the imidazole group H-bonds with protein groups and solvent. In order to form these H-bonds, however, important conformational rearrangements must occur. Helix I must slide toward the inhibitor while both Thr^{185} and Asp^{251} rotate inwards toward the inhibitor in order to form H-bonds with a new water molecule in the pocket. The various crystal structures demonstrate, but do not prove, that the range of motion available to Thr^{252} and helix I is sufficient for Thr^{252} to participate in protonation of, or at least stabilization of, the oxy complex via H-bonding with molecular oxygen. If so, Thr^{252} probably operates by rotating to form an H-bond with the distal O_2 oxygen atom.

Obviously Thr^{252} cannot play a universal role in P450 catalysis since $P450_{eryF}$ has an Ala at this position. Nevertheless, $P450_{eryF}$ could use water molecules for the same purpose. A superposition of the $P450_{cam}$ and $P450_{eryF}$ structures shows that $P450_{eryF}$ has a water molecule approximately 0.7 Å from where the Thr^{252} side chain OH would be. As illustrated in Fig. 9, the C-5 OH group of the 6DEB substrate H-bonds with another water molecule that is only 3.8 Å from the iron atom and, therefore, should contact the iron-bound O_2. Hence, these water molecules in $P450_{eryF}$ could serve the same role as the Thr^{252} OH group in other P450s. This could also explain why the C-5 OH group is so important since conversion of this OH group to a keto group results in a 100-fold decrease in activity.[11] This change may not be so much a steric mismatch but instead a disruption of an important H-bonded network required for efficient O–O bond cleavage. With $P450_{eryF}$ we may have the intriguing possibility of substrate-assisted acid catalysis.

The Thr^{252} proton shuttle is not the only possibility. Raag et al.[26] suggested that another highly conserved residue, Asp^{251}, might also operate as a proton donor. As shown in Fig. 15, Asp^{251} ion pairs with Lys^{178} and Arg^{186}. We have seen that Asp^{251} is capable of rotating inwards toward the substrate pocket which would be required if Asp^{251} operates as part of a proton relay. The new water molecule observed in the 2-phenylimidazole–$P450_{cam}$ structure is relevant here (Fig. 14). Using energetic consideration, Wade predicted that a solvent molecule ought to occupy this cavity in the native structure.[27] That we do not see electron density at this position in the native structure does not preclude the possibility of a highly mobile water. This water could also be part of a proton relay. In support of an important role for Asp^{251}, conversion of Asp^{251} to Asn dramatically lowers activity[18] (see Chapter 3).

If Asp^{251} is important, we might expect a similar set of interactions in other P450s. Both $P450_{BM-3}$ and $P450_{eryF}$ have Glu residues at this position although neither forms ion pairs with Arg and Lys residues as does Asp^{251} in $P450_{cam}$. Unfortunately, there is little hard evidence to allow a choice to be made between these possibilities as the primary site for a proton relay, although recent mutagenesis work indicates an important role for Asp^{251}.[28] Moreover, if either or both residues were absolutely essential for P450 function, they would be expected to be strictly conserved. Although a vast majority of P450s do have a Thr at position 252 and either Asp or Glu at position 251, there are exceptions. Although there are some intriguing guesses and possibilities, precisely how acid catalysis works in P450 remains an open question.

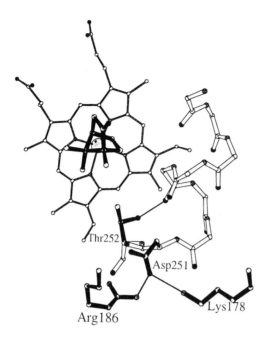

Thr252

Asp251

Arg186

Lys178

FIGURE 15. Ion pair formed between Asp[251] and Lys and Arg residues near the surface of P450$_{cam}$. This network is an alternate possibility for shuttling protons into the active site for the O–O bond cleavage reaction.[16,18] Glu[267] could play a similar role in P450$_{BM-3}$.[5]

9. Mutant Structure

The role of Thr[252] in the distal helix I has been examined using site-directed mutagenesis.[3,4] The primary effect of changing Thr[252] to Ala and other residues is to uncouple electron transfer from substrate hydroxylation. That is, only a small fraction of the reducing equivalents from NADH can be accounted for by hydroxylated product and the rest is used in water and peroxide formation.[3,4] The crystal structure of the Thr[252]Ala mutant provides a possible explanation[26] (Fig. 13). In the Ala[252] mutant, the local widening of helix I increases, allowing a new water molecule to move in closer to where oxygen is expected to bind. A more ready access of solvent to the oxy–P450 complex could be one explanation for uncoupling. Precisely how remains a problem. One possibility is that after cleavage of the O–O bond, the oxyferryl oxygen atom has ready access to protons so as the oxyferryl center is further reduced by incoming electrons, protonation by solvent leads to water release rather than hydroxylated product. Peroxide release could proceed by a similar mechanism. Another possibility is the loss of stereochemical control of the oxy complex in the mutants. In model metal–oxo complexes, there are two possible binding geometries, side-on or end-on[29] (Fig. 16). Under acid conditions the side-on complex generates peroxides.[29] Therefore, if, in the mutants, the oxygen binds side-on, reduction and protonation could lead to peroxide release without O–O bond cleavage although the side-on complex might be quite stable. An important unknown in all of this is the role played by the thiolate ligand and what effects it might have on water and peroxide production. In any event, one additional role of Thr[252] might be to control the stereochemistry of the oxy complex via H-bonding which ensures an end-on binding of oxygen. In

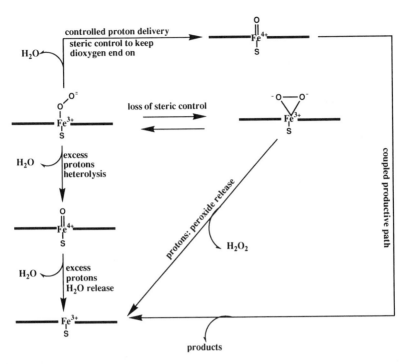

FIGURE 16. A possible explanation for the uncoupling of electron transfer from substrate hydroxylation observed in mutants of P450$_{cam}$ in which Thr^{252} is converted to other residues.[3,4] In this scheme, one possible role for Thr^{252} and the local distortion in helix I is to prevent O_2 from binding side-on. Side-on binding could promote peroxide release. The scheme shown was developed based on discussions with many colleagues, especially Drs. T. C. Bruice, S. Sligar, and J. Valentine.

P450$_{eryF}$, the substrate–water–protein H-bonded network would serve this role. A similar situation is present in peroxidases except, in this case, the H-bonding interactions between the distal catalytic histidine and arginine hold the peroxide in an end-on configuration. A summary of this view is provided in Fig. 16.

Although the crystal structure of the oxy complex has not been determined, the carbonmonoxy complex has[30] (Fig. 17). As shown in Fig. 17, the CO oxygen atom is within contact distances (3.2 and 3.4 Å) of Gly^{248} and Thr^{252}, the two key residues involved in forming the local helix I distortions. If we assume molecular oxygen makes similar contacts, then Thr^{252} would be expected to play an important role in oxygen binding.

10. Substrate Access

One problem in understanding how P450s work is how the substrate gains access to the oxyferryl center. As with many enzymes, the active site is complementary in shape and charge properties to the substrate, yet, in P450$_{cam}$, the active site opening is too small to accommodate a camphor molecule. As shown in Fig. 18, the access route in P450$_{cam}$ is

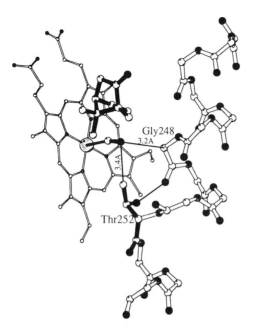

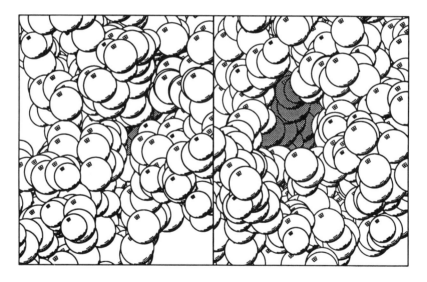

FIGURE 17. Active site region of the CO–camphor–P450$_{cam}$ ternary complex. The CO carbon atom contacts both Thr^{252} and Gly. Thr^{252} could be important for steric control of how O$_2$ binds (see Fig. 13).

FIGURE 18. Space-filled diagrams of P450$_{cam}$ and P450$_{BM-3}$ showing the substrate access routes. The hemes are shaded. Although this is a comparison between substrate-bound (P450$_{cam}$) and substrate-free (P450$_{BM-3}$) enzymes, there is no significant change in structure in the substrate-free form of P450$_{cam}$. This suggests that dynamic fluctuations are important for allowing substrate to enter the P450$_{cam}$ active site. In contrast, P450$_{BM-3}$ requires no such movement. This, however, does not preclude possible structural rearrangements in P450$_{BM-3}$ once substrate binds which might result in a "capping" of the access channel.

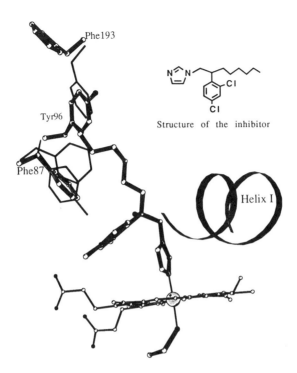

FIGURE 19. Crystal structure of the Pfizer inhibitor–P450$_{cam}$ complex (shaded thick bonds) superimposed on the camphor-bound structure (thin bonds). For clarity the camphor has been removed. Note the concerted movement of aromatic groups "up" and "out" toward the molecular surface in order to accommodate the large inhibitor.

closed while that of P450$_{BM-3}$ is open. Hence, there is no difficulty in envisioning how hydrophobic substrates diffuse from the molecular surface into the P450$_{BM-3}$ active site. With P450$_{cam}$ we must invoke one of two processes: either a conformational switch, one open and one closed, or a dynamic fluctuation. Most of the evidence favors dynamic fluctuations. In particular, this conclusion is based on the substrate-free crystal structure where it was found that those residues forming the opening to the active site (Fig. 18) exhibit significantly higher thermal motion when the substrate is not bound.[20] Similar differences were observed in those substrates that are smaller and less specific than camphor.[31,32]

Ideally one would like to trap the access route in the open conformation to better understand the range of motions available to the access route. Some progress has been made toward this goal using a large P450 inhibitor. The crystal structure of the complexes formed between both the (+) and (–) stereoisomers of the inhibitor and P450$_{cam}$ have been refined to 2.1 Å[33] (Fig. 19). The free imidazole nitrogen bonds with iron while the dichlorophenyl group occupies the space normally taken by camphor. The long alkyl side chain extends up and out toward the presumed route that camphor must take to enter the active site. This results in a large and concerted rearrangement of aromatic side chains (Fig.

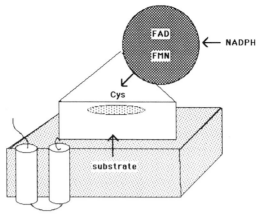

FIGURE 20. Schematic diagram showing the possible relationship between membrane-bound P450s and P450 reductase, adopted from Nelson and Strobel.[36] The substrate access channel is oriented down toward the lipid bilayer while the proximal, Cys ligand surface is oriented up where it is available for interaction with redox partners.

19). Phe[193] ends up protruding out on the surface. Although these data do not prove that such fluctuations occur in the uninhibited enzyme, these results do support the idea that this region of the structure is particularly sensitive to fluctuations. If Phe[193] does spend a significant fraction of its time on the surface, as we have seen in the inhibited structure, then this could serve as a surface hydrophobic patch for substrate binding. A similar surface hydrophobic patch has been seen in lignin peroxidase where a Phe residue is situated at the entrance to where small aromatic substrates are thought to bind.[13]

Since P450$_{BM-3}$ is generally considered to be the best available model system for structure/function relationships in microsomal P450s, it is of interest to speculate on the relationship between substrate access and the interaction of microsomal P450s with membranes. Modeling studies[34] together with site-directed mutagenesis work[35] indicate that the site of interaction between P450 and its redox partners is on the Cys-ligand side of the P450 (Fig. 20). By correlating a wealth of biochemical/biophysical data with the available sequence data, Nelson and Strobel[36] developed a model for how P450s interact with the membrane. In this model (Fig. 20), the P450 is anchored to the membrane via an N-terminal hydrophobic segment. The P450 is oriented such that the Cys-ligand surface predicted to be the main site of interaction with redox partners[35] is oriented away from the membrane leaving this region available for interactions with the soluble domain of the P450 reductase. This orientation places the substrate access channel facing toward the membrane. If the open access channel in P450$_{BM-3}$ is representative of those relatively nonspecific microsomal P450s, then the microsomal P450 access channel would be open and exposed to the lipid bilayer, raising the possibility that the bilayer itself forms part of the access channel. Lipophilic substrates, which concentrate in the membrane, would then have ready access to the P450 active center without ever leaving the membrane. If this view is correct, it raises some rather interesting limitations on determining the structure of microsomal P450s. Solubilization and removal of lipid may alter just those structural features that are most interesting. Indeed, the helices and loops that form part of the access channel in P450$_{BM-3}$ exhibit very high thermal motion as judged by the crystallographic temperature factors, suggesting that this region of P450 may be especially sensitive to the surrounding solvent.

11. Conformational Changes

P450$_{BM-3}$ presents a special problem in understanding substrate binding because the access channel actually is too open. Modeling how fatty acids might bind to P450$_{BM-3}$ indicates that there are few protein–substrate interactions as might be expected. Moreover, attempts to soak various fatty acid substrates into P450$_{BM-3}$ crystals lead to either uninterpretable difference Fouriers or crystal cracking.[6] To further model substrate binding, the P450$_{BM-3}$ heme domain structure was subjected to extensive energy minimization as a preamble to molecular dynamics simulation work.[6] Quite unexpectedly, the access channel was found to undergo a very large conformational change leading to a closure of the access channel (Fig. 21). Modeling the 14-carbon substrate, myristic acid, in the minimized model now reveals closer protein–substrate contacts: the hydrophobic walls of the access channel now are closer to the substrate. It also was found that if intermolecular crystal contacts are fixed in the energy minimization, the access channel remains in the open conformation.[6] Hence, it appears that the access channel is trapped in the open conformation by the crystalline lattice. These observations lend credence to the idea that P450s must undergo a rather large open/close motion which enables substrates to enter and bind.

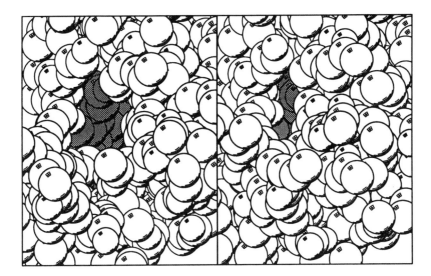

FIGURE 21. Spaced-filled diagrams of the P450$_{BM-3}$ substrate access channel before (left) and after (right) energy minimization.

12. Cation Effects on Substrate Binding

P450$_{cam}$ exhibits a unique relationship between metal ions and substrate binding. Peterson first showed that camphor binds eightfold more tightly to P450$_{cam}$ in the presence of various cations.[37] The spin equilibrium[37] and stability of the enzyme[38] also are affected by cations. The X-ray structure offers a possible explanation for these effects. One of the crystallographically well-defined solvent molecules is surrounded by six oxygen ligands suggestive of an octahedral cation ligand environment (Fig. 22). The precise identity of the cation remains unknown although the high concentrations of ammonium sulfate used for crystallization would indicate ammonium. Early on in the structural work, various cations like thallium were used in an attempt to make heavy atom derivatives. The lack of success in this approach indicates that the cation ion site has a relatively low affinity and

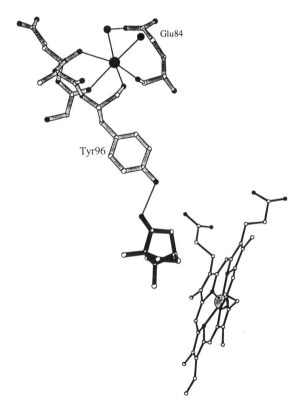

FIGURE 22. The postulated cation binding site in P450$_{cam}$ that is important for substrate binding. The ligands to the cation include four carbonyl oxygen atoms and one solvent molecule. The solvent molecule coordinated to the cation also H-bonds with Glu[84] so there is an indirect interaction between the cation and a negative charge. Note that one of the carbonyl oxygen atoms belongs to Tyr[96], a residue that also forms an important H-bond with the substrate. This Tyr[96]–cation-substrate link could be the structural basis for the observation that camphor binds more tightly in the presence of cations.[37]

cannot effectively compete with other solvent ions, as might be expected since the affinity of $P450_{cam}$ for cations is low.[37] The reason for suspecting that this electron density is a cation and not a water is based primarily on the oxygen ligand distances and geometry and the negative electrostatic potential in this region which should favor cation binding.

Four of the ligands to the cation are peptide carbonyl oxygen atoms and two are solvent molecules. One of the carbonyl oxygen atoms belongs to Tyr^{96}, a key active site residue that H-bonds with camphor (see Fig. 22). Tyr^{96} is at the C-terminal end of helix B′ where the helix abruptly ends with the peptide group of Tyr^{96} pointing nearly perpendicular to the helix axis. This is an unusual termination of a helix and is stabilized by the cation. As discussed earlier, Tyr^{96} and the surrounding protein are quite flexible and apparently cation binding aids in holding Tyr^{96} in position for interactions with the substrate. Thus, cation binding at this site enhances the interaction between the substrate and enzyme. Further support for Tyr^{96} being involved in the cation binding site stems from the mutagenesis studies. Di Promo *et al.*[39] showed that conversion of Tyr^{96} to Phe lowers the effect of cations on the substrate-induced spin-state conversion.

13. Substrate Control of Spin State and Redox Potential

Many P450s, including $P450_{cam}$, exhibit a pronounced change in the visible absorption spectrum on binding substrates indicative of a change from a predominantly low-spin heme iron to a high-spin iron. Moreover, with $P450_{cam}$ there also is an increase in redox potential for the Fe^{2+}/Fe^{3+} couple.[18,40] The shift in redox potential has functional significance. Reduction of $P450_{cam}$ in the substrate-free form by the $P450_{cam}$ redox partner, putidaredoxin, is thermodynamically unfavorable. When substrate binds leading to an increase in redox potential, reduction of the heme iron by putidaredoxin becomes thermodynamically favorable. Thus, substrate binding provides a redox switch which ensures that electron transfer occurs only when substrate is bound. A similar mechanism probably operates in $P450_{BM-3}$ since the heme iron is not reduced by NADPH unless a fatty acid substrate is present.[19]

The relationship between redox potential and spin equilibrium has been exhaustively characterized especially in the laboratory of S. Sligar (Chapter 3). These studies show that there is a strong correlation between redox potential and spin equilibrium in which the high-spin state correlates with the higher redox potential. Our laboratory[31,32] has solved the crystal structures of several substrate–$P450_{cam}$ complexes and a summary of both the crystallographic and biochemical results is provided in Table I. There are four structural parameters relevant to understanding how substrates regulate the redox potential and spin state: (1) solvent occupying the substrate pocket is displaced by substrate; (2) the presence or absence of an aqua ligand as the sixth ligand; (3) the mobility of the substrate as estimated by the crystallographic temperature factors; and (4) the presence or absence of an H-bond between the substrate and Tyr^{96}. Those substrates that do not lead to a full shift in spin equilibrium toward high spin have higher crystallographic temperature factors and hence higher mobility in the active site relative to camphor. In general, these more loosely fitting substrates allow an axial aqua ligand to remain bound to the iron while with camphor and other high-spin-state/high-redox-potential substrates, the aqua ligand is displaced. The

Table I
Summary of Crystallographic and Biochemical Data for Various Substrate–P450$_{cam}$ Complexes

Substrate	Camphor	Adamantone	Adamantane
Volume (Å^3)	315	300	293
H-bond Y96	yes[a]	yes[b]	no
No. of axial ligands	1[a]	1[b]	2
Redox potential (mV)	−170[c]	−175[c]	
% high spin	94–97[c,d,e]	96–98[c,d]	99[d]
Regiospecificity of hydroxylation	5-exo(100%)[d,e,f]	5-(100%)[a,d]	1-(100%)[d]
Substrate temp. factor (Å^2)	16.2	16.5	24.7
Hydroxylation efficiency	100%[e,f]		

Substrate	Thiocamphor	Norcamphor	Camphane
Volume (Å^3)	322	236	309
H-bond Y96	no	yes[b]	no
No. of axial ligands	2	2[b]	2
Redox potential (mV)		−206[c]	
% high spin	65[c]	46[c]	45[e]
Regiospecificity of hydroxylation	5-exo(100%)[e]	5-exo(45%)[e,f]	5-exo(100%)[f]
	6-exo(34%)	6-exo(47%)	6-exo(10%)
	3-exo(2%)	3-exo(8%)	
	9-(<1%)		
Substrate temp. factor (Å^2)	23.5	33.5	30.1
Hydroxylation efficiency	98%[e]	12%[f]	8%

[a]Ref. 2.
[b]Refs. 31, 32.
[c]Ref. 40.
[d]Ref. 42.
[e]Ref. 43.
[f]Ref. 44.

single exception is adamantane, which remains hexacoordinate by retaining the aqua ligand but is still high spin. However, the adamantane complex differs from the partially low-spin norcamphor and camphane complexes in that the iron–aqua ligand distance is about 2.0 Å in the adamantane complex and about 1.7 Å in the other two complexes. The longer Fe–O distance in the adamantane complex may account for why this complex converts to high spin. It also is clear that H-bonding between the substrate and Tyr[96] is another important factor contributing to rigidity of substrate binding and coordination state. This is further supported by site-directed mutagenesis where conversion of Tyr[96] to Phe gives an enzyme that does not undergo full spin conversion.[39] Not surprisingly, those substrates that bind more loosely also give multiply hydroxylated products. In these cases, the substrate is free to rotate in the active site, thereby presenting alternate carbon sites to the activated iron-linked oxygen atom.

The general picture that has emerged from these studies is that tight-fitting substrates displace a cluster of solvent present in the substrate-free enzyme[20,31,32] including the axial aqua ligand. The decreased dielectric environment plus a displacement of the aqua ligand promotes an increase in redox potential and a shift in spin state. One problem with this view is why the aqua ligand should favor a low-spin system. For example, metglobins are high spin yet these have a water coordinated to the iron. Based on this analogy, one would expect a water molecule ligand to favor the high-spin state. Theoretical work has provided some insight into this apparent paradox. Using *ab initio* and semiempirical quantum mechanical methods, Loew *et al.*[41] find that if there is substantial movement of the iron into the porphyrin core in the substrate-free enzyme and/or if the aqua ligand is hydroxide rather than water, the low-spin state will be favored.

14. Conclusions

While the structural work on P450s has provided critical information toward understanding how P450s work, two outstanding questions remain. First, how does the O–O bond cleave and how does acid catalysis work? Second, what is the mechanism of electron transfer? It is very likely that we now have sufficient structural information to answer at least one of these questions. This statement is based on the assumption that all P450s will operate by the same general mechanism so that a thorough understanding of one P450 should apply to the rest. This single P450 has, of course, been P450$_{cam}$. Given the recent results of Gerber and Sligar,[28] it may be possible to alter the rate-limiting step in P450 using site-directed mutagenesis. This may, in turn, allow heretofore unobserved intermediates to be studied. While P450$_{cam}$ is still the best system for such studies, what nature does or does not allow us to do in recombinant systems may dictate which P450 reveals the most about the O–O bond fission reaction. With regard to electron transfer, the structure of a P450 reductase is required in addition to the structure of a P450/reductase complex. P450$_{BM-3}$ may be the best system for unraveling this part of the P450 mechanism since the P450 heme and flavin reductase domains are tethered together, eliminating the "seek and bind" problem with other redox pairs. Therefore, P450$_{cam}$ and P450$_{BM-3}$ together should provide sufficient structural information for understanding the O–O bond cleavage reaction and the mechanism of electron transfer.

The more applied aspects of P450 research will require additional structural information. An obvious application of P450 structural data is to design novel and specific inhibitors of P450s for therapeutic and agrochemical uses. In addition, it would be highly desirable to alter substrate specificity and hence "create" P450s that catalyze the production of important hydroxylated intermediates of commercial interest. Do we have sufficient structural information to realistically attain these goals? The answer is no, as should be evident from the large structural variations between the P450s considered in this chapter. To achieve these goals we need to understand why $P450_{cam}$ and $P450_{BM-3}$ are more different from one another than the two peroxidases compared in this chapter. We need to develop the insight and homology modeling tools to be able to predict these changes. To realize these goals, we need additional structural information from each of the two classes of P450s: those with broad and those with narrow substrate specificities. Many microsomal P450s belong to the former class while the second, highly specific P450s are represented by $P450_{cam}$ and $P450_{eryF}$. The recent structural determination of new P450s should considerably aid in this effort. This additional structural information could take us a step closer to being able to correctly homology model P450s of either the "loose" or "tight" type for the purposes of rational inhibitor design and protein engineering for altering substrate specificity.

ACKNOWLEDGMENTS.   We would like to acknowledge various colleagues who have contributed to structural work over the years. In particular we thank Dr. Reetta Raag for crystallographic studies on $P450_{cam}$. Work in the authors' laboratory was supported by NIH Grant GM-33688.

References

 1. Raag, R., and Poulos, T. L., 1992, X-ray crystallographic studies of $p450_{cam}$: Factors controlling substrate metabolism, in: *Frontiers in Biotransformations*, Vol. 7 (K. Ruckpaul and H. Rein, eds.), Akademie Verlag, Weinheim, pp. 1–43. This review contains references to many of the pioneering works on $P450_{cam}$, especially from the laboratories of I. C. Gunsalus and S. Sligar.
 2. Poulos, T. L., Finzel, B. C., and Howard, A. J., 1987, High-resolution crystal structure of cytochrome $P450_{cam}$, *J. Mol. Biol.* **195:**697–700.
 3. Martinis, S. A., Atkins, W. A., Stayton, P. S., and Sligar, S. G., 1989, A conserved residue of cytochrome P450 is involved in heme-oxygen stability and activation, *J. Am. Chem. Soc.* **111:**9252–9253.
 4. Imai, M., Shimada, H., Watanabe, Y., Matsushima-Hibiya, Y., Makino, R., Koga, H., Horiuchi, T., and Ishimura, Y., 1989, Uncoupling of the cytochrome $P450_{cam}$ monooxygenase reaction by a single mutation, threonine-252 to alanine or valine: A possible role of the hydroxy amino acid in oxygen activation, *Proc. Natl. Acad. Sci. USA* **86:**7823–7827.
 5. Ravichandran, K. G., Boddupalli, S. S., Haserman, C. A., Peterson, J. A., and Deisenhofer, J., 1993, Crystal structure of hemoprotein domain of $P450_{BM-3}$, a prototype for microsomal P450s, *Science* **261:**731–736.
 6. Li, H., and Poulos, T. L., 1994, Modeling protein–substrate interactions in the heme domain of cytochrome $P450_{BM-3}$, *Acta Crystallogr.* **D51:**21–32.
 7. Hasemann, C. A., Ravichandran, K. G., Peterson, J. A., and Deisenhofer, J., 1994, Crystal structure and refinement of $P450_{terp}$ at 2.3 Å resolution, *J. Mol. Biol.* **236:**1169–1185.
 8. Cupp-Vickery, J., Li, H., and Poulos, T. L., 1994, Preliminary crystallographic analysis of an enzyme involved in erythromycin biosynthesis: Cytochrome $P450_{eryF}$, *Proteins* **20:**187–201.
 9. Cupp-Vickery, J., and Poulos, T. L., 1995, Structure of cytochrome P450 eryF: an enzyme involved in erythromycin biosynthesis, *Nat. Struct. Biol.* **2:**144–153.
10. Shaiffe, A., and Hutchinson, C. R., 1987, Macrolide antibiotic biosynthesis: Isolation and characterization of two forms of 6-deoxyerythronolide B hydroxylase from *Saccharopolyspora erythaea*, *Biochemistry* **26:**6204–6210.

11. Andersen, J. K., and Hutchinson, C. R., 1993, Substrate specificity of 6-deoxyerythronolide B hydroxylase, a bacterial cytochrome P450 of erythromycin biosynthesis, *Biochemistry* **32**:1905–1913.

12. Shaiffe, A., and Hutchinson, C. R., 1988, Purification and reconstitution of the electron transport components of 6-deoxyerythronolide B hydroxylase, a cytochrome P450 enzyme of macrolide antibiotic (erythromycin) biosynthesis, *J. Bacteriol.* **170**:1548–1553.

13. Poulos, T. L., Edwards, S. L., Wariishi, H., and Gold, M. H., 1993, Crystallographic refinement of lignin peroxidase at 2Å, *J. Biol. Chem.* **268**:4429–4440.

14. Kunishima, N., Fukuyama, K., Matsubara, H., Hatanaka, H., Shibano, Y., and Amachi, T., 1994, Crystal structure of the fungal peroxidase from *Arthromyces ramosus* at 1.9Å resolution, *J. Mol. Biol.* **235**:331–344.

15. Valli, K., Wariishi, H., and Gold, M. H., 1990, Oxidation of monoethoxylated aromatic compounds by lignin peroxidase: Role of veratryl alcohol in lignin biodegradation, *Biochemistry* **29**:8535–8539.

16. Ortiz de Montellano, P. R., 1992, Catalytic sites of hemoprotein peroxidases, *Annu. Rev. Pharmacol. Toxicol.* **32**:89–107.

17. Kraut, J., 1977, Serine proteases: Structure and mechanism of catalysis, *Annu. Rev. Biochem.* **46**:331–358.

18. Sligar, S. G., and Gunsalus, I. C., 1976, A thermodynamic model of regulation: Modulation of redox equilibria in camphor monooxygenase, *Proc. Natl. Acad. Sci. USA* **73**:1078–1082.

19. Li, H., Darwish, K., and Poulos, T. L., 1991, Characterization of recombinant *Bacillus megaterium* cytochrome P450$_{BM-3}$ and its two functional domains, *J. Biol. Chem.* **266**:11909–11914.

20. Poulos, T. L., Finzel, B. C., and Howard, A. J., 1986, Crystal structure of substrate-free *P. putida* cytochrome P450, *Biochemistry* **25**:5314–5322.

21. Erman, J. E., Vitello, L. B., Miller, M. A., Shaw, A., Brown, K. A., and Kraut, J., 1993, Histidine 52 is a critical residue for rapid formation of cytochrome c peroxidase compound I, *Biochemistry* **32**:9798–9806.

22. Choudhury, K., Sundaramoorthy, M., Mauro, J. M., and Poulos, T. L., 1992, Conversion of the proximal histidine ligand to glutamine restores activity to an inactive mutant of cytochrome c peroxidase, *J. Biol. Chem.* **267**:25656–25659.

23. Choudhury, K., Sundaramoorthy, M., Hickman, A., Yonetani, T., Woehl, E., Dunn, M. F., and Poulos, T. L., 1994, The role of the proximal ligand in peroxidase catalysis: Crystallographic, kinetic, and spectral studies of cytochrome c peroxidase proximal ligand mutants, *J. Biol. Chem.* **269**:20239–20249.

24. Dawson, J. H., 1988, Probing structure–function relations in heme-containing oxygenases and peroxidases, *Science* **240**:433–439.

25. Poulos, T. L., and Howard, A. J., 1987, Crystal structures of the metyrapone and phenylimidazole inhibited complexes of cytochrome P450$_{cam}$, *Biochemistry* **26**:8165–8174.

26. Raag, R., Martinis, S. A., Sligar, S. G., and Poulos, T. L., 1991, Crystal structure of the cytochrome P450$_{cam}$ active site mutant Thr252Ala, *Biochemistry* **30**:11420–11429.

27. Wade, R. C., 1990, Solvation at the active site of cytochrome P450$_{cam}$, *J. Comput. Aided Mol. Des.* **4**:199–204.

28. Gerber, N. C., and Sligar, S. G., 1992, Catalytic mechanism of cytochrome P450: Evidence for a distal charge relay, *J. Am. Chem. Soc.* **114**:8742–8743.

29. Klevan, L., Peone, J., and Madan, S. K., 1973, Molecular oxygen adducts of transition metal complexes, *J. Chem. Educ.* **50**:670–675.

30. Raag, R., and Poulos, T. L., 1989, Crystal structure of the carbon monoxide–substrate–cytochrome P450$_{cam}$ ternary complex, *Biochemistry* **28**:7586–7592.

31. Raag, R., and Poulos, T. L., 1989, The structural basis for substrate-induced changes in redox potential and spin equilibrium in cytochrome P450$_{cam}$, *Biochemistry* **28**:917–922.

32. Raag, R., and Poulos, T. L., 1991, Crystal structures of cytochrome P450$_{cam}$ complexed with camphane, thiocamphor, and adamantane: Factors controlling P450 substrate hydroxylation, *Biochemistry* **30**:2674–2684.

33. Raag, R., Li, H., Jones, B. C., and Poulos, T. L., 1993, Inhibitor-induced conformational change in cytochrome P450$_{cam}$, *Biochemistry* **32:**4571–4578.
34. Stayton, P. S., Poulos, T. L., and Sligar, S., 1989, Putidaredoxin competitively inhibits cytochrome b5–cytochrome P450$_{cam}$ association: A proposed model for a cytochrome 450cam electron-transfer complex, *Biochemistry* **28:**8201–8205.
35. Stayton, P. S., and Sligar, S. G., 1990, The cytochrome P450$_{cam}$ binding surface as defined by site-directed mutagenesis and electrostatic modeling, *Biochemistry* **29:**7381–7386.
36. Nelson, D. R., and Strobel, H. W., 1988, On the membrane topography of cytochrome P450 proteins, *J. Biol. Chem.* **263:**6038–6050.
37. Peterson, J. A., 1971, Camphor binding by *Pseudomonas putida* cytochrome P450, *Arch. Biochem. Biophys.* **144:**678–693.
38. Hoa, H. B., and Marden, M. C., 1982, The pressure dependence of the spin equilibrium in camphorbound ferric cytochrome P450, *Eur. J. Biochem.* **124:**311–315.
39. Di Promo, C., Hoa, H. B., Douzou, P., and Sligar, S., 1990, Mutagenesis of a single hydrogen bond in cytochrome P450 alters cation binding and heme solvation, *J. Biol. Chem.* **265:**5361–5363.
40. Fisher, M. T., and Sligar, S. G., 1983, Control of heme redox potential and reduction rate: A linear free energy relation between potential and ferric spin state equilibrium, *J. Am. Chem. Soc.* **107:**5018–5019.
41. Loew, G. H., Collins, J., Luke, B., Waleh, A., and Pudzanowski, K. A., 1986, Theoretical studies on cytochrome P450. Characterization of stable and transient active states, reaction mechanisms and substrate–enzyme interactions, *Enzyme* **36:**54–78.
42. White, R. E., McCarthy, M.-B., Egeberg, K. D., and Sligar, S. G., 1984, Regioselectivity in the cytochromes P450: Control by protein constraints and by chemical reactivities, *Arch. Biochem. Biophys.* **228:**493–502.
43. Atkins, W. M., and Sligar, S. G., 1988, The roles of active site hydrogen bonding in cytochrome P450$_{cam}$ as revealed by site-directed mutagenesis, *J. Biol. Chem.* **263:**18842–18849.
44. Atkins, W. M., and Sligar, S. G., 1989, Molecular recognition in cytochrome P450: Alteration of regioselective alkane hydroxylation via protein engineering, *J. Am. Chem. Soc.* **111:**2715–2717.
45. Kraulis, P., 1991, MOLSCRIPT: A program to produce both detailed and schematic plots of protein structures, *J. Appl. Crystallogr.* **24:**946–950.
46. Evans, S., 1993, SETOR: Hardware lighted three-dimensional solid model representation of macromolecules, *J. Mol. Graphics* **11:**134–138.

CHAPTER 5

Bacterial P450s

Structural Similarities and Functional Differences

JULIAN A. PETERSON and SANDRA E. GRAHAM-LORENCE

1. Introduction

Our understanding of the role in intermediary metabolism of the gene superfamily of proteins called P450 has changed with the cloning and sequencing of greater than 300 genes that encode unique forms of this protein.[1] While P450s have been thought of as being primarily located in the liver and endocrine organs of mammals where they catalyze the detoxification of drugs and the synthesis of steroid hormones,[2] P450s are found in essentially all mammalian tissues including brain,[3] skin,[4] and intestine.[5] Additionally, with the continued identification of unique P450s, which catalyze biosynthetic and biodegradative reactions in mammals as well as in plants, fungi, and bacteria, it has been realized that the role of P450s is much more diverse than had been previously suspected. (For example, the role of P450s in phytoalexin,[6] anthocyanidin,[7] isoflavonoid,[8] and jasmonic acid[9] synthesis in plants is well documented.) Thus, scientists have become interested in the mechanisms of disease states resulting from mutations in P450s, and in alternative substrates metabolized and new products formed by the engineering of known P450 proteins. Only through an understanding of the structures of these proteins can this be accomplished.

As shown in Eq. (1), the fundamental reaction that P450s catalyze is monooxygenation or mixed function oxidation; however, these enzymes also catalyze sulfur oxidation, epoxidation, cleavage of carbon–carbon bonds, alkyl group migration, and aromatization to name but a few different types of reactions.[10]

$$R-CH_3 + O_2 + NAD(P)H + H^+ \rightarrow R-CH_2-OH + NAD(P)^+ + H_2O \qquad (1)$$

JULIAN A. PETERSON and SANDRA E. GRAHAM-LORENCE • Department of Biochemistry, The University of Texas Southwestern Medical Center at Dallas, Dallas, Texas 75235.

Cytochrome P450: Structure, Mechanism, and Biochemistry (Second Edition), edited by Paul R. Ortiz de Montellano. Plenum Press, New York, 1995.

Monooxygenation reactions require the input of two electrons from reduced pyridine nucleotides. These electrons are transferred to the P450 via auxiliary electron-transfer proteins. The type of electron-transfer protein required for this reduction has been used as one means of dividing this large family of proteins. Class I P450s have two electron-transfer partners, an NAD(P)H flavoprotein reductase, and an iron sulfur protein with the iron sulfur protein serving as the proximal electron donor to the P450. These proteins are typically found in bacteria and mitochondria. Class II P450s have only one protein as their electron-transfer partner, an FAD/FMN-dependent, NADPH-P450 reductase with the electrons entering the reductase via the FAD domain while the FMN domain serves as the reductant of the P450. These P450s are, with only one exception, found in the endoplasmic reticulum of eukaryotes. The exception to this rule is the bacterial, fatty acid monooxy-genase P450$_{BM-3}$—a naturally occurring fusion protein containing both reductase and P450 domains on the same polypeptide chain[11] rendering this protein catalytically self-sufficient. The final group of P450s, Class III, contains those proteins that *do not* require an auxiliary electron-transfer partner since these enzymes do not catalyze a monooxygenation reaction as described by Eq. (1), rather they rearrange endoperoxides or hydroperoxides. Sequence alignments show that this group of proteins belongs to the P450 gene superfamily[1] and contains the conserved cysteinyl-loop which binds heme. Examples of this class of P450s are allene oxide synthase,[9] thromboxane synthase,[12] and prostacyclin synthase.[13]

1.1. The Fundamental Reaction Cycle of P450s

The most completely characterized P450 is the soluble, bacterial P450, P450$_{cam}$ (CYP101), which regio- and stereospecifically hydroxylates camphor to produce 5-*exo*-hydroxycamphor.[14] This protein was initially purified, characterized, cloned, and se-quenced by the Gunsalus group at Illinois.[14–16] P450$_{cam}$ is a Class I P450 that requires putidaredoxin, a 2Fe,2S ferredoxin, for efficient catalysis of the oxidation of camphor.

The fundamental catalytic cycle of P450s shown in Fig. 1 was originally proposed for microsomal P450s.[17] This cycle has been firmly established by studies of P450$_{cam}$. While there have been many elaborations on this cycle incorporating the interaction of P450s with alkyl peroxides, superoxide, etc.,[10] we will use this cycle for the description of the reactions of P450s. The first step of the reaction cycle is substrate binding to the low-spin ferric form of the enzyme resulting in the conversion of the heme iron to the high-spin ferric form.[18] Proton NMR studies of P450$_{cam}$ first indicated that the replaceable z-axis heme ligand was a solvent water molecule[19] which was displaced on camphor binding. Studies of the kinetics and thermodynamics of substrate binding to P450$_{cam}$ have clarified the changes that occur to the heme and their effect on the redox potential of the heme iron.[20–22] In the second step of the cycle, the first electron required for the reaction is transferred[23] followed by the binding of molecular oxygen shown in the third step to form the oxy complex.[24,25] Finally, in the fourth step, the addition of the second electron results in the formation of hydroxylated product and ferric substrate-free enzyme without the accumu-lation of measurable intermediates.[26] The demonstration that the oxy complex could be quantitatively converted to hydroxycamphor showed that it was an important intermediate in the cycle.[26] The ratio of putidaredoxin to P450$_{cam}$ determines which of the two electron-transfer

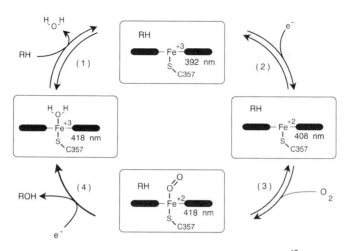

FIGURE 1. Fundamental cycle of P450s. The reaction cycle originally proposed[17] for microsomal P450s has been modified to include the cysteinyl ligation of the heme iron and the stepwise electron transfer reactions. The numbers under the heme ring in this figure are the wavelengths of maximum absorption of the Soret band of the P450 in each state. C357 represents the cysteinyl ligand from P450$_{cam}$.

steps is rate limiting in the overall cycle. In the case of a high ratio (>10:1), the first electron transfer is limiting, while at an equimolar ratio, the second step is limiting.[27]

Early in the studies of the mechanism of action of P450s it was realized that proton donation was important in the overall reaction cycle. However, the sequence determination of several P450s showed that there were only a few conserved residues among these proteins which might function in this capacity. With the subsequent determination of the atomic structure of P450$_{cam}$ at 1.6 Å resolution, Poulos and his colleagues[28,29] stated that there might be a unique role for a conserved threonine in oxygen binding, activation, and proton transfer during the catalytic cycle of P450s as is shown in Fig. 2. This figure is an elaboration of the original proposal for the mechanism of oxygen activation of Groves and McClusky.[30] In the X-ray crystal structure of P450$_{cam}$, T252 is in a position to hydrogen bond with iron-bound oxygen and to assist in its polarization. Input of the second electron is presumed to result in O–O bond cleavage with the formation of hydroxide ion and oxene bound to the heme iron (the oxene intermediate). This oxygen can abstract a hydrogen atom (a proton and an electron) from a carbon–hydrogen bond which is correctly positioned in the active site to allow the formation of an iron-bound hydroxyl radical and a carbon-centered radical. These two species react by radical recombination with retention of steric configuration around the carbon atom.[30] An important feature of the catalytic cycle of P450s is that completion of the cycle requires the input of protons to regenerate the hydroxyl group of threonine and to neutralize the hydroxyl group formed by O–O bond cleavage. Initially, Poulos, Sligar, and their colleagues identified a proton channel in P450$_{cam}$ which might serve to provide these protons during the reaction cycle.[31]

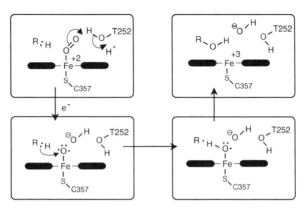

FIGURE 2. Proposed oxygen activation mechanism. The mechanism proposed by Groves[30] is shown with the residue numbering from P450$_{cam}$.

1.2. The Structure of P450$_{cam}$

With the publication of the atomic structure of P450$_{cam}$ at 1.6 Å resolution in 1987 by Poulos's group,[29] many questions were answered regarding how P450$_{cam}$ functioned. The structure heightened interest in the relationship between the structure of P450$_{cam}$ and that of distantly related eukaryotic P450s.[32–34] A ribbon diagram of the Cα backbone atoms of P450$_{cam}$ is shown in Fig. 3. P450$_{cam}$ is approximately triangular in shape with edge dimensions of 65 Å. Viewed from an edge, the molecule is domed on top (distal surface) and rather flat on the bottom (proximal surface) with a thickness of about 35 Å. The molecule can be divided into a β-sheet-rich region and an α-helical-rich region. The heme is in the center of the molecule and from this representation looks quite open and accessible from the surface. In the α-helical-rich region, the long I-helix runs from the top center to the bottom of the molecule and has 32 amino acid residues. In the middle of the I-helix, near the heme iron, is the conserved threonine mentioned above. In this representation, the cysteinyl thiolate of C357, which serves as the fixed axial ligand of the heme iron, is behind the heme.

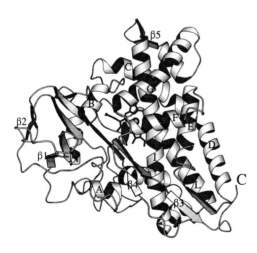

FIGURE 3. Ribbon drawing of P450$_{cam}$. The atomic coordinates of P450$_{cam}$[29] were used with the program MOLSCRIPT[98] to construct this figure. The α-helices are shown with coils and Roman letters while the β-sheets are shown with arrows and lowercase Greek letters. The helix designations are the same as those of Poulos et al.[28] while the β-sheets are named after those used for P450$_{BM-P}$ and P450$_{terp}$ by Ravichandran et al.[47] and Hasemann et al.,[48] respectively. The heme ring is shown in the middle of the figure with heavy dark lines under the intersection of the I- and F-helices. The NH$_2$- and COOH-terminal residues which can be observed in the crystals are indicated with a bold N and C, respectively.

Although many questions were answered about the structure and function of P450s by the determination of the structure of P450$_{cam}$, many questions remained unanswered. For example, in P450$_{cam}$ there is no obvious pathway from the surface into the substrate binding site/heme pocket which lies buried 20 Å from the surface.[28,35] Since the X-ray crystal structures of each of the various P450$_{cam}$ molecules and the respective ligands are essentially identical,[35] there have only been clues regarding the dynamics of substrate recognition, binding, and hydroxylation.

Because of the importance of P450s in intermediary metabolism, regulation, and carcinogenesis, a great deal of effort has been expended to extrapolate from the structure/function/mechanism studies on P450$_{cam}$ to the insoluble, membrane-bound P450s. We believe that the availability of additional P450s, to be used for detailed analyses, will assist us in better understanding their mechanism(s) and in designing inhibitors, activators, and P450s with wholly new specificities.

2. P450$_{BM-3}$ and the Reactions It Catalyzes

Several years ago, Fulco's group was able to purify three different P450s from *Bacillus megaterium* which had been grown in the presence of phenobarbital, all of which catalyze the oxidation of fatty acids.[36–38] One of these P450s, P450$_{BM-3}$(CYP102), is unique among P450s in that it is a self-sufficient monooxygenase with both the P450 and reductase domains on a single polypeptide chain having a molecular weight of approximately 120,000[11] that can be cleaved by trypsin into the two domains.[39] Using partially purified preparations of P450$_{BM-3}$, they were able to show that it would rapidly oxidize medium-chain-length fatty acids with the best substrate being pentadecanoic acid. The products of the oxidation of fatty acids by P450$_{BM-3}$ were ω-1, ω-2, and ω-3 hydroxy fatty acids with the ratio of the hydroxy derivatives being dependent on the length of the fatty acid.[40] They also showed that the enzyme would oxidize fatty acid amides and alcohols but not esters.[11,40]

Using homogeneous, recombinant enzyme purified from *E. coli*, we showed that the hydroxylation of fatty acids was tightly coupled: 1 mole of hydroxy fatty acid was produced per mole of oxygen, fatty acid, and NADPH consumed.[41] If the enzyme was incubated with a limiting concentration of fatty acid (<50 μM) and an excess of NADPH (>300 μM), other more polar products accumulated in the reaction mixture. These compounds were identified as dihydroxy and keto-hydroxy fatty acids which were formed by oxidation of the monohydroxy fatty acids from the initial round of oxidation.[42] To be able to accommodate these charged and uncharged fatty acids, as well as the hydroxy and keto-hydroxy fatty acids, the substrate binding site of this enzyme must be large and relatively polar. Because of the similarity between P450$_{BM-3}$ and microsomal P450s, this protein should provide a useful counterpoint to the studies of P450$_{cam}$.

3. P450$_{terp}$ and the Reaction It Catalyzes

To increase the range of purified P450s available for study, a microorganism was isolated from a wetlands area north of Dallas, Texas, for its ability to oxidize α-terpineol as its sole carbon source.[43] This microorganism was identified as a pseudomonad which contained an operon encoding a soluble, three-component $P450_{cam}$-like system.[44] From whole-cell DNA, the P450 gene was cloned and sequenced, and high-level expression was obtained for the encoded protein, $P450_{terp}$, in *E. coli*. $P450_{terp}$ (CYP108) is the closest homologue to $P450_{cam}$ with 27% sequence identity. The regio- and stereospecificity of monooxygenation by $P450_{terp}$ have been studied in collaboration with Ortiz de Montellano's group. The oxidation of α-terpineol produces 7-hydroxy-α-terpineol, but the enzyme will also catalyze the oxidation of thioanisoles and styrenes, in which case $P450_{terp}$ is even more stereospecific than $P450_{cam}$.[45]

4. Comparison of the Structures of $P450_{cam}$, $P450_{terp}$, and $P450_{BM-P}$

With the availability of large amounts of the hemoprotein domain of $P450_{BM-3}$ ($P450_{BM-P}$) and of $P450_{terp}$, we were able to crystallize both of these proteins. Large diffraction-quality crystals were grown by the hanging-drop, vapor diffusion technique using PEG 8000 as the primary precipitant.[46] Native X-ray diffraction data were collected with these crystals to 2.0 Å resolution with $P450_{BM-P}$ and 2.3 Å resolution with $P450_{terp}$. Initial molecular replacement calculations using a polyalanine model of $P450_{cam}$ for $P450_{BM-P}$ were unsuccessful[46] because the structures are not sufficiently alike. The structure of $P450_{BM-P}$ was solved using multiple isomorphous replacement with four heavy atom derivatives. The structure was refined to an *R* factor of 16.9% at 2.0 Å resolution.[47] In contrast to $P450_{BM-P}$, $P450_{terp}$ seemed to be a promising candidate for molecular replacement calculations using $P450_{cam}$ as the search model. Unfortunately, molecular replacement calculations based on $P450_{cam}$ did not permit us to complete the structure. Therefore, multiple isomorphous replacement phasing was performed with two heavy atom derivatives. The final structure, which was refined to 2.3 Å resolution, had an *R* factor of 18.9%.[48]

When the structures of $P450_{BM-P}$ and $P450_{terp}$ were available, we were struck by the dramatic similarity between them and $P450_{cam}$ (Figs. 3–5) even though these proteins share only approximately 7% sequence identity in their optimally aligned sequences.[49] In fact, in spite of the low sequence identity and their variable lengths, the topology diagram of

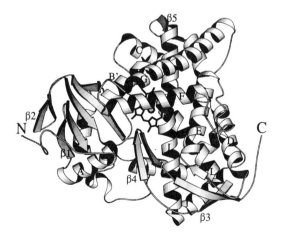

FIGURE 4. Ribbon drawing of $P450_{BM-P}$. The atomic coordinates of $P450_{BM-P}$[47] were used with MOLSCRIPT[98] to produce this figure. The α-helices, β-sheets, and NH_2- and COOH-termini which can be readily seen from this perspective are labeled.

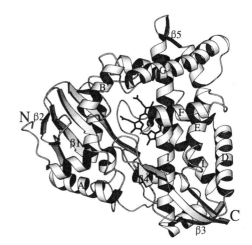

FIGURE 5. Ribbon drawing of P450$_{terp}$. The atomic coordinates of P450$_{terp}$[48] were used with MOLSCRIPT[98] to produce this figure. The α-helices, β-sheets, and NH$_2$- and COOH-termini which can be readily seen from this perspective are labeled.

the secondary structural elements of P450$_{BM-P}$ shown in Fig. 6 could be used for P450$_{cam}$ with only minor modification, or as discussed below, for P450$_{terp}$. When the secondary structural elements were compared in detail, it was found that additional residues were inserted in a small number of discrete locations.

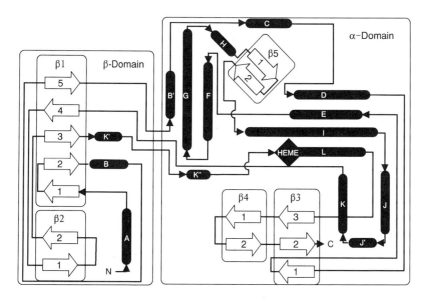

FIGURE 6. Topology drawing of P450$_{BM-P}$. The topology of P450$_{BM-P}$ is shown with helices represented by black bars. The length of each of the bars is in approximate proportion to the length of the helix. The strands of β-sheets are shown with arrows. The strands are grouped by the secondary structural elements which they comprise. The structural elements are grouped into the α-helical-rich domain and the β-sheet-rich domain. The heme is shown by the square at the NH$_2$-terminal end of the L-helix. With only minor modifications, this topology diagram could be used for all three of the P450s.

4.1. The Structural Core of P450s

The structural similarity of the three P450s shown in Figs. 3–5 is apparent, but to better evaluate the similarities and differences in the structures, the program "O"[50] was used to overlay the Cα backbone atoms of these molecules in three-dimensional space. Each of these molecules was superimposed in a pairwise fashion to obtain an "average" structure. Figure 7 is a difference plot of comparable Cα atoms obtained from this pairwise comparison demonstrating that there are regions of these molecules which are in similar positions in three-dimensional space. The pairwise comparison of the structural similarity between homologous proteins has been studied by Chothia and Lesk.[51] They found that the rms deviation between comparable positions in the structural core of homologous proteins was a function of the sequence identity. Their results predict a deviation of > 2.0 Å in the core structures of these P450s.[49] Poulos also pointed out, in his discussion of structure predictions based on P450$_{cam}$, that at best an rms deviation of approximately 2 Å might be obtained even in the structural core when P450s are superimposed.[35]

The drawing of the structure of P450$_{cam}$[29] provided an excellent view of the molecule with the heme ring, the I-helix, and the camphor molecule readily apparent. In fact, each of the three P450s for which the structure is known, when shown in this orientation, is triangular shaped with edge dimensions of approximately 65 Å, and look strikingly alike. However, if the structures are rotated by 90° as shown in Fig. 8, so that the heme is seen edge-on, the differences between the molecules become much more apparent. The heme ring is approximately in the middle of each of these drawings with the central I-helix being viewed end-on and located above the heme ring. The β-sheet-rich domain of these

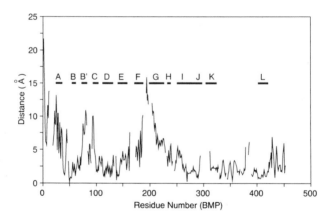

FIGURE 7. RMS deviation of Cα backbone between P450$_{cam}$ and P450$_{BM-P}$. The Cα backbone coordinates of P450$_{cam}$ and P450$_{BM-P}$ were overlaid in three-dimensional space[49] using the program "O"[50] and the distance between comparable Cα atoms calculated. P450$_{BM-P}$ has 472 amino acids while P450$_{cam}$ has 414 amino acids and this difference in length results in insertions having to be made in the sequence of P450$_{cam}$ to match the sequence of P450$_{BM-P}$. The abscissa is the number of the amino acid from this pairwise structural alignment of these P450s. The breaks in the data line in the figure are the sites where insertions had to be made in the P450$_{cam}$ sequence to align these proteins and therefore there were no comparable amino acids for the calculation. The helical secondary structural elements of P450$_{BM-P}$ are shown in this figure for reference.

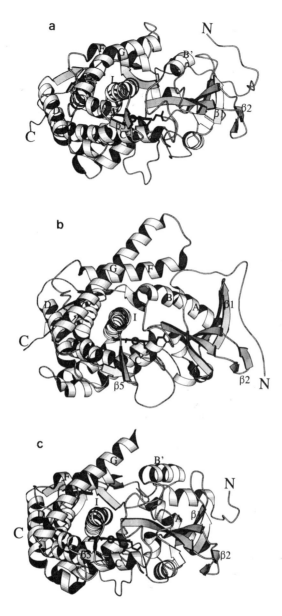

FIGURE 8. The structures of the P450s rotated. The coordinates from Figs. 3–5 were rotated by 90° in the x-axis and 180° in the y-axis and redrawn. The structural elements were drawn with MOLSCRIPT[98] as described in Fig. 3. The heme ring is seen in the middle of the figure as a dark line because it is viewed edge-on. $P450_{cam}$, $P450_{BM-P}$, and $P450_{terp}$ are seen in panels a, b, and c, respectively.

molecules is to the right in the drawings. In the loop between the F- and G-helices, the difference in Cα positions is approximately 15 Å as can be seen in Fig. 7. The 29-residue-long rodlike G-helix protrudes from the upper center of P450$_{BM-P}$ in this representation and extends down to the left edge. The G-helix is in approximately the same orientation in P450$_{terp}$. In contrast, the G-helix of P450$_{cam}$ is shorter by 7 residues and is markedly bowed over the top of the molecule. Overall, the G-helix in P450$_{terp}$ is more like that of P450$_{BM-P}$ in shape, but is 25 residues long. The F-helices are shorter and behind the G-helices in these representations. The differences in position and orientation of the B–B' loop, B'-helix, and B'–C loop are also apparent in Fig. 7.

In 1989, Presnell and Cohen[52] recognized that P450$_{cam}$ contained an unusual four-helical bundle in the core of the molecule. While most four-helical bundles contain two antiparallel helices, P450$_{cam}$ was recognized to have three parallel helices and an antiparallel helix.[52] This four-helical bundle is conserved in three-dimensional space in each of the P450s. The bundle is composed of the carboxy-terminal half of the I-helix, the D-, E-, and L-helices, with the E-helix being antiparallel to the other three helices. The conserved core of P450s also includes the amino-terminal portion of the J-helix, the K-helix, the cysteinyl heme-binding loop, the $\beta1$ and $\beta2$ sheets, and a segment of random coil which we have dubbed the "meander."

4.2. Heme Binding

Random probability would seem to dictate that the coordinating fixed ligand of the heme in myoglobin, hemoglobin, and P450s could be inserted from either face. Ortiz de Montellano in 1983 used the stereochemistry of heme alkylation to demonstrate that the heme of myoglobin, hemoglobin, and of the tested microsomal P450s is oriented stereospecifically with respect to the fixed axial ligand.[53] The high-resolution structural determination of three P450s showed that the heme is oriented in these proteins as predicted[53]: the heme is buried in the interior of the molecule with the charges from the propionate side chains of the heme being balanced by amino acid residues in the protein.

An examination of the alignment of Nelson *et al.*[1] shows that several of the residues involved in heme binding are either conserved or identical in all P450 molecules. The heme-binding pocket and the hydrogen bonds which hold the heme in place in P450$_{BM-P}$ are shown in Fig. 9. One of these residues is in the C-helix, one in $\beta1$–4 (the backbone nitrogen of S332), and one in the heme-binding loop (R393). Additionally, the heme-binding loop, which is the most highly conserved element of this family of proteins, contains the absolutely conserved cysteine which coordinates with the heme iron. Because of the importance of these residues in heme binding, each protein or gene sequence which is proposed for membership in this gene superfamily should have these or similar residues to bind the heme ring.

4.3. K-Helix and the Meander

As the structure of P450$_{BM-P}$ was being completed and structural elements were being assigned, it was observed that approximately 14 amino acids NH$_2$-terminal to the cysteinyl heme-binding loop seemed to take a "random walk" or "meander" on the proximal face of the molecule. Hence, this region was dubbed the "meander." Structural comparisons of

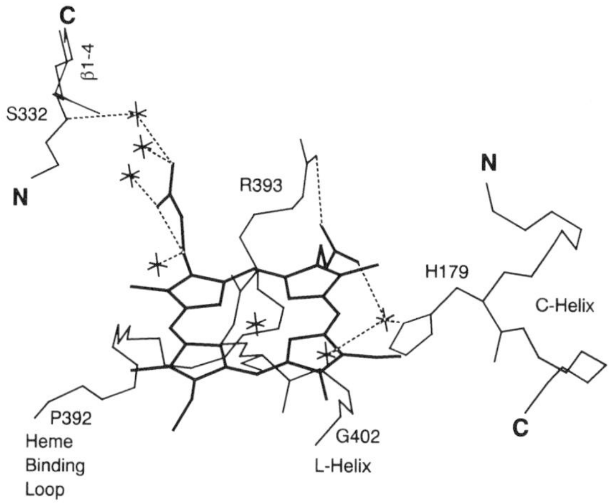

FIGURE 9. Heme propionate coordination in $P450_{BM-P}$. The backbone $C\alpha$, C, and N atoms of $P450_{BM-P}$ were displayed as light lines with the program InsightII running on a Silicon Graphics workstation. The heme ring atoms are shown in the center of the figure with the heavy lines. The residues which interact with the heme propionates were calculated as those residues which are within 3.0 Å of the specified atom. These interactions are shown with dashed lines. For example, the propionate side chain of pyrrole ring C is charge neutralized by R393 from the heme binding loop. The carbonyl oxygen of this propionate is also in hydrogen bond interaction distance of a crystallographically ordered water molecule which in turn is in hydrogen bond distance of another water molecule as well as H179 from the C-helix. The propionate side chain of pyrrole ring D is in hydrogen bond distance of a series of ordered water molecules which are directly connected to the proximal face of the molecule via a water channel. One of these water molecules, shown in this figure, is also within hydrogen bond distance of the backbone nitrogen of residue S332 from β-strand 1–4.

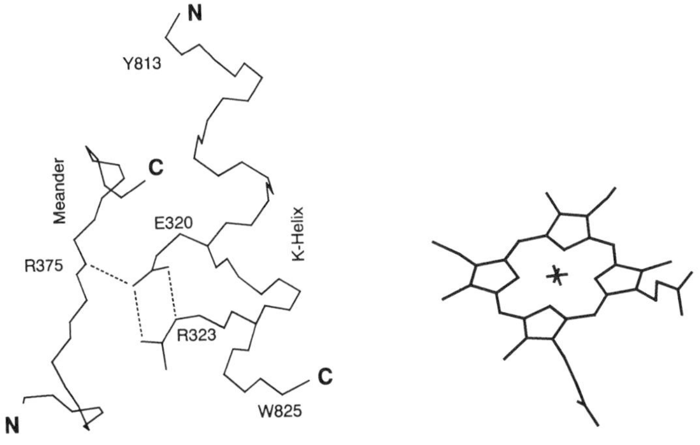

FIGURE 10. The hydrogen bond network of the K-helix and the meander. The atomic coordinates of $P450_{BM-P}$ were used to display the heme ring and the $C\alpha$, C, and N atoms of the backbone of the K-helix and the meander. The conserved E320 and R323 of this helix are shown by a stick model with the hydrogen bonds indicated by dashed lines. These two residues are also in hydrogen bond distance to the backbone nitrogen of a conserved residue from the meander, R375 in this case. The view in this figure is from the proximal face of the molecule with the heme ring displayed on the right for reference. The NH_2- and COOH-terminal ends of the meander are labeled **N** and **C**, respectively.

the three proteins[49] have shown that this "meander" is conserved in three-dimensional space. Figure 10 shows that it is held in place via a hydrogen bond network which is comprised of the ExxR pair from the K-helix and a backbone nitrogen of a conserved Arg, His, or Asn residue from the meander. The roles which the meander and the charged residues play in protein folding and structural stability of the P450 gene superfamily remain to be determined. In an effort to probe the role of these residues, the Arg of the ExxR pair in the K-helix of $P450_{arom}$ has been mutated, used to transform COS-1 cells, and found to be inactive.[54] Because of the low level of expression of P450s in COS-1 cells, the authors could not determine the reason for inactivity, i.e., whether it was lack of assembly of the protein, of heme insertion, or of stability of the protein once assembled.

4.4. Sequence Alignment of the P450s

Because of the importance of P450s in intermediary metabolism, structure/function studies have assumed a large role in the research community with models of various P450s being constructed based on homology modeling techniques.[33,55–60] These models have usually employed $P450_{cam}$ as their framework, but more recently research groups have begun to employ $P450_{BM-P}$ as the model of choice.[61]

In any homology modeling study, the first task is to align the amino acid sequence of the target protein with the sequence of the homologue for which the structure is known. The object of the alignment exercise is to enable the investigator to identify those residues which comprise secondary structural elements that can then be assembled into a whole molecule in some rational fashion. Some investigators employ secondary structure prediction algorithms to assist them in modifying the alignments which are produced by either pairwise or multiple sequence alignment algorithms. We utilized two secondary structure prediction algorithms and have found them to be generally useful in assigning secondary structure to the structural core of $P450_{cam}$, $P450_{BM-P}$ and $P450_{terp}$.[62–65] Outside of the structural core, the assignment of secondary structure was poor in one case[63–65] and very poor in the other case.[62,66] Unfortunately, those structural elements involved in substrate interaction or binding, which are of primary interest to most investigators, are outside of the structural core. Assisted by certain recognizable residues which are involved in heme contact as described above and hydrogen bond interactions in the structural core, we have been able to manually adjust the output of the program PILEUP from the GCG suite of programs[66] as shown in Fig. 11. Examples of the residues used to align these sequences are: the positively charged residue in the B-helix of most microsomal P450s (K122 for $P450_{BM-P}$ Fig. 11 numbering); heme propionate ligand in the C-helix (H179 for $P450_{BM-P}$ Fig. 11 numbering); ExxR pair in the K-helix (E477, R480, Fig. 11 numbering); heme propionate ligand from β1–4 in most P450s (R491 for $P450_{cam}$ and $P450_{terp}$, Fig. 11 numbering); and heme propionate ligand in the heme-binding loop (R562 for $P450_{BM-P}$ Fig. 11 numbering). While the initial alignment was adjusted to bring into register the secondary structural elements which are present in the three P450s, the alignment has been further modified over time as we have become more familiar with the composite structure of the proteins. We do not believe that this is the ultimate alignment of this protein superfamily; rather, individual protein sequences will need to be adjusted slightly to account for their structural peculiarities. The alignment in Fig. 11 is being offered in this review as our current effort at assisting other investigators in aligning their favorite P450. In instances where this information is to be utilized in a mutagenesis project, we would caution that close examination of the alignment along with some intuition (e.g., as to the beginning and end of helices) is necessary to adjust the alignment.

5. Substrate Binding to P450s

The publication of the structure of P450$_{cam}$ was exciting and provided clues to enable investigators to answer some but not all of the fundamental questions regarding the mechanism of action of the P450 gene superfamily of proteins. One of the questions in

```
          1                                                                  αA          β1-1
CYP101   .......... .......... .......... .......... ....MTTETI QSNANLAPLP PHVPEHLVFD FDMYNPSNLS A.GVQEAWAV LQESNV..PD
CYP108   .......... .......... .......... .......... ......MDAR ATIPEHIART VILPQGYA.. ..DDCEVIYP AFKWLRDEQP
CYP11A1  .......... MLARGLPLRS ALVKACPPIL STVGEGWGHH RVGTGEGAGI STKTPRPYSE IPSPGDNGWL NLYHFWREKG SQRIHFRHIE NFQKYG...P
CYP27    MAALGCARLR WALLGPRVAG CGLCPQGARA KAAIPTALPA DEAAQAPGAG PGDRRRRRSL EELPRLGQLR FFYQAFVQGY LLHLHKLQVL NKARYG...P
CYP21A2  .......... .......... MLLLGLLLLP LLAGARLLWN WWKLRS.LHL PPLAPGF... ..LHLLQP.. ..DLPIYLLG LTQKFG...P
CYP17A   .......... .......... ..MWELVGLL LLILAYFFWV KSKTPG.AKL PRSLPSLPLV GSLPFLPRRG ..HMHVNFFK LQEKYG...P
CYP1A2   .......... .......... MAFSQYISLA PELLLATAIF CLVFWVLRGT RTQVPKGLKS PPGPWGLPFI GHMLTLGK.. ..NPHLSLTK LSQQYG...D
CYP1A1   .......... .......MPS YYGFPAFTSA TELLLAVTTF CLGFWVVRVT RTWVPKGLKS PPGPWGLPFI GHVLTLGK.. ..NPHLSLTK LSQQYG...D
CYP2C11  .......... .......... .MDPVLVLVL TLSSLLLLSL WRQSFGRGKL PPGPTPLPII GNTLQIYMK. ..DIGQSIKK FSKVYG...P
CYP2C4   .......... .......... .MDPVAGLVL GLCCLLLLSL WKQNSGRGKL PPGPTPFPII GNILQIDVK. ..DISKSLTK FSERYG...P
CYP2E1   .......... .......MA VLGITVALLG WMVILLFISV WKQIHSSWNL PPGPFPLPII GNLLQLDLK. ..DIPKSFGR LAERFG...P
CYP2B1   .......... .......ME PSILLL..LA LLVGFLLLLV RGHPKSWGNF PPGPRPLPLL GNLLQLDRG. ..GLLNSFMQ LRKKYG...P
CYP2A4   .......... .......MLT SGLLLVAAVA FLSVLVLMSV WKQRKLSGKL PPGPTPLPFV GNFLQLNTE. ..QMYNSLMK ISQRYG...P
CYP3A4   .......... .......... ...MALIPDL AMETWLLLAV SLVLLYLYGT HSHGLFKKLG IPGPTPLPFL GNILSYHK.. ..GFCMFDME CHKKYG...K
CYP5A1   .......... .......MEA LGFLKLEVNG PMVTVALSVA LLALLKWYST SAFSRLEKLG LRHPKPSPFI GNLTFFRQ.. ..GFWESQME LRKLYG...P
CYP4A1   .......... MSVSALSSTR FTGSISGFLQ VASVLGLLLL LVKAVQFYLQ RQWLLKAFQQ FPSPPFHWFF GHK.QFGQD. ..KELQQIMT CVENFP...S
CYP102   .......... .......... .......... .......... ....MTIKE MPQPKTFGEL KNLPLLNTD. ..KPVQALMK IADELG....E
CYP19    .......... ....MVLEML NPIHYNITSI VPEAMPAATM SVLLLTGLFL LLWNYEGTSS IPGPGYCMGI GPLISHGRFL WMGSRSACNY YNRVYG...E
```

```
          101      β1-2      αB         β1-5      αB'              αC                          200
CYP101   LVWTRCNG.. GHWIATRGQL IREAYEDY.. RHFSSE..CP FIPREAGEAY .......... .DFIP.TSMD P.PEQRQFRA LANQVVGMPV VDK.......
CYP108   LAMAHIEGYD PMWIATKHAD VMQIGKQP.. GLFSNAEGSE ILYDQNNEAF MRSISGGCPH VIDSL.TSMD P.PTHTAYRG LTLNWFQPAS IRK.......
CYP11A1  IYREKLG.NL ESVYIIHPED VAHLFKFE.. GSYPER..ND MQLWKEHRDH QDLAY..... ..GVLFKKS. ..GTWKKDRV VLNTEVMAPE AIKN.....F
CYP27    MWVSYLG.PQ LFVNLASAPL VETVMRQE.. GKYPVR..ND MQLWKEHRDH QDLAY..... ..GVFTTDG. ..HDWYQLRQ ALNQRLLKPA EAAL.....Y
CYP21A2  IYRLHLG.LQ DVVVLNSKRT IEEAMVKKW. ADFAGR.... ..PEPLTYKL VSKNY..... PDLSL.GDYS ..LLWKAHKK LTRSAL.LLG IRDS......
CYP17A   IYSLRLG.TT TTVIIGHYQL AREVLIKKG. KEFSGR.... ..PQMVTQSL LSDQGK.... ..GVAFADAG ..SSWHLHRK LVFSTFSLFK DGQK......
CYP1A2   VLQIRIG.ST PVVVLSGLNT IKQALVKQG. DDFKGR.... ..PDLYSFTL ITNGK..... ..SMTFNPDS G.PVWAARRR LAQDALKSFS IASDPTSVSS
CYP1A1   VLQIRIG.ST PVVVLSGLNT IKQALVKQG. DDFKGR.... ..PDLYSFTL IANGQ..... ..SMTFNPDS G.PLWAARRR LAQNALKSFS IASDPTLASS
CYP2C11  IFTLYLG.MK PFVVLHGYEA VKEALVDLG. EEFSGR.... ..GSFPVSER VNKGL..... ..GVIFSNG. ..MQWKEIRR FSIMTLRTFG MGKRT.....
CYP2C4   VFTVYLG.MK PTVVLHGYKA VKEALVDLG. EEFAGR.... ..GHFPIAEK VNKGL..... ..GIVFTNA. ..NTWKEMRR FSLMTLRNFG MGKRS.....
CYP2E1   VFTVYLG.SR PVVVLHGYKA VREMLLNHK. NEFSGR.... ..GEIPAFREF KDK....... ..GIIFNNG. ..PTWKDTRR FSLTTLRDYG MGKQG.....
CYP2B1   VFTVHLG.PR PVVMLCGTDT IKEALVGQA. EDFSGR.... ..GTIAVIEPI FKE....... ..YGVIFANG. ..ERWKALRR FSLATMRDFG MGKRS.....
CYP2A4   VFTIYLG.SR RIVVLCGQEA VKEALVDQA. EEFSGR.... ..GEQATFDWL FKG....... ..YGIAFSSG. ..ERAKQLRS FSIATLRDFG VGKRG.....
CYP3A4   VWGFYDG.QQ PVLAITDPDM IKTVLVKECY SVFTNR.... ..RPFGPVGF MK........ ..SAISIAED. ..EEWKRLRS LLSPTFTSGK LKE.......
CYP5A1   LCGYYLG..R RMFIVISEPD MIKQVLVENF SNFTNR.... ...MASGLEF KSVAD..... ..SVLFL... RDKRWEEVRG ALMSAFSPEK LNE.......
CYP4A1   AFPRWFWGSK AYLIVYDPDY MKVIL..... GRSDPK.... ..ANGVYRL LAPWIG.... .YGLLLLNG. ..QPWFQHRR MLTPAFHYDI LKP.......
CYP102   IFKFEAPG.R VTRYLSSQRL IKEAC.DE.. SRFDKN.... ...LSQALKF VRDFAG.... ..DGLFTSWT HEKNWKKAHN ILLPSFSQQA MKG.......
CYP19    FMRVWISGEE TLIISKSSSM FHIMKHNHYS SRFGSK.... ..LGLQCIGM HEK....... ..GIIFNNNP ..ELWKTTRP FFMKALSGPG LVR.......
```

(continued)

FIGURE 11. Alignment of the sequences of some eukaryotic P450s with P450$_{cam}$, P450$_{BM-P}$, and P450$_{terp}$. This sequence alignment was constructed with the program PILEUP from the GCG suite of programs[66] and then manually adjusted as described in the text with the program LINEUP. The sequences were obtained from the composite table of Nelson *et al.*[1] and are here used with their family designations. The secondary structural elements of P450$_{BM-P}$, P450$_{terp}$, and P450$_{cam}$ are shown with α-helices and strands of β-sheets represented by single- and double-underlining, respectively.

P450 structure/function studies which remains is: how are hydrophobic substrates recognized and bound specifically in the active site? The structure of camphor-bound $P450_{cam}$ showed that in this enzyme Y96 is hydrogen-bonded to the carbonyl of camphor[28] while side chains of hydrophobic amino acids position the camphor molecule by steric/hydrophobic interactions. Although the atomic structures of many different complexes of $P450_{cam}$ have been reported,[28,29,31,35,67,68] there did not seem to be an unambiguous pathway for substrate entry and product release.

5.1. Substrate Access Channel

The atomic structures of $P450_{BM-P}$ and $P450_{terp}$ in combination with the structure of $P450_{cam}$ have provided a better understanding of the process of substrate recognition, access, and binding. In Poulos's review of the structure of $P450_{cam}$,[35] he proposed that substrate access should be through a region bounded by the loop between the F- and G-helices, β-sheets 1 and 4, and the B′ helix. As demonstrated in Fig. 12, which shows the Connolly surface of the residues around the substrate access channel in $P450_{BM-P}$ it is

```
        201     αD            β3-1     αE                                       αF          300
CYP101  ..LENRIQEL ACSLIESLRP ....QGQCNF TEDYAEPFPI RIFMLLAGL. .......... .......... .PEEDIPHLK YLTDQMTR.. ........PD
CYP108  ..LEENIRRI AQASVQRLLD F...DGECDF MTDCALYYPL HVVMTALGV. .......... .......... .PEDDEPLML KLTQDFFGVH EPDEQAVA..
CYP11A1 IPLLNPVSQD FVSLLHKRIK QQGSGKFVGD IKEDLFHFAF ESITNVMFGE RLGM...... .......... .LEETVNPEA QKFIDAVYKM FHTSVPLLNV
CYP27   TDALNEVIDS FVVRLDQLRA ESASGDQVPD MADLLYHFAL EAICYILFEK RIGC...... .......... .LEASIPKDT ENFIRSVGLM FQNSVYVT..
CYP21A2 .MEPVVEQL TQEFCERMRA Q..PGTPVAI EE.EFSLLTC SIICYLTFGD KI........ .......... KDDNLMPAYY KCIQEVLKTW SHWSIQIVDV
CYP17A  ..LEKLICQE AKSLCDMMLA HD..KESIDL ST.PIFMSVT NIICAICFNI SYE....... .......... KNDPKLTAIK TFTEGIVDAT GDRNLVDI..
CYP1A2  CYLEEHVSKE ANHLISKFQK LMAEVGHFEP VN.QVVESVA NVIGAMCFGK NFP....... .......... RKSEEMLNLV KSSKDFVENV TSGNAVDF..
CYP1A1  CYLEEHVSKE AEYLISKFQK LMAEVGHFDP FKYLVV.SVA NVICAICFGR RYD....... .......... HDDQELLSIV NLSNEFGEVT GSGYPADF..
CYP2C11 ..IEDRIQEE AQCLVEELRK ..SKGAPFDP TF.ILGCAPC NVICSIIFQN RFD....... .......... YKDPTFLNLM HRFNENFRLF SSPWLQVCNT
CYP2C4  ..IEDRVQEE ARCLVEELRK ..TNALPCDP TF.ILGCAPC NVICSVILHN RFD....... .......... YKDEEFLKLM ERLNENIRIL SSPWLQVYNN
CYP2E1  ..NEDRIQKE AHFLLEELRK ..TQGQPFDP TF.VIGCTPF NVIAKILFND RFD....... .......... YKDKQALRLM SLFNENFYLL STPWLQVYNN
CYP2B1  ..VEERIQEE AQCLVEELRK ..SQGAPLDP TF.LFQCITA NIICSIVFGE RFD....... .......... YTDRQFLRLL ELFYRTFSLL SSFSSQVFEF
CYP2A4  ..IEERIQEE AGFLIDSFRK ..TNGAFIDP TF.YLSRTVS NVISSIVFGD RFD....... .......... YEDKEFLSLL RMMLGSLQFT ATSMGQVYEM
CYP3A4  ..MVPIIAQY GDVLVRNLRR EAETGKPVTL KD.VFGAYSM DVITSTSFGV NID....... .......... .SLNNPQDPF VENTKKLLRF DFLDPFFLSI
CYP5A1  .MVPLISQA CDLLLAHLKR YAESGDAFDI QR.CYCNYTT DVVASVAFGT PVDSWQAPED PFVKHCKRFF EFCIPRPILV LLLSFPSIMV PLARILP...
CYP4A1  ..YVKNMADS IRLMLDKWEQ LAGQDSSIEI FQ.HISLMTL DTVMKCAFSH NGSVQ..... .......... VDGNYKSYIQ AIGNLNDLFH SRVRNIFHQN
CYP102  ..YHAMMVDI AVQLVQKWER L.NADEHIEV PE.DMTRLTL DTIGLCGFNY RFNSF..... .......... YRDQPHPFIT SMVRALDEAM NKLQR...AN
CYP19   ..MVTVCAES LKTHLDRLEE V.TNESGYVD VL.TLLRRVM LDTSNTLFLR IP........ .......... LDESAIVVKI QGYFDAWQAL LIKPDIFF..


        301      αG                                         αH         β5-1               400
CYP101  GSM....... ...TFAEAKE ALYDYLIPII EQRRQKP... .......... GTDAISIVAN .GQVNG... .......... .......... ........R
CYP108  ..APRQSADE AARRFHETIA TFYDYFNGFT VDRRSCP... .......... KDDVMSLLAN .SKLDG... .......... .......... ........N
CYP11A1 ..PPELYRLF RTKTWRDHVA AWDTIFNKAE KYTEIFYQD. ..LRRK..TE FRNYPGILY. ..CLLKS... .......... .......... .........E
CYP27   ..FLPKWTRP LLPFWKRYLD GWDTIFSFGK NLIDQKLQEV VAQLQSAGSD GVQVSGYLH. ..SLLTS... .......... .......... .........G
CYP21A2 IPFLRFFPNP GLRRLKQAIE KRDHIVEMQL RQHKESLVAG Q......... WRDMMDYMLQ ..GVA.QPSM EEGS...... .......... .........G
CYP17A  FPWLTIFPNK GLEVIKGYAK VRNEVLTGIF EKCREKFDSQ S......... ISSLTDILIQ AKMNSDNNNS CE........ .......... .....GRDPD
CYP1A2  FPVLRYLPNP ALKRFKNFND NFVLFLQKTV QEHYQDFNKN S......... IQDITGALFK .HSENY.... .......... ....KDNG.. .........G
CYP1A1  IPILRYLPNS SLDAFKDLNK KFYSFMKKLI KEHYRTFEKG H......... IRDITDSLIE .HCQDRRL.. .......... ....DENA. ........NV
CYP2C11 FPAIIDYFPG SHNQVLKNFF YIKNYVLEKV KEHQESLDKD N......... PRDFIDCFLN .KMEQE.... .......... ....KHNPQ .........S
CYP2C4  FPALLDYFPG IHKTLLKNAD YTKNFIMEKV KEHQKLLDVN N......... PRDFIDSFLI .KMEKENN.. .......... .......... .........L
CYP2E1  FSNYLQYMPG SHRKVIKNVS EIKEYTLARV KEHHKSLDPS C......... PRDFIDSLLI .EMEKD.... .......... .....KHST. ........EP
CYP2B1  FSGFLKYFPG AHRQISKNLQ EILDYIGHIV EKHRATLDPS A......... PRDFIDTYLL .RMEKE.... .......... ....KSMHH .........T
CYP2A4  FSSVMKHLPG PQQQAFKELQ GLEDFITKKV EHNQRTLDPN S......... PRDFIDSFLI .RMLEE.... .......... ....KKNPN .........T
CYP3A4  TVFPFLIPIL EVLNICVFPR EVTNFLRKSV KRMKESRLED T.....QKH RVDFLQLMID .SQNS..... .......... .....KETES ........HK
CYP5A1  ........NK NRDELNGFFN KLIRNVIALR ....DQQA.. .....AEERR R.DFLQMVLD ARHSASPMGV QDFDIVRDVF SSTGCKPNPS RQHQPSPMAR
CYP4A1  ..DTIYNFSS NGHLFNRACQ LAHDHTDGVI KLRKDQLQNA GELEKVKKKR RLDFLDILLL ARMEN..... .......... .......... .........GD
CYP102  PDDP..AYDE NKRQFQEDIK VMNDLVDKII ADRKASGEQ. .......... SDDLLTHMLN .GKDPET... .......... .......... ........GE
CYP19   ...KISWLYK KYEKSVKDLK DAIEVLIAEK RCRISTEEK. ..LEE..... CMDFATELIL .AEKR..... .......... .......... .........G
```

FIGURE 11. (*continued*)

```
            β5-2        αI              αJ                        αK       β1-4    β2-1
CYP101   PITSDEAKRM CGLLLVGGLD TVVNFLSFSM EFLAKSPEHR QELIERP... .......... .......ER IPAACEELLR RFSLV.A.DG RILTSDYEFH
CYP108   YIDDKYINAY YVAIATAGHD TTSSSGGAI IGLSRNPEQL ALAKSDP... .......... .......AL IPRLVDEAVR WTAPVKS.FM RTALADTEVR
CYP11A1  KMLLEDVKAN ITEMLAGGVN TTSMTLQWHL YEMARSLNVQ EMLREEVLNA RRQA...EGD ISKMLQMVPL LKASIKETLR LHPISVT.LQ RYPESDLVLQ
CYP27    QLSPREALGS LPELLLAGVD TTSNTLTWAL YHLSKNPEIQ AALRKEVVGV VAAG...QVP QHKDFAHMPL LKAVLKETLR LYPVIPA.NS RIIVDKEIEV
CYP21A2  QLLEGHVHMA AVDLLIGGTE TTANTLSWAV VFLLHHPEIQ QRLQEELDHE LGPGASSSRV PYKDRARLPL LNATIAEVLR LRPVVPLALP HRTTRPSSIS
CYP17A   VFSDRHILAT VGDIFGAGIE TTTTVLKWIL AFLVHNPEVK KKIQKEIDQY VG...FSRTP TFNDRSHLLM LEATIREVLR IRPVAPMLIP HKANVDSSIG
CYP1A2   LIPQEKIVNI VNDIFGAGFE TVTTAIFWSL LLLVTEPKVQ RKIHEELDTV IGRD...RQP RLSDRPQLPY LEAFILEIYR YTSFVPFTIP HSTTRDTSLN
CYP1A1   QLSDDKVITI VFDLFGAGFD TITTAISWSL MYLVTNPRIQ RKIQEELDTV IGRD...RQP RLSDRPQLPY LEAFILETFR HSSFVPFTIP HSTTRDTSLN
CYP2C11  EFTLESLVAT VTDMFGAGTE TTSTTLRYGL LLLKHVDVT AKVQEEIERV IGRN...RSP CMKDRSQMPY TDAVVHEIQR YIDLVPTNLP HLVTRDIKFR
CYP2C4   EFTLGSLVIA VFDLFGAGTE TTSTTLRYSL LLLKHPEVA ARVQEEIERV IGRH...RSP CMQDRSHMPY TDAVIHEIQR FIDLLPTNLP HAVTRDVKFR
CYP2E1   LYTLENIAVT VADMFFAGTE TTSTTLRYGL LILLKHPEIE EKLHEEIDRV IGPS...RMP SVRDRVQMPY MDAVVHEIQR FIDLVPSNLP HEATRDTTFQ
CYP2B1   EFHHENLMIS LLSLFFAGTE TSSTTLRYGF LLMLKYPHVA EKVQKEIDQV IGSH...RLP TLDDRSKMPY TDAVIHEIQR FSDLVPIGVP RVTKDTMFR
CYP2A4   EFYMKNLVLT TQNLFFAGTE TGSTTLRYGF LLLMKYPDIE AKVHEEIDRV IGRN...RQP KYEDRMKMPY TEAVIHEIQR FADLIPMGLA RRVTKDTKFR
CYP3A4   ALSDLELVAQ SIIFIFAGYE TTSSVLSFIM YELATHPDVQ QKLQEEIDAV LPNK...APP TYDTVLQMEY LDMVVNETLR LFPIAMR.LE RVCKKDVEIN
CYP5A1   PLTVDEIVGQ AFIFLIAGYE IITNTLSFAT YLLATNPDCQ EKLLREVD.V FKEK.HMAPE FCGLEEGLPY LDMVIAETLR MYPPAFR.FT REAAQDCEVL
CYP4A1   SLSDKDLRAE VDTFMFEGHD TTASGVSWIF YALATHPKHQ QRCREEVQSV LGDG...SSI TWDHLDQIPY TTMCIKEALR LYPPVPG.IV RELSTSVTFP
CYP102   PLDDENIRYQ IITFLIAGHE TTSGLLSFAL YFLVKNPHVL QKAAEEAARV LVDP....VP SYKQVKQLKY VGMVLNELLR LWPTAPA.FS LYAKEDTVLG
CYP19    DLTRENVNQC ILEMLIAAPD TMSVSLFFML FLIAKHPNVE EAIIKEIQTV IGE....RDI KIDDIQKLKV MENFIYESMR YQPVVDL.VM RKALEDDVID
```

```
          50β2-2  β1-3   αK'                                         αL                  β3-3    6β4-1
CYP101   G.VQLKKGDQ ILLPQMLSGL DERENA.CPM HVDFSRQKVS .......... ...HTTFGHG SHLCLGQHLA RREIIVTLKE WLTRIP..DF SIAPGAQIQH
CYP108   G.QNIKRGDR IMLSYPSANR DEEVFS.NPD EFDITRFPNR .......... ...HLGFGWG AHMCLGQHLA KLEMKIFFEE LLPKLK.... SVELSGPPRL
CYP11A1  D.YLIPAKTL VQVAIYAMGR DPAFFS.SPD KFDPTRWLSK DKDLIHFR.. ...NLGFGWG VRQCVGRRIA ELEMTLFLIH ILENFKV... EMQHIGDVDT
CYP27    GGFLFPKNTD FVFCHYVTSR DPSTFS.EPD TFWPYRWLRK GQPETSKTQH PFGSVPFGYG VRACLGRRIA ELLMAQLIQR YEL..... .MLAPETGEV
CYP21A2  G.YDIPEGTV IIPNLQGAHL DETVWE.RPH EFWPDRFLEP .GKNSR.... ...ALAFGCG ARVCLGEPLA RLELFVVLTR LLQAFTLLPS G.DALPSLQP
CYP17A   E.FTVPKDTH VVVNLWALHH DENEWD.QPD QFMPERFLDP TGSHL..ITP TQSYLPFGAG PRSCIGEALA RQELFVFTAL LLQRFDL... DVSSDKQLPR
CYP1A2   G.FHIPKECC IFINQWQVNH DEKQWK.DPF VFRPERFLTN DNTAIDKTL. SEKVMLFGLG KRRCIGEIPA KWEVFLFLAI LLHQLEF... TVPPGVKVDL
CYP1A1   G.FYIPKGHC VFVNQWQVNH DQELWG.DPN EFRPERFLTS SGT.LDKHL. SEKVILFGLG KRRCIGETIG RLKFLETIAR... NVSPQEKVDM
CYP2C11  N.YFIPKGTV VIVSLSSILH DDKEFPN.PE KFDPGHFLDE RGNFKK.... SDYFMPFSAG KRICAGEALA RTELFLFFTI ILQNFNL..K SLVDVKDIDT
CYP2C4   N.YFIPKGTD IITSLTSVLH DEKAFP.NPK VFDPGHFLDE SGNFKK.... SDYFMPFSAG KRMCVGEGLA RMELFLFLTS ILQNFKL..Q SLVEPKDLDI
CYP2E1   G.YVIPKGTV VIPTLDSLLY DKQEFP.DPE KFKPEHFLNE EGKFKY.... SDYFKPFSAG KRVCVGEGLA RMELFLLLSA ILQHFNL..K PLVDPEDIDL
CYP2B1   G.YLLPKNTQ VYPILSSALH DPQYFDH.PD SFNPEHFLDA NGALKK.... SEAFMPFSTG KRICLGEGIA RNELFLFFTT ILQNFSV..S SHLAPKDIDL
CYP2A4   D.FLLPKGTE VFPMLGSVLK DPKFFSN.PK DFNPKHFLDD KGQFKK.... SDAFVPFSIG KRYCFGEGLA RMELFLFLTN IMQNFHF..K STQAPQDIDV
CYP3A4   G.MFIPKGWV VMIPSYALHR DPKYWT.EPE KFLPERFSKK NKDNID.... PYIYTPFGSG PRNCIGMRFA LMNMKLALIR VLQNFSF... KPCKETQIPL
CYP5A1   G.QRIPAGAV LEMAVGALHH DPEHWP.SPE TFNPERFTAE AR....QQHR PFTYLPFGAG PRSCLGVRLG LLEVKLTLLH VLHKFRF..Q ACPETQVPLQ
CYP4A1   DGRSLPKGIQ VTLSIYGLHH NPKVWP.NPE VFDPSRFAPD SPRH...... SHSFLPFSGG ARNCIGKQFA MSEMKVIVAL TLLRFEL..L PDPTKVPIPL
CYP102   GEYPLEKGDE LMVLIPQLHR DKTIWGDDVE EFRPERFENP SAIP...... QHAFKPFGNG QRACIGQQFA LHEATLVLGM MLKHFDF... EDHTNYELDI
CYP19    G.YPVKKGTN IILNIGRMHR LE..FFPKPN EFTLENFAKN .VPYR..... ...YFQPFGFG PRGCAGKYIA MVMMKAILVT LLRRFHVKTL QGGCVESIQK
```

```
             601        β4-2    β3-2                 649
CYP101   KS...GIVSG VQ.ALPLVWD PATTKAV... .......... ........
CYP108   VATN..FVGG PK.NVPIRFT KA....... .......... ........
CYP11A1  IFN...LILT PD.KPIFLVF RPFNQDPPQA .......... ...
CYP27    QSVA..RIVL VP.NKKVGLR FLPTQR.... .......... ...
CYP21A2  LPHCS.VILK MQ.PFQVRLQ PRGMGAHSPG QNQ....... ..
CYP17A   LEGDPKVVFL ID.PFKVKIT VRQAWMDAQA EVST...... ....
CYP1A2   TPSY.GLTMK PRTCEHVQAW PRFSK...... .
CYP1A1   TPAY.GLTLK HARCEHFQVQ MRSSGPQHLQ A...
CYP2C11  TPAISGFGHL PP.FYEACFI PVQRADSLSS HL........ ..........
CYP2C4   TAVVNGFVSV PP.SYQLCFI PI....... .......... ...
CYP2E1   RNITVGFGRV PP.RYKLCVI PRS....... .......... ...
CYP2B1   TPKESGIGKI PP.TYQICFS AR....... .......... ...
CYP2A4   SPRLVGFVTI PP.TYTMSFL SR........ .......... ...
CYP3A4   KLSLGGLLQP EK.PVVLKVE SRDGTVSGA. .......... ...
CYP5A1   LES..KSALG PKNGVYIKIV SR........ .......... ...
CYP4A1   PR....LVLK SKNGIYLYLK KLH....... .......... .
CYP102   KET...LTLK PE.GFVVKAK SKKIPLGGIP SPSTEQSAKK VR
CYP19    IHD...LSLH PDETKNMLEM IFTPRNSDRC LEH....... .....
```

FIGURE 11. (*continued*)

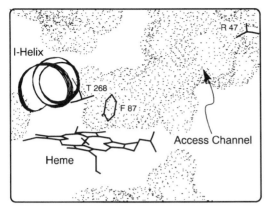

FIGURE 12. Substrate access channel in P450$_{BM-P}$ (molecule B). A Connolly surface with a 1.4-Å-diameter ball was constructed for P450$_{BM-P}$ with the program InsightII from Biosym, Inc. on a Silicon Graphics Indigo workstation. The representation of this "solvent accessible surface" was then constructed with a slab approximately 10 Å thick vertically through the molecule. The stippled areas are those parts of the molecule to which "solvent" has access while those areas of protein are shown as clear. Because of the lack of depth cueing in this figure, the access channel in the middle of the figure and the approach of solvent to the proximal face of this protein under the heme ring, appear to be quite close. In reality they are in slightly different planes and are separate. This solvent access channel to the heme propionate from the proximal face of the molecule is discussed under heme propionate binding above. The disclaimer regarding distance of approach of solvent is not true of solvent to the active site over the heme ring as indicated by the stippling over the I-helix. This is the region of the molecule into which E267 of P450$_{BM-P}$ protrudes and which is discussed later in this review under proton delivery to the active site. A ribbon drawing of the I-helix is shown at the left for reference. The heme ring is shown by the stick drawing at the lower left.

clear that there is a large cavity from the surface, in the upper right of this figure, down to the heme in the lower left. This cavity is approximately 10 Å in diameter at the mouth and extends for 20 Å to the heme iron. The access channel narrows as it approaches the heme surface such that at the heme iron there is space only for molecules the size of straight-chain aliphatic hydrocarbons (e.g., the ω-end of a fatty acid). Accommodation of other molecules in this substrate binding site will require movement of F87 from its position perpendicular to the heme ring to a position more parallel to the ring. The cavity is lined either with the side chains of hydrophobic amino acids or the carbon backbone of strands of the β1 sheet. The only polar residues present in the cavity are R47 seen at the mouth in the upper right and E52 which is just below the guanidino group of R47 and not seen in this representation.

Since fatty acids of about 20 Å in length are the optimal substrates for this enzyme,[40,41] it was tempting to propose that the carboxylate of the fatty acid would be anchored to R47 and the fatty acid then would extend down the access channel to the heme iron at the active site. Methylation of the fatty acid substrate has been shown to completely abolish the oxidation of either palmitic[40] or arachidonic acid lending further support to the hypothesis that the arginine residue should play an important role in orientation of the fatty acid in the substrate access channel. When the R47A and R47E mutants of P450$_{BM-3}$ were examined for their ability to oxidize either palmitic acid or arachidonic acid, we found that the activity was only decreased to 80 and 0%, respectively. Although R47 apparently does play a role in substrate binding as shown by these results, other factors must play important roles in substrate recognition and binding.

5.2. The Substrate Recognition/Docking Site

There is a hydrophobic region on the surface of $P450_{BM-P}$ (Fig. 13) which is immediately adjacent to the "mouth" of the substrate access channel described above. The residues which make up this hydrophobic surface are F11, L14, L17, P18, and L19. On one edge of this region, there are two lysine residues, K10 and K16. In Fulco's initial studies of the trypsinolysis of $P450_{BM-3}$[39] he noted that there were three sites of cleavage in the absence of myristic acid rather than the single site in the presence of the fatty acid at K472, which separated the P450 and reductase domains of the protein.[39] These additional cleavage sites were at K10 and K16. He noted also that the addition of fatty acid substrates to the purified hemoprotein domain did not give typical substrate binding spectra if the protein had been cleaved at K10 or K16. He stated that "1 or more residues of the first nine N-terminal amino acids of this protein are intimately involved in substrate binding."[39] As can be seen from Fig. 13, cleavage and removal of amino acids NH_2-terminal to either K10 or K16 would disrupt this substrate recognition/docking region and thus substrate binding. Therefore, we proposed a role for a hydrophobic patch on the surface of $P450_{BM-P}$ as a substrate recognition/docking region[47] assisted by the positive charges on these NH_2-terminal lysine residues. Fatty acids are probably attracted to the positive charges and will bind more readily to this region by the hydrophobic effect. The hydrophobic effect will then drive the ω-end of the fatty acid into the hydrophobic substrate access channel, thereby extruding water molecules from this region of the protein. This scenario for substrate binding is consistent with the experimental facts for $P450_{BM-P}$ regarding trypsinolysis, substrate specificity, and mutagenesis. We have also identified a hydrophobic patch at the mouth of the substrate access channel in $P450_{terp}$ which could function in substrate recognition/docking.[48] The identification of a hydrophobic substrate recognition/docking region on the surface of $P450_{BM-P}$ and $P450_{terp}$ has helped to clarify a long-standing question in P450 structure and function which was raised by the structure of $P450_{cam}$, that is, how is a hydrophobic substrate first "recognized" on the surface of a hydrophilic protein? Poulos has stated that $P450_{cam}$ must open up to allow access of camphor into the active site[29] and in a recent study of the atomic structure of inhibitor complexes, a

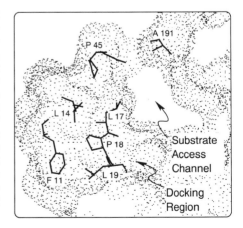

FIGURE 13. The substrate recognition/docking region of $P450_{BM-P}$. The Connolly solvent accessible surface of $P450_{BM-P}$ (molecule B) was constructed as described in Fig. 12. This is a 10-Å-thick representation of the molecule looking from the solvent back toward the molecule. The hydrophobic surface patch discussed in the text is shown at the left with the various amino acid structures shown by their stick figures. Because of the slab thickness, the substrate access channel is shown as a hole with no stippling. Again because of the thickness of the slab, we have sliced into the F–G loop at the upper right and there is no stippling because we are now inside the molecule. The solvent (outside) of the molecule is in the areas of no stippling at the upper and lower left.

hydrophobic patch at the mouth of the presumed substrate access channel was identified in this protein.[68]

5.3. Substrate Binding Site

In the structure of camphor-bound $P450_{cam}$ and in the models of the substrate-bound forms of both $P450_{BM-P}$ and $P450_{terp}$, substrate-contact residues are derived from amino acids in the B′ helix and from the loops on either side of it, from the I-helix, and from β4 sheet and β1–4 strand.[49] Initially modeling studies of eukaryotic P450s were based on $P450_{cam}$ and were focused on the I-helix in these comparisons primarily because this was one of the regions of highest homology between the proteins. A model for the I-helix region of $P450_{arom}$ has been constructed and tested by site-directed mutagenesis.[69] The data indicate that E302 plays an important role in the $P450_{arom}$ catalytic center. Other model-building efforts have been more global in nature with attempts to build a complete P450 based on the atomic coordinates of $P450_{cam}$.[56,70] Some studies have constrained the structure of their P450 to have essentially the same Cα coordinates as those of $P450_{cam}$.[58,71] Several investigators have focused their efforts to define the active sites of eukaryotic P450s on residues around the B′ helix.[72] However, in our examination of the location of these two loops (i.e., B–B′ loop and B′–C loop) and the B′-helix in the three proteins, as shown in Fig. 14, we have found that the position and orientation in three-dimensional space for these structural elements is different for each protein. Thus, one cannot state *a priori* which of these proteins is the most appropriate model for eukaryotic P450s in this region. The best model might be some composite which takes into account the position of the B′-helix in the alignment, the length of the helical element, and the number of residues on either side of the helix in the coil. We have been using this approach in our construction of a new model for $P450_{arom}$.[61] We cannot think of any way of assigning a residue to a position in space without this type of careful consideration in each and every instance.

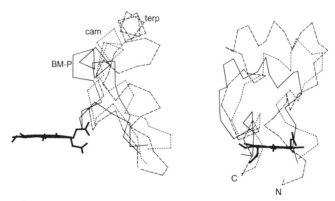

FIGURE 14. B′-helix region of $P450_{BM-P}$, $P450_{terp}$, and $P450_{cam}$. The atomic coordinates of the heme rings of $P450_{BM-P}$, $P450_{terp}$, and $P450_{cam}$ were superimposed and the lines connecting the Cα, C, and N atoms of the peptide backbones of these molecules were displayed from the end of the B-helix to the beginning of the C-helix in each case. The NH₂- and COOH-terminal residues are labeled **N** and **C**, respectively. The heme ring is shown edge-on by the heavy dark lines. The right panel was rotated by 90°. $P450_{BM-P}$, $P450_{cam}$, and $P450_{terp}$ are represented by solid, dotted, and dashed lines, respectively.

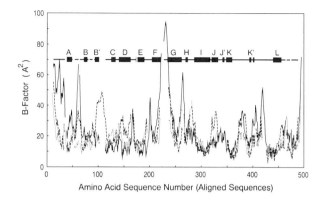

FIGURE 15. *B* factor plot for P450$_{BM-P}$. The temperature factors from the PDB file for molecule B of P450$_{BM-P}$,[47] P450$_{cam}$, and P450$_{terp}$ are plotted on the ordinate versus the number of the amino acid from the sequence of the protein plotted on the abscissa. P450$_{BM-P}$, P450$_{cam}$, and P450$_{terp}$ are indicated by solid, dotted, and dashed lines, respectively. The alignment of these three proteins is from the composite alignment.[49] For reference, the helical secondary structural elements of P450$_{BM-P}$ are given with bars in the upper portion of the figure.

5.4. B-Factor and Substrate Binding

Throughout this review, we have dwelt on the positions in three-dimensional space for atoms in each of these proteins, but it is well to remind ourselves that the assignment of coordinates for a particular atom contains a degree of uncertainty which is referred to as the structure factor or temperature factor *B*. This term can be thought of as being inversely related to the probability of an atom being located at the assigned coordinates at any given time. That is, as the value for *B* increases, the probability decreases. A plot of the structure factor *B* versus the linear sequence of these proteins is shown in Fig. 15. This uncertainty is most clearly reflected in P450$_{terp}$ in the region between the F- and G-helices in which the temperature factor increases at the carboxy-terminus of the F-helix and the amino-terminus of the G-helix, and in fact, there is no recorded *B* factor or coordinates for 18 amino acids between these helices. The molecule is so disordered in this region that there is no clear path for the α-carbon backbone through the observed electron density. Thus, as can be seen, the temperature factor of these molecules is important when considering the precise location of secondary structural elements. In those regions of the molecules which are in loops or on the surface, the structure is frequently very flexible with a high *B* factor. This flexibility becomes important when trying to model a new P450 based on the structure of an existing protein since those regions with high temperature factors are frequently implicated in substrate docking. Thus, not only are adequate sequence alignments difficult, but the position in three-dimensional space of the backbone of the model protein is uncertain.

6. Redox Partner Interactions

As discussed in the Introduction, there are three classes of P450s, two of which interact with specific redox partners. The third class does not require an external source of electrons for the reactions which they catalyze. How the interaction with redox partners occurs is a matter of some debate with the two major possibilities being either through electrostatic control[73,74] or hydrophobic interactions.[75]

Kinetic studies in our laboratory of the interaction of putidaredoxin with $P450_{cam}$ indicated that protein/protein interaction followed by electron transfer was at a minimum a two-phase process with the first phase of binding being electrostatic, and the second phase of binding as well as electron transfer probably being mediated by hydrophobic interactions.[23] With the availability of the structure of $P450_{cam}$, Sligar and his colleagues proposed that putidaredoxin approaches $P450_{cam}$ from the proximal face with the interaction being guided by charge coupling between arginine and lysine residues on the P450 and carboxylate groups on putidaredoxin.[76,77] Yet, in heterologous mixing experiments, the examination of various electron donors has shown that putidaredoxin cannot be substituted for linredoxin in the $P450_{lin}$ system,[78] while with $P450_{terp}$ putidaredoxin will substitute for terpredoxin but only with 10% of the maximal rate.[44] Thus, it is of interest to compare the charge distribution on the proximal faces of Class I P450s, i.e., $P450_{cam}$ and $P450_{terp}$, and to contrast them with the Class II P450, $P450_{BM-P}$. The charge distribution in the two Class I proteins, i.e., $P450_{cam}$ and $P450_{terp}$, is similar with four arginine residues (R72, R113, R364, and R365, for $P450_{cam}$) distributed around the hydrophobic patch immediately over the heme ring. In contrast, the hydrophobic patch immediately over the heme of $P450_{BM-P}$ is in a recessed groove on this face with the rim of the groove having numerous charged residues (K97, K113, K306, K309, K312, and K391) which may participate in docking the flavodoxin-like[79] region of the reductase domain to the P450. Two of these residues, K306 and K309, come from the J'-helix which is an insertion present in $P450_{BM-P}$ and which appears to be characteristic of eukaryotic P450s based on sequence alignments.

As was seen above, the distribution of charged residues on the proximal face of these P450s makes this face slightly positive, while the overall charge on these proteins is negative with isoelectric points of 5.05, 5.02, and 5.62 for $P450_{cam}$, $P450_{terp}$, and $P450_{BM-P}$ respectively. Calculation of the molecular dipole of these molecules using the program GRASP[80] indicates that there is a pronounced negative charge on the distal face.[49] While it will be very difficult to prove that this asymmetric charge distribution plays an important role in the catalytic cycle of P450s, the interaction with redox partners should be facilitated by this charge distribution. Putidaredoxin and terpredoxin both carry a strong negative charge with pI values of 4.16 and 3.38, respectively, and close approach of these molecules to the distal face will be disfavored by charge repulsion.

Thromboxane synthase CYP5A1, a Class III P450, is unusual among the P450 superfamily in that it catalyzes the synthesis of thromboxane A2 from the prostaglandin endoperoxide[81] without the need for an external source of electrons. With the determination of the nucleotide sequence encoding this protein,[82] it was found that the conserved ExxR pair in the K-helix and the residues of the cysteinyl heme-binding loop were present, thus indicating that this protein belonged to the P450 gene superfamily. Other characteristic

FIGURE 16. H-bond network in the I-helix of P450$_{BM-P}$. The Cα, C, and N atoms of the residues of the I-helix in the vicinity of the heme ring are represented with light lines. The atoms of the heme ring are displayed with heavy lines and potential hydrogen bond interactions are indicated by dotted lines. Some of the side chain atoms of residues in this region are also indicated. The crystallographically ordered water molecules immediately above the heme iron and in the I-helix are indicated with crosses. The conserved acidic amino acid E267 which immediately precedes the conserved threonine, T268, is shown at the top of the helix. E267 is exposed to bulk solvent in this molecule.

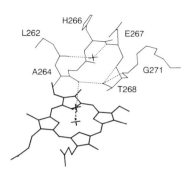

amino acid residues were also present, such as the arginine in the C-helix which serves to neutralize the charge of the heme propionate. In contrast, however, the conserved threonine of the I-helix is replaced by an isoleucine even though the adjacent residues, which cause the I-helix bulge, are conserved. As seen in Fig. 11, there is a large insertion in the amino acid sequence of CYP5A1 relative to the other P450s just NH$_2$-terminal of the I-helix. One might suppose that the insertion of approximately 30 amino acid residues could modify the structure of the protein on this face forming a large loop. We believe that this insertion would be in the region between the strands of the β5 sheet (see Fig. 8) and at the edge of their proximal faces. One can only conjecture as to how these amino acids might wind around and cover the proximal face, altering redox partner interaction.

7. Proton Channels and the Charge Relay

As shown in Eq. (1) and discussed above, the monooxygenation reaction requires the input of both electrons and protons. The precise pathway for protons into the active site has been the subject of some speculation with two different routes having been proposed for P450$_{cam}$.[31,83] There seems to be a possible route for protons into the active site in P450$_{terp}$ and P450$_{BM-P}$ which is somewhat different in detail than either of those proposed for P450$_{cam}$.[49] As can be seen in Fig. 16, there is a potential hydrogen-bond network in P450$_{BM-P}$ from T268 to the distal surface of the molecule. Although all three P450s have a bulge in the I-helix over the heme ring, the bulge in P450$_{cam}$ is caused by a disruption of the helical hydrogen-bond network with the insertion of a crystallographically ordered water molecule into the I-helix on the side distal to the heme ring.[47,48] In P450$_{terp}$ and P450$_{BM-P}$ an ordered water molecule is inserted into the I-helix on the side proximal to the heme ring. Whether this difference in position has any significance for the catalytic mechanism of these proteins is unclear at this time.

8. Extrapolation from Bacterial P450s to Eukaryotic Membrane-Bound P450s

The relationship between the structure of soluble, bacterial P450s and membrane-bound eukaryotic P450s has drawn an enormous amount of interest because of the crucial role which these enzymes play in steroid hormone metabolism, drug metabolism, and carcinogenesis. Subsequent to the membrane anchor, the sequences of eukaryotic P450s can be aligned with prokaryotic proteins starting with the A-helix as discussed above. Although the structural similarity among the bacterial P450s is not proof that eukaryotic P450s have similar structures, we believe that they do because of certain signature sequences and residues scattered throughout the length of the proteins. It is well to remember that the dendrogram of P450s indicates that $P450_{terp}$ and $P450_{cam}$ are members of different gene families with only 27% sequence identity, and that they are both very distantly related to $P450_{BM-P}$ with only 7% identity in optimum alignments. Alignments of eukaryotic P450s with these proteins are not appreciably more difficult than the alignment of $P450_{cam}$ and $P450_{terp}$ with $P450_{BM-P}$ as previously discussed.

8.1. Membrane Attachment and Substrate Access

When the amino acid sequences of the P450 gene superfamily are aligned, the eukaryotic P450s have an NH_2-terminal extension which is not present in any bacterial P450 sequenced to date[1] (see Fig. 11). In microsomal P450s, this extension is the membrane anchor and is 20 to 30 amino acids in length, composed largely of hydrophobic amino acids which terminate with a series of positive charges. Frequently this leader also has a conserved sequence, PPGP, at the carboxyl end of the hydrophobic leader[84] which may act as a hinge region between the leader and body of the protein. The hydrophobic leader sequence of eukaryotic P450s is attached at the NH_2-terminus of the triangular-shaped P450s adjacent to the β-sheet-rich region. If this leader is inserted into a membrane, as is illustrated in cartoon form in Fig. 17, the β-domain will then be adjacent to the surface of the membrane. This is important because the hydrophobic substrate docking region described above which is in the β-domain at the NH_2-terminus of the molecule would be positioned at the surface of the membrane or embedded in the membrane itself. Additionally as indicated in Fig. 17, the hydrophobic substrate access channel discussed above also would be positioned near the membrane surface so that hydrophobic compounds dissolved in the membrane could enter the P450 access channel directly from the membrane. This is consistent with the studies of the hydrophobic effect on a series of drug substrates which are oxidized by microsomal P450s[85–87] in which it was concluded that once dissolved in the membrane, hydrophobic substrates entered the active site of P450s where they were oxidized without having to enter the solvent. Additional evidence to support this arrange-

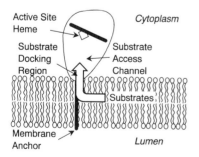

FIGURE 17. Model of eukaryotic microsomal p450s. The triangular shape of a hypothetical P450 is shown in this figure with the I-helix and the heme ring in the molecule for reference. The fatty acids of the membrane are shown by the wavy lines. The cytoplasmic and luminal faces of the membrane are indicated. The hydrophobic leader sequence (membrane anchor) is indicated by the bar which protrudes from the P450 into the membrane. The large arrow in the membrane into the P450 indicates the postulated path for hydrophobic substrates of these P450s.

ment of eukaryotic P450s on the membrane comes from the fluorescence energy transfer studies of Centeno and Gutierrez-Merino[88] and the antipeptide antibody studies of Boobis's group.[89,90] Centeno and Gutierrez-Merino measured the distance between the heme of P450 and the surface of the membrane. They calculated that the heme of P450 is approximately 50 Å from the surface of the membrane which is consistent with the cartoon structure based on soluble bacterial P450s shown in Fig. 17. Finally, Boobis's results indicated that the enzyme should be oriented so that the redox partner binding region was exposed and that the heme ring was perpendicular to the plane of the membrane.

8.2. Redox Partner Interaction

In addition to interacting with the membrane surface, P450s must interact with and receive electrons from their redox partners. As has been demonstrated by several groups, there seems to be only one NADPH-P450 reductase gene in any given organism.[91] Because there is only one gene for NADPH-P450 reductase, this protein must be able to interact with and reduce a variety of P450s. Thus, the affinity between NADPH-P450 reductase and a P450 may reflect the different number of charges depending on the particular P450 gene family.[92]

Because of the interest in the factors which control this interaction, i.e., whether they are electrostatic[73] or hydrophobic,[75] we have examined the charge distribution on the residues located on the proximal face of eukaryotic P450s. We used the composite alignment of Nelson et al.[1] for this purpose since it is quite good regarding those elements in the core structure. In ascertaining whether there was a charge signature in microsomal versus mitochondrial P450s, we selected only mammalian P450s for which we could find a definitive reference to a subcellular localization. As can be seen in Fig. 18, some residues have the same charge in both mitochondrial and microsomal P450s. Because of the evolutionary and functional diversity of these proteins, these conserved charges may indicate that these residues have a structural role in the P450 fold. Examples of these residues are 539 and 542, which are negatively and positively charged respectively, that is, the ExxR pair in the K-helix which was discussed above. Residues 220 and 223 in the C-helix are both positively charged. In P450$_{BM-P}$ these residues are His and Arg, respectively, and are hydrogen-bonded to one of the heme propionates. Residue 650 is positively charged and is an Arg which charge neutralizes the other heme propionate in some P450s.

Because these residues which appear to have a structural role in P450s were aligned and consistent with the structural analysis, the other charged residues on the proximal face were examined. Wada and Waterman mutated two lysine residues (K377 and K381) in P450$_{scc}$ (CYP11A), which is a Class I, mitochondrial protein.[93] In the composite sequence alignment, these residues are positively charged in the K-helix at positions 534 and 538 and are solvent exposed on the proximal face of the protein in the redox-partner interacting region. Mutations of K377 and K381 significantly decreased the K values for adrenodoxin binding to P450$_{scc}$ from 150- to 600-fold depending on the mutation. Shaw and Strobel chemically modified lysine residues in CYP1A1 which they believed might be involved in redox partner interactions.[74] One of the residues they identified corresponded to 156 in the B-helix, which is positively charged in microsomal P450s and negatively charged in mitochondrial P450s. There are several other interesting charged residues which stand out

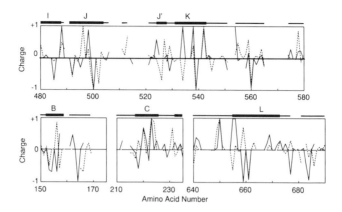

FIGURE 18. Charge distribution in microsomal and mitochondrial P450s. The charge distribution for approximately 100 microsomal and 17 mitochondrial P450s from the aligned sequences of Nelson *et al.*[1] was calculated by assigning a value of +1 to arginine and lysine residues, +0.5 to histidine residues, and –1 to aspartate and glutamate residues. All other residues were assigned a value of 0. The sum of all values for any position was divided by the number of sequences being used for the comparison, e.g., 17 for mitochondrial and 104 for microsomal P450s. Since many gaps have been inserted in this sequence alignment, the total number of amino acids is much greater than the expected 500 for a eukaryotic P450. To assist in recognizing regions of the molecules of interest, the helical secondary structural elements of P450$_{BM-P}$ are given as bars over the various panels. Mitochondrial P450s are shown by the solid lines while microsomal P450s are shown by the dashed lines.

in mitochondrial P450s such as the positively charged pair of residues at the beginning of the L-helix which are probably solvent exposed and very near the heme. We await the experimental test of the role of these and other residues in redox partner interaction.

In this analysis, we were struck by the consistency of the charge distribution on mitochondrial P450s and by the lack of consistency in microsomal P450s. To determine whether this inconsistency was a function of microsomal P450s in general or of differences in charged residues between different families which might modulate their interactions with the reductase,[92] we further subdivided the microsomal P450 family and examined several subfamilies for charge distribution as described above (Fig. 19). In P450$_{BM-P}$ the residues on either side of 650, which charge neutralizes one of the heme propionates, are solvent exposed and uncharged, and might be expected to be solvent exposed in eukaryotic P450s. In the CYP1A family, both of these residues are positively charged; in the CYP2A family, only one of these residues is charged; and in the CYP17 family, neither of these residues is charged. Also shown is residue 526 which is usually positively charged in these P450 families. This residue has been modified in CYP17 to an alanine with the surprising finding that while there was no difference in 17α-hydroxylase activity, the 17,20-lyase activity of the enzyme was abolished.[94] The relationship between this mutation and its effect on electron transfer and product formation by this enzyme may be important in unraveling the mode of action of the multifunctional P450s which catalyze several rounds of monooxygenation prior to product release.[95]

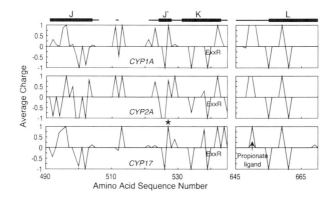

FIGURE 19. Charge distribution on microsomal P450 families. The charge distribution on selected P450 subfamilies which are indicated in the figure were calculated as described in Fig. 18. The helical secondary structural elements of P450$_{BM-P}$ are given over the panels for comparison. The positions of the conserved ExxR and heme propionate ligand are indicated in the panels. The asterisk above the CYP17 panel indicates the R346 residue which when mutated results in loss of 17,20-lyase activity with only nominal effects on 17-hydroxylase activity catalyzed by this enzyme.[94] We would predict that this arginine residue is solvent exposed and on the proximal face in the redox partner binding region of this family of proteins as discussed in the text.

9. Conclusions

In conclusion, we believe that all members of the P450 gene superfamily will have fundamentally the same secondary structural elements. This hypothesis will be tested further with the structural determination of the two P450s whose crystallization and preliminary X-ray structural analyses were reported recently.[96,97] Outside of the structural core, the spatial relationship between these elements will vary from one P450 to another depending on the substrate and redox partner specificity. Careful modeling of eukaryotic P450s based on the known structures should be helpful in identifying those residues which might be important in substrate and redox partner interactions. Although site-directed mutagenesis has been extensively applied to studies of the structure and function of P450s, we must remember that practicing site-directed mutagenesis can be likened to throwing hand grenades into a crowded area. Without determining the three-dimensional structure of the wild-type and mutant enzymes, it will be very difficult to draw firm conclusions regarding the structural origin of the changes in substrate specificity resulting from those mutations.

References

1. Nelson, D. R., Kamataki, T., Waxman, D. J., Guengerich, F. P., Estabrook, R. W., Feyereisen, R., Gonzalez, F. J., Coon, M. J., Gunsalus, I. C., Gotoh, O., Okuda, K., and Nebert, D. W., 1993, The P450 superfamily: Update on new sequences, gene mapping, accession numbers, early trivial names of enzymes, and nomenclature, *DNA Cell Biol.* **12**:1–51.
2. Guengerich, F. P., 1991, Reactions and significance of cytochrome P-450 enzymes, *J. Biol. Chem.* **266**:10019–10022.

3. Foidart, A., de Clerck, A., Harada, N., and Balthazart, J., 1994, Aromatase-immunoreactive cells in the quail brain: Effects of testosterone and sex dimorphism, *Physiol. Behav.* **55:**453–464.

4. Jugert, F. K., Agarwal, R., Kuhn, A., Bickers, D. R., Merk, H. F., and Mukhtar, H., 1994, Multiple cytochrome P450 isozymes in murine skin: Induction of P450 1A, 2B, 2E, and 3A by dexamethasone, *J. Invest. Dermatol.* **102:**970–975.

5. Macica, C., Balazy, M., Falck, J. R., Mioskowski, C., and Carroll, M. A., 1993, Characterization of cytochrome P-450-dependent arachidonic acid metabolism in rabbit intestine, *Am. J. Physiol.* **265:**G735–G741.

6. Kochs, G., Werck-Reichhart, D., and Grisebach, H., 1992, Further characterization of cytochrome P450 involved in phytoalexin synthesis in soybean: Cytochrome P450 cinnamate 4-hydroxylase and 3,9-dihydroxypterocarpan 6a-hydroxylase, *Arch. Biochem. Biophys.* **293:**187–194.

7. Holton, T. A., Brugliera, F., Lester, D. R., Tanaka, Y., Hyland, C. D., Menting, J. G., Lu, C.-Y., Farcy, E., Stevenson, T. W., and Cornish, E. C., 1993, Cloning and expression of cytochrome P450 genes controlling flower colour, *Nature* **366:**276–279.

8. Kochs, G., and Grisebach, H., 1989, Phytoalexin synthesis in soybean: Purification and reconstitution of cytochrome P450 3,9-dihydroxypterocarpan 6a-hydroxylase and separation from cytochrome P450 cinnamate 4-hydroxylase, *Arch. Biochem. Biophys.* **273:**543–553.

9. Song, W. C., Funk, C. D., and Brash, A. R., 1993, Molecular cloning of an allene oxide synthase: A cytochrome P450 specialized for the metabolism of fatty acid hydroperoxides, *Proc. Natl. Acad. Sci. USA* **90:**8519–8523.

10. White, R. E., and Coon, M. J., 1980, Oxygen activation by cytochrome P-450, *Annu. Rev. Biochem.* **49:**315–356.

11. Narhi, L. O., and Fulco, A. J., 1986, Characterization of a catalytically self-sufficient 119,000-dalton cytochrome P-450 monooxygenase induced by barbiturates in Bacillus megaterium, *J. Biol. Chem.* **261:**7160–7169.

12. Hecker, M., Haurand, M., Ullrich, V., and Terao, S., 1986, Spectral studies on structure–activity relationships of thromboxane synthase inhibitors, *Eur. J. Biochem.* **157:**217–223.

13. Miyata, A., Hara, S., Yokoyama, C., Inoue, H., Ullrich, V., and Tanabe, T., 1994, Molecular cloning and expression of human prostacyclin synthase, *Biochem. Biophys. Res. Commun.* **200:**1728–1734.

14. Katagiri, M., Ganguli, B. N., and Gunsalus, I. C., 1968, A soluble cytochrome P-450 functional in methylene hydroxylation, *J. Biol. Chem.* **243:**3543–3546.

15. Unger, B. P., Gunsalus, I. C., and Sligar, S. G., 1986, Nucleotide sequence of the Pseudomonas putida cytochrome P-450cam gene and its expression in Escherichia coli, *J. Biol. Chem.* **261:**1158–1163.

16. Gunsalus, I. C., and Sligar, S. G., 1978, Oxygen reduction by the P450 monooxygenase systems, *Adv. Enzymol. Relat. Areas Mol. Biol.* **47:**1–44.

17. Estabrook, R. W., Hildebrandt, A. G., Remmer, H., Schenkman, J. B., Rosenthal, O., and Cooper, D. Y., 1968, Role of cytochrome P-450 in microsomal mixed function oxidation reactions, in: *Biochemie des Sauerstoffs* (B. Hess and H. Staudinger, eds.), Springer-Verlag, Berlin, pp. 142–177.

18. Peterson, J. A., 1971, Camphor binding by Pseudomonas putida cytochrome 450, *Arch. Biochem. Biophys.* **42:**140–146.

19. Griffin, B. W., and Peterson, J. A., 1975, Pseudomonas putida cytochrome P-450. The effect of complexes of the ferric hemeprotein on the relaxation of solvent water protons, *J. Biol. Chem.* **250:**6445–6451.

20. Griffin, B. W., and Peterson, J. A., 1972, Camphor binding by Pseudomonas putida cytochrome P-450. Kinetics and thermodynamics of the reaction, *Biochemistry* **11:**4740–4746.

21. Sligar, S. G., and Gunsalus, I. C., 1976, A thermodynamic model of regulation: Modulation of redox equilibria in camphor monoxygenase, *Proc. Natl. Acad. Sci. USA* **73:**1078–1082.

22. Sligar, S. G., and Gunsalus, I. C., 1979, Proton coupling in the cytochrome P-450 spin and redox equilibria, *Biochemistry* **18:**2290–2295.

23. Hintz, M. J., Mock, D. M., Peterson, L. L., Tuttle, K., and Peterson, J. A., 1982, Equilibrium and kinetic studies of the interaction of cytochrome P-450cam and putidaredoxin, *J. Biol. Chem.* **257:**14324–14332.

24. Peterson, J. A., Ishimura, Y., and Griffin, B. W., 1972, Pseudomonas putida cytochrome P-450: Characterization of an oxygenated form of the hemoprotein, *Arch. Biochem. Biophys.* **149:**197–208.

25. Ishimura, Y., Ullrich, V., and Peterson, J. A., 1971, Oxygenated cytochrome P-450 and its possible role in enzymic hydroxylation, *Biochem. Biophys. Res. Commun.* **42:**140–146.

26. Brewer, C. B., and Peterson, J. A., 1988, Single turnover kinetics of the reaction between oxycytochrome P-450cam and reduced putidaredoxin, *J. Biol. Chem.* **263:**791–798.

27. Peterson, J. A., and Mock, D. M., 1975, Metabolic control of cytochrome P450cam, in: *Cytochromes P450 and b5* (D. Y. Cooper, O. Rosenthal, R. Snyder, and C. Witmer, eds.), Plenum Press, New York, pp. 311–324.

28. Poulos, T. L., Finzel, B. C., Gunsalus, I. C., Wagner, G. C., and Kraut, J., 1985, The 2.6Å crystal structure of Pseudomonas putida cytochrome P-450, *J. Biol. Chem.* **260:**16122–16130.

29. Poulos, T. L., Finzel, B. C., and Howard, A. J., 1987, High-resolution crystal structure of cytochrome P450cam, *J. Mol. Biol.* **195:**687–700.

30. Groves, J. T., and McClusky, G. A., 1978, Aliphatic hydroxylation by highly purified liver microsomal cytochrome P-450. Evidence for a carbon radical intermediate, *Biochem. Biophys. Res. Commun.* **81:**154–160.

31. Raag, R., Martinis, S. A., Sligar, S. G., and Poulos, T. L., 1991, Crystal structure of the cytochrome P-450CAM active site mutant Thr252Ala, *Biochemistry* **30:**11420–11429.

32. Nelson, D. R., and Strobel, H. W., 1988, On the membrane topology of vertebrate cytochrome P-450 proteins, *J. Biol. Chem.* **263:**6038–6050.

33. Tretiakov, V. E., Degtyarenko, K. N., Uvarov, V. Y., and Archakov, A. I., 1989, Secondary structure and membrane topology of cytochrome P450s, *Arch. Biochem. Biophys.* **275:**429–439.

34. Nelson, D. R., and Strobel, H. W., 1989, Secondary structure prediction of 52 membrane-bound cytochromes P450 shows a strong structural similarity to P450cam, *Biochemistry* **28:**656–660.

35. Poulos, T. L., 1991, Modeling of mammalian P450s on basis of P450cam X-ray structure, *Methods Enzymol.* **206:**11–30.

36. Fulco, A. J., Kim, B. H., Matson, R. S., Narhi, L. O., and Ruettinger, R. T., 1983, Nonsubstrate induction of a soluble bacterial cytochrome P-450 monooxygenase by phenobarbital and its analogs, *Mol. Cell. Biochem.* **53–54:**155–161.

37. Kim, B. H., and Fulco, A. J., 1983, Induction by barbiturates of a cytochrome P-450-dependent fatty acid monooxygenase in Bacillus megaterium: Relationship between barbiturate structure and inducer activity, *Biochem. Biophys. Res. Commun.* **116:**843–850.

38. Fulco, A. J., 1991, P450BM-3 and other inducible bacterial P450 cytochromes: Biochemistry and regulation, *Annu. Rev. Pharmacol. Toxicol.* **31:**177–203.

39. Narhi, L. O., and Fulco, A. J., 1987, Identification and characterization of two functional domains in cytochrome P-450BM-3, a catalytically self-sufficient monooxygenase induced by barbiturates in *Bacillus megaterium, J. Biol. Chem.* **262:**6683–6690.

40. Miura, Y., and Fulco, A. J., 1975, Omega-1, Omega-2 and Omega-3 hydroxylation of long-chain fatty acids, amides and alcohols by a soluble enzyme system from Bacillus megaterium, *Biochim. Biophys. Acta* **388:**305–317.

41. Boddupalli, S. S., Estabrook, R. W., and Peterson, J. A., 1990, Fatty acid monooxygenation by cytochrome P-450BM-3, *J. Biol. Chem.* **265:**4233–4239.

42. Boddupalli, S. S., Pramanik, B. C., Slaughter, C. A., Estabrook, R. W., and Peterson, J. A., 1992, Fatty acid monooxygenation by P450BM-3: Product identification and proposed mechanisms for the sequential hydroxylation reactions, *Arch. Biochem. Biophys.* **292:**20–28.

43. Peterson, J. A., and Lu, J. Y., 1991, Bacterial cytochromes P450: Isolation and identification, *Methods Enzymol.* **206:**612–620.

44. Peterson, J. A., Lu, J. Y., Geisselsoder, J., Graham-Lorence, S., Carmona, C., Witney, F., and Lorence, M. C., 1992, Cytochrome P-450terp. Isolation and purification of the protein and cloning and sequencing of its operon, *J. Biol. Chem.* **267**:14193–14203.

45. Fruetel, J. A., Mackman, R. L., Peterson, J. A., and Ortiz de Montellano, P. R., 1994, Relationship of active site topology to substrate specificity for cytochrome P450terp (CYP108), *J. Biol. Chem.* **269**:28815–28821.

46. Boddupalli, S. S., Hasemann, C. A., Ravichandran, K. G., Lu, J. Y., Goldsmith, E. J., Deisenhofer, J., and Peterson, J. A., 1992, Crystallization and preliminary x-ray diffraction analysis of P450terp and the hemoprotein domain of P450BM-3, enzymes belonging to two distinct classes of the cytochrome P450 superfamily, *Proc. Natl. Acad. Sci. USA* **89**:5567–5571.

47. Ravichandran, K. G., Boddupalli, S. S., Hasemann, C. A., Peterson, J. A., and Deisenhofer, J., 1993, Crystal structure of hemoprotein domain of P450BM-3, a prototype for microsomal P450's, *Science* **261**:731–736.

48. Hasemann, C. A., Ravichandran, K. G., Peterson, J. A., and Deisenhofer, J., 1994, Crystal structure and refinement of cytochrome P450terp at 2.3Å resolution, *J. Mol. Biol.* **236**:1169–1185.

49. Hasemann, C. A., Ravichandran, K. G., Boddupalli, S. S., Peterson, J. A., and Deisenhofer, J., 1995, Structure and function of cytochromes P450: A comparative analysis of the three-dimensional structures of P450terp, P450cam, and the hemoprotein domain of P450BM-3, *Structure* **3**:41–62.

50. Jones, T. A., Bergdoll, M., and Kjeldgaard, M., 1990, O: A macromolecule modeling environment, in: *Crystallographic and Modeling Methods in Molecular Design* (C. E. Bugg and S. E. Ealick, eds.), Springer-Verlag, Berlin, pp. 189–199.

51. Chothia, C., and Lesk, A. M., 1987, The evolution of protein structures, *Cold Spring Harbor Symp. Quant. Biol.* **52**:399–405.

52. Presnell, S. R., and Cohen, F. E., 1989, Topological distribution of four-alpha-helix bundles, *Proc. Natl. Acad. Sci. USA* **86**:6592–6596.

53. Ortiz de Montellano, P. R., Kunze, K. L., and Beilan, H. S., 1983, Chiral orientation of prosthetic heme in the cytochrome P-450 active site, *J. Biol. Chem.* **258**:45–47.

54. Chen, S., and Zhou, D., 1992, Functional domains of aromatase cytochrome P450 inferred from comparative analyses of amino acid sequences and substantiated by site-directed mutagenesis experiments, *J. Biol. Chem.* **267**:22587–22594.

55. Edwards, R. J., Murray, B. P., Boobis, A. R., and Davies, D. S., 1989, Identification and location of alpha-helices in mammalian cytochromes P450, *Biochemistry* **28**:3762–3770.

56. Laughton, C. A., Neidle, S., Zvelebil, M. J., and Sternberg, M. J., 1990, A molecular model for the enzyme cytochrome P450(17 alpha), a major target for the chemotherapy of prostatic cancer, *Biochem. Biophys. Res. Commun.* **171**:1160–1167.

57. Laughton, C. A., and Neidle, S., 1990, Inhibitors of the P450 enzymes aromatase and lyase. Crystallographic and molecular modeling studies suggest structural features of pyridylacetic acid derivatives responsible for differences in enzyme inhibitory activity, *J. Med. Chem.* **33**:3055–3060.

58. Zvelebil, M. J., Wolf, C. R., and Sternberg, M. J., 1991, A predicted three-dimensional structure of human cytochrome P450: Implications for substrate specificity, *Protein Eng.* **4**:271–282.

59. Lewis, D. F., and Moereels, H., 1992, The sequence homologies of cytochromes P-450 and active-site geometries, *J. Comput. Aided Mol. Des.* **6**:235–252.

60. Zhou, D. J., Korzekwa, K. R., Poulos, T., and Chen, S. A., 1992, A site-directed mutagenesis study of human placental aromatase, *J. Biol. Chem.* **267**:762–768.

61. Amarneh, B., Corbin, C. J., Peterson, J. A., Simpson, E. R., and Graham-Lorence, S., 1993, Functional domains of human aromatase cytochrome P450 characterized by linear alignment and site-directed mutagenesis, *Mol. Endocrinol.* **7**:1617–1624.

62. Garnier, J., Osguthorpe, D. J., and Robson, B., 1978, Analysis of the accuracy and implications of simple methods for predicting the secondary structure of globular proteins, *J. Mol. Biol.* **120**:97–120.

63. Rost, B., Sander, C., and Schneider, R., 1994, PHD—An automatic mail server for protein secondary structure prediction, *Comput. Appl. Biosci.* **10**:53–60.

64. Rost, B., Schneider, R., and Sander, C., 1993, Progress in protein structure prediction? *Trends Biochem. Sci.* **18**:120–123.

65. Rost, B., and Sander, C., 1993, Prediction of protein secondary structure at better than 70% accuracy, *J. Mol. Biol.* **232**:584–599.

66. (1991) GCG Package. 7. 575 Science Drive, Madison, WI 53711.

67. Poulos, T. L., Finzel, B. C., and Howard, A. J., 1986, Crystal structure of substrate-free Pseudomonas putida cytochrome P-450, *Biochemistry* **25**:5314–5322.

68. Raag, R., Li, H., Jones, B. C., and Poulos, T. L., 1993, Inhibitor-induced conformational change in cytochrome P-450CAM, *Biochemistry* **32**:4571–4578.

69. Graham-Lorence, S., Khalil, M. W., Lorence, M. C., Mendelson, C. R., and Simpson, E. R., 1991, Structure–function relationships of human aromatase cytochrome P-450 using molecular modeling and site-directed mutagenesis, *J. Biol. Chem.* **266**:11939–11946.

70. Laughton, C. A., McKenna, R., Neidle, S., Jarman, M., McCague, R., and Rowlands, M. G., 1990, Crystallographic and molecular modeling studies on 3-ethyl-3-(4-pyridyl)piperidine-2,6-dione and its butyl analogue, inhibitors of mammalian aromatase. Comparison with natural substrates: Prediction of enantioselectivity for N-alkyl derivatives, *J. Med. Chem.* **33**:2673–2679.

71. Vijayakumar, S., and Salerno, J. C., 1992, Molecular modeling of the 3-D structure of cytochrome P-450scc, *Biochim. Biophys. Acta* **1160**:281–286.

72. Straub, P., Lloyd, M., Johnson, E. F., and Kemper, B., 1994, Differential effects of mutations in substrate recognition site 1 of cytochrome P450 2C2 on lauric acid and progesterone hydroxylation, *Biochemistry* **33**:8029–8034.

73. Shen, S., and Strobel, H. W., 1993, Role of lysine and arginine residues of cytochrome P450 in the interaction between cytochrome P4502B1 and NADPH-cytochrome P450 reductase, *Arch. Biochem. Biophys.* **304**:257–265.

74. Shen, S., and Strobel, H. W., 1992, The role of cytochrome P450 lysine residues in the interaction between cytochrome P450IA1 and NADPH-cytochrome P450 reductase, *Arch. Biochem. Biophys.* **294**:83–90.

75. Voznesensky, A. I., and Schenkman, J. B., 1992, The cytochrome P450 2B4–NADPH cytochrome P450 reductase electron transfer complex is not formed by charge-pairing, *J. Biol. Chem.* **267**:14669–14676.

76. Davies, M. D., and Sligar, S. G., 1992, Genetic variants in the putidaredoxin–cytochrome P-450cam electron-transfer complex: Identification of the residue responsible for redox-state-dependent conformers, *Biochemistry* **31**:11383–11389.

77. Stayton, P. S., Poulos, T. L., and Sligar, S. G., 1989, Putidaredoxin competitively inhibits cytochrome b5–cytochrome P-450cam association: A proposed molecular model for a cytochrome P-450cam electron-transfer complex, *Biochemistry* **28**:8201–8205.

78. Ullah, A. J., Murray, R. I., Bhattacharyya, P. K., Wagner, G. C., and Gunsalus, I. C., 1990, Protein components of a cytochrome P-450 linalool 8-methyl hydroxylase, *J. Biol. Chem.* **265**:1345–1351.

79. Porter, T. D., 1991, An unusual yet strongly conserved flavoprotein reductase in bacteria and mammals, *Trends Biochem. Sci.* **16**:154–158.

80. Nicholls, A., and Honig, B., 1993, *GRASP: Graphical Representation and Analysis of Surface Properties*, Columbia University, New York.

81. Hecker, M., and Ullrich, V., 1989, On the mechanism of prostacyclin and thromboxane A2 biosynthesis, *J. Biol. Chem.* **264**:141–150.

82. Yokoyama, C., Miyata, A., Ihara, H., Ullrich, V., and Tanabe, T., 1991, Molecular cloning of human platelet thromboxane A synthase, *Biochem. Biophys. Res. Commun.* **178**:1479–1484.

83. Gerber, N. C., and Sligar, S. G., 1992, Catalytic mechanism of cytochrome P450: Evidence for a distal charge relay, *J. Am. Chem. Soc.* **114**:8742–8743.

84. Yamazaki, S., Sato, K., Suhara, K., Sakaguchi, M., Mihara, K., and Omura, T., 1993, Importance of the proline-rich region following signal-anchor sequence in the formation of correct conformation of microsomal cytochrome P-450s, *J. Biochem.* **114**:652–657.

85. Al-Gailany, K. A., Houston, J. B., and Bridges, J. W., 1978, The role of substrate lipophilicity in determining type 1 microsomal P450 binding characteristics, *Biochem. Pharmacol.* **27**:783–788.

86. Ebel, R. E., O'Keeffe, D. H., and Peterson, J. A., 1978, Substrate binding to hepatic microsomal cytochrome P-450. Influence of the microsomal membrane, *J. Biol. Chem.* **253**:3888–3897.

87. Parry, G., Palmer, D. N., and Williams, D. J., 1976, Ligand partitioning into membranes: Its significance in determining Km and Ks values for cytochrome P-450 and other membrane bound receptors and enzymes, *FEBS Lett.* **67**:123–129.

88. Centeno, F., and Gutierrez-Merino, C., 1992, Location of functional centers in the microsomal cytochrome P450 system, *Biochemistry* **31**:8473–8481.

89. Edwards, R. J., Murray, B. P., Singleton, A. M., and Boobis, A. R., 1991, Orientation of cytochromes P450 in the endoplasmic reticulum, *Biochemistry* **30**:71–76.

90. Edwards, R. J., Singleton, A. M., Murray, B. P., Murray, S., Boobis, A. R., and Davies, D. S., 1991, Identification of a functionally conserved surface region of rat cytochromes P450IA, *Biochem. J.* **278**:749–757.

91. Porter, T. D., Beck, T. W., and Kasper, C. B., 1990, NADPH-cytochrome P-450 oxidoreductase gene organization correlates with structural domains of the protein, *Biochemistry* **29**:9814–9818.

92. Voznesensky, A. I., and Schenkman, J. B., 1994, Quantitative analyses of electrostatic interactions between NADPH-cytochrome P450 reductase and cytochrome P450 enzymes, *J. Biol. Chem.* **269**:15724–15731.

93. Wada, A., and Waterman, M. R., 1992, Identification by site-directed mutagenesis of two lysine residues in cholesterol side chain cleavage cytochrome P450 that are essential for adrenodoxin binding, *J. Biol. Chem.* **267**:22877–22882.

94. Kitamura, M., Buczko, E., and Dufau, M. L., 1991, Dissociation of hydroxylase and lyase activities by site-directed mutagenesis of the rat P45017 alpha, *Mol. Endocrinol.* **5**:1373–1380.

95. Grogan, J., Shou, M., Zhou, D., Chen, S., and Korzekwa, K. R., 1993, Use of aromatase (CYP19) metabolite ratios to characterize electron transfer from NADPH-cytochrome P450 reductase, *Biochemistry* **32**:12007–12012.

96. Nakahara, K., Shoun, H., Adachi, S., Iizuka, T., and Shiro, Y., 1994, Crystallization and preliminary X-ray diffraction studies of nitric oxide reductase cytochrome P450nor from Fusarium oxysporum, *J. Mol. Biol.* **239**:158–159.

97. Cupp-Vickery, J. R., Li, H., and Poulos, T. L., 1994, Preliminary crystallographic analysis of an enzyme involved in erythromycin biosynthesis: Cytochrome P450eryF, *Proteins* **20**:197–201.

98. Kraulis, J., 1991, MOLSCRIPT: A program to produce both detailed and schematic plots of protein structures, *J. Appl. Crystallogr.* **24**:946–950.

Part III

Structures and Mechanisms of Membrane-Bound P450 Enzymes

Structures of Eukaryotic Cytochrome P450 Enzymes

CLAES VON WACHENFELDT and ERIC F. JOHNSON

1. Introduction

A three-dimensional structure will facilitate our understanding of the relationship between structure and substrate specificity within and between different eukaryotic members of the cytochrome P450 superfamily. Such information would provide insights into the structural arrangement of these proteins that would assist in the rational design of therapeutic enzyme inhibitors or in the engineering of P450s as biotechnological tools.

Although the substrate binding sites of the individual cytochrome P450 monooxygenases are likely to be distinct, these enzymes probably exhibit a common structural framework. One of the axial ligands for the heme prosthetic group is provided by the thiol moiety of a conserved cysteine residue that coordinates the heme iron. The unique properties of the thiolate ligand contribute to the spectral and catalytic properties of P450 enzymes. P450s exhibit a conserved amino acid sequence motif (FXXGXXXCXG) in the C-terminal region of the protein that surrounds this cysteine residue.[1-3] Oxygen reduction and bond scission are thought to be catalyzed solely by the heme prosthetic group which is likely to be bound by each enzyme in a similar fashion.[4] In addition, the protoporphyrin ring is probably sequestered in the interior of P450 proteins, which limits its participation in peroxidase reactions.[5] This arrangement distinguishes P450s from other heme enzymes which normally have a more exposed heme group relative to the surface of the protein. Moreover, it has been suggested, based on amino acid sequence comparisons, that all P450 genes have diverged from a common ancestral gene.[6] This, together with the basic structural similarities exhibited by the bacterial P450 enzymes 101, 102, and 108[7] (see Chapters 4 and 5) for which the three-dimensional structures are known, suggests that a

CLAES VON WACHENFELDT and ERIC F. JOHNSON • Division of Biochemistry, Department of Molecular and Experimental Medicine, The Scripps Research Institute, La Jolla, California 92037.

Cytochrome P450: Structure, Mechanism, and Biochemistry (Second Edition), edited by Paul R. Ortiz de Montellano. Plenum Press, New York, 1995.

common structural framework is highly probable for the members of the P450 superfamily as a whole.

The eukaryotic P450s differ significantly in one respect from the prokaryotic enzymes in that, with a few exceptions, they are bound either to the endoplasmic reticulum or to the inner membrane of mitochondria. The general difficulties involved in producing well-ordered three-dimensional crystals of membrane proteins have precluded the use of standard X-ray crystallographic methods. However, other indirect methods have made substantial contributions to our understanding of the structure of these proteins. A considerable body of indirect evidence derived from comparative sequence analysis, studies of chimeric proteins, site-directed mutagenesis, and three-dimensional homology modeling suggests that these enzymes share several structural features with the prokaryotic enzymes.

In this chapter, we will review not only the experimental evidence that suggests a common architecture for the eukaryotic P450s but also studies that have defined the structural features of the eukaryotic P450s that contribute to membrane binding.

2. Substrate Binding Sites

2.1. Identification of Amino Acid Determinants of Substrate Metabolism by Site-Directed Mutagenesis

A number of investigators have identified amino acid differences that are principally responsible for characteristic functional differences between two closely related enzymes. The experimental objective of these studies was the identification of the smallest number of amino acid changes to the target enzyme corresponding to the source enzyme that confer the catalytic properties of the source enzyme to the target. These homology-based amino acid substitutions suggest that P450s within the largest and most catalytically diverse family of P450s, family 2, share a common structural organization and that these features are shared with the prokaryotic enzymes.

For example, P450s 2A4 and 2A5 differ at 11 of 494 amino acids. P450 2A4 is an androgen 15α-hydroxylase, whereas P450 2A5 is a coumarin 7-hydroxylase. Neither enzyme catalyzes the other reaction at an appreciable rate. By examining the effects of single substitutions of the residues found in P450 2A4 for those in P450 2A5, Lindberg and Negishi[8] demonstrated that the mutation F209L[*] conferred steroid 15α-hydroxylation activity to P450 2A5 which also retained coumarin hydroxylase activity. In contrast, any of three single mutations, V117A, L209F, or L365M, conferred coumarin hydroxylase activity to P450 2A4. The triple, reciprocal mutants at these sites effectively exchanged the enzymatic phenotypes between the two enzymes.[8,9]

As illustrated by this example, the design of the mutation and the predicted outcome are based on the phenotype of the related enzyme. P450 enzymes of large and catalytically diverse families, such as family 2, are particularly well suited for this approach as these enzymes can often exhibit very similar amino acid sequences but distinct catalytic differences. The latter include substrate specificity, regio- and stereospecificity of oxida-

---

*Mutations will be designated by the one-letter code of the residue to be mutated followed by the residue number and the one-letter code for the new amino acid.

tion, and kinetic properties. These distinctions are all likely to reflect alterations of the active site resulting from differences of amino acid sequence that affect substrate binding. The identification of determinants of substrate specificity by homology-based site-directed mutagenesis is an unbiased, empirical approach that is not based on structural models for the enzymes. However, as will be evident, a pattern for the distribution of key residues that were identified in this way has emerged that is consistent with proposed structures for the mammalian enzymes that are based on the experimentally determined structures for prokaryotic P450s.

Concurrent studies of other enzymes demonstrated that residues that align with the V117A difference between P450s 2A4 and 2A5 were determinants of catalytic differences between P450s 2C4 and 2C5 as well as between allelic variants of P450 2B2 (Fig. 1). P450s 2C4 and 2C5 differ at 24 of 487 amino acid residues and display a 10-fold difference in apparent K_m for progesterone 21-hydroxylation. A difference between 2C4 and 2C5 at this alignment position, V113A, underlies the lower apparent K_m exhibited by P450 2C5.[10] In the case of P450 2B2, the critical difference, I114F, is one of three amino acid differences found in an allozyme exhibiting a distinct regiospecificity for steroid metabolism when compared to P450 2B2.[11] This mutation together with another (L58F) produced the same phenotype in P450 2B1.[*] The I114F substitution in P450 2B2 was also shown to substantially diminish the turnover number for the 12-hydroxylation of 7,12-dimethylbenz[a]anthracene.[12] Thus, key differences were found by empirical approaches at the same alignment positions in pairs of enzymes from three distinct subfamilies. Interestingly, a V117A difference occurs between two allelic forms of P450 2A5 leading to lower rates of liver microsomal coumarin hydroxylation in inbred strains of mice bearing the Ala[117] allele.[13]

A cluster of key residues were also identified for allelic variants of P450s 2D1, 2C9, and 2C3 that align in close proximity to the L364M difference between P450s 2A4 and 2A5 (Fig. 1). An I380F difference, one of four amino acid differences between allelic variants of P450 2D1, underlies a selective loss of bufuralol but not debrisoquine hydroxylase activity.[14] A Ser/Thr difference between P450 2C3 and a variant, P450 2C3v, respectively, at residue 364 enables P450 2C3v to catalyze steroid 6β-hydroxylation.[15] This is also a selective effect as both variants catalyze steroid 16α-hydroxylation as well as the metabolism of a number of other substrates. A second of the five differences between 2C3 and 2C3v, I178M, was also shown to contribute to differences in the apparent K_m for progesterone 16α- and 6β-hydroxylation. In addition, an Ile/Leu difference at residue 359 of P450 2C9 allozymes also maps to this cluster of key residues. The Ile[359] variant catalyzes the 7-hydroxylation of (S)-warfarin, whereas the Leu[359] allele catalyzes the 4′-hydroxylation of (R)-warfarin.[16] The Leu substitution for Ile[359] also lowers the ratio of phenytoin to tolbutamide hydroxylase activity[17] and the turnover number for tolbutamide.[18]

*P450 2B1 rather than P450 2B2 was chosen as the host for site-directed mutagenesis as it generally exhibits a higher catalytic activity than P450 2B2 while displaying the same regiospecificity for steroid hydroxylation. P450 2B1 differs from P450 2B2 by only 12 amino acid differences, and it is identical to 2B2 in the region where P450 2B2 differs from P450 2B2v.

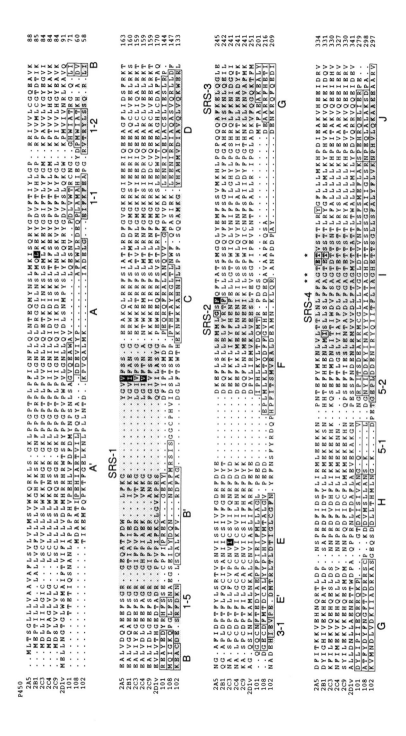

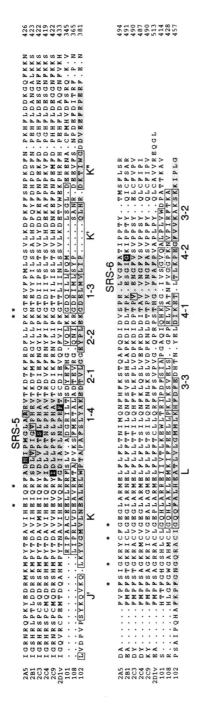

FIGURE 1. Alignment of the amino acid sequences of family 2 enzymes used in homology-based substitution studies with the prokaryotic P450s of known structure. Key residues identified by empirical approaches are designated by reverse lettering. Additional, residues mutated in the enzyme shown or in another member of the same subfamily where the mutation was found to alter catalytic activities are boxed and shown in bold. The alignment of family 2 sequences was produced using Pileup in the GCG software suite.[190] The alignment of the family 2 sequences with that of P450 101 is essentially that of Gotoh[29] with minor modifications. The proposed SRSs[29] are defined by the shaded regions for the family 2 sequences. The alignments of P450s 102 and 108 with P450 101 correspond to that of Graham-Lorence and Peterson.[7] The regions where the Cα backbone of P450 102 can be superimposed on that of P450s 101 and 108 are shaded in the sequence of P450 102. The α-helices and β-sheets are boxed and designated by a letter or number, respectively.[7] Highly conserved residues are indicated by an asterisk above the aligned sequences. The sequences were obtained from the Swiss protein data base. Accession numbers: P20852 (2A5), P00176 (2B1), P00182 (2C3), P11371 (2C4), P11712 (2C9), P10633 (2D1v), P00183 (101), P33006 (108) and P14779 (102).

2.2. Targeted Homologous Substitutions Based on Empirical Studies

Further studies have shown that the alteration of residues that align with key residues can lead to predictable changes in the properties of a more distantly related target enzyme. Kronbach *et al.*[19] demonstrated that the introduction of a single mutation, V113A, corresponding to the key residue identified in P450 2C5, conferred the progesterone 21-hydroxylase activity of P450 2C5 to P450 2C1. P450 2C1 exhibits no progesterone hydroxylase activity and only 75% amino acid sequence identity with P450 2C5. The introduction of a V113A mutation into P450 2C3v also confers progesterone 21-hydroxylase activity to 2C3v.[20] In contrast to 2C1, 2C3v exhibits progesterone 6β- and 16α-hydroxylase activity. The V113A mutation had little effect on 6β-hydroxylase activity but greatly diminished the rate of 16α-hydroxylation. Hydroxylation at these two sites on the substrate probably occurs from two distinct binding orientations of the substrate, and the V113A mutation appears to selectively decrease the regioselectivity of hydroxylation for the orientation that places the α-face and the D-ring of the steroid in proximity to the catalytic site.[20]

Further evidence for the importance of this key residue was obtained by modifying residue 114 of P450 2B1, which aligns with residue 113 of P450 2C5. Changes at this position alter the regiospecificity of androgen hydroxylation by P450 2B1. Halpert and He[21] characterized L114V and L114A mutants of P450 2B1. They found that decreasing the volume of the side chain of the residue at 114 increased the ratio of 15α- to 16α-hydroxylation and decreased the ratio of 16β- to 16α-hydroxylation. Halpert and colleagues also characterized an allelic variant of P450 2B1 exhibiting a single amino acid substitution, A478G, that displays a reduced ratio of 16β- to 16α-hydroxylation.[22] They demonstrated by site-directed mutagenesis that increasing the volume of the side chain of residue 478 increased the ratio of 15α- to 16-hydroxylation as well as of 16β- to 16α-hydroxylation with a serine exhibiting the highest ratios of the mutants examined.[23] A double mutant of P450 2B1, I114A, G478S, was found to exhibit a 1000-fold higher ratio of 15α- to 16-hydroxylation than the wild-type P450 2B1.[21] Thus, the combined effect of mutations at the two sites produced the expected alteration in regiospecificity of the enzyme. These mutations were also found to affect the regiospecificity of the epoxidation of arachidonic acid catalyzed by P450 2B1.[24]

The resulting changes in the regiospecificity of androgen hydroxylation effected by the I114A and G478S mutations in P450 2B1 produced an enzyme that resembles mouse P450 2A4 in catalyzing androgen 15α-hydroxylation. However, the double mutant did not catalyze the 15α-hydroxylation of progesterone. The latter activity could be introduced to the enzyme by making the double mutant I114A and F206L.[25] The F206L mutation was selected based on the conversion of P450 2A5 to a steroid 15α-hydroxylase by the F209L mutation at the same alignment position.[8]

These studies indicate that the residues identified as determinants of substrate specificity for one family 2 enzyme by empirical approaches are predictive for determinants of other family 2 enzymes. Moreover, the effects created by changes at one site can be combined with those at another to produce additive and predictable outcomes.

Mutations at key residues also produce novel catalytic specificities. Negishi and co-workers demonstrated that mutants of P450 2A enzymes at residues 117 and 209 acquire new catalytic activities.[26,27] For instance, the introduction of an F209N mutation in P450 2A5 results in the acquisition of 15α-hydroxylase activity for corticosterone in addition to its capacity to hydroxylate 11-deoxysteroids as seen for the F209L mutant.[26] Also, the P450 2A4 A117V mutant exhibits a dehydroepiandrosterone (DHEA) 7α-hydroxylase activity not seen for the wild-type enzyme.[27] DHEA is also rapidly metabolized by the P450 2A5 F209N mutant, and this activity is greatly diminished by a V117A substitution.[27]

2.3. Correspondence of Key Residues Identified by Empirical Approaches to Substrate Binding Sites in Prokarytic P450s

The key residues defined by the studies described in the preceding sections generally map to regions containing substrate-contacting amino acid residues of P450 101 when the amino acid sequences of the family 2 enzymes are aligned with those of P450 101 by different approaches.[28] Gotoh[29] suggested that amino acids that align within three residues of the substrate contact residues identified in P450 101 be considered potential substrate recognition sites (SRSs). SRS-1 was lengthened further to include helix B′ as well as regions that align with substrate-contacting residues found in the loops connecting the ends of helix B′ to the B and C helices (see Chapter 4). Almost all of the key residues mapped by homology-based substitutions fall within one of the six SRS regions defined by Gotoh[29] (Fig. 1).

The importance of SRSs has been borne out by more complex comparisons between enzymes. He et al.[30] targeted 5 residues in 2B1 that differ from 2B2 and that occur in SRS regions, residues 303, 360, 363, 367, and 473. These residues were also found to differ among other 2B enzymes from other species. Of these residues, only those homologous substitutions at positions 363 and 367 of P450 2B1 that occur in SRS-5 altered the regiospecificity of steroid hydroxylation. Increasing the volume of the side chain by substitutions for Val[367] led to a relative suppression of 16β-hydroxylation whereas decreasing the volume of the side chain by substitutions for Val[363] had the same effect. The V367A substitution was also observed to confer androgen 6β-hydroxylase activity to the enzyme. In a similar approach, Hasler et al.[31] mutated 11 residues in 2B11 that reside in SRSs and were unique to 2B11 when compared to P450s 2B1, 2B2, 2B4, 2B5, and 2B10. These residues were changed to the corresponding residues found in either P450 2B1 or 2B5. Of these mutants, the V114I, D290I, and L363V mutations were found to alter the regiospecificity of steroid hydroxylation. The author's alignment with P450 101 juxtaposed these residues with active site residues of P450 101, whereas the other residues which also fall within the SRS boundaries but that did not alter activity were aligned with residues of P450 101 that are found at more distant positions from the active site or that were predicted to be oriented away from the active site.

Some aspects of these sequence alignments are controversial, however, reflecting the low sequence similarity and differences in length exhibited by P450 101 when compared to the mammalian enzymes. Poulos[32] noted that early sequence alignments of family 2 enzymes with P450 101[33,34] mapped the key L209F difference between P450s 2A4 and 2A5 to helix E which is far from the active site of P450 101. A subsequent alignment of the mammalian P450s with P450 101 by Zvelebil et al.[35] that was based on multiple sequence alignments and secondary structure predictions placed residue 209 of P450s

2A4/5 in the region corresponding to the loop between the F and G helices of P450 101 that contains a substrate-contacting residue. However, the three-dimensional model[35] generated from this alignment using the coordinates for the Cα backbone of P450 101 placed residue 209 of P450 2A4/5 at the surface of the enzyme pointing outward. As discussed by the latter authors, homology-based models depend critically on the alignment of the sequences with P450 101, on conservation of the packing of α and β elements, and on the modeling of loop structures. As discussed in Chapter 5 by Peterson and Graham-Lorence, the F and G helices and the intervening loop differ greatly between P450s 101, 102, and 108, suggesting that predictions for this region are very likely to be uncertain. The alignment of Gotoh,[29] Fig. 1, also placed the L209F difference between P450s 2A4 and 2A5 in a site in P450 101 that corresponds to the loop between helices F and G that was designated as SRS-2. The alignment of Gotoh was based on comparisons of multiple sequences within family 2, hydropathy profiles, and secondary structure predictions.

The characterization of a variety of mutants at residue 209 of P450 2A5 suggests that this residue is likely to directly affect substrate binding. For instance, K_m and V_{max} values increased as the volume of the amino acid side chain of residue 209 decreased in the hydrophobic amino acids Leu, Val, Ala, and Gly.[36] Dissociation constants for the substrate determined from type I spectral changes paralleled the changes in K_m, indicating that substrate binding was affected.[37] The mutations also affected the relative proportion of high- versus low-spin spectral characteristics of the ferric cytochrome P450.[37] As the volume of the hydrophobic side chain diminished or its polarity increased, the low-spin character increased, suggesting that water gained increasing access to the sixth coordination site of the heme. This did not generally lead to an increase in uncoupling as judged by hydrogen peroxide formation during substrate metabolism.[38] Although the formation of hydrogen peroxide was elevated for some mutants, there was no clear dependence on the volume, hydrophobicity, or charge of the amino acid for this effect. It was also found that mutation of the adjacent Gly[207] to a Pro produced a low-spin enzyme. This mutation also increased the K_m for coumarin greatly, whereas an Ala substitution had little effect on either spin state or K_m. The authors suggested that the Pro substitution resulted in a larger substrate binding site.[39]

Of the mutations studied, only the F209V, F209N, and F209L substitutions conferred significant steroid 15α-hydroxylase activity to P450 2A4.[37] As noted earlier, the F209N mutant of P450 2A5 acquired the capacity to hydroxylate the 15α-position of corticosterone, which is not normally a substrate for this enzyme.[26] The authors suggested that the acquisition of corticosterone as a substrate might reflect a direct interaction of Asn[209] with the 11β-hydroxyl group of corticosterone as it was not a substrate for any of the other mutants. Whether or not residue 209 directly contacts the substrate, these results indicate that the identity of the amino acid residue at position 209 can determine the substrate specificity of the enzyme and that progressive changes in the volume and polarity of the substituted residues can elicit a graded alteration in the kinetic parameters and binding constants for coumarin.

Site-directed mutagenesis has also been used to better characterize the portion of SRS-1 corresponding to the B′–C loop of the mammalian family 2 enzymes. The SRS-1 region encompasses two substrate-contacting loops in P450 101 that connect the ends of the B' helix of P450 101 to the B and C helices (see Chapter 4). Both of the loops invaginate the active site of P450 101, and a substrate-contacting residue also resides at the C-terminal end of the B' helix. In contrast, the B' helices of P450s 102 and 108 reside at a greater distance from the active site, and in the case of P450 102, the B–B' loop does not invaginate the active site. On the other hand, the B'–C loop is a common feature of the active sites of structures of P450s 101, 102, and 108. As can be seen from the sequence alignment in Fig. 1, the length of this loop differs between the three prokaryotic enzymes, and a one-to-one correspondence between the prokaryotic enzymes and the mammalian enzymes does not occur.

A series of mutations in the SRS-1 region[40–42] were examined in P450 2C2, a lauric acid ω-1 hydroxylase, and in a chimera between P450s C2 and C1, C2MstC1. The C2MstC1 chimera comprises the 184 N-terminal amino acids of P450 2C2 and the 306 C-terminal amino acids of P450 2C1, and it exhibits progesterone 21-hydroxylase activity in addition to lauric acid ω-1 hydroxylase activity.[40] In contrast to P450 2C1, replacement of the SRS-1 region of P450 2C2 with that of P450 2C5 does not confer 21-hydroxylase activity to P450 2C2. However, a chimera, C2HincC1, where the C-terminus of P450 2C1 was substituted for that of P450 2C2 beginning at residue 388, was found to exhibit 21-hydroxylase activity although neither source enzyme exhibits this activity. The C-terminal region of P450 2C2 is highly divergent from 2C1 and other 2C enzymes, and it has been shown by Imai and co-workers[43] that this segment of the sequence is an important determinant of the catalytic differences between 2C2 and 2C14. Examination of the differences in this region by site-directed mutagenesis indicates that progesterone 21-hydroxylation can be conferred by substitution of the Val found in P450 2C1 for the Ser of P450 2C2 at residue 473.* This residue maps to SRS-6. The 21-hydroxylase activity of the C2MstC1 chimera is greater than that of the C2HincC1 chimera, reflecting additional, complementary differences in SRS-5 and the β-sheet 1 region.[44]

A number of mutants bearing alternative hydrophobic amino acid residues at residue 113 were characterized in P450 2C2 and the C2MstC1 chimera. The laurate ω-1 hydroxylase activity of P450 2C2 was not greatly affected by the substitution of Ala, Val, Leu, or Phe for Ile[113], whereas the activities of the Gly, Tyr, and Cys mutants were very low.[40] The C2MstC1 chimera appeared to be more sensitive to the substitution of Ala, Val, Leu, or Phe for Ile[113], and this effect was greater for progesterone hydroxylation than for laurate hydroxylation. However, these effects were small. Interestingly, the Ala substitution did not enhance progesterone 21-hydroxylation of the C2MstC1 chimera as was seen for P450 2C4,[10] 2C1,[45] and 2C3v.[20] However, the apparent K_m was not assessed in these studies, and under the conditions of the assays, the apparent rate might not be very sensitive to changes in K_m of < 10-fold. These results indicate that a number of hydrophobic residues of different volume could be accommodated in the two enzymes at this alignment position. In addition, hydroxylation of the more flexible and less bulky substrate, lauric acid, was affected to a lesser extent than that of the rigid, planar progesterone molecule by amino acid side chains of increasing size.[40,42]

*B. Kemper, University of Illinois (personal communication).

A degenerate cassette was also synthesized, and at least one mutant for each position between residues 107 and 120 was obtained for P450 2C2[41] and for the C2MstC1 chimera.[42] In some cases, the resulting mutants exhibited relatively conservative amino acid changes whereas in others the changes were nonconservative. For the mutants obtained, the alteration of amino acids from 107 to 110 and from 116 to 119 did not greatly affect lauric acid ω-1 hydroxylase activity of P450 2C2. A relatively nonconservative G109E mutation had little affect on activity, but the G111V and G117V mutations diminished the lauric acid ω-1 hydroxylase activity of P450 2C2[41] by 50- and 7-fold, respectively, and affected the activity of the C2MstC1 chimera similarly.[42] The S115R mutants of both enzymes also exhibited low activity. In addition, as was seen for the substitutions at position 113, the V112F and F114L mutations also reduced activity, and in the case of the C2MstC1 chimera, the progesterone 21-hydroxylase activity appeared to be more sensitive to the substitutions at residue 112. These results are consistent with the likelihood that the short stretch of hydrophobic amino acids that is flanked by the two highly conserved glycine residues forms the B′–C loop. Gly[111] may serve to terminate the putative B′ helix in the SRS-1 region of mammalian P450s as it does in P450 102 (Fig. 1).[46] However, these experiments did not identify which of these residues are likely to contact the substrate directly.

SRS-4 and SRS-5 are the only SRSs that occur in segments of the amino acid sequences that can be aligned rather unambiguously with P450 101, 102, and 108 (Fig. 1). As predicted by Poulos,[47] SRS-4 and SRS-5 constitute part of the core structure of P450s 101, 102, and 108 where the positions of the Cα backbones of the three proteins are relatively well conserved in space.[46,48] SRS-4 resides in helix I, and SRS-5 spans from the end of helix K into a portion of the β-sheet designated as 1–4 in P450 102. These topological elements form two sides of the substrate binding sites of P450s 101, 102, and 108. Key residues identified in homology-based substitution studies are spread across each region, and only some of these residues are likely to be in direct contact with the substrate (Fig. 2). One of these, the T364S difference identified in variants of P450 2C3, underlies a phenotypic difference in the capacity to hydroxylate progesterone at the 6β-position. Both P450 2C3 variants catalyze progesterone 16α-hydroxylation although the efficiency is lower for the Ser variant.[15] This residue corresponds by alignment to Phe[331] of P450 102, Phe[317] of P450 108, and Asp[297] of P450 101 (Fig. 1). Although the α-carbons of these residues are superimposable when the structures are compared, the orientations of the side chains differ in the three enzymes. In P450 102, the side chain of Phe[331] is oriented down and away from the active site and the substrate access channel (Fig. 2), whereas the side chain of Phe[317] is directed into the active site for P450 108. Asp[297] of P450 101 is also directed into the active site, where it hydrogen bonds to one of the heme propionates. In contrast, the corresponding propionate of P450 108 is directed down and away from Phe[317]. Thus, although these alignments are highly suggestive that residue 364 of P450 2C3 occurs at the juncture of the substrate access channel and the active site, it is difficult to predict whether this side chain is in contact with the substrate.

A number of amino acid substitutions have been examined for Thr[364] of P450 2C3v.[20] Substitution of Gly, Asp, or Asn produced the 6β-hydroxylase-deficient phenotype of the Ser variant, whereas substitutions of the more hydrophobic, bulky residues Val, Ile, and

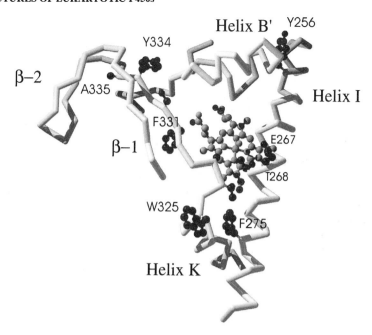

FIGURE 2. Topological features of P450 102 that correspond to SRS-1, SRS-4, and SRS-5. The Cα backbone of P450 102 corresponding to SRS-1 is rendered as a tube for residues from the end of helix B to the beginning of helix C, E64–E93. The Cα backbone of helix I, L249–Q283, is rendered similarly. The region spanning SRS-5 is also rendered as a tube beginning in helix K, residue V314, and extending through β-sheets 1 and 2 terminating at I357. Side chains are depicted as dark balls and sticks for those residues that align with key residues identified for mammalian enzymes in SRS-4 and SRS-5, Fig. 1. The heme is rendered as a ball and stick model with a lighter hue. The figure was generated from the coordinates deposited in the Brookhaven protein data base, 2HPD for molecule A, using Molw and rendered by ray tracing using Showcase (US Science, San Diego, CA).

Leu exhibited the characteristic steroid 6β-hydroxylase activity of the Thr variant.[20] Little variation was evident between the catalytic activity of mutants in each phenotypic group, suggesting that although the alignments indicate that these residues would potentially contact the substrate, the reaction was not greatly affected by differences in the volume of the side chain at this position. Interestingly, the opposite trend was observed for mutations at residue 367 of P450 2B1 that aligns with Ser[332] of P450 102 (Fig. 1). Substitutions that decreased the bulk of residue 367 of P450 2B1 conferred steroid 6β-hydroxylation to 2B1, and a similar effect was seen at residue 363 of P450 2B1.[30] If the β-sheet structure is conserved in the mammalian enzymes, the side chains of residues 367 of P450 2B1 and 364 of P450 2C3 would be predicted to be oriented in different directions. This β-sheet in P450 102 forms the base of the proposed substrate access channel. Thus, some mutations in this region may alter the docking of the steroid substrates by limiting access of the steroid to the active site in an orientation that places the β-face and B-ring toward the heme-iron, whereas access for the steroid in the opposite orientation which place the face of the D-ring of the steroid at the heme-iron is not affected.

Additional studies have identified homologous substitutions in SRS-4 that alter substrate metabolism. Studies of P450s 2A1 and 2A2 have revealed differences at residues 302, 303, and 310 that alter the regiospecificity of steroid hydroxylation.[49] Residues 302 and 303 reside in SRS-4 whereas residue 310 does not (Fig. 1). When mapped to the structure of P450 102 (Fig. 2, residue F275), residue 310 of P450 2A1/2 does not align with the active site, but it does correspond to a position in opposition to the site where the key Ile/Leu difference in SRS-5 identified for residue 359 of 2C9[16-18] maps to the structure of P450 102. These residues correspond by alignment (Fig. 1), with F275 and W325, respectively, of P450 102 (Fig. 2). This suggests that the effects seen for mutations at these two positions could reflect alterations in the contacts between these two structural elements that form sides of the active site and thereby perturb substrate binding indirectly. Similarly, the Ile/Phe difference between P450 2D1 allozymes at residue 380[14] and the Arg/His difference at residue 372 of P450s 2A11 and 2A10[50] are both predicted to occur at or near contacts between SRS-5 and SRS-1. The corresponding residues of A335 and Y334, respectively, in P450 102 are shown in Fig. 2. In addition, the key residue at 290 in P450 2B1[31] appears to reside near the contact of the B'–C loop of SRS-1 with helix I. This corresponds to Y256 of P450 102 (Fig. 2). Thus, some of the residues identified by homologous substitution studies that occur in SRSs 4 and 5 may form critical contacts between the topological elements of the active site rather than directly contacting the substrate.

2.4. Targeted Mutagenesis

The recognition of a conserved sequence motif, (A/G)GxxT, that is found in helix I of P450 101 led to early predictions of the corresponding helix in mammalian P450s (Fig. 1).[51] This motif resides in SRS-4 and there are no gaps when sequences from different families are aligned across this region. A highly conserved Thr, residue 252 in P450 101, was hypothesized to play a role in deformation of the I helix to facilitate oxygen binding. Hydrogen bond donation from the Thr side chain to the carbonyl oxygen of Gly[249] was postulated to contribute to the stabilization of this deformation.[51] However, site-directed mutagenesis of the corresponding residue in P450 2C2 demonstrated that a variety of side chains were compatible with retention of catalytic activity.[52,53] In addition, a T319A mutant of P450 1A2 retained activity for ethoxycoumarin as substrate whereas metabolism of benzphetamine was lost.[54] Similarly, T252S, V, and A mutants of P450 101 retain their capacity to hydroxylate camphor. However, substrate hydroxylation became uncoupled from oxygen reduction, yielding increasing amounts of hydrogen peroxide.[55] It was suggested that this might reflect the loss of threonine's function as a proton donor and/or an acid–base catalyst in oxygen scission.

Determination of the structure of the P450 101 T252A[56] mutant indicated that this mutation increased the space in the vicinity of the axial ligation site and was likely to permit greater access for water to the activated oxygen. A similar effect was suggested for the T319A mutant in P450 1A2 as it increases the likelihood that aryl hydrazines will attack the nitrogen of the pyrrole ring of the heme group that is predicted to reside most closely to the conserved Thr.[57] The oxidation of heme iron-phenyldiazene complexes leads to a phenyl iron complex that rearranges to form N-phenylprotoporphyrin IX regioisomers.

The distribution of these isomers differs from that of heme in solution and is dependent on the protein environment, and a topological model of the active site with reference to the heme can be deduced from these types of experiments. Examination of this reaction for P450s 101, 102, and 108[58] indicates that the ratios of the isomers formed differ for each P450. This reaction is highly selective for the pyrrole nitrogen of the D-ring, and products derived from the B-ring are not detected, for P450s 101 and 108. For P450 102, the principal product is formed with the A-ring nitrogen. However, each of the other rings is also attacked to a minor extent, suggesting that active site is more open above all four pyrrole rings of the heme, in agreement with the experimentally determined structures. In the case of P450s 1A1, 2B1, 2B2, and 2E1, the products represent the D- and A-ring pyrroles with slightly different ratios seen for each enzyme.[59] For P450s 2B4, 2B10, and 2B11, products are also derived from pyrrole ring C, although to a lesser extent.[60] With the exception of P450 102, the B-ring nitrogen is protected in all of the P450s. The regiospecificity of these reactions suggests a common orientation for the heme, and analysis of the products formed with other suicide inhibitors that attack the heme leads to a similar conclusion.[61] It is interesting that the Pro[329] of P450 102 aligns with a conserved Pro in family 2 enzymes in SRS-5 (Fig. 1). The deformation of the $C\alpha$ backbone by the Pro may contribute to the greater accessibility of the A-ring nitrogen in both P450 102 and the family 2 enzymes. The D-ring is relatively accessible in all of the enzymes studied to date and in P450 102 it resides near Phe[331] that forms the base of the proposed substrate access channel (Figs. 1 and 2).

The effects of a T319A mutation in rat 1A2 on coupling were evaluated for the metabolism of 7-ethoxycoumarin[62] and methanol.[63] Although the efficiency of oxygen incorporation in these substrates is low (< 10%), it was not diminished but rather enhanced when the Thr was replaced by Ala or Asp. These substitutions had little effect on the kinetic parameters characterizing the metabolism of 7-ethoxycoumarin when the mutant and wild-type enzyme were reconstituted with NADPH and NADPH P450 oxidoreductase.[64] However, these mutations did affect the reaction supported by *tert*-butylhydroperoxide (TBHP). It is possible that the mutations affect the binding of TBHP and/or oxygen transfer to the heme-iron as the K_m for TBHP was altered. In contrast, mutations at the adjacent Glu[318] of rat P450 1A2 affect the kinetic parameters characterizing the NADPH P450 oxidoreductase-supported reaction, whereas the TBHP-supported reaction was largely unaffected. In fact, the E318D mutant exhibited a much higher rate for the NADPH P450 oxidoreductase-supported reaction than the wild-type.[64] This residue has also been mutated in other enzymes leading to positive and negative alterations in catalytic activity as well as differences in regiospecificity that depend on the substrate, enzyme, and substituted amino acid.[53,65-72] Thus, the results of these mutagenesis experiments do not indicate a necessary or extraordinary role in P450-catalyzed reactions for this highly conserved threonine. The effects that are seen are consistent with the proximity of this and the adjacent residue to the binding sites for oxygen, organic substrate, and alternative oxygen donors. The corresponding residues of P450 102, T268 and E267, are shown in Fig. 2. It is interesting to note that the substitution of a Lys for Thr[301] of P450 2C2 or Thr[303] of P450 2E1 produces an enzyme exhibiting visible absorption spectra consistent with the binding of the epsilon amino group of the Lys side chain to the open coordination site of the heme-iron[73,74] suggesting a close proximity to the heme-iron.

In contrast to the empirical approach described earlier, targeted mutagenesis is based on a model that influences the choice of residues selected for mutagenesis. For the mammalian P450s, these studies are largely based on the premise that a correspondence exists between the structures of the eukaryotic enzymes and the prokaryotic structures, and mutagenesis experiments have been employed to seek support for the proposed models. However, the predicted effects of these mutations are often a loss of function that could arise by additional mechanisms such as disruption of fundamental aspects of catalysis or loss of structural integrity. More often, the informative experiments are those that either produce an unexpected outcome, such as those indicating that conservation of the Thr/Ser was not essential for residues corresponding to Thr^{252} of P450 101, or that produce a distinct, new phenotype, such as the Lys substitutions for Thr^{303} of P450 2E1.[73,74]

In many cases, the choice of target depends critically on the precise alignment of the target sequence with the model. The examples described in this section are some of the few where these alignments are very likely to define a correspondence of the residues in mammalian P450s with those in the prokaryotic P450s and where the three-dimensional models are likely to be highly predictive. As discussed in this and earlier chapters, the correspondence between P450s in the structures of other topological elements that form the active site, such as SRSs-1, 2, 3, and 6, are less clear and more likely to vary between families of enzymes.

2.5. Active Site Probes

Active site residues can, in principle, be identified by chemical approaches such as the use of photoaffinity probes. There are only a few reports of the use of photoaffinity probes for the active sites of P450s that have identified specific targets. 2-Ethynyl-naphthalene was found to react with P450s 2B1 and 2B4 and inactivate these enzymes.[75] Peptide cleavage results indicated that the probe bound to peptides that correspond to portions of helix I that span SRS-4. P450 2B4 has also been investigated with spiro[adamantane-2,2′-diazirine] as a photoaffinity probe.[76] Following mild acid cleavage, the probe was found associated with a 2-kDa peptide. Although the peptide was not identified by sequence analysis, the size corresponds to a predicted cleavage product comprising residues 258–278 that by alignment spans helix H (Fig. 1) and would not be predicted on the basis of the prokaryotic structures to occur near the active site. In contrast, 4-azidobiphenyl was found to label a peptide in rabbit P450 1A2, residues 212–220, that would align near SRS-2.[77] 2-Ethynylnaphthalene has also been employed with rat and rabbit P450 1A2.[77] The label was localized to residues 67–78 and to 175–184 for the rat and rabbit enzymes, respectively. The basis for the distinct targets in these two similar enzymes is unclear. The two regions would be predicted to correspond to helices A and D, respectively, of the prokaryotic enzymes. The D helix lies outside the substrate binding sites of the prokaryotic enzymes, but the A helix may lie close to the proposed substrate access channel. A similar region of the bovine mitochondrial side chain cleavage enzyme, P450 11A, predicted to occur close to the substrate access channel, residues 8 to 28, was labeled by the suicide substrate methoxychlor.[78] However, a region of P450 11B that corresponds to the loop preceding helix L was labeled with the photoaffinity probe, methyltrienolone.[79] Overall, these studies have not revealed a pattern of labeling suggesting that targets are restricted to specific portions of the sequence. The target regions identified in these studies are inconsistent, in some cases, with models for the mammalian enzymes based on the structures of the prokaryotic enzymes. However, these compounds are, in general, hydrophobic and may bind to sites on the enzymes that are distinct from the active site. The primary evidence for the binding of the probe at the active site is inactivation of the enzyme and protection by substrate. However, binding may occur at other sites, and competitive binding for these alternative sites by substrates and inhibitors of the enzyme could also occur. Thus, peptides lying outside the active site could be labeled. The inconsistencies with models for the mammalian P450s based on the prokaryotic enzymes are not sufficient to reject these models but should be considered in relation to future experiments.

2.6. Relation between Catalytic Diversity and Genetic Diversity

In addition to suggesting common structural features for the eukaryotic and prokaryotic P450s, the pattern of key residues revealed by homology-based substitutions indicates the location of genetic differences that underlie catalytic diversity. Kronbach et al.[80] noted that the key residue Val[113] of P450 2C5 that determines 21-hydroxylase activity resides in a portion of the amino acid sequence (corresponding to SRS-1) that displays a high degree of sequence divergence among P450 families that exhibit catalytic diversity. This key residue lies in a surface loop that might tolerate a significant range of amino acid substitutions without altering the overall structure of the enzyme. As this loop also forms a portion of the substrate binding site, this region could readily accumulate changes that lead to alterations of substrate specificity.[80] In support of this topology, Kronbach and Johnson[10] demonstrated that residues immediately adjacent to the key amino acid formed an epitope recognized by an inhibitory monoclonal antibody to P450 2C5, confirming that this segment is surface exposed. Antibodies generated to peptides corresponding to this region of 2B1 also recognize the native enzyme (see Table II).[81,82]

This suggested a framework model for the catalytic and genetic diversity of family 2 enzymes[28] in which segments that determine substrate specificity occur on structural elements that can tolerate changes, whereas more highly conserved regions generally preserve the helices and sheets as well as the spatial architecture found for P450 101. This is also evident when genetic variations within SRSs are compared to those of other regions.[29] Each of the SRS regions exhibits very high rates of nonsynonymous nucleotide substitutions in family 2 when compared to other regions,[29] suggesting that this might reflect a selection for enzymatic diversity within this family of xenobiotic metabolizing enzymes.

The subsequent determination of the structures for P450s 102[46] and 108[48] indicates that the topological features of P450 101 that form the substrate binding site are also present in P450s 102 and 108. However, there is a considerable range of spatial variation in the regions that correspond to SRSs 1, 2, 3, and 5, in which a high degree of genetic variation is seen for family 2 enzymes. The more highly conserved regions of the family 2 enzymes correspond to the topological elements of the prokaryotic enzymes that are spatially conserved and that can be considered as the structural framework of these enzymes that underlies their basic architecture.

3. Membrane Topology

In contrast to the prokaryotic cytochromes P450, all eukaryotic P450s analyzed to date with the exception of the fungal P450 55A1[83] can be classified as membrane proteins. Detailed knowledge on how these proteins are bound to, and interact with, biological membranes is of fundamental importance for our understanding of this group of proteins. Moreover, insights concerning the region(s) of the P450 molecule involved in membrane binding might guide the engineering of less hydrophobic and possibly water-soluble variants more amenable to crystallization or industrial applications. As discussed in the previous sections there is considerable evidence that the mammalian P450 enzymes share substantial overall three-dimensional similarities with the three water-soluble bacterial P450s for which three-dimensional structures are known. These similarities provide useful information regarding the overall structure of the P450 molecule, but do not address the question of how the membrane-bound forms are anchored to, or associate with, the lipid bilayer of membranes. A wide range of experimental methods employing, for example, protease digestions, site-specific antibodies, biophysical techniques, chemical modifications, sequence analysis, and genetic engineering have been used to study the P450 membrane topology. In the following section the results of some of these studies will be discussed. The reader is also referred to other previously published reviews on this topic.[84–87]

3.1. Role of the N-Terminal Region of Eukaryotic P450s

The different forms of P450 are each encoded in the nucleus and cytoplasmically synthesized, but they are incorporated in either mitochondria or the endoplasmic reticulum (ER). Hence, the signals required for targeting of the polypeptide to the correct compartment within the cell as well as signals that direct protein folding are expected to be determined by the amino acid sequence of each polypeptide.

3.1.1. Microsomal P450 Enzymes

Haugen *et al.*[88] first pointed out that the amino acid sequence of the N-terminal portion of rabbit liver P450 2B1 resembles signal sequences of secretory proteins. Using an mRNA-dependent *in vitro* synthesis system, Bar-Nun *et al.*[89] found that P450 2B1 or 2B2 had the same size and N-terminal amino acid sequence whether microsomal membranes were present or absent in the reaction mixture, indicating that these microsomal P450s contained a noncleavable membrane insertion signal. Although similar to known secretory signal sequences, the various N-terminal regions of microsomal P450 enzymes show significant differences relative to this group of targeting sequences in that they are generally longer and contain an apolar segment comprised of 16 to 20 residues that is more hydrophobic than the corresponding region of cleaved signal sequences. In this respect, the N-terminal region of microsomal P450s resembles membrane-spanning anchor domains more than typical secretory signals. Acidic residues usually precede the first apolar segment and basic residues are rarely found. In addition, a cluster of basic residues is usually present after the hydrophobic segment in most microsomal P450s. This type of signal sequence is referred to as a signal-anchor (SA) sequence.[90]

Microsomal P450s are synthesized on membrane-bound polysomes, and the hydrophobic N-terminus directly interacts with the signal recognition particle which mediates the insertion into the ER membrane during translation, as has been shown for rabbit P450 2B4.[91] A number of studies employing chimeric proteins in which the signal sequence for a secretory protein has been replaced by the SA sequence of a P450 have shown that the P450 SA sequence can function to direct normally secreted proteins to the ER membrane and stop their translocation across the membrane.[92–97] Recently, Ahn et al.[96] have shown that the N-terminal 29 amino acid residues of P450 2C1, when fused to either a secreted protein or a soluble cytoplasmic protein, are sufficient for retention in the ER, suggesting that the globular domain of the P450 molecule is not required for ER retention.

It is clear that the hydrophobicity of the SA sequence is important for membrane insertion, but the final membrane orientation is also strongly influenced by the hydrophilic flanking regions. For example, by exchanging two N-terminal residues (Asp^2, Leu^3) of rabbit P450 2C2 for basic residues (Lys and Arg, respectively) the orientation of the protein changes from being exposed on the cytoplasmic side of the ER membrane to facing the luminal side of the membrane.[98] Sakaguchi et al.[99] have proposed that the balance between the length of the hydrophobic segment and the number of flanking charged residues determines whether a sequence functions as an SA sequence or a cleaved secretory signal sequence.

In a recent study,[100] the effect of deleting increasing portions of the N-terminal region of the steroid 21-hydroxylase (P450 21B) was analyzed. When six P450 mutants were synthesized in vitro using a coupled transcription/translation system in the presence of microsomal membranes, it was found that a variant lacking approximately one-third of the hydrophobic N-terminal region still bound to membranes. However, when more than half of the hydrophobic segment was deleted, membrane binding was abolished and the majority of the P450 was found in the soluble fraction. Using a similar experimental approach, Monier et al.[92] showed that amino acid residues 2 to 42 of P450 2B1 are required for binding to microsomal membranes. These results strongly indicate that the N-terminal hydrophobic segment is required for proper membrane binding in vitro.

Several studies have addressed the effect of truncations of N-terminal amino acid residues on the in vivo synthesis, cellular localization, and catalytic activity of microsomal P450s. In these studies, P450s have been heterologously expressed either in eukaryotic cells or in the enterobacterium Escherichia coli. There are at least two important differences between expression of a microsomal P450 in bacterial and eukaryotic systems. First, bacteria lack an intracellular membrane equivalent to the ER membrane of eukaryotic cells. Second, although recent data[101] suggest that the basic processes of protein insertion into the ER membrane and the inner membrane of E. coli are similar, insertion into the bacterial membrane takes place posttranslationally, whereas in eukaryotes, it is a cotranslational process. These differences could explain why truncations of N-terminal amino acid residues may have different effects on the properties of the P450 molecule expressed in eukaryotic and prokaryotic cells.

When N-terminally truncated variants of P450 21B were synthesized in vivo in Rat-1 or COS-1 cells, it was found that the steady-state levels and specific activities decreased in correlation with the length of the truncations.[100] N-terminally truncated variants of both rat and mouse P450 1A1 behaved in a similar way when expressed in yeast cells.[102,103] The expression level of the holo form of the truncated rat P450 1A1, lacking residues 2–30 of the mature protein, was decreased 3-fold and there was a 4-fold decrease in specific activity relative to the wild-type 1A1. Ultracentrifugation in sucrose density gradients showed that the truncated P450 1A1, like the full-length protein, co-sedimented with the microsomes, indicating that both forms were associated with the microsomal membranes.[104] Incubation of the microsomes in a buffer containing 800 mM NaCl resulted only in a minor release of the truncated P450 1A1 into the soluble fraction showing that the protein was not simply electrostatically bound to the microsomes. Clark and Waterman[105] expressed wild-type and two N-terminally modified variants of the bovine 17α-hydroxylase (P450 17A) in COS-1 cells and showed by subcellular fractionation that the expressed polypeptides were present in the microsomal fraction. Deletion of amino acid residues 2 to 17 did not abolish membrane binding. However, the amount of protein that was recovered in the microsomal fraction decreased about 5-fold compared with the native enzyme. Unlike the modified P450 21B or 1A1, the truncated P450 17 could not be synthesized as the holo form and consequently showed no enzymatic activity when expressed in COS-1 cells. Even when the N-terminal deletion was extended to include residues 2–34, the polypeptide remained associated with the ER membrane. In a subsequent study,[106] it was shown that the enzymatic activity could be restored when chimeric proteins combining the truncated P450 17 with the SA sequence from either bovine 21-hydroxylase (P450 21) or the rat liver NADPH P450 oxidoreductase were expressed. Using a yeast expression system, it was found that deleting the first 22 N-terminal residues of human P450 2D6 resulted in an enzymatically inactive protein that did not insert properly into the ER membrane.[107] The expression level of the truncated 2D6 was reduced almost 10-fold compared to the wild-type indicating that the polypeptide could not fold properly and hence was degraded in the yeast cells. Taken together, these data suggest that the P450 SA sequence, in addition to the functions previously discussed, may be important for *in vivo* protein stability. Furthermore, these results indicate that proper localization of the P450 protein, via the signal-recognition particle pathway in the ER membrane, might be a required intermediate step for efficient folding of the polypeptide to form an active enzyme when expressed in eukaryotic cells.

In contrast to the results obtained in eukaryotic cells, expression of truncated P450s in *E. coli* have resulted in active proteins that can be found either bound to the inner membrane or as soluble polypeptides in the cytoplasm. Some of these data are summarized in Table I. Generally, microsomal P450s expressed in *E. coli* with an intact SA sequence are present in the membrane fraction of bacterial cells. However, this does not seem to be the case for the housefly P450 6A1 that was found to be present in almost equal amounts in the soluble and the membrane fraction when expressed in *E. coli*.[108] The N-terminally truncated, but still membrane-bound forms of P450 2E1 (Δ3–29)[109] and P450 17 (Δ2–17)[110] retained similar catalytic properties as the respective full-length enzymes, suggesting that the N-terminal SA sequence is not required for catalytic activity. This is in agreement with findings that a proteolytically cleaved variant of rat P450 1A1 lacking the first 31 amino acid residues of the mature protein had similar enzymatic properties as the intact enzyme.[111]

TABLE I

Subcellular Localization of Full-Length and N-Terminally Modified Microsomal P450 Enzymes Expressed in *E. coli*

Protein	Extent of deletion	Distribution % of total P450		Ref.
		Membrane	Cytosol	
2B4	wt[a]	73±5	30±5	181
2B4	2–20	33±6	67±6	181
2B4	2–27	32±5	68±5	181
102:2B4	[b]	<30	>70	182
2C3	wt	>99	<1	[d]
2C3	2–20	<60	>40	[d]
102:2C3	[c]	<40	>60	[d]
2C5	wt	>95	<5	[d]
102:2C5	[c]	<20	>80	[d]
2C8	wt	>99	<1	[d]
2C8	2–20	80	20	[d]
2E1	wt	70±5	30±5	181
2E1	3–29	65±5	35±5	181
2E1	3–48	45±5	55±5	181
102:2E1	[b]	20	80	182
6A1	wt	45	55	108
7	wt	~100	—	183
7	2–24	~15	~85	183
17	wt	~100	—	110
17	2–17	>90	<10	110

[a] wt, wild-type N-terminal sequence or an N-terminal sequence that was slightly modified in order to increase the expression levels.
[b] A chimeric protein in which the 17 N-terminal amino acid residues of 2B4 (Δ2–27) or 2E1 (Δ3–29) have been replaced with the 19 N-terminal residues of P450 102.
[c] A chimeric protein composed of the first 20 N-terminal amino acid residues of P450 102 fused to the first N-terminal proline residue (approximately at residue 30) of the respective mammalian P450.
[d] von Wachenfeldt and Johnson (unpublished data).

The truncated forms of P450 2E1 and P450 17 were both tightly bound to the bacterial membrane as indicated by the inability to extract any significant amount of P450 by treatments that perturb electrostatic interactions, such as high ionic strength or alkaline pH.[109,110] These findings indicate that the N-terminal SA sequence might, for most eukaryotic P450s, not be the only region of the molecule that anchors the protein to the membrane, as will be further discussed in Section 3.3.

3.1.2. Mitochondrial P450 Enzymes

Mitochondrial P450 enzymes have an amphipathic N-terminal signal sequence (presequence) rich in positively charged amino acid residues that allows the protein to bind to the mitochondrial surface at contact sites between the inner and outer membranes.[112] The size of the signal sequence ranges from 24 amino acid residues for P450 11B to 39 residues for the cholesterol side-chain cleavage enzyme, P450 11A. The signal peptide is removed during the import process by a specific protease that resides in the matrix, and subsequently the mature protein is incorporated into the mitochondrial inner membrane. The mechanism by which the polypeptide becomes associated with the inner membrane is not fully understood. The mature polypeptides of mitochondrial P450 enzymes lack the highly hydrophobic SA sequence found in microsomal P450s. Hence, other regions of the molecule must be responsible for anchoring the protein to the membrane. This is demonstrated by the mature form of P450 11A (P450$_{scc}$) that, when heterologously expressed in *E. coli*, was found exclusively in the membrane fraction of the bacterial cells.[113]

Sakaki *et al.*[114] have shown that by replacing the presequence of rat P450 27A with the SA sequence of bovine P450 17 it is possible to convert a mitochondrial P450 to a variant that is located in the ER membrane and that is enzymatically active. In contrast, Black *et al.*[115] found, in a similar type of experiment, that P450 11A is only active when expressed in mitochondria. The reason for the discrepancy between these studies is not clear.

3.2. Cellular Localization of the N-Terminus of Microsomal P450 Enzymes

Cumulative recent data suggest that the N-terminus of microsomal P450s has a luminal orientation. Chemical labeling using fluorescein isothiocyanate (FITC) has been employed in these studies. FITC is a membrane-impermeable reagent that has been shown to react specifically and stoichiometrically with the N-terminal α-amino group of P450 2B4.[116] Using this method Vergères *et al.*[117] have shown that the N-terminal methionine of P450 2B1 was accessible to FITC labeling only when the protein was solubilized from the microsomes, which is in agreement with the idea that the N-terminus faces the lumen of the ER. Recently, Shumyantseva *et al.*[118] used a similar approach to study the topology of P450 1A2 and 2B4 and concluded that the N-termini of these proteins are exposed on the luminal side of the ER membrane. In contrast, Bernhardt *et al.*[119] found that FITC could label the N-terminus of P450 2B4 even when the protein was bound to microsomes. However, in this latter study no information on the integrity of the microsomal vesicles or on the efficiency of the labeling was reported.

The core glycosylation enzyme that is responsible for the *N*-glycosylation of proteins is located in the ER membrane and has its catalytic site facing the luminal compartment. Two studies have taken advantage of this fact to probe P450 membrane topology. The isolated human aromatase (P450 19) has been shown to be a glycoprotein.[120] Recently, Shimozawa *et al.*[121] demonstrated that Asp^{12} is the site of glycosylation, clearly indicating a luminal localization of the N-terminus of P450 19. In addition, when a peptide containing a glycosylation site was fused to the N-terminus of P450 2C1, the protein became glycosylated both *in vitro* and *in vivo*.[122]

In this context, it is interesting to note that the human thromboxane synthase appears to have the opposite N-terminal topology. Thromboxane synthase belongs to family 5 of the P450 superfamily and is present in microsomes where it catalyzes the isomerization of prostaglandin H$_2$ to thromboxane A$_2$. This enzyme does not show any monooxygenase activity and is not dependent on the P450 oxidoreductase for activity. From immunocytochemical data, using antibodies raised against two peptides that correspond to the first 10 or 15 residues of the enzyme, it was concluded that the N-terminal region is exposed to the cytoplasmic side of the ER membrane.[123] It is possible that thromboxane synthase has its catalytic domain facing the luminal side of the ER membrane, although experimental evidence is lacking.

3.3. Overall Membrane Topology

From the foregoing discussion it is clear that the N-terminal hydrophobic region of microsomal P450 enzymes spans the lipid bilayer of the ER membrane with the N-terminus facing the luminal side of the membrane. Based on physiochemical criteria, it is reasonable to assume that the transmembrane segment is in an α-helical conformation. The thickness of the nonpolar portion of the ER membrane is between 30 and 40 Å. A typical straight α-helix rises 1.5 Å per residue along the helix axis. Hence, the N-terminal transmembrane segment must have a minimum length of 20 predominantly hydrophobic amino acid residues to be able to span the hydrocarbon region of a 30-Å-thick fluid bilayer.

Using antibodies raised against the whole molecule of different P450 enzymes, it has been established that the major antigenic determinants of microsomal P450s are located on the cytoplasmic side of the ER membrane.[124-127] Both of the electron donors to P450, cytochrome b_5 and NADPH P450 oxidoreductase, are anchored to the ER membrane by a single transmembrane helix, and the bulk of these proteins is exposed in the cytoplasm.[128,129] This indicates that at least the portion of the P450 molecule that interacts with these electron donors must be exposed on the cytoplasmic side.

3.3.1. Topology Predictions from Primary Structure

Hydrophobicity analysis of primary structure can be used to identify amino acid sequences that are sufficiently long and hydrophobic to imply the existence of a transmembrane helix. There are a number of different hydropathy scales for estimating the relative hydrophobicity of amino acid residues (for review see White[130]) and that of Kyte and Doolittle[131] is one of the most used. It is based on thermodynamic considerations and on the extent to which a given type of residue has been found buried in the nonpolar interior of globular, water-soluble proteins. In Fig. 3, hydropathy profiles are shown for a microsomal, a mitochondrial, and a bacterial P450. From the hydropathy profiles it is clear that the region corresponding to the proposed SA sequence of the microsomal P450s is by far the most hydrophobic segment. No corresponding hydrophobic region is present in the bacterial or the mitochondrial P450s. Also, most microsomal P450s exhibit a stretch of hydrophobic residues in the region aligning with helix E'/E that is not found in the water-soluble bacterial enzymes (see Fig. 1).

Based on hydropathy analysis of the polypeptide sequence of the bovine P450 11A1, Degli Esposti et al.[132] concluded that there was no evidence for the presence of transmembrane segments in mitochondrial P450s. Hence, mitochondrial P450s could be considered as extrinsic membrane proteins. A three-dimensional model of P450 11A1 based on the crystallographic coordinates for P450 101 has been presented.[133] Based on this structural model and on the sequence alignment with P450 101, an inserted amphipathic sequence between the amino acid residues that align with helices E and F of P450 101 was suggested

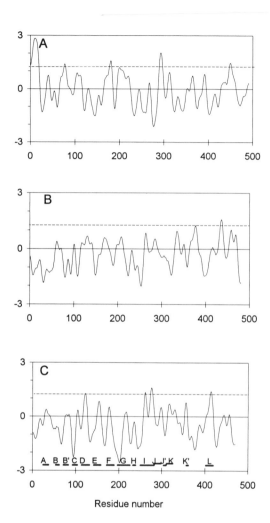

FIGURE 3. Hydropathy profiles of three types of cytochromes P450. (A) P450 2B1 (microsomal); (B) P450 11A1 (mitochondrial); (C) heme domain of P450 102 (bacterial). The amino acid sequences were obtained from the Swiss protein data base. Accession numbers: P00176 (P450 2B1), P00189 (P450 11A1), and P14779 (P450 102). The software package pSAAM (Protein Sequence Analysis and Modeling) written by A. R. Crofts (University of Illinois) was employed to generate the hydropathy profiles using the recommended settings of a window of 7 followed by smoothing through two cycles with a window of 7. Positive values represent stretches of mainly hydrophobic residues. All plots use the hydropathy index of Kyte and Doolittle[131] and have the same scale and abscissa length to facilitate comparisons. In plot C, the locations of the α-helical elements of the heme domain of P450 102 are indicated by heavy horizontal lines at the bottom of the plot. The dotted line is set at 1.25, which represents the lowest value for a known α-helical transmembrane segment.

as a likely candidate for membrane binding.[133] However, it should be noted that Usanov *et al.*[134] found that antibodies to P450 11A1 could recognize parts of the protein on both sides of the inner mitochondrial membrane, suggesting that the protein did span the inner mitochondrial membrane.

From the comparison of hydropathy profiles of 34 aligned P450 sequences, Nelson and Strobel[135] proposed a model for the microsomal P450s in which the protein is anchored to the membrane by two N-terminal transmembrane helices. In this model, the pair of N-terminal helices forms a transmembrane hairpin with the N-terminus at the cytoplasmic surface and a small region of the molecule facing the luminal side of the membrane. This model for the anchor region is less likely in light of the studies described above that show the N-terminus is most likely located in the lumen.

Recent evaluation of hydropathy profiles of the primary sequences of microsomal P450s supports the idea that these proteins have a simple membrane topology in which only the hydrophobic region at the N-terminus traverses the membrane.[86,132,136] However, the lack of an N-terminal hydrophobic sequence in membrane-associated mitochondrial P450s, together with results obtained with N-terminally truncated P450s, argues for the presence of hydrophobic regions other than the SA sequence that are involved in membrane binding.

3.3.2. Topology Based on Immunolocalization Studies Using Antibodies

Both polyclonal and monoclonal antibodies have been used as tools to study the membrane topology of P450s. Antibodies have been raised either against synthetic peptides corresponding to specified segments on a given P450 or by immunization with the intact enzyme. Binding of an antibody to membrane-bound P450s provides evidence for an exposed location of the epitope, provided that the epitope to which the antibody binds has been correctly mapped.

De Lemos-Chiarandini *et al.*[82] studied the reactivity of site-specific polyclonal antibodies raised to peptides corresponding to 15 different regions of rat P450 2B1. Immunocytochemical experiments using colloidal gold-labeled antibodies showed that several of these antibodies bound to the cytoplasmic surface of the microsomal membranes of intact isolated rat liver microsomes (see Table II and Fig. 4). Antibodies raised against the peptide corresponding to parts of the SA sequence (amino acid residues 1–30) of P450 2B1 showed poor binding to microsomes containing P450 2B1 but bound relatively strongly to the purified enzyme, which is concordant with the hypothesis that the major portion of the SA sequence is embedded in the ER membrane. None of the antibodies were found to bind to epitopes localized on the luminal side of the ER membrane, providing further evidence that the majority of the P450 is cytoplasmic.

Edwards *et al.*[137] used a similar approach to study the orientation of mammalian P450s with respect to the membrane. They built a generalized model of eukaryotic P450s using the crystallographic coordinates for residues 10–414 of P450 101 to which a membrane anchoring helix was added onto the N-terminus. The predicted binding sites of ten antipeptide antibodies were mapped on the structural model. Based on the exposure of the antibody binding sites on the structural model it was concluded that the P450 molecule most likely is oriented such that the heme plane lies perpendicular to the plane of the lipid

TABLE II

Location of Antipeptide Antibody Binding Sites and of Mapped Epitopes of Monoclonal
Antibodies on the Primary Structure of P450 2B1[a]

Predicted position on P450 2B1[b]	Position in various P450s	Inhibitory[c]	Binding to microsomes[d]	Ref.
18–29	2B4 18–29	nr[e]	yes	138
24–38	2B1 24–38	–	no	82
40–48	2B1 40–48	–	no	82
61–72	2B1 61–72	+	yes	81, 82
93–98	2B1 93–98	–	no	82
108–116	2B1 108–116	+	yes	81, 82
116–119	2C5 115–118[f]	+	yes	10
122–131	2B1 122–131	+	yes	82
153–161	1A1 174–182	nr	yes	184
186–193	2B1 186–193	–	yes	82
211–223	2B1 211–223	–	no	82
225–232	2B1 225–232	–	yes	82
226–233	2E1 227–234	nr	yes	137
251–263	2C5 250–262[f]	+	yes	[g]
256–263	2D6 262–270[f]	+	yes	185, 186
272–279	1A2 290–296	+	yes	145, 187
315–323	2B1 315–323	—[h]	yes	82
333–340	1A2 350–357	–	yes	137, 188
357–362	1A1 380–385[f]	–	yes	189
398–408	2B1 398–408	–	yes	82

[a]The sequence was obtained from the Swiss protein data base. Accession No. P00176.
[b]The sequence alignment of Nelson was used (see Appendix A by Nelson for details).
[c]Antibodies that reduced the activity by more than 50% were considered as inhibitory.
[d]The data concerning the binding to P450 2B1 were taken from the study of De Lemos-Chiarandini.[82] Antipeptide antibody numbers 1 and 8 were not included. Only the antibodies that showed maximum binding (+ + +) to microsomes by employing protein A–colloidal gold as a marker were considered to bind to microsomes.
[e]nr, not reported.
[f]Monoclonal antibody for which the epitope has been mapped.
[g]Richardson and Johnson (unpublished data).
[h]Part of an epitope for an inhibitory monoclonal antibody has been mapped to this region.[188] However, the antibody raised against the corresponding peptide did not show any inhibitory effect. It was suggested that the inhibition of enzymatic activity of P450 1A1 by the monoclonal antibody involves binding to a discontinuous epitope.

bilayer. Furthermore, it was suggested that the entrance to the active site is positioned away from the membrane.

Figure 4 shows the localization of the predicted binding sites for antipeptide antibodies and known epitopes of monoclonal antibodies on a three-dimensional homology model of P450 2B1. The predicted epitopes and some of the properties of the corresponding antibodies are shown in Table II. As can be seen in Fig. 4, a large portion of the P450 molecule appears to be recognized by different antibodies, suggesting that these epitopes are surface accessible. The central parts of the surface on the distal and proximal sides of the heme, Fig. 4, panels A and C, respectively, are not represented by any of the proposed antibody epitopes listed in Table II that generate antibodies reactive with the membrane-bound enzyme. A majority of the mapped antibody epitopes seem to cluster around the edges of the molecule. Most of the antibodies that have been shown to exhibit an inhibitory

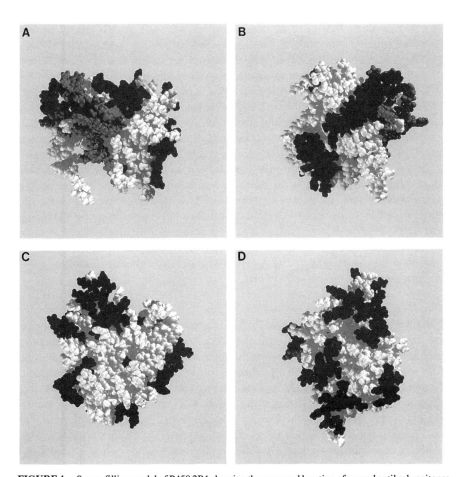

FIGURE 4.   Space-filling model of P450 2B1 showing the proposed location of several antibody epitopes. The three-dimensional structure was modeled based on the crystallographic coordinates for P450 101[170] and further refined using the coordinates for P450s 102 and 108 (Grazyna D. Szklarz, personal communication). The modeled protein (A) was rotated around the y-axis by 90° (B), 180° (C), and 270° (D). The view in A is from the distal side of the heme and the view in C is from the proximal side. The NADPH P450 oxidoreductase has been suggested to bind to the proximal side of the molecule.[7,47] The proposed binding sites for the antibodies (see Table II for further information) that recognize membrane-bound P450 are shown in black. The midgray area represents epitopes that appear to be exposed only after the P450 molecule has been solubilized from the microsomal membrane. Antibodies to the peptide corresponding to amino acid residues 24–38 of P450 2B1 showed intermediate binding to microsomes.[82] This region partially overlaps with the peptide (residues 18–29) that was used to generate antibodies that could recognize P450 2B4 in microsomes.[138] In the model of P450 2B1, residues 18–29 are shown in black and residues 30–38 in midgray. The images were created using the Molw molecular visualization program and rendered by ray tracing using Showcase (US Science, San Diego, CA).

effect on enzymatic activity seem to bind at the top edges of the molecule as oriented in Fig. 4. Surfaces of the P450 that either face the membrane or participate in protein–protein interactions between P450 molecules are likely to be parts of the surface on membrane-bound P450s to which no antibody binding sites have been mapped or to which antibodies only bind after the P450 has been solubilized from the membrane.

Recently, an immunochemical approach was used to examine whether microsomal P450s are bound to the membrane via a transmembranous hairpin as proposed by Nelson and Strobel.[135] Site-specific monoclonal antibodies, raised against a synthetic peptide corresponding to the C-terminal portion (amino acid residues 18–29) of the SA sequence of rabbit P450 2B4, were shown to readily react with P450 2B4 in isolated microsomes.[138] If P450 2B4 was anchored to the membrane by two transmembrane helices, then this epitope would be exposed on the luminal side, whereas if only a single transmembrane segment was involved, the epitope would be found on the cytoplasmic side of the ER membrane. When the microsomes were solubilized with detergent, no enhanced binding by the antibodies was detected, suggesting that exposure of a putative epitope hidden in the lumen did not occur after solubilization. This further indicates that the epitope is localized on the cytoplasmic side of the membrane and that only one N-terminal transmembrane segment is present in P450 2B4.

3.3.3. Topology Probed by Proteases

Proteolytic enzymes are commonly utilized to address questions regarding membrane protein topology. For example, trypsin has been used to determine which parts of the P450 molecule are exposed on the outer surface of microsomal vesicles. Those regions of the P450 polypeptide chain that are likely to be accessible to proteolytic cleavage are exposed loops within domains or linkage regions between domains. It has been shown that cytochrome P450 is sensitive to proteolytic digestion in intact microsomes.[127,139] Unlike cytochrome b_5 or the NADPH P450 oxidoreductase, limited proteolysis using trypsin or other proteases has not been shown to liberate any hydrophilic domain(s) with catalytic activity or that contain(s) heme. However, in a recent study[140] using a genetically engineered variant of yeast P450 52A3, in which a recognition site for endoproteinase Factor Xa had been introduced at position 63–66 in the primary sequence, it was demonstrated that proteolysis resulted in a water-soluble, enzymatically active form of the protein.

Brown and Black[141] treated isolated rabbit liver microsomes with trypsin and purified the resulting peptides and subjected them to N-terminal amino acid microsequencing. The rabbits were treated with phenobarbital prior to the isolation of microsomes in order to induce the levels of P450 2B so that it was the predominant P450. From these experiments it was concluded that most of the P450 2B4, except for the N-terminal region up to about residue 50, was accessible to proteolytic digestion with trypsin. However, a portion of this N-terminal region is reactive with antipeptide antibodies raised to residues 18–29 of P450 2B4 which include two arginines and one lysine residue that are potential cleavage sites for trypsin. It is not clear why trypsin-sensitive sites could not be identified in this region of the molecule as it seems to be relatively well exposed for antibody binding.

Vergères et al.[142] have presented additional data in favor of the single membrane-spanning model. They found that when isolated rat P450 2B1 was incorporated into phospholipid vesicles and then exposed to extensive trypsinolysis, only one peptide corresponding to the SA sequence of 2B1 (amino acid residues Met[1] to Arg[21]) remained bound to the liposomes.[142] In contrast, exhaustive trypsinolysis of P450 17 or P450 21 incorporated into liposomes did not release any peptides from the vesicles.[143] SDS-PAGE analysis revealed that the P450 17 molecule was almost completely resistant to cleavage by trypsin. On the other hand, trypsinolysis of P450 21 resulted in the formation of four major fragments with molecular masses of about 30, 25, 20, and 6 kDa. A similar result was observed when isolated bovine adrenocortical microsomes were subjected to trypsin treatment, followed by SDS-PAGE and analyzed by Western blotting using antibodies raised against P450 21,[144] indicating that a large portion of the molecule may be embedded in the membrane and hence protected from proteolysis. The site of cleavage was identified by N-terminal amino acid sequence analysis of the three largest fragments obtained after trypsin digestion. Based on sequence alignments of the three bacterial P450s with known three-dimensional structures, the two major cleavage sites fall in putative loop regions between helices H and I and between helices J and K, respectively, indicating that these regions are exposed to the aqueous phase. The major sites of trypsinolysis of rat liver P450 forms 1A1, 2B1, and 2E1 have also been mapped to the area corresponding to the connecting region between helices H and I of P450 101. As discussed above, antibodies that recognize a continuous epitope have been mapped to this region of P450 1A1 and P450 1A2,[145,146] providing further evidence that this portion of the molecule is exposed at the surface.

Interestingly, the trypsin cleavage of P450 21 incorporated into liposomes did not appear to affect the enzymatic activity or the spectroscopic properties of the protein.[144] This may be because portions of the P450 molecule are embedded in the lipid phase of the liposomes and that hydrophobic interactions with the membrane somehow stabilize the active site. Similar results were obtained with isolated mitochondrial P450 11B1 incorporated into phospholipid vesicles and treated with trypsin.[147] Even though trypsin was shown to cleave P450 11B1 into several fragments, it did not affect the binding of heme to the protein and the enzyme retained some catalytic activity.

3.3.4. Biophysical Studies Related to Topology

The bacterial P450s with known structure and the models of eukaryotic P450s derived from these structures have an asymmetric shape resembling a triangular prism (e.g., see Fig. 4). Thus, the distance between the heme and the membrane depends on the orientation of the molecule with respect to the membrane plane. Data from fluorescence quenching and energy-transfer studies using P450 isolated from rat microsomes incorporated into lipid vesicles, suggest that the heme is located approximately 70 Å from the center of the lipid bilayer.[148]

Electron paramagnetic resonance (EPR) spectroscopy on oriented membrane multi-layers has been used to study the orientation of the heme group of several hemoproteins with respect to the lipid/water interface. These studies indicate that the heme-plane orientation of P450 11A1 and P450 11B1, unlike most integral membrane hemoproteins, is largely parallel to the membrane plane.[149,150] This orientation implies that the heme group is relatively close to the membrane surface. In sharp contrast to the distance derived from the fluorescence quenching and energy-transfer experiments for the microsomal P450, EPR studies suggest that the distance of the heme from the membrane surface is less than 20 Å for the two mitochondrial P450s. The reason for this difference is not clear.

Several biophysical studies on the rotational diffusion of P450 in the plane of the bilayer of liposomes indicate that part of the enzyme is deeply embedded in the lipid bilayer.[143,151,152]The rotational mobilities of both mitochondrial (P450 11A1, 11B1) and microsomal (P450 1A1, 1A2, 2B1, 2B2, 17, 21) P450 enzymes have been studied. The slow rotation of the P450 molecule in membranes is similar to that of integral membrane proteins having multiple membrane-spanning segments, such as cytochrome oxidase[153] and bacteriorhodopsin.[154] Studies on P450 1A2 and 2B4 indicate that these P450s exist in a hexameric quaternary structure when bound to membranes.[155,156]Based on the mobility of P450 2B4 in lipid vesicles, Schwarz et al.[157] suggested that the slow rotation is mainly the result of interaction of the oligomeric form of P450 2B4 with the membrane surface, or if the molecule is partly embedded, with the hydrophobic core of the membrane. In a recent study,[104] the mobility in yeast microsomes of the truncated rat P450 1A1 described earlier, was shown to be similar to that of the full-length protein. These data indicate that the truncated P450 1A1, even though it lacks the hydrophobic N-terminal tail, is firmly bound to the membrane via at least two hydrophobic surfaces which restrict the freedom of rotation of the molecule in the membrane. Alternatively, if the truncated 1A1 exists in an oligomeric conformation in the membrane, only one hydrophobic surface per monomer would be required to limit rotational freedom.

3.3.5. Membrane-Binding Regions

Before discussing other portions of the P450 molecule that, in addition to the N-terminal SA region of the microsomal P450s, interact with the lipid bilayer, we will briefly describe the salient features of the few membrane proteins for which structures are known. High-resolution structures of membrane proteins exist for the bacterial photosynthetic reaction centers,[158,159]the bacteriorhodopsin of Halobacterium halobium,[160]the pea light-harvesting complex II,[161] the porins that are present in the outer membrane of gram-negative bacteria,[162,163] and the ovine prostaglandin H synthase.[164] All of these membrane proteins, except the porins, ligate one or several chromophoric cofactors. Based on their folding motifs, revealed by structural analysis, they can be divided into three groups. The first group comprises polytopic membrane proteins in which the transmembrane domains are composed of a bundle of α-helices that are oriented more or less perpendicular to the plane of the membrane. The porins define the second group in which the membrane domain is formed by a large antiparallel β-sheet. The recent structural analysis of the prostaglandin H synthase (PGHS), a hemoprotein that is bound to the ER membrane, revealed that this enzyme has a unique membrane-binding motif and hence could be classified into a third group. In contrast to the members of group 1 and 2, PGHS is a monotopic membrane protein that has a large hydrophilic domain that most likely faces the luminal side of the ER membrane. The protein does not span the membrane but is thought to be anchored to the lipid bilayer by a membrane-binding motif consisting of three amphipathic helices (A, B, and C) and the N-terminal portion of a fourth helix (D). The structure of PGHS predicts that the A, B, and C helices lie parallel to the plane of the membrane such that the hydrophobic side chains can be buried in the hydrophobic core of the membrane. The lipophilic substrate is proposed to enter the active site through a long, narrow, hydrophobic channel that faces the membrane. A similar hydrophobic channel is present in P450 102.[46] In the crystallized form, PGHS exists as a dimer in which the proposed membrane anchor domain forms an extensive hydrophobic surface on one side of the enzyme. It is interesting to note that current algorithms designed to produce hydropathy plots cannot be used to identify the membrane anchor region of PGHS. In fact, several models based on hydropathy profile analyses of PGHS have predicted one or more transmembrane segments.[165] These models are incompatible with the X-ray structure of PGHS that shows that the hydrophobic region suggested to span the membrane corresponds to parts of the catalytic domain.

Eukaryotic P450s could have a similar membrane-binding motif as that found in PGHS with the substrate access channel facing the membrane. The hydrophobicity of the membrane-anchoring domain would depend on the number of amino acid residues involved, and their relative hydrophobicity. The more hydrophobic the membrane-binding region is, the deeper the molecule could be embedded in the core of the membrane. This type of membrane interaction would allow different members of the eukaryotic P450s to have similar overall three-dimensional structure and at the same time be more or less strongly bound and inserted into lipid bilayers. The way in which P450s interact with membranes may be of importance for regulating substrate specificity as has been proposed by several groups.[12,166,167] A P450 molecule that has the substrate access channel deeply embedded in the membrane would have a preference for very hydrophobic substrates, whereas a less buried molecule could accept moderately lipophilic substrates. Furthermore, a PGHS type of membrane interaction could also explain the sometimes contradictory results obtained concerning membrane binding of different P450 enzymes.

Recently, Uvarov et al.[168] attempted to identify possible membrane-binding regions of P450 2B4 using three different radioactively labeled photoreactive phospholipids. The phospholipids had the photoreactive group either close to the apolar end or at the polar side of the fatty acid chain. After chemical cleavage of the polypeptide with cyanogen bromide, three peptides (42–65 residues long) were identified as having reacted with the radiolabeled phospholipids. The phospholipids that had the reactive group at the apolar end only reacted with a peptide that corresponds to the region that covers the SA sequence of P450 2B4. The identified peptides mapped to the same side of a three-dimensional model of P450 2B4 generated by a computer-aided molecular modeling package using the coordinates for P450 101 as a framework. Two regions apart from the SA region were suggested as candidates for membrane binding. The first aligns with the beginning of helix I (residues 283–292) and the second is very close to the C-terminus of the protein (residues 471–480). However, it should be noted that if the substrate access channel opens into the membrane bilayer, the photoreactive phospholipids may gain access to the active site. Hence, regions that are not directly in contact with the membrane could become labeled.

Recently, Peterson and colleagues[7] constructed a detailed molecular model for the human aromatase (P450 19) based on the three-dimensional structure of the heme domain of P450 102. From this structural model of P450 19 it is seen that the region at the surface of the molecule that is most hydrophobic is found in the region surrounding the proposed entrance to the substrate access channel. The region that potentially could be partly buried in the lipid bilayer includes parts of the β1 and β2 sheets, the B' helix, and the F–G loop. Apart from the area of the B' helix, no antibodies have been shown to bind to these regions of membrane-bound P450s.

Taken together, the data presented above indicate that portions of the distal face (Fig. 4A) of eukaryotic P450s are likely to provide the amino acid side chains that in addition to the SA sequence bind the P450 molecule to the membrane in a rather rigid way. However, it still remains to be more precisely established which regions are responsible for the peripheral membrane binding.

4. Summary and Future Directions

In this chapter, we have reviewed the current information on the membrane topology and structure of various eukaryotic P450 isozymes. The majority of the eukaryotic P450s are associated with membranes; most are found in the ER membrane while others are present in the inner membrane of mitochondria. The microsomal enzymes are most likely anchored to the ER membrane by a single N-terminal transmembrane-spanning α-helix. In addition, several microsomal P450s appear to have other regions that are inserted in and interact with the interfacial region of the lipid bilayer. The mitochondrial P450 enzymes do not contain a hydrophobic N-terminal region equivalent to that present in the microsomal enzymes and are likely to be anchored by hydrophobic regions that are partially inserted in the membrane. From a number of studies utilizing diverse techniques, it has become apparent that the major bulk of the eukaryotic P450s is solvent exposed and that the active site is located at the external face of the membrane.

X-ray crystallography is one of the most powerful techniques to gain information on the three-dimensional structure of proteins. Unfortunately, because of the general difficulties involved in growing large and sufficiently well-ordered crystals of membrane proteins, high-resolution structural data for the membrane-associated P450s are still lacking. Therefore, a number of noncrystallographic methods have been used to probe the structures of the eukaryotic P450s. Studies involving site-directed mutagenesis have identified a number of key residues that are important for substrate specificity and catalytic activity. These studies have shown that relatively conservative single amino acid residue substitutions can confer new substrate specificities onto P450s or alter the stereo- and regioselectivity of these enzymes. In general, the above-mentioned key residues have been shown by sequence comparisons to map to substrate binding residues of the bacterial P450s, P450 101, 102, and 108, for which the structures are known. Accumulating data from sequence analysis and other studies suggest that there exists substantial conservation of structure between most P450s. Indeed, the crystallographic studies on the three bacterial P450s have shown that despite low sequence similarities, these enzymes share an overall similar fold (see Chapters 4 and 5).

During the past few years molecular modeling has become an increasingly powerful method for predicting three-dimensional protein structures. Homology model building based on the crystal coordinates of P450 101 have been reported for P450 1A1,[35,169] P450 2B1,[170] P450 2C9,[167] P450 2D6,[171] P450 2E1,[172] P450 3A4,[173] P450 11A1,[133] P450

17A,[174] P450 19,[69,175] and P450 51.[176,177] The availability of the coordinates for the recently determined P450 structures and of others that will follow, including that of P450$_{eryF}$ (P450 107),[178] are likely to increase the accuracy of P450 tertiary structure predictions. Recently, the structure of the human thromboxane synthase (P450 5A) was modeled using the coordinates for P450 102 as a template.[179] The primary structure of thromboxane synthase is more similar to P450 102 than to P450 101. The accuracy of the alignment with P450 101 is therefore likely to be lower and hence it was concluded that P450 102 provides a better template for homology modeling of the thromboxane synthase than P450 101.

The low sequence identity, 10 to 30%, between prokaryotic and eukaryotic P450s makes it extremely difficult to generate an accurate high-resolution model based on the prokaryotic P450 structures. In addition, the diversity in primary structure and physico-chemical properties such as substrate specificity and membrane binding among the P450s is likely to be reflected by a variation in helical topography and overall structure of these enzymes. Homology-based models would be greatly aided by experimentally determined high-resolution structures of eukaryotic P450s. The elucidation of the three-dimensional structure for P450s belonging to different subfamilies will ultimately have to be achieved. The development of heterologous expression systems for high-level expression[180] together with the possibilities of obtaining functional eukaryotic P450s without membrane-binding regions increase the potential for the determination of a high-resolution structure for these P450s in the not too distant future.

ACKNOWLEDGMENTS.   We wish to thank Dr. Grazna Szklarz and Dr. James Halpert for providing us with the coordinates for their revised model of P450 2B1 based on P450s 101, 102, and 108. The helpful comments of Dr. Toby R. Richardson and Keith J. Griffin during the preparation of this manuscript were greatly appreciated. The authors' work is supported by USPHS grant GM31001 and a long-term fellowship to C.v.W. from the European Molecular Biology Organization.

References

1. Gotoh, O., and Fujii-Kuriyama, Y., 1989, Evolution, structure, and gene regulation of cytochrome P-450, in: *Frontiers in Biotransformation. Basis and Mechanisms of Regulation of Cytochrome P-450* (K. Ruckpaul and H. Rein, eds.), Taylor & Francis, London, pp. 195–243.
2. Koymans, L., Donné-Op den Kelder, G. M., Koppele Te, J. M., and Vermeulen, N. P. E., 1993, Cytochromes P450: Their active-site structure and mechanism of oxidation, *Drug Metab. Rev.* **25**:325–387.
3. Kalb, V. F., and Loper, J. C., 1988, Proteins from eight eukaryotic cytochrome P-450 families share a segmented region of sequence similarity, *Proc. Natl. Acad. Sci. USA* **85**:7221–7225.
4. White, R. E., 1991, The involvement of free radicals in the mechanisms of monooxygenases, *Pharmacol. Ther.* **49**:21–42.
5. Ator, M. A., and Ortiz de Montellano, P. R., 1987, Protein control of prosthetic heme reactivity. Reaction of substrates with the heme edge of horseradish peroxidase, *J. Biol. Chem.* **262**:1542–1551.
6. Nelson, D. R., Kamataki, T., Waxman, D. J., Guengerich, F. P., Estabrook, R. W., Feyereisen, R., Gonzalez, F. J., Coon, M. J., Gunsalus, I. C., Gotoh, O., Okuda, K., and Nebert, D. W., 1993, The P450 superfamily: Update on new sequences, gene mapping, accession numbers, early trivial names of enzymes, and nomenclature, *DNA Cell Biol.* **12**:1–51.

7. Graham-Lorence, S. E., and Peterson, J. A., 1995, The molecular structure of P450s: The conserved and the variable elements, in: *Advances in Molecular and Cell Biology* (C. Jefcoate, ed.), JAI Press, Greenwich, CT.

8. Lindberg, R. L. P., and Negishi, M., 1989, Alteration of mouse cytochrome P450coh substrate specificity by mutation of a single amino-acid residue, *Nature* **339:**632–634.

9. Lindberg, R. L. P., and Negishi, M., 1991, Modulation of specificity and activity in mammalian cytochrome P-450, *Methods Enzymol.* **202:**741–752.

10. Kronbach, T., and Johnson, E. F., 1991, An inhibitory monoclonal antibody binds in close proximity to a determinant for substrate binding in cytochrome P450IIC5, *J. Biol. Chem.* **266:**6215–6220.

11. Aoyama, T., Korzekwa, K., Nagata, K., Adesnik, M., Reiss, A., Lapenson, D. P., Gillette, J., Gelboin, H. V., Waxman, D. J., and Gonzalez, F. J., 1989, Sequence requirements for cytochrome P-450IIB1 catalytic activity. Alteration of the stereospecificity and regioselectivity of steroid hydroxylation by a simultaneous change of two hydrophobic amino acid residues to phenylalanine, *J. Biol. Chem.* **264:**21327–21333.

12. Christou, M., Mitchell, M. J., Aoyama, T., Gelboin, H. V., Gonzalez, F. J., and Jefcoate, C. R., 1992, Selective suppression of the catalytic activity of cDNA-expressed cytochrome P4502B1 toward polycyclic hydrocarbons in the microsomal membrane: Modification of this effect by specific amino acid substitutions, *Biochemistry* **31:**2835–2841.

13. Lindberg, R. L. P., Juvonen, R., and Negishi, M., 1992, Molecular characterization of the murine Coh locus: An amino acid difference at position 117 confers high and low coumarin 7-hydroxylase activity in P450coh, *Pharmacogenetics* **2:**32–37.

14. Matsunaga, E., Zeugin, T., Zanger, U. M., Aoyama, T., Meyer, U. A., and Gonzalez, F. J., 1990, Sequence requirements for cytochrome P-450IID1 catalytic activity: A single amino acid change (ILE^{380}PHE) specifically decreases V_{max} of the enzyme for bufuralol but not debrisoquine hydroxylation, *J. Biol. Chem.* **265:**17197–17201.

15. Hsu, M.-H., Griffin, K. J., Wang, Y., Kemper, B., and Johnson, E. F., 1993, A single amino acid substitution confers progesterone 6β-hydroxylase activity to rabbit cytochrome P450 2C3, *J. Biol. Chem.* **268:**6939–6944.

16. Kaminsky, L. S., De Morais, S. M. F., Faletto, M. B., Dunbar, D. A., and Goldstein, J. A., 1993, Correlation of human cytochrome P4502C substrate specificities with primary structure: Warfarin as a probe, *Mol. Pharmacol.* **43:**234–239.

17. Veronese, M. E., Doecke, C. J., Mackenzie, P. I., McManus, M. E., Miners, J. O., Rees, D. L. P., Gasser, R., Meyer, U. A., and Birkett, D. J., 1993, Site-directed mutation studies of human liver cytochrome P-450 isoenzymes in the CYP2C subfamily, *Biochem. J.* **289:**533–538.

18. Goldstein, J. A., Faletto, M. B., Romkes-Sparks, M., Sullivan, T., Kitareewan, S., Raucy, J. L., Lasker, J. M., and Ghanayem, B. I., 1994, Evidence that CYP2C19 is the major (S)-mephenytoin 4'-hydroxylase in humans, *Biochemistry* **33:**1743–1752.

19. Kronbach, T., Kemper, B., and Johnson, E. F., 1990, Multiple determinants for substrate specificities in cytochrome P450 isozymes, in: *Current Research in Protein Chemistry: Techniques, Structure, and Function* (J. J. Villafranca, ed.), Academic Press, New York, pp. 481–488.

20. Richardson, T. H., and Johnson, E. F., 1994, Alterations of the regiospecificity of progesterone metabolism by the mutagenesis of two key amino acid residues in rabbit cytochrome P450 2C3v, *J. Biol. Chem.* **269:**23937–23943.

21. Halpert, J. R., and He, Y., 1993, Engineering of cytochrome P450 2B1 specificity. Conversion of an androgen 16β-hydroxylase to a 15α-hydroxylase, *J. Biol. Chem.* **268:**4453–4457.

22. Kedzie, K. M., Balfour, C. A., Escobar, G. Y., Grimm, S. W., He, Y., Pepperl, D. J., Regan, J. W., Stevens, J. C., and Halpert, J. R., 1991, Molecular basis for a functionally unique cytochrome P450IIB1 variant, *J. Biol. Chem.* **266:**22515–22521.

23. He, Y., Balfour, C. A., Kedzie, K. M., and Halpert, J. R., 1992, Role of residue 478 as a determinant of the substrate specificity of cytochrome P450 2B1, *Biochemistry* **31:**9220–9226.

24. Laethem, R. M., Halpert, J. R., and Koop, D. R., 1994, Epoxidation of arachidonic acid as an active-site probe of cytochrome P-450 2B isoforms, *Biochim. Biophys. Acta Protein Struct. Mol. Enzymol.* **1206**:42–48.

25. Luo, Z., He, Y., and Halpert, J. R., 1994, Role of residues 363 and 206 in conversion of cytochrome P450 2B1 from a steroid 16-hydroxylase to a 15α-hydroxylase, *Arch. Biochem. Biophys.* **309**:52–57.

26. Iwasaki, M., Darden, T. A., Pedersen, L. G., Davis, D. G., Juvonen, R. O., Sueyoshi, T., and Negishi, M., 1993, Engineering mouse P450coh to a novel corticosterone 15α-hydroxylase and modeling steroid-binding orientation in the substrate pocket, *J. Biol. Chem.* **268**:759–762.

27. Iwasaki, M., Darden, T. A., Parker, C. E., Tomer, K. B., Pedersen, L. G., and Negishi, M., 1994, Inherent versatility of P-450 oxygenase. Conferring dehydroepiandrosterone hydroxylase activity to P-450 2a-4 by a single amino acid mutation at position 117, *J. Biol. Chem.* **269**:9079–9083.

28. Johnson, E. F., 1992, Mapping determinants of the substrate selectivities of P450 enzymes by site-directed mutagenesis, *Trends Pharmacol. Sci.* **13**:122–126.

29. Gotoh, O., 1992, Substrate recognition sites in cytochrome P450 family 2 (CYP2) proteins inferred from comparative analyses of amino acid and coding nucleotide sequences, *J. Biol. Chem.* **267**:83–90.

30. He, Y., Luo, Z., Klekotka, P. A., Burnett, V. L., and Halpert, J. R., 1994, Structural determinants of cytochrome P450 2B1 specificity: Evidence for five substrate recognition sites, *Biochemistry* **33**:4419–4424.

31. Hasler, J. A., Harlow, G. R., Szklarz, G. D., John, G. H., Kedzie, K. M., Burnett, V. L., He, Y.-A., Kaminsky, L. S., and Halpert, J. R., 1994, Site-directed mutagenesis of putative substrate recognition sites in cytochrome P450 2B11: Importance of amino acid residues 114, 290, and 363 for substrate specificity, *Mol. Pharmacol.* **46**:338–345.

32. Poulos, T. L., 1989, Site-directed mutagenesis: Reversing enzyme specificity, *Nature* **339**:580–581.

33. Nelson, D. R., and Strobel, H. W., 1989, Secondary structure prediction of 52 membrane-bound cytochromes P450 shows a strong structural similarity to P450cam, *Biochemistry* **28**:656–660.

34. Edwards, R. J., Murray, B. P., Boobis, A. R., and Davies, D. S., 1989, Identification and location of α-helices in mammalian cytochrome P450, *Biochemistry* **28**:3762–3770.

35. Zvelebil, M. J. J. M., Wolf, C. R., and Sternberg, M. J. E., 1991, A predicted three-dimensional structure of human cytochrome P450: Implications for substrate specificity, *Protein Eng.* **4**:271–282.

36. Juvonen, R. O., Iwasaki, M., and Negishi, M., 1991, Structural function of residue-209 in coumarin 7-hydroxylase (P450coh). Enzyme-kinetic studies and site-directed mutagenesis, *J. Biol. Chem.* **266**:16431–16435.

37. Iwasaki, M., Juvonen, R., Lindberg, R., and Negishi, M., 1991, Alteration of high and low spin equilibrium by a single mutation of amino acid 209 in mouse cytochromes P450, *J. Biol. Chem.* **266**:3380–3382.

38. Juvonen, R. O., Iwasaki, M., and Negishi, M., 1992, Roles of residues 129 and 209 in the alteration by cytochrome *b*5 of hydroxylase activities in mouse 2A P450S, *Biochemistry* **31**:11519–11523.

39. Juvonen, R. O., Iwasaki, M., Sueyoshi, T., and Negishi, M., 1993, Structural alteration of mouse P450coh by mutation of glycine-207 to proline: Spin equilibrium, enzyme kinetics, and heat sensitivity, *Biochem. J.* **294**:31–34.

40. Straub, P., Johnson, E. F., and Kemper, B., 1993, Hydrophobic side chain requirements for lauric acid and progesterone hydroxylation at amino acid 113 in cytochrome P450 2C2, a potential determinant of substrate specificity, *Arch. Biochem. Biophys.* **306**:521–527.

41. Straub, P., Lloyd, M., Johnson, E. F., and Kemper, B., 1993, Cassette-mutagenesis of a potential substrate recognition region of cytochrome P450 2C2, *J. Biol. Chem.* **268**:21997–22003.

42. Straub, P., Lloyd, M., Johnson, E. F., and Kemper, B., 1994, Differential effects of mutations in substrate recognition site 1 of cytochrome P450 2C2 on lauric acid and progesterone hydroxylation, *Biochemistry* **33**:8029–8034.

43. Uno, T., Yokota, H., and Imai, Y., 1990, Replacing the carboxy-terminal 28 residues of rabbit liver P-450 (laurate (omega-1)-hydroxylase) with those of P-450 (testosterone 16α-hydroxylase) produces a new stereospecific hydroxylase activity, *Biochem. Biophys. Res. Commun.* **167**:498–503.

44. Ramarao, M. K., Straub, P., and Kemper, B., 1995, Identification by in vitro mutagenesis of the interaction of two segments of C2MSTC1, a chimera of cytochromes P450 2C2 and P450 2C1, *J. Biol. Chem.* **270**:1873–1880.

45. Kronbach, T., Kemper, B., and Johnson, E. F., 1991, A hypervariable region of P450IIC5 confers progesterone 21-hydroxylase activity to P450IIC1, *Biochemistry* **30**:6097–6102.

46. Ravichandran, K. G., Boddupalli, S. S., Hasemann, C. A., Peterson, J. A., and Deisenhofer, J., 1993, Crystal structure of hemoprotein domain of P450BM-3, a prototype for microsomal P450's, *Science* **261**:731–736.

47. Poulos, T. L., 1991, Modeling of mammalian P450s on basis of P450cam X-ray structure, *Methods Enzymol.* **206**:11–30.

48. Hasemann, C. A., Ravichandran, K. G., Peterson, J. A., and Deisenhofer, J., 1994, Crystal structure and refinement of cytochrome P450$_{terp}$ at 2·3 Å resolution, *J. Mol. Biol.* **236**:1169–1185.

49. Hanioka, N., Gonzalez, F. J., Lindberg, N. A., Liu, G., Gelboin, H. V., and Korzekwa, K. R., 1992, Site-directed mutagenesis of cytochrome P450s CYP2A1 and CYP2A2: Influence of the distal helix on the kinetics of testosterone hydroxylation, *Biochemistry* **31**:3364–3370.

50. Ding, X., Peng, H.-M., and Coon, M. J., 1994, Structure–function analysis of CYP2A10 and CYP2A11, P450 cytochromes that differ in only eight amino acids but have strikingly different activities toward testosterone and coumarin, *Biochem. Biophys. Res. Commun.* **203**:373–378.

51. Poulos, T. L., Finzel, B. C., and Howard, A. J., 1987, High-resolution crystal structure of cytochrome P450cam, *J. Mol. Biol.* **195**:687–700.

52. Imai, Y., and Nakamura, M., 1988, The importance of threonine-301 from cytochromes P-450 (laurate (w-1)-hydroxylase and testosterone 16α-hydroxylase) in substrate binding as demonstrated by site-directed mutagenesis, *FEBS Lett.* **234**:313–315.

53. Imai, Y., and Nakamura, M., 1989, Point mutations at threonine-301 modify substrate specificity of rabbit liver microsomal cytochromes P-450 (laurate (omega-1)-hydroxylase and testosterone 16α-hydroxylase), *Biochem. Biophys. Res. Commun.* **158**:717–722.

54. Furuya, H., Shimizu, T., Hirano, K., Hatano, M., Fujii-Kuriyama, Y., Raag, R., and Poulos, T. L., 1989, Site-directed mutageneses of rat liver cytochrome P-450d: Catalytic activities toward benzphetamine and 7-ethoxycoumarin, *Biochemistry* **28**:6848–6857.

55. Martinis, S. A., Atkins, W. M., Stayton, P. S., and Sligar, S. G., 1989, A conserved residue of cytochrome P-450 is involved in heme-oxygen stability and activation, *J. Am. Chem. Soc.* **111**:9252–9253.

56. Raag, R., Martinis, S. A., Sligar, S. G., and Poulos, T. L., 1991, Crystal structure of the cytochrome P-450$_{CAM}$ active site mutant Thr252Ala, *Biochemistry* **30**:11420–11429.

57. Tuck, S. F., Hiroya, K., Shimizu, T., Hatano, M., and Ortiz de Montellano, P. R., 1993, The cytochrome P450 1A2 active site: Topology and perturbations caused by glutamic acid-318 and threonine-319 mutations, *Biochemistry* **32**:2548–2553.

58. Tuck, S. F., Peterson, J. A., and Ortiz de Montellano, P. R., 1992, Active site topologies of bacterial cytochromes P450101 (P450*cam*), P450108 (P450*terp*), and P450102 (P450$_{BM-3}$). *In situ* rearrangement of their phenyl-iron complexes, *J. Biol. Chem.* **267**:5614–5620.

59. Swanson, B. A., Dutton, D. R., Lunetta, J. M., Yang, C. S., and Ortiz de Montellano, P. R., 1991, The active sites of cytochromes P450 IA1, IIB1, IIB2, and IIE1. Topological analysis by *in situ* rearrangement of phenyl-iron complexes, *J. Biol. Chem.* **266**:19258–19264.

60. Swanson, B. A., Halpert, J. R., Bornheim, L. M., and Ortiz de Montellano, P. R., 1992, Topological analysis of the active sites of cytochromes P450IIB4 (rabbit), P450IIB10 (mouse), and P450IIB11 (dog) by *in situ* rearrangement of phenyl-iron complexes, *Arch. Biochem. Biophys.* **292**:42–46.

61. Kunze, K. L., Mangold, B. L. K., Wheeler, C., Beilan, H. S., and Ortiz de Montellano, P. R., 1983, The cytochrome P-450 active site, *J. Biol. Chem.* **258**:4202–4207.

62. Ishigooka, M., Shimizu, T., Hiroya, K., and Hatano, M., 1992, Role of Glu318 at the putative distal site in the catalytic function of cytochrome P450$_d$, *Biochemistry* **31**:1528–1531.

63. Hiroya, K., Ishigooka, M., Shimizu, T., and Hatano, M., 1992, Role of Glu318 and Thr319 in the catalytic function of cytochrome P450$_d$ (P4501A2): Effects of mutations on the methanol hydroxylation, *FASEB J.* **6:**749–751.

64. Hiroya, K., Murakami, Y., Shimizu, T., Hatano, M., and Ortiz de Montellano, P. R., 1994, Differential roles of Glu318 and Thr319 in cytochrome P450 1A2 catalysis supported by NADPH-cytochrome P450 reductase and *tert*-butyl hydroperoxide, *Arch. Biochem. Biophys.* **310:**397–401.

65. Chen, S., and Zhou, D., 1992, Functional domains of aromatase cytochrome P450 inferred from comparative analyses of amino acid sequences and substantiated by site-directed mutagenesis experiments, *J. Biol. Chem.* **267:**22587–22594.

66. Amarneh, B., Corbin, C. J., Peterson, J. A., Simpson, E. R., and Graham-Lorence, S., 1993, Functional domains of human aromatase cytochrome P450 characterized by linear alignment and site-directed mutagenesis, *Mol. Endocrinol.* **7:**1617–1624.

67. Zhou, D., Korzekwa, K. R., Poulos, T., and Chen, S., 1992, A site-directed mutagenesis study of human placental aromatase, *J. Biol. Chem.* **267:**762–768.

68. Furuya, H., Shimizu, T., Hatano, M., and Fujii-Kuriyama, Y., 1989, Mutations at the distal and proximal sites of cytochrome P-450$_d$ changed regio-specificity of acetanilide hydroxylations, *Biochem. Biophys. Res. Commun.* **160:**669–676.

69. Graham-Lorence, S., Khalil, M. W., Lorence, M. C., Mendelson, C. R., and Simpson, E. R., 1991, Structure–function relationships of human aromatase cytochrome P-450 using molecular modeling and site-directed mutagenesis, *J. Biol. Chem.* **266:**11939–11946.

70. Kadohama, N., Zhou, D., Chen, S., and Osawa, Y., 1993, Catalytic efficiency of expressed aromatase following site-directed mutagenesis, *Biochim. Biophys. Acta Protein Struct. Mol. Enzymol.* **1163:**195–200.

71. Kadohama, N., Yarborough, C., Zhou, D., Chen, S., and Osawa, Y., 1992, Kinetic properties of aromatase mutants Pro308Phe, Asp309Asn, and Asp309Ala and their interactions with aromatase inhibitors, *J. Steroid Biochem. Mol. Biol.* **43:**693–701.

72. Chen, S., Zhou, D., Swiderek, K. M., Nobuyuki, K., Osawa, Y., and Hall, P. F., 1993, Structure–function studies of human aromatase, *J. Steroid Biochem. Mol. Biol.* **44:**347–356.

73. Imai, Y., and Nakamura, M., 1991, Nitrogenous ligation at the sixth coordination position of the Thr-301 to Lys-mutated P450IIC2 heme iron, *J. Biochem.* **110:**884–888.

74. Fukuda, T., Imai, Y., Komori, M., Nakamura, M., Kusunose, E., Satouchi, K., and Kusunose, M., 1993, Replacement of Thr-303 of P450 2E1[1] with serine modifies the regioselectivity of its fatty acid hydroxylase activity, *J. Biochem.* **113:**7–12.

75. Roberts, E. S., Hopkins, N. E., Zaluzec, E. J., Gage, D. A., Alworth, W. L., and Hollenberg, P. F., 1994, Identification of active-site peptides from ^{3}H-labeled 2-ethynylnaphthalene-inactivated P450 2B1 and 2B4 using amino acid sequencing and mass spectrometry, *Biochemistry* **33:**3766–3771.

76. Miller, J. P., and White, R. E., 1994, Photoaffinity labeling of cytochrome P450 2B4: Capture of active site heme ligands by a photocarbene, *Biochemistry* **33:**807–817.

77. Yun, C.-H., Hammons, G. J., Jones, G., Martin, M. V., Eddy Hopkins, N., Alworth, W. L., and Guengerich, F. P., 1992, Modification of cytochrome P450 1A2 enzymes by the mechanism-based inactivator 2-ethynylnaphthalene and the photoaffinity label 4-azidobiphenyl, *Biochemistry* **31:**10556–10563.

78. Tsujita, M., and Ichikawa, Y., 1993, Substrate-binding region of cytochrome *P*-450$_{scc}$ (*P*-450 XIA1). Identification and primary structure of the cholesterol binding region in cytochrome *P*-450$_{scc}$, *Biochim. Biophys. Acta Protein Struct. Mol. Enzymol.* **1161:**124–130.

79. Ohnishi, T., Miura, S., and Ichikawa, Y., 1993, Photoaffinity labeling of cytochrome *P*-450$_{11\beta}$ with methyltrienolone as a probe for the substrate binding region, *Biochim. Biophys. Acta Protein Struct. Mol. Enzymol.* **1161:**257–264.

80. Kronbach, T., Larabee, T. M., and Johnson, E. F., 1989, Hybrid cytochromes P-450 identify a substrate binding domain in P-450IIC5 and P-450IIC4, *Proc. Natl. Acad. Sci. USA* **86:**8262–8265.

81. Frey, A. B., Waxman, D. J., and Kreibich, G., 1985, The structure of phenobarbital-inducible rat liver cytochrome P-450 isoenzyme PB-4: Production and characterization of site-specific antibodies, *J. Biol. Chem.* **260:**15253–15265.

82. De Lemos-Chiarandini, C., Frey, A. B., Sabatini, D. D., and Kreibich, G., 1987, Determination of the membrane topology of the phenobarbital-inducible rat liver cytochrome P-450 isoenzyme PB-4 using site-specific antibodies, *J. Cell Biol.* **104:**209–219.

83. Shoun, H., Suyama, W., and Yasui, T., 1989, Soluble, nitrate/nitrite-inducible cytochrome P-450 of the fungus, *Fusarium oxysporum*, *FEBS Lett.* **244:**11–14.

84. Black, S. D., 1992, Membrane topology of the mammalian P450 cytochromes, *FASEB J.* **6:**680–685.

85. Vergères, G., Winterhalter, K. H., and Richter, C., 1989, Microsomal cytochrome P-450: Substrate binding, membrane interactions, and topology, *Mutat. Res.* **213:**83–90.

86. Tretiakov, V. E., Degtyarenko, K. N., Uvarov, V. Y., and Archakov, A. I., 1989, Secondary structure and membrane topology of cytochrome P450s, *Arch. Biochem. Biophys.* **275:**429–439.

87. Sakaguchi, M., and Omura, T., 1993, Topology and biogenesis of microsomal cytochrome P-450s, in: *Medicinal Implications in Cytochrome P-450 Catalyzed Biotransformations* (K. Ruckpaul and H. Rein, eds.), Akademie Verlag, Berlin, pp. 59–73.

88. Haugen, D. A., Armes, L. G., Yasunobu, K. T., and Coon, M. J., 1977, Amino-terminal sequence of phenobarbital-inducible cytochrome P-450 from rabbit liver microsomes: Similarity to hydrophobic amino-terminal segments of preproteins, *Biochem. Biophys. Res. Commun.* **77:**967–973.

89. Bar-Nun, S., Kreibich, G., Adesnik, M., Alterman, L., Negishi, M., and Sabatini, D. D., 1980, Synthesis and insertion of cytochrome P-450 into endoplasmic reticulum membranes, *Proc. Natl. Acad. Sci. USA* **77:**965–969.

90. High, S., and Dobberstein, B., 1992, Mechanisms that determine the transmembrane disposition of proteins, *Curr. Opin. Cell Biol.* **4:**581–586.

91. Sakaguchi, M., Katsuyoshi, M., and Sato, R., 1984, Signal recognition particle is required for co-translational insertion of cytochrome P-450 into microsomal membranes, *Proc. Natl. Acad. Sci. USA* **81:**3361–3364.

92. Monier, S., Van Luc, P., Kreibich, G., Sabatini, D. D., and Adesnik, M., 1988, Signals for the incorporation and orientation of cytochrome P450 in the endoplasmic reticulum membrane, *J. Cell Biol.* **107:**457–470.

93. Sakaguchi, M., Mihara, K., and Sato, R., 1987, A short amino-terminal segment of microsomal cytochrome P-450 functions both as an insertion signal and as a stop-transfer sequence, *EMBO J.* **6:**2425–2431.

94. Szczesna-Skorupa, E., Browne, N., Mead, D., and Kemper, B., 1988, Positive charges at the NH_2 terminus convert the membrane-anchor signal peptide of cytochrome P-450 to a secretory signal peptide, *Proc. Natl. Acad. Sci. USA* **85:**738–742.

95. Sato, T., Sakaguchi, M., Mihara, K., and Omura, T., 1990, The amino-terminal structures that determine topological orientation of cytochrome P-450 in microsomal membrane, *EMBO J.* **9:**2391–2397.

96. Ahn, K., Szczesna-Skorupa, E., and Kemper, B., 1993, The amino-terminal 29 amino acids of cytochrome P450 2C1 are sufficient for retention in the endoplasmic reticulum, *J. Biol. Chem.* **268:**18726–18733.

97. Murakami, K., Mihara, K., and Omura, T., 1994, The transmembrane region of microsomal cytochrome P450 identified as the endoplasmic reticulum retention signal, *J. Biochem.* **116:**164–175.

98. Szczesna-Skorupa, E., and Kemper, B., 1989, NH_2-terminal substitutions of basic amino acids induce translocation across the microsomal membrane and glycosylation of rabbit cytochrome P450IIC2, *J. Cell Biol.* **108:**1237–1243.

99. Sakaguchi, M., Tomiyoshi, R., Kuroiwa, T., Mihara, K., and Omura, T., 1992, Functions of signal and signal-anchor sequences are determined by the balance between the hydrophobic segment and the N-terminal charge, *Proc. Natl. Acad. Sci. USA* **89:**16–19.

100. Hsu, L.-C., Hu, M.-C., Cheng, H.-C., Lu, J.-C., and Chung, B., 1993, The N-terminal hydrophobic domain of P450c21 is required for membrane insertion and enzyme stability, *J. Biol. Chem.* **268:**14682–14686.

101. Wolin, S. L., 1994, From the elephant to E. coli: SRP-dependent protein targeting, *Cell* **77:**787–790.

102. Yabusaki, Y., Murakami, H., Sakaki, T., Shibata, M., and Ohkawa, H., 1988, Genetically engineered modification of P450 monooxygenases: Functional analysis of the amino-terminal hydrophobic region and hinge region of the P450/reductase fused enzyme, *DNA* **7:**701–711.

103. Cullin, C., 1992, Two distinct sequences control the targeting and anchoring of the mouse P450 1A1 into the yeast endoplasmic reticulum membrane, *Biochem. Biophys. Res. Commun.* **184:**1490–1495.

104. Ohta, Y., Sakaki, T., Yabusaki, Y., Ohkawa, H., and Kawato, S., 1994, Rotation and membrane topology of genetically expressed methylcholanthrene-inducible cytochrome P-450IA1 lacking the N-terminal hydrophobic segment in yeast microsomes, *J. Biol. Chem.* **269:**15597–15600.

105. Clark, B. J., and Waterman, M. R., 1991, The hydrophobic amino-terminal sequence of bovine 17α-hydroxylase is required for the expression of a functional hemoprotein in COS 1 cells, *J. Biol. Chem.* **266:**5898–5904.

106. Clark, B. J., and Waterman, M. R., 1992, Functional expression of bovine 17α-hydroxylase in COS 1 cells is dependent upon the presence of an amino-terminal signal anchor sequence, *J. Biol. Chem.* **267:**24568–24574.

107. Krynetsky, E. Y., Drutsa, V. L., Kovaleva, I. E., Luzikov, V. N., and Uvarov, V. Y., 1993, Effects of amino-terminus truncation in human cytochrome P450IID6 on its insertion into the endoplasmic reticulum membrane of *Saccharomyces cerevisiae*, *FEBS Lett.* **336:**87–89.

108. Andersen, J. F., Utermohlen, J. G., and Feyereisen, R., 1994, Expression of house fly CYP6A1 and NADPH-cytochrome P450 reductase in *Escherichia coli* and reconstitution of an insecticide-metabolizing P450 system, *Biochemistry* **33:**2171–2177.

109. Larson, J. R., Coon, M. J., and Porter, T. D., 1991, Purification and properties of a shortened form of cytochrome P-450 2E1: Deletion of the NH_2-terminal membrane-insertion signal peptide does not alter the catalytic activities, *Proc. Natl. Acad. Sci. USA* **88:**9141–9145.

110. Sagara, Y., Barnes, H. J., and Waterman, M. R., 1993, Expression in *Escherichia coli* of functional cytochrome P450$_{c17}$ lacking its hydrophobic amino-terminal signal anchor, *Arch. Biochem. Biophys.* **304:**272–278.

111. Sakaki, T., Oeda, K., Miyoshi, M., and Ohkawa, H., 1985, Characterization of rat cytochrome P-450mc synthesized in Saccharomyces cerevisiae, *J. Biochem.* **98:**167–175.

112. Omura, T., and Ito, A., 1991, Biosynthesis and intracellular sorting of mitochondrial forms of cytochrome P450, *Methods Enzymol.* **206:**75–81.

113. Wada, A., Mathew, P. A., Barnes, H. J., Sanders, D., Estabrook, R. W., and Waterman, M. R., 1991, Expression of functional bovine cholesterol side chain cleavage cytochrome P450 (P450scc) in *Escherichia coli, Arch. Biochem. Biophys.* **290:**376–380.

114. Sakaki, T., Akiyoshi-Shibata, M., Yabusaki, Y., and Ohkawa, H., 1992, Organella-targeted expression of rat liver cytochrome P450c27 in yeast. Genetically engineered alteration of mitochondrial P450 into a microsomal form creates a novel functional electron transport chain, *J. Biol. Chem.* **267:**16497–16502.

115. Black, S. M., Harikrishna, J. A., Szklarz, G. D., and Miller, W. L., 1994, The mitochondrial environment is required for activity of the cholesterol side-chain cleavage enzyme, cytochrome P450scc, *Proc. Natl. Acad. Sci. USA* **91:**7247–7251.

116. Bernhardt, R., Ngoc Dao, N. T., Stiel, H., Schwarze, W., Friedrich, J., Janig, G.-R., and Ruckpaul, K., 1983, Modification of cytochrome P-450 with fluorescein isothiocyanate, *Biochim. Biophys. Acta* **745:**140–148.

117. Vergères, G., Winterhalter, K. H., and Richter, C., 1991, Localization of the N-terminal methionine of rat liver cytochrome P-450 in the lumen of the endoplasmic reticulum, *Biochim. Biophys. Acta Bio-Membr.* **1063:**235–241.

118. Shumyantseva, V. V., Kuznetsova, G. P., Yu, V., Archakov, U., and Archakov, A. I., 1994, Membrane topology of N-terminal residues of cytochromes P-450 2B4 and 1A2, *Biochem. Mol. Biol. Int.* **34:**183–190.

119. Bernhardt, R., Kraft, R., and Ruckpaul, K., 1988, A simple determination of the sideness of the NH$_2$-terminus in the membrane bound cytochrome P-450 LM2, *Biochem. Int.* **17:**1143–1150.

120. Sethumadhavan, K., Bellino, F. L., and Thotakura, N. R., 1991, Estrogen synthetase (aromatase). The cytochrome P-450 component of the human placental enzyme is a glycoprotein, *Mol. Cell. Endocrinol.* **78:**25–32.

121. Shimozawa, O., Sakaguchi, M., Ogawa, H., Harada, N., Mihara, K., and Omura, T., 1993, Core glycosylation of cytochrome P-450(arom). Evidence for localization of N terminus of microsomal cytochrome P-450 in the lumen, *J. Biol. Chem.* **268:**21399–21402.

122. Szczesna-Skorupa, E., and Kemper, B., 1993, An N-terminal glycosylation signal on cytochrome P450 is restricted to the endoplasmic reticulum in a luminal orientation, *J. Biol. Chem.* **268:**1757–1762.

123. Ruan, K.-H., Wang, L.-H., Wu, K. K., and Kulmacz, R. J., 1993, Amino-terminal topology of thromboxane synthase in endoplasmic reticulum, *J. Biol. Chem.* **268:**19483–19490.

124. Thomas, P. E., Lu, A. Y. H., West, S. B., Ryan, D., Miwa, G. T., and Levin, W., 1977, Accessibility of cytochrome P-450 in microsomal membranes: Inhibition of metabolism by antibodies to cytochrome P-450, *Mol. Pharmacol.* **13:**819–831.

125. Matsuura, S., Fujii-Kuriyama, Y., and Tashiro, Y., 1979, Quantitative immunoelectron-microscopic analyses of the distribution of cytochrome P-450 molecules on rat liver microsomes, *J. Cell Sci.* **36:**413–435.

126. Matsuura, S., Fujii-Kuriyama, Y., and Tashiro, Y., 1978, Immunoelectron microscope localization of cytochrome P-450 on microsomes and other membrane structures of rat hepatocytes, *J. Cell Biol.* **78:**503–519.

127. Nilsson, O. S., DePierre, J. W., and Dallner, G., 1978, Investigation of the transverse topology of the microsomal membrane using combinations of proteases and the non-penetrating reagent diazobenzene sulfonate, *Biochim. Biophys. Acta* **511:**93–104.

128. Black, S. D., and Coon, M. J., 1982, Structural features of liver microsomal NADPH-cytochrome P-450 reductase. Hydrophobic domain, hydrophilic domain, and connecting region, *J. Biol. Chem.* **257:**5929–5938.

129. Ozols, J., 1989, Structure of cytochrome b5 and its topology in the microsomal membrane, *Biochim. Biophys. Acta* **997:**121–130.

130. White, S. H., 1994, Hydropathy plots and the prediction of membrane protein topology, in: *Membrane Protein Structure: Experimental Approaches* (S. H. White, ed.), Oxford University Press, London, pp. 97–124.

131. Kyte, J., and Doolittle, R. F., 1982, A simple method for displaying the hydropathic character of a protein, *J. Mol. Biol.* **157:**105–132.

132. Degli Esposti, M., Crimi, M., and Venturoli, G., 1990, A critical evaluation of the hydropathy profile of membrane proteins, *Eur. J. Biochem.* **190:**207–219.

133. Vijayakumar, S., and Salerno, J. C., 1992, Molecular modeling of the 3-D structure of cytochrome P-450$_{scc}$, *Biochim. Biophys. Acta Protein Struct. Mol. Enzymol.* **1160:**281–286.

134. Usanov, S. A., Chernogolov, A. A., and Chashchin, V. L., 1990, Is cytochrome P-450$_{scc}$ a transmembrane protein? *FEBS Lett.* **275:**33–35.

135. Nelson, D. R., and Strobel, H. W., 1988, On the membrane topology of vertebrate cytochrome P-450 proteins, *J. Biol. Chem.* **263:**6038–6050.

136. Edelman, J., 1993, Quadratic minimization of predictors for protein secondary structure. Application to transmembrane alpha-helices, *J. Mol. Biol.* **232:**165–191.

137. Edwards, R. J., Murray, B. P., Singleton, A. M., and Boobis, A. R., 1991, Orientation of cytochromes P450 in the endoplasmic reticulum, *Biochemistry* **30:**71–76.

138. Black, S. D., Martin, S. T., and Smith, C. A., 1994, Membrane topology of liver microsomal cytochrome P450 2B4 determined via monoclonal antibodies directed to the halt-transfer signal, *Biochemistry* **33**:6945–6951.

139. Vlasuk, G. P., Ghrayeb, J., Ryan, D. E., Reik, L., Thomas, P. E., Levin, W., and Walz, F. G., Jr., 1982, Multiplicity, strain differences, and topology of phenobarbital-induced cytochromes P-450 in rat liver microsomes, *Biochemistry* **21**:789–798.

140. Scheller, U., Kraft, R., Schröder, K.-L., and Schunck, W.-H., 1994, Generation of the soluble and functional cytosolic domain of microsomal cytochrome P450 52A3, *J. Biol. Chem.* **269**:12779–12783.

141. Brown, C. A., and Black, S. D., 1989, Membrane topology of mammalian cytochromes P-450 from liver endoplasmic reticulum. Determination by trypsinolysis of phenobarbital-treated microsomes, *J. Biol. Chem.* **264**:4442–4449.

142. Vergères, G., Winterhalter, K. H., and Richter, C., 1989, Identification of the membrane anchor of microsomal rat liver cytochrome P-450, *Biochemistry* **28**:3650–3655.

143. Ohta, Y., Kawato, S., Tagashira, H., Takemori, S., and Kominami, S., 1992, Dynamic structures of adrenocortical cytochrome P-450 in proteoliposomes and microsomes: Protein rotation study, *Biochemistry* **31**:12680–12687.

144. Kominami, S., Tagashira, H., Ohta, Y., Yamada, M., Kawato, S., and Takemori, S., 1993, Membrane topology of bovine adrenocortical cytochrome P-450$_{C21}$: Structural studies by trypsin digestion in vesicle membranes, *Biochemistry* **32**:12935–12940.

145. Edwards, R. J., Singleton, A. M., Murray, B. P., Murray, S., Boobis, A. R., and Davies, D. S., 1991, Identification of a functionally conserved surface region of rat cytochromes P450IA, *Biochem. J.* **278**:749–757.

146. Murray, B. P., Edwards, R. J., Davies, D. S., and Boobis, A. R., 1993, Conservation of a functionally important surface region between two families of the cytochrome P-450 superfamily, *Biochem. J.* **292**:309–310.

147. Lombardo, A., Laine, M., Defaye, G., Monnier, N., Guidicelli, C., and Chambaz, E. M., 1986, Molecular organization (topography) of cytochrome P-450(11)beta in mitochondrial membrane and phospholipid vesicles as studied by trypsinolysis, *Biochim. Biophys. Acta* **863**:71–81.

148. Centeno, F., and Gutiérrez-Merino, C., 1992, Location of functional centers in the microsomal cytochrome P450 system, *Biochemistry* **31**:8473–8481.

149. Blum, H., Leigh, J. S., Salerno, J. C., and Ohnishi, T., 1978, The orientation of bovine adrenal cortex cytochrome P-450 in submitochondrial particle multilayers, *Arch. Biochem. Biophys.* **187**:153–157.

150. Kamin, H., Batie, C., Lambeth, J. D., Lancaster, J., Graham, L., and Salerno, J. C., 1985, Paramagnetic probes of multicomponent electron-transfer systems, *Biochem. Soc. Trans.* **13**:615–618.

151. Gut, J., Richter, C., Cherry, R. J., Winterhalter, K. H., and Kawato, S., 1983, Rotation of cytochrome P-450: Complex formation of cytochrome P-450 with NADPH-cytochrome P-450 reductase in liposomes demonstrated by combining protein rotation with antibody-induced cross-linking, *J. Biol. Chem.* **258**:8588–8594.

152. Kawato, S., Gut, J., Cherry, R. J., Winterhalter, K. H., and Richter, C., 1982, Rotation of cytochrome P-450. I. Investigations of protein–protein interactions of cytochrome P-450 in phospholipid vesicles and liver microsomes, *J. Biol. Chem.* **257**:7023–7029.

153. Kawato, S., Sigel, E., Carafoli, E., and Cherry, R. J., 1981, Rotation of cytochrome oxidase in phospholipid vesicles. Investigations of interactions between cytochrome oxidases and between cytochrome oxidase and cytochrome bc1 complex, *J. Biol. Chem.* **256**:7518–7527.

154. Cherry, R. J., Heyn, M. P., and Oesterhelt, D., 1977, Rotational diffusion and exciton coupling of bacteriorhodopsin in the cell membrane of Halobacterium halobium, *FEBS Lett.* **78**:25–30.

155. Tsuprun, V. L., Myasoedova, K. N., Berndt, P., Sograf, O. N., Orlova, E. V., Chernyak, V. Y., Archakov, A. I., and Skulachev, V. P., 1986, Quaternary structure of the liver microsomal cytochrome P-450, *FEBS Lett.* **205**:35–40.

156. Myasoedova, K. N., and Tsuprun, V. L., 1993, Cytochrome P-450: Hexameric structure of the purified LM4 form, *FEBS Lett.* **325**:251–254.

157. Schwarz, D., Pirrwitz, J., Meyer, H. W., Coon, M. J., and Ruckpaul, K., 1990, Membrane topology of microsomal cytochrome P-450: Saturation transfer EPR and freeze-fracture electron microscopy studies, *Biochem. Biophys. Res. Commun.* **171**:175–181.

158. Deisenhofer, J., and Michel, H., 1991, High-resolution structures of photosynthetic reaction centers, *Annu. Rev. Biophys. Biophys. Chem.* **20**:247–266.

159. Rees, D. C., Komiya, H., Yeates, T. O., Allen, J. P., and Feher, G., 1989, The bacterial photosynthetic reaction center as a model for membrane proteins, *Annu. Rev. Biochem.* **58**:607–633.

160. Henderson, R., Baldwin, J. M., Ceska, T. A., Zemlin, F., Beckmann, E., and Downing, K. H., 1990, Model for the structure of bacteriorhodopsin based on high-resolution electron cryo-microscopy, *J. Mol. Biol.* **213**:899–929.

161. Kuhlbrandt, W., Wang, D. N., and Fujiyoshi, Y., 1994, Atomic model of plant light-harvesting complex by electron crystallography, *Nature* **367**:614–621.

162. Weiss, M. S., and Schulz, G. E., 1992, Structure of porin refined at 1.8 A resolution, *J. Mol. Biol.* **227**:493–509.

163. Cowan, S. W., Schirmer, T., Rummel, G., Steiert, M., Ghosh, R., Pauptit, R. A., Jansonius, J. N., and Rosenbusch, J. P., 1992, Crystal structures explain functional properties of two E. coli porins, *Nature* **358**:727–733.

164. Picot, D., Loll, P. J., and Garavito, R. M., 1994, The X-ray crystal structure of the membrane protein prostaglandin H_2 synthase-1, *Nature* **367**:243–249.

165. Merlie, J. P., Fagan, D., Mudd, J., and Needleman, P., 1988, Isolation and characterization of the complementary DNA for sheep seminal vesicle prostaglandin endoperoxide synthase (cyclooxygenase), *J. Biol. Chem.* **263**:3550–3553.

166. Kominami, S., Itoh, Y., and Takemori, S., 1986, Studies on the interaction of steroid substrates with adrenal microsomal cytochrome P-450 (P-450 C21) in liposome membranes, *J. Biol. Chem.* **261**:2077–2083.

167. Korzekwa, K. R., and Jones, J. P., 1993, Predicting the cytochrome P450 mediated metabolism of xenobiotics, *Pharmacogenetics* **3**:1–18.

168. Uvarov, V. Y., Sotnichenko, A. I., Vodovozova, E. L., Molotkovsky, J. G., Kolesanova, E. F., Lyulkin, Y. A., Stier, A., Krueger, V., and Archakov, A. I., 1994, Determination of membrane-bound fragments of cytochrome P-450 2B4, *Eur. J. Biochem.* **222**:483–489.

169. Lewis, D. F. V., Ioannides, C., and Parke, D. V., 1994, Molecular modelling of cytochrome CYP1A1: A putative access channel explains differences in induction potency between the isomers benzo(a)pyrene and benzo(e)pyrene, and 2- and 4-acetylaminofluorene, *Toxicol. Lett.* **71**:235–243.

170. Szklarz, G. D., Ornstein, R. L., and Halpert, J. R., 1994, Application of 3-dimensional homology modeling of cytochrome P450 2B1 for interpretation of site-directed mutagenesis results, *J. Biomol. Struct. Dynam.* **12**:61–78.

171. Koymans, L. M., Vermeulen, N. P., Baarslag, A., and Donne-Op den Kelder, G. M., 1993, A preliminary 3D model for cytochrome P450 2D6 constructed by homology model building, *J. Comput. Aided Mol. Des.* **7**:281–289.

172. Lewis, D. F. V., 1987, Quantitative structure–activity relationships in a series of alcohols exhibiting inhibition of cytochrome P-450-mediated aniline hydroxylation, *Chem. Biol. Int.* **62**:271–280.

173. Ferenczy, G. G., and Morris, G. M., 1989, The active site of cytochrome P-450 nifedipine oxidase: A model-building study, *J. Mol. Graph.* **7**:206–211.

174. Laughton, C. A., Neidle, S., Zvelebil, M. J. J. M., and Sternberg, M. J. E., 1990, A molecular model for the enzyme cytochrome $P450_{17\alpha}$, a major target for the chemotherapy of prostatic cancer, *Biochem. Biophys. Res. Commun.* **171**:1160–1167.

175. Laughton, C. A., Zvelebil, M. J. J. M., and Neidle, S., 1993, A detailed molecular model for human aromatase, *J. Steroid Biochem. Mol. Biol.* **44**:399–407.

176. Morris, G. M., and Richards, W. G., 1991, Molecular modelling of the sterol C-14 demethylase of *Saccharomyces cerevisiae*, *Biochem. Soc. Trans.* **19**:793–795.

177. Boscott, P. E., and Grant, G. H., 1994, Modeling cytochrome P450 14α demethylase (*Candida albicans*) from P450cam, *J. Mol. Graph.* **12**:185–200.

178. Cupp-Vickery, J. R., and Poulos, T. L., 1995, Structure of cytochrome P450eryF involved in erythromycin biosynthesis, *Nature Struct. Biol.* **2**:144–153.

179. Ruan, K.-H., Milfeld, K., Kulmacz, R. J., and Wu, K. K., 1994, Comparison of the construction of a 3-D model for human thromboxane synthase using P450cam and BM-3 as templates: Implications for the substrate binding pocket, *Protein Eng.* **7**:1345–1351.

180. Waterman, M. R., 1993, Heterologous expression of cytochrome *P*-450 in *Escherichia coli, Biochem. Soc. Trans.* **21**:1081–1085.

181. Pernecky, S. J., Larson, J. R., Philpot, R. M., and Coon, M. J., 1993, Expression of truncated forms of liver microsomal P450 cytochromes 2B4 and 2E1 in *Escherichia coli*: Influence of NH_2-terminal region on localization in cytosol and membranes, *Proc. Natl. Acad. Sci. USA* **90**:2651–2655.

182. Pernecky, S. J., Larson, J. R., and Coon, M. J., 1994, Cytosolic localization of NH_2-terminal-modified microsomal P450s expressed in E. coli, *FASEB J.* **7**:A1200.

183. Li, Y. C., and Chiang, J. Y. L., 1991, The expression of a catalytically active cholesterol 7alpha-hydroxylase cytochrome P450 in Escherichia coli, *J. Biol. Chem.* **266**:19186–19191.

184. Edwards, R. J., Singleton, A. M., Sesardic, D., Boobis, A. R., and Davies, D. S., 1988, Antibodies to a synthetic peptide that react specifically with a common surface region on two hydrocarbon-inducible isoenzymes of cytochrome P-450 in the rat, *Biochem. Pharmacol.* **37**:3735–3741.

185. Manns, M. P., Griffin, K. J., Sullivan, K. F., and Johnson, E. F., 1991, LKM-1 autoantibodies recognize a short linear sequence in P450IID6, a cytochrome P-450 monooxygenase, *J. Clin. Invest.* **88**:1370–1378.

186. Yamamoto, A. M., Cresteil, D., Boniface, O., Clerc, F. F., and Alvarez, F., 1993, Identification and analysis of cytochrome P450IID6 antigenic sites recognized by anti-liver-kidney microsome type-1 antibodies (LKM1), *Eur. J. Immunol.* **23**:1105–1111.

187. Edwards, R. J., Singleton, A. M., Murray, B. P., Sesardic, D., Rich, K. J., Davies, D. S., and Boobis, A. R., 1990, An anti-peptide antibody targeted to a specific region of rat cytochrome *P*-450IA2 inhibits enzyme activity, *Biochem. J.* **266**:497–504.

188. Edwards, R. J., Murray, B. P., Murray, S., Singleton, A. M., Davies, D. S., and Boobis, A. R., 1993, An inhibitory monoclonal anti-protein antibody and an anti-peptide antibody share an epitope on rat cytochrome P-450 enzymes CYP1A1 and CYP1A2, *Biochim. Biophys. Acta* **1161**:38–46.

189. Edwards, R. J., Sesardic, D., Murray, B. P., Singleton, A. M., Davies, D. S., and Boobis, A. R., 1992, Identification of the epitope of a monoclonal antibody which binds to several cytochromes P450 in the CYP1A subfamily, *Biochem. Pharmacol.* **43**:1737–1746.

190. Devereux, J., Haeberli, P., and Smithies, O., 1984, A comprehensive set of sequence analysis programs for the VAX, *Nucleic Acids Res.* **12**:387–395.

CHAPTER 7

NADPH Cytochrome P450 Reductase and Its Structural and Functional Domains

HENRY W. STROBEL, ANNE V. HODGSON, and SIJIU SHEN

1. Introduction

NADPH cytochrome P450 reductase has been a topic of interest and study since Horecker[1] first reported the isolation from pig liver after acetone extraction and trypsin treatment of a protein that catalyzed the reduction of cytochromes *c*. Since that initial report, much information has been revealed and reported about the nature, reactivities, structure, and regulation of the reductase that we have come to recognize as a component of the cytochrome P450-dependent drug metabolism system. The development of knowledge about the role and mechanism of NADPH cytochrome P450 reductase has been periodically discussed and reviewed.[2–5] The most recent review (by Backes[5]) in 1993 focused on the function of cytochrome P450 reductase, summarizing and evaluating very elegantly a number of studies of the mechanism of electron transfer to the flavin centers of the reductase, interflavin electron transfer and transfer of electrons from the flavin centers to oxygen and/or the redox partners of cytochrome P450 reductase. This present review, therefore, will not cover mechanism in any more than a cursory fashion, but will attempt to focus on a summary of the salient features of the study of structural/functional regions or domains of cytochrome P450 reductase.

2. Background

The cytochrome P450-dependent drug metabolism system was resolved into its two protein components cytochrome P450 and NADPH cytochrome P450 reductase by Lu and Coon.[6] The system was further purified and characterized[7] and the heat-stable lipid

HENRY W. STROBEL, ANNE V. HODGSON, and SIJIU SHEN • Department of Biochemistry and Molecular Biology, The University of Texas Medical School at Houston, Houston, Texas 77225.

Cytochrome P450: Structure, Mechanism, and Biochemistry (Second Edition), edited by Paul R. Ortiz de Montellano. Plenum Press, New York, 1995.

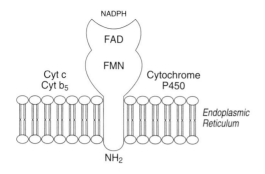

FIGURE 1. Representation of cytochrome P450 reductase illustrating the functional domains. Indicated are regions for binding NADPH, for binding to the endoplasmic reticulum, for interacting with cytochrome P450, and for interacting with other substrates, as well as regions for binding both of the flavin cofactors of cytochrome P450 reductase, FAD and FMN.

component was identified as phosphatidylcholine.[8] The reconstituted system consisting of cytochrome P450, cytochrome P450 reductase, and phosphatidylcholine was able to catalyze the hydroxylation of a variety of drug hydrocarbon and fatty acid substrates.[9,10] The general reaction for hydroxylation catalyzed by the resolved reconstituted system, as well as by the cytochrome P450 system in the endoplasmic reticulum, is:

$$RH + NADPH + H^+ + O_2 \rightarrow ROH + NADP^+ + H_2O$$

where RH represents a large variety of compounds including N- and O-alkyl drugs, polycyclic aromatic hydrocarbons, alkanes, fatty acids, pesticides, and chemical carcinogens.[11,12] The hydroxylated product, ROH, may be more or less pharmacologically active than the initial substrate, RH. Thus, the cytochrome P450 system can inactivate and clear therapeutic agents, as well as activate chemicals, such as environmental toxicants, to their carcinogenic active metabolites.

The cytochrome P450-dependent drug metabolism system is present in a wide range of animal and plant species, as well as in many microorganisms. In mammals the highest concentration seems to be in liver, but expression of the P450 system has been reported in almost all tissues including lung,[13] small intestine,[14] kidney,[15] colon,[16] and brain.[17,18] In these and other tissues, one form of cytochrome P450 reductase, as depicted in Fig. 1, has been found, as evidenced by purification techniques,[18,19] immunochemical techniques,[20,21] and expression techniques.[17,22] On the other hand, to date more than 400 forms of cytochrome P450 are known to exist[23] and although the number of mammalian forms in any given species is much less, not every tissue expresses the same spectrum of forms. Expression analysis by polymerase chain reaction can demonstrate which forms may be expressed,[17] but a clear definition of which forms are expressed must be made at the protein level, either by protein purification or immunochemical means.[22]

2.1. Purification

After the resolution of the hepatic microsomal mixed-function oxidase system into its protein components by Lu *et al.*[7] in 1969, purification of those components began. Cytochrome P450 was recognized as existing in multiple forms by purification and definition of catalytic activities.[24–27] Since those initial descriptions of multiple forms of cytochrome P450 were published, a vast array of reports on forms of cytochrome P450 have been made and summarized.[3] To date over 400 forms of cytochrome P450 have been sequenced and assigned to families and subfamilies based on their sequence homologies.[23] Moreover, some of the bacterial forms have been crystallized and their crystal structures defined by X-ray crystallography. The first such structure to be published was that for P450 101 (P450$_{cam}$) from *Pseudomonas putida*.[28,29] The crystal structure for P450 102 (P450$_{BM-3}$) has been reported[30] and more recently the structure for P450 103 (P450$_{terp}$) has been solved.[31] These advances certainly have made and will make great contributions to our understanding of cytochrome P450 mechanism of action and to the study of the interaction of cytochromes P450 with substrates and redox partners.

Cytochrome P450 reductase holoenzyme was purified to homogeneity by standard chromatographic techniques in 1975 and shown to competently catalyze the reduction of cytochrome P450 supporting cytochrome P450-catalyzed substrate hydroxylation.[32] Prior to purification to homogeneity, many of the salient features of the reductase were known: its content of FMN and FAD in an equimolar ratio and its redox behavior,[33,34] its ability to support reduction of multiple electron acceptors,[35–37] and its attachment to the endoplasmic reticulum as an integral membrane protein as evidenced by the requirement for its solubilization by proteases[35,36,38] or ionic detergents such as deoxycholate[6] or nonionic detergents such as Renex 690.[32] Yasukochi and Masters[39] were the first to develop an efficient purification procedure for the reductase utilizing biospecific affinity chromatography with 2′, 5′-ADP as the affinity ligand. Dignam and Strobel[40] utilized NADP Sepharose column chromatography to achieve 60% recoveries of homogeneous holoenzyme.

2.2. Characterization

The purified enzyme was utilized to establish the equimolar content of FMN and FAD and the apparent molecular weight by SDS gel electrophoresis as between 76,000 and 80,000.[39,40] An estimate of α-helix, β-sheet, and random-coil structure obtained through circular dichroism studies was reported by Knapp *et al.*[41] These workers also reported that purified cytochrome P450 reductase holoenzyme exhibited the same free mobility in SDS as many soluble proteins, suggesting no remarkably hydrophobic character even though the reductase is a membrane-bound protein. Redox characteristics of the enzyme were established by spectrophotometric and stopped-flow kinetic studies.[42] In an elegant series of studies, Vermilion, Coon, and their co-workers established that electron flow through the reductase follows the pathway from NADPH to FAD and FMN to electron acceptors and suggested that FMN alone produced the air-stable semiquinone and was the high-potential flavin.[43–45] These results were extended and the conclusions confirmed by studies with FAD-free reductase preparations.[46–48] These aspects are discussed in the following section.

3. Mechanism

The mechanism of electron transfer through the flavin centers of the reductase was the focus of the most recent review of cytochrome P450 reductase.[5] The reader is referred to the Backes review for a more comprehensive discussion of electron transfer than is summarized here.[5]

It had been observed in many laboratories that during the course of purification flavin was lost from the enzyme, resulting in lowered specific activity. This loss could be redressed by adding FMN to the enzyme[18] and, further, the loss could be substantially prevented by adding low concentrations of FMN to buffers used during the course of enzyme preparation.

The tendency to lose FMN was enhanced by treatment with KBr to produce a preparation that lacked FMN but had almost all of its complement of FAD.[44] The FMN-depleted enzyme was unable to reduce cytochrome P450 or cytochrome c, but was able to transfer electrons from NADPH to ferricyanide leading to the postulate that electron flow from NADPH passed through FAD to FMN and then to cytochrome c or cytochrome P450. The FMN-depleted reductase, however, did not produce an air-stable semiquinone,[43] as the holoreductase did.[44] These data suggested that FAD is the low-potential group that interacts with NADPH. ^{31}P NMR studies proved that the unpaired electrons in the semiquinone state (air-stable) reside exclusively on FMN.[49] This agrees with data from several earlier light absorption studies.[33,34,43]

Kurzban and Strobel[46,47] prepared an FAD-depleted reductase that was able to transfer electrons to ferricyanide and exhibit an air-stable semiquinone.[48] With these data, the pathway of electron transfer from NADPH to FAD, then to FMN, and finally to redox partners was established as predicted earlier. Further, the identity of FAD as the low-potential flavin and the identity of FMN as the source of the air-stable semiquinone were confirmed.

Data from time-resolved fluorescence spectroscopic studies have demonstrated that energy is transferred between FMN and FAD and that the distance between the flavins in reductase is between 2.0 and 2.5 nm.[50] Kuki and Wolynes[51] have shown that two redox centers up to a distance of 2.8 nm can efficiently transfer electrons. Energy transfer between FAD and FMN was determined to be about 31% efficient,[52] which is consistent with a distance in redox centers of 2 nm. Work by Centeno and Gutierrez-Merino[52] also measured the probable distance between the flavins in reductase as 2 nm using steady-state fluorescence intensity and anisotropy data. This juxtaposition of flavins may be of importance in interflavin electron transfer. The fluorescence results of Centeno and Gutierrez-Merino[52] indicate that FAD is more deeply buried within the structure of the reductase than is FMN, which is consistent with the more ready discharge of FMN than FAD from the reductase and with the evidence that FMN is the site of electron egress to the redox partners of the reductase.

4. Domains of Reductase: Structural and Functional

Like most proteins, the cytochrome P450 polypeptide chain is oriented in three dimensions defining functional domains, the concerted action of which enables the overall function of the protein as shown in Fig. 2. In this relatively simple model, the reductase has a membrane binding domain, domains that bind each of the two flavin cofactors of the reductase, a domain that binds the electron donor NADPH, and a perhaps more loosely constructed domain that mediates interaction with the substrate cytochrome c or the cytochrome P450 component of the mixed-function oxidase reaction. The evidence that

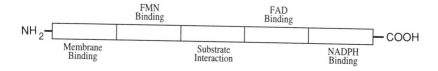

FIGURE 2. Model of functional domains arrayed schematically in cytochrome P450 reductase.

permits this general statement has been developed through the efforts of many laboratories over time in a domain-by-domain fashion. For instance, the demonstration of the existence of a membrane binding domain for cytochrome P450 reductase was provided through the efforts of Gum and Strobel,[53,54] and Black and Coon[55,56] by selective proteolysis of the intact reductase and purification of the membrane binding peptide. The impetus for this work derived in part from the failure of the purified protease-solubilized cytochrome *c* reductase to support cytochrome P450-catalyzed substrate hydroxylation in a solubilized reconstructed system.[7] Thus, identification and characterization of the domain structure and function has occurred as a result of, and has been stimulated by, studies of the mechanism of the cytochrome P450 mixed-function oxidase system.

Much information about the domain structure of the reductase has been gained through the determination of the primary amino acid sequence of the reductase[57] and from the study of the structure of the reductase gene made possible by the successful preparation of the reductase genomic clone.[58] This clone is composed of 16 exons spanning 20 kilobases of DNA and reductase exon 1 is a completely untranslated segment of the DNA.[58] Traut[59] has reported that functional domains of proteins are often encoded by discrete exons. This appears true for the reductase as well. Reductase exon 2 contains the start codon, 13 nucleotides of untranslated sequence, and the complete coding sequence for the N-terminal hydrophobic domain. The relationship between discrete exons and functional domains is illustrated in other reductase functional domains. The connecting sequence between the hydrophobic (membrane binding) domain and the catalytic domains is encoded by reductase exon 3. The proposed binding site for the FMN ribitoyl phosphate group is encoded by reductase exon 4. Reductase exons 7 and 8 probably encode a segment between the two flavin-binding groups, whereas amino acids binding the pyrophosphate group of FAD are encoded by exon 9.

Traut also reported that many other protein functional domains cross exon boundaries.[59] In the reductase protein, the FMN isoalloxazine ring putatively binds amino acid residues encoded by exons 5 and 6. The substrate recognition regions for cytochrome P450 and cytochrome *c* are encoded by exons 6 and 7. Likewise, the FAD pyrophosphate group apparently binds to amino acid residues encoded by exons 9 and 10. The FAD and NADPH binding domains encompass three intron/exon borders encoded by exons 13 to 16. As with studies of protein structure, development of a base of information on gene structure has provided an abundant area for further investigation of domain structure.

4.1. Probes of Domain Structure and Function

A number of methods have been used to define and characterize the structural/functional domains of cytochrome P450 reductase. The list of methods includes the isolation and purification of particular domains by classical protein chemistry procedures, the use of amino acid residue modification and cross-linking studies, the use of polyclonal antibodies to the whole protein, as well as antibodies to isolated or synthesized peptide regions of the protein, and the use of molecular biological approaches either to overexpress the whole protein or truncated regions of the protein for functional analysis.

4.1.1. Isolation and Purification of Defined Regions

Isolation of functional/structural domains has been a useful technique in the definition of the domain structure of cytochrome P450 reductase. One example of the utility of these approaches is found in the dissection and definition of the membrane binding peptide of the reductase. The use of proteases[35,36] and of detergents[7] to release the reductase from the endoplasmic reticulum led to the isolation of two polypeptides of differing molecular weights and differing activities. Steapsin or trypsin solubilization of reductase from microsomal membranes gave rise to a polypeptide of 68 kDa that was active in reduction of cytochrome c and other artificial electron acceptors, but not cytochrome P450.[7] Solubilization of the reductase from the microsomal membrane with ionic detergents such as deoxycholate[6] or nonionic detergents such as Renex 690[32] resulted in reductase preparations of 76 kDa with activity toward cytochrome c and artificial electron acceptors, as well as cytochrome P450.[32] Gum and Strobel[53] compared these two preparations for their abilities to interact with phospholipid vesicles and reductase-depleted microsomes and showed that only the detergent-solubilized larger molecular mass form was able to bind to phospholipid vesicles or microsomes and support cytochrome P450-catalyzed drug metabolism. Taking advantage of the ability of the detergent-solubilized residue to insert in phospholipid vesicles, Gum and Strobel[53] and Black et al.[55] were able to use proteases to separate the phospholipid vesicles containing the membrane binding domain from the hydrophilic catalytic portion of the reductase. After resolution of these portions of the reductase by high-speed ultracentrifugation, the membrane binding peptide was purified and characterized.[54,56]

Purification techniques also led to the definition of the functional roles of the FMN and FAD binding domains. For both flavin binding sites a strategy of discharging one of the flavins, but not the other, and purifying the one-flavin depleted form was followed. Vermilion and Coon[44] used this strategy to purify FMN-depleted reductase and to propose a sequence of electron flow from NADPH $\rightarrow$ FAD $\rightarrow$ FMN $\rightarrow$ acceptor. Kurzban and Strobel[46] developed conditions that discharged FAD but retained FMN in its site and purified FAD-less reductase. The characterization of this FAD-depleted enzyme supported the proposed pathway of electron transit through the reductase.[48]

4.1.2. Modification of Specific Amino Acid Residues

Treatment of purified cytochrome P450 reductase with reagents that modify specific amino acid residues has been employed to identify domains within the reductase that interact with the redox partners of cytochrome P450 reductase. Tamburini and Schenkman[60] modified carboxyl residues with 1-ethyl-3-(3-dimethylaminopropyl) carbodiimide (EDC) and showed that the modifications affected functional interaction with various cytochromes P450. Nadler and Strobel[61] used acetic anhydride to modify specific reductase lysine residues and EDC with an exogeneous nucleophile (e.g., methylamine) to modify carboxyl groups. The resultant net charge changes increased the efficacy of electron transfer between reductase and cytochrome P450 as judged by product formation when specific reductase lysines were modified and decreased product formation when carboxyl groups on the reductase were modified. These findings were interpreted as indicative of a role for electrostatic charge-pairing in the interaction of reductase with its redox partners. Further studies with radiolabeled modifying reagents followed by proteolysis of the labeled protein and amino acid sequencing of the tryptic peptides led to the proposal of a region of the reductase molecule as important in interaction with cytochromes P450.[62] Modification of negatively charged residues such as aspartate or glutamate and positively charged residues such as lysine or arginine have led to the recognition of the role of electrostatic charge in protein–protein interaction with reductase

Modification of amino acid residues with cross-linking reagents has been used as a probe for reductase amino acid residues that interact with specific amino acid residues of redox partner proteins. Nisimoto[63] using this approach has demonstrated that certain reductase carboxyl residues can be cross-linked with specific cytochrome c lysine residues. Other studies have probed the sites of cross-linking between cytochrome P450 reductase and cytochrome b_5.[64]

4.1.3. Use of Antibodies to Specific Peptides

Several strategies have been utilized to probe cytochrome P450 reductase functionality with antibodies. Polyclonal antibodies to isolated cytochrome P450 reductase have been prepared and used in various studies of protein purity, or in concentration of the protein using Western blotting techniques.[65] Wada et al.[66] used immunochemical approaches to demonstrate the role of cytochrome P450 reductase in cytochrome P450-dependent reactions. Masters et al.[67] developed an antibody to cytochrome c reductase (a preparation of cytochrome P450 reductase missing the amino-terminal membrane binding peptide) that was capable of inhibiting reduction of cytochrome c by the cytochrome c reductase preparation. Moreover, this antibody was able to inhibit microsomal cytochrome P450-mediated drug hydroxylation,[67] as well as laurate hydroxylation catalyzed by a partially purified and reconstituted cytochrome P450 reductase and cytochrome P450 system in the presence of lipid.[7] These data provided evidence that the electron transfer site(s) were located on that part of the reductase exposed to the aqueous environment and, therefore, accessible to the antibody. Polyclonal antibodies have been used to identify the presence of reductase in other tissues such as hepatoma,[19] colon,[20] and brain.[18]

More recently, antipeptide antibodies have been used as probes of functional regions of cytochrome P450 reductase. Shen and Strobel[68] developed a series of antipeptide antibodies to regions of the reductase previously proposed to be involved in interaction with cytochrome c[63] or cytochrome P450.[62] Antipeptide antibodies were designed empirically to react with specific regions of cytochrome P450 reductase shown by other techniques such as site-specific mutagenesis or amino acid modification to be involved in interaction with cytochrome P450 reductase redox partners. This technique has proven useful in assigning specific structural regions to particular functions. For example, Shen and Strobel[68] have suggested that two regions (110–119 and 204–218) of rat reductase seem to be involved in the interaction of reductase with cytochrome P450 and, thus, may define at least a portion of the protein–protein interaction domain of the reductase for cytochrome P450.

4.1.4. Site-Directed Mutagenesis and Cloning

A dramatic step forward in the study of functional domains of cytochrome P450 reductase was made through the cloning and sequencing of the reductase by Porter and Kasper.[69] This made possible the identification of numerous domain sequences by virtue of commonalities in sequence with known proteins containing FAD, FMN, and/or NADPH binding sites.[69,70]

Site-specific mutagenesis studies have also been utilized to examine the amino acid composition of particular domains. For instance, Shen *et al.*[71] altered specific tyrosines to phenylalanines or aspartates in reductase to probe the involvement of these residues in FMN binding. Alternatively, truncated portions of the reductase have been prepared that contain the putative FAD and NADPH binding regions but not the FMN or membrane binding regions as a way to study the discrete partial reactivities of reductase domains.[72] These truncated polypeptides actively catalyze the transfer of electrons from NADPH to ferricyanide.

In addition, cloning the reductase has enabled overexpression of the enzyme and the preparation of large quantities of pure cytochrome P450 reductase for NMR and X-ray crystallographic studies, both of which consume great amounts of enzyme. The resultant data have been, and will continue to be, of great value in defining the domain structure of cytochrome P450 reductase.

4.2. FMN Site

The FMN[43,47] and FAD[46,47] domains have been enzymatically characterized by selectively removing the flavin from the native protein. Both flavins are reported to be embedded deeply inside the protein, residing greater than 5 nm from the lipid/water interface or the border of the hydrophobic domain.[52] Although the three-dimensional structure of the reductase is not yet completed, putative binding sites for prosthetic groups have been elucidated by comparison of the cDNA sequence of the reductase to those of flavoproteins with known crystal structures.[73] Chemical modification studies suggest that sulfhydryl[74–77] and lysine[77,78] residues are involved in reductase catalysis and in binding of FAD, FMN, and NADPH to reductase. The cloning and sequencing of the reductase by Porter and Kasper[69] have made possible the identification of numerous domain sequences by virtue of commonalities with sequences of known proteins containing FAD, FMN, and/or NADPH binding sites.[69,70] Based on their cDNA sequence of reductase, Porter and Kasper[73] have proposed that the two flavin binding domains of cytochrome P450 reductase arose from two separate ancestral genes, each of which encoded a single binding site for FAD or FMN.

This postulate led Strobel and co-workers to use both classical protein chemical and molecular biological techniques to isolate the peptide proposed to bind FAD. Using protein chemical techniques, tryptic peptides containing amino acids 54–388 which contain the FMN binding region, and a peptide consisting of amino acids 389 to the C-terminus which contains the FAD and NADPH binding regions, were prepared and resolved on SDS gels.[79] Using PCR and subcloning techniques, truncated clones of the reductase cDNA were expressed and purified. These clones which contain the putative FAD and NADPH binding regions but not the FMN or membrane binding regions, provide a way to study the discrete partial reactivities of the reductase domains.[72] Both the cDNA-generated truncated reductase peptide and the proteolytically produced FAD-containing peptide catalyze reduction of ferricyanide but not cytochrome c or DCIP, giving more evidence for the dual ancestral gene hypothesis as well as the proposed electron pathway stipulating that electrons flow from NADPH to FAD, then to FMN, and ultimately to substrate.

A goodly amount of detail regarding specific amino acid interactions has been uncovered for the FMN binding domain. Chemical modification and fluorescence studies suggested that FMN binding involves tryptophan stacking.[76] In these studies, oxidation of tryptophan indole rings reduced the ability of FMN to bind to FMN-depleted reductase and caused the loss of cytochrome c reduction, but not NADPH to ferricyanide, activity. The same studies also showed that release of FMN from reductase caused an increase in tryptophan fluorescence consistent with FMN binding-induced quenching of tryptophan fluorescence. Photochemically induced dynamic nuclear polarization studies also indicate that an interactive tryosine residue is located in the vicinity of the FMN binding domain.[80] Predictions for the FMN binding domain were confirmed using site-directed mutants. Specifically, a $Tyr^{178}Asp$ mutation abolished FMN binding and cytochrome c reductase activity.[71] The $Tyr^{140}Asp$ reductase mutant maintained the ability to bind FMN, but the cytochrome c reductase activity was decreased by fivefold compared to the wild type. Substitution of a Phe for either Tyr^{178} or Tyr^{140} did not alter FMN binding or cytochrome c activity.[71]

4.3. FAD Site

Many studies have begun to clarify and define the FAD binding domain structure of reductase. The terminal half of the reductase, which is proposed to bind FAD and NADPH, is structurally homologous to a family of novel flavoenzymes which also bind FAD and nicotinamide dinucleotide or nicotinamide dinucleotide phosphate. This family includes ferredoxin-$NADP^+$, NADPH-sulfite reductase, NADH-cytochrome b_5 reductase, glutathione reductase, and NADPH cytochrome P450 reductase. The three-dimensional structure of ferredoxin has recently been solved to 2.6 Å by X-ray diffraction[81] and has been used as a model for the study of the FAD binding domain of reductase. Glutathione reductase has also served as a model since the amino acid residues within 5 Å of FAD are known.[82,83] Our laboratory has cleaved reductase with *Staphylococcus aureus* V_8 protease into three fragments. Microsequencing has shown that the protease cleaves the reductase at residues 54 and 388. The fragment encompassing residues 388 to the C-terminus retains the ability to reduce ferricyanide, indicating binding of FAD and NADPH. These studies provided evidence that the FAD/NADPH binding domain may be expressed, purified, and characterized as a stable and enzymatically active fragment.

The FAD binding domain has been more difficult to study than the FMN binding domain because the FAD moiety binds much more tightly to reductase than FMN. Studies of the FMN-depleted reductase[43–45] demonstrated that the FAD accepts electrons directly from NADPH, but left open the possibility that the FMN could also accept electrons from NADPH at catalytic rates. Kurzban and Strobel[47] developed a technique to remove selectively the FAD moiety from the FAD site of the reductase and determined that the FAD-less reductase had very little ability to donate electrons to one-electron acceptors such as DCIP, NBT, menadione, and cytochrome c or to accept electrons from electron donors such as the physiological donor NADPH. These data were consistent with FAD serving as the direct acceptor of electrons from NADPH, interflavin electron transfer of one electron at a time, and FMN as the sole electron donor to substrates such as cytochrome c, NBT, and DCIP. The work of Kurzban and Strobel[47] and Kurzban et al.[48] confirmed the role of FAD in the proposed electron transport sequence.

The selective removal of specific flavins from reductase supported the proposed electron transport studies, yet some important considerations remained unaddressed. For each flavin removal study, there is evidence that some of the flavin to be removed remained bound to the reductase. Additionally, there is evidence that either FAD or FMN may bind to the reductase in the binding domain for either one of the two flavins. These two complications lead to undesired flavin contamination of the aporeductases, presumably generating background electron transport activities, spectral peaks, and flavin contents.

In order to extend these FAD studies, improve the flavin background problem, and confirm the predicted FAD binding site, Hodgson and Strobel[72] isolated two cDNA clones for the FAD and NADPH binding domains to study the FAD binding domain without

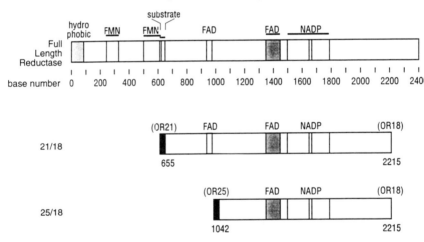

FIGURE 3. FAD binding domain constructs. A schematic diagram of the full-length rat liver cytochrome P450 reductase is shown at the top. Putative binding sites for the flavin cofactors, substrates, NADPH, and the hydrophobic domain are labeled. FAD binding domain constructs are shown and labeled 21/18 and 25/18 according to the two oligonucleotide primers used to amplify them. Clone 21/18 was amplified from oligonucleotide primer OR21 and OR18 and spans from nucleic acid base pair 655 to 2215. This fragment includes all of the putative FAD and NADPH binding sites. Clone 25/18 was amplified from oligonucleotide primer OR 25 and OR 18 and spans from nucleic acid base pair 1042 to 2215. This fragment includes all of the putative NADPH binding sites and the larger of the FAD binding domains from nucleic acid base pair 1358 to 1411.

influence of the FMN binding domain. Both clones, #21/18 and #25/18 as shown in Fig. 3, have been expressed in *E. coli* under the control of the *trc* promoter. The expressed truncated proteins have been purified to apparent homogeneity and are stable and enzymatically functional, catalyzing the reduction of ferricyanide at rates of 54% (#25/18) and 104% (#21/18) of the rate of FMN-less reductase on a per mole of flavin basis. Consistent with the proposed electron transfer sequence, the FAD binding fragments have very little or no activity toward DCIP or cytochrome *c*, but have 41% (#25/18) and 65% (#21/18) of the FMN-less reductase activity toward 3 AcPyADP. Flavin analysis clearly demonstrates that the proposed FAD binding domain is indeed the FAD binding domain, binding 0.71 mole of FAD and 0.03 mole of FMN per mole of #21/18 protein and 0.66 mole of FAD and 0.02 mole of FMN per mole of #25/18 protein.

4.4. NADPH Site

When the cytochrome P450-dependent drug metabolism system was first resolved, it was demonstrated that NADPH was the preferred electron donor for the system. Lu *et al.*[7] reported that the K_m of the partially purified reductase was 2.0×10^{-5} M for NADPH and almost an order of magnitude higher for NADH, 6.3×10^{-4} M. For the purified reductase the K_m for NADPH is reported as 5 μM when either cytochrome *c* or cytochrome P450 is used as electron acceptor.[40] The NADPH binding site is located in the carboxy-terminal portion of the reductase.[57,69,70]

Chemical modification studies demonstrated that two sulfhydryl groups that are protected by both NADP$^+$ and 2′-AMP will abolish cytochrome *c* or DCIP reductase activity when modified. These studies suggest that two cysteine residues are involved in binding NADPH to reductase.[74] Similar conclusions were reached by later chemical modification studies[75] showing that NADPH blocked all of the reactive cysteine residues from chemical modification. A lysine residue required for catalysis was proposed by separate chemical modification studies[78] to be located near the 2′ phosphate group on the adenosine ribose of NADPH.

4.5. Substrate Interaction Site

Substrate interaction with the reductase has been a closely investigated subject over the years. A wide range of substrates were known for the proteolytically solubilized form of the reductase before the physiological acceptor cytochrome P450 was known.[35–37] The fact that the proteolytically solubilized reductase competently catalyzed the reduction of nonphysiological acceptors such as cytochrome *c* and was incapable of supporting catalysis of the physiological acceptor cytochrome P450 suggests both that the cytosolic portion of the reductase contained sites important in electron flow to cytochrome *c* or other exogenous acceptors and that the membrane binding portion cleaved off the N-terminus by proteolytic solubilization of the reductase either contained an alternate site of electron transfer to cytochrome P450 or played a role in facilitating the interaction of cytochrome P450 with the cytochrome P450 reductase site located in the cytosolic portion of the reductase.

Evidence showing the importance of an electron transfer site in the cytosolic portion of the reductase has suggested a role for complementary charge pairing in this process of electron transfer. First, in the reconstituted system containing reductase and cytochrome

P450 2B4 (P450$_{LM2}$), high ionic strength was shown to inhibit the demethylation of benzphetamine.[84] On the other hand, Voznesensky and Schenkman[85,86] have reported an increase in activity by changing from low to intermediate ionic strengths, consistent with earlier observations on the effects of low and intermediate ionic strength on the reduction of artificial electron acceptors by the protease-solubilized reductase,[35–37] and with ionic strength effects on the detergent-solubilized enzyme.[39] While variations in ionic strength may or may not have physiological parallels, these results may suggest that some looseness in the interaction is required in order to achieve the most effective mode for electron transfer. A second line of evidence derives from alteration of charge on specific amino acid residues. The chemical modification of Lys[384] of cytochrome P450 2B4 was shown to inhibit electron transfer from the reductase to cytochrome P450.[87,88]

Direct evidence for a substrate interaction site in the cytosolic portion of cytochrome P450 reductase was provided by the work of Tamburini and Schenkman[60] and Tamburini et al.[64] who showed the importance of complementary charge pairing and the formation of a binary complex between cytochrome P450 reductase and cytochrome b_5 by cross-linking studies. Nisimoto[63] using cross-linking studies showed that cytochrome c forms a covalent complex with cytochrome P450 reductase, further demonstrating a substrate interaction site on the cytosolic portion of cytochrome P450 reductase centered around residues 204–208. Nadler and Strobel[61] modified carboxyl residues of cytochrome P450 reductase and demonstrated dramatic changes in the K_m and V_{max} values of a substrate metabolizing system reconstituted with either cytochrome P450 1A1 or 2B1. In subsequent studies Nadler and Strobel[62] identified a peptide containing residues 109–130 as the locus of carboxyl-containing residues which on modification resulted in the previously documented dramatic loss in the ability of reductase to support the cytochrome P450-dependent substrate hydroxylation.[61]

More recently, antipeptide antibodies have been used as probes for locating the functional regions of cytochrome P450 reductase. Shen and Strobel[68] developed three antipeptide antibodies to regions of the reductase previously proposed to be involved in the interaction with cytochrome c[63] or cytochrome P450.[62] Two of the specific antipeptide antibodies against reductase were shown to be inhibitory to reductase-supported cytochrome P450 1A1 ethoxycoumarin and P450 2B1 benzphetamine demethylation activity in an $in\ vitro$ reconstituted system. These antipeptide antibodies did not have any effect on the ability of reductase to reduce cytochrome c. These data confirmed the important role of regions 110–119 and 204–208 of reductase in the association with cytochrome P450, and the difference in the manner of interaction between cytochrome P450 reductase with cytochrome P450 and cytochrome c. Interestingly, these sites seem to occur around the sites proposed to be involved in FMN binding. This is consistent with the demonstration that FMN is the electron egress site for cytochrome P450 and for cytochrome c. It would be reasonable to expect substrate interaction residues to occur in the vicinity of the electron egress site but this reasoning must withstand the test of three-dimensional analysis when the complete crystal structure of the reductase is known.

4.6. Membrane Binding Site

The membrane binding domain is comprised of the N-terminal end of the reductase molecule. It is made up of the N-terminal 45–50 amino acids, depending on the protease used to cleave the membrane binding domain[54,56] or choices made in evaluating the amino acid sequence.[69] The size of the membrane binding peptide is about 6400 Da.

The hydrophobic domain has been purified and characterized by this lab[54] and others.[56] This domain is responsible for anchoring the enzyme to the endoplasmic reticulum membrane.[53] The isolated domain is about 32% hydrophobic amino acids as determined from the amino acid composition of the total isolated peptide.[54] Data from ^{31}P NMR and TLC analysis of the native reductase and the reductase cleaved away from the hydrophobic domain demonstrated that phospholipids are associated with the hydrophobic domain rather than the cytosolic portion.[49] The interaction of P450 and reductase is limited by the degrees of freedom granted to membrane-bound proteins. Ironically, the limited movement dictated by the membrane binding domain may also increase the interaction between proteins. One requirement for interaction between reductase and P450 is the membrane-bound domain of each protein,[53,89,90] though this requirement is curtailed in single polypeptide chain P450:reductase fusion proteins, e.g., P450 102 from *Bacillus megaterium*.[91]

5. Crystal Structure and Primary Sequence

The primary protein sequence of cytochrome P450 reductase was determined for different species in different laboratories. The complete amino acid sequence for pig liver cytochrome P450 reductase was determined by Shively and colleagues in 1986.[57] At the same time, Katagiri *et al.*[70] published the cloning and sequence analysis of the full-length rabbit liver reductase cDNA. These two publications followed the solution of the rat reductase cDNA sequence in 1985.[69] Since that time, interest in the crystal structure of the reductase has increased. The crystallization of cytochrome P450 reductase and preliminary X-ray studies have been reported by Masters and colleagues.[92] When the crystal structure is finished, the domain structure studies will be clarified and can be extended based on actual three-dimensional interrelations of the various structural elements of the reductase.

6. Distribution

That the cytochrome P450-dependent drug metabolism system is very widely distributed in mammalian species and in the various tissues of individual mammalian species was recognized early in the study of drug metabolizing activities.[93] The description of the character and control of drug metabolism activities in various tissues and species has expanded since its first demonstration in extrahepatic tissues, just as the specifics of regulation and enzyme induction in hepatic tissues of various species developed and expanded. The presence and unique properties of the drug metabolizing systems of lung, kidney, small intestine, heart, adrenals, mammary glands, male reproductive organs including testis and prostate, skin, pancreas, brain, and other tissues have been described.[94] In all systems, cytochrome P450 reductase has been shown by one or more techniques to be present. For instance, the first direct demonstration of cytochrome P450 reductase in brain was accomplished using immunohistochemical techniques.[95] Later, Bergh and Strobel[18] purified rat brain cytochrome P450 reductase to homogeneity and characterized brain reductase as essentially identical to hepatic cytochrome P450 reductase as far as its immunologic, physical, kinetic, and catalytic properties are concerned. Another demonstration of cytochrome P450 reductase identity was provided by Fennell and Strobel who purified the reductase from the Novikoff hepatoma to homogeneity and demonstrated its identity with hepatic reductase.[19]

In summary, abundant evidence indicates that both cytochrome P450 and cytochrome P450 reductase are widely distributed throughout the constituent tissues of most mammals and also in other species. The weight of the evidence shows that while the distribution of cytochromes P450 may vary from tissue to tissue, cytochrome P450 reductase seems to exist in one form that is widely distributed in tissues with no isoforms of reductase being identified to date.

7. Chimeric Proteins/Fused Proteins

While cytochrome P450 reductase is considered and has been treated as a single individual protein, in some instances it exists as part of a longer, single polypeptide that also includes a heme-containing cytochrome P450-like domain. Fulco and his colleagues[91] reported the isolation of a cytochrome P450-dependent fatty acid hydroxylase from B. megaterium which contained two functional domains, a P450 domain and a domain containing FMN, FAD, and an NADPH binding site, within a single polypeptide chain of 119 kDa. This protein has been studied extensively over the intervening years. Recently its heme domain was crystallized and its crystal structure solved by X-ray diffraction studies, becoming the second P450 crystal structure to be determined.[30] Mammalian nitric oxide synthase has recently been recognized as a natural fusion protein containing a flavoprotein domain similar to cytochrome P450 reductase, as well as a cytochrome P450-like heme domain and a calmodulin binding site.[96] These natural fusion proteins containing a reductase-like domain may function to facilitate catalysis. This is certainly true for the B. megaterium polypeptide,[97] which has an exceptionally high turnover number (greater than 1500 nmole of palmitic acid per minute per mole of cytochrome P450) in comparison to reconstituted systems composed of cytochrome P450 and cytochrome P450 reductase.[10]

The natural fusion proteins have inspired the preparation of chimeric forms containing most or all of the sequence of reductase and various cytochromes P450. Murakami et al.[98] reported the cloning and expression of a functional fusion protein containing rat cytochrome P450 reductase and rat cytochrome P450 1A1. Fisher et al.[99] achieved a high level of expression in E. coli of catalytically active fusions of reductase and bovine adrenal cytochrome P450 17A or P450 4A1. Both of these fusion proteins catalyze their constituent reactions efficiently. In subsequent studies these investigators showed that the fusion proteins behave as tight complexes functioning as self-contained biocatalytic units.[97] Thus, a very efficient fused protein construct could have very important uses in synthetic processes for biologically important compounds (e.g., hormones). Clearly the reductase retains throughout these constructions sufficient structural integrity to enable catalysis to proceed at least as fast as the optimized reconstituted system.

8. Prospective

Work over the last three decades has taken our knowledge of cytochrome P450 reductase from a flavoprotein without a physiological function solubilized from the endoplasmic reticulum by proteases[35,36] to a well-characterized pure protein with a clearly defined physiological acceptor functioning in a well-characterized catalytic process. Furthermore, the reductase has been sequenced, cloned, expressed, and mutated to provide many molecular details of its structure and function. What lies ahead for study of the reductase surely includes solution of the crystal structure since preliminary studies have already been reported.[92] This advance will provide answers to the present questions of physical relationships of domains to one another. It will not only allow the definition of relationships but will also permit the devising of experiments based on physical relationships to test hypotheses of which amino acid residues are actually directly involved in binding flavin molecules.

Also anticipated is greater detail about the mechanism of electron transfer, and clarity about the kinds of interactions the reductase has with its redox partners. Regulation of expression of cytochrome P450 reductase in liver as well as in other tissues appears to be another area of likely advance in knowledge concerning the reductase. Whatever the area or time scale, it is likely that future studies will uncover as many surprises and new insights as the prior studies, which have provided such a good scientific base for further study of NADPH cytochrome P450 reductase.

ACKNOWLEDGMENTS.   The authors acknowledge the generous support of Grant CA 53191 from the National Cancer Institute DHHW. The authors express their thanks to Ms. Karyn Zamojski for her expert editorial assistance and to Drs. Hidenori Kawashimi, Petr Hodek, David Sequeira, and Xuan Chuan Yu and to Hernan Vasquez, Huamin Wang, Jun Geng, Tomas Cvrk, and Laura Bankey for their comments and assistance.

References

1. Horecker, B. L., 1950, Triphosphopyridine nucleotide-cytochrome *c* reductase in liver, *J. Biol. Chem.* **183**:593–605.

2. Strobel, H. W., Dignam, J. D., and Gum, J. R., 1980, NADPH cytochrome P450 reductase and its role in the mixed function oxidase reaction, *Pharmacol. Ther.* **8**:525–537.

3. Guengerich, F. P., 1987, *Mammalian Cytochromes P450*, Vols. 1 and 2, CRC Press, Boca Raton, FL.

4. Porter, T. D., and Coon, M. J., 1991, Cytochrome P450: Multiplicity of isoforms, substrates and catalytic and regulatory mechanisms, *J. Biol. Chem.* **266**:13469–13472.

5. Backes, W. L., 1993, NADPH cytochrome P450 reductase: Function, in: *Handbook of Experimental Pharmacology*, Vol. 105, *Cytochrome P450* (J. B. Schenkman and H. Greim, eds.), Springer-Verlag, Berlin, pp. 15–34.

6. Lu, A. Y. H., and Coon, M. J., 1968, Role of hemoprotein P450 in fatty acid ω-hydroxylation in a soluble enzyme system from liver microsomes, *J. Biol. Chem.* **243**:1331–1332.

7. Lu, A. Y. H., Junk, K. W., and Coon, M. J., 1969, Resolution of the cytochrome P450 containing ω-hydroxylation system of liver microsomes into three components, *J. Biol. Chem.* **244**:3714–3721.

8. Strobel, H. W., Lu, A. Y. H., Heidema, J., and Coon, M. J., 1970, Phosphatidylcholine requirement in the enzymatic reduction of hemoprotein P450 and in fatty acid, hydrocarbon and drug hydroxylation, *J. Biol. Chem.* **245**:4851–4854.

9. Lu, A. Y. H., Strobel, H. W., and Coon, M. J., 1969, Hydroxylation of benzphetamine and other drugs by a solubilized form of cytochrome P450 from liver microsomes lipid: Requirement for drug demethylation, *Biochem. Biophys. Res. Commun.* **36**:545–551.

10. Lu, A. Y. H., Strobel, H. W., and Coon, M. J., 1970, Properties of a solubilized form of the cytochrome P450-containing mixed-function oxidase of liver microsomes, *Mol. Pharmacol.* **6**:213–220.

11. Gillette, J. R., 1979, Effects of induction of cytochrome P450 enzymes on the concentrations of foreign compounds and their metabolites and on the toxicological effects of the compounds, *Drug Metab. Rev.* **10**:59–87.

12. Conney, A. H., 1982, Induction of microsomal enzymes by foreign compounds and carcinogenesis by polycyclic aromatic hydrocarbons, *Cancer Res.* **42**:4875–4917.

13. Guengerich, F. P., 1977, Metabolism of vinyl chloride: Destruction of the heme of highly purified liver microsomal cytochrome P450 by a metabolite, *Mol. Pharmacol.* **13**:911–923.

14. Stohs, S. J., Grafstrom, R. C., Burke, M. D., Moldeus, P. W., and Orrenius, S., 1976, The isolation of rat intestinal microsomes with stable cytochrome P450 and their metabolism of benzo[a]pyrene, *Arch. Biochem. Biophys.* **177**:105–116.

15. Ellin, A., Jakobsson, S. B., Schenkman, J. B., and Orrenius, S., 1971, P450$_k$ of rat kidney cortex microsomes: Its involvement in fatty acid ω- and (ω-1) hydroxylation, *Arch. Biochem. Biophys.* **150**:64–71.

16. Fang, W. F., and Strobel, H. W., 1978, The drug and carcinogen metabolism system of rat colon microsomes, *Arch. Biochem. Biophys.* **186**:128–138.

17. Hodgson, A. V., White, T. B., White, J. W., and Strobel, H. W., 1993, Expression analysis of the mixed-function oxidase system in rat brain by the polymerase chain reaction, *Mol. Cell. Biochem.* **121**:171–174.

18. Bergh, A. F., and Strobel, H. W., 1992, Reconstitution of the brain mixed-function oxidase system: Purification of NADPH-cytochrome P450 reductase and partial purification of cytochrome P450 from whole rat brain, *J. Neurochem.* **59**:575–581.

19. Fennell, P. M., and Strobel, H. W., 1982, Preparation of homogeneous NADPH-cytochrome P450 reductase from rat hepatoma, *Biochim. Biophys. Acta* **709**:173–177.

20. Oshinsky, R. J., and Strobel, H. W., 1987, Distribution and properties of cytochromes P450 and cytochrome P450 reductase from rat colon mucosal cells, *Int. J. Biochem.* **19**:575–588.

21. Hammond, D. K., and Strobel, H. W., 1990, Human colon cell line LS174T drug metabolizing system, *Mol. Cell. Biochem.* **93**:95–105.

22. White, T. B., Hammond, D. K., Vasquez, H., and Strobel, H. W., 1991, Expression of two cytochromes P450 involved in carcinogen activation in a human colon cell line, *Mol. Cell. Biochem.* **102**:61–69.

23. Nelson, D. R., Kamataki, T., Waxman, D. J., Guengerich, F. P., Estabrook, R. W., Feyereisen, R., Gonzales, F. J., Coon, M. J., Gunsalus, I. C., Gotoh, O., Okuda, K., and Nebert, D. W., 1993, The P450 superfamily: Update on new sequences, gene mapping, accession numbers, early trivial names and nomenclature, *DNA Cell Biol.* **12**:1–51.

24. Lu, A. Y. H., Kuntzman, R., West, S., Jacobson, M., and Conney, A. H., 1972, I. Reconstituted liver microsomal system that hydroxylates drugs, other foreign compounds, and endogenous substrates. II. Role of the cytochrome P450 and P448 fractions in drug and steroid hydroxylations, *J. Biol. Chem.* **247**:1724–1734.

25. van der Hoeven, T. A., Haugen, D. A., and Coon, M. J., 1974, Cytochrome P450 purified to apparent homogeneity from phenobarbital-induced rabbit liver microsomes: Catalytic activity and other properties, *Biochem. Biophys. Res. Commun.* **60**:569–575.

26. Levin, W., Ryan, D., West, S., and Lu, A. V. H., 1974, Preparation of partially purified, lipid-depleted cytochrome P450 and reduced nicotinamide adenine dinucleotide phosphate-cytochrome c reductase from rat liver microsomes, *J. Biol. Chem.* **249:**1747–1754.

27. Ryan, D., Lu, A. Y. H., West, S. B., and Levin, W., 1975, Multiple forms of cytochrome P450 in phenobarbital- and 3-methylcholanthrene-treated rats: Separation and spectral properties, *J. Biol. Chem.* **250:**2157–2163.

28. Poulos, T. L., Finzel, B. C., Gunsalus, I. C., Wagner, G. C., and Kraut, J., 1985, The 2.6 Å crystal structure of *Pseudomonas putida* cytochrome P450, *J. Biol. Chem.* **260:**16122–16130.

29. Poulos, T. L., Finzel, B. C., and Howard, A. J., 1987, High resolution crystal structure of cytochrome P450$_{cam}$, *J. Mol. Biol.* **195:**687–700.

30. Ravichandran, K. G., Boddupalli, S. S., Hasemann, C. A., Peterson, J. A., and Deisenhofer, J., 1993, Crystal structure of hemoprotein domain of P450 BM-3. A prototype for microsomal P450s, *Science* **261:**731–736.

31. Hasemann, C. A., Ravichandran, K. G., Peterson, J. A., and Deisenhofer, J., 1994, Crystal structure and refinement of cytochrome P450$_{terp}$ at 2.3 Å resolution, *J. Mol. Biol.* **236:**1169–1185.

32. Dignam, J. D., and Strobel, H. W., 1975, Preparation of homogeneous NADPH cytochrome P450 reductase from rat liver, *Biochem. Biophys. Res. Commun.* **63:**845–852.

33. Iyanagi, T., and Mason, H. S., 1973, Some properties of hepatic reduced nicotinamide adenine dinucleotide phosphate-cytochrome c reductase, *Biochemistry* **12:**2297–2308.

34. Iyanagi, T., Makino, N., and Mason, H. S., 1974, Redox properties of the reduced nicotinamide adenine dinucleotide phosphate-cytochrome P450 and reduced nicotinamide adenine dinucleotide-cyto-chrome b$_5$ reductases, *Biochemistry* **13:**1701–1710.

35. Williams, C. H., and Kamin, H., 1962, Microsomal triphosphopyridine nucleotide cytochrome c reductase of liver, *J. Biol. Chem.* **237:**587–595.

36. Phillips, A. H., and Langdon, R. G., 1962, Hepatic triphosphopyridine nucleotide cytochrome c reductase: Isolation, characterization and kinetic studies, *J. Biol. Chem.* **237:**2652–2660.

37. Masters, B. S. S., Bilimoria, M. H., and Kamin, H., 1965, The mechanism of 1- and 2- electron transfers catalyzed by reduced triphosphopyridine nucleotide-cytochrome c reductase, *J. Biol. Chem.* **240:**4081–4088.

38. Pederson, T. C., Buege, J. A., and Aust, S. D., 1973, Microsomal electron transport: The role of reduced nicotinamide adenine dinucleotide-phosphate-cytochrome c reductase in liver microsomal peroxida-tion, *J. Biol. Chem.* **248:**7134–7141.

39. Yasukochi, Y., and Masters, B. S. S., 1976, Some properties of a detergent solubilized NADPH-cyto-chrome c (cytochrome P450) reductase purified by biospecific affinity chromatography, *J. Biol. Chem.* **251:**5337–5344.

40. Dignam, J. D., and Strobel, H. W., 1977, NADPH-cytochrome P450 reductase from rat liver: Purification by affinity chromatography and characterization, *Biochemistry* **16:**1116–1123.

41. Knapp, J. A., Dignam, J. D., and Strobel, H. W., 1977, NADPH cytochrome P450 reductase: Circular dichroism and physical studies, *J. Biol. Chem.* **252:**437–443.

42. Yasukochi, Y., Peterson, J. A., and Masters, B. S. S., 1979, NADPH cytochrome c (cytochrome P450) reductase: Spectrophotometric and stopped flow kinetic studies on the formation of reduced flavopro-tein intermediates, *J. Biol. Chem.* **254:**7097–7104.

43. Vermilion, J. L., and Coon, M. J., 1978, Identification of the high and low potential flavins of liver microsomal NADPH-cytochrome P450 reductase, *J. Biol. Chem.* **253:**8812–8819.

44. Vermilion, J. L., and Coon, M. J., 1978, Purified liver microsomal NADPH-cytochrome P450 reductase: Spectral characterization of oxidation reduction states, *J. Biol. Chem.* **253:**2694–2704.

45. Vermilion, J. L., Ballou, D. P., Massey, V., and Coon, M. J., 1981, Separate roles for FMN and FAD in catalysis by liver microsomal NADPH-cytochrome P450 reductase, *J. Biol. Chem.* **256:**266–277.

46. Kurzban, G. P., and Strobel, H. W., 1986, Preparation and characterization of FAD-dependent NADPH cytochrome P450 reductase, *J. Biol. Chem.* **261:**7824–7830.

47. Kurzban, G. P., and Strobel, H. W., 1986, Purification of flavin mononucleotide-dependent and flavin-adenine dinucleotide-dependent reduced nicotinamide-adenine dinucleotide phosphate-cytochrome P450 reductase by high-performance liquid chromatography on hydroxylapatite, *J. Chromatogr.* **358**:296–301.

48. Kurzban, G. P., Howarth, J., Palmer, G., and Strobel, H. W., 1990, NADPH-cytochrome P450 reductase: Physical properties and redox behavior in the absence of the FAD moiety, *J. Biol. Chem.* **265**:12272–12279.

49. Narayanasami, R., Otvos, J. D., Kasper, C. B., Shen, A., Rajagopalan, J., McCabe, T. J., Okita, J. R., Hanahan, D. J., and Masters, B. S. S., 1992, ^{31}P NMR spectroscopic studies on purified, native and cloned, expressed forms of NADPH-cytochrome P450 reductase, *Biochemistry* **31**:4210–4218.

50. Bastiaens, P. I. H., Bonants, P. J. M., Muller, F., and Visser, A. J. W. G., 1989, Time resolved fluorescence spectroscopy of NADPH cytochrome P450 reductase: Demonstration of energy transfer between the two prosthetic groups, *Biochemistry* **28**:8416–8425.

51. Kuki, A., and Wolynes, P. G., 1987, Electron tunnelling paths in proteins, *Science* **236**:1642–1652.

52. Centeno, F., and Gutierrez-Merino, C., 1992, Location of functional centers in the microsomal cytochrome P450 system, *Biochemistry* **31**:8473–8481.

53. Gum, J. R., and Strobel, H. W., 1979, Purified NADPH cytochrome P450 reductase: Interaction with hepatic microsomes and phospholipid vesicles, *J. Biol. Chem.* **254**:4177–4185.

54. Gum, J. R., and Strobel, H. W., 1985, Isolation of the membrane-binding peptide of NADPH cytochrome P450 reductase: Characterization of the peptide and its role in the interaction of reductase with cytochrome P450, *J. Biol. Chem.* **256**:7478–7486.

55. Black, S. D., French, J. S., Williams, C. H., and Coon, M. J., 1979, Role of hydrophobic polypeptide in the N-terminal region of NADPH-cytochrome P450 reductase in complex formation with P450$_{LM}$, *Biochem. Biophys. Res. Commun.* **91**:1528–1535.

56. Black, S. D., and Coon, M. J., 1982, Structural features of liver microsomal NADPH cytochrome P450 reductase: Hydrophobic domain, hydrophilic domain and connecting region, *J. Biol. Chem.* **257**:5929–5938.

57. Haniu, M., Iyanagi, T., Miller, P., Lee, T. D., and Shively, J. E., 1986, Complete amino sequence of NADPH cytochrome P450 reductase from porcine hepatic microsomes, *Biochemistry* **25**:7906–7911.

58. Porter, T. D., Beck, T. W., and Kasper, C. B., 1990, NADPH cytochrome P450 oxidoreductase gene organization correlates with structural domains of the protein, *Biochemistry* **29**:9814–9818.

59. Traut, T. W., 1988, Do exons code for structural or functional units in proteins? *Proc. Natl. Acad. Sci. USA* **85**:2944–2948.

60. Tamburini, P. P., and Schenkman, J. B., 1986, Differences in the mechanism of functional interaction between NADPH cytochrome P450 reductase and its redox partners, *Mol. Pharmacol.* **30**:178–185.

61. Nadler, S. G., and Strobel, H. W., 1988, Role of electrostatic interactions in the reaction of NADPH cytochrome P450 reductase with cytochromes P450, *Arch. Biochem. Biophys.* **261**:418–429.

62. Nadler, S. G., and Strobel, H. W., 1991, Identification and characterization of an NADPH-cytochrome P450 reductase-derived peptide involved in binding to cytochrome P450, *Arch. Biochem. Biophys.* **290**:277–284.

63. Nisimoto, Y., 1986, Localization of cytochrome *c*-binding domain on NADPH cytochrome P450 reductase, *J. Biol. Chem.* **261**:14232–14239.

64. Tamburini, P. P., MacFarquhar, S., and Schenkman, J. B., 1986, Evidence of binary complex formation between cytochrome P450, cytochrome b$_5$ and NADPH-cytochrome P450 reductase, *Biochem. Biophys. Res. Commun.* **134**:519–526.

65. Towbin, H., Staehelin, T., and Gordin, J., 1979, Electrophoretic transfer of proteins from polyacrylamide gels to nitrocellulose sheets: Procedure and some applications, *Proc. Natl. Acad. Sci. USA* **76**:4350–4354.

66. Wada, F., Shibata, H., Goto, M., and Sakamoto, Y., 1968, Participation of the microsomal electron transport system involving cytochrome P450 in ω-oxidation of fatty acids, *Biochim. Biophys. Acta* **162**:518–524.

67. Masters, B. S. S., Baron, J., Taylor, W. E., Isaacson, E. L., and LoSpalluto, J., 1971, Immunochemical studies on electron transport chains involving cytochrome P450, *J. Biol. Chem.* **246**:4143–4150.
68. Shen, S., and Strobel, H. W., 1994, Probing the putative cytochrome P450- and cytochrome *c*-binding sites on NADPH cytochrome P450 reductase by antipeptide antibodies, *Biochemistry* **33**:8807–8812.
69. Porter, T. D., and Kasper, C. B., 1985, Coding nucleotide sequence of rat NADPH cytochrome P450 oxidoreductase cDNA and identification of flavin-binding domains, *Proc. Natl. Acad. Sci. USA* **82**:973–977.
70. Katagiri, M., Murakami, H., Yabusaki, Y., Sugigama, T., Okamoto, M., Yamano, T., and Ohkawa, H., 1986, Molecular cloning and sequence analyses of full-length cDNA for rabbit liver NADPH-cytochrome P450 reductase mRNA, *J. Biochem.* **100**:945–954.
71. Shen, A. L., Porter, T. D., Wilson, T. E., and Kasper, C. B., 1989, Structural analysis of the FMN-binding domain of NADPH-cytochrome P450 oxidoreductase by site-directed mutagenesis, *J. Biol. Chem.* **264**:7584–7590.
72. Hodgson, A. V., and Strobel, H. W., 1994, Polymerase chain reaction cloning, expression and purification of two FAD-binding domain fragments of cytochrome P450 reductase, *FASEB J.* **8**:A1422.
73. Porter, J. D., and Kasper, C. B., 1985, NADPH-cytochrome P450 oxidoreductase: Flavin mononucleotide and flavin adenine dinucleotide domains evolved from different flavoproteins, *Biochemistry* **25**:1682–1687.
74. Nisimoto, Y., and Shibata, Y., 1981, Location of functional -SH groups in NADPH cytochrome P450 reductase from rabbit liver microsomes, *Biochim. Biophys. Acta* **662**:291–299.
75. Lee, J. J., and Kaminsky, L. S., 1986, Fluorescence probing of the function-specific cysteines of rat microsomal NADPH-cytochrome P450 reductase, *Biochem. Biophys. Res. Commun.* **134**:393–399.
76. Nisimoto, Y., and Shibata, Y., 1982, Studies in FAD and FMN-binding domains in NADPH cytochrome P450 reductase from rabbit liver microsomes, *J. Biol. Chem.* **257**:12532–12539.
77. Inano, H., Kurihara, S., and Tamaoki, B., 1988, Inactivation of rat testicular NADPH-cytochrome P450 reductase by 2,4,6-trinitrobenzene sulfonate, *J. Steroid Biochem.* **29**:227–232.
78. Inano, H., and Tamaoki, B., 1986, Chemical modification of NADPH-cytochrome P450 reductase: Presence of a lysine residue in the rat hepatic enzyme as the recognition site of the 2′ phosphate moiety of the cofactor, *Eur. J. Biochem.* **155**:485–489.
79. Strobel, H. W., Nadler, S. G., and Nelson, D. R., 1989, Cytochrome P450: Cytochrome P450 reductase interactions, *Drug Metab. Rev.* **20**:519–533.
80. Nisimoto, Y., Hayashi, R., Akutsu, H., Kyogoku, Y., and Shibata, Y., 1984, Photochemically induced dynamic nuclear polarization study on microsomal NADPH-cytochrome P450 reductase, *J. Biol. Chem.* **259**:2480–2483.
81. Karplus, P. A., Daniels, M. K., and Herriott, J. R., 1991, Atomic structure of ferredoxin NADPH reductase: Prototype for a structurally novel family, *Science* **251**:60–66.
82. Karplus, P. A., and Schulz, G. E., 1987, Refined structure of glutathione reductase at 1.5 Å resolution, *J. Mol. Biol.* **195**:701–729.
83. Karplus, P. A., and Schulz, G. E., 1989, Substrate binding and catalysis by glutathione reductase as derived from refined enzyme: Substrate crystal structures at 2 Å resolution, *J. Mol. Biol.* **210**:163–180.
84. Bosterling, B., and Trudell, J. R., 1982, Association of cytochrome b5 and cytochrome P450 reductase with cytochrome P450 in the membrane and reconstituted vesicles, *J. Biol. Chem.* **257**:4783–4787.
85. Voznesensky, A. I., and Schenkman, J. B., 1992, The cytochrome P450 2B4–NADPH cytochrome P450 reductase electron transfer complex is not formed by charge pairing, *J. Biol. Chem.* **267**:14669–14676.
86. Voznesensky, A. I., and Schenkman, J. B., 1992, Inhibition of cytochrome P450 reductase by polyols has an electrostatic nature, *Eur. J. Biochem.* **210**:741–746.
87. Makower, A., Bernhardt, R., Rabe, H., Janig, G.-R., and Ruckpaul, K., 1984, Identification of lysine (384) in cytochrome P450$_{LM2}$ as functionally-linked residue, *Biomed. Biochim. Acta* **43**:1333–1341.
88. Bernhardt, R., Makower, A., Janig, G.-R., and Ruckpaul, K., 1984, Selective chemical modification of a functionally-linked lysine in cytochrome P450$_{LM2}$, *Biochim. Biophys. Acta* **785**:186–190.

89. Nelson, D. R., and Strobel, H. W., 1988, On the membrane topology of vertebrate cytochrome P450 proteins, *J. Biol. Chem.* **263:**6038–6050.

90. Nelson, D. R., and Strobel, H. W., 1989, Secondary structure prediction of 52 membrane-bound cytochromes P450 shows a strong structural similarity to P450$_{cam}$, *Biochemistry* **28:**656–660.

91. Narhi, L. D., and Fulco, A. J., 1986, Characterization of a catalytically self-sufficient 119,000 dalton cytochrome P450 monooxygenase induced by barbiturates in *Bacillus megaterium*, *J. Biol. Chem.* **261:**7160–7169.

92. Djordjevic, S., Wang, M., Shea, T., Roberts, D., Camitta, M., Masters, B. S. S., and Kim, J. J. P., 1994, Crystallization and preliminary x-ray studies of NADPH cytochrome P450 reductase, *FASEB J.* **8:**A1244.

93. Gillette, J. R., 1966, Biochemistry of drug oxidation and reduction by enzymes in hepatic endoplasmic reticulum, *Adv. Pharmacol.* **4:**219–261.

94. Rydstrom, J., Montelius, J., and Bengtsson, M., 1983, *Extrahepatic Drug Metabolism and Chemical Carcinogenesis*, Elsevier, Amsterdam.

95. Hagland, L., Kohler, C., Haaparanta, T., Goldstein, M., and Gustafsson, J.-A., 1983, Immunohisto-chemical evidence for a heterogeneous distribution of NADPH-cytochrome P450 reductase in the rat and monkey brain, in: *Extrahepatic Drug Metabolism and Chemical Carcinogenesis* (J. Rydström, J. Montelius, and M. Bengtsson, eds.), Elsevier, Amsterdam, pp. 89–93.

96. McMillan, K., Bredt, D. S., Hirsch, D. J., Snyder, S. H., Clark, J. E., and Masters, B. S. S., 1992, Cloned expressed rat cerebellar nitric acid synthase contains stoichiometric amounts of heme which binds carbon monoxide, *Proc. Natl. Acad. Sci. USA* **89:**11141–11145.

97. Shet, M. S., Fisher, C. W., Arlotto, M. P., Shackleton, C. H. L., Holmans, P. L., Martin-Wixtrom, C. A., Saeki, Y., and Estabrook, R. W., 1994, Purification and enzymatic properties of a recombinant fusion protein expressed in *Escherichia coli* containing the domains of bovine P450 17A and rat NADPH cytochrome P450 reductase, *Arch. Biochem. Biophys.* **311:**402–417.

98. Murakami, H., Yabusaki, Y., Sakaki, T., Shibata, M., and Ohkawa, H., 1987, A genetically engineered P450 monooxygenase: Construction of the functional used enzyme between rat cytochrome P450 *c* and NADPH cytochrome P450 reductase, *DNA* **6:**189–197.

99. Fisher, C. W., Shet, M. S., Caudle, D. C., Martin-Wixtrom, C. A., and Estabrook, R. W., 1992, High-level expression in *Escherichia coli* of enzymatically active fusion proteins containing the domains of mammalian cytochromes P450 and NADPH-cytochrome P450 reductase flavoprotein, *Proc. Natl. Acad. Sci. USA* **89:**10817–10821.

CHAPTER 8

Oxygen Activation and Reactivity

PAUL R. ORTIZ de MONTELLANO

1. Introduction

The cytochrome P450-catalyzed insertion of an oxygen into a substrate culminates a process that reduces molecular oxygen to a species equivalent, in terms of formal electron count and reactivity, to an oxygen atom.[1–6] The cytochrome P450 catalytic cycle traverses the following steps (Fig. 1): (1) reversible substrate binding, (2) reduction of cytochrome P450 from the ferric to the ferrous state by auxiliary electron transport proteins, (3) binding of molecular oxygen to give the ferrous dioxygen complex, (4) transfer of a second electron from the auxiliary electron transport proteins, (5) cleavage of the oxygen–oxygen bond to give a molecule of water and an oxidizing species in which the second oxygen is bound to the iron, (6) insertion of the iron-bound oxygen into the substrate, and (7) product dissociation. In the case of microsomal cytochrome P450 enzymes, the auxiliary electron transport protein is cytochrome P450 reductase, although cytochrome b_5 is able in some instances to provide the second electron. In the case of most bacterial and mitochondrial cytochrome P450 enzymes the electron transfer partners are a flavoprotein (e.g., adreno-doxin reductase) and an iron–sulfur protein (e.g., adrenodoxin) (see Chapters 3 and 11). Uncoupling of catalytic turnover from substrate oxidation can divert the consumption of reducing equivalents toward the production of superoxide, H_2O_2, or water rather than substrate-derived products (see Chapter 3).[6] Although the physiological turnover of cytochrome P450 generally adheres to the above sequence, alternative oxygen donors such as peroxides can react with cytochrome P450 via "shunt" mechanisms that produce the activated oxidizing species without recourse to reducing equivalents or molecular oxygen. The early parts of this catalytic cycle involving reductive activation of molecular oxygen are discussed in detail in Chapters 3 and 7. This chapter concentrates on the nature of the activated oxidizing species, the shunt pathways for its formation, and its reactions with organic substrates.

PAUL R. ORTIZ de MONTELLANO • Department of Pharmaceutical Chemistry, School of Pharmacy, University of California, San Francisco, San Francisco, California 94143.

Cytochrome P450: Structure, Mechanism, and Biochemistry (Second Edition), edited by Paul R. Ortiz de Montellano. Plenum Press, New York, 1995.

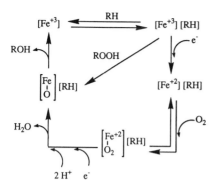

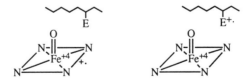

FIGURE 1. The catalytic cycle of cytochrome P450. The prosthetic heme group of the enzyme is represented in this figure by an iron in brackets, but is often represented in subsequent figures as an iron in a square of nitrogen atoms.

2. Formation of Activated Oxygen

2.1. Oxygen–Oxygen Bond Cleavage

The delivery of an electron to the ferrous dioxygen complex (Chapter 7) initiates dioxygen bond cleavage and unmasks the actual oxidizing species. The details of the dioxygen bond cleavage, the structure of the activated species, and the mechanisms by which the activated oxygen is transferred to substrates continue to pose a challenge to our understanding of cytochrome P450 catalysis because the steps after introduction of the second electron occur too rapidly for reaction intermediates to be observed. The information provided by X-ray crystallography on the active sites of several cytochrome P450 enzymes (see Chapters 4 and 5) has shed considerable light on these problems, but our view of the final steps of the catalytic sequence still relies heavily on the results of studies with artificial oxygen donors and of hemoprotein and metalloporphyrin models (see Chapter 1).

The available evidence strongly suggests that a single oxygen atom is bound to the prosthetic heme iron in the catalytically active species. The electron inventory for such a complex shows that the oxygen bears six valence electrons if the iron is held in the ferric state. No metalloporphyrin or hemoprotein complex is known, however, that actually has this specific electron distribution. Formally equivalent complexes in which two electrons are transferred from the iron to the oxygen to give an $[Fe^{5+}][O^{2-}]$ species, or in which transfer of one electron yields an $[Fe^{4+}][O^-]$ complex with seven electrons on the oxygen, are also unknown. The two-electron oxidized hemoproteins that have been characterized have a tetravalent iron and a full electron octet on oxygen. The electron required to complete the oxygen octet is drawn from the porphyrin ring or from an active-site amino

FIGURE 2. The known electron distributions for the two electron-oxidized species (Compound I) of peroxidases and related hemoproteins.

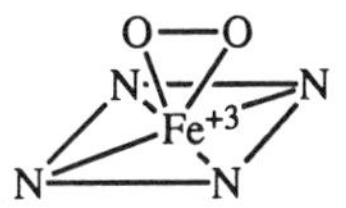

FIGURE 3. The cyclic peroxoiron structure.

acid residue (Fig. 2). A porphyrin radical cation is therefore present in the former and a protein radical or radical cation in the latter. The oxoiron complexes (Compounds I) generated when horseradish peroxidase or catalase react with hydrogen peroxide are examples of the former electron distribution (Chapter 2).[7-10] Porphyrin radical cations are also observed with model metalloporphyrin complexes (Chapter 1). Compound I of cytochrome c peroxidase, on the other hand, is an example of a structure with a protein-centered radical.[11] However, the electronic properties of a thiolate ligand differ markedly from those of the imidazole in horseradish and cytochrome c peroxidases or the phenoxide in catalase, a fact that may radically alter the electronic structure of the catalytic complex. The extent of unpaired electron density on the oxygen is governed by the degree of orbital mixing and the extent of electron transfer from the iron and thiolate sulfur to the oxygen. The extent of this electron delocalization is important because it influences the propensity of the activated oxygen to enter into unpaired electron (radical) reactions.

The catalytic species is normally generated by reductive cleavage of the dioxygen bond in the ferrous dioxygen complex, but the mechanism by which the enzyme facilitates this bond cleavage is not yet fully understood (Chapter 3). The finding that the reaction of superoxide with ferrous porphyrins and the electrochemical reduction of ferroporphyrin dioxygen complexes yield cyclic iron dioxygen complexes (Fig. 3) suggests the possible involvement of such a species in dioxygen bond cleavage.[12-14] Although a cyclic peroxo complex such as this may be an intermediate in oxygen activation, it is unlikely to be the actual oxidizing species because structurally related iron cycloperoxides fail to oxidize substrates as reactive as styrene.[15-17] Unless the thiolate ligand of P450 markedly alters their chemistry, the chemical inertness of iron cycloperoxides would appear to preclude their involvement as the immediate oxidizing species in P450 catalysis.

2.2. Peroxides as Oxygen Donors

The catalytic choreography of three membrane-bound proteins (cytochrome P450, cytochrome P450 reductase, and cytochrome b_5) and two cosubstrates (NADPH and molecular oxygen) in normal P450 function obscures the molecular details of the catalytic process. The discovery that hydroperoxides and other artificial oxygen donors support catalytic turnover of P450 in the absence of electron transfer proteins or cosubstrates has therefore been important for mechanistic studies. It is necessary, however, to carefully evaluate all inferences based on work with alternative oxygen donors such as peroxides to determine their relevance to the normal function of the enzyme.

The decomposition of linoleic acid hydroperoxide by hepatic microsomes freed of hemoproteins other than P450 by trypsin treatment is inhibited by cyanide or alternative cytochrome P450 substrates, and is elevated by phenobarbital pretreatment, to the same extent as the P450 catalytic activity.[18,19] This decomposition of linoleic acid hydroperoxide can be coupled to the oxidation of N,N,N',N'-tetramethyl-p-phenylenediamine,[18] NADPH,[20] diaminobenzidine,[21] alcohols,[22] phenols,[21] and other peroxidase substrates. P450 similarly catalyzes the decomposition of cumene hydroperoxide,[18] cholesterol hydroperoxides,[19] pregnenolone 17α-hydroperoxide,[19] progesterone 17α-hydroperoxide,[19] *tert*-butylhydroperoxide,[23] and H_2O_2.[18] A role for this activity in the hydroperoxide-dependent peroxidation of microsomal lipids[24–26] is suggested by the fact that reduced pyridine nucleotides attenuate lipid peroxidation, the inhibition is enhanced by substrates that promote the reduction of P450, and carbon monoxide inhibits the peroxidative process.[26,27]

The mechanistic interest in reactions of P450 with peroxides, however, stems from the occurrence of peroxygenative reactions in which an oxygen of the peroxide is incorporated into the substrate rather than from the coupling of peroxide reduction to one-electron oxidation of electron donors. The cumene hydroperoxide-supported microsomal N-demethylations of dimethylaniline, aminopyrine, benzphetamine, propoxyphene, ethylmorphine, and methamphetamine are reportedly reactions of this type.[28] The role of P450 in these N-demethylation reactions is confirmed by the fact that N-methylaniline, formaldehyde, and cumyl alcohol are formed in stoichiometric amounts in the reaction of reconstituted P450 2B4 with cumene hydroperoxide and N,N-dimethylaniline.[29] N-demethylation, however, is a well-established peroxidative as well as peroxygenative reaction and therefore is a fragile measure of peroxygenase reactivity.[30,31] More persuasive evidence for peroxygenative reactions is provided by the hydroperoxide-dependent O-dealkylation of alkyl ethers[32,33] and the hydroxylation of aromatic rings[33,34] and unactivated hydrocarbons.[35–39] Lauric acid,[36,37] androstenedione, testosterone, progesterone, 17β-estradiol, and 5β-cholestane-3α,7α,12α-triol[35,37–39] are among the substrates shown to be hydroxylated by a hydroperoxide-dependent mechanism. The hydroxylation of aromatic substrates, first demonstrated for biphenyl, benzpyrene, coumarin, and aniline,[33,34] occurs with the NIH shift characteristic of normal NADPH-dependent reactions.[40] The oxygen incorporated into the epoxide in the cumene hydroperoxide-supported oxidation of phenanthrene by hepatic microsomes derives from a source other than the medium but was not specifically shown to derive from the peroxide.[40] The cumene hydroperoxide-dependent hydroxylation of cyclohexane by reconstituted CYP2B4 also results in incorporation of an oxygen from a source other than the medium.[29] More definitive experiments with isotopically labeled oxygen have been carried out and are described below.

The ferryl species produced by reaction of a hydroperoxide with cytochrome P450 sometimes reacts with the alcohol produced from the hydroperoxide before it diffuses out of the active site, resulting in net enzyme-catalyzed conversion of the hydroperoxide to a diol. The first reaction of this type reported was the P450$_{scc}$-catalyzed conversion of 20α-hydroperoxycholesterol to 20α,22R-dihydroxycholesterol and of 20β-hydroperoxycholesterol to 20β,21-dihydroxy-2-isocholesterol (Fig. 4).[41–43] The oxygen atom incorporated into these diols was shown by ^{18}O studies to derive from the hydroperoxide moiety. Furthermore, radiolabeled cholesterol added to the incubation at the same time as the hydroperoxide was not oxidized, demonstrating that diol formation occurred by intramolecular oxygen transfer. In contrast, the reactions of P450$_{scc}$ with (20S)-hydroperoxy-22-keto-, 23-hydroperoxy-22-keto-, or 25-hydroperoxy-24-keto-cholesterol result in

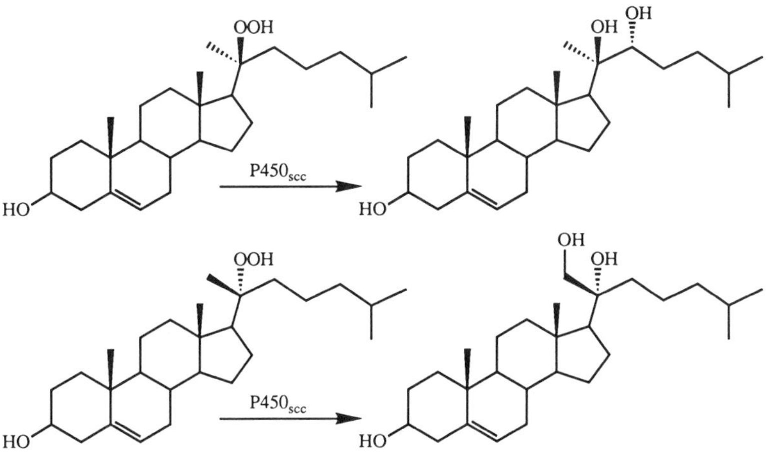

FIGURE 4. Conversion of 20α-hydroperoxycholesterol to 20α,22R-dihydroxycholesterol and 20β-hydroperoxycholesterol to 20β,21-dihydroxy-20-isocholesterol by P450$_{scc}$, two examples of intramolecular hydroxylation by oxygen transfer from a peroxide.

cleavage of the carbon–carbon bond between the oxygenated carbons.[43] EPR studies have detected a radical species in the reaction of (20R)-20-hydroperoxycholesterol with P450$_{scc}$ that has been attributed to an Fe^{IV} porphyrin radical cation by comparison with the spectra of horseradish peroxidase Compound I.[44] However, the EPR detectable species corresponds to oxidation of only 0.7% of the total hemoprotein. The signal could be low because of the reactive nature of the proposed ferryl intermediate but could very well result from a minor, aberrant reaction other than that which yields (20R)-20,21-dihydroxycholesterol.

The reactions of 2,6-di-*tert*-butyl-4-hydroperoxy-4-methyl-2,5-cyclohexadienone, a hydroperoxide derived from butylated hydroxytoluene, have been used to investigate the nature of hemoprotein-catalyzed oxygen–oxygen bond cleavage.[45] Reaction of this hydroperoxide with microsomal cytochrome P450 results in conversion of the hydroperoxide to the alcohol with concomitant hydroxylation of one of the *tert*-butyl groups. Isotopic labeling experiments show the reaction occurs by an intramolecular mechanism. These results are consistent with heterolytic cleavage of the peroxide to give the ferryl species and the alcohol, followed by normal hydroxylation of one of the *tert*-butyl groups by the ferryl species.[45] A complete analysis of the products formed in the reaction shows that the hydroperoxide also undergoes homolytic scission.[46] Dioxygen bond homolysis produces both a quinone as a result of extrusion of the 4-methyl group from the alkoxy radical intermediate and a variety of rearranged products (Fig. 5). The proportion of quinone to rearranged products formed by the homolytic pathway can be altered by changing the 4-alkyl group to make it more easily eliminated as a carbon radical.[46] The 4-benzyl compound, for example, primarily gives the quinone product. Analysis of the fate of the 4-benzyl group shows that it is converted into benzyl alcohol, benzaldehyde, and toluene.[47] Studies with ^{18}O-labeled substrate indicate that most of the benzyl alcohol is produced by recombination of the benzyl radical with the homolytically generated ferryl oxygen. These results mirror the earlier finding that P450 2B4 catalyzes the conversion of 2-phenylper-

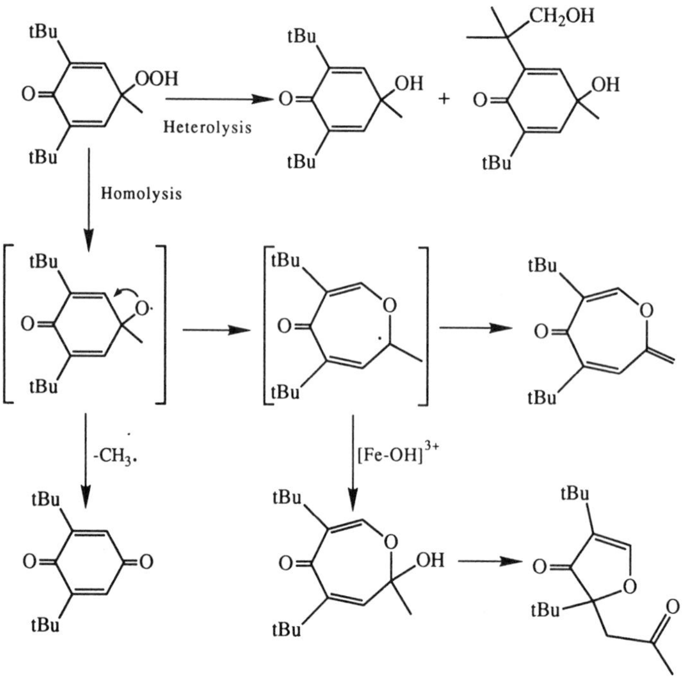

FIGURE 5. Products formed in the reaction of 2,6-di-*tert*-butyl-4-hydroperoxy-4-methyl-2,5-cyclo-hexadienone with cytochrome P450 and other hemoproteins. Heterolysis of the dioxygen bond results in formation of the alcohol and partial intramolecular hydroxylation to give a diol, whereas homolysis results in extrusion of an alkyl group or rearrangement of the cyclohexadienone skeleton.

acetic acid to benzyl alcohol via homolytic oxygen transfer to the iron, decarboxylation, and return of the iron-bound oxygen to the benzyl radical (Fig. 6).[48]

In the same vein, cumene hydroperoxide [PhC(CH$_3$)$_2$OOH] undergoes both homolytic cleavage to give benzophenone[45,49] and heterolytic cleavage followed by intramolecular hydroxylation to give 2-phenyl-1,2-propanediol and 2(*p*-hydroxyphenyl)-2-propanol.[45,50] Presumably the intramolecular hydroxylation competes with the hydroxylation of other substrates shown to be supported by cumene hydroperoxide (see above). Groves has also reported conversion of 1-hexylhydroperoxide to 1,2-hexanediol (see Chapter 1).[51] It is clear from these results that hydroperoxides undergo concurrent homolytic and heterolytic cleavage by cytochrome P450 and that intramolecular rearrangement of peroxides to diol products is likely to be fairly common.

The peroxygenase activities of rabbit liver microsomes[52] and reconstituted P450 2B4[53–55] have been analyzed by kinetic methods. An "ordered bi bi" kinetic sequence in which the substrate binds before the peroxide and the catalytic event occurs in a ternary enzyme/substrate/peroxide complex is suggested by (1) the observation that the *O*-demethylation of *p*-nitroanisole by rabbit liver microsomes and *tert*-butylhydroperoxide gives rise to convergent double reciprocal plots when the substrate or peroxide concentra-

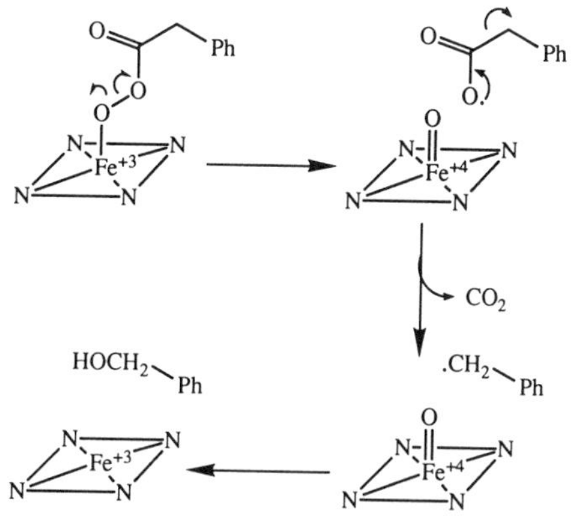

FIGURE 6. Homolytic scission of the dioxygen bond in 2-phenylperacetic acid followed by decarboxylation and transfer of the iron-bound oxygen to the benzyl radical to give benzyl alcohol.

tion is varied and (2) the finding that cyanide inhibits the reaction competitively with respect to the peroxide but noncompetitively with respect to the substrate.[53] The same conclusion has been reached in a kinetic analysis of the hydroperoxide-dependent hydroxylation of toluene by reconstituted P450 2B4.[53–55] Two reversibly formed ternary complexes were observed with the purified enzyme. The first of these (C) is directly related to the hydroxylation event but has a low affinity for the peroxide and is difficult to detect, whereas the second (D) is readily detected but is a dead-end complex:

$$[\text{enzyme–substrate}] + \text{peroxide} \rightleftharpoons C \rightleftharpoons D$$

The difference spectrum of complex D, with a trough at 416 nm and a maximum at approximately 436 nm, is essentially identical to that observed when cumene hydroperoxide is added to hepatic microsomes from phenobarbital-pretreated rabbits.[56] The magnitude and direction of the P450 2B4 shift are comparable to those incurred in the conversion of ferric horseradish peroxidase to Compound II.[57] In contrast to horseradish peroxidase, however, the magnitude and position of the Soret bands of complexes C and D depend on the peroxide used to generate the complex. The implication that the organic moiety of the peracid is retained in the complexes is supported by the following observations: (1) the rate of conversion of complex C to complex D depends on the electron-withdrawing ability of the substituents on the peracid, (2) the rate of substrate hydroxylation is limited by the decay of complex C, and (3) the hydroxylation rate depends on the structure of both the substrate and the peracid. The demonstration that alkyl hydroperoxides are retained in the active site long enough to become substrates (see above) is consistent with these observations.

At least four transient intermediates have been detected spectroscopically in the reaction of P450$_\text{cam}$ with *meta*-chloroperbenzoic acid.[58–60] The first intermediate appears within 10 msec and has an absorption spectrum (λ_max at 367 and 694 nm) essentially identical to that of Compound I of chloroperoxidase (λ_max at 367 and 688 nm), a hemoprotein in which the iron, as in P450, is coordinated to a cysteine thiolate. The first intermediate detected in the reaction thus appears to be the Compound I equivalent of P450$_\text{cam}$. A spectrum (λ_max at 380 and 440 nm) similar to that of Compound II of chloroperoxidase (λ_{max} of 370 and 470 nm) is reportedly obtained if the ferrous dioxy complex of P450$_{cam}$ is reduced by pulse radiolysis.[61] The heme in the enzyme was replaced for this experiment by iron 2,4-diacetyldeuteroporphyrin because the ferrous dioxy complex of the modified enzyme is much more stable than that of the unmodified enzyme. The correlation between the absorption maxima is therefore closer than the numbers suggest if the effect of the electron-withdrawing acetyl groups on the position of the P450$_{cam}$ maxima is taken into account.[61] The reaction of ferrous P450$_{cam}$ with superoxide in theory provides an alternative route to the catalytically active species, but the reaction gives an intermediate with a spectrum similar to that of the ferrous dioxy complex rather than to a Compound I or II species.[62] The collective results suggest that P450$_{cam}$ reacts with peroxides to give Compound I and Compound II species analogous to those of chloroperoxidase. It has definitely not been established, however, that these species are formed under normal, NADPH-dependent catalytic turnover.

The reaction of n- or $tert$-butylhydroperoxide with purified CYP2C11 gives rise at low temperature to the EPR spectrum of a ferric, low-spin species with parameters ($g_1 = 2.29$, $g_2 = 2.24$, and $g_3 = 1.96$) very similar to those ($g_1 = 2.285$, $g_2 = 2.198$, and $g_3 = 1.959$) of a model thiolate–iron–peroxide (R-S-Fe-OOR') complex.[63] As the reaction of $tert$-butylhydroperoxide with liver microsomes gives rise to peroxyl and alkoxyl radicals, it is likely that formation of the peroxide complex is followed, in part, by homolytic scission to give the $tert$-butoxy radical.[23] A similar conclusion is suggested by the observation of transient EPR signals at approximately 2.01 g in incubations of hepatic microsomes with cumene hydroperoxide.[56] The Fe-OOR species undoubtedly is also the precursor, via heterolytic scission, of the ferryl species involved in peroxygenative catalysis.

The parallel operation of peroxidative and peroxygenative pathways readily rationalizes the differences between the metabolite profiles in cytochrome P450 reactions supported by hydroperoxides and NADPH. Differences in the kinetic parameters and the sensitivity to inhibitors of the CYP2B4-catalyzed decarboxylation of phenylperacetic acid and concurrent hydroxylation of cyclohexane support the view that the two reactions are products of different catalytic trajectories.[64] The metabolism of benzo[a]pyrene to phenols with NADPH but to quinones with cumene hydroperoxide,[65] and of 4-chloroaniline to (4-chlorophenyl)hydroxylamine with NADPH but 4-chloro-1-nitrobenzene with cumene hydroperoxide,[66] can be explained by secondary peroxidative oxidation of the initial phenol and hydroxylamine metabolites. The observation by EPR of nitrogen radicals in the hydroperoxide- but not NADPH-dependent N-demethylation of aminopyrine,[30,31,67] and of the nitroxide radical in the hydroperoxide but not NADPH-supported oxidation of N-hydroxy-2-acetylaminofluorene,[68–70] is consistent with such an explanation. The formation of peroxyl and alkoxyl radicals from $tert$-butylhydroperoxide in incubations with liver microsomes, established by spin trapping experiments,[23] supports the idea that the peroxidative reactions are triggered by homolytically generated oxygen radicals. Differences may exist, however, that require a more complex explanation than the superposition of peroxidative (homolytic) and peroxygenative (heterolytic) processes.

Hydrogen peroxide supports the peroxygenase activity of P450 less effectively than cumene hydroperoxide[28,29,35,67,71–74] and supports the peroxidase activity very poorly.[18] The product distribution for the hydrogen peroxide-supported oxidation of aniline[71] and benzo[a]pyrene[72] resembles that of the NADPH- rather than cumene hydroperoxide-supported process. The P450 2B4-catalyzed N-demethylation of aminopyrine supported by hydrogen peroxide proceeds without the detectable formation of aminopyrine radicals.[65] Differences are observed in the regiochemistry of warfarin metabolism when cumene hydroperoxide is substituted for NADPH and oxygen.[75] Cumene hydroperoxide supports ω-1- but not ω-hydroxylation of lauric acid,[36] and does not support the aromatization of sterols by placental P450.[76] Hydrogen peroxide supports the hydroxylation of prostaglandins at different positions by liver microsomes from differentially induced rats.[74] The discrepancies between the hydrogen peroxide- and NADPH-dependent N-demethylation rates for 15 secondary and tertiary amines,[73] and the differences in the products obtained with NADPH and H_2O_2 in other instances (e.g., prostaglandin hydroxylation),[74] indicate that differences exist in the abilities of different isoforms to turn over with H_2O_2.

In sum, alkyl hydroperoxides undergo heterolysis to produce an oxidizing species similar if not identical to that generated with NADPH and cytochrome P450 reductase. However, the hydroperoxides simultaneously undergo homolytic reaction, forming alkoxy radicals that diffuse from the active site and initiate enzyme-independent peroxidative reactions. The reactions supported by alkyl hydroperoxides differ, furthermore, in that the alkyl group can remain in the active site long enough to be hydroxylated and/or to alter the metabolism of other substrates. It is not yet clear whether H_2O_2 is homolytically as well as heterolytically cleaved by cytochrome P450. Although H_2O_2 eliminates differences associated with the retention of an organic fragment in the active site, the differences observed in product distributions, the inability of some enzymes to turn over with H_2O_2, and the spectroscopic observation of Compound I-like intermediates in peroxide- but not NADPH-supported turnover, indicate that the activation of P450 by peroxides differs in some essential manner from the activation by NADPH and oxygen. This difference is supported by the efficient degradation of the prosthetic heme group observed with peroxides but not with NADPH and oxygen (see Chapter 9). The difference is likely to be subtle and may be no more than a higher rate of enzyme oxidation by the peroxides, or the absence of the protection provided by the availability of reducing equivalents.

2.3. Iodosobenzene and Other Single Oxygen Donors

The delivery of a single oxygen atom in the appropriate oxidation state simplifies the catalytic system even further and circumvents peroxidative processes associated with homolytic peroxide scission. The initial substitution of $NaIO_4$ and $NaClO_2$ for NADPH in the hydroxylation of steroids[35,37–39,77] and fatty acids[78] yielded to the use of iodosobenzene,[77–80] a lipophilic agent that more readily interacts with the membrane-bound enzymes. Iodosobenzene supports the catalytic turnover of P450 in the absence of NADPH or molecular oxygen but the regioselectivities of the reactions supported by iodosobenzene and NADPH are not identical.[77,80] These differences could result from oxygen transfer in a ternary complex analogous to that implicated in at least some hydroperoxide-dependent reactions but could also result, in microsomal incubations, from preferential interaction of the oxygen donor with a subset of the P450 isozymes.

The strong parallels between the iodosobenzene- and NADPH-dependent reactions of P450 and the iodosobenzene-supported model reactions of metalloporphyrins (Chapter

1) provide important support for the view that a ferryl complex is the activated catalytic species. The finding that the oxygen from $H_2^{18}O$ is incorporated into the products obtained from camphor,[81] cyclohexane,[82] and parathion[83] in the iodosobenzene- but not NAD(P)H- or peroxide-supported reactions appears, at first glance, to be inconsistent with formation of the same activated species. Thus, oxygen from the medium is not incorporated into the products during (1) the NADPH-dependent microsomal hydroxylation of sterols,[84,85] (2) the NADPH- or hydroperoxide-supported hydroxylation of camphor by $P450_{cam}$[81] (3) the NADPH- or hydroperoxide-dependent sulfoxidation of p-methylthioanisole by P450 2B4,[86,87] (4) the microsomal oxidation of sulfur compounds with either NADPH or cumene hydroperoxide,[83] or, except for a trace, (5) the hydroxylation of cyclohexane catalyzed by P450 2B4 and cumene hydroperoxide.[88] The dichotomy between the iodosobenzene and NADPH reactions is clarified, however, by the observation that the oxygen of iodosobenzene is stable to uncatalyzed exchange[89] but exchanges with the medium in the presence of intact microsomes.[82] It appears likely that the oxygen of *iodosobenzene* is exchanged in a reaction catalyzed by P450 prior to actual formation of the ferryl intermediate. The catalytic involvement of P450 and the high incorporation (>98%) of oxygen from the medium into metabolites can be readily explained if water coordinated to the iron adds to iodosobenzene to yield a trisubstituted iodine species that dissociates from the enzyme, resulting in exchange of the oxygen in iodosobenzene, or decomposes to the catalytically active species. The sixth iron ligand in the P450 enzymes for which crystal structures are available is, in fact, a water molecule (Chapters 4 and 5). A hydrolytic mechanism can also be invoked to rationalize the finding that the tosylimine analogue of iodosobenzene ($MeC_6H_5SO_2N=IC_6H_5$) supports the P450 2B4-catalyzed hydroxylation rather than tosy-lamination of cyclohexane,[90,91] although P450-catalyzed *intramolecular* insertion of a closely related nitrogen species into a C–H bond of the same molecule has been success-fully accomplished.[92]

Aryldimethylamine N-oxides also activate cytochrome P450 but their significance and utility are compromised by the fact that they only support reactions in which the organic framework of the aryl N-oxide is itself the substrate. The N-oxide of N,N-dimethylaniline is thus converted by $P450_{cam}$ to N-methylaniline but camphor is not simultaneously hydroxylated.[93] Comparable results have been obtained with CYP2B1.[94] Transfer of the oxygen to P450 possibly produces the nitrogen radical cation and an $Fe^{IV}=O$ species equivalent to the Compound II rather than Compound I intermediate of a peroxidase. Deprotonation of the methyl and addition of the carbon radical to the $Fe^{IV}-OH$ species would then yield N-dealkylated products without allowing alternative substrate oxidation.

3. Mechanisms of Oxygen Transfer Reactions

3.1. Introduction

The oxygen of the activated ferryl species in cytochrome P450, as implied by its monooxygenase function, is ordinarily transferred to a substrate. The structurally related ferryl species found in Compound I of the peroxidases differs from that of P450 because it functions as an electron sink rather than as an oxygen donor. This functional differentia-tion does not, however, preclude P450 mechanisms in which the two oxidation equivalents

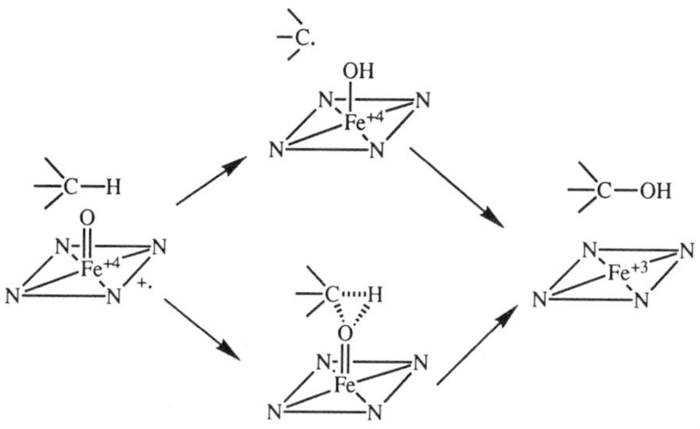

FIGURE 7. Schematic illustration of concerted versus nonconcerted hydrocarbon hydroxylation mechanisms. In order to illustrate the changes in redox state of the P450, the ferryl species is arbitrarily portrayed with a porphyrin radical cation rather than a protein radical.

are committed individually rather than simultaneously, or in which electron transfer precedes oxygen transfer. The difference between a concerted mechanism with one transition state and a nonconcerted mechanism with two or more transition states is illustrated for hydrocarbon hydroxylation in Fig. 7. The accumulated evidence suggests, in fact, that transfer of the oxygen from P450 to substrates often proceeds via a nonconcerted mechanism.

3.2. Hydrocarbon Hydroxylation

The 7α- and 11α-hydroxylation of steroids,[95,96] the benzylic hydroxylation of ethyl benzene,[97] the ω- and ω-1-hydroxylation of lauric acid,[98] the terminal hydroxylation of $(1R)$- and $(1S)$-[1-^{3}H,^{2}H,^{1}H]octane,[99] and the terminal hydroxylation of geraniol[100] have been reported to occur with retention of stereochemistry. Other studies have shown, however, that P450-catalyzed carbon hydroxylation reactions proceed with loss of stereochemistry. Stereochemical scrambling was first observed in the hydroxylation of *exo*-tetradeuterated norbornane by rabbit liver microsomes, which yielded, among other products, the *endo*-alcohol with three rather than four deuterium atoms and the *exo*-alcohol with four deuterium atoms (see Chapter 1).[101] Related examples of stereochemical scrambling are provided by the observation that P450$_{cam}$ removes the *endo* or *exo* hydrogen from camphor but only delivers the hydroxyl to the *exo* position,[102] and comparable results obtained for the hydroxylation of camphor derivatives by the fungus *Beauveria sulfurescens*.[103] Even the hydroxylation of ethylbenzene, a reaction initially considered to proceed with retention of configuration,[97] proceeds with 23–40% loss of stereochemistry.[104] These results suggest that retention of stereochemistry is not an intrinsic property of carbon hydroxylation reactions but rather is imposed on the reaction process by the structure of the active site. This follows from the fact that retention of stereochemistry is only a valid measure of the timing of the reaction if stereochemical fidelity is *not* imposed by the active site, whereas *loss* of stereochemistry requires a nonconcerted mechanism.

Independent evidence for a nonconcerted mechanism is provided by the allylic rearrangement of a double bond observed during the hydroxylation of 3,4,5,6-tetrachloro-

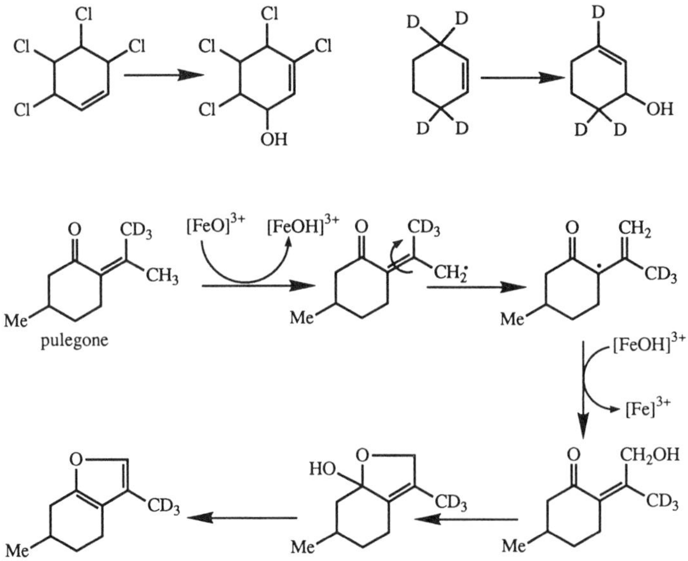

FIGURE 8. Hydroxylation reactions that demonstrably proceed via nonconcerted mechanisms include the hydroxylation of 3,4,5,6-tetrachlorocyclohexene and 3,3,6,6-tetradeuteriocyclohexene, and the topomerization of deuterated pulegone. The mechanism proposed for the latter reaction is shown.

cyclohexene and related compounds by rat or housefly microsomes (Fig. 8),[105] 3,3,6,6-tetradeuterated cyclohexene (Fig. 8), methylenecyclohexane, and β-pinene by CYP2B4,[106] linoleic acid,[107] and (R)-(+)-pulegone by rat liver microsomes (Fig. 8).[108] The oxidation of pulegone is of unique interest because the allylic rearrangement is accompanied by extensive isomerization about the double bond prior to oxygen recombination with the radical (Fig. 8). The reason for the high degree of topomerization is not known but it is likely that the reaction involves removal of a hydrogen from the (E)-methyl group, isomerization of the resulting allylic radical, recombination of the radical with the ferryl oxygen to give an alcohol, and chemical cyclization/dehydration to give the furan. In contrast, the allylic transposition involved in the conversion of linoleic acid to 9-hydroxyoctadeca-10E,12Z-dienoic acid or 13-hydroxyoctadeca-9Z,11E-dienoic acid involves stereospecific suprafacial hydrogen abstraction from C-11 and addition of the ferryl oxygen at either C-9 or C-13.[107] The allylic transposition of functionality observed in all of the isomerization reactions requires the formation of a delocalized intermediate. The intermediate is presumably a free radical produced by hydrogen abstraction, although the intermediate could, in principle, be generated by sequential abstraction of an electron from the double bond followed by loss of a proton from the adjacent carbon. However, the magnitudes of the isotope effects (k_H/k_D = 4–6) for allylic hydroxylation and the absence of evidence for radical cation intermediates in π-bond oxidations (see Section 3.3) support a direct hydrogen abstraction mechanism for these allylic rearrangement reactions.

The rate of recombination of the carbon radical and iron-bound hydroxyl formed in the first step of a nonconcerted hydroxylation reaction (Fig. 7) has been explored with the

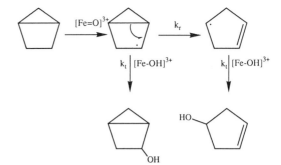

FIGURE 9. Radical clock timing of the rate of the hydrocarbon hydroxylation "rebound" step with bicyclo[2.1.0]pentane.

help of "radical clock" probes. The carbon radical formed from such a probe undergoes a rearrangement reaction at a known rate (k_r) that competes with the rate (k_t) of the radical recombination reaction.[109] Efforts to detect the radical intermediate by incorporating a simple cyclopropyl moiety adjacent to the hydroxylated position, an approach intended to exploit the rearrangement of the cyclopropylmethylene radical to the 3-butenyl radical (k_r = 1.3×10^8 sec^{-1} at 25°C), have only given rise to unrearranged hydroxylated products.[110–115] This failure to detect rearranged products indicates either that the reaction does not proceed via a radical intermediate or that combination of the carbon radical with the ferryl oxygen occurs substantially faster than the rearrangement. To address this question, substrates that unmask radicals that rearrange at even faster rate than the simple cyclopropylmethylene radical have been constructed. The first such substrate was bicyclo[2.1.0]pentane.[116] The radical obtained on removal of a hydrogen from one of the carbons adjacent to the cyclopropyl ring in this strained system rearranges to the ring-opened 3-cyclopentenyl radical at a rate of k_r = 2.2×10^{10} sec^{-1}.[117] Oxidation of bicyclo[2.1.0]pentane by microsomal cytochrome P450 produces a 7:1 mixture of unrearranged and rearranged alcohol products (Fig. 9).[116] This indicates that the rate of the combination reaction between the iron-bound oxygen and the carbon radical is on the order of k_t = 10^{10} sec^{-1}. Extension of these studies to a family of other rapid radical clock systems based on the substituted cyclopropane ring yields values for the recombination rate (k_t) to a primary carbon of 2.4×10^{11} sec^{-1} and to a secondary carbon of 1.4×10^{10} sec^{-1}.[117,118]

The kinetic isotope effects for P450-catalyzed hydrocarbon hydroxylations calculated from the rates of product formation from deuterated and undeuterated substrates are generally small,[119] although V_{max} isotope effects greater than 5 have occasionally been observed.[120] These small isotope effects provided early support for a concerted hydroxylation mechanism because the bent transition state required by such a process, as illustrated by the isotope effects for carbene insertions into C–H bonds (k_H/k_D = 0.9–2.5),[121,122] are theoretically predicted to exhibit small isotope effects.[123] However, the isotope effects calculated from the rates of metabolite formation are inaccurate measures of the isotope effects intrinsic to the hydroxylation step if the catalytic rate is significantly determined by enzymatic steps other than the hydroxylation itself. The intrinsic isotope effects determined from the ratio of oxygen insertion into C–H and C–D bonds in equivalent sites within a single molecule circumvent the ambiguities related to multiple rate-limiting steps and are in fact quite large: (1) k_H/k_D = 11.5 for the hydroxylation of tetradeuterated norbornane (see Chapter 1),[101] (2) $k_H/k_D = 11$ for the benzylic hydroxylation of $[1,1-^2H]$-1,3-diphenylpropane $(PhCD_2CH_2CH_2Ph)$,[124] and (3) $k_H/k_D = 10$ for the O-demethylation of p-trideuteromethoxyanisole.[125] The intrinsic isotope effect for the O-dealkylation of 7-ethoxycoumarin, calculated from the measured deuterium and tritium isotope effects on V_{max}, yields an equally high value $(k_H/k_D = 13.5)$.[126] The intrinsic isotope effect can be unmasked not only by intramolecular competition between two equivalent sites on a substrate but also by intramolecular competition between unidentical sites. Thus, an intramolecular k_H/k_D of 9.5–9.8 for ω-hydroxylation of octane is unmasked by competition for hydroxylation of the terminal methyl versus the internal methylenes of the hydrocarbon chain.[127] Analysis of this reaction has dissected the measured isotope effect into a primary isotope effect of 7.9 and a secondary isotope effect of 1.14.[128,129] The isotope effects for the reaction are essentially the same for the ω-hydroxylations catalyzed by CYP2B1, 1A1, and 2B4, suggesting that the hydroxylation mechanism is the same for all three enzymes.[129] The hydroxylation of deuterated xylenes based on a competition between the hydroxylation of CH_3 and CD_3 on the same molecule yields a primary isotope effect of 7.5–9.5 and a secondary isotope effect of 1.09–1.19.

An interesting isotope effect is unmasked for the hydroxylation of toluene when the effect is unmasked by considering the shift from methyl to aromatic ring oxidation in the total products formed. The intermolecular isotope effect determined by comparing the yield of benzyl alcohol from $PhCH_3$ and $PhCD_3$ is 1.92 whereas the corresponding isotope effect is 0.67 when the yield of both benzyl alcohol and phenol products is considered.[130,131] The *inverse* isotope effect on the total product yield has been rationalized by invoking slower product release of the benzyl alcohol than phenol from the active site, resulting in an increase in enzyme turnover as deuterium shifts the hydroxylation from the methyl group to the aromatic ring. Similar phenomena, rather than the postulated change in mechanism from hydrogen to electron abstraction, could possibly explain the substituent-dependent isotope effects observed in the hydroxylation of substituted toluenes by the fungus *Mortierella isabellina*.[132]

Kinetic analysis of the hydroxylation of (R)- and (S)-2-phenylpropane-$1,1,1$-d_3 by uninduced rat liver microsomes shows that the intramolecular isotope effect for the (R)-isomer is 15.6 but that for the (S)-isomer is only 4.2.[133] The authors provide evidence that the observed intramolecular isotope effect depends not only on the intrinsic isotope effect, which must be the same for both enantiomers, but also on the rates of reorientation of the substrate within the active site and the rate of its dissociation from the enzyme.[133] The large intrinsic isotope effects associated with carbon hydroxylation reactions are consistent with a direct correlation, in the absence of overriding steric or substrate positioning factors,[133] between the strength of a C–H bond and its susceptibility to hydroxylation.[134,135] Calculations based on the susceptibility of a hydrogen to abstraction by the p-nitrosophenoxy radical have been proposed as a model that can be used to predict the relative reactivities of C–H bonds to hydroxylation.[136] These results indicate that the hydrogen is transferred from the carbon to the ferryl oxygen via a roughly symmetrical transition state, and that the isotope effects for the reaction are large but are usually masked in kinetic terms by rate-determining steps other than actual oxygen insertion.

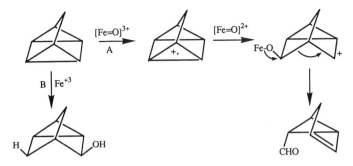

FIGURE 10. A: Electron transfer in the P450-catalyzed oxidation of quadricyclane to give a bicyclic aldehyde. B: In contrast, the simple autooxidation of quadricyclane catalyzed by ferric ions yields an unrearranged bicyclic alcohol in which water has simply added across one of the carbon–carbon bonds.

Nonconcerted hydroxylation mechanisms generally invoke removal of the hydrogen by the activated oxygen followed by rapid collapse of the carbon intermediate with the activated hydroxyl (Fig. 7). However, the first step of a hydroxylation reaction could, in special cases, be electron rather than hydrogen atom abstraction. Oxidation of a hydrocarbon by initial removal of an electron has been observed in the case of quadricyclane,[137] a strained hydrocarbon with a very low oxidation potential ($E_{1/2} = 0.92$ V).[138] The result of this reaction is insertion of oxygen into a *carbon–carbon* rather than carbon–hydrogen bond. Quadricyclane slowly autooxidizes in the presence of unreduced liver microsomes or trace metals to nortricyclanol. The NADPH-dependent microsomal oxidation of quadricyclane, however, yields the rearranged aldehyde that is also obtained as the principal product in the enzymatic oxidation of norbornadiene (Fig. 10). Control experiments have shown that norbornadiene is not an intermediate in the quadricyclane oxidation but rather that a common intermediate is probably generated from both substrates. Oxidation of quadricyclane to the radical cation, followed by capture of the carbon radical by the enzymatically activated oxygen, rationalizes these results because the resulting cationic intermediate is known to rearrange to the observed aldehyde. The radical cation is also generated in the autooxidative reaction but it is trapped by water rather than by the activated oxygen and thus gives a radical rather than cationic intermediate that does not rearrange.

In sum, the scrambling of stereochemistry in the hydroxylation of norbornane and camphor, the allylic rearrangements observed in the hydroxylation of unsaturated hydrocarbons, the results of radical clock experiments, the correlation of reactivity with bond strength, the large intrinsic isotope effects, and the dissociation of electron from oxygen transfer in the oxidation of quadricyclane provide firm evidence for nonconcerted hydroxylation mechanisms. The retention of stereochemistry observed in most hydroxylations stems from the structural constraints of the active site rather than from the intrinsic nature of the hydroxylation mechanism. The intermediates in hydroxylation reactions are tentatively identified as free radicals by the chemistry observed in model systems (see Chapter 1), the radical clock data, and the general absence of the skeletal rearrangements expected from cationic species.[101] The mechanism of P450-catalyzed hydroxylation reactions is best represented by the nonconcerted sequence in Fig. 7, in which the carbon radical obtained by hydrogen atom transfer to the activated oxygen complex is captured, in a second step, by the equivalent of an iron-bound hydroxyl radical at a rate in excess of 10^{10} sec^{-1}. The evidence for a free radical intermediate does not, however, preclude the formation of cationic intermediates in at least some reaction pathways (see Section 3.6).

3.3. Olefin Epoxidation

The transfer of oxygen to carbon–carbon π-bonds, as specifically shown for *cis*-stilbene,[139] oleic acid,[140] and *trans*-[1-^{2}H]-1-octene,[141] proceeds with retention of the olefin stereochemistry. However, as noted in Section 3.2, retention of stereochemistry does not unambiguously differentiate between concerted and nonconcerted mechanisms. Asynchronous transfer of oxygen to the π-bond of styrene during its enzymatic epoxidation is suggested by the observation of an inverse isotope effect ($k_H/k_D = 0.93$) for deuterium on the internal but not terminal carbon of the double bond.[142] The converse, no isotope effect for an internal deuterium but a large inverse isotope effect for a terminal deuterium, has been reported for the epoxidation carried out with *m*-chloroperbenzoic acid.[143] Similar isotope effects are expected for deuterium substitution at both ends of the π-bond if the two carbon–oxygen bonds are formed simultaneously and at comparable rates. The observation of differential isotope effects at the two carbons, however, does not require a nonconcerted mechanism if the two carbon–oxygen bonds are formed simultaneously but at different rates.

A nonconcerted π-bond oxidation mechanism is suggested by two phenomena: (1) alkylation of the prosthetic heme group associated with catalytic turnover of certain terminal olefins and (2) direct formation of carbonyl products during the oxidation of certain olefins. Although the structure of the *N*-alkyl heme adducts obtained during the oxidation of terminal olefins is that expected from addition of one of the porphyrin pyrrole nitrogens to the epoxide metabolite (Fig. 11), control experiments clearly establish that the adducts do not arise by such a mechanism (see Chapter 9). *N*-alkylation of the prosthetic heme group clearly requires oxidation of the double bond of the terminal olefin but diverges from the formation of epoxide metabolites prior to closure of the epoxide ring.

The oxidation of trichloroethylene to trichloroethylene epoxide results in the simultaneous formation of trichloroacetaldehyde. Although the aldehyde could conceivably be

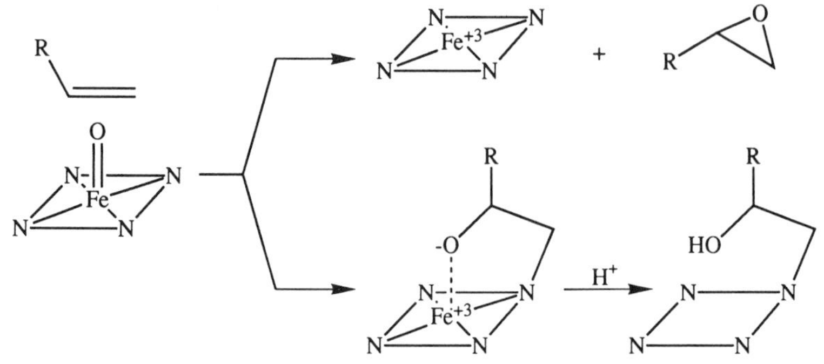

FIGURE 11. Heme alkylation during the cytochrome P450-catalyzed oxidation of terminal olefins to epoxides. The *N*-alkylated porphyrins isolated by acidic demetallation of the initial heme adducts have been fully characterized by spectroscopic methods.

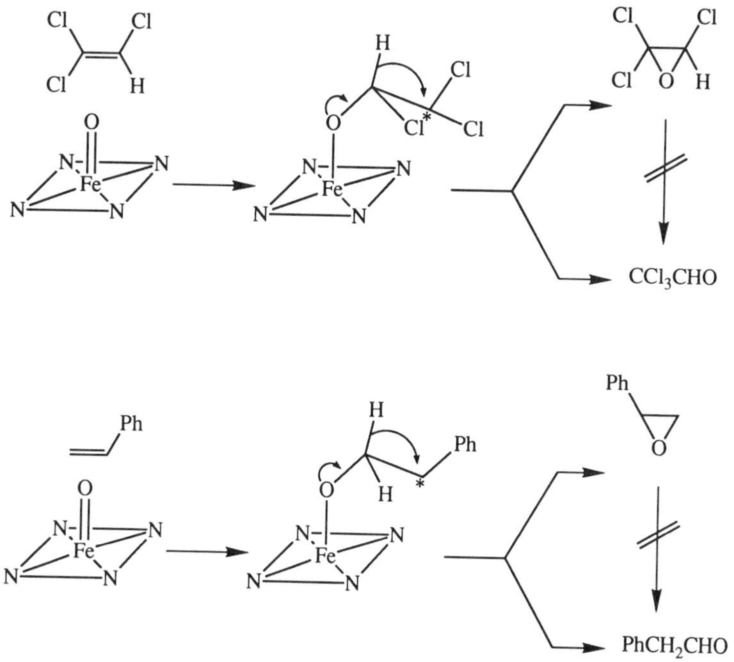

FIGURE 12. Formation of carbonyl products during the oxidation of olefins by mechanisms that bypass the epoxide metabolites. The asterisk denotes a radical or a cation. Oxidation of a radical species in the epoxidation pathway to a cation could divert the catalytic process toward the formation of rearranged products.

formed by acid-catalyzed rearrangement of the epoxide, the demonstration that synthetic trichloroethylene oxide does not rearrange to trichloroacetaldehyde under physiological conditions indicates that the rearrangement occurs prior to epoxide formation (Fig. 12).[144,145] Similarly, metabolic formation of chloro- and dichloroacetic acids from 1,1-dichloroethylene but not from the synthetically prepared epoxide suggests that chlorine migration also occurs here prior to epoxide formation.[146] Direct formation of a carbonyl function during the oxidation of a double bond has also been observed to a small extent with some unchlorinated olefins. Thus, *trans*-1-phenylbutene yields 1-phenyl-1-butanone and 1-phenyl-2-butanone as minor products,[146] and styrene provides a trace of 2-phenyl-acetaldehyde.[147] These rearranged products reportedly are not obtained from the corresponding epoxides under physiological conditions. These rearrangements argue for the formation of cationic intermediates that do not derive from the epoxides. The cations may be intermediates in the epoxidation pathway, in which case the shifts compete with epoxide formation, but are more likely to arise by leakage from the epoxidation pathway into a secondary and independent reaction pathway. Partitioning between the two reactions could occur very early in the reaction or could involve a later step, for example, oxidation of a radical in the epoxidation pathway to a cation (Fig. 12). Efforts to detect radical or radical

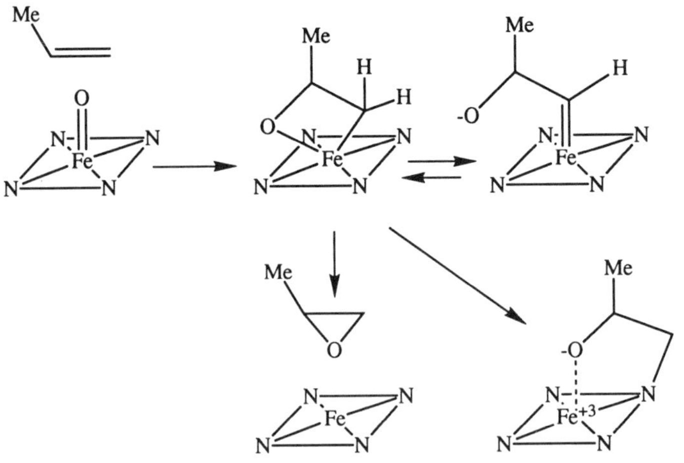

FIGURE 13. Mechanism proposed for the exchange of deuterium for hydrogen during the cytochrome P450-catalyzed conversion of *trans*-1-deuteriopropene to propene oxide.

cation intermediates in olefin oxidation by looking for radical rearrangement products have been unsuccessful. Thus, *trans*-1-phenyl-2-vinylcyclopropane is oxidized to the epoxide without the formation of rearranged products, either because the cyclopropane ring opening is too slow to compete with closure of the epoxide ring or because a radical intermediate is not formed.[148] *Trans*-1-phenyl-2-vinylcyclopropane is not an ideal probe for these experiments, however, because rearrangement products must be measured above a background provided by rearrangement of the epoxide under physiological conditions.

The remarkable observation that *trans*-1-deuteriopropene is epoxidized by cytochrome P450 with significant loss of the deuterium label, and that the corresponding oxidation of unlabeled propylene in deuterated water yields *trans*-1-deuteriopropene, clearly requires the intervention of an intermediate in the epoxidation of this substrate.[149] Groves *et al.* rationalize the exchange by invoking the formation of a metallaoxetane intermediate that equilibrates with a carbene species via a mechanism that results in exchange of one of the hydrogens with a proton from the medium (Fig. 13; see also Chapter 1). Computational studies of the insertion of metal atoms into the C–O bond of ethylene oxide suggest that metallaoxetanes are thermodynamically reasonable intermediates in metalloporphyrin-catalyzed epoxidation reactions, although the calculation cannot be taken as evidence that such intermediates are actually formed.[150] One caveat concerning this observation is that hydrogen exchange is seen when the enzymatic reaction is supported by cytochrome P450 reductase and NADPH but not when it is supported by iodosobenzene. Furthermore, it is not seen in the epoxidation catalyzed by metalloporphyrin models, nor is it observed during the enzyme-catalyzed epoxidation of hexene[149] or *trans*-[1-²H]-1-octene.[141] This intriguing result excludes epoxidation mechanisms that do not rationalize the hydrogen exchange, but the circumscribed conditions under which the exchange is observed leave open the possibility that it is part of an anomalous rather than universal reaction pathway.

The P450-catalyzed oxidation of terminal acetylenes is accompanied by *quantitative* rearrangement of the acetylenic hydrogen to the vicinal carbon.[151–153] The hydrogen shift in the case of acetylenes is not prima facie evidence for a nonconcerted oxidation mechanism, however, because the expected unsaturated epoxide (oxirene) metabolite would itself rearrange rapidly to the observed products. However, the hydrogen migration must occur in the rate-determining step of the catalytic process because the oxidation of aryl acetylenes to arylacetic acid metabolites is subject to a large isotope effect when the acetylenic hydrogen is replaced by deuterium.[153,154] The oxidation of aryl acetylenes by *m*-chloroperbenzoic acid yields the same products[151,154] and is subject to the same isotope effect as the enzymatic reaction.[151] This suggests that the hydrogen migration and the oxygen transfer occur simultaneously in both the chemical and enzymatic reactions. The enzymatic oxidation of terminal acetylenes, as of terminal olefins, results in alkylation of the prosthetic heme group of the enzyme (see Chapter 9). It is clear from the structures of the heme adducts that the heme is alkylated by a species produced by delivery of the activated oxygen to the internal carbon of the π-bond, whereas formation of the metabolites follows from delivery of the oxygen to the terminal carbon. In agreement with this regiochemical difference, the destruction of P450 by aryl acetylenes is not subject to a detectable isotope effect when deuterium is substituted for the acetylenic hydrogen. These results require the oxidative sequence leading to metabolites to diverge from that resulting in heme alkylation prior to, or concurrent with, transfer of the oxygen to the π-bond.

The evidence for nonconcerted oxidation of π-bonds is more indirect and tenuous than that marshaled in support of a nonconcerted hydroxylation mechanism. P450 enzymes under appropriate conditions clearly catalyze nonconcerted double bond oxidations that lead to heme alkylation or rearranged products. The rearranged products require a cationic intermediate, but these products are relatively rare and may arise by an aberrant pathway distinct from that which results in epoxide formation. It is possible that heme alkylation is also the result of an aberrant process, although heme alkylation by terminal olefins and acetylenes is quite common and occurs as frequently as once in every two substrate turnovers (see Chapter 9). The precise nature of the olefin epoxidation mechanism is therefore still ambiguous. Model studies with metalloporphyrins suggest that the olefin epoxidation manifold is entered via the rate-limiting formation of a charge transfer complex that subsequently decomposes via multiple pathways into the observed products (see Chapter 1).[155] The epoxidation could actually occur by a concerted mechanism subsequent to charge transfer formation, or could arise by one of the pathways indicated in Fig. 14. Although charge transfer complex formation is likely, oxidation of the olefin to a full-fledged π-radical cation is unlikely except in the special situation of highly oxidizable substrates. Two radicals can potentially arise from the charge transfer complex by addition of the oxygen to the two ends of the π-bond. These radicals, if formed, may lead to the epoxide, to heme alkylation, or, after oxidation to the cation, to the observed rearrangement products. Although the involvement of a metallaoxetane intermediate is suggested by the deuterium exchange experiments with propene (Fig. 13), it is difficult at this time to determine whether such intermediates play a general role in the olefin oxidation manifold.

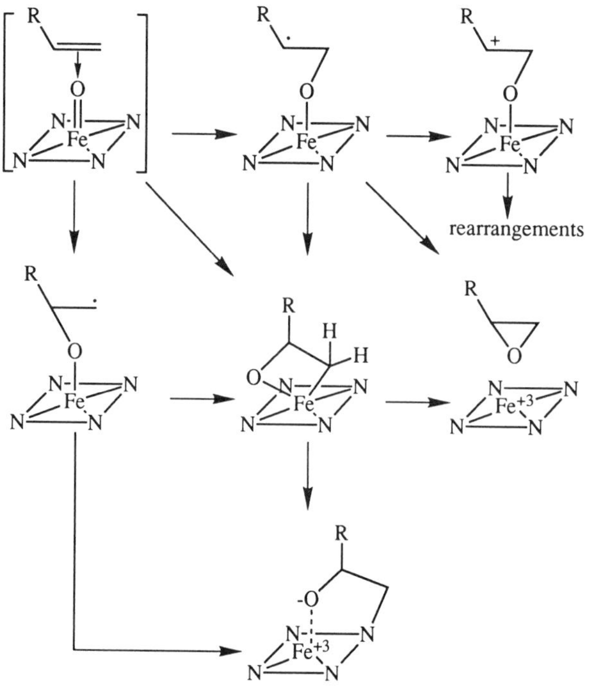

FIGURE 14. Mechanistic alternatives for the oxidation of a carbon–carbon π-bond. The first intermediate in the process is shown as a charge transfer complex. The rearrangement reaction certainly comes from the indicated cationic intermediate, but the relationship of this intermediate and species such as metallaoxetanes to olefin epoxidation and heme alkylation remains unclear.

3.4. Aromatic Oxidation

The introduction of a hydroxyl group into an aromatic ring by P450 commonly involves epoxidation of the aromatic ring followed by epoxide ring opening, migration of a hydride to the carbocation thus generated, and tautomerization of the ketone product thus formed (the so-called "NIH shift") (Fig. 15).[156] A fraction of the hydrogen that shifts is retained in the hydroxylated product because it becomes equivalent to the hydrogen already on the receiving carbon and only one hydrogen is lost in the tautomerization step. Deuterium substitution does not markedly alter the hydroxylation rate because the deuterium-sensitive tautomerization occurs after the rate-limiting enzymatic epoxidation step. Nevertheless, an inverse secondary isotope effect (0.83–0.94) has been measured for hydroxylation of the aromatic ring in *ortho-* and *para-*xylene.[131] This is consistent with rate-limiting addition of the oxygen to a π-bond of the aromatic ring, as the transition state for the reaction then involves partial rehybridization of the carbon from the sp^2 to the sp^3 state. The inverse isotope effect provides no support for mechanisms in which the rate-determining step is oxidation of the aromatic ring to a π-cation radical. In this context, it should be noted that the oxidation of cyclopropylbenzene by cytochrome P450 gives

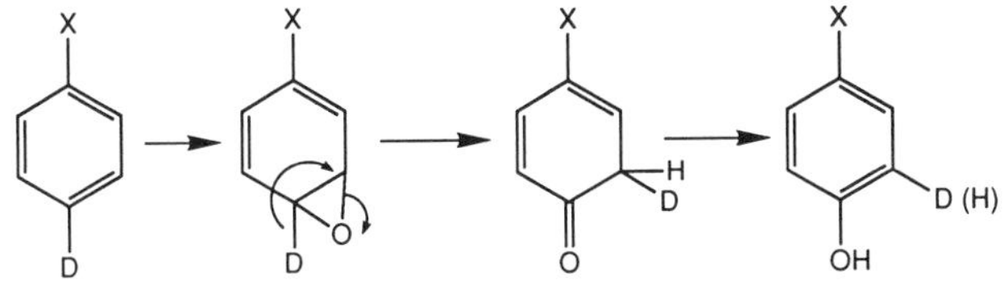

FIGURE 15. Cytochrome P450-catalyzed hydroxylation of an aromatic ring via the NIH shift mechanism.

1-phenylcyclopropanol and phenol metabolites without the detectable formation of ring-opened products suggestive of electron abstraction from the aromatic ring.[157]

Quantitative loss of the hydrogen on the hydroxylated carbon and a small deuterium kinetic isotope effect are sometimes observed, particularly in hydroxylations *meta* to a halide substituent.[158,159] These hydroxylations could result from direct oxygen insertion into the C–H bond, as in a true "hydroxylation" reaction, but could also result from oxidation of the aromatic ring by a nonconcerted mechanism that does not pass through a discrete epoxide metabolite. Detailed studies of the isotope effects associated with the hydroxylation of deuterated benzenes with a variety of substituents have been carried out in an effort to deconvolute the process and to determine if mechanisms other than that involving an epoxide intermediate operate in aromatic hydroxylation.[160,161] A key finding of these studies is that a small, normal isotope effect is observed for *meta*-hydroxylation when a deuterium is located *meta*- to the halogen in chlorobenzene but a small, inverse isotope effect is observed for *ortho*- or *para*-hydroxylation when the deuterium is *ortho* or *para*, respectively, to the halide (Fig. 16).[161] Concerted epoxide formation would be expected to give a small, normal isotope effect at both of the carbons involved in epoxide formation, so these isotope effects have been used to argue against such a mechanism. It is difficult, however, to completely rule out the possibility that the two carbon–oxygen bonds of the epoxide are formed in a concerted, asynchronous manner that gives rise to different isotope effects at the two carbon atoms. In the limiting case, addition of the oxygen to one carbon of the π-system would be followed by ring closure to the epoxide. The regioselectivity of hydroxylation of 1,2-difluorobenzene, 1,3-difluorobenzene, 1,2,3-trifluorobenzene, and 1,2,4-trifluorobenzene has been correlated with the frontier orbital characteristics of the substrate with respect to attack by an electrophilic atom.[162] These results exclude initial electron transfer to give a radical cation intermediate and have been advanced as evidence against epoxide intermediates. The latter argument may be invalid,

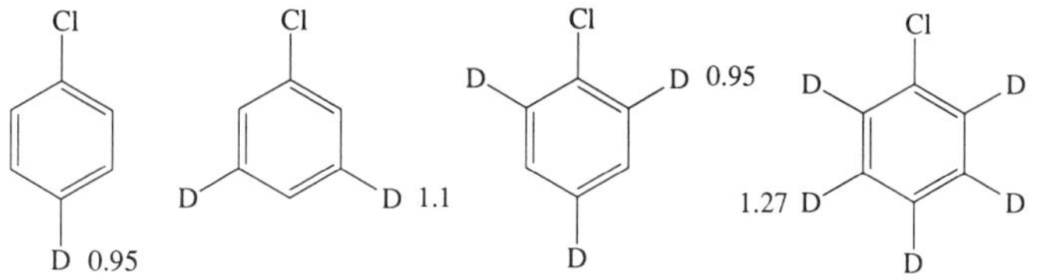

FIGURE 16. Hydrogen isotope effects for the cytochrome P450-catalyzed hydroxylation of deuterated chlorobenzenes.

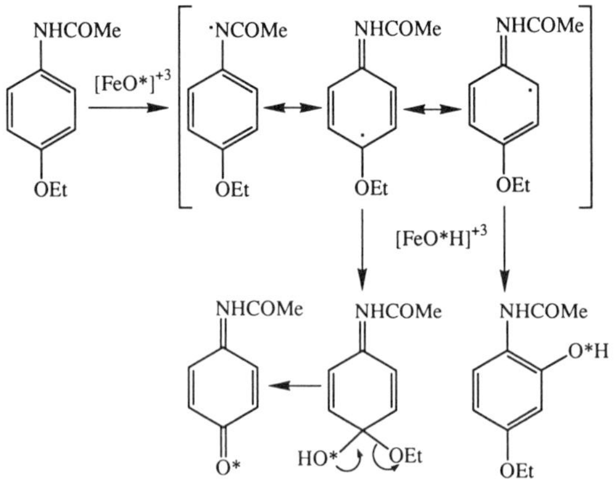

FIGURE 17. Mechanism proposed for oxidation of phenacetin to N-acetyl-p-benzoquinoneimine with partial incorporation of an atom of molecular oxygen into the carbonyl group. The asterisk indicates ^{18}O-labeled oxygen.

however, as the direction of ring opening of the epoxides may also correlate with the orbital characteristics because the tetrahedral species formed by epoxide ring opening is very similar to that obtained by addition of the oxygen to one carbon of the aromatic ring.

The aromatic rings of anilines and phenols are the most likely to be oxidized by pathways that circumvent epoxide intermediates due to the strong electron-donating effects of the substituents. The first evidence for such a mechanism was provided by the demonstration that oxidation of p-ethoxyacetanilide (phenacetin) to N-acetyl-p-benzoquinoneimine results in partial incorporation of an atom of molecular oxygen into the quinone.[163,164] This suggests that the reaction proceeds, in part, by cleavage of the aryl–oxygen rather than alkyl–oxygen bond. This can be explained if the initial step in the P450-catalyzed reaction is hydrogen abstraction from the nitrogen to give a radical that recombines with the iron-bound oxygen at the carbon bearing the ethoxy group (Fig. 17).[165] Concurrent formation of the 2-hydroxy derivative of phenacetin as a minor metabolite, as observed, is consistent with such a mechanism because the radical density would be expected to reside on both the *ortho*- and *para*-carbons of the aniline ring.[166] The formation of quinoneimines accompanied by elimination of fluoride anion in the oxidation of 4-fluoroanilines or pentafluorophenol,[167,168] and the oxidative defluorination of substrates such as 2-fluoro-17α-ethynylestradiol,[169] is readily rationalized by similar mechanisms.

In a similar vein, the oxidation of p-aryloxyphenols by either rat liver microsomes or an iron porphyrin system cleaves the C–O bond between the aryloxy substituent and the phenol ring. The oxidation of p-(p-nitrophenoxy)phenol thus yields p-nitrophenol and p-benzoquinone as metabolites (Fig. 18).[170] ^{18}O$_2$ studies demonstrate incorporation of an atom of molecular oxygen into one of the quinone carbonyl groups. Structure–activity studies indicate that diaryl ether cleavage only occurs when one of the aryl rings is a phenol. Phenol cleavage, like the reactions of phenacetin and fluorinated phenols, thus appears to result from one-electron oxidation of the phenol to the phenoxy radical followed by *ipso*-recombination of the phenoxy radical to give a *para*-hydroxylated species. Elimina-

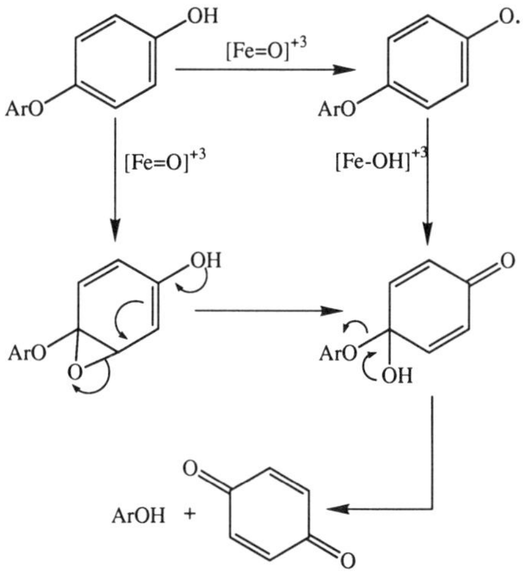

FIGURE 18. Mechanisms that rationalize the oxidative cleavage of *p*-(*p*-nitrophenoxy)phenol to quinone and *p*-nitrophenol (Ar = *p*-nitrophenyl).

tion of the aryloxy group then gives the observed quinone and phenol metabolites. Although epoxidation of substituted aromatic π-bonds is not common, it must be kept in mind that the same products are expected from epoxidation of the substituted double bond because the epoxide would open exclusively in the appropriate direction as a result of electron donation from the hydroxyl (or amino) function (Fig. 18).

The oxidation of pentafluorochlorobenzene by cytochrome P450 to tetrafluorochlorophenol has been suggested to proceed by a related mechanism in which the ferryl oxygen adds to the aromatic ring at the fluorine-substituted carbon *para* to the chloride. Electron donation by the chloride is then postulated to produce a positively charged species with a double bond to the chlorine that is finally reduced to the phenol (or hydrolyzed to tetrafluoroquinone).[171] The oxidation of a hexahalogenated benzene substrate is difficult to rationalize by an epoxidation mechanism.

A cytochrome P450 enzyme isolated from *Berberis stolonifera* cell cultures catalyzes the cross-linking of phenol rings in the biosynthesis of bibenzylisoquinoline alkaloids (Fig. 19).[172] It has been proposed that this reaction, like the cross-linking of phenols in solution, involves the coupling of two phenoxy radicals. Thus, oxidation of one of two phenol moieties to a phenoxy radical would be followed not by *ipso*-hydroxylation but by electron transfer from the second phenol to give a second phenoxy radical. The two electrons provided by the phenols reduce the ferryl enzyme to the ferric state, while the concomitantly formed phenoxy radicals couple to yield the observed product. The related transformation of reticuline to salutaridine is similarly catalyzed by microsomal hog and plant cytochrome P450 enzymes.[173,174] Mechanistic studies are required to confirm the proposed mechanisms of these reactions, but if they are valid they provide independent evidence for the stepwise one-electron oxidation of phenols by cytochrome P450 enzymes. It is important to note that the phenoxy radicals invoked in these reactions and in the elimination reactions above are trapped within the active site either by *ipso*-hydroxylation or, in the plant enzyme, by *in situ* coupling with a second phenoxy radical. There is no evidence that aryloxy radicals escape from P450 active sites into the surrounding medium.

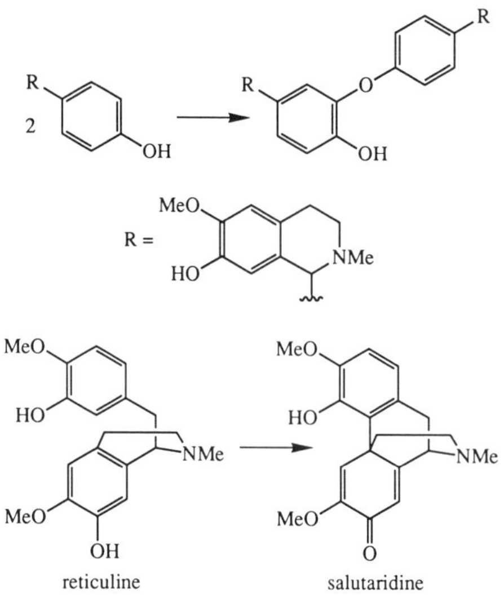

FIGURE 19. Reaction catalyzed by an isolated cytochrome P450 enzyme from *Berberis stolonifera* and microsomal cytochrome P450-catalyzed conversion of reticuline to salutaridine.

Cavalieri and co-workers have argued for some years that the covalent binding of polycyclic aromatic hydrocarbons to DNA involves the radical cations produced by the catalytic action of cytochrome P450 or peroxidases.[175] Their evidence suggests that polycyclic aromatic hydrocarbons with ionization potentials below 7.35 eV are susceptible to oxidation to radical cations by horseradish peroxidase and prostaglandin synthase.[176,177] The presence of cytochrome P450 enzymes in the nuclear membrane,[178] and the formation of a benzo[a]pyrene–DNA adduct consistent with activation of the hydrocarbon to a radical cation by rat liver microsomes and rat skin,[179,180] fuels the proposal that cytochrome P450 likewise activates polycyclic aromatic hydrocarbons to the corresponding radical cations.[175] The oxidation of 6-fluorobenzo[a]pyrene by liver microsomes to 6-hydroxy-benzo[a]pyrene has been interpreted as evidence for cytochrome P450-catalyzed radical cation formation, although the reaction could arise, as postulated for pentafluorochlorobenzene,[171] by direct addition of the activated oxygen to the aromatic system.[181] Direct addition would be expected to occur at the 6-position because that is the position that is most activated toward electrophilic attack. There is as yet no unambiguous evidence for the direct formation of a diffusible radical cation in the cytochrome P450-catalyzed oxidation of polycyclic aromatic hydrocarbons. The question is therefore still open as to whether the normal (as distinguished from peroxide-supported) cytochrome P450-catalyzed oxidation of polycyclic aromatic hydrocarbons produces diffusible radical cations.

Incubation of 9-ethyl-10-methylanthracene for 1 hr with rat liver microsomes produces two products, the 9-ethyl-10-hydroxymethyl derivative (3%) and a deethylated product identified as 10-methylanthrone (83%).[182] The same two products are obtained when the substrate is incubated for 20 hr with rat liver cytosol, but in this case the hydroxymethyl product predominates (86%) over 10-methylanthrone (9%). Oxidation of 9,10-diethylanthracene by microsomes exclusively produces 10-ethylanthrone. To explain the formation of the 10-methyl- and 10-ethylanthrones, it has been postulated that the anthracene skeleton is oxidized to a radical cation to which a molecule of water adds to

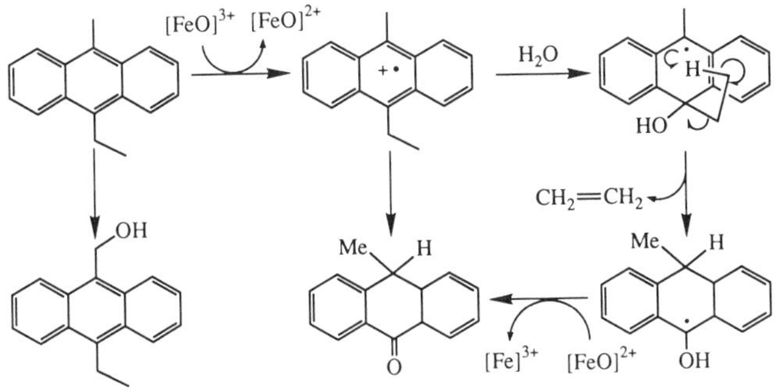

FIGURE 20. Oxidation of 9-ethyl-10-methylanthracene by rat liver microsomes.

give a ring-hydroxylated radical species. It is then postulated that hydrogen transfer from the terminal carbon of the ethyl group to the aromatic ring concomitant with elimination of ethylene gas produces a radical that is readily oxidized to anthrone (Fig. 20). Chemical oxidation studies using *tris*(phenanthroline)iron(III) *tris*(hexafluorophosphate) in aqueous acetonitrile give the same products. The chemical mechanism is supported by evidence that ethylene gas is formed and that the reaction is subject to a deuterium isotope effect of 5.7 when deuterium is placed on the ethyl group. Although the experimental procedures used for the initial biological work do not rule out the possibility that the reaction is catalyzed by adventitious iron,[182] recent work reported in a communication indicates that the initial findings relative to the dealkylation process have been confirmed with better defined enzyme preparations.[183] These results, if confirmed, provide considerable support for the existence of a radical cation pathway in polycyclic aromatic oxidation.

In sum, the weight of evidence for oxidation of some aromatic rings by mechanisms other than direct epoxide formation is increasing but the evidence is not yet unambiguous. The conventional pathway for aromatic oxidation is epoxidation followed, in some cases, by the NIH shift. It appears, however, that addition of the ferryl oxygen to a single carbon in some cases produces a tetrahedral intermediate that decays by one of several pathways, including closure to an epoxide, extrusion of a substituent from the tetrahedral carbon, or electron transfer followed by addition of a nucleophile. This type of reaction is particularly prevalent with phenols and anilines. The evidence for the formation of radical cations by direct electron abstraction from polycyclic aromatic hydrocarbons is still ambiguous. Although it appears that radical cations are involved in some DNA alkylation reactions, the role of cytochrome P450 as the actual oxidation catalyst rather than as a source of H_2O_2 or other activated oxygen species must yet be firmly established.

3.5. Heteroatom Oxidation and Dealkylation

Cytochrome P450 oxidizes nitrogen and sulfur to the corresponding oxides but does not similarly oxidize the more electronegative (and consequently less reactive) oxygen atom. P450 also catalyzes the addition of a hydroxyl group to the carbon adjacent to heteroatoms in a reaction that ultimately eliminates the heteroatom, but the electronegativity of oxygen again sets it apart. The O-dealkylation of ethers and other oxygen

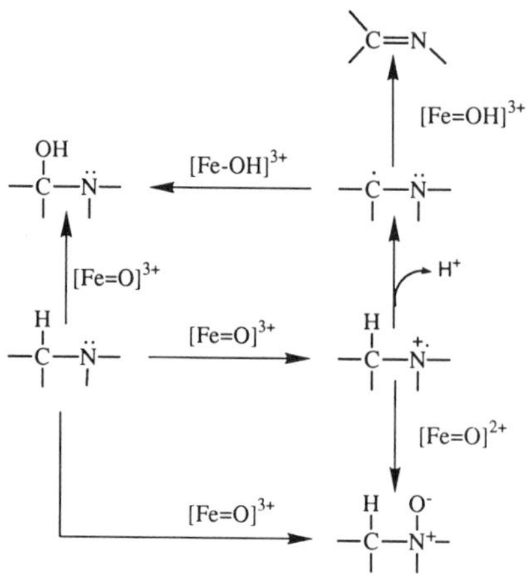

$$\text{C}=\text{N}$$

$$[\text{Fe}=\text{OH}]^{3+}$$

$$\begin{array}{c} \text{OH} \\ | \\ -\text{C}-\text{N}- \\ | \quad | \end{array} \quad \xleftarrow{[\text{Fe-OH}]^{3+}} \quad \begin{array}{c} \\ -\text{C}-\text{N}- \\ | \quad | \end{array}$$

$$[\text{Fe}=\text{O}]^{3+} \qquad \qquad \text{H}^+$$

$$\begin{array}{c} \text{H} \\ | \\ -\text{C}-\text{N}- \\ | \quad | \end{array} \quad \xrightarrow{[\text{Fe}=\text{O}]^{3+}} \quad \begin{array}{c} \text{H} \\ | \quad +. \\ -\text{C}-\text{N}- \\ | \quad | \end{array}$$

$$[\text{Fe}=\text{O}]^{2+}$$

$$[\text{Fe}=\text{O}]^{3+} \qquad \begin{array}{cc} \text{H} & \text{O}^- \\ | & | \\ -\text{C}-\text{N}^+- \\ | & | \end{array}$$

FIGURE 21. Mechanisms for the P450-catalyzed oxidation, N-dealkylation, and desaturation of nitrogen compounds. Similar mechanisms can be written for sulfur, although the role of radical cations in sulfur oxidations is less well established and the desaturation pathway has not been reported for sulfur compounds.

derivatives appears to involve direct hydroxylation of the carbon and is therefore most usefully viewed in the context of hydrocarbon hydroxylations rather than heteroatom oxidations. On the other hand, the dealkylation of alkylamines and probably of alkylsulfides, like the oxidation of nitrogen or sulfur atoms, is apparently initiated by electron abstraction from the heteroatom (Fig. 21). The nitrogen or sulfur radical cation thus formed generally either collapses with the activated oxygen to give the corresponding oxide or loses a proton from the adjacent carbon to give a delocalized radical that is trapped by addition of the activated oxygen to the carbon atom. The resulting α-hydroxylated product, identical to that expected from a conventional carbon hydroxylation reaction, subsequently decomposes to the dealkylation products.

As indicated above, the oxidation of alkylamines can result in N-oxygenation or N-dealkylation. The cytochrome P450-catalyzed formation of alkylamine N-oxides is not a common reaction but the formation of N-oxides has been observed and sensitive assays indicate that they are formed as minor products in the oxidation of compounds as simple as N,N-dialkylanilines.[94,184,185] Thus, the ratio of N-dealkylation to N-oxygenation catalyzed by CYP2B1 has been found to range from 940 for N,N-dimethylaniline or 1020 for N,N-diethylaniline to 6 for N,N-dimethyl-2-aminofluorene.[94] N-Oxides become major products when the nitrogen is locked in a ring system that does not permit full conjugation of the nitrogen with the p-orbital that would be generated by deprotonation of the adjacent carbon atom.[185]

Partitioning between N-hydroxylation and N-dealkylation has been demonstrated during the microsomal oxidation of cyclohexylamine.[186] Thus, introduction of a deuterium adjacent to the amine group results in metabolic switching from deamination (N-dealkylation) to N-oxygenation with a V_H/V_D isotope effect of 1.75. This result does not discriminate, however, between a mechanism with a common radical cation intermediate and the alternative of two different mechanisms. Hlavica and Künzel-Mulas have observed that N-oxide formation from N,N-dimethylaniline catalyzed by purified CYP2B4 is strongly inhibited by superoxide dismutase and can be supported by H_2O_2.[184] They suggest that N-oxide formation may be catalyzed by the ferric peroxide (Fe^{3+}-OOH) P450 complex

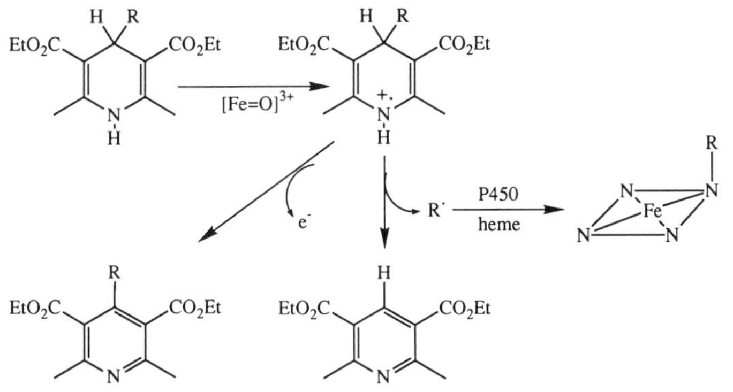

FIGURE 22. Mechanisms for the formation of aromatic metabolites and alkylation of the heme group during the oxidative metabolism of 4-alkyl-1,4-dihydropyridines.

rather than by the ferryl species. Although there is now considerable precedent for nucleophilic addition of the ferric peroxide to aldehydes (see Sections 3.7 and 4), there is as yet no clear evidence for reaction of the ferric peroxide as an electrophilic species. Nevertheless, the possibility cannot be ruled out at this time that N-oxygenation is partially or fully mediated by the ferric peroxide complex rather than by the ferryl species that catalyzes N-dealkylation.

The oxidation of 3,5-(bis)carbethoxy-2,6-dimethyl-4-ethyl-1,4-dihydropyridine (Fig. 22) by hepatic microsomes results in transfer of the 4-ethyl moiety from the substrate to a nitrogen of the prosthetic heme group.[187] Alkylation of the heme group in the reaction inactivates the cytochrome P450 enzyme(s) involved in the reaction (see Chapter 9). The isolation of 4-dealkylated metabolites and spin trapping of the ethyl radical indicate that the 4-ethyl group is eliminated as a free radical. Both heme alkylation and metabolite formation require one-electron oxidation of the dihydropyridine to a radical or radical cation that aromatizes by extruding the ethyl radical. Although heme alkylation requires formation of the alkylating radical within the enzyme active site, the ethyl radical that is spin trapped is produced, at least in part, by oxidative mechanisms other than direct cytochrome P450-catalyzed oxidation because the spin-trapping reaction is inhibited by catalase and by prewashing the microsomes with metal chelating agents.[187,188]

The isotope effects for the dealkylation of alkyl amines provide useful information on the mechanisms by which these heteroatoms are oxidized. The O-dealkylation of 7-ethoxycoumarin, as expected for a carbon hydroxylation, is characterized by a large intrinsic isotope effect ($k_H/k_D = 13$–14).[126] In contrast, the intrinsic isotope effects for N-dealkylation of alkylamines, estimated from the ratio of deuterated products obtained from aryl amines bearing both deuterated and undeuterated N-methyl groups, fall in the range $k_H/k_D = 1.3$–3.0.[189–191] These small isotope effects are comparable to those measured for the dealkylation of amines by photochemical[192] or electrochemical[193] methods, reactions in which one-electron oxidation of the nitrogen is known to be the first step (however, see below). The photochemical dealkylation of a mixture of dimethylaniline and di(trideuteriomethyl)aniline thus proceeds without a measurable isotope effect, as does the metabolic reaction, whereas an isotope effect between 2 and 3 is observed for both the photochemical and enzymatic dealkylation of N-methyl-N-trideuteriomethylaniline.[192] The product ratios for the electrochemical and enzymatic N-dealkylation of amines bearing two different N-alkyl groups are also very similar.[193]

Substituent effects on the rate of N-dealkylation of dimethylanilines are consistent with a nitrogen radical cation mechanism but do not exclude alternative mechanisms, including direct hydrogen abstraction, because positive charge is also likely to accrue in the transition states in alternative mechanisms.[194,195] Thus, the V_{max} and K_m values for N-demethylation of 12 *para-substituted* N,N-dimethylanilines by rat liver microsomes have been determined and the log V_{max} values shown to fit a linear free energy equation that includes terms for lipophilicity, electronic factors, and steric bulk: log $V_{max} = 0.41\pi - 1.02\sigma - 0.023MR + 1.72$ ($r = 0.953$).[194] The electronic term (-1.02σ) shows that the oxidation is strongly facilitated by electron-donating groups, as expected if positive charge accumulates on the substrate during the reaction. In contrast, N-hydroxylation of substituted anilines does not appear to depend on the electronic nature of the substituent, suggesting that the transition state for aniline N-hydroxylation differs from that for N-demethylation of N,N-dimethylaniline.[195] An analysis of the oxidation of substituted N,N-dimethylanilines by P450 2B1 supported by both NADPH/P450 reductase and iodosobenzene shows that the rates correlate with the substrate redox potentials. Theoretical treatment of the results has led to the conclusion that the oxidation potential for P450 2B1 is in the range of 1.7–2.0 V (versus the saturated calomel electrode).[196,197]

In contrast to the N-dealkylation of amines, the N-demethylation of amides with methyl and trideuteriomethyl groups on the nitrogen [RCON(CH$_3$)CD$_3$] is subject to a relatively large intramolecular isotope effect ($k_H/k_D = 4$–7).[198,199] The corresponding intermolecular isotope effects, however, are quite small ($k_H/k_D < 2$), indicating a dominant effect of factors other than actual amide oxidation in determining the rate. These results suggest that N-dealkylation of amides involves direct hydrogen atom abstraction rather than oxidation of the nitrogen to the radical cation, a change in mechanism warranted by the strong electron-withdrawing effect of the carbonyl group on the nitrogen. In agreement with this conclusion, the kinetic isotope effect for the demethylation of N,N-dimethylbenzamide by an iron porphyrin model system is 5.6,[200] whereas the isotope effect for the corresponding electrochemical demethylation, a radical cation reaction, is only 1.4–1.7.[201]

It has been argued that isotope effects may sometimes be misleading indicators of the oxidative mechanism.[202] This caveat is supported by the finding that the deprotonation of chemically generated tertiary amine radical cations can give isotope effects as large as 6–9, so that dealkylation via a radical cation need not be characterized by a low isotope effect.[203] This is consistent with the demonstration that the pK_a for a trialkylammonium cation is ~15 and for N,N-dimethylaniline is ~9.[204] Comparison of the enzymatic demethylation of *para*-substituted N-methyl-N-(trideuteriomethyl)anilines [XPhN(CD$_3$)CH$_3$] and N,N-bis(dideuteriomethylanilines) [XPhN(CHD$_2$)$_2$] shows that the intramolecular isotope effects determined from the two types of substrates sometimes differ (e.g., 2.0 versus 2.7 for X = H).[202] This suggests that intramolecular isotope effects can also be masked, leaving open the possibility that the true isotope effect for a dealkylation reaction is larger than that which is measured. Perhaps the strongest caution concerning the interpretation of

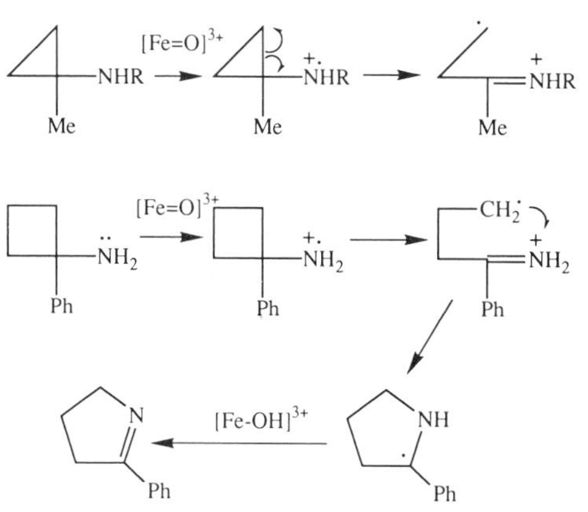

FIGURE 23. Ring opening caused by the one-electron oxidation of cyclopropyl- and cyclobutylamines.

isotope effects is provided by the early finding of Miwa *et al.*, who showed that *N*-dealkylation by horseradish peroxidase, a reaction that almost certainly proceeds via a nitrogen radical cation, exhibits an intramolecular isotope effect of 10.[191] This high isotope effect, which is close to those observed for chemically generated radical cations,[203] suggests a symmetrical transition state for transfer of a proton from the radical cation to an acceptor atom in the protein. Evidence supporting this inference has been reported.[205] Suppression of the intramolecular isotope effect by the intrinsic enantioselectivity of the enzyme has been observed during the *O*-demethylation of methoxychlor with one deuterated and one undeuterated methoxy group.[206] The apparent intramolecular isotope effect for *O*-demethylation of the (*S*)-[d_3] substrate is 13.6 but that for the corresponding (*R*)-[d_3] substrate is only 1.06. These results reinforce the age-old dictum that isotope effects must be analyzed with caution.

Cyclopropylamines inactivate P450 even when there is no hydrogen on the cyclopropyl carbon adjacent to the nitrogen. The inactivation therefore cannot be simply explained by oxidation of the cyclopropyl amine to an iminium cation that alkylates the enzyme.[207–211] It has been proposed that one-electron oxidation of the amine to the radical cation is followed by ring opening to the β-iminium radical and alkylation of the protein (Fig. 23). This theoretically attractive mechanism is supported by the recent demonstration that the oxidation potentials of a series of cyclopropylamines and other cyclopropyl derivatives correlate reasonably well with the rate of enzyme inactivation.[210] In a similar vein, *N*-cyclobutylbenzylamine also causes turnover-dependent inactivation of cytochrome P450.[211] The metabolites isolated from the oxidation of 1-phenylcyclobutylamine, 2-phenyl-1-pyrroline and 2-phenylpyrrolidine, support the postulate that oxidation of the cyclobutylamine to a nitrogen radical cation is followed by ring opening to a γ-iminium radical that recyclizes to the observed products (Fig. 23).[211]

The intrinsic isotope effects for *S*-dealkylation reactions are not known, but a correlation exists between the acidity of the protons vicinal to the sulfur and the ratio of dealkylation to *S*-oxide formation.[87] A linear free energy correlation ($\rho^+ = -0.16$) has been reported between the V_{max} values for the P450-catalyzed sulfoxidation of four substituted thioanisoles and the one-electron potentials of the substrates, although a fifth value, that for *para*-chlorothioanisole, did not fit the correlation.[86] An analogous correlation ($\rho^+ =$ –0.2) has been reported for oxidation of the four sulfoxides to the corresponding sulfones.[212] Correlations of reaction rates with oxidation potentials, however, particularly when based on a very limited number of compounds, are suggestive but do not require enzymatic oxidation of the sulfur to a radical cation because two-electron oxidation mechanisms also depend on the substrate oxidation potentials. An example of the coincidence of substituent effects for one- and two-electron oxidation mechanisms is provided by work on the chemical oxidation of NAD analogues.[213] Direct competition between a sulfide and a sulfoxide in a symmetric molecule results in oxidation of the sulfide, as expected if electron availability is a major determinant of sulfur oxidizability.[214] The oxidation of phenyl cyclopropyl sulfide by the fungus *Mortierella isabellina* has been investigated in the hope that cyclopropyl ring opening would provide evidence for a sulfur radical cation intermediate.[115] However, sulfoxidation occurred without detectable opening of the cyclopropane ring.

The intrinsic isotope effects and product ratios for N-dealkylation reactions, the much lower intrinsic isotope effects for N-dealkylations than for carbon hydroxylations or O-dealkylations, and the results obtained with radical probes support the thesis that N-dealkylations are initiated by oxidation of the heteroatom to the corresponding radical cation (Fig. 21). Although the evidence is limited, a similar mechanism can be postulated for sulfur oxidation and dealkylation.

3.6. Dehydrogenation Reactions

Cytochrome P450 enzymes occasionally catalyze dehydrogenation rather than oxygenation reactions. This recently identified class of reactions includes the oxidation of saturated to unsaturated hydrocarbons, alcohols to ketones, and nitrogen compounds to more unsaturated products. The most impressive and convincing of these transformations is the desaturation of hydrocarbons, among which the desaturation of valproic acid to 2-n-propyl-2(E)-pentenoic acid is the best understood (Fig. 24).[215–218] The oxidation of valproic acid to the $\Delta^{4,5}$-unsaturated product is catalyzed by rat, rabbit, mouse, and human liver microsomes and by purified, reconstituted cytochrome P450 2B1 in a reaction that depends on NADPH and oxygen and is depressed by inhibitors of the hemoprotein.[215,216] The desaturation reaction is accompanied by the formation of 3- and 4-hydroxyvalproic acid but the hydroxylated products are not precursors of the Δ^4-unsaturated compound.[215] Oxidation of the two enantiomers of stereospecifically [3-^{13}C]-labeled valproic acid by cultured hepatocytes shows that the pro-(R)-side chain is preferentially desaturated.[217] The $\Delta^{2,3}$-unsaturated analogue of valproic acid, 2-n-propyl-2(E)-pentenoic acid, is similarly desaturated to give the corresponding $\Delta^{2,3},\Delta^{4,5}$-diene.[218]

The intramolecular isotope effects obtained with valproic acid bearing two C-4 deuterium atoms on one of the two propyl side chains demonstrate that 4-hydroxylation ($k_H/k_D = 5.05$) and $\Delta^{4,5}$-desaturation ($k_H/k_D = 5.58$) are sensitive to deuterium substitution.[216] In contrast, only minor intramolecular isotope effects are observed for 4-hydroxylation ($k_H/k_D = 1.09$) or $\Delta^{4,5}$-desaturation ($k_H/k_D = 1.62$) when the terminal methyl group

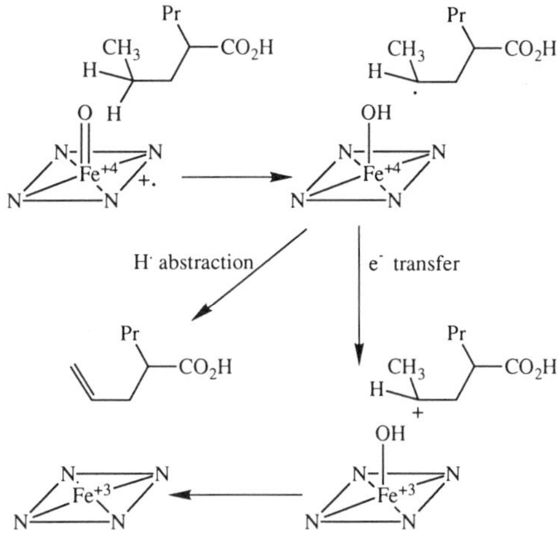

FIGURE 24. Cytochrome P450-catalyzed desaturation of valproic acid to 2-*n*-propyl-2(*E*)-pentenoic acid. As shown, the second step of the reaction involves either electron transfer followed by deprotonation or direct hydrogen radical abstraction. In order to illustrate the changes in redox state of the P450, the ferryl species is arbitrarily portrayed with a porphyrin radical cation rather than a protein radical.

of one of the side chains is trideuterated. These results suggest that removal of a C-4 hydrogen triggers both 4-hydroxylation and desaturation, whereas loss of the hydrogen at C-5 is not rate determining. A mechanism consistent with these results involves removal of a C-4 hydrogen followed by either oxygen rebound to give the 4-hydroxy product or transfer of a hydrogen atom from the terminal methyl to the iron-bound oxygen to give the desaturated derivative. A similar mechanism can be written in which electron transfer to the ferryl species from the carbon radical intermediate precedes transfer of a proton from the terminal methyl to the iron-bound oxygen (Fig. 24).

7-Hydroxylation of testosterone by CYP2A1 is accompanied as minor reactions by 6-hydroxylation and $\Delta^{6,7}$-desaturation of the substrate to give 17β-hydroxy-4,6-androstadiene-3-one (Fig. 25).[219,220] The ratio of the three products is, respectively, 38:1:1.[220] The double bond is introduced in this reaction at an allylically activated position. A primary intermolecular isotope effect is observed for 7-hydroxylation when deuterium is introduced at C-7 but primary intermolecular deuterium isotope effects are only observed for 6-hydroxylation and $\Delta^{6,7}$-desaturation when the deuterium is at C-6.[221] In agreement with the valproic acid results, hydrogen abstraction of only one of the two carbons involved in the desaturation reaction is important in kinetic terms. A mechanism similar to that proposed for valproic acid is suggested by these results (Fig. 25).

Hydrocarbon desaturation appears to be a minor but not necessarily uncommon pathway. The first reported example of P450-mediated hydrocarbon desaturation appears to be the conversion of lindane (1,2,3,4,5,6-hexachlorocyclohexane) to 1,2,3,4,5,6-hexachlorocyclohexene,[222] but the P450-mediated desaturations so far reported include $\Delta^{22,23}$-desaturation of ergosterol by yeast microsomes,[223] $\Delta^{6,7}$-desaturation of androstenedione and deoxycorticosterone by adrenal mitochondria,[224] oxidation of dihydronaphthalene to naphthalene,[225] conversion of warfarin to dehydrowarfarin,[226] and desaturation of lovostatin and simvastatin to the 6-*exo*-methylene derivatives.[227–229] In a study of mechanisms of possible relevance to the catalytic action of aromatase (Section 4.3), Watanabe and Ishimura have shown that cytochrome $P450_{cam}$ oxidizes 1-trimethylsilyloxy-2-methyl-3,4-dihydronaphthalene to the aromatic naphthalene product.[230]

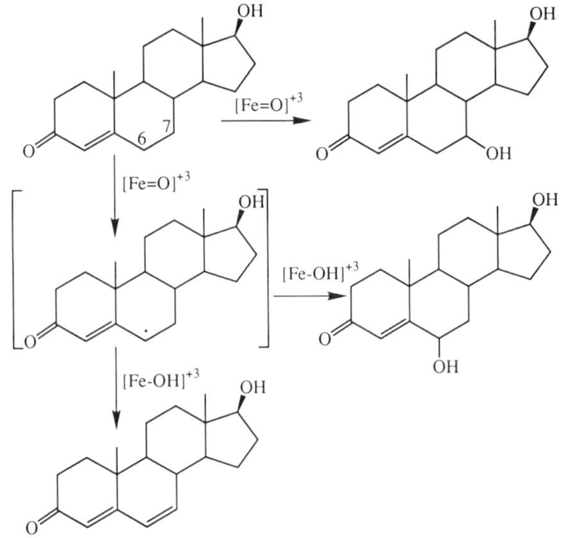

FIGURE 25. CYP2A1-catalyzed oxidation of testosterone to 7α-hydroxytestosterone, 6α-hydroxytestosterone, and 17β-hydroxy-4,6-androstadiene-3-one in a 38:1:1 ratio, respectively. The probable radical precursor of the 6α- and 6,7-desaturated products is indicated.

Functionalities other than carbon–carbon bonds are subject to P450-catalyzed dehydrogenation reactions. The aromatization of 4-alkyl- and 4-aryl-1,4-dihydropyridines (Section 3.5) is one example of a P450-catalyzed dehydrogenation reaction involving a nitrogen atom.[231,232] A second example is provided by the oxidation of acetaminophen to an iminoquinone without the intervention of a hydroxylated intermediate (Fig. 26).[233] Oxidation of the 17-hydroxy to the 17-keto function in the P450 2B1-catalyzed conversion of testosterone to androstenedione may be an example of a dehydrogenation reaction involving an oxygen. Only 5–8% of the oxygen in the keto group derives from molecular oxygen when testosterone is the substrate, whereas 84% of the oxygen is from molecular oxygen when the substrate is epitestosterone.[234] A similar observation has been made for the P450-catalyzed oxidation of 6-hydroxy- to 6-keto-progesterone by CYP2C13.[235] Either there is strong stereospecificity in the elimination of one of the two hydroxyls of a *gem*-diol intermediate, or the reaction involves hydrogen abstraction followed by electron abstraction without formation of the *gem*-diol:

$$RR'CHOH + [FeO]^{3+} \rightarrow RR'C \cdot (OH) + [FeOH]^{3+} \rightarrow RR'C=O + [Fe]^{3+} + H_2O$$

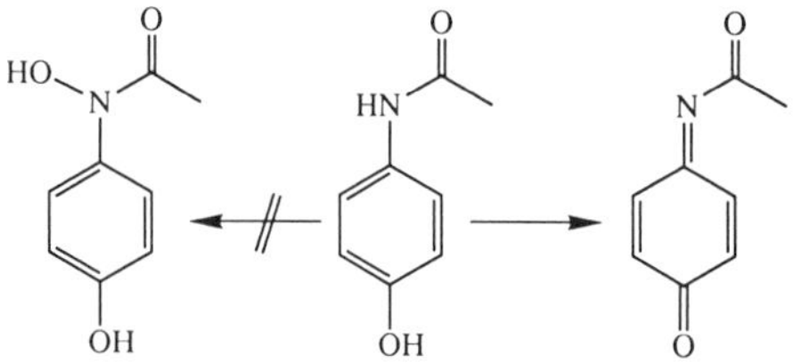

FIGURE 26. Dehydrogenation of acetaminophen to the reactive iminoquinone responsible for the protein arylation and hepatic necrosis caused by high doses of this agent. Desaturation does not proceed via the *N*-hydroxy derivative.

3.7. Aldehyde Oxidation

The oxidation of aldehydes by cytochrome P450 normally yields the corresponding carboxylic acids. Examples include the oxidation of acetaldehyde to acetic acid by CYP2E1,[236] retinal to retinoic acid by rabbit liver microsomes,[237] 5β-cholestane-3α,7α,12α-triol to 3α,7α,12α-trihydroxy-5β-cholestanoic acid by the sterol 26-hydroxylase of rabbit liver mitochondria,[238] 11-oxo-Δ^8-tetrahydrocannabinol to Δ^8-tetrahydrocannabinol-11-oic acid by CYP2C29,[239,240] aliphatic α,β-unsaturated aldehydes and anthraldehyde to the corresponding acids by the latter enzyme,[240–242] and 15-oxopentadecanoic acid to the corresponding diacid by P450$_{BM-3}$(CYP102).[243] Examination of the isozyme specificity for the oxidation of anthraldehyde to 9-anthracenecarboxylic acid indicates that this reaction is catalyzed by CYP2A1, 2B2, 2C6, 2C11, and 3A2, with CYP2C11 having the highest activity.[244] The oxidation of aldehydes to carboxylic acids can thus be catalyzed by a diversity of cytochrome P450 isozymes. Mechanistic studies indicate that one atom of molecular oxygen is incorporated into the carboxylic acid product when the reaction is carried out under an atmosphere of $^{18}O_2$.[237,239] This provides evidence for a normal hydrogen abstraction–oxygen rebound hydroxylation mechanism (Fig. 27, Path A).

A second reaction process has been identified in the oxidation of aldehydes by cytochrome P450.[245,246] This alternative process is illustrated by the CYP2B4-catalyzed oxidation of cyclohexanecarboxaldehyde to cyclohexene and formic acid, a reaction that results in loss of the carbonyl group with concomitant introduction of a double bond (Fig. 27, Path B). Decarbonylation also occurs with isobutyraldehyde, trimethylacetaldehyde, isovaleraldehyde, 2-methylbutyraldehyde, trimethylacetaldehyde, and citronellal, but not with the unbranched compounds propionaldehyde or valeraldehyde.[246] Branching of the

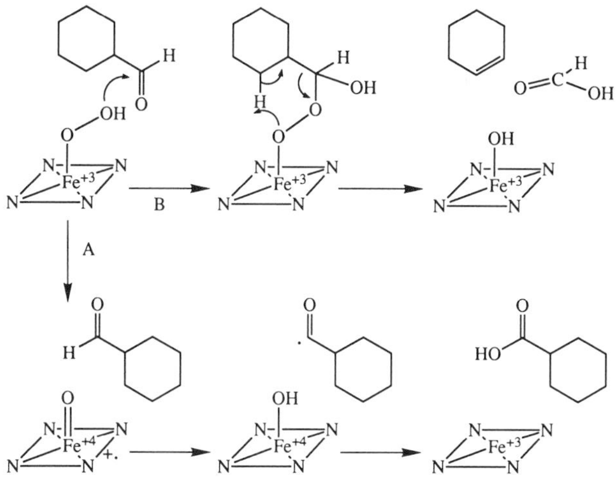

FIGURE 27. The two pathways proposed for the oxidation of aldehydes by cytochrome P450. In order to illustrate the changes in redox state of the P450, the ferryl species is arbitrarily portrayed with a porphyrin radical cation rather than a protein radical. The reaction of Path B could occur by a concerted mechanism, as shown, but is more likely to involve a stepwise mechanism triggered by dioxygen bond homolysis.

hydrocarbon chain thus appears to be important for the decarbonylation reaction. The reaction is catalyzed to different extents by several P450 enzymes, including CYP2B4, 1A2, 2E1, 2C3, and 3A6, and is therefore not isoform-limited. The reaction is not universal, however, as studies of the oxidation of fatty acids with an ω-aldehyde group by CYP102 (P450$_{BM-3}$) show that they are converted to the diacid without detectable decarbonylation.[243] Mechanistic studies have demonstrated that the oxidation of cyclohexanecarboxaldehyde by CYP2B4 is supported by NADPH/cytochrome P450 reductase or H_2O_2, but not by iodosobenzene, *meta*-chloroperbenzoic acid, or cumene hydroperoxide.[245] The reaction is not inhibited by catalase or superoxide dismutase. The ability of H_2O_2 but no other source of activated oxygen to substitute for the normal oxygen activation mechanism supports the proposal that the reaction is catalyzed by the ferric peroxide complex of cytochrome P450 rather than by the ferryl species obtained after cleavage of the oxygen–oxygen bond (Fig. 27).

The oxidation of a 3-oxodecalin-4-ene-10-carboxaldehyde (Fig. 28) by CYP2B4 has been investigated as a direct model for the reaction catalyzed by aromatase (see Section 4.3).[247] The bicyclic aldehyde is decarboxylated to the aromatic product in a reaction that produces formic acid. Deuterium isotope studies show that the reaction occurs with retention of the formyl hydrogen in the formic acid, specific loss of the 1β-hydrogen, and no specificity for loss of the 2α- or 2β-hydrogen. These results faithfully mimic the properties of the aromatase-catalyzed reaction.

A variety of aldehydes, including substituted benzaldehydes, *trans*-cinnamaldehyde, 10-undecenal, and 2-phenylpropionaldehyde are oxidized by a system consisting of an iron porphyrin and *meta*-chloroperbenzoic acid to the corresponding acids.[248] Oxidation of 2-phenylpropionaldehyde to the corresponding acid was paralleled by the formation of a trace of the decarbonylated product 1-phenylethanol. This decarbonylated product accounted for up to 10% of the product when a relatively unreactive iron porphyrin complex was used. The formation of the decarbonylated product in this reaction was attributed to a competition between the normal oxygen rebound and extrusion of carbon monoxide from the carbonyl radical intermediate in the hydroxylation process.[248] The chemical model thus reproduces the oxidation of aldehydes to acids but does not mimic the biological decarbonylation process because it is supported by *meta*-chloroperbenzoic acid, yields a saturated rather than olefinic product, and may proceed via a different mechanism.

Substantial evidence now exists for two distinct aldehyde oxidation mechanisms, one of which (decarboxylation) strongly resembles the mechanisms invoked for the decarboxylation reactions catalyzed by aromatase, lanosterol 14-demethylase, and the sterol 17,20-lyase (Sections 4.3–4.5). The hypothesis that decarbonylation is catalyzed by the iron peroxide species (Fe^{3+}-OOH) and carboxylic acid formation by the ferryl species provides

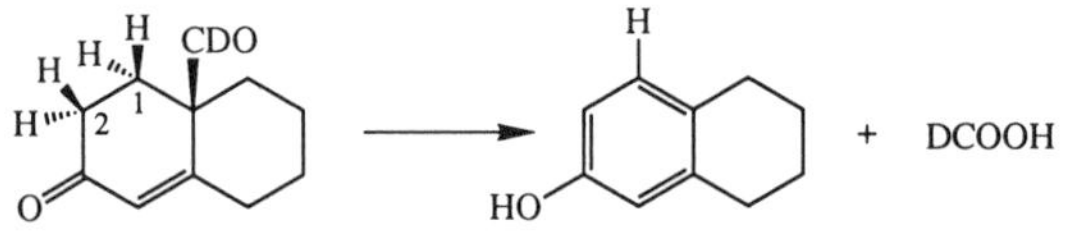

FIGURE 28. Oxidation of 3-oxodecalin-4-ene-10-carboxaldehyde by CYP2B4 as a model for the catalytic action of aromatase (Section 4.3).

an attractive explanation for the formation of two types of products. It is unclear, however, why some enzymes favor carboxylic acid formation and others decarbonylation, although some of this apparent divergence may be an artifact of the different focus of different research groups. The oxidation of cyclohexanecarboxaldehyde, for example, apparently yields the acid as well as cyclohexene but the relative importance of the two pathways has not been reported.[245] If the mechanistic proposals are correct, the product ratio presumably will reflect isoform- and substrate-dependent differences in the lifetime or reactivity of the FeOOH complex.

4. Mechanisms of Biosynthetic P450 Enzymes

4.1. Introduction

Cytochrome P450 enzymes that are integrated into biosynthetic pathways are generally more substrate-specific than enzymes primarily devoted to xenobiotic metabolism. Furthermore, several biosynthetic enzymes catalyze a sequence of two or three oxidative reactions that terminate in cleavage of a carbon–carbon bond. Nevertheless, the oxygen activation and substrate hydroxylation mechanisms appear to be similar for all forms of cytochrome P450. Enzymes that catalyze straightforward reactions, such as the 7α-hydroxylation of cholesterol during bile acid synthesis, or consecutive hydroxylations without unusual consequences, such as the double hydroxylation of the 18-methyl group in aldosterone biosynthesis,[249] therefore require no further mechanistic elaboration. However, the information available on the biosynthetic enzymes that catalyze carbon–carbon bond cleavage reactions is reviewed in this section.

4.2. Cholesterol Side-Chain Cleavage

The C-20–C-22 bond of cholesterol is cleaved by a mitochondrial P450 enzyme that catalyzes three sequential oxidative steps, each of which consumes one molecule of oxygen and one of NADPH. The three steps are 22(R)-hydroxylation, 20(S)-hydroxylation, and severance of the C-20–C-22 bond (Fig. 29). The efficient conversion of cholesterol to pregnenolone is ensured by the fact that the hydroxylated intermediates bind to the enzyme

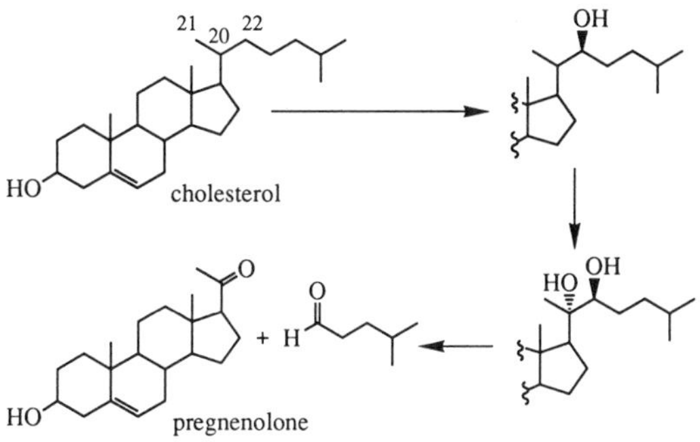

FIGURE 29. The intermediates in the triple turnover of P450$_{scc}$ that cleaves the side chain of cholesterol to give pregnenolone and 4-methylpentanal.

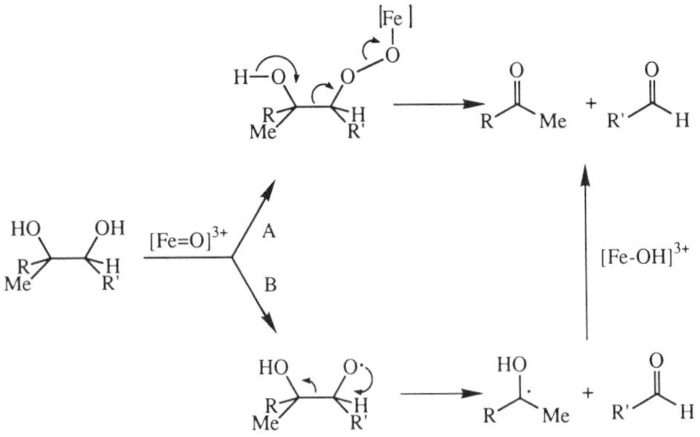

FIGURE 30. Possible mechanisms for the final step in the cholesterol side-chain cleavage reaction, where R = the sterol nucleus and R' = $CH_2CH_2CH(CH_3)_2$.

up to 300 times more tightly than cholesterol[250] and by the increased stability of the ferrous dioxygen complex in each successive turnover.[251,252] The first two hydroxylations, which proceed with retention of configuration,[253,254] are unexceptional reactions but the final step, the oxidative carbon–carbon bond cleavage, is of considerable mechanistic interest.

Mechanisms that couple carbon–carbon bond cleavage to introduction of a third hydroxyl group at C-22 are ruled out by the demonstration that the 22(S) hydrogen is retained in the 4-methylpentanal fragment that is produced in the reaction.[253] Two mechanistic approaches for the carbon–carbon bond scission are plausible. In one, the activated oxygen complex generated in the third turnover is intercepted by addition of a hydroxyl group of 20(R),22(R)-dihydroxycholesterol. Proton removal from the hydroxyl adjacent to the resulting hydroperoxide then initiates carbon–carbon bond cleavage (Fig. 30, Path A). A variant of this mechanism that does not involve a hydroperoxide has been proposed to rationalize kinetic evidence suggesting the formation of a complex prior to carbon–carbon bond scission in the superoxide and dihydropyridine-dependent cleavage of diols by an iron porphyrin model system.[255] The authors propose replacement of the ferryl oxygen by the diol hydroxyl group to give an Fe^{IV}-OCH_2CH_2OH species that subsequently undergoes carbon–carbon bond cleavage. In an alternative mechanistic approach, abstraction of the hydrogen from one of the two side-chain hydroxyls by the activated oxygen complex could produce an alkoxy radical. Homolytic scission of the carbon–carbon bond in this species and electron transfer from the resulting carbon radical to the protonated iron–oxygen complex completes the reaction process (Fig. 30, Path B). The two reactions required in the second mechanism are well established for alkyl peroxides, but their relevance to the side-chain cleavage reaction is not known.

4.3. Aromatase

Aromatase, like $P450_{scc}$, catalyzes a sequence of two carbon hydroxylations followed by an oxidative carbon–carbon bond cleavage (Fig. 31).[256] Three molecules of oxygen and three of NADPH are required for this catalytic sequence.[257] The 19-methyl group is

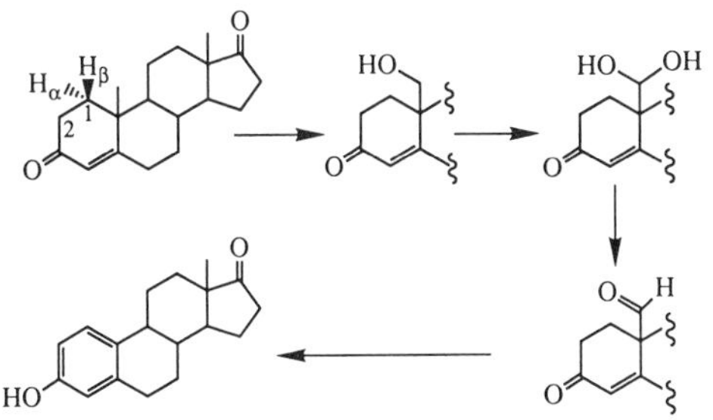

FIGURE 31. Intermediates in the catalytic turnover of aromatase.

eliminated in the final oxidative step, which also results in aromatization of the A-ring of androst-4-ene-3,17-dione and related substrates. The mechanism has primarily been studied with placental preparations, although the purified enzyme is available (see Chapter 12).[258–260] The aromatization reaction is not supported by cumene hydroperoxide or iodosobenzene,[76] in agreement with the common failure of alternative oxygen donors to replace the NAD(P)H and oxygen requirements of biosynthetic P450 enzymes.

The first 19-methyl hydroxylation catalyzed by human placental aromatase proceeds, as expected, with retention of configuration.[261,262] The second hydroxylation reaction, which removes the 19-pro-*R* hydrogen,[263,264] yields a *gem*-diol that chemically eliminates water to give the 19-aldehyde. The small tritium isotope effect reported for the aromatization of [19-³H]androst-4-ene-3,17-dione[265] is exclusively associated with the first hydroxylation step.[266] The second hydroxylation of the 19-methyl group proceeds without a detectable isotope effect. A large tritium isotope effect is also observed in a substrate analogue for the first but not second oxidation of the 19-methyl group.[267] This difference in isotope effects is readily explained by the fact that the first hydroxylation can discriminate between the tritium and the hydrogen in a given methyl group, whereas the second hydroxylation stereospecifically removes the pro-*R* hydrogen and thus is constrained to react at that position whether it is occupied by a hydrogen or a tritium. The first isotope effect thus results from an internal competition, whereas an isotope effect in the second hydroxylation would have to result from the kind of *inter*molecular discrimination that is commonly suppressed in P450 reactions. The final oxidative reaction, which stereospecifically removes the 1β- and 2β-hydrogens,[268–273] has been suggested to be a 2β-hydroxylation. The expected 2β-hydroxy-19-oxo sterol has been synthesized and shown to rapidly aromatize in the absence of the enzyme.[274] The 2β-hydroxy sterol has furthermore been detected in incubations carried out at low pH, conditions that slow down the aromatization step.[275] The importance of stereochemistry and the order of the hydroxylation reactions is demonstrated by the finding that the 2α-hydroxy-19-oxo sterol and analogues with a 19-hydroxyl rather than 19-oxo function are not aromatized chemically or enzymatically. Additional evidence for introduction of a 2β-hydroxyl group is provided by studies with antibodies that specifically recognize sterols with a 2β-hydroxyl function.[276] The antibodies inhibit the aromatization of androst-4-ene-3,17-dione but this inhibition is relieved when the antibody is presaturated with 2β,19-dihydroxyandrost-4-ene-3,17-dione. These

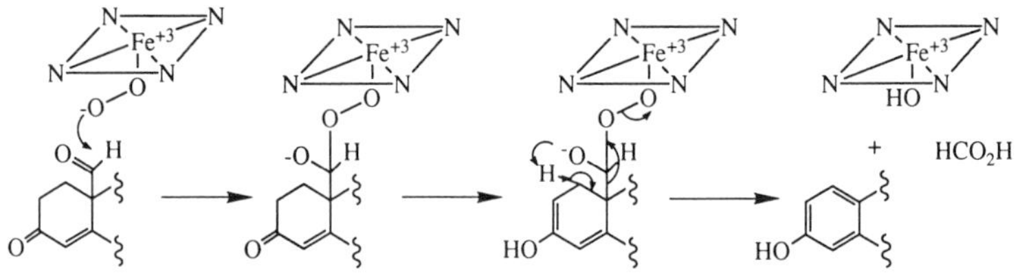

FIGURE 32. The favored mechanism for the final step in the aromatase-catalyzed reaction. The exact sequence of electron transfer within the peroxo complex and the order in which the C–C, H_β–C, and O–O bonds are ruptured is unclear. It is likely that the peroxo intermediate will fragment stepwise rather than by the concerted process shown in the figure.

results, and the finding that the formic acid eliminated in the aromatization step contains the first and third oxygen atoms consumed in the catalytic sequence,[277,278] have led to the postulate that intramolecular addition of the 2β-hydroxyl to the 19-aldehyde moiety to give a hemiacetal precedes rupture of the carbon–carbon bond.[277] However, the report that the 2β-hydroxyl is not incorporated into the formic acid when the synthetic intermediate is incubated with placental microsomes conflicts with the finding that the third oxygen *is incorporated* into formic acid in the normal catalytic reaction.[278] 2β-Hydroxylation thus appears not to be an obligatory step in estradiol biosynthesis. It appears, furthermore, that the stereochemistry of the hydrogen abstraction from C-2 differs for testosterone and androstenedione and thus may be substrate-dependent.[279] Alternative mechanisms in which the third oxidation introduces a 4,5-epoxide,[280] a 1β-hydroxyl,[272] or a C-19-peroxide[277,281] have been excluded, as has the possibility that the substrate is covalently bound to the enzyme by Schiff base formation through the 3-keto function prior to catalytic turnover.[282] The exact nature of the third oxidative step therefore remains uncertain but a mechanism involving addition of the iron-bound dioxygen species to the 19-aldehyde, followed by oxygen–oxygen bond cleavage, carbon loss, and aromatization has received strong support (Fig. 32),[283–286] including support by model studies and the observation of comparable reactions with other cytochrome P450 enzymes.[246,287]

4.4. Lanosterol 14α-Demethylation

Removal of the 14α-methyl group of lanosterol as formic acid in the biosynthesis of cholesterol[288,289] involves sequential oxidation of the methyl group to the alcohol and aldehyde moiety followed by oxidative elimination of the methyl as formic acid with concomitant introduction of a $\Delta^{14,15}$-double bond into the sterol framework (Fig. 33).[290–300] Purification of the enzyme from yeast, rat, pig, and human sources,[299–302] and cloning and functional expression of the rat enzyme,[303] have clearly established that this sequence of oxidative reactions is catalyzed by a single cytochrome P450 enzyme. The opening 14-methyl hydroxylation is an unexceptional P450-catalyzed hydroxylation.[304–306] Conversion of the alcohol to the aldehyde involves a second hydroxylation to give a *gem*-diol that eliminates water to give the aldehyde. These first two steps of the reaction thus closely parallel the first two steps of the reaction catalyzed by aromatase.

Conversion of the 14-aldehyde with stereospecific removal of the 15α-hydrogen to give the $\Delta^{14,15}$-unsaturated steroid[291,307,308] is suggested to proceed via a Baeyer–Villiger

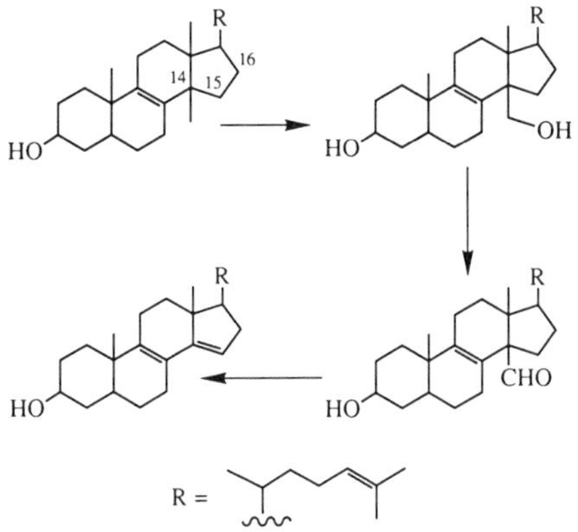

FIGURE 33. The stable intermediates in the 14-demethylation of lanosterol.

reaction by the recent isolation of a new reaction intermediate (Fig. 34).[309] The intermediate isolated from the reaction was identified as the 14-formyl derivative by spectroscopic
methods and was shown to be converted by the enzyme to the final unsaturated product.
The Baeyer–Villiger reaction may be initiated by nucleophilic addition of the ferrous
dioxygen species to the aldehyde to give an iron-coordinated alkyl peroxide species that
rearranges to the 14-formyl product. The mechanism of the final elimination reaction in
which the 14-formyl and 15α-proton are eliminated has not been elucidated but, in
principle, requires nothing more than conventional acid–base catalysis. The possibility
remains that the normal reaction intermediate decays to a small extent to the Baeyer–
Villiger product but primarily undergoes a reaction in which the 15α-hydrogen is simultaneously eliminated. The present results provide strong support for the involvement of a
ferric peroxide intermediate in the demethylation reaction. In contrast, the 14-aldehyde
with a 15α-hydroxyl group is an inhibitor rather than a substrate for the enzyme.[309] The
earlier proposal that 15-hydroxylation might precede loss of the aldehyde moiety and
introduction of the 14,15-double bond is therefore not tenable. It is interesting in this
context that 14α-methylcholest-7-ene-3β,15β-diol, in which the 15-hydroxyl has the
wrong stereochemistry, is converted to cholesterol by rat liver homogenates.[310] This well

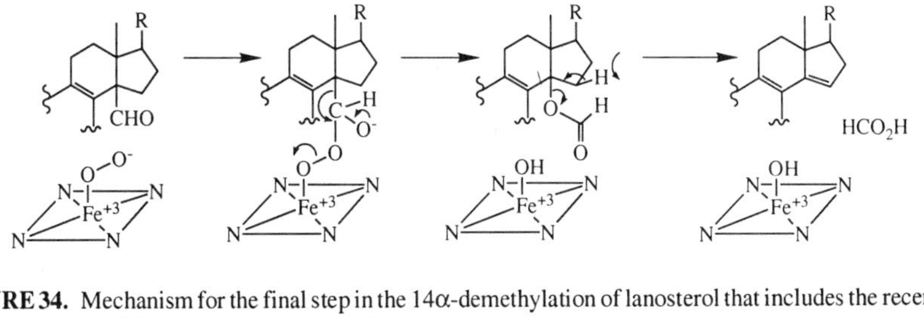

FIGURE 34. Mechanism for the final step in the 14α-demethylation of lanosterol that includes the recently
isolated Baeyer–Villiger rearrangement product.

illustrates the caution that must be exercised in interpreting studies of the fate of chemically synthesized hypothetical intermediates.

4.5. CYP17A: Progesterone 17α-Hydroxylase/17,20-lyase

The 17α-hydroxylation of pregnenolone (or progesterone) and cleavage of the C-17–C-20 carbon–carbon bond in the 17α-hydroxylated steroid to give dehydroepiandrosterone (or androstenedione), as shown by heterologous expression of the recombinant protein, are catalyzed by a single enzyme (see Chapter 12).[311,312] The mechanism proposed for these reactions closely resembles the mechanisms proposed for the lanosterol 14α-demethylation and aromatase reactions. Detailed studies of the reaction have shown that the dominant pathway involving 17-hydroxylation and side chain elimination to give the 17-keto sterol is paralleled by minor pathways which result in cleavage of pregnenolone to the 17α-hydroxyandrost-5-en-3β-ol (note inversion of stereochemistry) and $\Delta^{16,17}$-olefin (Fig. 35).[313–315] The unexceptional 17α-hydroxylation of pregnenolone is followed by a carbon–carbon bond cleavage in which the 17α-hydroxyl group becomes the carbonyl group of the product and the two-carbon side chain is converted to acetic acid. ^{18}O isotopic labeling studies indicate that the acetic acid that is formed incorporates one atom of oxygen from molecular oxygen.[316] One atom of molecular oxygen is similarly incorporated into the acetic acid that is eliminated in the oxidations that directly transform pregnenolone into either a $\Delta^{16,17}$-sterol or androst-5-ene-3β,17α-diol.[315–317] An atom of molecular oxygen is also incorporated into the 3β-17α-diol at position 17.[317] These results are consistent with a mechanism in which the iron dioxygen species adds to the carbonyl group of pregnenolone and then fragments to give an alkoxy radical. Fragmentation of this alkoxy radical produces a carbon radical, acetic acid, and a one-electron oxidized ferryl species (Fig. 36). According to this mechanism, the principal product is formed by rebound delivery of the ferryl oxygen to the carbon radical formed from 17α-hydroxypregnenolone to give a *gem*-diol that decays to the ketone group of androstenedione.[316,317] Similar reaction of pregnenolone itself produces the 17α-hydroxy product if the ferryl oxygen is transferred to the carbon radical or the olefin if the carbon radical is oxidized to a cation

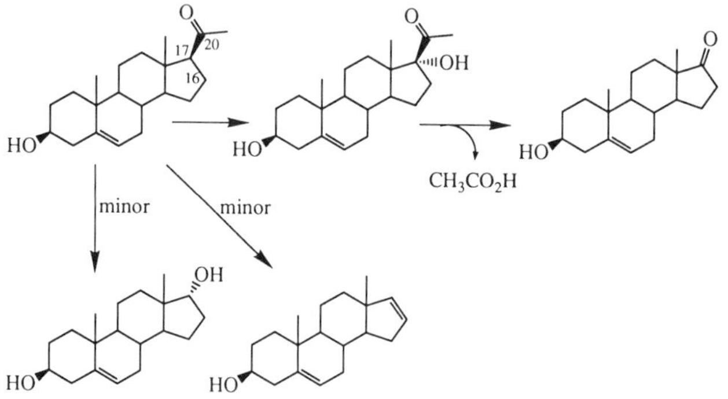

FIGURE 35. Steps in the C-17–C-20 bond cleavage catalyzed by CYP17. Two minor nonhormonal steroids, the $\Delta^{16,17}$-olefin and the 17α-hydroxylated steroid, are also produced by the enzyme as a result of competing divergent reactions.

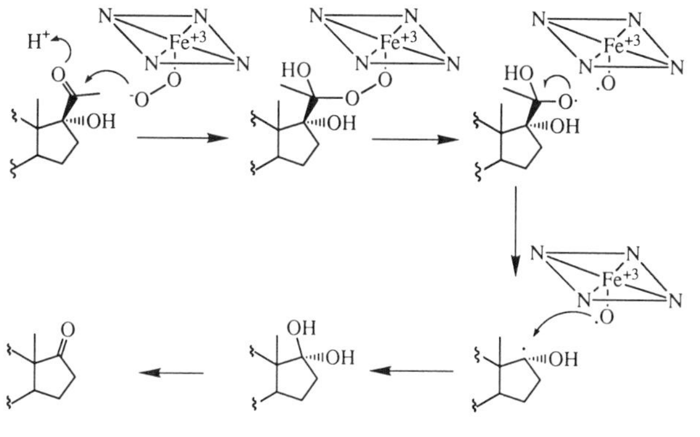

FIGURE 36. Mechanism proposed for the C-17–C-20 lyase reaction involving addition of a cytochrome P450 ferric peroxide intermediate to the C-20 carbonyl group followed by free radical fragmentation of the peroxide adduct.

by electron transfer to the ferryl species. The oxidation of progesterone by microsomal or purified CYP17 has also been shown to produce 17-O-acetyltestosterone.[318] The formation of this product is more readily rationalized by a Baeyer–Villiger rearrangement similar to that postulated for lanosterol 14-demethylase (Fig. 37). There is some ambiguity, therefore, concerning the mechanism of decomposition of the adduct formed by addition of the ferric peroxide to the 20-keto group (i.e., radical or Baeyer–Villiger). Both mechanisms, however, require initial formation of the peroxide adduct and may compete with each other. A study of the deuterium solvent isotope effect on the reaction as a function of pH has led to the proposal that protonation of the FeOO⁻ intermediate governs whether the reaction proceeds via an oxene (17α-hydroxylation) or peroxide (Baeyer–Villiger or alkoxy radical) pathway.[319]

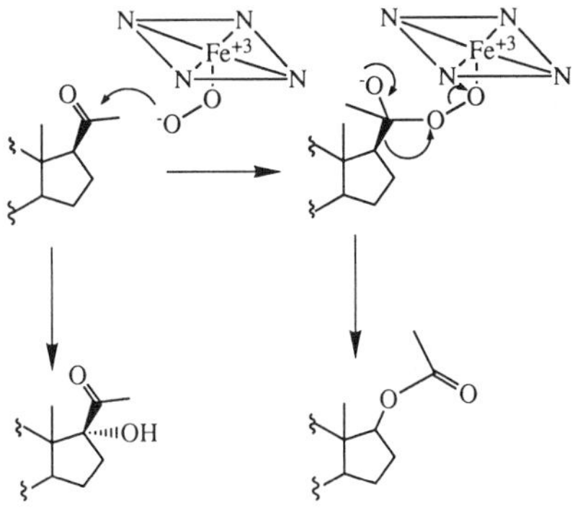

FIGURE 37. Mechanism proposed for the CYP17-catalyzed oxidation of pregnenolone in which there is a partitioning between addition of the cytochrome P450 ferric peroxide intermediate to the 20-ketone followed by Baeyer–Villiger rearrangement and decay of the ferric peroxide intermediate to the usual ferryl species that gives rise to the 17α-hydroxylated product.

4.6. Other Biosynthetic Cytochrome P450-Like Enzymes

In the last few years, a growing number of enzymes have been discovered that have a thiolate ligand and, in some cases, sufficient sequence identity to the cytochrome P450 class of enzymes to be considered as part of that class. The enzymes, however, are distinct in that they catalyze the rearrangement of lipid hydroperoxide substrates. These enzymes include prostacyclin, thromboxane, and allene oxide synthases. Nitric oxide synthase, on the other hand, is a thiolate protein with very little, if any, sequence identity to cytochrome P450. Nevertheless, the reactions catalyzed by nitric oxide synthase depend on interaction with a two-flavin domain very similar to cytochrome P450 reductase and the chemistry catalyzed by the enzyme is very similar to that expected for a P450 enzyme. The mechanisms of these novel hemoprotein catalysts are discussed in Chapter 15.

5. Summary

The catalytic cycle of cytochrome P450 culminates in the production of an activated species two oxidation equivalents above the ferric state. The activated species appears to be a ferryl complex ($Fe^{IV}=O$), but it is not known if the second oxidation equivalent is present as a porphyrin or protein radical. The ferryl species presumably arises by heterolytic cleavage of a ferric peroxide complex (Fe^{3+}-OOH). This is consistent with the fact that catalytic turnover of many P450 enzymes is supported by H_2O_2, although the normal and H_2O_2-derived oxidizing species may not be entirely identical because differences are seen in product ratios and in inactivation of the enzyme. The ferric peroxide intermediate may also play a limited direct catalytic role because it appears to act as the actual oxidizing species in some of the reactions of cytochrome P450 with aldehydes and other carbonyl compounds. The ferric peroxide species may be of particular importance in the catalytic action of aromatase, lanosterol 14-demethylase, and the adrenal sterol 17-20 lyase.

As previously stated,[1] the support for a concerted reaction mechanism is generally the absence of evidence for a nonconcerted process. The accumulated evidence that cytochrome P450-catalyzed hydroxylations occur by nonconcerted oxygen-transfer mechanisms is sufficiently strong to rule out concerted mechanisms for this process. Strong evidence also argues that nitrogen oxidation and dealkylation reactions are also achieved by nonconcerted mechanisms. The oxidation and dealkylation of sulfur moieties is also probably mediated by nonconcerted mechanisms but only limited data are available on these reactions. The greatest mechanistic ambiguities are in the area of π-bond oxidation. Although some reaction products are obtained that implicate nonconcerted mechanisms, it is not yet clear whether these products arise by the same pathway as epoxidation or via pathways that diverge prior to oxygen transfer to the π-bond.

In sum, no evidence is available that *requires* a concerted mechanism, but considerable evidence exists that can only be explained by nonconcerted mechanisms.

ACKNOWLEDGMENT.   The preparation of this chapter was aided by grant GM25515 from the National Institutes of Health.

References

1. Ortiz de Montellano, P. R., 1986, Oxygen activation and transfer, in: *Cytochrome P-450: Structure, Mechanism, and Biochemistry* (P. R. Ortiz de Montellano, ed.), Plenum Press, New York, pp. 217–271.
2. White, R. E., and Coon, M. J., 1980, Oxygen activation by cytochrome P450, *Annu. Rev. Biochem.* **49**:315–356.
3. Dawson, J. H., 1988, Probing structure–function relationships in heme-containing oxygenases and peroxidases, *Science* **240**:433–439.
4. Ishimura, Y., 1993, Oxygen activation and transfer, in: *Cytochrome P-450*, 2nd ed. (T. Omura, Y. Ishimura, and Y. Fujii-Kuriyama, eds.), VCH, Weinheim, pp. 80–91.
5. Koymans, L., Donné-op Den Kelder, G. M., Koppele Te, J. M., and Vermeulen, N. P. E., 1993, Cytochromes P450: Their active-site structure and mechanism of oxidation, *Drug Metab. Rev.* **25**:325–387.
6. Archakov, A. I., and Bachmanova, G. I., 1990, *Cytochrome P450 and Active Oxygen*, Taylor & Francis, London.
7. Dolphin, D., Forman, A., Borg, D. C., Fajer, J., and Felton, R. H., 1971, Compounds I of catalase and horse radish peroxidase: π-cation radicals, *Proc. Natl. Acad. Sci. USA* **68**:614–618.
8. Morishima, I., Takamuki, Y., and Shiro, Y., 1984, Nuclear magnetic resonance studies of metalloporphyrin π-cation radicals as models for compound I of peroxidases, *J. Am. Chem. Soc.* **106**:7666–7672.
9. Roberts, J. E., Hoffman, B. M., Rutter, R., and Hager, L. P., 1981, Electron–nuclear double resonance of horseradish peroxidase compound I, *J. Biol. Chem.* **256**:2118–2121.
10. La Mar, G. N., de Ropp, J. S., Latos-Grazynski, L., Balch, A. L., Johnson, R. B., Smith, K. M., Parish, D. W., and Cheng, R.-J., 1983, Proton NMR characterization of the ferryl group in model heme complexes and hemoproteins: Evidence for the Fe^{IV}=O group in ferryl myoglobin and compound II of horseradish peroxidase, *J. Am. Chem. Soc.* **105**:782–787.
11. Hoffman, B. M., Roberts, J. E., Kang, C. H., and Margoliash, E., 1981, Electron paramagnetic and electron nuclear double resonance of the hydrogen peroxide compound of cytochrome c peroxidase, *J. Biol. Chem.* **256**:6556–6564.
12. McCandlish, E., Miksztal, A. R., Nappa, M., Sprenger, A. Q., Valentine, J. S., Stong, J. D., and Spiro, T. G., 1980, Reactions of superoxide with iron porphyrins in aprotic solvents: A high spin ferric porphyrin peroxo complex, *J. Am. Chem. Soc.* **102**:4268–4271.
13. Ataollah, S., and Goff, H. M., 1982, Characterization of superoxide–metalloporphyrin reaction products: Effective use of deuterium NMR spectroscopy, *J. Am. Chem. Soc.* **104**:6318–6322.
14. Burstyn, J. N., Roe, J. A., Miksztal, A. R., Shaevitz, B. A., Lang, G., and Valentine, J. S., 1988, Magnetic and spectroscopic characterization of an iron porphyrin peroxide complex. Peroxoferrioctaethylporphyrin (1-), *J. Am. Chem. Soc.* **110**:1382–1388.
15. Welborn, C. H., Dolphin, D., and James, B. R., 1981, One-electron electrochemical reduction of a ferrous porphyrin dioxygen complex, *J. Am. Chem. Soc.* **103**:2869–2871.
16. Miksztal, A. R., and Valentine, J. S., 1984, Reactivity of the peroxo ligand in metalloporphyrin complexes. Reaction of sulfur dioxide with iron and titanium porphyrin peroxo complexes to give sulfato complexes or sulfate, *Inorg. Chem.* **23**:3548–3552.
17. Valentine, J. S., Burstyn, J. N., and Margerum, L. D., 1988, *Mechanisms of Dioxygen Activation in Metal-Containing Monooxygenases: Enzymes and Model Systems*, Plenum Press, New York, pp. 175–187.
18. Hrycay, E. G., and O'Brien, P. J., 1971, Cytochrome P450 as a microsomal peroxidase utilizing a lipid peroxide substrate, *Arch. Biochem. Biophys.* **147**:14–27.
19. Hrycay, E. G., and O'Brien, P. J., 1972, Cytochrome P450 as a microsomal peroxidase in steroid hydroperoxide reduction, *Arch. Biochem. Biophys.* **153**:480–494.
20. Hrycay, E. G., and O'Brien, P. J., 1974, Microsomal electron transport. II. Reduced nicotinamide adenine dinucleotide-cytochrome b₅ reductase and cytochrome P450 as electron carriers in microsomal NADH-peroxidase activity, *Arch. Biochem. Biophys.* **160**:230–245.

21. O'Brien, P. J., 1978, Hydroperoxides and superoxides in microsomal oxidations, *Pharmacol. Ther. A* **2:**517–536.

22. Rahimtula, A. D., and O'Brien, P. J., 1977, The hydroperoxide dependent microsomal oxidation of alcohols, *Eur. J. Biochem.* **77:**210–211.

23. Davies, M. J., 1989, Detection of peroxyl and alkoxyl radicals produced by reaction of hydroperoxides with rat liver microsomal fractions, *Biochem. J.* **257:**603–606.

24. O'Brien, P. J., and Rahimtula, A., 1975, Involvement of cytochrome P450 in the intracellular formation of lipid peroxides, *J. Agr. Food Chem.* **23:**154–158.

25. Weiss, R. H., and Estabrook, R. W., 1986, The mechanism of cumene hydroperoxide-dependent lipid peroxidation: The function of cytochrome P-450, *Arch. Biochem. Biophys.* **251:**348–360.

26. Lindstrom, T. D., and Aust, S. D., 1984, Studies on cytochrome P450-dependent lipid hydroperoxide reduction, *Arch. Biochem. Biophys.* **233:**80–87.

27. Cavallini, L., Valente, M., and Bindoli, A., 1983, NADH and NADPH inhibit lipid peroxidation promoted by hydroperoxides in rat liver microsomes, *Biochim. Biophys. Acta* **752:**339–345.

28. Kadlubar, F. F., Morton, K. C., and Ziegler, D. M., 1973, Microsomal-catalyzed hydroperoxide-dependent C-oxidation of amines, *Biochem. Biophys. Res. Commun.* **54:**1255–1261.

29. Nordblom, G. D., White, R. E., and Coon, M. J., 1976, Studies on hydroperoxide-dependent substrate hydroxylation by purified liver microsomal cytochrome P450, *Arch. Biochem. Biophys.* **175:**524–533.

30. Ashley, P. L., and Griffin, B. W., 1981, Involvement of radical species in the oxidation of aminopyrine and 4-aminoantipyrine by cumene hydroperoxide in rat liver microsomes, *Mol. Pharmacol.* **19:**146–152.

31. Griffin, B. W., 1982, Use of spin traps to elucidate radical mechanisms of oxidations by hydroperoxides catalyzed by hemoproteins, *Can. J. Chem.* **60:**1463–1473.

32. Rahimtula, A. D., and O'Brien, P. J., 1975, Hydroperoxide-dependent O-dealkylation reactions catalyzed by liver microsomal cytochrome P450, *Biochem. Biophys. Res. Commun.* **62:**268–275.

33. Burke, D. M., and Mayer, R. T., 1975, Inherent specificities of purified cytochrome P450 and P-448 toward biphenyl hydroxylation and ethoxyresorufin deethylation, *Drug Metab. Dispos.* **3:**245–253.

34. Rahimtula, A. D., and O'Brien, P. J., 1974, Hydroperoxide catalyzed liver microsomal aromatic hydroxylation reactions involving cytochrome P450, *Biochem. Biophys. Res. Commun.* **60:**440–447.

35. Hrycay, E. G., Gustafsson, J.-A., Ingelman-Sundberg, M., and Ernster, L., 1975, Sodium periodate, sodium chlorite, organic hydroperoxides, and H_2O_2 as hydroxylating agents in steroid hydroxylation reactions catalyzed by partially purified cytochrome P450, *Biochem. Biophys. Res. Commun.* **66:**209–216.

36. Ellin, A., and Orrenius, S., 1975, Hydroperoxide-supported cytochrome P450-linked fatty acid hydroxylation in liver microsomes, *FEBS Lett.* **50:**378–381.

37. Danielsson, H., and Wikvall, K., 1976, On the ability of cumene hydroperoxide and $NaIO_4$ to support microsomal hydroxylations in biosynthesis and metabolism of bile acids, *FEBS Lett.* **66:**299–302.

38. Gustafsson, J.-A., Hrycay, E. G., and Ernster, L., 1976, Sodium periodate, sodium chlorite, and organic hydroperoxides as hydroxylating agents in steroid hydroxylation reactions catalyzed by adrenocortical microsomal and mitochondrial cytochrome P450, *Arch. Biochem. Biophys.* **174:**440–453.

39. Hrycay, E. G., Gustafsson, J.-A., Ingelman-Sundberg, M., and Ernster, L., 1976, The involvement of cytochrome P450 in hepatic microsomal steroid hydroxylation reactions supported by sodium periodate, sodium chlorite, and organic hydroperoxides, *Eur. J. Biochem.* **61:**43–52.

40. Rahimtula, A. D., O'Brien, P. J., Seifried, H. E., and Jerina, D. M., 1978, The mechanism of action of cytochrome P450: Occurrence of the "NIH shift" during hydroperoxide-dependent aromatic hydroxylations, *Eur. J. Biochem.* **89:**133–141.

41. van Lier, J. E., and Rousseau, J., 1976, Mechanism of cholesterol side-chain cleavage: Enzymic rearrangement of 20β-hydroperoxy-20-isocholesterol to 20β,21-dihydroxy-20-isocholesterol, *FEBS Lett.* **70:**23–27.

42. Larroque, C., and van Lier, J. E., 1986, Hydroperoxysterols as a probe for the mechanism of cytochrome P-450$_{scc}$-mediated hydroxylation: Homolytic versus heterolytic oxygen–oxygen bond scission, *J. Biol. Chem.* **261**:1083–1087.

43. Larroque, C., and van Lier, J. E., 1983, Spectroscopic evidence for the formation of a transient species during cytochrome P450$_{scc}$ induced hydroperoxysterol–glycol conversions, *Biochem. Biophys. Res. Commun.* **112**:655–662.

44. Larroque, C., Lange, R., Maurin, L., Bienvenue, A., and van Lier, J. E., 1990, On the nature of the cytochrome P450scc "ultimate oxidant": Characterization of a productive radical intermediate, *Arch. Biochem. Biophys.* **282**:198–201.

45. Thompson, J. A., and Wand, M. D., 1985, Interaction of cytochrome P-450 with a hydroperoxide derived from butylated hydroxytoluene. Mechanism of isomerization, *J. Biol. Chem.* **260**:10637–10644.

46. Wand, M. D., and Thompson, J. A., 1986, Cytochrome P-450-catalyzed rearrangement of a peroxyquinol derived from butylated hydroxytoluene. Involvement of radical and cationic intermediates, *J. Biol. Chem.* **261**:14049–14056.

47. Yumibe, N. P., and Thompson, J. A., 1988, Fate of free radicals generated during one-electron reductions of 4-alkyl-1,4-peroxyquinols by cytochrome P-450, *Chem. Res. Toxicol.* **1**:385–390.

48. White, R. E., Sligar, S. G., and Coon, M. J., 1980, Evidence for a homolytic mechanism of peroxide oxygen–oxygen bond cleavage during substrate hydroxylation by cytochrome P450, *J. Biol. Chem.* **255**:11108–11111.

49. Vaz, A. D. N., and Coon, M. J., 1987, Hydrocarbon formation in the reductive cleavage of hydroperoxides by cytochrome P-450, *Proc. Natl. Acad. Sci. USA* **84**:1172–1176.

50. Thompson, J. A., and Yumibe, N. P., 1989, Mechanistic aspects of cytochrome P-450–hydroperoxide interactions: Substituent effects on degradative pathways, *Drug. Metab. Rev.* **20**:365–378.

51. Fish, K. M., Avaria, G. E., and Groves, J. T., 1988, Rearrangement of alkyl hydroperoxides mediated by cytochrome P-450: Evidence for the oxygen rebound mechanism, in: *Microsomes and Drug Oxidations* (J. O. Miners, D. J. Birkett, R. Drew, B. K. May, and M. E. McManus, eds.), Taylor & Francis, London, pp. 176–183.

52. Koop, D. R., and Hollenberg, P. F., 1980, Kinetics of the hydroperoxide-dependent dealkylation reactions catalyzed by rabbit liver microsomal cytochrome P450, *J. Biol. Chem.* **255**:9685–9692.

53. Blake, R. C., and Coon, M. J., 1980, On the mechanism of action of cytochrome P450: Spectral intermediates in the reactions of P450$_{LM2}$ with peroxy compounds, *J. Biol. Chem.* **255**:4100–4111.

54. Blake, R. C., and Coon, M. J., 1981, On the mechanism of action of cytochrome P450: Role of peroxy spectral intermediates in substrate hydroxylation, *J. Biol. Chem.* **256**:5755–5763.

55. Blake, R. C., and Coon, M. J., 1981, On the mechanism of action of cytochrome P450: Evaluation of homolytic and heterolytic mechanisms of oxygen–oxygen bond cleavage during substrate hydroxylation by peroxides, *J. Biol. Chem.* **256**:12127–12133.

56. Rahimtula, A. D., O'Brien, P. J., Hrycay, E. G., Peterson, J. A., and Estabrook, R. W., 1974, Possible higher valence states of cytochrome P450 during oxidative reactions, *Biochem. Biophys. Res. Commun.* **60**:695–702.

57. Dunford, H. B., 1982, Peroxidases, in: *Advances in Inorganic Biochemistry* (G. Eichhorn and L. G. Marzilli, eds.), Elsevier, Amsterdam, pp. 41–68.

58. Wagner, G. C., Palcic, M. M., and Dunford, H. B., 1983, Absorption spectra of cytochrome P450$_{cam}$ in the reaction with peroxy acids, *FEBS Lett.* **156**:244–248.

59. Egawa, T., Shimada, H., and Ishimura, Y., 1994, Evidence for compound I formation in the reaction of cytochrome P450$_{cam}$ with *m*-chloroperbenzoic acid, *Biochem. Biophys. Res. Commun.* **201**:1464–1469.

60. Wagner, G. C., and Gunsalus, I. C., 1982, Cytochrome P450: Structure and states, in: *Biology and Chemistry of Iron, NATO Adv. Study Inst. Ser. C* **89**:405–412.

61. Kobayashi, K., Amano, M., Kanbara, Y., and Hayashi, K., 1987, One-electron reduction of the oxyform of 2,4-diacetyldeuterocytochrome P-450$_{cam}$, *J. Biol. Chem.* **262**:5445–5447.

62. Kobayashi, K., Iwamoto, T., and Honda, K., 1994, Spectral intermediates in the reaction of ferrous cytochrome $P450_{cam}$ with superoxide anion, *Biochem. Biophys. Res. Commun.* **201**:1348–1355.

63. Tajima, K., Edo, T., Ishzu, K., Imaoka, S., Funae, Y., Oka, S., and Sakurai, H., 1993, Cytochrome P-450–butyl peroxide complex detected by ESR, *Biochem. Biophys. Res. Commun.* **191**:157–164.

64. McCarthy, M.-B., and White, R. E., 1983, Competing modes of peroxyacid flux through cytochrome P450, *J. Biol. Chem.* **258**:11610–11616.

65. Capdevila, J., Estabrook, R. W., and Prough, R. A., 1980, Differences in the mechanism of NADPH- and cumene hydroperoxide-supported reactions of cytochrome P450, *Arch. Biochem. Biophys.* **200**:186–195.

66. Hlavica, P., Golly, I., and Mietaschk, J., 1983, Comparative studies on the cumene hydroperoxide- and NADPH-supported N-oxidation of 4-chloroaniline by cytochrome P450, *Biochem. J.* **212**:539–547.

67. Renneberg, R., Damerau, W., Jung, C., Ebert, B., and Scheller, F., 1983, Study of H_2O_2-supported N-demethylations catalyzed by cytochrome P450 and horseradish peroxidase, *Biochem. Biophys. Res. Commun.* **113**:332–339.

68. Bartsch, H., and Hecker, E., 1971, On the metabolic activation of the carcinogen N-hydroxy-N-2-ace-tylaminofluorene. III. Oxidation with horseradish peroxidase to yield 2-nitrosofluorene and N-ace-toxy-N-2-acetylaminofluorene, *Biochim. Biophys. Acta* **237**:567–578.

69. Reigh, D. L., and Floyd, R. A., 1981, Evidence for a cytochrome P-420 catalyzed mechanism of activation of N-hydroxy-2-acetylaminofluorene, *Cancer Biochem. Biophys.* **5**:213–217.

70. Stier, A., Reitz, I., and Sackmann, E., 1972, Radical accumulation in liver microsomal membranes during biotransformation of aromatic amines and nitro compounds, *Naunyn-Schmiedebergs Arch. Pharmacol.* **274**:189–191.

71. Renneberg, R., Scheller, F., Ruckpaul, K., Pirrwitz, J., and Mohr, P., 1978, NADPH and H_2O_2-de-pendent reactions of cytochrome $P450_{LM}$ compared with peroxidase catalysis, *FEBS Lett.* **96**:349–353.

72. Renneberg, R., Capdevila, J., Chacos, N., Estabrook, R. W., and Prough, R. A., 1981, Hydrogen peroxide-supported oxidation of benzo[a]pyrene by rat liver microsomal fractions, *Biochem. Pharmacol.* **30**:843–848.

73. Estabrook, R. W., Martin-Wixtrom, C., Saeki, Y., Renneberg, R., Hildebrandt, A., and Werringloer, J., 1984, The peroxidatic function of liver microsomal cytochrome P450: Comparison of hydrogen peroxide and NADPH-catalyzed N-demethylation reactions, *Xenobiotica* **14**:87–104.

74. Holm, K. A., Engell, R. J., and Kupfer, D., 1985, Regioselectivity of hydroxylation of prostaglandins by liver microsomes supported by NADPH versus H_2O_2 in methylcholanthrene-treated and control rats: Formation of novel prostaglandin metabolites, *Arch. Biochem. Biophys.* **237**:477–489.

75. Fasco, M. J., Piper, L. J., and Kaminsky, L. S., 1979, Cumene hydroperoxide-supported microsomal hydroxylations of warfarin—A probe of cytochrome P450 multiplicity and specificity, *Biochem. Pharmacol.* **28**:97–103.

76. Kelly, W. G., and Stolee, A. H., 1978, Stabilization of placental aromatase by dithiothreitol in the presence of oxidizing agents, *Steroids* **31**:533–539.

77. Gustafsson, J.-A., Rondahl, L., and Bergman, J., 1979, Iodosylbenzene derivatives as oxygen donors in cytochrome P450 catalyzed steroid hydroxylations, *Biochemistry* **18**:865–870.

78. Gustafsson, J.-A., and Bergman, J., 1976, Iodine- and chlorine-containing oxidation agents as hydroxylating catalysts in cytochrome P450-dependent fatty acid hydroxylation reactions in rat liver microsomes, *FEBS Lett.* **70**:276–280.

79. Lichtenberger, F., Nastainczyk, W., and Ullrich, V., 1976, Cytochrome P450 as an oxene transferase, *Biochem. Biophys. Res. Commun.* **70**:939–946.

80. Berg, A., Ingelman-Sundberg, M., and Gustafsson, J.-A., 1979, Purification and characterization of cytochrome $P450_{meg}$, *J. Biol. Chem.* **254**:5264–5271.

81. Heimbrook, D. C., and Sligar, S. G., 1981, Multiple mechanisms of cytochrome P450-catalyzed substrate hydroxylation, *Biochem. Biophys. Res. Commun.* **99**:530–535.

82. Macdonald, T. L., Burka, L. T., Wright, S. T., and Guengerich, F. P., 1982, Mechanisms of hydroxylation by cytochrome P450: Exchange of iron–oxygen intermediates with water, *Biochem. Biophys. Res. Commun.* **104**:620–625.

83. Kexel, H., Schmelz, E., and Schmidt, H.-L., 1977, Oxygen transfer in microsomal oxidative desulfuration, in: *Microsomes and Drug Oxidations* (V. Ullrich, I. Roots, A. Hildebrandt, R. W. Estabrook, and A. H. Conney, eds.), Pergamon Press, Elmsford, NY, pp. 269–274.

84. Hayano, M., Lindberg, M. C., Dorfman, R. I., Hancock, J. E. H., and von Doering, W. E., 1955, On the mechanism of the C-1 1β-hydroxylation of steroids: A study with H_2O^{18} and O_2^{18}, *Arch. Biochem. Biophys.* **59**:529–532.

85. Hayano, M., Saito, A., Stone, D., and Dorfman, R. I., 1956, Hydroxylation of steroids by microorganisms in the presence of $^{18}O_2$, *Biochim. Biophys. Acta* **21**:380–381.

86. Watanabe, Y., Iyanagi, T., and Oae, S., 1980, Kinetic study on enzymatic S-oxygenation promoted by a reconstituted system with purified cytochrome P450, *Tetrahedron Lett.* **21**:3685–3688.

87. Watanabe, Y., Numata, T., Iyanagi, T., and Oae, S., 1981, Enzymatic oxidation of alkyl sulfides by cytochrome P450 and hydroxyl radical, *Bull. Chem. Soc. Jpn.* **54**:1163–1170.

88. Nordblom, G. D., White, R. E., and Coon, M. J., 1976, Studies on hydroperoxide-dependent substrate hydroxylation by purified liver microsomal cytochrome P450, *Arch. Biochem. Biophys.* **175**:524–533.

89. Schardt, B. C., and Hill, C. L., 1983, Preparation of iodobenzene dimethoxide: A new synthesis of [^{18}O]iodosylbenzene and a reexamination of its infrared spectrum, *Inorg. Chem.* **22**:1563–1565.

90. White, R. E., and McCarthy, M.-B., 1984, Aliphatic hydroxylation by cytochrome P450: Evidence for rapid hydrolysis of an intermediate iron–nitrene complex, *J. Am. Chem. Soc.* **106**:4922–4926.

91. White, R. E., 1987, Methanolysis of ((tosylimino)iodo)benzene, *Inorg. Chem.* **26**:3916–3919.

92. Svastits, E. W., Dawson, J. H., Breslow, R., and Gellman, S. H., 1985, Functionalized nitrogen atom transfer catalyzed by cytochrome P-450, *J. Am. Chem. Soc.* **107**:6427–6428.

93. Heimbrook, D. C., Murray, R. E., Egeberg, K. D., Sligar, S. G., Nee, M. W., and Bruice, T. C., 1984, Demethylation of N,N-dimethylaniline and p-cyano-N,N-dimethylaniline and their N-oxides by cytochromes $P450_{LM2}$ and $P450_{cam}$, *J. Am. Chem. Soc.* **106**:1514–1515.

94. Seto, Y., and Guengerich, F. P., 1993, Partitioning between N-dealkylation and N-oxygenation in the oxidation of N,N-dialkylarylamines catalyzed by cytochrome P450 2B1, *J. Biol. Chem.* **268**:9986–9997.

95. Bergstrom, S., Lindstredt, S., Samuelson, B., Corey, E. J., and Gregoriou, G. A., 1958, The stereochemistry of 7α-hydroxylation in the biosynthesis of cholic acid from cholesterol, *J. Am. Chem. Soc.* **80**:2337–2338.

96. Corey, E. J., Gregoriou, G. A., and Peterson, D. H., 1958, The stereochemistry of 11α-hydroxylation of steroids, *J. Am. Chem. Soc.* **80**:2338.

97. McMahon, R. E., Sullivan, H. R., Craig, J. C., and Pereira, W. E., 1969, The microsomal oxygenation of ethyl benzene: Isotopic, stereochemical, and induction studies, *Arch. Biochem. Biophys.* **132**:575–577.

98. Hamberg, M., and Bjorkhem, I., 1971, ω-Oxidation of fatty acids. I. Mechanism of microsomal ω1- and ω2-hydroxylation, *J. Biol. Chem.* **246**:7411–7416.

99. Shapiro, S., Piper, J. U., and Caspi, E., 1982, Steric course of hydroxylation at primary carbon atoms: Biosynthesis of 1-octanol from (1*R*)- and (1*S*)-[1-^{3}H,^{2}H,^{1}H; 1-^{14}C]octane by rat liver microsomes, *J. Am. Chem. Soc.* **104**:2301–2305.

100. Fretz, H., Woggon, W.-D., and Voges, R., 1989, The allylic oxidation of geraniol catalyzed by cytochrome $P450_{Cath}$, proceeding with retention of configuration, *Helv. Chim. Acta* **72**:391–400.

101. Groves, J. T., McClusky, G. A., White, R. E., and Coon, M. J., 1978, Aliphatic hydroxylation by highly purified liver microsomal cytochrome P450: Evidence for a carbon radical intermediate, *Biochem. Biophys. Res. Commun.* **81**:154–160.

102. Gelb, M. H., Heimbrook, D. C., Malkonen, P., and Sligar, S. G., 1982, Stereochemistry and deuterium isotope effects in camphor hydroxylation by the cytochrome $P450_{cam}$ monooxygenase system, *Biochemistry* **21**:370–377.

103. Fourneron, J. D., Archelas, A., and Furstoss, R., 1989, Microbial transformations. 10. Evidence for a carbon-radical intermediate in the biohydroxylations achieved by the fungus *Beauveria sulfurescens*, *J. Org. Chem.* **54**:2478–2483.

104. White, R. E., Miller, J. P., Favreau, L. V., and Bhattacharyaa, A., 1986, Stereochemical dynamics of aliphatic hydroxylation by cytochrome P450, *J. Am. Chem. Soc.* **108**:6024–6031.

105. Tanaka, K., Kurihara, N., and Nakajima, M., 1979, Oxidative metabolism of tetrachlorocyclohexenes, pentachlorocyclohexenes, and hexachlorocyclohexenes with microsomes from rat liver and house fly abdomen, *Pestic. Biochem. Physiol.* **10**:79–95.

106. Groves, J. T., and Subramanian, D. V., 1984, Hydroxylation by cytochrome P450 and metalloporphyrin models: Evidence for allylic rearrangement, *J. Am. Chem. Soc.* **106**:2177–2181.

107. Oliw, E. H., Brodowsky, I. D., Hörnsten, L., and Hamberg, M., 1993, Bis-allylic hydroxylation of polyunsaturated fatty acids by hepatic monooxygenases and its relation to the enzymatic and nonenzymatic formation of conjugated hydroxy fatty acids, *Arch. Biochem. Biophys.* **300**:434–439.

108. McClanahan, R. H., Huitric, A. C., Pearson, P. G., Desper, J. C., and Nelson, S. D., 1988, Evidence for a cytochrome P450 catalyzed allylic rearrangement with double bond topomerization, *J. Am. Chem. Soc.* **110**:1979–1981.

109. Griller, D., and Ingold, K. U., 1980, Free-radical clocks, *Acc. Chem. Res.* **13**:317–323.

110. White, R. E., Groves, J. T., and McClusky, G. A., 1979, Electronic and steric factors in regioselective hydroxylation catalyzed by purified cytochrome P450, *Acta Biol. Med. Ger.* **38**:475–482.

111. Sligar, S. G., Gelb, M. H., and Heimbrook, D. C., 1984, Bio-organic chemistry and cytochrome P450-dependent catalysis, *Xenobiotica* **14**:63–86.

112. Houghton, J. D., Beddows, S. E., Suckling, K. E., Brown, L., and Suckling, C. J., 1986, 5α,6α-Methanocholestan-3β-ol as a probe of the mechanism of action of cholesterol 7α-hydroxylase, *Tetrahedron Lett.* **27**:4655–4658.

113. Hoyte, R. M., and Hochberg, R. B., 1979, Enzymatic side chain cleavage of C-20 alkyl and aryl analogs of (20-*S*)-20-hydroxycholesterol. Implications for the biosynthesis of pregnenolone, *J. Biol. Chem.* **254**:2278–2286.

114. Strugnell, S., Calverley, M. J., and Jones, G., 1990, Metabolism of a cyclopropane-ring-containing analog of 1α-hydroxyvitamin D_3 in a hepatocyte cell model. Identification of 24-oxidized metabolites, *Biochem. Pharmacol.* **40**:333–341.

115. Holland, H. L., Chernishenko, M. J., Conn, M., Munoz, A., Manoharan, T. S., and Zawadski, M. A., 1990, Enzymic hydroxylation and sulfoxidation of cyclopropyl compounds by fungal biotransformation, *Can. J. Chem.* **68**:696–700.

116. Ortiz de Montellano, P. R., and Stearns, R. A., 1987, Timing of the radical recombination step in cytochrome P450 catalysis with ring-strained probes, *J. Am. Chem. Soc.* **109**:3415–3420.

117. Bowry, V. W., and Ingold, K. U., 1991, A radical clock investigation of microsomal cytochrome P450 hydroxylation of hydrocarbons. Rate of oxygen rebound, *J. Am. Chem. Soc.* **113**:5699–5707.

118. Atkinson, J. K., and Ingold, K. U., 1993, Cytochrome P450 hydroxylation of hydrocarbons: Variation in the rate of oxygen rebound using cyclopropyl radical clocks including two new ultrafast probes, *Biochemistry* **32**:9209–9214.

119. Foster, A. B., 1985, Deuterium isotope effects in the metabolism of drugs and xenobiotics: Implications for drug design, *Adv. Drug Res.* **14**:2–40.

120. Lu, A. Y. H., Harada, N., and Miwa, G. T., 1984, Rate-limiting steps in cytochrome P450-catalyzed reactions: Studies on isotope effects in the O-deethylation of 7-ethoxycoumarin, *Xenobiotica* **14**:19–26.

121. Simons, J. W., and Rabinovitch, B. S., 1963, Deuterium isotope effects in rates of methylene radical insertion into carbon–hydrogen bonds and across carbon–carbon double bonds, *J. Am. Chem. Soc.* **85**:1023–1024.

122. Goldstein, M. J., and Dolbier, W. R., 1965, The intramolecular insertion mechanism of α-haloneopentyl lithium, *J. Am. Chem. Soc.* **87**:2293–2295.

123. O'Ferrall, R. A. M., 1970, Model calculations of hydrogen isotope effects for nonlinear transition states, *J. Chem. Soc. B* **1970**:785–790.

124. Hjelmeland, L. M., Aronow, L., and Trudell, J. R., 1977, Intramolecular determination of primary kinetic isotope effects in hydroxylations catalyzed by cytochrome P450, *Biochem. Biophys. Res. Commun.* **76**:541–549.

125. Foster, A. B., Jarman, M., Stevens, J. D., Thomas, P., and Westwood, J. H., 1974, Isotope effects in O- and N-demethylations mediated by rat liver microsomes: An application of direct insertion electron impact mass spectrometry, *Chem. Biol. Interact.* **9**:327–340.

126. Miwa, G. T., Walsh, J. S., and Lu, A. Y. H., 1984, Kinetic isotope effects on cytochrome P450-catalyzed oxidation reactions: The oxidative O-dealkylation of 7-ethoxycoumarin, *J. Biol. Chem.* **259**:3000–3004.

127. Jones, J. P., Korzekwa, K. R., Rettie, A. E., and Trager, W. F., 1986, Isotopically sensitive branching and its effect on the observed intramolecular isotope effects in cytochrome P-450 catalyzed reactions: A new method for the estimation of intrinsic isotope effects, *J. Am. Chem. Soc.* **108**:7074–7078.

128. Jones, J. P., and Trager, W. F., 1987, The separation of the intramolecular isotope effect for the cytochrome P-450 catalyzed hydroxylation of *n*-octane into its primary and secondary components, *J. Am. Chem. Soc.* **109**:2171–2173.

129. Jones, J. P., Rettie, A. E., and Trager, W. F., 1990, Intrinsic isotope effects suggest that the reaction coordinate symmetry for the cytochrome P-450 catalyzed hydroxylation of octane is isozyme independent, *J. Med. Chem.* **33**:1242–1246.

130. Hanzlik, R. P., and Ling, K.-H. J., 1990, Active site dynamics of toluene hydroxylation by cytochrome P-450, *J. Org. Chem.* **55**:3992–3997.

131. Hanzlik, R. P., and Ling, K.-H. J., 1993, Active site dynamics of xylene hydroxylation by cytochrome P-450 as revealed by kinetic deuterium isotope effects, *J. Am. Chem. Soc.* **115**:9363–9370.

132. Holland, H. L., Brown, F. M., and Conn, M., 1990, Side chain hydroxylation of aromatic compounds by fungi. Part 4. Influence of the para substituent on kinetic isotope effects during benzylic hydroxylation of Mortierella isabellina, *J. Chem. Soc. Perkin Trans. 2* **1990**:1651–1665.

133. Sugiyama, K., and Trager, W. F., 1986, Prochiral selectivity and intramolecular isotope effects in the cytochrome P450 catalyzed ω-hydroxylation of cumene, *Biochemistry* **25**:7336–7343.

134. Frommer, U., Ullrich, V., and Staudinger, H., 1970, Hydroxylation of aliphatic compounds by liver microsomes. I. The distribution of isomeric alcohols, *Hoppe-Seylers Z. Physiol. Chem.* **351**:903–912.

135. White, R. E., McCarthy, M.-B., Egeberg, K. D., and Sligar, S. G., 1984, Regioselectivity in the cytochromes P450: Control by protein constraints and by chemical reactivities, *Arch. Biochem. Biophys.* **228**:493–502.

136. Korzekwa, K. R., Jones, J. P., and Gillette, J. R., 1990, Theoretical studies on cytochrome P450 mediated hydroxylation: A predictive model for hydrogen atom abstractions, *J. Am. Chem. Soc.* **112**:7042–7046.

137. Stearns, R. A., and Ortiz de Montellano, P. R., 1985, Cytochrome P450-catalyzed oxidation of quadricyclane: Evidence for a radical cation intermediate, *J. Am. Chem. Soc.* **107**:4081–4082.

138. Gassman, P. G., and Yamaguchi, R., 1982, Electron transfer from highly strained polycyclic molecules, *Tetrahedron* **38**:1113–1122.

139. Watabe, T., and Akamatsu, K., 1974, Microsomal epoxidation of *cis*-stilbene: Decrease in epoxidase activity related to lipid peroxidation, *Biochem. Pharmacol.* **23**:1079–1085.

140. Watabe, T., Ueno, Y., and Imazumi, J., 1971, Conversion of oleic acid into *threo*-dihydroxystearic acid by rat liver microsomes, *Biochem. Pharmacol.* **20**:912–913.

141. Ortiz de Montellano, P. R., Mangold, B. L. K., Wheeler, C., Kunze, K. L., and Reich, N. O., 1983, Stereochemistry of cytochrome P450-catalyzed epoxidation and prosthetic heme alkylation, *J. Biol. Chem.* **258**:4202–4207.

142. Hanzlik, R. P., and Shearer, G. O., 1978, Secondary deuterium isotope effects on olefin epoxidation by cytochrome P450, *Biochem. Pharmacol.* **27**:1441–1444.

143. Hanzlik, R. P., and Shearer, G. O., 1975, Transition state structure for peracid epoxidation: Secondary deuterium isotope effects, *J. Am. Chem. Soc.* **97:**5231–5233.

144. Henschler, D., Hoos, W. R., Fetz, H., Dallmeier, E., and Metzler, M., 1979, Reactions of trichloroethylene epoxide in aqueous systems, *Biochem. Pharmacol.* **28:**543–548.

145. Miller, R. E., and Guengerich, F. P., 1982, Oxidation of trichloroethylene by liver microsomal cytochrome P450: Evidence for chlorine migration in a transition state not involving trichloroethylene oxide, *Biochemistry* **21:**1090–1097.

146. Liebler, D. C., and Guengerich, F. P., 1983, Olefin oxidation by cytochrome P450: Evidence for group migration in catalytic intermediates formed with vinylidene chloride and trans-1-phenyl-1-butene, *Biochemistry* **22:**5482–5489.

147. Mansuy, D., Leclaire, J., Fontecave, M., and Momenteau, M., 1984, Oxidation of monosubstituted olefins by cytochromes P450 and heme models: Evidence for the formation of aldehydes in addition to epoxides and allylic alcohols, *Biochem. Biophys. Res. Commun.* **119:**319–325.

148. Miller, V. P., Fruetel, J. A., and Ortiz de Montellano, P. R., 1992, Cytochrome P450$_{cam}$-catalyzed oxidation of a hypersensitive radical probe, *Arch. Biochem. Biophys.* **298:**697–702.

149. Groves, J. T., Avaria-Neisser, G. E., Fish, K. M., Imachi, M., and Kuczkowski, R. L., 1986, Hydrogen–deuterium exchange during propylene epoxidation by cytochrome P-450, *J. Am. Chem. Soc.* **108:**3837–3838.

150. Bäckvall, J.-E., Bökman, F., and Blomberg, M. R. A., 1992, Metallaoxetanes as possible intermediates in metal-promoted deoxygenation of epoxides and epoxidation of olefins, *J. Am. Chem. Soc.* **114:**534–538.

151. Ortiz de Montellano, P. R., and Kunze, K. L., 1981, Shift of the acetylenic hydrogen during chemical and enzymatic oxidation of the biphenylacetylene triple bond, *Arch. Biochem. Biophys.* **209:**710–712.

152. Ortiz de Montellano, P. R., 1985, Alkenes and alkynes, in: *Bioactivation of Foreign Compounds* (M. W. Anders, ed.), Academic Press, New York, pp. 121–155.

153. McMahon, R. E., Turner, J. C., Whitaker, G. W., and Sullivan, H. R., 1981, Deuterium isotope effect in the biotransformation of 4-ethynylbiphenyls to 4-biphenylacetic acids by rat hepatic microsomes, *Biochem. Biophys. Res. Commun.* **99:**662–667.

154. Ortiz de Montellano, P. R., and Komives, E. A., 1985, Branchpoint for heme alkylation and metabolite formation in the oxidation of aryl acetylenes by cytochrome P450, *J. Biol. Chem.* **260:**3330–3336.

155. Ostovic, D., and Bruice, T. C., 1992, Mechanism of alkene epoxidation by iron, chromium, and manganese higher valent oxo-metalloporphyrins, *Acc. Chem. Res.* **25:**314–320.

156. Jerina, D. M., and Daly, J. W., 1974, Arene oxides: A new aspect of drug metabolism, *Science* **185:**573–582.

157. Riley, P., and Hanzlik, R. P., 1994, Electron transfer in P450 mechanisms. Microsomal metabolism of cyclopropylbenzene and *p*-cyclopropylanisole, *Xenobiotica* **24:**1–16.

158. Tomaszewski, J. E., Jerina, D. M., and Daly, J. W., 1975, Deuterium isotope effects during formation of phenols by hepatic monooxygenases: Evidence for an alternative to the arene oxide pathway, *Biochemistry* **14:**2024–2030.

159. Preston, B. D., Miller, J. A., and Miller, E. C., 1983, Non-arene oxide aromatic ring hydroxylation of 2,2′,5,5′-tetrachlorobiphenyl as the major metabolic pathway catalyzed by phenobarbital-induced rat liver microsomes, *J. Biol. Chem.* **258:**8304–8311.

160. Hanzlik, R. P., Hogberg, K., and Judson, C. M., 1984, Microsomal hydroxylation of specifically deuterated monosubstituted benzenes: Evidence for direct aromatic hydroxylation, *Biochemistry* **23:**3048–3055.

161. Korzekwa, K. R., Swinney, D. C., and Trager, W. F., 1989, Isotopically labeled chlorobenzenes as probes for the mechanism of cytochrome P-450 catalyzed aromatic hydroxylation, *Biochemistry* **28:**9019–9027.

162. Rietjens, I. M. C. M., Soffers, A. E. M. F., Veeger, C., and Vervoort, J., 1993, Regioselectivity of cytochrome P-450 catalyzed hydroxylation of fluorobenzenes predicted by calculated frontier orbital substrate characteristics, *Biochemistry* **32:**4801–4812.

163. Hinson, J. A., Nelson, S. D., and Mitchell, J. R., 1977, Studies on the microsomal formation of arylating metabolites of acetaminophen and phenacetin, *Mol. Pharmacol.* **13**:625–633.

164. Hinson, J. A., Nelson, S. D., and Gillette, J. R., 1979, Metabolism of [p-^{18}O]-phenacetin: The mechanism of activation of phenacetin to reactive metabolites in hamsters, *Mol. Pharmacol.* **15**:419–427.

165. Koymans, L., Lenthe, J. H. V., Den Kelder, G. M. D., and Vermeulen, N. P. E., 1990, Mechanisms of activation of phenacetin to reactive metabolites by cytochrome P-450: A theoretical study involving radical intermediates, *Mol. Pharmacol.* **37**:452–460.

166. Veronese, M. E., McLean, S., D'Souze, C. A., and Davies, N. W., 1985, Formation of reactive metabolites of phenacetin in humans and rats, *Xenobiotica* **15**:929–940.

167. Rietjens, I. M. C. M., Tyrakowska, B., Veeger, C., and Vervoort, J., 1990, Reaction pathways for biodehalogenation of fluorinated anilines, *Eur. J. Biochem.* **194**:945–954.

168. den Besten, C., van Bladeren, P. J., Duizer, E., Vervoort, J., and Rietjens, I. M. C. M., 1993, Cytochrome P450-mediated-oxidation of pentafluorophenol to tetrafluorobenzoquinone as the primary reaction product, *Chem. Res. Toxicol.* **6**:674–680.

169. Morgan, P., Maggs, J. L., Page, P. C. B., and Park, B. K., 1992, Oxidative dehalogenation of 2-fluoro-17α-ethynyloestradiol in vivo. A distal structure–metabolism relationship of 17α-ethynyla-tion, *Biochem. Pharmacol.* **44**:1717–1724.

170. Ohe, T., Mashino, T., and Hirobe, M., 1994, Novel metabolic pathway of arylethers by cytochrome P450: Cleavage of the oxygen–aromatic ring bond accompanying *ipso*-substitution by the oxygen atom of the active species in cytochrome P450 models and cytochrome P450, *Arch. Biochem. Biophys.* **310**:402–409.

171. Rietjens, I. M. C., and Vervoort, J., 1992, A new hypothesis for the mechanism for cytochrome P-450 dependent aerobic conversion of hexahalogenated benzenes to pentahalogenated phenols, *Chem. Res. Toxicol.* **5**:10–19.

172. Stadler, R., and Zenk, M. H., 1993, The purification and characterization of a unique cytochrome P-450 enzyme from *Berberis stolonifera* plant cell cultures, *J. Biol. Chem.* **268**:823–831.

173. Amann, T., and Zenk, M. H., 1991, Formation of the morphine precursor salutaridine is catalyzed by a cytochrome P-450 enzyme in mammalian liver, *Tetrahedron Lett.* **32**:3675–3678.

174. Zenk, M. H., Gerardy, R., and Stadler, R., 1989, Phenol oxidative coupling of benzylisoquinoline alkaloids is catalyzed by regio- and stereo-selective cytochrome P-450 linked plant enzymes: Salutaridine and berbamunine, *J. Chem. Soc. Chem. Commun.* **1989**:1725–1727.

175. Cavalieri, E. L., and Rogan, E. G., 1992, The approach to understanding aromatic hydrocarbon carcinogenesis. The central role of radical cations in metabolic activation, *Pharmacol. Ther.* **55**:183–199.

176. Cavalieri, E., Rogan, E., Roth, R. W., Saugier, R. K., and Hakam, A., 1983, The relationship between ionization potential and horseradish peroxidase/hydrogen peroxide-catalyzed binding of aromatic hydrocarbons to DNA, *Chem. Biol. Interact.* **47**:87–109.

177. Devanesan, P., Rogan, E., and Cavalieri, E., 1987, The relationship between ionization potential and prostaglandin H synthase-catalyzed binding of aromatic hydrocarbons to DNA, *Chem. Biol. Interact.* **61**:89–95.

178. Khandwala, A. S., and Kasper, C. B., 1973, Preferential induction of aryl hydroxylase activity in rat liver nuclear envelope by 3-methylcholanthrene, *Biochem. Biophys. Res. Commun.* **54**:1241–1246.

179. Cavalieri, E. L., Rogan, E. G., Devanesan, P. D., Cremonesi, P., Cerny, R. L., Gross, M. L., and Bodell, W. J., 1990, Binding of benzo[a]pyrene to DNA by cytochrome P450-catalyzed one-electron oxidation in rat liver microsomes and nuclei, *Biochemistry* **29**:4820–4827.

180. Rogan, E. G., Devanesan, P. D., RamaKrishna, N. V. S., Higginbotham, S., Padmavathi, N. S., Chapman, K., Cavalieri, E. L., Jeong, H., Jankowiak, R., and Small, G. J., 1993, Identification and quantitation of benzo[a]pyrene–DNA adducts formed in mouse skin, *Chem. Res. Toxicol.* **6**:356–363.

181. Cavalieri, E. L., Rogan, E. G., Cremonesi, P., and Devanesan, P. D., 1988, Radical cations as precursors in the metabolic formation of quinones from benzo[a]pyrene and 6-fluorobenzo[a]pyrene. Fluoro

substitution as a probe for one-electron oxidation in aromatic substrates, *Biochem. Pharmacol.* **37:**2173–2182.

182. Tolbert, L. M., Khanna, R. K., Popp, A. E., Gelbaum, L., and Bottomley, L. A., 1990, Stereoelectronic effects in the deprotonation of arylalkyl radical cations: *meso*-ethylanthracenes, *J. Am. Chem. Soc.* **112:**2373–2378.

183. Guengerich, F. P., Niwa, T., Anzenbacher, P., Sirrimane, S., and Tolbert, L. M., 1994, Oxidation of 9-alkyl anthracene derivatives by cytochrome P450 and peroxidases, *FASEB J.* **8:**A1380.

184. Hlavica, P., and Künzel-Mulas, U., 1993, Metabolic N-oxide formation by rabbit liver microsomal cytochrome P-4502B4: Involvement of superoxide in the NADPH-dependent N-oxygenation of N,N-dimethylaniline, *Biochim. Biophys. Acta* **1158:**83–90.

185. Guengerich, F. P., 1984, Oxidation of sparteines by cytochrome P-450: Evidence against the formation of N-oxides, *J. Med. Chem.* **27:**1101–1103.

186. Kurebayashi, H., 1989, Kinetic deuterium isotope effects on deamination and N-hydroxylation of cyclohexylamine by rabbit liver microsomes, *Arch. Biochem. Biophys.* **270:**320–329.

187. Augusto, O., Beilan, H. S., and Ortiz de Montellano, P. R., 1982, The catalytic mechanism of cytochrome P450: Spin-trapping evidence for one electron substrate oxidation, *J. Biol. Chem.* **257:**11288–11295.

188. Kennedy, C. H., and Mason, R. P., 1990, A reexamination of the cytochrome P-450-catalyzed free radical production from dihydropyridine: Evidence of trace transition metal catalysis, *J. Biol. Chem.* **265:**11425–11428.

189. Abdel-Monem, M. M., 1975, Isotope effects in enzymatic N-demethylation of tertiary amines, *J. Med. Chem.* **18:**427–430.

190. Miwa, G. T., Garland, W. A., Hodshon, B. J., Lu, A. Y. H., and Northrop, D. B., 1980, Kinetic isotope effects in cytochrome P450-catalyzed oxidation reactions: Intermolecular and intramolecular deuterium isotope effects during the N-demethylation of N,N-dimethylphentermine, *J. Biol. Chem.* **255:**6049–6054.

191. Miwa, G. T., Walsh, J. S., Kedderis, G. L., and Hollenberg, P. F., 1983, The use of intramolecular isotope effects to distinguish between deprotonation and hydrogen atom abstraction mechanisms in cytochrome P450- and peroxidase-catalyzed N-demethylation reactions, *J. Biol. Chem.* **258:**14445–14449.

192. Dopp, D., and Heufer, J., 1982, N-Demethylation of N,N-dimethylaniline by photoexcited 3-nitrochlorobenzene, *Tetrahedron Lett.* **23:**1553–1556.

193. Shono, T., Toda, T., and Oshino, N., 1982, Electron transfer from nitrogen in microsomal oxidation of amine and amide: Simulation of microsomal oxidation by anodic oxidation, *J. Am. Chem. Soc.* **104:**2639–2641.

194. Galliani, G., Nali, M., Rindone, B., Tollari, S., Rocchetti, M., and Salmona, M., 1986, The rate of N-demethylation of N,N-dimethylanilines and N-methylanilines by rat-liver microsomes is related to their ionization potential, their lipophilicity and to a steric bulk factor, *Xenobiotica* **16:**511–517.

195. Burstyn, J. N., Iskandar, M., Brady, J. F., Fukuto, J. M., and Cho, A. K., 1991, Comparative studies of N-hydroxylation and N-demethylation by microsomal cytochrome P-450, *Chem. Res. Toxicol.* **4:**70–76.

196. Macdonald, T. L., Gutheim, W. G., Martin, R. B., and Guengerich, F. P., 1989, Oxidation of substituted N,N-dimethylanilines by cytochrome P-450: Estimation of the effective oxidation–reduction potential of cytochrome P-450, *Biochemistry* **28:**2071–2077.

197. Burka, L. T., Guengerich, F. P., Willard, R. J., and Macdonald, T. L., 1985, Mechanism of cytochrome P-450 catalysis. Mechanism of N-dealkylation and amine oxide deoxygenation, *J. Am. Chem. Soc.* **107:**2549–2551.

198. Hall, L. R., and Hanzlik, R. P., 1989, Kinetic deuterium isotope effects on the N-demethylation of tertiary amides by cytochrome P-450, *J. Biol. Chem.* **264:**12349–12355.

199. Constantino, L., Rosa, E., and Iley, J., 1992, The microsomal demethylation of N,N-dimethylbenzamides. Substituent and kinetic deuterium isotope effects, *Biochem. Pharmacol.* **44:**651–658.

200. Iley, J., Constantino, L., Norberto, F., and Rosa, E., 1990, Oxidation of the methyl groups of N,N-dimethylbenzamides by a cytochrome P450 mono-oxygenase model system, *Tetrahedron Lett.* **31**:4921–4922.

201. Hall, L. R., Iwamoto, R. T., and Hanzlik, R. P., 1989, Electrochemical models for cytochrome P-450. N-Demethylation of tertiary amides by anodic oxidation, *J. Org. Chem.* **54**:2446–2451.

202. Dinnocenzo, J. P., Karki, S. B., and Jones, J. P., 1993, On isotope effects for the cytochrome P-450 oxidation of substituted N,N-dimethylanilines, *J. Am. Chem. Soc.* **115**:7111–7116.

203. Dinnocenzo, J. P., and Banach, T. E., 1989, Deprotonation of tertiary amine cation radicals. A direct experimental approach, *J. Am. Chem. Soc.* **111**:8646–8653.

204. Nelsen, S. F., and Ippoliti, J. T., 1986, On the deprotonation of trialkylamine cation radicals by amines, *J. Am. Chem. Soc.* **108**:4879–4881.

205. Okazaki, O., and Guengerich, F. P., 1993, Evidence for specific base catalysis in dealkylation reactions catalyzed by cytochrome P450 and chloroperoxidase. Differences in rates of deprotonation of aminium radicals as an explanation for high kinetic hydrogen isotope effects observed with peroxidases, *J. Biol. Chem.* **268**:1546–1552.

206. Ichinose, R., and Kurihara, N., 1987, Intramolecular deuterium isotope effect and enantiotopic differentiation in oxidative demethylation of chiral [monomethyl-D3]methoxychlor in rat liver microsomes, *Biochem. Pharmacol.* **36**:3751–3756.

207. Hanzlik, R. P., and Tullman, R. H., 1982, Suicidal inactivation of cytochrome P450 by cyclopropylamines: Evidence for cation-radical intermediates, *J. Am. Chem. Soc.* **104**:2048–2050.

208. Macdonald, T. L., Zirvi, K., Burka, L. T., Peyman, P., and Guengerich, F. P., 1982, Mechanism of cytochrome P450 inhibition by cyclopropylamines, *J. Am. Chem. Soc.* **104**:2050–2052.

209. Tullman, R. H., and Hanzlik, R. P., 1984, Inactivation of cytochrome P450 and monoamine oxidase by cyclopropylamines, *Drug Metab. Dispos.* **15**:1163–1182.

210. Guengerich, F. P., Willard, R. J., Shea, J. P., Richards, L. E., and Macdonald, T. L., 1984, Mechanism-based inactivation of cytochrome P450 by heteroatom-substituted cyclopropanes and formation of ring opened products, *J. Am. Chem. Soc.* **106**:6446–6447.

211. Bondon, A., Macdonald, T. L., Harris, T. M., and Guengerich, F. P., 1989, Oxidation of cycloalkylamines by cytochrome P-450. Mechanism-based inactivation, adduct formation, ring expansion, and nitrone formation, *J. Biol. Chem.* **264**:1988–1997.

212. Watanabe, Y., Iyanagi, T., and Oae, S., 1982, One electron transfer mechanism in the enzymatic oxygenation of sulfoxide to sulfone promoted by a reconstituted system with purified cytochrome P450, *Tetrahedron Lett.* **23**:533–536.

213. Powell, M. F., Wu, J. C., and Bruice, T. C., 1984, Ferricyanide oxidation of dihydropyridines and analogues, *J. Am. Chem. Soc.* **106**:3850–3856.

214. Alvarez, J. C., and Ortiz de Montellano, P. R., 1992, Thianthrene-5-oxide as a probe of the electrophilicity of hemoprotein oxidizing species, *Biochemistry* **31**:8315–8322.

215. Rettie, A. E., Rettenmeier, A. W., Howald, W. N., and Baillie, T. A., 1987, Cytochrome P-450-catalyzed formation of Δ^4-VPA, a toxic metabolite of valproic acid, *Science* **235**:890–893.

216. Rettie, A. E., Boberg, M., Rettenmeier, A. W., and Baillie, T. A., 1988, Cytochrome P-450-catalyzed desaturation of valproic acid in vitro. Species differences, induction effects, and mechanistic studies, *J. Biol. Chem.* **263**:13733–13738.

217. Porubeck, D. J., Barnes, H., Meier, G. P., Theodore, L. J., and Baillie, T. A., 1989, Enantiotopic differentiation during the biotransformation of valproic acid to the hepatotoxic olefin 2-*n*-propyl-4-pentenoic acid, *Chem. Res. Toxicol.* **2**:35–40.

218. Kassahun, K., and Baillie, T. A., 1993, Cytochrome P-450-mediated dehydrogenation of 2-n-propyl-2(E)-pentenoic acid, a pharmacologically-active metabolite of valproic acid, in rat liver microsomal preparations, *Drug Metab. Dispos.* **21**:242–248.

219. Nagata, K., Liberato, D. J., Gillette, J. R., and Sasame, H. A., 1986, An unusual metabolite of testosterone: 17β-hydroxy-4,6-androstadiene-3-one, *Drug Metab. Dispos.* **14**:559–565.

220. Aoyama, T., Korzekwa, K., Nagata, K., Gillette, J., Gelboin, H. V., and Gonzalez, F. J., 1989, cDNA-directed expression of rat testosterone 7α-hydroxylase using the modified vaccinia virus, T7-RNA-polymerase system and evidence for 6α-hydroxylation and Δ^6-testosterone formation, *Eur. J. Biochem.* **181:**331–336.

221. Korzekwa, K. R., Trager, W. F., Nagata, K., Parkinson, A., and Gillette, J. R., 1990, Isotope effect studies on the mechanism of the cytochrome P-450IIA1-catalyzed formation of Δ^6-testosterone from testosterone, *Drug Metab. Dispos.* **18:**974–979.

222. Chadwick, R. W., Chuang, L. T., and Williams, K., 1975, Dehydrogenation: A previously unreported pathway of lindane metabolism in mammals, *Pestic. Biochem. Physiol.* **5:**575–586.

223. Hata, S., Nishino, T., Katsuki, H., Aoyama, Y., and Yoshida, Y., 1983, Two species of cytochrome P-450 involved in ergosterol biosynthesis of yeast, *Biochem. Biophys. Res. Commun.* **116:**162–166.

224. Mochizuki, H., Suhara, K., and Katagiri, M., 1992, Steroid 6β-hydroxylase and 6-desaturase reactions catalyzed by adrenocortical mitochondrial P-450, *J. Steroid Biochem. Mol. Biol.* **42:**95–101.

225. Boyd, D. R., Sharma, N. D., Agarwal, R., McMordie, R. A. S., Bessems, J. G. M., van Ommen, B., and van Bladeren, P. J., 1993, Biotransformation of 1,2-dihydronaphthalene and 1,2-dihydroanthracene by rat liver microsomes and purified cytochromes P-450. Formation of arene hydrates of naphthalene and anthracene, *Chem. Res. Toxicol.* **6:**808–812.

226. Kaminsky, L. S., Fasco, M. J., and Guengerich, F. P., 1980, Comparison of different forms of purified cytochrome P-450 from rat liver by immunological inhibition of regio- and stereoselective metabolism of warfarin, *J. Biol. Chem.* **255:**85–91.

227. Vyas, K. P., Kari, P. H., Prakash, S. R., and Duggan, D. E., 1990, Biotransformation of lovastatin. II. *In vitro* metabolism by rat and mouse liver microsomes and involvement of cytochrome P-450 in dehydrogenation of lovastatin, *Drug Metab. Dispos.* **18:**218–222.

228. Wang, R. W., Kari, P. H., Lu, A. Y. H., Thomas, P. E., Guengerich, F. P., and Vyas, K. P., 1991, Biotransformation of lovostatin. IV. Identification of cytochrome P450 3A proteins as the major enzymes responsible for the oxidative metabolism of lovostatin in rat and human liver microsomes, *Arch. Biochem. Biophys.* **290:**355–361.

229. Vickers, S., and Duncan, C. A., 1991, Studies on the metabolic inversion of the 6′ chiral center of simvastatin, *Biochem. Biophys. Res. Commun.* **181:**1508–1515.

230. Watanabe, Y., and Ishimura, Y., 1989, Aromatization of tetralone derivatives by FeIIIPFP(Cl)/PhIO and cytochrome P-450$_{cam}$: A model study on aromatase cytochrome P-450 reaction, *J. Am. Chem. Soc.* **111:**410–411.

231. Lee, J. S., Jacobsen, N. E., and Ortiz de Montellano, P. R., 1988, 4-Alkyl radical extrusion in the cytochrome P-450-catalyzed oxidation of 4-alkyl-1,4-dihydropyridines, *Biochemistry* **27:**7703–7710.

232. Guengerich, F. P., Brian, W. R., Iwasaki, M., Sari, M.-A., Bäärnhielm, C., and Berntsson, P., 1991, Oxidation of dihydropyridine calcium channel blockers and analogues by human liver cytochrome P-450 IIIA4, *J. Med. Chem.* **34:**1838–1844.

233. Nelson, S. D., Forte, A. J., and Dhalin, D. C., 1980, Lack of evidence for N-hydroxyacetaminophen as a reactive metabolite of acetaminophen *in vitro*, *Biochem. Pharmacol.* **29:**1617–1620.

234. Wood, A. W., Swinney, D. C., Thomas, P. E., Ryan, D. E., Hall, P. F., Levin, W., and Garland, W. A., 1988, Mechanism of androstenedione formation from testosterone and epitestosterone catalyzed by purified cytochrome P-450b, *J. Biol. Chem.* **263:**17322–17332.

235. Swinney, D. C., Ryan, D. E., Thomas, P. E., and Levin, W., 1988, Evidence for concerted kinetic oxidation of progesterone by purified rat hepatic cytochrome P-450g, *Biochemistry* **27:**5461–5470.

236. Terelius, Y., Norsten-Höög, C., Cronholm, T., and Ingelman-Sundberg, M., 1991, Acetaldehyde as a substrate for ethanol-inducible cytochrome P450 (CYP2E1), *Biochem. Biophys. Res. Commun.* **179:**689–694.

237. Tomita, S., Tsujita, M., Matsuo, Y., Yubisui, T., and Ichikawa, Y., 1993, Identification of a microsomal retinoic acid synthase as a microsomal cytochrome P-450-linked monooxygenase system, *Int. J. Biochem.* **25:**1775–1784.

238. Dahlbäck, H., and Holmberg, I., 1990, Oxidation of 5β-cholestane-3α,7α,12α-triol into 3α,7α,12α-trihydroxy-5β-cholestanoic acid by cytochrome P-450$_{26}$ from rabbit liver mitochondria, *Biochem. Biophys. Res. Commun.* **167**:391–395.

239. Watanabe, K., Narimatsu, S., Yamamoto, I., and Yoshimura, H., 1991, Oxygenation mechanism in conversion of aldehyde to carboxylic acid catalyzed by a cytochrome P-450 isozyme, *J. Biol. Chem.* **266**:2709–2711.

240. Matsunaga, T., Watanabe, K., Yamamoto, I., Negishi, M., Gonzalez, F. J., and Yoshimura, H., 1994, cDNA cloning and sequence of CYP2C29 encoding P-450 MUT-2, a microsomal aldehyde oxygenase, *Biochim. Biophys. Acta* **1184**:299–301.

241. Watanabe, K., Matsunaga, T., Narimatsu, S., Yamamoto, I., and Yoshimura, H., 1992, Mouse hepatic microsomal oxidation of aliphatic aldehydes (C_8 to C_{11}) to carboxylic acids, *Biochem. Biophys. Res. Commun.* **188**:114–119.

242. Watanabe, K., Narimatsu, S., Matsunaga, T., Yamamoto, I., and Yoshimura, H., 1993, A cytochrome P450 isozyme having aldehyde oxygenase activity plays a major role in metabolizing cannabinoids by mouse hepatic microsomes, *Biochem. Pharmacol.* **46**:405–411.

243. Davis, S. C., Sui, Z., Peterson, J. A., and Ortiz de Montellano, P. R., 1995, Oxidation of fatty acid aldehydes by cytochrome P450$_{BM-3}$ (CYP102), *FASEB J.* **9**:A1490.

244. Matsunaga, T., Iwawaki, Y., Watanabe, K., Narimatsu, S., Yamamoto, I., Imaoka, S., Funae, Y., and Yoshimura, H., 1993, Cytochrome P450 isozymes catalyzing the hepatic microsomal oxidation of 9-anthraldehyde to 9-anthracene carboxylic acid in adult male rats, *Biol. Pharm. Bull.* **16**:866–869.

245. Vaz, A. D. N., Roberts, E. S., and Coon, M. J., 1991, Olefin formation in the oxidative deformylation of aldehydes by cytochrome P-450. Mechanistic implications for catalysis by oxygen-derived peroxide, *J. Am. Chem. Soc.* **113**:5886–5887.

246. Roberts, E. S., Vaz, A. D. N., and Coon, M. J., 1991, Catalysis by cytochrome P-450 of an oxidative reaction in xenobiotic aldehyde metabolism: Deformylation with olefin formation, *Proc. Natl. Acad. Sci. USA* **88**:8963–8966.

247. Vaz, A. D. N., Kessell, K. J., and Coon, M. J., 1994, Aromatization of a bicyclic steroid analog, 3-oxodecalin-4-ene-10-carboxaldehyde, by liver microsomal cytochrome P450 2B4, *Biochemistry* **33**:13651–13661.

248. Watanabe, Y., Takehira, K., Shimizu, M., Hayakawa, T., and Orita, H., 1990, Oxidation of aldehydes by an iron(III) porphyrin complex–*m*-chloroperbenzoic acid system, *J. Chem. Soc. Chem. Commun.* **1990**:927–928.

249. Wada, A., Okamoto, M., Nonaka, Y., and Yamano, T., 1984, Aldosterone biosynthesis by a reconstituted cytochrome P450$_{11β}$ system, *Biochem. Biophys. Res. Commun.* **119**:365–371.

250. Lambeth, J. D., Kitchen, S. E., Farooqui, A. A., Tuckey, R., and Kamin, H., 1982, Cytochrome P450$_{scc}$–substrate interactions: Studies of binding and catalytic activity using hydroxycholesterols, *J. Biol. Chem.* **257**:1876–1884.

251. Tuckey, R. C., and Kamin, H., 1983, Kinetics of O_2 and CO binding to adrenal cytochrome P450scc: Effect of cholesterol, intermediates, and phosphatidylcholine vesicles, *J. Biol. Chem.* **258**:4232–4237.

252. Tuckey, R. C., and Kamin, H., 1982, The oxyferro complex of adrenal cytochrome P450$_{scc}$: Effect of cholesterol and intermediates on its stability and optical characteristics, *J. Biol. Chem.* **257**:9309–9314.

253. Byon, C.-Y., and Gut, M., 1980, Steric considerations regarding the biodegradation of cholesterol to pregnenolone: Exclusion of (22S)-22-hydroxycholesterol and 22-ketocholesterol as intermediates, *Biochem. Biophys. Res. Commun.* **94**:549–552.

254. Burstein, S., Middleditch, B. S., and Gut, M., 1975, Mass spectrometric study of the enzymatic conversion of cholesterol to (22R)-22-hydroxycholesterol, (20R,22R)-20,22-dihydroxycholesterol, and pregnenolone, and of (22R)-22-hydroxycholesterol to the glycol and pregnenolone in bovine adrenocortical preparations, *J. Biol. Chem.* **250**:9028–9037.

255. Okamoto, T., Sasaki, K., and Oka, S., 1988, Biomimetic oxidation with molecular oxygen. Selective carbon–carbon bond cleavage of 1,2-diols by molecular oxygen and dihydropyridine in the presence of iron–porphyrin catalysts, *J. Am. Chem. Soc.* **110:**1187–1196.

256. Fishman, J., 1982, Biochemical mechanisms of aromatization, *Cancer Res.* **42:**3277s–3280s.

257. Thompson, E. A., and Siiteri, P. K., 1974, Utilization of oxygen and reduced nicotinamide adenine dinucleotide phosphate by human placental microsomes during aromatization of androstenedione, *J. Biol. Chem.* **249:**5364–5372.

258. Hagerman, D. D., 1987, Human placenta estrogen synthetase (aromatase) purified by affinity chromatography, *J. Biol. Chem.* **262:**2398–2400.

259. Nakajin, S., Shinoda, M., and Hall, P. F., 1986, Purification to homogeneity of aromatase from human placenta, *Biochem. Biophys. Res. Commun.* **134:**704–710.

260. Kellis, J. T., and Vickery, L. E., 1987, Purification and characterization of human placental aromatase cytochrome P-450, *J. Biol. Chem.* **262:**4413–4420.

261. Caspi, E., Arunachalam, T., and Nelson, P. A., 1983, Biosynthesis of estrogens: The steric mode of the initial C-19 hydroxylation of androgens by human placental aromatase, *J. Am. Chem. Soc.* **105:**6987–6989.

262. Caspi, E., Arunachalam, T., and Nelson, P. A., 1986, Biosynthesis of estrogens: Aromatization of (19R)-, (19S)-, and (19RS)-[19-^{3}H,^{2}H,^{1}H]-3β-hydroxyandrost-5-en-17-ones by human placental aromatase, *J. Am. Chem. Soc.* **108:**1847–1852.

263. Osawa, Y., Shibata, K., Rohrer, D., Weeks, C., and Duax, W. L., 1975, Reassignment of the absolute configuration of 19-substituted 19-hydroxysteroids and stereomechanism of estrogen biosynthesis, *J. Am. Chem. Soc.* **97:**4400–4402.

264. Arigoni, D., Battaglia, R., Akhtar, M., and Smith, T., 1975, Stereospecificity of oxidation at C-19 in oestrogen biosynthesis, *J. Chem. Soc. Chem. Commun.* **1975:**185–186.

265. Miyairi, S., and Fishman, J., 1983, Novel method of evaluating biological 19-hydroxylation and aromatization of androgens, *Biochem. Biophys. Res. Commun.* **117:**392–398.

266. Miyairi, S., and Fishman, J., 1985, Radiometric analysis of oxidative reactions in aromatization by placental microsomes: Presence of differential isotope effects, *J. Biol. Chem.* **260:**320–325.

267. Numazawa, M., Midzuhashi, K., and Nagaoka, M., 1994, Metabolic aspects of the 1β-proton and the 19-methyl group of androst-4-ene-3,6,17-trione during aromatization by placental microsomes and inactivation of aromatase, *Biochem. Pharmacol.* **47:**717–726.

268. Brodie, H. J., Kripalani, K. J., and Possanza, G., 1969, Studies on the mechanisms of estrogen biosynthesis. VI. The stereochemistry of hydrogen elimination at C-2 during aromatization, *J. Am. Chem. Soc.* **91:**1241–1242.

269. Fishman, J., and Guzik, H., 1969, Stereochemistry of estrogen biosynthesis, *J. Am. Chem. Soc.* **91:**2805–2806.

270. Fishman, J., Guzik, H., and Dixon, D., 1969, Stereochemistry of estrogen biosynthesis, *Biochemistry* **8:**4304–4309.

271. Fishman, J., and Raju, M. S., 1981, Mechanism of estrogen biosynthesis: Stereochemistry of C-1 hydrogen elimination in the aromatization of 2β-hydroxy-19-oxoandrostenedione, *J. Biol. Chem.* **256:**4472–4477.

272. Townsley, J. D., and Brodie, H. J., 1968, Studies on the mechanism of estrogen biosynthesis. III. The stereochemistry of aromatization of C_{19} and C_{18} steroids, *Biochemistry* **7:**33–40.

273. Osawa, Y., Yoshida, N., Fronckowiak, M., and Kitawaki, J., 1987, Immunoaffinity purification of aromatase cytochrome P450 from human placental microsomes, metabolic switching from aromatization to 1β and 2β-monohydroxylation, and recognition of aromatase isozymes, *Steroids* **50:**11–28.

274. Hosoda, H., and Fishman, J., 1974, Usually facile aromatization of 2β-hydroxy-19-oxo-4-androstene-3,17-dione to estrone: Implications in estrogen biosynthesis, *J. Am. Chem. Soc.* **96:**7325–7329.

275. Goto, J., and Fishman, J., 1977, Participation of a nonenzymatic transformation in the biosynthesis of estrogens from androgens, *Science* **195:**80–81.

276. Hahn, E. F., and Fishman, J., 1984, Immunological probe of estrogen biosynthesis: Evidence for the 2β-hydroxylative pathway in aromatization of androgens, *J. Biol. Chem.* **259**:1689–1694.

277. Akhtar, M., Calder, M. R., Corina, D. L., and Wright, J. N., 1982, Mechanistic studies on C-19 demethylation in oestrogen biosynthesis, *Biochem. J.* **201**:569–580.

278. Caspi, E., Wicha, J., Arunachalam, T., Nelson, P., and Spiteller, G., 1984, Estrogen biosynthesis: Concerning the obligatory intermediacy of 2β-hydroxy-10β-formylandrost-4-ene-3,17-dione, *J. Am. Chem. Soc.* **106**:7282–7283.

279. Cole, P. A., and Robinson, C. H., 1990, Conversion of 19-oxo[2β-^{2}H]androgens into oestrogens by human placental aromatase. An unexpected stereochemical outcome, *Biochem. J.* **268**:553–561.

280. Morand, P., Williamson, D. G., Layne, D. S., Lompa-Krzymien, L., and Salvador, J., 1975, Conversion of an androgen epoxide into 17β-estradiol by human placental microsomes, *Biochemistry* **14**:635–638.

281. Covey, D. F., and Hood, W. F., 1982, A new hypothesis based on suicide substrate inhibitor studies for the mechanism of action of aromatase, *Cancer Res.* **42**:3327s–3333s.

282. Beusen, D. D., and Covey, D. F., 1984, Study of the role of Schiff base formation in the aromatization of androgen substrates by human placenta, *Fed. Proc.* **43**:330.

283. Akhtar, M., Corina, D., Pratt, J., and Smith, T., 1976, Studies on the removal of C-19 in oestrogen biosynthesis using ^{18}O₂, *J. Chem. Soc. Chem. Commun.* **1976**:854–856.

284. Stevenson, D. E., Wright, J. N., and Akhtar, M., 1988, Mechanistic consideration of P450 dependent enzymic reactions: Studies on oestriol biosynthesis, *J. Chem. Soc. Perkin Trans. I* **1988**:2043–2052.

285. Akhtar, M., Njar, V. C. O., and Wright, J. N., 1993, Mechanistic studies on aromatase and related C–C bond cleaving enzymes, *J. Steroid Biochem. Mol. Biol.* **44**:375–387.

286. Akhtar, M., Corina, D., Miller, S., Shyadehi, A. Z., and Wright, J. N., 1994, Mechanism of the acyl-carbon cleavage and related reactions catalyzed by multifunctional P450s: Studies on cytochrome P45017α, *Biochemistry* **33**:4410–4418.

287. Cole, P. A., and Robinson, C. H., 1988, A peroxide model reaction for placental aromatase, *J. Am. Chem. Soc.* **110**:1284–1285.

288. Alexander, K., Akhtar, M., Boar, R. B., McGhie, J. F., and Barton, D. H. R., 1972, The removal of the 32-carbon atom as formic acid in cholesterol biosynthesis, *J. Chem. Soc. Chem. Commun.* **1972**:383–385.

289. Mitropoulos, K. A., Gibbons, G. F., and Reeves, E. A., 1976, Lanosterol 14α-demethylase: Similarity of the enzyme system from yeast and rat liver, *Steroids* **27**:821–829.

290. Canonica, L., Fiecchi, A., Galli Kienle, M., Scala, A., Galli, G., Grossi Paoletti, E., and Paoletti, R., 1968, Evidence for the biological conversion of Δ^{8,14} sterol dienes into cholesterol, *J. Am. Chem. Soc.* **90**:6532–6534.

291. Gibbons, G. F., Goad, L. J., and Goodwin, T. W., 1968, The stereochemistry of hydrogen elimination from C-15 during cholesterol biosynthesis, *J. Chem. Soc. Chem. Commun.* **1968**:1458–1460.

292. Watkinson, I. A., Wilton, D. C., Munday, K. A., and Akhtar, M., 1971, The formation and reduction of the 14,15-double bond in cholesterol biosynthesis, *Biochem. J.* **121**:131–137.

293. Shafiee, A., Trzaskos, J. M., Paik, Y.-K., and Gaylor, J. L., 1986, Oxidative demethylation of lanosterol in cholesterol biosynthesis: Accumulation of sterol intermediates, *J. Lipid Res.* **27**:1–10.

294. Trzaskos, J. M., Fischer, R. T., and Favata, M. F., 1986, Mechanistic studies of lanosterol C-32 demethylation. Conditions which promote oxysterol intermediate accumulation during the demethylation process, *J. Biol. Chem.* **261**:16937–16942.

295. Saucier, S. E., Kandutsch, A. A., Phirwa, S., and Spencer, T. A., 1987, Accumulation of regulatory oxysterols, 32-oxolanosterol and 32-hydroxylanosterol in mevalonate-treated cell cultures, *J. Biol. Chem.* **262**:14056–14062.

296. Aoyama, Y., Yoshida, Y., Sonoda, Y., and Sato, Y., 1987, Metabolism of 32-hydroxy-24,25-dihydro-lanosterol by purified cytochrome P45014DM from yeast. Evidence for contribution of the cytochrome to whole process of lanosterol demethylation, *J. Biol. Chem.* **262**:1239–1243.

297. Aoyama, Y., Yoshida, Y., Sonoda, Y., and Sato, Y., 1989, Deformylation of 32-oxo-24,25-dihydro-lanosterol by the purified cytochrome P45014DM (lanosterol 14α-demethylase) from yeast. Evidence confirming the intermediate step of lanosterol 14α-demethylation, *J. Biol. Chem.* **264:**18502–18505.

298. Trzaskos, J., Kawata, S., and Gaylor, J. L., 1986, Microsomal enzymes of cholesterol biosynthesis. Purification of lanosterol 14α-methyl demethylase cytochrome P450 from hepatic microsomes, *J. Biol. Chem.* **261:**14651–14657.

299. Sono, H., Sonoda, Y., and Sato, Y., 1991, Purification and characterization of cytochrome P45014DM (lanosterol 14α-demethylase) from pig liver microsomes, *Biochim. Biophys. Acta* **1078:**388–394.

300. Sonoda, Y., Endo, M., Ishida, K., Sato, Y., Fukusen, N., and Fukuhara, M., 1993, Purification of a human cytochrome P450 isozyme catalyzing lanosterol 14α-demethylation, *Biochim. Biophys. Acta* **1170:**92–97.

301. Yoshida, Y., and Aoyama, Y., 1984, Yeast cytochrome P450 catalyzing lanosterol 14α-demethylation. I. Purification and spectral properties, *J. Biol. Chem.* **259:**1655–1660.

302. Aoyama, Y., Yoshida, Y., and Sato, R., 1984, Yeast cytochrome P450 catalyzing lanosterol 14α-de-methylation. II. Lanosterol metabolism by purified P450₁₄DM and by intact microsomes, *J. Biol. Chem.* **259:**1661–1666.

303. Sloane, D. L., So, O.-Y., Leung, R., Scarafia, L. E., Saldou, N., Jarnagin, K., and Swinney, D. C., 1995, Molecular cloning and functional expression of rat lanosterol 14α-demethylase, *Gene*, in press.

304. Alexander, K. T. W., Akhtar, M., Boar, R. B., McGhie, J. F., and Barton, D. H. R., 1971, The pathway for the removal of C-32 in cholesterol biosynthesis, *J. Chem. Soc. Chem. Commun.* **1971:**1479–1481.

305. Akhtar, M., Freeman, C. W., Wilton, D. C., Boar, R. B., and Copsey, D. B., 1977, The pathway for the removal of the 14α-methyl group of lanosterol: The role of lanost-8-ene-3β,32-diol in cholesterol biosynthesis, *Bioorg. Chem.* **6:**473–481.

306. Akhtar, M., Alexander, K., Boar, R. B., McGhie, J. F., and Barton, D. H. R., 1978, Chemical and enzymic studies on the characterization of intermediates during the removal of the 14α-methyl group in cholesterol biosynthesis: The use of 32-functionalized lanostan derivatives, *Biochem. J.* **169:**449–463.

307. Ramm, P. J., and Caspi, E., 1969, The stereochemistry of tritium at carbon atoms 1, 7, and 15 in cholesterol derived from (3R,2R)-(2-³H)-mevalonic acid, *J. Biol. Chem.* **244:**6064–6073.

308. Akhtar, M., Rahimtula, A. D., Watkinson, I. A., Wilton, D. C., and Munday, K. A., 1969, The status of C-6, C-7, C-15, and C-16 hydrogen atoms in cholesterol biosynthesis, *Eur. J. Biochem.* **9:**107–111.

309. Fischer, R. T., Trzaskos, J. M., Magolda, R. L., Lo, S. S., Brosz, C. S., and Larsen, B., 1991, Lanosterol 14α-methyl demethylase: Isolation and characterization of the third metabolically generated oxidative demethylation intermediate, *J. Biol. Chem.* **266:**6124–6132.

310. Spike, T. E., Wang, A. H.-J., Paul, I. C., and Schroepfer, G. J., 1974, Structure of a potential intermediate in cholesterol biosynthesis, *J. Chem. Soc. Chem. Commun.* **1974:**477–478.

311. Zuber, M. X., Simpson, E. R., and Waterman, M. R., 1986, Expression of bovine 17α-hydroxylase cytochrome P450 cDNA in nonsteroidogenic (COS 1) cells, *Science* **234:**1258–1261.

312. Barnes, H. J., Arlotto, M. P., and Waterman, M. R., 1991, Expression and enzymatic activity of recombinant cytochrome P450 17α-hydroxylase in *Escherichia coli*, *Proc. Natl. Acad. Sci. USA* **88:**5597–5601.

313. Nakajin, S., Takahashi, M., Shinoda, M., and Hall, P. F., 1985, Cytochrome b₅ promotes the synthesis of Δ¹⁶-C₁₉ steroids by homogeneous cytochrome P-450 C₂₁ side-chain cleavage enzyme from pig testis, *Biochem. Biophys. Res. Commun.* **132:**708–713.

314. Shimizu, K., 1978, Formation of S-[17β-²H]androstene-3β,17α-diol from 3β-hydroxy-S-[17,21,21,21-²H]pregnen-20-one by the microsomal fraction of boar testis, *J. Biol. Chem.* **253:**4237–4241.

315. Corina, D. L., Miller, S. L., Wright, J. N., and Akhtar, M., 1991, The mechanism of cytochrome P-450 dependent C–C bond cleavage: Studies on 17a-hydroxylase-17,20-lyase, *J. Chem. Soc. Chem. Commun.* **1991:**782–783.

316. Akhtar, M., Corina, D. L., Miller, S. L., Shyadehi, A. Z., and Wright, J. N., 1994, Incorporation of label from $^{18}O_2$ into acetate during side-chain cleavage catalyzed by cytochrome P450$_{17}\alpha$ (17α-hydroxylase-17,20-lyase), *J. Chem. Soc. Perkin Trans. I* **1994**:263–267.

317. Akhtar, M., Corina, D., Miller, S., Shyadehi, A. Z., and Wright, J. N., 1994, Mechanism of the acyl-carbon cleavage and related reactions catalyzed by multifunctional P450s: Studies on cytochrome P45017α, *Biochemistry* **33**:4410–4418.

318. Mak, A. Y., and Swinney, D. C., 1992, 17-O-Acetyltestosterone formation from progesterone in microsomes from pig testes; evidence for the Baeyer–Villiger rearrangement in androgen formation catalyzed by CYP17, *J. Am. Chem. Soc.* **114**:8309–8310.

319. Swinney, D. C., and Mak, A. Y., 1994, Androgen formation by cytochrome CYP17. Solvent isotope effect and pH studies suggest a role for protons in the regulation of oxene versus peroxide chemistry, *Biochemistry* **33**:2185–2190.

Inhibition of Cytochrome P450 Enzymes

PAUL R. ORTIZ de MONTELLANO and
MARIA ALMIRA CORREIA

1. Introduction

The catalytic cycle of cytochrome P450 (see Chapters 3 and 8) traverses three steps that are particularly vulnerable to inhibition: (1) the binding of substrates, (2) the binding of molecular oxygen subsequent to the first electron transfer, and (3) the catalytic step in which the substrate is actually oxidized. This chapter focuses on inhibitors that act at one of these three steps. Inhibitors that act at other steps in the catalytic cycle, such as agents that interfere with the electron supply to the hemoprotein by accepting electrons directly from cytochrome P450 reductase,[1-3] are not discussed here.

Cytochrome P450 inhibitors can be divided into three categories that differ in mechanism: (1) agents that bind reversibly, (2) agents that form quasi-irreversible complexes with the heme iron atom, and (3) agents that bind irreversibly to the protein or the prosthetic heme group, or that accelerate degradation of the prosthetic heme group. For the most part, inhibitors that interfere in the catalytic cycle prior to the actual oxidative event are reversible competitive or noncompetitive inhibitors. Agents that act during or subsequent to the oxygen transfer step are generally irreversible or quasi-irreversible inhibitors and, in many instances, fall into the category of mechanism-based (or suicide) inhibitors. Extended lists of P450 inhibitors are available in various reviews.[4-8] The emphasis in this chapter is on the mechanisms of inhibition; thus, most of the chapter is devoted to a discussion of agents that require catalytic turnover of the enzyme because the mechanisms of reversible competitive and noncompetitive inhibitors, despite their practical importance, are relatively straightforward.

PAUL R. ORTIZ de MONTELLANO • Department of Pharmaceutical Chemistry, School of Pharmacy, University of California, San Francisco, San Francisco, California 94143. *MARIA ALMIRA CORREIA* • Department of Cellular and Molecular Pharmacology, University of California, San Francisco, San Francisco, California 94143.

Cytochrome P450: Structure, Mechanism, and Biochemistry (Second Edition), edited by Paul R. Ortiz de Montellano. Plenum Press, New York, 1995.

2. Reversible Inhibitors

Inhibitors that compete reversibly with substrates for occupancy of the active site include substances that (1) bind to its hydrophobic domain, (2) coordinate to the prosthetic heme iron atom, or (3) participate in specific hydrogen bonding or ionic interactions with specific active-site residues.[4-8] The first mechanism, simple competition for binding to the lipophilic domain of the active site, is evidenced by the competition that exists between the substrates of a given P450 isozyme. A clear-cut example of such inhibition is provided by the mutual *in vitro* and *in vivo* inhibition of benzene and toluene metabolism.[9] This type of inhibition is optimal when the inhibitory substance is bound tightly but is a poor substrate. Inhibition by this mechanism is usually not particularly effective but, in appropriate situations, can cause physiologically relevant metabolic changes and clinically significant interactions.[10]

2.1. Coordination to Ferric Heme

The coordination of a strong ligand to the pentacoordinated iron, or the displacement of a weak ligand from the hexacoordinated heme by a strong ligand, shifts the enzyme from the high- to the low-spin form and gives rise to a "type II" binding spectrum with a Soret maximum at 425–435 nm and a trough at 390–405 nm.[11-13] This spin state change is accompanied by a change in the redox potential of the enzyme that makes its reduction by cytochrome P450 reductase more difficult (see Chapter 3).[14,15] This change in reduction potential, as much as physical occupation of the sixth coordination site, is responsible for the inhibition associated with the binding of strong iron ligands.

Ionic ligands such as cyanide bind preferentially to the ferric form of P450.[16,17] The triple ferric positive charge is matched in the enzyme by the negative charges of its three ligands (the two porphyrin nitrogens and the thiolate), but the negative charges exceed the two positive charges of the ferrous iron. The cyanide binds more readily to the neutral (ferric) than the negative (ferrous) enzyme. In fact, cyanide binds more weakly to ferric P450 than to ferric myoglobin because the thiolate ligand of P450 places a higher electron density on the iron than does the imidazole ligand of myoglobin.[18] The chelation of ionic ligands is disfavored, in addition, by the lipophilic nature of the P450 active site.[19]

Nitric oxide, a molecule of great interest because of the role it plays in diverse physiological and pathological processes, inhibits P450 enzymes. Inhibition initially involves reversible coordination of the nitrogen to the iron but subsequent time-dependent, irreversible inactivation of the enzyme by an undefined mechanism has been observed.[20-22]

2.2. Coordination to Ferrous Heme

Carbon monoxide, the simplest uncharged ligand, binds exclusively to the ferrous (reduced) form of P450. Binding of carbon monoxide to the ferrous prosthetic heme group entails donation of electrons from the carbon to the iron through a σ-bond as well as back-donation of electrons from the occupied iron d-orbitals to the empty antibonding π-orbitals of carbon monoxide.[23] P450 enzymes are so named because their carbon monoxide complexes have absorption maxima at approximately 450 nm.[24] The finding that the 450-nm absorption can be reproduced with model ferroporphyrins only with a thiolate ligand *trans* to the carbon monoxide provided key early evidence for a thiolate fifth ligand in P450.[25] Inhibition by carbon monoxide is one of the hallmarks of processes catalyzed by P450, although different isozymes have different sensitivities to carbon monoxide[26] and a number of the reactions catalyzed by biosynthetic P450 enzymes are relatively resistant to inhibition by carbon monoxide.[27–29] In particular, the sensitivity of aromatase[30,31] and $P450_{scc}$[32] to inhibition by carbon monoxide decreases drastically as the enzymes traverse the conformational and ligand states inherent in their multistep catalytic sequences. Among human liver enzymes, the sensitivity of different families to carbon monoxide appears to decrease in the order 2D > 2C > 3A.[26]

2.3. Heme Coordination and Lipophilic Binding

Inhibitors that bind to lipophilic regions of the protein, and simultaneously bind to the prosthetic heme iron (Fig. 1), are inherently more effective than agents that depend on only one of these binding interactions. The activity of such agents as inhibitors of P450 is governed both by their hydrophobic character and by the strength of the bond between their heteroatomic lone pair and the prosthetic heme iron. The less effective agents, including alcohols, ethers, ketones, lactones, and other structures in which the coordinating atom is an oxygen, only coordinate weakly to the prosthetic heme iron.[13,33–37] The Soret band of such complexes is found at approximately 415 nm.[11,13] The most effective reversible inhibitors, in contrast, interact strongly with both the protein and the prosthetic heme iron.[4–8] The binding of inhibitors that are strong iron ligands, as already noted for cyanide, gives rise to what is termed a type II difference spectrum with a Soret maximum at 430 nm.[11,13,38,39] For the most part, these powerful inhibitors are nitrogen-containing aliphatic and aromatic compounds.

Pyridine and imidazole derivatives have found particularly widespread utility as P450 inhibitors.[4] Metyrapone (Fig. 1), historically one of the most frequently employed P450 inhibitors, first gained prominence as an inhibitor of the 11β-hydroxylase that catalyzes the final step in cortisol biosynthesis.[40] This activity led to its use in the diagnosis and treatment of hypercortisolism (Cushing's syndrome) and other hormonal disorders.[41] The factors that determine the inhibitory potency of metyrapone and other nitrogen heterocycles are valid for most reversible inhibitors: (1) the intrinsic affinity of the ligand electron

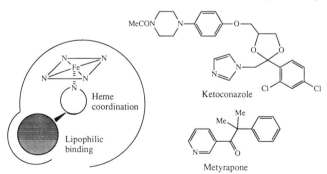

FIGURE 1. Schematic diagram of the two-point binding of agents with a lipophilic domain and a nitrogen function that coordinates to the heme iron atom. The structures of two such agents, metyrapone and ketoconazole, are shown.

pair for the prosthetic heme iron, (2) the degree to which the intrinsic affinity of the ligand for the iron is moderated by steric interactions with substituents on the inhibitor,[42,43] (3) the lipophilicity of the nonligating portion of the inhibitor,[19,44] and, naturally, (4) the congruence between the geometry of the inhibitor and that of the active site. The synergism between binding to the lipophilic domain and coordination with the heme iron is illustrated by the fact that imidazole and benzene individually are weak inhibitors, but when joined in phenylimidazole form a powerful inhibitor.[42] Optimization of these features in keto-conazole (Fig. 1) has led to its clinical use as both an antifungal and cancer chemotherapeutic agent.[5] On the other hand, the inadvertent effectiveness of similar features in cimetidine that result in inhibition of the metabolism of coadministered drugs led to a search for non-imidazole-containing H2 antagonists devoid of this undesirable side effect.[34] Improvement of the specificity of metyrapone, aminoglutethimide, and other classical inhibitors by structural modification continues to be of therapeutic interest. This consideration has prompted the development and testing of pyridylaminoglutethimide, CGS 16949A {4-(5,6,7,8-tetrahydroimidazo[1,5α]pyridin-5-yl) benzonitrile}, CGS 18320B bis-(p-cyanophenyl)imidazo-1-yl-methane hemisuccinate, and R-76713 [6-(4-chloro-phenyl)(1H-1,2,4-triazol-1-yl)methyl]-1-methyl-1H-benzotriazole as aromatase inhibitors in breast cancer chemotherapy (see Section 5.2, Fig. 22).[5] Efforts to improve the properties of ketoconazole have similarly fostered the rational design and use in fungal therapy of 14α-demethylase inhibitors such as miconazole, fluconazole, saperconazole, and terconazole (see also Section 4).[5]

3. Catalysis-Dependent Inhibition

Several classes of inhibitors are known that are catalytically activated by the enzyme to transient intermediates that irreversibly or quasi-irreversibly inhibit the enzyme. The inhibitory activity of the catalytically generated species is superimposed on the reversible inhibition associated with binding of the parent structure to the ferric enzyme. Mechanism-based[45,46] (catalysis-dependent) inhibitors are potentially more enzyme-specific than reversible inhibitors because: (1) the inhibitor must first bind to the enzyme and therefore must satisfy the constraints imposed on classical inhibitors, (2) the inhibitor must then be catalytically activated and therefore must be acceptable as a substrate, and, finally, (3) reactive species produced by catalytic turnover irreversibly alter the enzyme and remove it permanently from the catalytic pool. Four classes of catalysis-dependent irreversible inhibitors of P450 are known: (1) agents that bind covalently to the protein, (2) agents that quasi-irreversibly coordinate to the prosthetic heme iron, (3) agents that alkylate or arylate the prosthetic heme group, and (4) agents that degrade the prosthetic heme group, in some cases to products that modify the protein.

3.1. Covalent Binding to the Protein

3.1.1. Sulfur and Halogenated Compounds

Agents that inactivate P450 by binding covalently to the protein after they are oxidatively activated by the enzyme include (1) a variety of sulfur compounds (e.g., carbon disulfide,[47–49] parathion,[50,51] diethyldithiocarbamate,[52] isothiocyanates,[53] thioureas,[54] thiophenes,[55] tienilic acid,[55,56] and possibly mercaptosteroids,[57,58] (2) halogenated structures such as chloramphenicol,[59–62] N-monosubstituted dichloroacetamides,[63] and N-(2-p-nitrophenethyl)dichloroacetamide,[64] and (3) terminal alkyl and aryl olefins and acetylenes[65–69] such as 10-undecynoic acid,[65,70] 1-ethynylpyrene,[68,70] 17β-ethynylprogesterone,[71,72] 9- and 2-ethynylnaphthalene,[67,69,73] and secobarbital.[74] It is likely that other classes of compounds will be found that act by this mechanism. For instance, the photoactive linear furocoumarin, 8-methoxypsoralen (methoxsalen), is activated by P450 enzymes to reactive electrophilic intermediates that bind covalently to the protein.[75–77] Studies with P450 inducers and P450-selective functional markers reveal that methoxypsoralen targets P450s 2B1/2, 1A, 3A, and 2C11 for inactivation.[75,76] Inclusion of cysteine or glutathione in incubations of 8-methoxypsoralen with NADPH-supplemented rodent liver microsomes did not prevent P450 inactivation but suppressed nonspecific covalent binding, resulting in a residual covalent binding of methoxypsoralen that was nearly stoichiometric with the loss of microsomal P450.[77] This finding, together with observation of a high-spin type I spectrum in incubations of rat liver microsomes with methoxypsoralen in the presence but not absence of NADPH,[76] suggests an active site-directed inactivation mechanism. Furthermore, because cysteine markedly quenches the covalent binding of label from [^{3}H]methoxy- but not [4-^{14}C]-ring-labeled methoxypsoralen, the reactive intermediate has been proposed to derive from oxidation of the furan ring.[77] This finding is consistent with the fact that trioxsalen (trimethylpsoralen), in which the furan ring is methyl-substituted, fails to destroy P450.

Similar findings have been obtained with cannabidiol, which inactivates mouse P450 isoforms 2C and 3A via a mechanism that involves stoichiometric covalent binding of the inhibitor and loss of cannabidiol oxidizing activity.[78] Limited structure–activity studies show that a free phenolic group in the resorcinol moiety of cannabidiol is essential for inactivation.[78] More recently, 2-phenylphenanthridinone, a representative of a new class of P450 1A1 suicide inactivators, has been proposed to inactivate the enzyme by binding covalently to the protein.[79] Many details of the mechanisms by which these compounds inactivate P450 remain to be defined but major progress has been made in elucidating the mechanisms of inhibition of certain compounds.

It has been shown for parathion that the protein is radiolabeled when [^{35}S]parathion is used but not when the radioactivity is present in the ethyl groups as ^{14}C.[50,51] Ninety percent of the ^{35}S label bound covalently to microsomal proteins is precipitated by antibodies to P450 enzymes. Approximately 75% of the prosthetic heme of P450 is lost in incubations of the enzyme with parathion, but no information is available on its fate. The bulk (50–75%) of the sulfur radiolabel is removed from the protein by cyanide or dithiothreitol, a fact consistent with the suggestion that it is present in the form of hydrodisulfides (RSSH), but the activity of the enzyme is not restored by these treatments. Approximately 4 nmole of radiolabeled sulfur binds covalently to the protein for each nanomole of heme chromophore that is lost. Catalytic turnover and sulfur activation apparently continue despite covalent attachment of sulfur to the protein until the heme itself is damaged[50] or is released from the protein as a consequence of protein damage. The covalent binding of sulfur, the well-established oxidation of sulfur compounds to

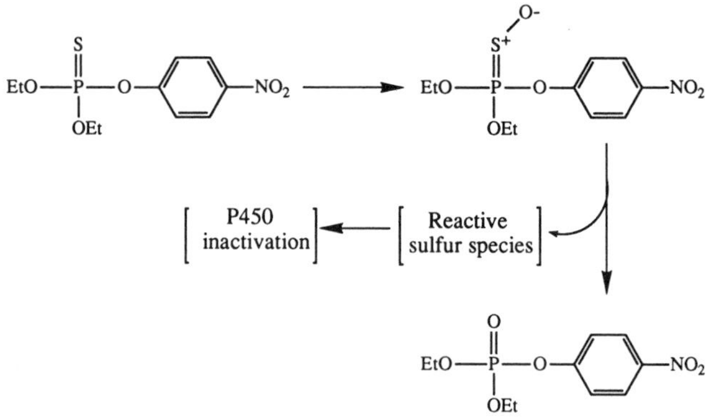

FIGURE 2. Proposed mechanism for the cytochrome P450-catalyzed activation of parathion. A sulfur species formally equivalent to atomic sulfur is thought to be produced that binds covalently to the protein.

S-oxides, and the formation of metabolites in which the sulfur is replaced by an oxygen, suggest the oxidative activation mechanism in Fig. 2.

Tienilic acid (Fig. 3), a substituted thiophene, is oxidized by cytochromes P450 2C9 and 2C10 to a product that causes irreversible inactivation of these enzymes ($K_I = 4.3$ µM, $k_{inact} = 0.22$ min^{-1}, and partition ratio = 11.6).[55] Tienilic acid is oxidized by the enzymes that are inactivated to 5-hydroxytienilic acid and a product of tienilic acid oxidation is covalently bound to the protein. Covalent labeling of the protein is partially prevented by glutathione but glutathione does not protect the enzyme from inactivation. In the presence of glutathione, approximately 0.9 equivalent of label is bound to the protein before catalytic activity is completely suppressed. A mechanism has been proposed for this reaction involving the formation of a thiophene sulfoxide that reacts with water to give the metabolite or with a protein nucleophile to inactivate the enzyme.

Spironolactone (Fig. 4), an aldosterone antagonist employed as a diuretic and antihypertensive,[80] provides a related example. Spironolactone inactivates P450 enzymes in both

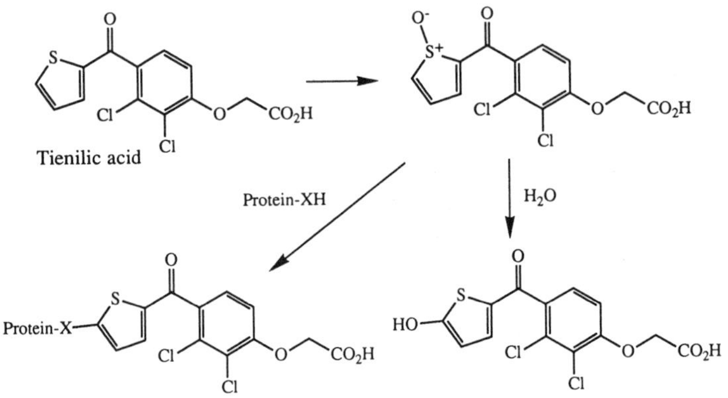

FIGURE 3. Proposed mechanism for the P450-catalyzed activation of tienilic acid to a reactive, protein-arylating intermediate. The protein nucleophile involved in the reaction is denoted as Protein–XH.

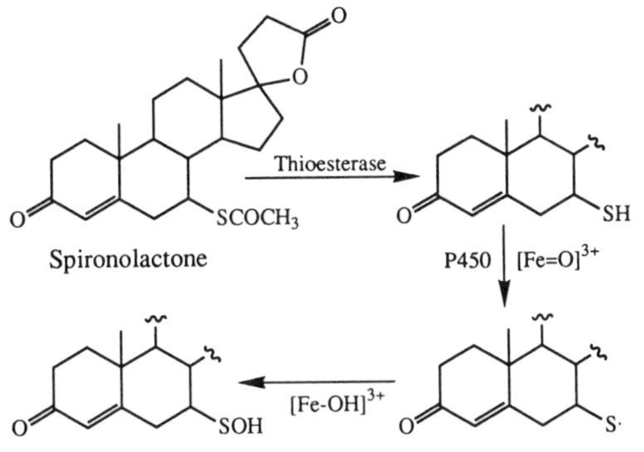

FIGURE 4. Possible mechanisms for the activation of spironolactone to products that inactivate the adrenal sterol 17α-hydroxylase by covalent attachment to the protein, and hepatic P450 3A by degradation of the heme group to fragments that bind to the protein.

Thioesterase

Spironolactone

P450 $[Fe=O]^{3+}$

$[Fe\text{-}OH]^{3+}$

the liver and steroidogenic tissues[57,81]: the enzyme inactivated in the adrenals is the steroid 17α-hydroxylase,[58,81] while members of the CYP3A and CYP2C families are inactivated in the liver.[82,83] P450 inactivation in the liver by spironolactone occurs after hydrolysis of the thioester function to give the free thiol.[58,84,85] P450 then oxidizes the thiol group to a reactive species that binds covalently to the protein and/or modifies the heme group.[84] P450 inactivation in the liver involves degradation of the prosthetic heme group to products that bind covalently to the protein (see Section 3.4).[83] In contrast, the inactivation of adrenal P450 appears to involve direct covalent binding of the thiosteroid to the protein.[81,86] A role for oxidation of the thiol in enzyme inactivation is supported by the demonstration that hepatic microsomes oxidize the thiol group ($-SH$) to the sulfinic ($-SO_2H$) and sulfonic ($-SO_3H$) metabolites,[83] and give rise to a disulfide adduct with glutathione.[87] However, the disulfide adduct arises, at least in part, by a flavin monooxygenase-catalyzed rather than P450-catalyzed reaction.[87] The two reactive intermediates that can be envisioned to arise by thiol oxidation are the sulfhydryl radical ($-S\cdot$) and the sulfenic acid ($-SOH$), either or both of which could be involved in P450 inactivation. One attractive possibility is that loss of the heme group involves reaction of the heme with the sulfhydryl radical whereas protein modification is likely to arise by reaction of amino acid side chains with the sulfenic acid metabolite (Fig. 4).

In the case of chloramphenicol, not only has binding of [^{14}C]chloramphenicol to the apoprotein been correlated with loss of ethoxycoumarin deethylase activity, but proteolytic digestion of the protein has been shown to yield a single radiolabeled amino acid.[59–62] Hydrolysis of the modified amino acid yielded lysine and the chloramphenicol fragment shown in Fig. 5. Chloramphenicol is thus oxidized to an oxamyl chloride intermediate that either is hydrolyzed to the oxamic acid metabolite or acylates a critical lysine in the protein. Acylation of the lysine apparently interferes with electron transfer from the reductase to the heme because the inactivated hemoprotein is still able to catalyze the deethylation of ethoxycoumarin supported by cumene hydroperoxide or iodosobenzene.[62] Furthermore, because ethoxycoumarin is still O-deethylated with substitute oxygen donors, it appears that chloramphenicol does not bind covalently at the substrate binding site of the enzyme.

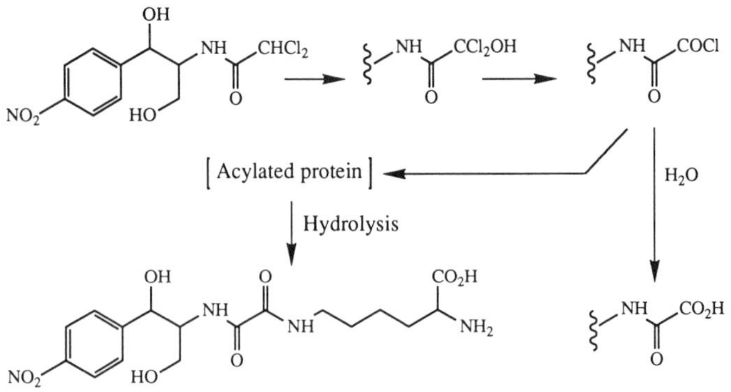

FIGURE 5. Mechanism proposed for the catalysis-dependent inactivation of P450 by chloramphenicol. The acylated protein residue has been isolated and characterized.

3.1.2. Olefins and Acetylenes

Although terminal acetylenes have been known for some time to alkylate the prosthetic heme group (see Section 3.3.2), it is now apparent that terminal acetylenic compounds such as 10-undecynoic acid, 1-ethynylpyrene, 17β-ethynylprogesterone, and 9- and 2-ethynylnaphthalene inactivate P450 by binding covalently to the protein with little loss of the prosthetic heme group.[67–73] Thus, near-stoichiometric binding of 2-ethynyl-naphthalene and 1-ethynylpyrene to P450s 1A1 and 1A2, and of 10-undecynoic acid to purified rat liver CYP4A1 (ω-hydroxylase) have been reported.[65–70,73] Isolation of the expected acid metabolites during the P450 inactivation reactions mediated by 10-undecynoic acid and 1-ethynylpyrene argues for a mechanism in which oxygen transfer from P450 to the terminal acetylenic carbon occurs at the same time as migration of the terminal hydrogen to the vicinal carbon (Fig. 6). The reactive ketene species produced by this migration, which has been confirmed by deuterium labeling experiments, either acylates the protein or is hydrolyzed to give the carboxylic acid metabolite.[70] Analogous ketene

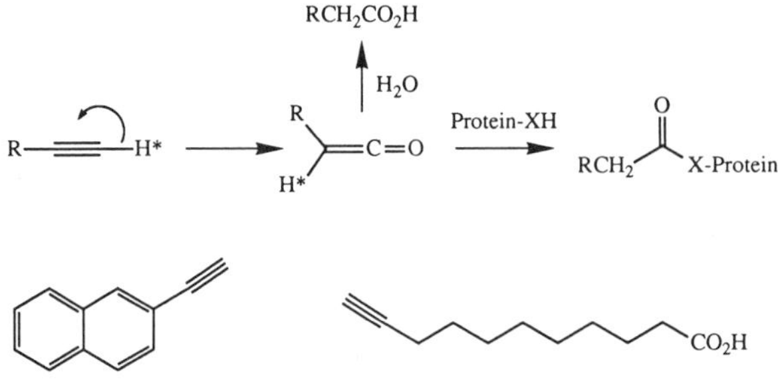

FIGURE 6. Terminal acetylenes are oxidized via ketene intermediates to the carboxylic acid metabolites (see Chapter 8). The ketene intermediates are presumed to be the activated species that acylate the protein during the catalytic turnover of some acetylenes. The hydrogen that migrates is indicated by an asterisk. The structures of 2-ethynylnaphthalene and 10-undecynoic acid are shown.

intermediates have been invoked in acylation of the bovine adrenal CYP21 protein by 17β-ethynylprogesterone[72] and of P450s 1A2, 2B1, and 2B4 by 2-ethynylnaphthalene (Fig. 6).[67,69,73] As expected, CYP2B1 produces the substituted acetic acid metabolite by oxidation of the triple bond of 2-ethynylnaphthalene, a reaction that presumably also produces the reactive species that modifies the protein. The mechanism-based inactivation of CYP2B1 mediated by 2-ethynylnaphthalene is characterized by a $K_I = 0.08$ μM and $k_{inact} = 0.83$ min^{-1}.[69] The partition ratio between metabolite formation and protein alkylation is approximately 4–5 moles of acid formed per inactivation event.

In contrast to inactivation by 2-ethynylnaphthalene, P450 2B1-catalyzed addition of the oxygen to the internal rather than terminal triple bond carbon of phenylacetylene results in N-alkylation of the heme rather than protein acylation (see below).[70,88] Predominant inactivation of P450 2B1 by phenylacetylene via heme alkylation[88] and by 2-ethynyl-naphthalene via protein acylation[67,69,73] provides mechanistic insight into the influence of substrate fit within the enzyme active site as a determinant of the final course of the inactivation process. Both of these terminal acetylenic compounds generate reactive ketene species but only 2-ethynylnaphthalene acylates the 2B1 protein. Protein acylation by 2-ethynylnaphthalene clearly establishes that the failure of phenylacetylene to acylate the 2B1 protein is not related to the absence of appropriate nucleophilic residues. Furthermore, oxidation of phenylacetylene to the acetic acid metabolite demonstrates that the ketene intermediate is actually formed. In fact, the differential inactivation of P450 2B1 by 2-ethynylnaphthalene and phenylacetylene suggests that: (1) 2-ethynylnaphthalene, but not phenylacetylene, binds in a manner that suppresses delivery of the ferryl oxygen to the internal carbon, and (2) heme alkylation by phenylacetylene is sufficiently efficient with respect to acylation by phenylketene that the enzyme is inactivated before protein acylation becomes significant. These differences may stem from differences in the binding affinities or the orientations of the two agents within the P450 2B1 active site.

Considerable progress has been made in elucidating the specific protein sites that are modified. Peptide mapping and amino acid sequence analyses of radiolabeled peptides show that 2-ethynylnaphthalene covalently binds to amino acid residues 67–78 and 175–184, respectively, of rat and rabbit CYP1A2.[67] However, the instability of the adducts has prevented definitive identification of the residue that is actually modified or the nature of the covalent link to the inhibitor. Alignment of the sequences of the modified peptides with the sequence of P450$_{cam}$ (CYP101) suggests that the 1A2 peptides correspond to helices A and D of P450$_{cam}$ (see Chapter 4). This correlation suggests that the rat 1A2 peptide may contain residues that are part of the substrate binding site.[89,90] The 2-ethynyl-naphthalene-modified peptides from rat and rabbit CYP2B1 and CYP2B4 have also been characterized.[69,73] Oxidation of the radiolabeled agent by P450 2B1 has led to the isolation of a radiolabeled peptide (ISLLSLFFAGTETSSTTLRYGFLLM) comprising residues 290–314 of the protein.[69,73] Similar studies with P450 2B4 reveal that an analogous peptide is alkylated in that protein.[69,73] The two alkylated peptides correspond in sequence to the highly conserved distal I helix of P450$_{cam}$ that has contacts with both the substrate and the heme group (see Chapters 4 and 5).[89,90] Again, the actual residue alkylated in the peptides was not established but several nucleophilic residues, including serine, threonine, and tyrosine, are present. Protein alkylation presumably involves the formation of a ketene intermediate that partitions between reaction with a protein nucleophile and reaction with water to give the acetic acid metabolite (Fig. 6). If the protein nucleophile is a hydroxyl group such as that of serine or tyrosine, the resulting adduct would be an ester.

Studies with the olefin secobarbital reveal that it completely inactivates P450 2B1 with only partial loss of the heme chromophore.[74,91] Isolation of the N-alkylated porphyrins (see Section 3.3.1) and of the modified 2B1 protein reveal that the compound partitions between heme N-alkylation, 2B1 protein modification, and epoxidation, in a 1:1:16 ratio.[45,74] The N-modified porphyrin has been isolated and characterized by mass spectrometric analysis as the N-adduct of protoporphyrin IX and hydroxysecobarbital (He and Correia, preliminary findings). The P450 2B1 peptide that is modified by the drug has also been isolated and shown to correspond to the distal I helix in P450$_{cam}$,[91] a result suggesting that inactivation occurs at the active site.

These collective results clearly show that P450 enzymes can generate reactive species that alkylate, acylate, or otherwise modify the protein skeleton. It is possible, furthermore, that protein alkylation has gone undetected in instances where enzyme inactivation has been attributed solely to heme modification. Conversely, the data on parathion suggest that heme destruction is required for enzyme inactivation in some instances where protein modification clearly occurs. Furthermore, the fact that P450 2B1 is inactivated by secobarbital by heme alkylation and protein modification, by phenylacetylene predominantly by heme alkylation, and by 2-ethynylnaphthalene predominantly by protein acylation, underscores the critical role of substrate–protein interactions in determining the nature of the inactivating event. Protein–inhibitor interactions presumably also explain why N-phenyl or N-octyl 2,2-dichloroacetamides predominantly inactivate P450 2B1 via protein acylation whereas the corresponding N-hexyl, N-butyl, or N-methyl dichloroacetamides do so via heme destruction.[63,64] Similar factors should also explain the differential modes of inactivation observed with terminal acetylenes (e.g., N-alkylation of P450 2B1 by phenylacetylene versus protein acylation of P450 4A1 by 10-undecynoic acid).[65,70]

3.2. Quasi-irreversible Coordination to the Prosthetic Heme

Agents that are catalytically oxidized to intermediates that coordinate so tightly to the prosthetic heme of P450 that they can only be displaced under special experimental conditions are considered in this section. The two major classes of such inhibitors are compounds with a dioxymethylene function and nitrogen compounds, usually amines, that are converted *in situ* to nitroso metabolites. 1,1-Disubstituted hydrazines and acyl hydrazines also appear to inhibit P450 to some extent by a related mechanism. Reductive coordination of halocarbons to the prosthetic heme under anaerobic conditions is covered in Section 3.4 because the reaction is closely associated with heme destruction.

3.2.1. Methylenedioxy Compounds

Aryl and alkyl methylenedioxy compounds, some of which are commercially employed as insecticide synergists,[92,93] are oxidized by P450 to species that coordinate tightly to the prosthetic heme iron.[94] The catalytic role of the enzyme in unmasking the inhibitory species is confirmed by the time, NADPH, oxygen, and concentration dependence of the process, as well as by the finding that cumene hydroperoxide can substitute for the NADPH and oxygen requirement.[92–97] The ferrous complex is characterized by a difference absorption spectrum with maxima at 427 and 455 nm, whereas the ferric complex has a single absorption maximum at 437 nm.[94,96,97] The peaks at 427 and 455 nm are related to distinct complexes, although the structural relationship between the two complexes remains obscure.[94] The quasi-irreversible nature of the ferrous complex is evidenced by the fact that it can be isolated intact from animals treated with isosafrole. The catalytically active enzyme can be regenerated from the much less stable ferric complex, however, by incubation with lipophilic compounds that displace the inhibitor from the active site.[98,99] The ferrous complex is not disrupted by incubation with lipophilic compounds but can be broken by irradiation at 400–500 nm.[100,101] Structure–activity studies show that the size and lipophilicity of the alkyl group in 4-alkoxymethylenedioxybenzene are an important determinant of the stability of the complexes, analogues with alkyl chains of one to three carbons giving unstable complexes while those with longer alkyl groups are stable.[102] This indicates that the stability of the ferrous complex stems, as already described for reversible inhibitors (Section 2.3), from concurrent binding to the lipophilic active site[103] and to the prosthetic heme group. The weakening of the complex associated with the ferrous-to-ferric transition indicates that the activated species, like carbon monoxide, only strongly coordinates to the ferrous heme.

The evidence suggests the catalysis-dependent formation of a carbene–iron complex (Fig. 7), although the actual structure of the hemoprotein complex has yet to be established. Circumstantial evidence for a carbene complex is provided by the synthesis and characterization of model complexes.[104,105] The structural resemblance between carbenes and carbon monoxide readily rationalizes the 455-nm absorption maximum of the complexes. The absorbance maximum at 427 nm presumably is related to a different complex, perhaps to a carbene complex in which the *trans* ligand, as in P420, is not a thiolate.[106] The carbene formulation rationalizes the incorporation of oxygen from water into the carbon monoxide derived from the dioxymethylene bridge (see below), and the observation that electron-withdrawing substituents increase the proportion of the carbon monoxide metabolite.[107] Addition of a hydroxyl to the iron-coordinated carbene would yield an iron-coordinated anion that could readily fragment into the observed catechol and carbon monoxide

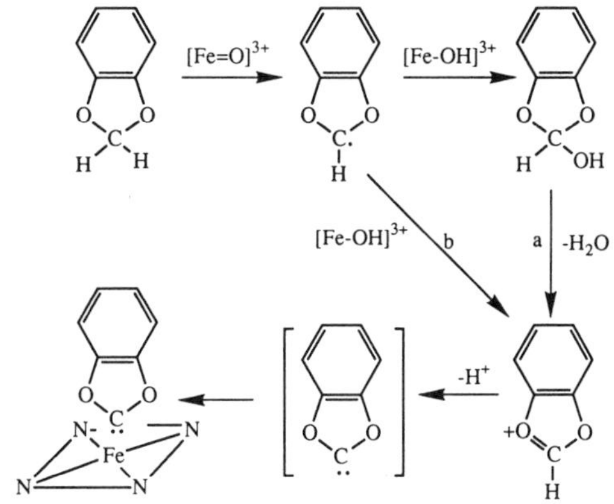

FIGURE 7. Structure proposed for the quasi-irreversible complex formed during the catalytic turnover of methylenedioxy phenyl compounds and possible mechanisms for P450-catalyzed formation of the putative methylenedioxy carbene function.

metabolites. A different but undefined mechanism must be postulated to explain the observation that an atom of molecular oxygen is incorporated into a fraction of the carbon monoxide.[107]

Association of the inhibitory activity with the dioxymethylene function, the importance of enzymatic turnover for inhibition, and enzymatic oxidation of the dioxymethylene group indicate that oxidation of that group is essential for quasi-irreversible inhibition. Free radical,[108] carbocation,[109] and carbanion[100] intermediates have been suggested, but the results are most consistent with formation of the carbene from the 2-hydroxylated metabolite or from a radical precursor of it (Fig. 7). Substituents on the dioxymethylene group, except for an alkoxy group, block complex formation.[93,101,110] The anomalous activity of an ethoxy-substituted compound is readily explained because O-dealkylation would yield the 2-hydroxylated metabolite obtained by oxidation of the unsubstituted compound.[101] The metabolism of aryldioxymethylene compounds to catechols, carbon monoxide, carbon dioxide, and formic acid is consistent with such a hydroxylation reaction,[93,107,111–113] particularly in view of the fact that carbon monoxide is evolved more slowly when the dioxymethylene hydrogens are replaced by deuteriums ($k_H/k_D = 1.7$–2.0). The observation of a similar isotope effect on the *in vivo* activity of the compounds as insecticide synergists supports the hypothesis that carbon monoxide and complex formation are mechanistically linked.[114]

Oxidation of the dioxymethylene bridge to the putative carbene involved in complex formation could occur by one of three mechanisms. Hydroxylation of the dioxymethylene bridge, followed by elimination of a molecule of water, would yield an oxonium ion (Fig. 7, path a). Deprotonation of this acidic oxonium intermediate would provide the desired carbene. The oxonium intermediate could be obtained without actually passing through the 2-hydroxylated metabolite if the radical generated by removal of a hydrogen during the hydroxylation process is oxidized by the ferryl species rather than collapsing with it (Fig. 7, path b). Finally, the radical formed by hydrogen abstraction could collapse with the resulting $[Fe-OH]^{3+}$ species to give a carbon–iron complex in which the hydroxyl and the carbon are simultaneously bound to the iron.[104] Deprotonation and transfer of the oxygen from the iron to the carbon would then yield an intermediate that could give the carbene complex by eliminating a molecule of water or could decompose to the carbon monoxide metabolite. It is not possible at this time to say which of these mechanisms is operative.

3.2.2. Amines

The second, quite large, class of agents that form quasi-irreversible complexes [metabolic intermediate (MI) complexes][115] with the heme of P450 is composed of alkyl and aromatic amines, including a number of clinically useful amine antibiotics such as troleandomycin (TAO) and erythromycin (Fig. 8).[4,115–117] Oxidation of these amines yields intermediates that coordinate tightly to the ferrous heme and give rise to a spectrum with an absorbance maximum in the region of 445–455 nm.[115] Primary amines are required for complex formation, but secondary and tertiary amines, as in the case of TAO, are suitable precursors of the P450 complexes if they are N-dealkylated *in situ* to the primary amines. The complexes formed with aromatic amines differ from those obtained with alkyl amines

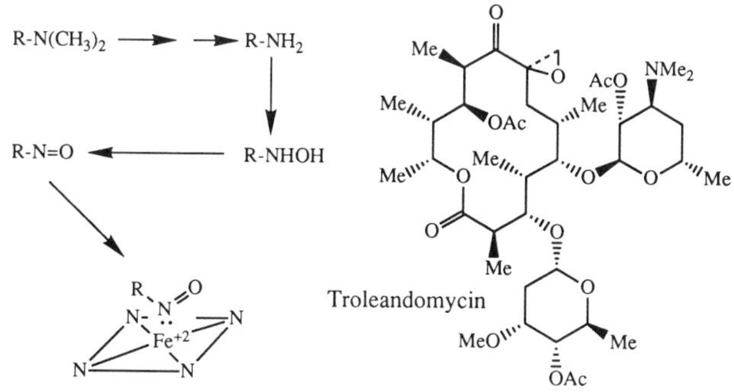

FIGURE 8. Structure assigned to the metabolic intermediate (MI) heme complexes formed during the P450-catalyzed oxidation of some primary amines. As shown, the reaction sequence that generates the complex often involves initial N-dealkylation of an alkylamine group to unmask the primary amine function. The structure of troleandomycin (TAO), a compound that forms a strong MI complex with P450s of the 3A family, is shown (AcO in the structure stands for CH_3CO_2).

in that they are unstable to reduction by dithionite.[118] The normal competitive inhibition associated with the binding of amines does not, of course, require catalytic activation of the inhibitor, but activation is essential for formation of the tight, quasi-irreversible complexes.[115,118,119] The primary amines appear to first be hydroxylated because the corresponding hydroxylamines also yield the complexes,[120] but the functionality that coordinates to the iron lies beyond the hydroxylamine in the oxidative scale because the hydroxylamines must also be oxidatively activated.[118,120] The chelated function appears, in fact, to be the nitroso group obtained by two-electron oxidation of the hydroxylamine (Fig. 8).[119–121] The final oxidation may not actually require catalytic turnover of the enzyme because hydroxylamines readily autooxidize.[122] The conclusion that the nitroso function is involved in iron chelation is supported by the demonstration that apparently identical complexes are obtained from nitro compounds under reductive conditions.[123] The crystal structure of the complex between a nitroso compound and a model iron porphyrin indicates that, as expected, the iron binds to the nitrogen rather than the oxygen.[104]

It is noteworthy that *in vivo* complexation of TAO to the heme of P450 results in a stabilization of the enzyme and prolongation of its lifetime in the hepatocyte.[124–126] A consequence of this stabilization is an increase in the concentration of the protein in the cell (i.e., a form of induction). It is not yet clear whether this elevation in protein levels is related to sustained substrate-induced conformational stabilization or whether inactivation of the enzyme by formation of a quasi-irreversible heme complex suppresses normal damage to the protein caused by reactive O_2 species produced by futile turnover of the enzyme in the absence of tightly coupled substrates.

3.2.3. 1,1-Disubstituted and Acyl Hydrazines

1,1-Disubstituted hydrazines, in contrast to monosubstituted hydrazines (see Section 3.3.4), are oxidized by P450s to intermediates that chelate tightly to the prosthetic heme iron. The complexes, formed in a time-, NADPH-, and oxygen-dependent manner, are characterized by a ferric absorption maximum at approximately 438 nm and a ferrous

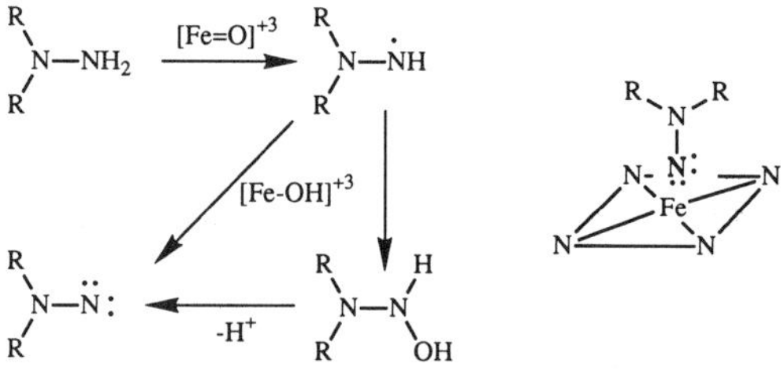

FIGURE 9. Structure of the nitrene–iron structure proposed for the complexes formed during the metabolism of 1,1-dialkylhydrazines and possible metabolic routes to formation of the nitrene function.

maximum at 449 nm.[127] A similar transient complex with an absorbance maximum of 449 nm has been detected during the microsomal oxidation of isoniazid and other acyl hydrazines,[128,129] but the isoniazid complex falls apart on addition of ferricyanide and thus is only stable in the ferrous state.[130] Chemical model studies indicate that oxidation of 1,1-dialkylhydrazines yields disubstituted nitrenes that form end-on complexes with the iron of metalloporphyrins. Specifically, nitrene complexes derived from 1-amino-2,2,6,6-tetra-methyl-piperidine and several iron tetraarylporphyrins have been isolated and characterized by NMR, Mössbauer, and X-ray methods.[131,132] It therefore appears likely that the P450 complexes formed during the metabolism of 1,1-disubstituted hydrazines, and possibly acyl hydrazines, are aminonitrene–iron complexes (Fig. 9). Oxidation of the dialkylhydrazines to the required aminonitrenes is readily rationalized by hydroxylation of the hydrazine or, more probably, by stepwise electron removal from the hydrazine (Fig. 9).

3.3. Covalent Binding to the Prosthetic Heme

It is now well documented that irreversible inactivation of P450 often reflects covalent attachment of the inhibitor, or a fragment derived from it, to the prosthetic heme group. The evidence for a heme alkylation mechanism includes, in many instances, the demonstration of equimolar enzyme and heme loss and isolation and structural characterization of the resulting modified heme. It is important to note that an equimolar loss of enzyme and heme is not sufficient, in the absence of explicit evidence for heme adduct formation, to conclude that heme alkylation is responsible for enzyme inactivation because alternative mechanisms exist for catalysis-dependent destruction of the prosthetic group (see Section 3.4). The possibility also exists, of course, that a heme adduct is formed but is too unstable to be isolated. The technical difficulties inherent in quantitating heme adducts continue to hinder the quantitative correlation of heme adduct formation with enzyme inactivation. In the absence of such data, it is possible that mechanisms other than heme modification operate even when heme alkylation is conclusively demonstrated. In particular, it is now clear that protein alkylation competes in some instances with heme alkylation.

3.3.1. Terminal Olefins

The P450-catalyzed epoxidation of terminal olefins is paralleled, in many instances, by N-alkylation of the prosthetic heme group and inactivation of the enzyme (see Chapter 8, Fig. 11).[133] The realization that self-catalyzed heme alkylation is a relatively common phenomenon evolved from early studies with 2-isopropyl-4-pentenamide (AIA) and

5-allyl-substituted barbiturates such as secobarbital,[134,135] which established that the oxidative metabolism of homoallylic amides results in (1) equimolar loss of P450 and microsomal heme, (2) accumulation of green (red-fluorescing) porphyrins, and (3) derangement of the heme biosynthetic pathway.

A double bond with no more than one substituent is the only structural prerequisite for prosthetic heme alkylation by olefins. This is clearly demonstrated by the ability of ethylene, but not ethane, to destroy the prosthetic heme group, and by the inactivity of structures such as 3-hexene, cyclohexene, and 2-methyl-1-heptene.[136] Even monosubstituted olefins do not inactivate the enzyme if they are not accepted as substrates by the enzyme, if a group other than the double bond is oxidized, or if the double bond, as in styrene, is part of a conjugated system.[136] Indeed, studies of the oxidation of styrene by a model iron porphyrin system indicate that heme alkylation only occurs once in 10,000 turnovers,[137] a ratio of product formation to heme alkylation much higher than is commonly observed with unconjugated terminal olefins, where usually less than 300 substrate molecules are oxidized per heme alkylation event.[7] These observations suggest that heme alkylation by olefins is subject to strong steric constraints and is suppressed by substituents that delocalize charge or electron density from the double bond.

The structures of the N-alkylated porphyrins isolated from the livers of rats treated with a variety of olefins (ethylene, propene, octene, fluroxene, 2,2-diethyl-4-pentenamide, 2-isopropyl-4-pentenamide, and vinyl fluoride) have been unambiguously elucidated by spectroscopic methods. Analogous products are probably formed in inactivation mediated by other olefins, such as the inactivation of P450 2E1 by the garlic components diallyl sulfide and diallylsulfone,[138] but the adducts have not been isolated. In all of the biological adducts that have been characterized, a porphyrin nitrogen is bound to the terminal carbon of the double bond and an oxygen to the internal carbon (Fig. 10).[135,139] The oxygen in the ethylene and AIA adducts has been shown by ^{18}O studies to derive from molecular oxygen

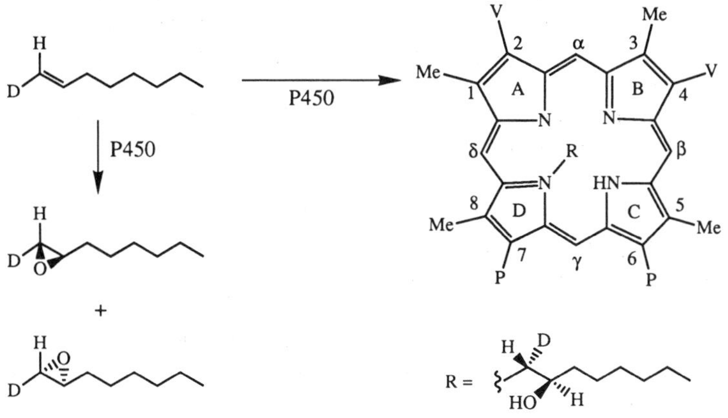

FIGURE 10. Structures of the heme adducts and the epoxide metabolites isolated from incubations of trans-1-[1-^{2}H]octene with liver microsomes from phenobarbital-induced rats. The heme substituents are: Me, CH_3; V, $CH=CH_2$; P, $CH_2CH_2CO_2H$. The peripheral carbons and the meso positions of the porphyrin ring are labeled.

and therefore is presumed to be the catalytically activated oxygen.[139,140] The not unreasonable hypothesis that the heme adduct is formed by nucleophilic addition of the porphyrin nitrogen to the terminal carbon of the epoxide metabolite is untenable because (1) the enzyme is not inactivated by the epoxides of olefins that destroy the enzyme,[133,136] (2) the nitrogen and the oxygen add across the double bond in a *cis* fashion rather than in the *trans* fashion expected for addition of a nucleophile to an epoxide,[140] (3) the nitrogen reacts with the terminal rather than internal carbon of vinyl ethers even though the internal (oxygen-substituted) carbon is more reactive in the corresponding epoxides,[141] and (4) the pyrrole nitrogens of the heme are very poor nucleophiles and do not react with epoxides even under harsh chemical conditions. These results, in conjunction with the clear demonstration that enzyme turnover is required for enzyme inactivation, indicate that enzyme inactivation is initiated by catalytic oxygen transfer to the double bond but is *not* mediated by the epoxide metabolite.

Linear olefins (ethylene, propene, octene) only detectably alkylate pyrrole ring D of the prosthetic group of the phenobarbital-inducible P450 enzymes from rat liver (Fig. 10), but heme alkylation by two "globular" olefins (2-isopropyl-4-pentenamide and 2,2-diethyl-4-pentenamide) is less regiospecific.[139,142] A detailed study of the regiochemistry and stereochemistry of heme alkylation by *trans*-$[1\text{-}^2H]$-1-octene shows that (1) the olefin stereochemistry is preserved during the alkylation reaction and (2) heme alkylation only occurs when the oxygen is delivered to the *re* face of the double bond even though stereochemical analysis of the epoxide metabolite indicates that the oxygen is delivered almost equally to *both* faces of the π-bond.[140] Heme alkylation is thus a highly regio- and stereospecific process.

Chemical models have recently succeeded in reproducing the heme alkylation and, in the process, in confirming and extending the mechanistic information first gleaned through enzymatic studies.[143–150] These studies have shown that the oxidation of terminal olefins by iron porphyrins results in alkylation of the porphyrin nitrogens to give the same types of adducts as are obtained during the oxidation of terminal olefins by P450.[143–146] As found for the biological reaction, the oxygen is added to the olefin from the same side as the porphyrin nitrogen, is not mediated by the epoxide or aldehyde, and is subject to steric effects from substituents on the olefin.[143–145] However, in the case of the model systems, adducts are obtained from disubstituted olefins and to some extent from primary olefins with the porphyrin nitrogen attached to the internal carbon and the oxygen to the terminal carbon.[143,147] Spectroscopic studies suggest that norbornene gives a transient species tentatively identified as the N-alkylated porphyrin.[148,149] This addition appears to be reversible and has led to the suggestion that secondary N-alkyl adducts may reverse to eventually result in the observed primary adducts.[147,150,151] Reversibility of N-alkyl adduct formation has not been detected with P450. However, recent work has shown that the catalytic oxidation of terminal olefins by chloroperoxidase/H_2O_2 results in N-alkylation of the heme group. Furthermore, time course studies using absorption spectroscopy and mass spectrometry indicate that N-alkyl adduct formation is reversible in this protein, with up to 80% of the enzyme activity being recovered over a period of several hours at 25 °C.[152]

A number of heme alkylation mechanisms are consistent with the available data, none of which involves single-step transfer of the oxygen to the π-bond. Subsequent to the proposed formation of a charge transfer complex between the ferryl species and the olefin π-bond (see Chapter 8), the oxygen could add to the π-bond to give a transient carbon radical that alkylates the heme, closes to the epoxide, or transfers the unpaired electron to the heme before alkylating the heme. It is probable that the partition between metabolite formation and heme alkylation is determined by the regiochemistry (i.e., inner or outer carbon) of oxygen addition to the π-bond. The oxidation of olefins by P450 can also be explained by initial addition of the oxoiron complex to the π-bond, directly or via a radical intermediate, to give the two possible metallacyclobutane intermediates (see Chapter 8, Fig. 14). The partitioning between epoxide formation and heme alkylation then could be determined by the relative importance of the metallacyclobutane with the oxygen bound to the internal carbon versus the metallacyclobutane with the oxygen bound to the terminal carbon. However, the details of the heme alkylation mechanism, the parameters that govern partitioning between epoxidation and heme alkylation, and the extent to which the heme alkylation mechanism mirrors that which results in epoxide formation, remain to be clarified. Differentiation of the mechanistic possibilities is complicated by the finding from model studies that disfavored *N*-alkyl adducts can be reversibly formed.[143,147] If this is true for the enzymatic reaction, information in addition to the structures of the isolated heme adducts will be required to determine the details of the mechanism of heme *N*-alkylation during the catalytic oxidation of olefins.

3.3.2. Terminal Acetylenes

The oxidation of terminal acetylenes to substituted acetic acids by P450 (Chapter 8) is even more frequently accompanied than in the case of terminal olefins by alkylation of the prosthetic heme group. The relationship between structure and activity resembles that for olefins, except that there are fewer exceptions to the rule that oxidation of triple bonds results in enzyme inactivation by either heme or protein modification. This is evident in the fact that phenylacetylene, but not styrene, inactivates the enzyme,[88] and that internal acetylenes inactivate the enzyme, albeit without the formation of detectable heme adducts, whereas internal olefins do not.[136,153] Enzyme inactivation requires catalytic turnover of the inhibitor and yields, in the case of terminal acetylenes, heme adducts similar to those obtained with terminal olefins.[141,142] The primary difference in the structure of the adducts is that addition of a hydroxyl group and a porphyrin nitrogen across the triple bond yields an *enol* structure that is isolated in the more stable keto form.[141,142] The other difference is that linear acetylenes react with the P450 enzymes of phenobarbital-induced rat liver almost exclusively to give the adduct with the nitrogen of pyrrole ring A whereas, as noted before, linear olefins react with the nitrogen of pyrrole ring D.[142] This difference in the alkylation regiochemistry has been used to propose an active site topography.[142]

The mechanisms written for the inactivation of P450 by terminal olefins also apply to its inactivation by terminal acetylenes, with the proviso that an additional double bond is carried by all of the reaction intermediates (see Chapter 8) and that the oxidation of an acetylene is relatively difficult. The oxidation of a triple bond is furthermore distinguished by the fact that the carbon to which the oxygen is *initially* attached can be discerned, whereas this information is lost when an olefin is oxidized because of the symmetry of the final epoxide metabolite. The acetylene data demonstrate that the oxygen is added to the

FIGURE 11. Partitioning of terminal acetylene oxidation between ketene production and heme alkylation is determined by the carbon to which the P450 activated oxygen is added. The broad arrows indicate to which carbon the oxygen is added for the indicated

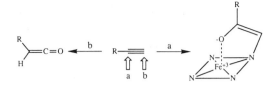

pathway. Removal of the iron from the adduct results in tautomerization of the N-alkyl enol to the ketone.

terminal carbon in the metabolite but the internal carbon in the heme adduct. Addition of the oxygen to the terminal carbon gives a ketene that can be hydrolyzed to the isolated carboxylic acid metabolites (Fig. 6). It is now also clear, as already discussed, that the ketene or a closely related species can also react with the protein, as opposed to the heme, to cause enzyme inactivation (see Section 3.1.2).[65,68] This difference in the regiochemistry of oxygen addition, and the finding that replacement of the acetylenic hydrogen in phenylacetylene by deuterium decreases the rate of metabolite formation without altering the rate of heme alkylation, indicate that commitment of a given catalytic turnover to metabolite formation rather than heme alkylation occurs prior to the oxygen transfer step (Fig. 11).[88] The factors that determine whether the oxygen is added to the internal or external carbon of the triple bond, and therefore whether heme or protein modification is observed, remain to be elucidated. The fact that 1-ethynylpyrene and phenylacetylene inactivate CYP1A2 by, respectively, protein and heme modification clearly shows that the reaction regiochemistry is controlled by specific substrate–enzyme interactions rather than simply being a property of the enzyme in question.[70]

The acetylenic function is particularly useful for the construction of P450 form-selective or specific irreversible inhibitors. It has been used (Section 4) as the enabling moiety in isoform-selective or specific inhibitors of P450$_{scc}$, aromatase, and the P450 enzymes that oxidize saturated fatty acids, arachidonic acid, and leukotriene B$_4$ (Section 5). It may also play a role in the alterations of oxidative metabolism observed in individuals treated with ethynyl sterol drugs such as gestodene.[154]

3.3.3. Dihydropyridines and Dihydroquinolines

The perturbation of heme biosynthesis and the decrease in hepatic P450 content that follow administration of 3,5-bis(carbethoxy)-2,4,6-trimethyl-1,4-dihydropyridine (DDC)[155–157] have been traced to alkylation of the prosthetic heme of P450 by this agent.[135,158,159] Heme alkylation is detected if the substituent at position 4 of the dihydropyridine ring is a primary, unconjugated moiety (methyl, ethyl, propyl, *sec*-butyl, nonyl), but not if it is an aryl (phenyl), secondary (isopropyl), or conjugated (benzyl) group.[160–163] The 4-aryl-substituted dihydropyridines do not inactivate the enzyme at all, but those with secondary or conjugated substituents inactivate the enzyme but do not yield detectable N-alkyl heme adducts. Dihydropyridines with simple 4-alkyl groups cause not only heme N-alkylation but inactivation of the enzyme by a mechanism that results in covalent binding of the porphyrin or its degradation products to the protein (Section 3.4).

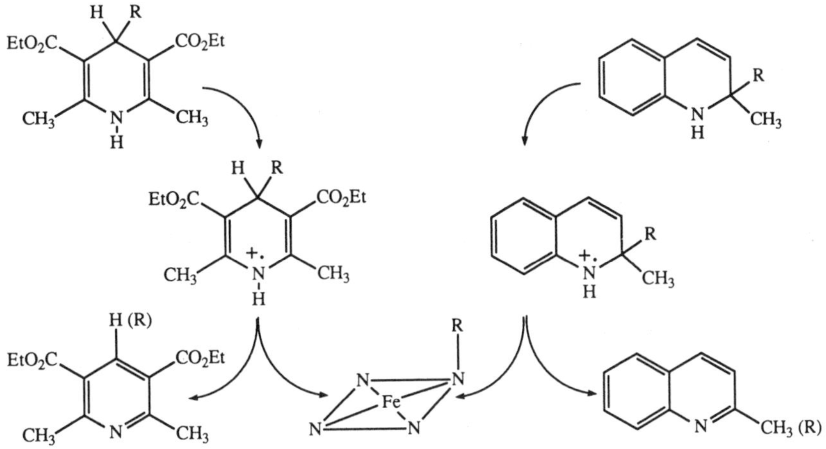

FIGURE 12. Oxidative activation of 4-alkyl-1,4-dihydropyridines and 2,2-dialkyl-1,2-dihydroquinolines to putative radical cation intermediates that either lose a second electron and a proton to give an aromatic metabolite or eliminate an alkyl radical that N-alkylates the prosthetic heme group of P450. Elimination of the alkyl radical simultaneously results in formation of the dealkylated aromatic metabolite.

The mechanisms of enzyme inactivation and heme destruction by analogues that do not give identifiable heme adducts remain obscure (see Section 3.4), but the mechanism of analogues that alkylate the heme group is better understood. The adducts consist of protoporphyrin IX with the 4-alkyl group of the parent substrate covalently attached to one of the nitrogen atoms (Fig. 12).[160,164,165] Evidence exists that different nitrogens are alkylated in different P450 enzymes,[166,167] and that the dihydropyridines cause isoform-selective inactivation.[167–169] Oxidation of the 4-alkyl-1,4-dihydropyridines is therefore accompanied by transfer of the 4-alkyl group to the prosthetic heme. The nature of the enzyme inactivation is further clarified by the following observations: (1) oxidation of the 4-alkyldihydropyridines, but not 4-aryldihydropyridines, results in pyridine formation with partial loss of the 4-alkyl group,[170,171] (2) substitution of the dihydropyridine nitrogen with an N-methyl or N-ethyl moiety does not prevent inactivation although inactivation may follow N-dealkylation because the preferential reaction with those substrates is N-dealkylation rather than dihydropyridine aromatization,[170] (3) there is no primary isotope effect on enzyme inactivation when the hydrogen at position 4 is replaced by deuterium,[170] (4) the heme adducts that are formed are chiral and thus are clearly formed within the active site,[172] and (5) incubation of the 4-ethyldihydropyridine analogue with hepatic microsomes in the presence of a spin trap results in turnover-dependent accumulation of the ethyl radical spin adduct.[160] Studies with desferoxamine-washed microsomes suggest, however, that most if not all of the radical that can be spin-trapped is formed by transition metal-catalyzed oxidation of the dihydropyridine.[173] However, as neither glutathione nor the radical trap blocks enzyme inactivation, radicals formed in the medium are apparently not involved in the heme alkylation reaction. Conversely, it is likely that radicals formed in the active site are not readily detected in the medium. The results strongly suggest that electron abstraction from the dihydropyridine generates a radical cation that aromatizes either by extruding the 4-alkyl group as a radical (see Chapter 8, Fig. 22) or by directly transferring the alkyl group to the heme. The 4-alkyl group may react directly with the porphyrin nitrogen atom but model studies suggest that it probably reacts first with the iron atom to form an alkyl–iron complex from which it migrates to the porphyrin nitrogen atom (see Section 3.3.4). The absence of heme adducts in the inactivation caused by the 4-isopropyl and 4-benzyl analogues is readily rationalized by such a mechanism, not only because the iron–nitrogen shift is sensitive to steric effects,[174] but also because the more oxidizable secondary or benzylic moieties may be converted to the corresponding cations by electron loss to the iron in preference to undergoing the oxidative iron–nitrogen shift.

The oxidation of dihydropyridines to radical cations that aromatize by radical extrusion suggests that other structures may behave similarly. This possibility is confirmed by the demonstration that the P450-catalyzed oxidation of 2,2-dialkyl-1,2-dihydroquinolines results in enzyme inactivation and heme adduct formation. One of the 2-alkyl substituents of the dihydroquinoline is found covalently attached in the heme adduct to a nitrogen of protoporphyrin IX.[175] The analogy with the action of 4-alkyl-1,4-dihydropyridines suggests the operation of a similar mechanism (Fig. 12).

3.3.4. Alkyl- and Arylhydrazines and Hydrazones

Phenelzine, an alkylhydrazine for which the P450-destructive mechanism is relatively well defined, causes an approximately equimolar loss of enzyme and heme when incubated with hepatic microsomes.[176] The enzyme and heme losses are paralleled by the formation of a prosthetic heme adduct identified as N-(2-phenylethyl)protoporphyrin IX (Fig. 13).[177]

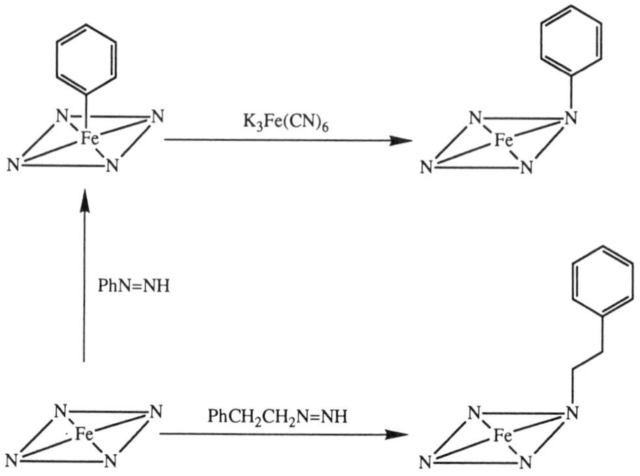

FIGURE 13. Alkylation of the prosthetic heme group of cytochrome P450 during the oxidative metabolism of phenelzine ($PhCH_2CH_2NHNH_2$) or phenylhydrazines ($PhNHNH_2$). The reaction with phenylhydrazine yields an iron–phenyl complex. The phenyl group can be induced to migrate from the iron to the porphyrin nitrogens within the intact active site by treatment with ferricyanide. A comparable iron complex is not readily detected for phenelzine or other alkylhydrazines. Two-electron oxidation of the hydrazines to the diazenes (RN=NH) is a prerequisite to heme modification.

A role for the 2-phenylethyl radical in this enzyme inactivation is suggested by formation in a hepatic microsomal system of products that implicate the 2-phenylethyl radical as a major metabolic species.[178] Spin-trapping experiments confirm that the 2-phenylethyl radical is generated in such incubations but much of the radical that can be spin-trapped is formed by transition-metal- rather than P450-catalyzed reactions.[177,179] The results suggest that the N-(2-phenylethyl) adduct is formed by oxidation of phenelzine to the 2-phenylethyl radical within the P450 active site where it is captured by the heme (Fig. 13). The 2-phenylethyl radical formed in solution by autooxidation is almost certainly too short-lived to diffuse into the enzyme active site and participate in enzyme inactivation. The 2-phenylethyl group could react directly with the porphyrin nitrogens, but by analogy with the reactions of hemoproteins with arylhydrazines (see below) it is likely that the alkyl radical is trapped by reaction with the iron atom to give an unstable alkyl–iron complex that subsequently rearranges to the isolated N-alkyl adduct.

The reaction of P450 with phenylhydrazine and N-phenylhydrazones yields a complex with an absorbance maximum at 480 nm. Formation of this complex inactivates the enzyme and precedes irreversible destruction of the prosthetic heme group.[180–183] A complex with an absorbance maximum at ~430 nm is similarly formed in the reactions of myoglobin, hemoglobin, and catalase with phenylhydrazine.[184–188] The complexes formed with myoglobin and P450$_{cam}$ have specifically been shown by X-ray crystallography to have the phenyl group bound end-on to the iron, as required for a σ-aryl–iron complex (Fig. 13).[183,189] Extraction of the heme complex from P450$_{cam}$ under oxidative, denaturing conditions, as found for the complexes from myoglobin, hemoglobin, and catalase, yields an equal mixture of the four possible N-phenylprotoporphyrin IX regioisomers.[183,190] Extraction of the prosthetic group under *anaerobic* conditions, in contrast, provides the intact phenyl–iron heme complex which has been characterized by absorption spectroscopy and NMR.[183,187] If the phenyl–iron heme complex obtained by anaerobic extraction is exposed to oxygen or other oxidizing agents under acidic conditions, the phenyl migrates from the iron to the porphyrin nitrogen atoms (Fig. 13).[187] This migration is sensitive to steric effects because the aryl moiety in aryl–iron complexes obtained from *ortho*-substituted phenylhydrazines does not undergo the oxidative shift.[174] A variety of arylhydrazines has been shown to work with specific P450 enzymes, among which the most common are phenyl-, 2-naphthyl-, and *p*-biphenylhydrazine.[191] Finally, addition of ferricyanide to the intact P450 complexes has been shown to result in migration within the intact active site of the phenyl group from the iron to the porphyrin nitrogens.[192–194] The distribution of the four regioisomers of protoporphyrin IX thus produced is a function of the active site topology and varies markedly from enzyme to enzyme. The *in situ* migration of aryl groups within the active sites of P450 enzymes provides a tool for the determination of active site topology because the regioselectivity of the migration is controlled by the degree to which the active site is sterically unencumbered above each of the four pyrrole ring nitrogens.[191–197] *In situ* migration works for most P450 iron–aryl complexes but does not work for the corresponding complexes of proteins such as myoglobin in which the fifth iron ligand is an imidazole rather than a cysteine thiolate.

The possibility that alkyl radicals, like their aryl counterparts, bind to the iron before shifting to the nitrogen is supported by the observation that the type II complexes formed between alkyldiazenes and P450 in the absence of oxygen are converted, when limited amounts of oxygen are introduced, to complexes with the absorption maximum in the vicinity of 480 nm characteristic of complexes with an iron–carbon σ-bond.[181,198] Model alkyl diazene–iron tetraphenylporphyrin complexes have been shown to exist under anaerobic conditions.[199] Alkyl–iron complexes, however, are much less stable and less well characterized than aryl–iron complexes and their role in heme N-alkylation reactions remains unproven.

3.3.5. Other N–N Functions

In addition to the reactions with alkyl- and arylhydrazines, the prosthetic heme of P450 is N-alkylated or N-arylated by reactive intermediates produced during the oxidation of 1-aminoaryltriazoles, 2,3-bis(carbethoxy)-2,3-diazabicyclo[2.2.0]hex-5-ene, and the sydnones.

1-Aminobenzotriazole (ABT) is oxidized by chemical reagents to benzyne, an exceedingly reactive species, and two molecules of nitrogen.[200] The P450-catalyzed oxidation of ABT apparently follows a similar reaction course because benzyne, or its equivalent, has been shown to add across two of the nitrogens of the prosthetic heme group.[201,202] The benzyne may add directly to the two nitrogens, generating an N,N-bridged species that autooxidizes to the isolated bridged porphyrin, or may first bridge the iron and a nitrogen of the heme and subsequently rearrange to the N,N-bridged species (Fig. 14). A broad range of P450 enzymes are inactivated by ABT without detectable toxic effects.[201–205] Destructive activity is retained if substituents are placed on the phenyl ring or on the exocyclic nitrogen, or if the phenyl framework is replaced by other aryl moieties.[202,206] P450 isoform and tissue selectivity is conveyed by placing substituents on the exocyclic nitrogen of the aminotriazole function.[206–208] It is not known if the oxidation of ABT to benzyne involves hydroxylation of the exocyclic nitrogen or initial oxidation to a radical or radical cation (Fig. 15). The similarity between the activation mechanism proposed for ABT (Fig. 15) and other 1,1-disubstituted hydrazines (Fig. 8) is to be noted. ABT is a highly effective

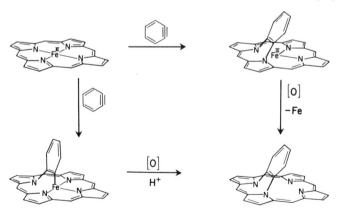

FIGURE 14. The two alternatives for formation of the heme adduct in the reaction of heme with benzyne generated enzymatically from ABT. The peripheral porphyrin substituents have been omitted for clarity.

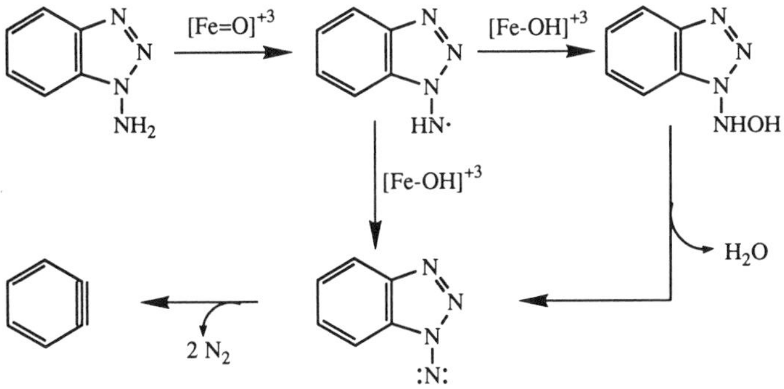

FIGURE 15. Mechanistic alternatives for the P450-catalyzed generation of benzyne from ABT.

agent for the *in vivo* inactivation of P450 enzymes in plants,[209,210] insects,[211] and animals.[203,212–214]

Cyclobutadiene, which may exist as a rectangular structure with a singlet electronic state or as a square structure with a triplet electronic state, can be generated by chemical oxidation of 2,3-diazabicyclo[2.2.0]hex-5-ene.[215] Bis(carbethoxy)-2,3-diazabicyclo-[2.2.0]hex-5-ene (DDBCH) is a mechanism-based irreversible inhibitor of P450 that exploits the basic reactivity of the parent bicyclic system.[216] The bis(carbethoxy) derivative was selected for enzymatic work because the parent bicyclic hydrazine autooxidizes too readily to be of biological utility. The prosthetic heme group of the enzyme is converted by the inactivation reaction into the *N*-2-cyclobutenyl derivative (Fig. 16). The secondary, allylic, *N*-alkyl moiety in this adduct makes it much less stable than other adducts, which bear primary, unactivated, *N*-alkyl groups. The failure of internal olefins and acetylenes to detectably alkylate the prosthetic heme group suggests, in fact, that secondary carbons are generally too sterically encumbered to react with the heme. The 2-cyclobutenyl adduct implicates cyclobutadiene, or a closely related species, in heme alkylation, although the precise nature of the reactive species remains undefined. The observed adduct is readily explained by addition of cyclobutadiene to a nitrogen of the heme to give a transient intermediate that abstracts a hydrogen from an active-site residue. The transient interme-

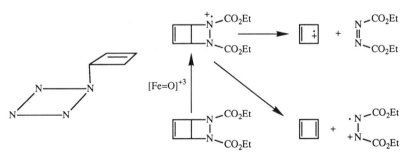

FIGURE 16. Alternative mechanisms for the oxidative generation of a cyclobutadienoid species that alkylates the prosthetic heme group of P450. The structure of the porphyrin obtained by removal of the iron from the heme adduct is also shown.

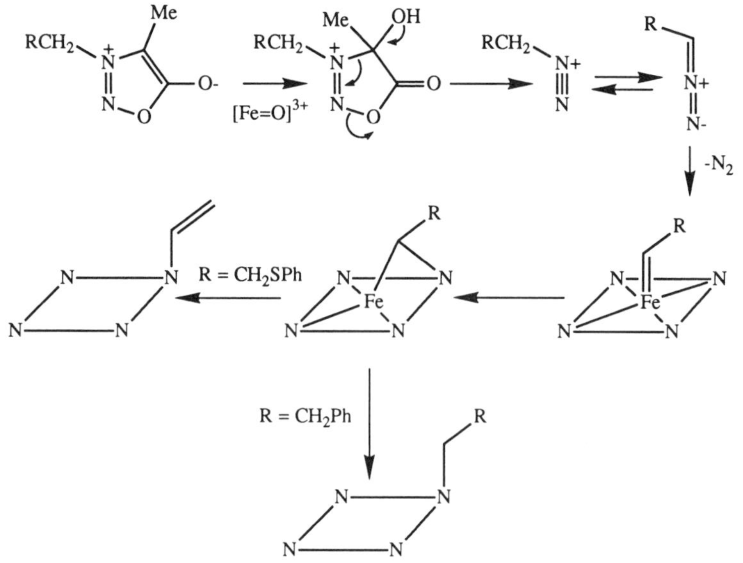

FIGURE 17. Mechanism proposed for the P450-catalyzed oxidation of sydnones to species that alkylate the prosthetic heme, and structures of the N-alkylporphyrins isolated from rats treated with the 3-(2-phenylthioethyl) and 3-(2-phenylethyl) sydnones.

diate could be stabilized by bond formation between the iron of the heme and the carbon that eventually abstracts the active-site hydrogen. As shown in Fig. 16, electron abstraction from DDBCH can lead to the observed adduct by several pathways that differ in whether the cyclobutadiene is neutral, cationic, or anionic.

The report that a fluorescent hepatic pigment accumulates in the livers of dogs and rats administered a sydnone derivative[217] led to the discovery that the sydnone is catalytically activated by P450 to a species that alkylates the prosthetic heme group.[218] The heme adduct isolated from rats treated with the sydnone has been identified as N-vinylprotoporphyrin IX (Fig. 17). Activation of the sydnone by hydroxylation of the electron-rich zwitterionic carbon, followed by ring opening and elimination of the carboxylic fragment to give the diazo species (Fig. 17), is suggested by this finding. The chemical oxidation of sydnones has been shown to proceed by such a mechanism.[219] The diazoalkane, in turn, reacts with the heme, possibly via an initial carbene complex, to give a nitrogen–iron bridged species. The formation of such bridged nitrogen–iron species has been documented in model porphyrin systems.[220,221] The negative charge on the carbon in the bridged intermediate finally eliminates the thiophenyl moiety and generates the N-vinyl adduct.[218] Further light is shed on the mechanism by the observation that the inactivation of P450 by 3-(2-phenylethyl)-4-methylsydnone is associated with the formation of both N-(2-phenylethyl)- and N-(2-phenylethenyl)protoporphyrin IX.[222] These results suggest that oxidation of the sydnone generates the (2-phenylethyl)diazonium cation that can react with the heme in two ways. In the mechanism analogous to that discussed above, deprotonation of the diazonium intermediate results in a carbene-like addition. In the absence of a β-leaving group, the carbanion intermediate thus formed is apparently oxidized to a cation that is deprotonated to introduce the double bond into the N-alkyl group.[222] On the other hand, reduction of the diazo intermediate prior to deprotonation presumably yields a phenyldiazenyl radical that adds to the heme nitrogen atom in a reaction entirely analogous to that proposed for the 2-phenylethyl radical produced by oxidation of phenylethyldiazene (Fig. 13).[177] This mechanism is supported by the finding that the phenylethyl adduct obtained from the 1,1-dideuterated substrate retains both deuteriums and therefore arises by a mechanism that does not involve deprotonation of the diazo intermediate.[222]

Diethylnitrosamine ($Et_2N-N=O$), when administered *in vivo* to mice, has been reported to cause the formation of an alkylated porphyrin tentatively identified by mass spectrometry as N-(2-hydroxyethyl)protoporphyrin IX.[223] *In vitro* studies with rabbit liver microsomes and purified P450 enzymes have independently shown that diethylnitrosamine can be converted to ethylene by a P450-mediated reaction.[224] Although not explicitly demonstrated, it is likely that the proposed N-(2-hydroxyethyl) porphyrin adduct is derived from P450 enzymes that are inactivated by the ethylene metabolically generated from diethylnitrosamine (see Section 3.3.1).

3.3.6. Other Functionalities

The herbicide 1-[4-(3-acetyl-2,4,6-trimethylphenyl)-2,6-cyclohexanedionyl]-O-ethyl propionaldehyde oxime (ATMP; see Fig. 18) has been found to cause hepatic protoporphyria in the mouse, albeit not in other species, and the evidence suggests that derangement of the porphyrin biosynthetic pathway is linked to inactivation of P450 via a heme-alkylation mechanism.[225] A pigment with the HPLC properties of N-methylprotoporphyrin IX has been isolated from mice treated with ATMP and, in accord with a P450-dependent process, its formation is decreased by preadministration of the P450 inhibitors SKF 525A or piperonyl butoxide.[225] Replacement of the ethyl on the oxime carbon by a propyl suppresses the porphyrinogenic activity of the analogue, presumably

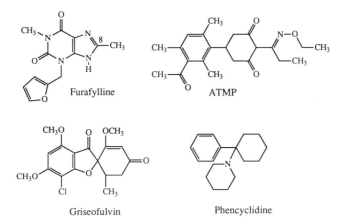

FIGURE 18. Structures of ATMP, griseofulvin, furafylline, and phencyclidine, all of which are reported to cause heme alkylation but for which the products and/or mechanisms are unclear.

by preventing the heme alkylation reaction. Apart from this clue, nothing is known about the mechanism by which ATMP mediates the P450-dependent formation of an N-alkyl (possibly N-methyl) protoporphyrin IX adduct.

Griseofulvin, an agent long known to cause hepatic porphyrias, has been reported to cause the catalysis-dependent destruction of P450 and the formation of a green pigment in mice.[226] Characterization of the pigment by copper-mediated transfer of the N-alkyl group to an amine followed by mass spectrometric analysis of the alkylated amine suggests that N-methylprotoporphyrin IX is formed as a minor product and a porphyrin with the entire griseofulvin structure bound to the nitrogen as the major product.[227,228] The detailed structures of the adducts, however, have not been established and the reaction may be species-selective because much lower levels of pigment formation are reported to occur in rats,[229,230] and efforts in some laboratories to detect such products in rats have not been successful. The mechanism for the formation of the proposed N-methyl heme adduct during the metabolism of griseofulvin is particularly unclear because a pigment identical to it is reportedly present in lower amounts in the livers of control mice.[229,230] Although the *in vivo* formation of an N-alkylporphyrin in untreated mice would be of considerable interest, more definitive evidence on the structures of the adducts and their origin is required to evaluate the significance of these results and to postulate mechanisms for formation of the two types of adducts.

Furafylline, a potent inhibitor selective for human CYP1A2, causes time- and NADPH-dependent inactivation of the enzyme.[231,232] CYP 1A1, 2A6, 2B6, 2C9, 2C19, 2D6, 3A4, and 2E1 are not similarly inhibited but evidence suggests that other enzymes can be inhibited (see Chapter 14).[232] The activity loss is paralleled by a corresponding loss of the heme chromophore, with $K_I = 23$ μM, $k_{inact} = 0.87$ min^{-1}, and a partition ratio of 3–6 for the average number of substrate molecules oxidized by an enzyme molecule before it is inactivated.[231] Although the mechanism of the inactivation reaction is not known, replacement of the C-8 methyl hydrogens by deuteriums produces an isotope effect of approximately 2.0 on k_{inact} but not on K_I, and removal of the methyl group results in an inactive agent.[231] These results suggest that oxidation of the C-8 methyl contributes to the inactivation process but the mechanism of the inactivation is unclear.

Phencyclidine is oxidized to an iminium derivative that binds to proteins but also appears to undergo further NADPH-dependent metabolism by some P450 enzymes that results in inactivation of those enzymes.[233–235] The inactivation reaction is reported to involve parallel loss of the activity and the heme chromophore.[234] Analogues of the phencyclidine imine in which the size of the heterocyclic ring is varied cause similar activity and heme loss.[236] The phencyclidine imine reaction is reported to result in the accumulation of modified porphyrins, but the nature of these porphyrins was not established.[234] It is therefore not possible to determine whether heme modification or degradation is causally or incidentally connected with the loss of P450 activity.

The inactivation mechanism remains unclear with some agents. For example, the inactivation of CYP2E1 by 3-amino-1,2,4-triazole does not result in incorporation of radioactivity into the protein, the formation of P420, or the loss of heme even though the inactivation is time- and NADPH-dependent.[237]

3.4. Covalent Binding of Modified Heme to the Protein

In the course of the catalytic oxidation of some substrates, certain P450s (i.e., CYP3A, CYP2E) undergo mechanism-based inactivation as a result of conversion of their prosthetic heme groups to products that irreversibly bind to the protein. Examples of such inactivators include CCl_4, spironolactone, 3,5-dicarbethoxy-2,6-dimethyl-4-ethyl-1,4-dihydropyridine (DDEP) and its 4-isopropyl and 4-isobutyl analogues.[81,238–244] A related process is mediated by peroxides such as H_2O_2 and cumene hydroperoxide (Chapter 8). These peroxides partially degrade the prosthetic heme group to monopyrrole and dipyrrole fragments,[238,245–247] but the bulk of the heme moiety is converted to as yet uncharacterized fragments that irreversibly bind to the protein. Several peptides have been isolated from cumene hydroperoxide-inactivated CYP3A and CYP2B1 and active site peptides that correspond to the distal I and proximal L helices of $P450_{cam}$ have been shown to be covalently bound to structurally uncharacterized heme fragment(s).[248,249]

Myoglobin and hemoglobin have been utilized as hemoprotein models in efforts to elucidate this unusual heme degradation process, although it is not yet clear that these hemoproteins are good models for the P450 reaction.[250–252] Oxidation of myoglobin with H_2O_2 results in covalent binding of the heme group via either the α or β-*meso* carbon or one of the vinyl groups to Tyr^{103}.[250] In contrast, the reductive metabolism of CCl_4 or $CBrCl_3$ by myoglobin results in covalent attachment of the prosthetic group to His^{93} via one of its vinyl groups.[252,253] In both of these processes, however, as well as during the hemoglobin-mediated reductive metabolism of $CBrCl_3$,[254] the cross-linked heme retains an absorption maximum at approximately 405 nm and is thus bound to the protein without any modification that alters the chromophore. Moreover, the myoglobin-mediated oxidative metabolism of alkylhydrazines yields γ-*meso* alkylated heme adducts without appreciable heme-protein cross-linking.[251] In contrast, binding of the prosthetic group to the protein in the reactions of P450 with H_2O_2, cumene hydroperoxide, DDEP, or spironolactone destroys the heme chromophore and is therefore associated with major degradation of the heme group.[83,238,241–244] This is also true if the cumene hydroperoxide-mediated inactivation is carried out under anaerobic conditions.[248]

The features that predispose an enzyme to the covalent attachment of a prosthetic heme fragment to the protein remain obscure. Studies with several substrates suggest that the generation of free radical products is important, but this alone is not enough because this covalent attachment does not occur with all of the P450 enzymes that produce such radicals. Thus, one-electron oxidation of DDEP by CYP2C6 and CYP2C11 results in *N*-alkylation of their heme groups, whereas oxidation of DDEP by CYP3A results predominantly in heme-protein cross-linking.[244] Furthermore, studies with DDEP and its analogues in which a secondary carbon is attached to the 4-position (4-isopropyl and 4-isobutyl) reveal that the course of the reaction is largely dictated by the specific P450 active site structure rather than by the inability of the inactivator to *N*-alkylate heme.[243] Thus, the relative extents of heme destruction and heme-protein binding observed with DDEP, which can form *N*-alkyl porphyrins, and with the 4-isopropyl and 4-isobutyl analogues, which cannot, are comparable.[243] This indicates that the heme-protein cross-linking pathway is not merely the result of a defective or inefficient heme *N*-alkylation process. Further support for the proposal that binding of heme fragments to the protein is isoform-determined is provided by the fact that spironolactone inactivates hepatic CYP3A via heme-protein cross-linking[83] but inactivates adrenal P450s by direct protein modification.[58] It is possible that the propensity of the CYP3A family of enzymes to heme-protein cross-linking is related to an unusual active site flexibility or polarity of these proteins. The CYP3A enzymes apparently accommodate not only substrates as large as cyclosporin, macrolide antibiotics, and FK506, but also smaller substrates such as DDEP and therefore may undergo a relatively large breathing motion or may have an unusually large active site water content.[255] Regardless of the mechanism, it is clear that P450 3A active sites are particularly susceptible to inactivation by this cross-linking process.

3.5. Other Modes of P450 Heme Degradation and Protein Denaturation

P450 is sometimes inactivated by mechanisms that involve destruction of the prosthetic heme group but that are not accompanied by the detectable formation of heme adducts. In some instances these reactions are associated with the binding of heme fragments to the protein (Section 3.4), but in most instances the incidence of heme-protein cross-linking has not been investigated. The destructive mechanisms of most peroxides,[245–247] halocarbons (CCl_4),[256] internal acetylenes (3-hexyne),[153] allenes (1,1-dimethylallene),[257] cyclopropylamines (N-methyl-N-benzylcyclopropylamine),[258,259] and benzothiadiazoles (5,6-dichloro-1,2,3-benzothiadiazole)[260] remain poorly characterized but are members of this somewhat indeterminate class. The internal acetylenes are probably enzymatically oxidized to reactive species analogous to those obtained from the terminal acetylenes but, unlike the terminal analogues, do not detectably N-alkylate the prosthetic heme. The heme adducts expected from N-alkylation by disubstituted acetylenes have been sought but have not been found despite the fact that they should be at least as stable as the adduct obtained with cyclobutadiene (Section 3.3.4). It is possible, of course, that oxidation of internal acetylenes results in covalent binding to the protein, but this possibility has not been investigated. Similar ambiguities cloud the destructive mechanisms of allenes ($RCH=C=CH_2$), cyclopropyl amines (Chapter 8), and benzothiadiazoles. A hypothetical activation mechanism can be formulated for each of these functionalities but experimental evidence to support the mechanisms is not available. Thus, for 5,6-dichlorobenzothiadiazoles, it is likely that oxidation of the heteroatomic ring systems results in loss of nitrogen and formation of an alkylating radical or cationic species (Fig. 19).

The destructive action of halocarbons, one of the most studied members of this class, was believed at one time to result from the secondary action of the lipid peroxides that are concomitantly formed. It now appears, however, that substances like CCl_4 destroy the prosthetic heme group directly rather than by a process mediated by lipid peroxides.[261–264] The reductive metabolism of halocarbons, including CCl_4, gives rise to semistable complexes with Soret maxima between 450 and 500 nm.[265–268] Model studies, including the isolation and detailed characterization of a dichlorocarbene–metalloporphyrin complex,[269] suggest that the long-wavelength Soret bands are related to complexes of halocarbenes with the reduced heme. The finding that carbon monoxide is a product of the reduction of CCl_4 by P450, a reaction that presumably occurs by a mechanism analogous to that proposed for the generation of carbon monoxide from methylenedioxyphenyl complexes (Section 3.2.1), supports this hypothesis.[270] Model porphyrin dichlorocarbene–iron com-

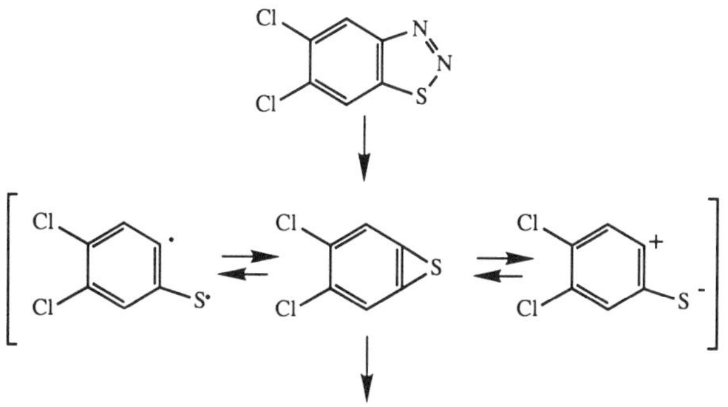

Enzyme inactivation

FIGURE 19. Activation of 5,6-dichlorobenzothiadiazole to a species that inactivates cytochrome P450.

plexes do, in fact, react with water to give carbon monoxide and with primary amines to give isonitriles.[270,271] Work with halothane suggests that, in some instances, complexes in which the halocarbon is σ-bonded to the ferric prosthetic heme group are formed in preference to carbene complexes.[272,273] It has been argued that halocarbons *only* give unstable σ-bonded complexes, but the data invoked in support of this view are not convincing because they were obtained in an experimental system where the complexes with Soret maxima at long wavelengths were not observed.[274]

The causal steps that link iron–alkyl complex formation to irreversible destruction of the enzyme and its heme moiety are not known, although the results of model studies with diaryl- and carbethoxy-substituted carbene complexes suggest that the halogenated carbenes may migrate to the nitrogens of the porphyrin.[275–278] The N-haloalkyl adduct obtained by migration of a dichlorocarbene would probably be unstable toward reaction with water and would therefore not be detected by the methods that have been used to isolate other N-alkyl porphyrins. It is not possible at this time, however, to rule out release of the halogenated carbenes or alkyl moieties as reactive species that attack the protein or the heme in an undetermined manner. Furthermore, the relationship of halocarbon reactions with the heme iron to the finding that reaction of CCl_4 with P450 2E1 results in cross-linking of heme fragment(s) to the protein is not clear at this time (see above).[239]

Denaturation of P450 by high (1–5 mM) concentrations of indomethacin and other nonsteroidal anti-inflammatory agents has been reported and evidence provided that denaturation is related to the surfactant properties of these molecules.[279] The loss of P450 seen when indomethacin is added to liver microsomes is paralleled by essentially stoichiometric appearance of a P420 peak. Although it is likely that other agents cause denaturation of P450, it is unclear to what extent the process is physiologically important because of the relatively high drug concentrations that are required.

4. P450 Enzyme Specificity

Isoform-specific P450 inhibitors are potentially of substantial importance not only as therapeutic, insecticidal, or herbicidal agents, but also as probes of the structures, mechanisms, and biological roles of specific P450 enzymes. Efforts to develop P450 form-specific inhibitors have focused on biosynthetic enzymes because their inhibition is potentially of greater practical utility. In addition, their relatively high substrate specificity makes them more amenable to specific inhibition. The broad, overlapping, substrate specificities of the P450s that primarily metabolize xenobiotics make it difficult to obtain isoform-specific rather than isoform-selective inhibitors.[280,281] Selective inhibitors of hepatic P450 enzymes, examples of which are provided by the amphetamines,[115] troleandomycin,[116,117] gestodene,[154] furafylline,[231] and 1-ethynylpyrene,[67,68] are fairly common (see Appendix B). A note of caution is appropriate with respect to the inhibitor specificities listed above and reported in the literature. Inhibitors can only be said to be specific to the extent that they have been tested against the diversity of P450 isoforms. The claim for specificity of inhibitors tested against only two or three isoforms is necessarily limited.

5. Inhibitors of Biosynthetic Enzymes

The inhibition of biosynthetic P450 enzymes and the potential clinical applications of such inhibitors have recently been reviewed.[5,282–284] The discussion below is therefore intended to delineate the diversity of inhibitors and their mechanisms rather than to provide a comprehensive review of the available inhibitors.

5.1. P450scc

The three oxidative steps required to cleave the side chain of cholesterol are catalyzed by a single P450 enzyme (P450scc) (see Chapter 8, Fig. 29 and Chapter 12). In a rational approach to the development of such inhibitors, amino[285–288] and thiol[289] functions have been placed on the cholesterol side chain at positions that favor chelation to the prosthetic heme iron (Fig. 20). These agents are potent reversible inhibitors (K_I = 25 to 700 nM) of P450scc. The most potent of these inhibitors is (22R)-22-aminocholesterol, in which the amine function is substituted for the first hydroxyl group inserted into the cholesterol side chain by the catalytic action of the enzyme.[288] (22S)-22-aminocholesterol, in which the amino function is located on the correct carbon but with the incorrect stereochemistry, is bound ~1000 times more weakly (K_I = 13 μM).

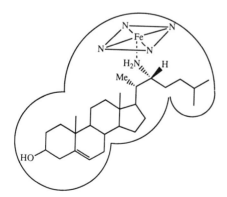

FIGURE 20. Most reversible inhibitors of P450scc and other biosynthetic P450 enzymes are constructed by the classic approach of coupling a hydrophobic ligand that binds well to the protein to a nitrogen or sulfur that coordinates to the heme iron (see Fig. 1). For P450scc, inhibitors of this type have been rationally obtained by placing an amino group at the position normally occupied by the first hydroxyl inserted into the cholesterol side chain by P450scc.

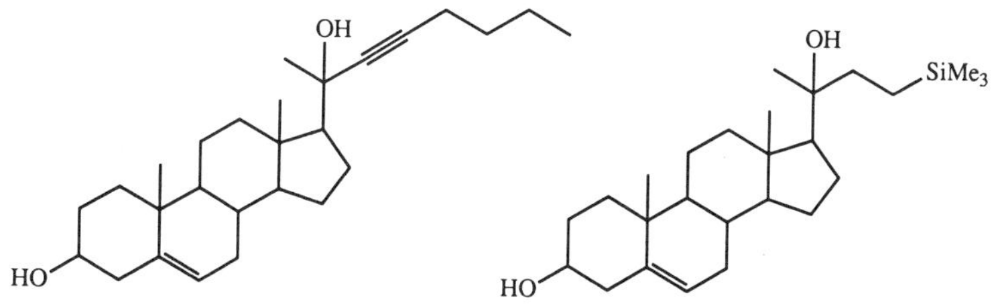

FIGURE 21. Mechanism-based inhibitors of $P450_{scc}$.

A number of mechanism-based, irreversible inhibitors of $P450_{scc}$ have been developed. Analogues of pregnenediol with an acetylenic group grafted into the side chain (Fig. 21) inactivate the enzyme in a time-, concentration-, and NADPH-dependent manner.[290-292] The heme chromophore is destroyed in the inactivation reaction but an alkylated heme, in agreement with the data on the inactivation of hepatic enzymes by internal acetylenes (see Section 3.4), is not detected. A mechanism-based inhibitor of $P450_{scc}$ is also obtained if the side-chain carbons beyond C-23 in 20-hydroxycholesterol are replaced by a trimethylsilyl group (Fig. 21).[293] One-electron chemical oxidation of 1-substituted 3-trimethylsilyl-1-propanols has been shown to yield ethylene, the trimethylsilyl radical, and an aldehyde:[294]

$$RCH(OH)CH_2CH_2Si(CH_3)_3 \rightarrow RCHO + CH_2{=}CH_2 + (CH_3)_3Si\cdot$$

The inactivation of $P450_{scc}$, if the chemical model is relevant, may therefore result from reaction of the enzyme with the trimethylsilyl radical, or from secondary oxidation of the ethylene liberated in the first catalytic turnover. An interesting variant of a mechanism-based inhibitor is provided by (20S)-22-nor-22-thiacholesterol, in which the sulfur that replaces the carbon at position 22 is oxidized by $P450_{scc}$ to a sulfoxide that is a potent but not irreversible inhibitor of the enzyme.[292,295,296]

5.2. Aromatase

The three-step transformation catalyzed by aromatase, the enzyme that controls the conversion of androgens to estrogens, is discussed in Chapters 8 and 12. Inhibitors of aromatase are potentially useful agents for the control of estrogen-dependent mammary tumors,[6,283,284,297-302] benign prostatic hyperplasia,[303,304] and, possibly, for the control of coronary heart disease.[305] Indeed, some of the more promising agents are now either in clinical use or in clinical trials. Aminoglutethimide, an inhibitor of aromatase, has been used to treat hormone-dependent metastatic breast carcinoma, but its poor specificity, particularly its inhibition of $P450_{scc}$, causes side effects that compromise its utility.[306] Replacement of the aminophenyl group in aminoglutethimide by a pyridine moiety has been reported to yield an agent [pyridoaminoglutethimide (3-ethyl-3-(4-pyridyl)piperidine-2,6-dione)] that inhibits aromatase without inhibiting $P450_{scc}$.[307] The basis for the enhanced specificity is not known, but the pyridine nitrogen is probably so positioned that it can coordinate with the heme of aromatase but not with that of $P450_{scc}$. It is interesting, in this context, that an aminoglutethimide analogue with a nitrogen at the opposite end is

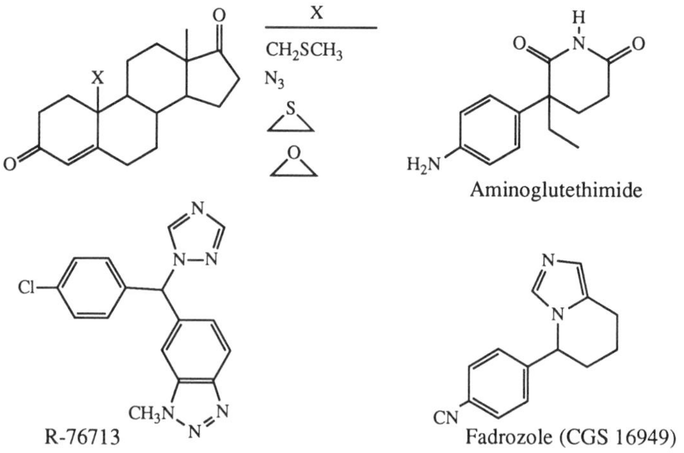

FIGURE 22. Structures of some competitive aromatase inhibitors.

more potent than glutethimide as an inhibitor of $P450_{scc}$ but has little or no activity against aromatase.[308] The 19-methyl group and carbon 22 of the sterol side chain are separated by a distance roughly equal to the length of the aminoglutethimide structure. It is therefore tempting to speculate that aminoglutethimide is bound in roughly the same orientation relative to the sterol substrate in the active sites of $P450_{scc}$ and $P450_{arom}$, and that the differential inhibitory activity of the two analogues derives from positioning of the nitrogen so that it can only coordinate with the $P450_{scc}$ prosthetic heme.

More recently, other potent inhibitors of aromatase that incorporate an imidazole function capable of coordinating tightly with the aromatase heme-iron have also been developed. Of these, fadrozole (CGS 16949) {4-(5,6,7,8-tetrahydroimidazo[1,5α]pyridin-5-yl)benzonitrile monochloride} (Fig. 22), and its congener CGS 18320, [bis-(p-cyano-phenyl)imidazo-1-yl-methane hemisuccinate], are the most promising breast cancer therapeutic agents.[309] Both CGS 16949 and CGS 18320 are relatively more selective inhibitors of aromatase than of $P450_{scc}$, P450 21, or P450 11B.[309,310] However, although CGS 16949 appears to inhibit aldosterone production (18-hydroxylase activity) in rats, this is much less pronounced with CGS 18320.[309] Preliminary findings in Phase I clinical trials of CGS 16949 indicate that it is a potent inhibitor of estrogen biosynthesis in postmenopausal women with advanced breast cancer.[310–313] In addition to being well tolerated by these patients, at the maximal effective dose it did not significantly alter cortisol, androstenedione, testosterone, or aldosterone levels, confirming its relative selectivity for inhibition of aromatase at the doses utilized.[310–313]

R-76713 [6-(4-chlorophenyl)(1H-1,2,4-triazol-1-yl)methyl]-1-methyl-1H-ben-zotriazole (Fig. 22) is a new, relatively selective, and very potent inhibitor of human placental aromatase.[314–316] Its (+) enantiomer (R 83842) exhibits the lower IC_{50} values for human placental aromatase activity, and is the most potent of the two enantiomers with no appreciable inhibition of other steroidogenic enzymes or liver microsomal P450s until concentrations exceed 1000 times the IC_{50} value for inhibition of aromatase.[315,316]

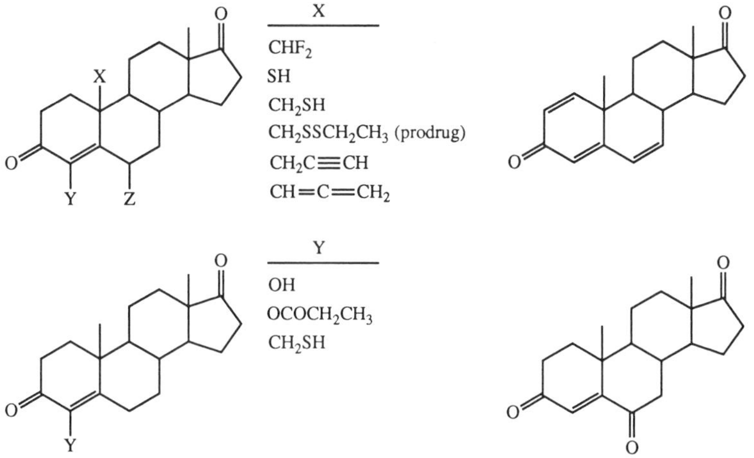

X
CHF$_2$
SH
CH$_2$SH
CH$_2$SSCH$_2$CH$_3$ (prodrug)
CH$_2$C$\equiv$CH
CH$=$C$=$CH$_2$

Y
OH
OCOCH$_2$CH$_3$
CH$_2$SH

FIGURE 23. Structures of some of the mechanism-based inactivators of aromatase.

As already described for P450$_{scc}$, powerful competitive inhibitors of aromatase have been rationally prepared by replacing the C-19 methyl with sulfur- or nitrogen-containing functions that can coordinate to the heme iron (Fig. 22).[317–321] The functionalities that have been used to replace the 19-methyl include CH$_3$SCH$_2$,[317–319] CH$_3$SCH$_2$CH$_2$,[320] HSCH$_2$,[321,322] RSSCH$_2$ (Org-30958, R = ethyl is best),[322] SH,[321] and NH$_2$.[323] The 19-methyl has also been replaced with the oxygen (oxiranyl), sulfur (thiiranyl), and nitrogen (aziridinyl) three-membered ring structures (Fig. 23).[324–328] The best of these 19-heteroatom-substituted steroids have K_I values in the 1 nM range. The epoxide, episulfide, and aziridine functions inhibit the enzyme by stereoselective coordination of the heteroatom to the iron atom but, despite their reactivity, apparently do not inactivate the enzyme. The steroids in which the 19-methyl group is replaced by a sulfhydryl or HSCH$_2$ group (Fig. 23), however, are irreversible mechanism-based inhibitors rather than simple competitive inhibitors.[321]

The development of mechanism-based inhibitors for aromatase reflects, to some extent, the strategies developed for inactivation of hepatic P450 enzymes. Replacement of the normally hydroxylated methyl group (C-19) by a propargylic or allenic moiety (Fig. 23) converts the sterol into an irreversible inhibitor of aromatase.[329–335] The details of the mechanisms by which these acetylenic and allenic agents inactivate the enzyme have not been clearly defined but are likely to involve, as demonstrated for the inactivation of hepatic P450 enzymes by acetylenic agents (see Sections 3.3.2 and 3.4), either heme or protein alkylation. A mechanism-based inhibitor is also obtained if the C-19 methyl is replaced by a difluoromethyl group (Fig. 23).[334–339] Tritium release studies with the tritium-labeled difluoromethyl derivative establish that C-19 hydroxylation is required for enzyme inactivation.[340] The difluoromethylalcohol thus produced presumably eliminates fluoride to give the reactive acyl fluoride that acylates a protein nucleophile.

The 19-substituted analogue of androst-4-ene-3,17-dione steroid inhibitors, Org-30958 [19-(ethyldithio)androst-4-ene-3,17-dione], is currently in Phase I clinical trials for estrogen-dependent breast cancer chemotherapy.[322] The ethyldithio substitution apparently renders the steroid more stable extracellularly than the free thiol Org-30365 (19-mercapto-androst-4-ene-3,17-dione), with 8-fold greater activity than either 4-OHA or SH-489 in *in vivo* assays of aromatase inhibition in several animal models. Its *in vivo* activity is apparently due to the intracellular reduction of its disulfide releasing the 19-mercapto analogue Org-30365, a potent mechanism-based aromatase inactivator,[321] that is a 16- or 8-fold more potent *in vitro* irreversible inhibitor of human placental microsomal aromatase, respectively, than 4-OHA or SH-489.[322]

Substitution of the 4- or 6-positions of the sterol substrate can also produce mechanism-based inactivating agents. 4-Acetoxy- and 4-hydroxy-4-androstene-3,17-dione (4-OHA) (Fig. 23) irreversibly inactivate placental aromatase by catalysis-dependent mechanisms.[341,342] The 19-methyl group is required for the activity of these agents. The mechanism for the inhibitory activity of the 4-substituted analogues remains elusive but a possible mechanism for the action of 4-OHA is shown in Fig. 24. 4-OHA is currently in clinical use for the treatment of estrogen-dependent breast cancer.[302,343] Of a series of $\Delta^{1,2}$, $\Delta^{4,5}$, and $\Delta^{6,7}$ analogues evaluated as prospective aromatase inhibitors in preclinical trials, FC 24928 (4-aminoandrostan-1,4,6-triene-3,17-dione) is the most promising candidate because it inactivates human placental aromatase activity as potently as 4-OHA and FCE-24304 (6-methylene-androstan-1,4-diene-3,17-dione), but unlike both of these compounds it has little intrinsic androgenic activity and it did not affect 5α-reductase or $P450_{scc}$.[344-346]

Conjugation of the 4-hydroxyandrostene nucleus as in 1,4,6-androstatriene-3,17-dione (ATD) conveys aromatase inhibitory and marked tumor regression activities ($\approx 80\%$).[347,348] On the other hand, the introduction of a C-1-methyl into 1,4-androstadiene-3,17-dione as in Atamestane (1-methylandrosta-1,4-diene-3,17-dione, SH-489) apparently

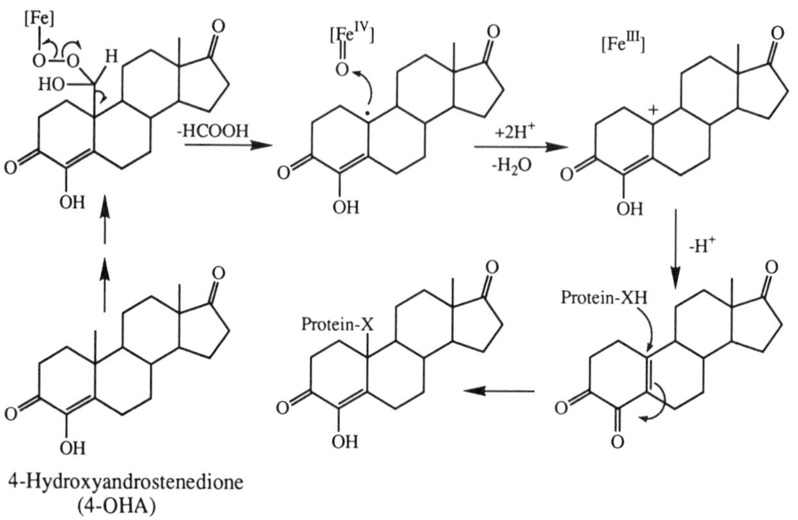

4-Hydroxyandrostenedione
(4-OHA)

FIGURE 24. Possible mechanism for the inactivation of aromatase by 4-OHA.

enhances its affinity ($K_I \approx 2$ nM versus K_I of 29 nM for 4-OHA) for the human placental aromatase while considerably slowing its inactivation of the enzyme, thereby reducing the production of estrogenic products.[346,349] The compound along with its 1,2-methylene-substituted congeners is currently being evaluated in Phase I clinical trials for possible therapy of estrogen-dependent conditions (breast cancer and benign prostatic hypertrophy). More recently, androst-5-ene-7,17-dione and its 19-hydroxy derivative have been shown to represent another class of suicide substrates of aromatase.[350]

The presence of a 6-keto group (Fig. 24) also causes turnover-dependent irreversible inactivation of the enzyme.[351–353] Radiolabeling experiments indicate that inactivation involves covalent attachment of the inhibitor to the enzyme.[352] Measurements of the ^{3}H:^{14}C ratio with the inhibitor double-labeled on C-19 indicate that the C-19 methyl, one of the C-19 hydrogens, and, from a separate double-label experiment, the 1β-hydrogen are retained in the covalently bound species.[353] These results do not uniquely define an inactivation mechanism but do clearly show C-19 demethylation and aromatization are not involved, although the concurrent formation of 6-oxoestrone and 6-oxoestradiol indicate that the normal aromatization reaction can take place.[353]

FCE-24304 (6-methylene-androsta-1,4-diene-3,17-dione) is an aromatase inhibitor that is comparable to 4-OHA in its IC_{50} for human placental aromatase.[344,345] The relative K_I (nM) and $t_{1/2}$ (min) values for the inactivation processes were 26 ± 1.4 and 29.0 ± 7.5, and 13.9 ± 0.7 and 2.1 ± 0.2 for FCE-24304 and 4-OHA, respectively.[344,345] In spite of its relatively slow inactivation of aromatase both orally and subcutaneously, FCE-24304 is relatively more potent in experimental animals,[344] and also much more effective than either MDL 18962 or SH-489 in producing regression of chemically induced mammary tumors in rats.[346]

MDL 18,962 [10-(2-propynyl)estr-4-ene-3,17-dione] is clearly one of the most potent of the "rationally" designed suicide substrates of aromatase, with K_I values on the order of 3–4 nM, and $t_{1/2}$ of 9 and 6 min, respectively, for the human and baboon placental aromatases.[329–332,334–337,354] MDL 18,962 is currently in Phase I clinical trials, but preclinical findings in normal human volunteers indicate that MDL 18,962 is indeed effective in lowering estrogen levels.[355]

The time-dependent inactivation of aromatase by 10-hydroperoxy-4-estrene-3,3-dione has been reported.[356,357] The inactivation, which is inhibited by NADPH or substrate, is partially reversed by dithiothreitol. Other high-affinity 10-substituted analogues, such as the mechanism-based inhibitor 10β-mercapto-estr-4-ene-3,17-dione, and the competitive inhibitors (19-R)-10-oxiranyl and 10-thiiranyl-estr-4-ene-3,17-diones and 10β-MeSCH$_2$-estr-4-ene-3,17-dione, have been reported.[318,319,322,323,358,359] C-10, C-2 hydroxyethyl bridged steroids synthesized as stable carbon analogues of the 2β-hydroxylated 19-oxoandrostenedione, a putative intermediate in the aromatization reaction, have been found to be potent competitive inhibitors of the human placental aromatase.[360] In contrast, related halohydrin analogues are potent mechanism-based aromatase inhibitors.[360]

5.3. Lanosterol 14-Demethylation

A key step in the biosynthesis of cholesterol is the P450-catalyzed 14-demethylation of lanosterol (Chapters 8 and 12). The preferential inhibition of this enzyme by a number

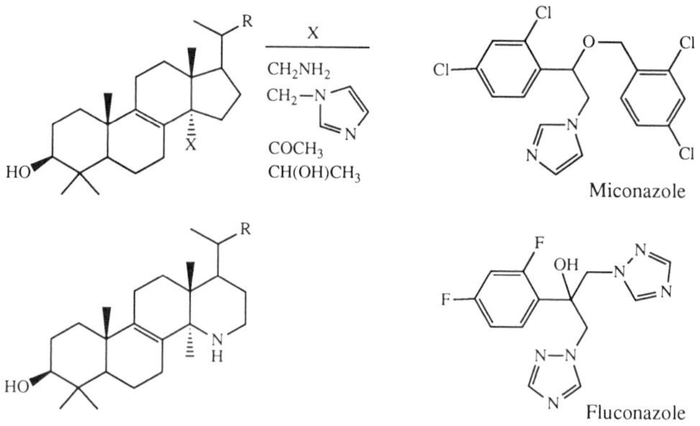

X
CH$_2$NH$_2$
CH$_2$—N
COCH$_3$
CH(OH)CH$_3$

Miconazole

Fluconazole

FIGURE 25. Competitive lanosterol 14-demethylase inhibitors. The group R on the sterols is either C(CH$_3$)CH$_2$CH$_2$CH$_2$C(CH$_3$)$_2$ or a closely related variant.

of substituted imidazoles, pyridines, pyrimidines, and other lipophilic heterocycles has been exploited successfully in the construction of clinically important antifungal agents.[6,361–363] Miconazole (Fig. 25), fluconazole (Fig. 25), ketoconazole (Fig. 1), and related azole structures inhibit the 14-demethylase activity of fungi at extremely low (nM) concentrations,[364,365] but only inhibit the 14-demethylase activity of the mammalian host[366] or other P450 activities (aminopyrine N-demethylase)[367] at higher (up to 100 μM) concentrations. Low doses of ketoconazole, however, appear to also inhibit the C$_{17-20}$ lyase in man.[368] The substituted imidazoles and other nitrogen-based heterocyclic antifungal agents bind tightly to P450 and give rise to type II binding spectra, as expected if the inhibition results from coordination of the inhibitors with the prosthetic heme group.[369] The 14-methyl sterols that accumulate in the membranes of susceptible fungi when the 14-demethylase is inhibited are believed to cause the changes in membrane permeability responsible for the antifungal action of the inhibitors.[370–372]

Similar structural considerations have prompted the development of 14α-methyl-15-aza-D-homosterols (Fig. 25), which exhibit relatively high *in vitro* fungistatic potencies and high fungicidal activities as well as respectable *in vivo* efficacies in a murine model of candidiasis.[373] Introduction of an oxime functionality (=NOH) at position 15 of the sterol group, giving a compound in which the geometry of the nitrogen is not dissimilar from that in the 15-azahomosterols, also conveys inhibitory activity.[374] Similarly, 7-oxo-24,25-dihydrolanosterols, which uncouple catalytic turnover of CYP51 by competitively blocking transfer of a second electron to the oxyferrocomplex, have also been considered as prospective inhibitors.[375]

Steroid analogues that inhibit lanosterol 14α-demethylation have been obtained by attaching amino and imidazole functions to the 14-methyl carbon (Fig. 25).[376] Steroids with the 14-methyl replaced by a 1-hydroxyethyl, 1-oxoethyl, or a carboxyl group are competitive inhibitors of the enzyme (Fig. 25).[377–379] Lanosterol analogues with modified 14α-methyl groups have also been synthesized as potential mechanism-based inhibitors of lanosterol 14α-demethylase. These include sterols in which the methyl group (X in Fig.

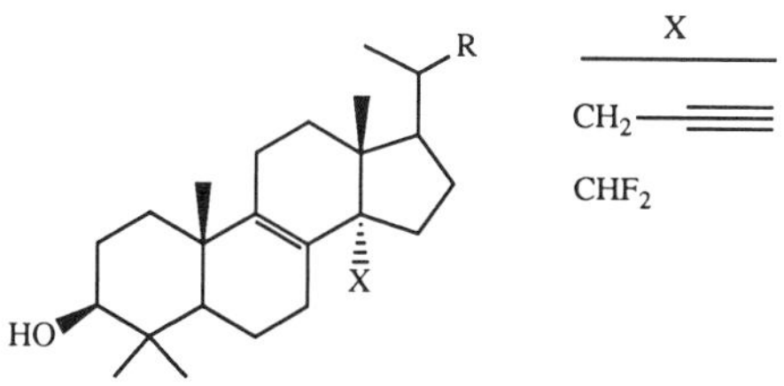

FIGURE 26. Some irreversible, mechanism-based inhibitors of lanosterol 14-demethylase.

26) has been replaced by vinyl, ethynyl, allyl, propargyl, 1-hydroxypropargyl, 1-keto-propargyl, difluoromethyl, epoxide, or episulfide moieties.[379–381] Although most of these compounds inhibit the enzyme, some of them act as simple competitive rather than irreversible inhibitors (e.g., the epoxide and episulfide analogues).[381] In the case of the ethynyl sterols, there is clear evidence for mechanism-based inactivation of the enzyme.[382] 15α-Fluorolanost-7-en-3β-ol, in which the 15-hydrogen normally lost during catalysis is replaced by a fluorine, is oxidized by the enzyme to the 14-aldehyde but proceeds no further because of the fluorine substitution.[383] The compound is therefore an inhibitor that is activated by the catalytic action of the enzyme but is, in fact, a latent form of the actual inhibitor rather than a true mechanism-based inactivating agent.

5.4. Other Biosynthetic Sterol Hydroxylases

17β-(Cyclopropylamino)-androst-5-en-3β-ol is a mechanism-based inactivating agent for CYP17, which has both 17-hydroxylase and C_{17-20}lyase activity.[384,385] It appears to have no inhibitory activity against $P450_{scc}$ or the sterol 21-hydroxylase. In contrast, sterols bearing a difluoromethyl group at position 17 selectively inactivate the C-21 hydroxylase, but sterols bearing a dichloromethyl, vinyl, or ethynyl function at the same position inactivate both the C-17- and C-21-hydroxylases.[72] The same approach has been used to synthesize aldosterone analogues that target the C-18 hydroxylation involved in the biosynthesis of aldosterone.[386–388] The functionalities that have been investigated at that position include the iodomethyl, chloromethyl, allyl, propargyl, vinyl, and methylthiomethyl, some of which appear to cause irreversible inactivation. A rationally designed active-site-directed 18-acetylenic deoxycorticosterone [21-hydroxy-13(-2-propynyl)-18-nor-preg-4-ene-3,20-dione, MDL 19,347] has been found to be a promising suicide inactivator of the adrenal corticosterone 18-hydroxylase ($K_I \approx 38$ nM; $t_{1/2}$, 4.6 min) from rats and rhesus monkeys that was also effective in reducing plasma aldosterone levels *in vivo*.[330,389] Furthermore, this compound apparently is a relatively selective inhibitor of aldosterone biosynthesis, since it fails to inhibit the 11β-hydroxylation of corticosterone and DOC.[330,389] Rationally developed inhibitors such as these may be useful in the management of conditions associated with hyperaldosteronism such as hypertension, hypokalemia, and edema.

5.5. Fatty Acid and Leukotriene Hydroxylases

Fatty acids, including arachidonic acid and its derivatives, are hydroxylated at the ω and ω-1 positions by P450 enzymes in liver, kidney, intestine, adrenal, polymorphonuclear leukocyte, and lung microsomes.[390–392] The analogous enzymes of plants and bacteria introduce hydroxyl groups at the ω, ω-1, ω-2, ω-3, and ω-4 positions.[392,393] 11-Dodecynoic acid, the terminal acetylenic analogue of lauric acid, and 10-undecynoic acid inactivate the hepatic P450 enzymes that hydroxylate lauric acid while minimally altering the spectroscopically measured concentration of P450 or the benzphetamine or N-methyl-p-chloroaniline N-demethylase activities.[65,394] The acetylenic fatty acids similarly inactivate the plant P450 enzyme responsible for in-chain hydroxylation of fatty acids.[395] The *in vivo* utility of the acetylenic fatty acids, however, is unfortunately compromised by their toxicity. This difficulty has been partially circumvented by modifying the inhibitors so that they are not subject to β-oxidation. The results indicate that 10-undecynyl sulfate, the acetylenic analogue of 10-undecynoic acid in which the carboxyl group is replaced by a sulfate, is sufficiently less toxic to be used *in vivo* as a mechanism-based inactivator of the lauric acid hydroxylases.[396]

The P450-catalyzed oxidation of arachidonic acid to hydroxyeicosatetraenoic acids and ω-hydroxylated metabolites may be involved in the regulation of physiological processes such as kidney sodium excretion and hypertension.[397,398] The arachidonic acid ω-hydroxylase, presumably an enzyme of the 4A family, and the enzymes that oxidize the double bonds are selectively inactivated by 17-octadecynoic acid (17-ODYA).[399–401] This same agent inactivates P450$_{BM-3}$, an enzyme that hydroxylates fatty acids at positions other than the terminal (ω) methyl group, by a heme alkylation mechanism.[402]

Leukotriene B$_4$, an important mediator of inflammation in humans, is primarily metabolized in polymorphonuclear leukocytes by ω-hydroxylation.[390,403] The ω-hydroxylated LTB$_4$ may be as important as the parent compound in mediating the inflammation response.[404] The LTB$_4$ ω-hydroxylase is inactivated by 15-hexadecynoic and 17-octadecynoic acids in whole cells and cell lysates.[400] The saturated analogues of these long-chain acetylenic acids are inactive, while 10-undecynoic acid, the short-chain acid, is much less effective. The long-chain acetylenic fatty acids also inactivate the lung isozymes responsible for ω-hydroxylation of prostaglandins.[401]

6. Summary

The development of inhibitors of P450 enzymes, and our understanding of the mechanisms of action of such inhibitors, have flowered over the past few years. Efforts to increase the potency and specificity of reversible inhibitors by modifying the lipophilic framework and the heme-coordinating nitrogen function have yielded a number of important clinical and agricultural agents. The more recent development of mechanism-based irreversible inhibitors (suicide substrates) for P450 enzymes greatly enhances the potential specificity and utility of P450 inhibitors. A variety of inactivator functionalities are now available for the construction of such mechanism-based inhibitors, each of which irreversibly inactivates the enzyme that catalyzes its oxidation.

Although agents such as chloramphenicol inactivate P450 by reacting with its protein structure, by far the most vulnerable site for the action of mechanism-based inhibitors appears to be the prosthetic heme group. Identification of the prosthetic heme as the Achilles' heel of P450 enzymes is of practical importance for the design of appropriate inhibitors. As borne out by experience, the functionalities that inactivate P450 with the highest specificity and efficiency are generally oxidized to radical or neutral reactive species rather than to cationic alkylating intermediates. The catalytic involvement of P450 in its own inactivation, and the detailed reconstruction of the destructive event made possible by isolation and structural characterization of the prosthetic heme adducts, make the irreversible inhibitors important mechanistic and topological probes. The practical utility of mechanism-based P450 inhibitors, which is only now being explored, is likely to surpass that of the more classical reversible agents.

ACKNOWLEDGMENT. The preparation of this review was assisted by grants from the National Institutes of Health.

References

1. Rahimtula, A. D., and O'Brien, P. J., 1977, The peroxidase nature of cytochrome P450, in: *Microsomes and Drug Oxidations* (V. Ullrich, I. Roots, A. Hildebrandt, R. W. Estabrook, and A. H. Conney, eds.), Pergamon Press, Elmsford, NY, pp. 210–217.
2. Rodrigues, A. D., Fernandez, D., Nosarzewski, M. A., Pierce, W. M., and Prough, R. A., 1991, Inhibition of hepatic microsomal cytochrome P-450 dependent monooxygenation activity by the antioxidant 3-*tert*-butyl-4-hydroxyanisole, *Chem. Res. Toxicol.* **4:**281–289.
3. Kharasch, E. D., Wendel, N. K., and Novak, R. F., 1987, Anthracenedione antineoplastic agent effects on drug metabolism *in vitro* and *in vivo*: Relationship between structure and mechanism of inhibition, *Fundam. Appl. Toxicol.* **9:**18–25.
4. Testa, B., and Jenner, P., 1981, Inhibitors of cytochrome P-450s and their mechanism of action, *Drug. Metab. Rev.* **12:**1–117.
5. Correia, M. A., and Ortiz de Montellano, P. R., 1993, Inhibitors of cytochrome P450 and possibilities for their therapeutic application, in: *Frontiers in Biotransformation* (K. Ruckpaul, ed.), Akademie-Verlag, Berlin, pp. 74–146.
6. Murray, M., and Reidy, G. F., 1990, Selectivity in the inhibition of mammalian cytochromes P-450 by chemical agents, *Pharmacol. Rev.* **42:**85–101.
7. Ortiz de Montellano, P. R., 1988, Suicide substrates for drug metabolizing enzymes: Mechanism and biological consequences, in: *Progress in Drug Metabolism* (G. G. Gibson, ed.), Taylor & Francis, London, pp. 99–148.
8. Vanden Bossche, H., 1992, Inhibitors of P450-dependent steroid biosynthesis: From research to medical treatment, *J. Steroid Biochem. Mol. Biol.* **43:**1003–1021.
9. Sato, A., and Nakajima, T., 1979, Dose-dependent metabolic interaction between benzene and toluene *in vivo* and *in vitro*, *Toxicol. Appl. Pharmacol.* **48:**249–256.
10. Watkins, P. B., 1990, Role of cytochromes P450 in drug metabolism and hepatotoxicity, *Semin. Liver Dis.* **10:**235–250.
11. Jefcoate, C. R., 1978, Measurement of substrate and inhibitor binding to microsomal cytochrome P-450 by optical-difference spectroscopy, *Methods Enzymol.* **52:**258–279.
12. Kumaki, K., Sato, M., Kon, H., and Nebert, D. W., 1978, Correlation of type I, type II, and reverse type I difference spectra with absolute changes in spin state of hepatic microsomal cytochrome P-450 iron from five mammalian species, *J. Biol. Chem.* **253:**1048–1058.

13. Schenkman, J. B., Sligar, S. G., and Cinti, D. L., 1981, Substrate interactions with cytochrome P-450, *Pharmacol. Ther.* **12:**43–71.

14. Sligar, S. G., Cinti, D. L., Gibson, G. G., and Schenkman, J. B., 1979, Spin state control of the hepatic cytochrome P-450 redox potential, *Biochem. Biophys. Res. Commun.* **90:**925–932.

15. Guengerich, F. P., 1983, Oxidation–reduction properties of rat liver cytochromes P450 and NADPH-cytochrome P-450 reductase related to catalysis in reconstituted systems, *Biochemistry* **22:**2811–2820.

16. Kitada, M., Chiba, K., Kamataki, T., and Kitagawa, H., 1977, Inhibition by cyanide of drug oxidations in rat liver microsomes, *Jpn. J. Pharmacol.* **27:**601–608.

17. Ho, B., and Castagnoli, N., 1980, Trapping of metabolically generated electrophilic species with cyanide ion: Metabolism of 1-benzylpyrrolidine, *J. Med. Chem.* **23:**133–139.

18. Sono, M., and Dawson, J. H., 1982, Formation of low spin complexes of ferric cytochrome P-450-CAM with anionic ligands: Spin state and ligand affinity comparison to myoglobin, *J. Biol. Chem.* **257:**5496–5502.

19. Backes, W. L., Hogaboom, M., and Canady, W. J., 1982, The true hydrophobicity of microsomal cytochrome P-450 in the rat: Size dependence of the free energy of binding of a series of hydrocarbon substrates from the aqueous phase to the enzyme and to the membrane as derived from spectral binding data, *J. Biol. Chem.* **257:**4063–4070.

20. Wink, D. A., Osawa, Y., Darbyshe, J. F., Jones, C. R., Eshenaur, S. C., and Nims, R. W., 1993, Inhibition of cytochromes P450 by nitric oxide and a nitric oxide-releasing agent, *Arch. Biochem. Biophys.* **300:**115–123.

21. Khatsenko, O. G., Gross, S. S., Rifkind, A. B., and Vane, J. R., 1993, Nitric oxide is a mediator of the decrease in cytochrome P450-dependent metabolism caused by immunostimulants, *Proc. Natl. Acad. Sci. USA* **90:**11147–11151.

22. Griscavage, J. M., Fukuto, J. M., Komori, Y., and Ignarro, L. J., 1994, Nitric oxide inhibits neuronal nitric oxide synthase by interacting with the heme prosthetic group. Role of tetrahydrobiopterin in modulating the inhibitory action of nitric oxide, *J. Biol. Chem.* **269:**21644–21649.

23. Hanson, L. K., Eaton, W. A., Sligar, S. G., Gunsalus, I. C., Gouterman, M., and Connell, C. R., 1976, Origin of the anomalous Soret spectra of carboxycytochrome P450, *J. Am. Chem. Soc.* **98:**2672–2674.

24. Omura, T., and Sato, R., 1964, The carbon monoxide-binding pigment of liver microsomes. 1. Evidence for its hemoprotein nature, *J. Biol. Chem.* **239:**2370–2378.

25. Collman, J. P., and Sorrell, T. N., 1975, A model for the carbonyl adduct of ferrous cytochrome P-450, *J. Am. Chem. Soc.* **97:**4133–4134.

26. Leeman, T., Bonnabry, P., and Dayer, P., 1994, Selective inhibition of major drug metabolizing cytochrome P450 isozymes in human liver microsomes by carbon monoxide, *Life Sci.* **54:**951–956.

27. Canick, J. A., and Ryan, K. J., 1976, Cytochrome P-450 and the aromatization of 16-alpha-hydroxytestosterone and androstenedione by human placental microsomes, *Mol. Cell. Endocrinol.* **6:**105–115.

28. Gibbons, G. F., Pullinger, C. R., and Mitropoulos, K. A., 1979, Studies on the mechanism of lanosterol 14-alpha-demethylation: A requirement for two distinct types of mixed-function-oxidase systems, *Biochem. J.* **183:**309–315.

29. Hansson, R., and Wikvall, K., 1982, Hydroxylations in biosynthesis of bile acids: Cytochrome P-450 LM$_4$ and 12α-hydroxylation of 5β-cholestane-3α,7α-diol, *Eur. J. Biochem.* **125:**423–429.

30. Meigs, R. A., and Ryan, K. J., 1971, Enzymatic aromatization of steroids. I. Effects of oxygen and carbon monoxide on the intermediate steps of estrogen biosynthesis, *J. Biol. Chem.* **246:**83–87.

31. Zachariah, P. K., and Juchau, M. R., 1975, Interactions of steroids with human placental cytochrome P-450 in the presence of carbon monoxide, *Life Sci.* **16:**1689–1692.

32. Tuckey, R. C., and Kamin, H., 1983, Kinetics of O$_2$ and CO binding to adrenal cytochrome P-450scc: Effect of cholesterol, intermediates, and phosphatidylcholine vesicles, *J. Biol. Chem.* **258:**4232–4237.

33. Cohen, G. M., and Mannering, G. J., 1972, Involvement of a hydrophobic site in the inhibition of the microsomal para-hydroxylation of aniline by alcohols, *Mol. Pharmacol.* **8:**383–397.

34. Gerber, M. C., Tejwani, G. A., Gerber, N., and Bianchine, J. R., 1985, Drug interactions with cimetidine: An update, *Pharmacol. Ther.* **27**:353–370.

35. Testa, B., 1981, Structural and electronic factors influencing the inhibition of aniline hydroxylation by alcohols and their binding to cytochrome P-450, *Chem. Biol. Interact.* **34**:287–300.

36. Wattenberg, L. W., Lam, L. K. T., and Fladmoe, A. V., 1979, Inhibition of chemical carcinogen-induced neoplasia by coumarins and alpha-angelicalactone, *Cancer Res.* **39**:1651–1654.

37. Remmer, H., Schenkman, J., Estabrook, R. W., Sasame, H., Gillette, J., Narasimhulu, S., Cooper, D. Y., and Rosenthal, O., 1966, Drug interaction with hepatic microsomal cytochrome, *Mol. Pharmacol.* **2**:187–190.

38. Jefcoate, C. R., Gaylor, J. L., and Callabrese, R. L., 1969, Ligand interactions with cytochrome P-450. 1. Binding of primary amines, *Biochemistry* **8**:3455–3463.

39. Schenkman, J. B., Remmer, H., and Estabrook, R. W., 1967, Spectral studies of drug interaction with hepatic microsomal cytochrome P-450, *Mol. Pharmacol.* **3**:113–123.

40. Dominguez, O. V., and Samuels, L. T., 1963, Mechanism of inhibition of adrenal steroid 11-beta-hydroxylase by methopyrapone (metopirone), *Endocrinology* **73**:304–309.

41. Temple, T. E., and Liddle, G. W., 1970, Inhibitors of adrenal steroid biosynthesis, *Annu. Rev. Pharmacol.* **10**:199–218.

42. Rogerson, T. D., Wilkinson, C. F., and Hetarski, K., 1977, Steric factors in the inhibitory interaction of imidazoles with microsomal enzymes, *Biochem. Pharmacol.* **26**:1039–1042.

43. Wilkinson, C. F., Hetarski, K., Cantwell, G. P., and DiCarlo, F. J., 1974, Structure–activity relationships in the effects of 1-alkylimidazoles on microsomal oxidation in vitro and in vivo, *Biochem. Pharmacol.* **23**:2377–2386.

44. Duquette, P. H., Erickson, R. R., and Holtzman, J. L., 1983, Role of substrate lipophilicity on the N-demethylation and type I binding of 3-O-alkylmorphine analogues, *J. Med. Chem.* **26**:1343–1348.

45. Ator, M. A., and Ortiz de Montellano, P. R., 1990, Mechanism-based (suicide) enzyme inactivation, in: *The Enzymes: Mechanisms of Catalysis*, Vol. 19, 3rd ed. (D. S. Sigman and P. D. Boyer, eds.), Academic Press, New York, pp. 214–282.

46. Silverman, R. B., 1988, *Mechanism-Based Enzyme Inactivation: Chemistry and Enzymology*, CRC Press, Boca Raton, FL.

47. Dalvi, R. R., 1987, Cytochrome P-450-dependent covalent binding of carbon disulfide to rat liver microsomal protein in vitro and its prevention by reduced glutathione, *Arch. Toxicol.* **61**:155–157.

48. De Matteis, F. A., and Seawright, A. A., 1973, Oxidative metabolism of carbon disulphide by the rat: Effect of treatments which modify the liver toxicity of carbon disulphide, *Chem. Biol. Interact.* **7**:375–388.

49. Bond, E. J., and De Matteis, F. A., 1969, Biochemical changes in rat liver after administration of carbon disulphide, with particular reference to microsomal changes, *Biochem. Pharmacol.* **18**:2531–2549.

50. Halpert, J., Hammond, D., and Neal, R. A., 1980, Inactivation of purified rat liver cytochrome P-450 during the metabolism of parathion (diethyl *p*-nitrophenyl phosphorothionate), *J. Biol. Chem.* **255**:1080–1089.

51. Neal, R. A., Kamataki, T., Lin, M., Ptashne, K. A., Dalvi, R., and Poore, R. Y., 1977, Studies of the formation of reactive intermediates of parathion, in: *Biological Reactive Intermediates* (D. J. Jollow, J. J. Koesis, R. Snyder, and H. Vaino, eds.), Plenum Press, New York, pp. 320–332.

52. Miller, G. E., Zemaitis, M. A., and Greene, F. E., 1983, Mechanisms of diethyldithiocarbamate-induced loss of cytochrome P-450 from rat liver, *Biochem. Pharmacol.* **32**:2433–2442.

53. El-hawari, A. M., and Plaa, G. L., 1979, Impairment of hepatic mixed-function oxidase activity by alpha- and beta-naphthylisothiocyanate: Relationship to hepatotoxicity, *Toxicol. Appl. Pharmacol.* **48**:445–458.

54. Lee, P. W., Arnau, T., and Neal, R. A., 1980, Metabolism of alpha-naphthylthiourea by rat liver and rat lung microsomes, *Toxicol. Appl. Pharmacol.* **53**:164–173.

55. Lopez-Garcia, M. P., Dansette, P. M., and Mansuy, D., 1993, Thiophene derivatives as new mechanism-based inhibitors of cytochromes P450: Inactivation of yeast-expressed human liver P450 2C9 by tienilic acid, *Biochemistry* **33:**166–175.

56. Lopez-Garcia, M. P., Dansette, P. M., Valadon, P., Amar, C., Beaune, P. H., Guengerich, F. P., and Mansuy, D., 1993, Human liver P450s expressed in yeast as tools for reactive metabolite formation studies: Oxidative activation of tienilic acid by P450 2C9 and P450 2C10, *Eur. J. Biochem.* **213:**223–232.

57. Menard, R. H., Guenthner, T. M., Taburet, A. M., Kon, H., Pohl, L. R., Gillette, J. R., Gelboin, H. V., and Trager, W. F., 1979, Specificity of the in vitro destruction of adrenal and hepatic microsomal steroid hydroxylases by thiosterols, *Mol. Pharmacol.* **16:**997–1010.

58. Kossor, D. C., Kominami, S., Takemori, S., and Colby, H. D., 1991, Role of the steroid 17α-hydroxylase in spironolactone-mediated destruction of adrenal cytochrome P-450, *Mol. Pharmacol.* **40:**321–325.

59. Halpert, J., and Neal, R. A., 1980, Inactivation of purified rat liver cytochrome P-450 by chloramphenicol, *Mol. Pharmacol.* **17:**427–434.

60. Halpert, J., 1982, Further studies of the suicide inactivation of purified rat liver cytochrome P-450 by chloramphenicol, *Mol. Pharmacol.* **21:**166–172.

61. Halpert, J., 1981, Covalent modification of lysine during the suicide inactivation of rat liver cytochrome P-450 by chloramphenicol, *Biochem. Pharmacol.* **30:**875–881.

62. Halpert, J., Naslund, B., and Betner, I., 1983, Suicide inactivation of rat liver cytochrome P-450 by chloramphenicol in vivo and in vitro, *Mol. Pharmacol.* **23:**445–452.

63. Halpert, J., Balfour, C., Miller, N. E., and Kaminsky, L. S., 1986, Dichloromethyl compounds as mechanism-based inactivators of rat liver cytochromes P450 in vitro, *Mol. Pharmacol.* **30:**19–24.

64. Halpert, J., Jaw, J.-Y., Balfour, C., and Kaminsky, L. S., 1990, Selective inactivation by chlorofluoroacetamides of the major phenobarbital-inducible form(s) of rat liver cytochrome P-450, *Drug Metab. Dispos.* **18:**168–174.

65. CaJacob, C. A., Chan, W., Shephard, E., and Ortiz de Montellano, P. R., 1988, The catalytic site of rat hepatic lauric acid ω-hydroxylase. Protein vs prosthetic heme alkylation in the ω-hydroxylation of acetylenic fatty acids, *J. Biol. Chem.* **263:**18640–18649.

66. Hammons, G. J., Alworth, W. L., Hopkins, N. E., Guengerich, F. P., and Kadlubar, F. F., 1989, 2-Ethynylnaphthalene as a mechanism-based inactivator of the cytochrome P-450-catalyzed N-oxidation of 2-naphthylamine, *Chem. Res. Toxicol.* **2:**367–374.

67. Yun, C.-H., Martin, M. V., Hopkins, N. E., Alworth, W. L., Hammons, G. J., and Guengerich, F. P., 1992, Modification of cytochrome P4501A2 enzymes by the mechanism-based inactivator 2-ethynylnaphthalene, *Biochemistry* **31:**10556–10563.

68. Gan, L.-S. L., Acebo, A. L., and Alworth, W. L., 1984, 1-Ethynylpyrene, a suicide inhibitor of cytochrome P-450 dependent benzo(a)pyrene hydroxylase activity in liver microsomes, *Biochemistry* **23:**3827–3836.

69. Roberts, E. S., Hopkins, N. E., Alworth, W. L., and Hollenberg, P. F., 1993, Mechanism-based inactivation of cytochrome P450 2B1 by 2-ethynylnaphthalene: Identification of an active-site peptide, *Chem. Res. Toxicol.* **6:**470–479.

70. Chan, W. K., Sui, Z., and Ortiz de Montellano, P. R., 1993, Determinants of protein modification versus heme alkylation: Inactivation of cytochrome P450 1A1 by 1-ethynylpyrene and phenylacetylene, *Chem. Res. Toxicol.* **6:**38–45.

71. Halpert, J., Jaw, J.-Y., and Balfour, C., 1989, Specific inactivation by 17β-substituted steroids of rabbit and rat liver cytochromes P-450 responsible for progesterone 21-hydroxylation, *Mol. Pharmacol.* **34:**148–156.

72. Stevens, J. C., Jaw, J.-Y., Peng, C.-T., and Halpert, J., 1991, Mechanism-based inactivation of bovine adrenal cytochromes P450 C-21 and P450 17α by 17β-substituted steroids, *Biochemistry* **30:**3649–3658.

73. Roberts, E. S., Hopkins, N. E., Zalulec, E. J., Gage, D. A., Alworth, W. L., and Hollenberg, P. F., 1994, Identification of active site peptides from ^{3}H-labeled 2-ethynylnaphthalene-inactivated P450 2B1 and 2B4 using amino acid sequencing and mass spectrometry, *Biochemistry* **33:**3766–3771.

74. Lunetta, J. M., Sugiyama, K., and Correia, M. A., 1989, Secobarbital-mediated inactivation of rat liver cytochrome P-450b: A mechanistic reappraisal, *Mol. Pharmacol.* **35:**10–17.

75. Fouin-Fortunet, H., Tinel, M., Descatoire, V., Letteron, P., Larrey, D., Geneve, J., and Pessayre, D., 1986, Inactivation of cytochrome P450 by the drug methoxsalen, *J. Pharmacol. Exp. Ther.* **236:**237–247.

76. Labbe, G., Descatoire, V., Beaune, P., Letteron, P., Larrey, D., and Pessayre, D., 1989, Suicide inactivation of cytochrome P450 by methoxsalen. Evidence for the covalent binding of a reactive intermediate to the protein moiety, *J. Pharmacol. Exp. Ther.* **250:**1034–1042.

77. Mays, D. C., Hilliard, J. B., Wong, D. D., Chambers, M. A., Park, S. S., Gelboin, H. V., and Gerber, N., 1990, Bioactivation of 8-methoxypsoralen and irreversible inactivation of cytochrome P450 in mouse liver microsomes: Modification by monoclonal antibodies, inhibition of drug metabolism and distribution of covalent adducts, *J. Pharmacol. Exp. Ther.* **254:**720–731.

78. Bornheim, L. M., Everhart, E. T., Li, J., and Correia, M. A., 1993, Characterization of cannabidiol-mediated cytochrome P450 inactivation, *Biochem. Pharmacol.* **45:**1323–1331.

79. Liu, H., Santostefano, M., and Safe, S., 1994, 2-Phenylphenanthridinone and related compounds: Aryl hydrocarbon receptor agonists and suicide inactivators of P4501A1, *Arch. Biochem. Biophys.* **313:**206–214.

80. Saunders, F. J., and Alberti, R. L., 1978, *Aldactone: Spironolactone: A Comprehensive Review*, Searle, New York.

81. Menard, R. H., Guenthner, T. M., Taburet, A. M., Kon, H., Pohl, L. R., Gillette, J. R., Gelboin, H. V., and Trager, W. F., 1979, Specificity of the in vitro destruction of adrenal and hepatic microsomal steroid hydroxylases by thiosteroids, *Mol. Pharmacol.* **16:**997–1010.

82. Decker, C., Sugiyama, K., Underwood, M., and Correia, M. A., 1986, Inactivation of rat hepatic cytochrome P-450 by spironolactone, *Biochem. Biophys. Res. Commun.* **136:**1162–1169.

83. Decker, C. J., Rashed, M. S., Baillie, T. A., Maltby, D., and Correia, M. A., 1989, Oxidative metabolism of spironolactone: Evidence for the involvement of electrophilic thiosteroid species in drug-mediated destruction of rat hepatic cytochrome P450, *Biochemistry* **28:**5128–5136.

84. Menard, R. H., Guenthner, T. M., Kon, H., and Gillette, J. R., 1979, Studies on the destruction of adrenal and testicular cytochrome P-450 by spironolactone: Requirement for the 7-alpha-thio group and evidence for the loss of the heme and apoproteins of cytochrome P-450, *J. Biol. Chem.* **254:**1726–1733.

85. Sherry, J. H., O'Donnell, J. P., Flowers, L., Lacagnin, L. B., and Colby, H. D., 1986, Metabolism of spironolactone by adrenocortical and hepatic microsomes: Relationship to cytochrome P-450 destruction, *J. Pharmacol. Exp. Ther.* **236:**675–680.

86. Colby, H. D., O'Donnell, J. P., Lynn, N., Kossor, D. C., Johnson, P. B., and Levitt, M., 1991, Relationship between covalent binding to microsomal protein and the destruction of adrenal cytochrome P-450 by spironolactone, *Toxicology* **67:**143–154.

87. Decker, C. J., Cashman, J. R., Sugiyama, K., Maltby, D., and Correia, M. A., 1991, Formation of glutathionyl-spironolactone disulfide by rat liver cytochromes P450 or hog liver flavin-containing monooxygenases: A functional probe of two-electron oxidations of the thiosteroid? *Chem. Res. Toxicol.* **4:**669–677.

88. Ortiz de Montellano, P. R., and Komives, E. A., 1985, Branchpoint for heme alkylation and metabolite formation in the oxidation of aryl acetylenes, *J. Biol. Chem.* **260:**3330–3336.

89. Poulos, T. L., Finzel, B. C., and Howard, A. J., 1987, High-resolution crystal structure of cytochrome P450$_{cam}$, *J. Mol. Biol.* **195:**687–700.

90. Nelson, D. R., and Strobel, H. W., 1988, On the membrane topology of vertebrate cytochrome P-450 proteins, *J. Biol. Chem.* **263:**6038–6050.

91. He, K., Chen, B., Falick, A. M., and Correia, M. A., 1994, Identification of an active site peptide modified during mechanism-based inactivation of cytochrome P450 2B1 by secobarbital, *Abstracts, 10th International Symposium on Microsomes & Drug Oxidations*, p. 558.

92. Casida, J. E., 1970, Mixed function oxidase involvement in the biochemistry of insecticide synergists, *J. Agric. Food Chem.* **18:**753–772.

93. Hodgson, E., and Philpot, R. M., 1974, Interaction of methylene dioxyphenol (1,3-benzodioxole) compounds with enzymes and their effects on mammals, *Drug Metab. Rev.* **3:**231–301.

94. Wilkinson, C. F., Murray, M., and Marcus, C. B., 1984, Interactions of methylenedioxyphenyl compounds with cytochrome P-450 and effects on microsomal oxidation, in: *Reviews in Biochemical Toxicology*, Vol. 6 (E. Hodgson, J. R. Bend, and R. M. Philpot, eds.), Elsevier, Amsterdam, pp. 27–63.

95. Kulkarni, A. P., and Hodgson, E., 1978, Cumene hydroperoxide-generated spectral interactions of piperonyl butoxide and other synergists with microsomes from mammals and insects, *Pestic. Biochem. Physiol.* **9:**75–83.

96. Franklin, M. R., 1971, The enzymic formation of a methylene dioxyphenyl derivative exhibiting an isocyanide-like spectrum with reduced cytochrome P-450 in hepatic microsomes, *Xenobiotica* **1:**581–591.

97. Elcombe, C. R., Bridges, J. W., Nimmo-Smith, R. H., and Werringloer, J., 1975, Cumene hydroperoxide-mediated formation of inhibited complexes of methylenedioxyphenyl compounds with cytochrome P-450, *Biochem. Soc. Trans.* **3:**967–970.

98. Elcombe, C. R., Bridges, J. W., Gray, T. J. B., Nimmo-Smith, R. H., and Netter, K. J., 1975, Studies on the interaction of safrole with rat hepatic microsomes, *Biochem. Pharmacol.* **24:**1427–1433.

99. Dickins, M., Elcombe, C. R., Moloney, S. J., Netter, K. J., and Bridges, J. W., 1979, Further studies on the dissociation of the isosafrole metabolite–cytochrome P-450 complex, *Biochem. Pharmacol.* **28:**231–238.

100. Ullrich, V., and Schnabel, K. H., 1973, Formation and binding of carbanions by cytochrome P-450 of liver microsomes, *Drug Metab. Dispos.* **1:**176–183.

101. Ullrich, V., 1977, Mechanism of microsomal monooxygenases and drug toxicity, in: *Biological Reactive Intermediates* (D. J. Jollow, J. Kocsis, R. Snyder, and H. Vaino, eds.), Plenum Press, New York, pp. 65–82.

102. Murray, M., Hetnarski, K., and Wilkinson, C. F., 1985, Selective inhibitory interactions of alkoxymethylenedioxybenzenes towards mono-oxygenase activity in rat-hepatic microsomes, *Xenobiotica* **15:**369–379.

103. Murray, M., Wilkinson, C. F., Marcus, C., and Dube, C. E., 1983, Structure–activity relationships in the interactions of alkoxymethylenedioxybenzene derivatives with rat hepatic microsomal mixed-function oxidases in vivo, *Mol. Pharmacol.* **24:**129–136.

104. Mansuy, D., 1981, Use of model systems in biochemical toxicology: Heme models, in: *Reviews in Biochemical Toxicology*, Vol. 3 (E. Hodgson, J. R. Bend, and R. M. Philpot, eds.), Elsevier, Amsterdam, pp. 283–320.

105. Mansuy, D., Battioni, J. P., Chottard, J. C., and Ullrich, V., 1979, Preparation of a porphyrin-iron-carbene model for the cytochrome P-450 complexes obtained upon metabolic oxidation of the insecticide synergists of the 1,3-benzodioxole series, *J. Am. Chem. Soc.* **101:**3971–3973.

106. Dahl, A. R., and Hodgson, E., 1979, The interaction of aliphatic analogs of methylenedioxyphenyl compounds with cytochromes P-450 and P-420, *Chem. Biol. Interact.* **27:**163–175.

107. Anders, M. W., Sunram, J. M., and Wilkinson, C. F., 1984, Mechanism of the metabolism of 1,3-benzodioxoles to carbon monoxide, *Biochem. Pharmacol.* **33:**577–580.

108. Hansch, C., 1968, The use of homolytic, steric, and hydrophobic constants in a structure–activity study of 1,3-benzodioxole synergists, *J. Med. Chem.* **11:**920–924.

109. Hennessy, D. J., 1965, Hydride-transferring ability of methylene dioxybenzenes as a basis of synergistic activity, *J. Agric. Food Chem.* **13:**218–231.

110. Cook, J. C., and Hodgson, E., 1983, Induction of cytochrome P-450 by methylenedioxyphenyl compounds: Importance of the methylene carbon, *Toxicol. Appl. Pharmacol.* **68:**131–139.

111. Casida, J. E., Engel, J. L., Essac, E. G., Kamienski, F. X., and Kuwatsuka, S., 1966, Methylene-^{14}C-dioxyphenyl compounds: Metabolism in relation to their synergistic action, *Science* **153**:1130–1133.

112. Kamienski, F. X., and Casida, J. E., 1970, Importance of methylenation in the metabolism *in vivo* and *in vitro* of methylenedioxyphenyl synergists and related compounds in mammals, *Biochem. Pharmacol.* **19**:91–112.

113. Yu, L.-S., Wilkinson, C. F., and Anders, M. W., 1980, Generation of carbon monoxide during the microsomal metabolism of methylenedioxyphenyl compounds, *Biochem. Pharmacol.* **29**:1113–1122.

114. Metcalf, R. L., Fukuto, C. W., Fahmy, S., El-Azis, S., and Metcalf, E. R., 1966, Mode of action of carbamate synergists, *J. Agric. Food. Chem.* **14**:555–562.

115. Franklin, M. R., 1977, Inhibition of mixed-function oxidations by substrates forming reduced cytochrome P-450 metabolic-intermediate complexes, *Pharmacol. Ther. A* **2**:227–245.

116. Larrey, D., Tinel, M., and Pessayre, D., 1983, Formation of inactive cytochrome P450 Fe(II)–metabolite complexes with several erythromycin derivatives but not with josamycin and midecamycin in rats, *Biochem. Pharmacol.* **32**:1487–1493.

117. Delaforge, M., Jaquen, M., and Mansuy, D., 1983, Dual effects of macrolide antibiotics on rat liver cytochrome P-450. Induction and formation of metabolite-complexes: A structure–activity relationship, *Biochem. Pharmacol.* **32**:2309–2318.

118. Mansuy, D., Beaune, P., Cresteil, T., Bacot, C., Chottard, J. C., and Gans, P., 1978, Formation of complexes between microsomal cytochrome P-450-Fe(II) and nitrosoarenes obtained by oxidation of arylhydroxylamines or reduction of nitroarenes in situ, *Eur. J. Biochem.* **86**:573–579.

119. Jonsson, J., and Lindeke, B., 1976, On the formation of cytochrome P-450 product complexes during the metabolism of phenylalkylamines, *Acta Pharm. Suec.* **13**:313–320.

120. Franklin, M. R., 1974, The formation of a 455 nm complex during cytochrome P-450-dependent N-hydroxylamphetamine metabolism, *Mol. Pharmacol.* **10**:975–985.

121. Mansuy, D., 1978, Coordination chemistry of cytochromes P-450 and iron-porphyrins: Relevance to pharmacology and toxicology, *Biochimie* **60**:969–977.

122. Lindeke, B., Anderson, E., Lundkvist, G., Jonsson, H., and Eriksson, S.-O., 1975, Autoxidation of N-hydroxyamphetamine and N-hydroxyphentermine: The formation of 2-nitroso-1-phenylpropanes and 1-phenyl-2-propanone oxime, *Acta Pharm. Suec.* **12**:183–198.

123. Mansuy, D., Gans, P., Chottard, J.-C., and Bartoli, J.-F., 1977, Nitrosoalkanes as Fe(II) ligands in the 455-nm-absorbing cytochrome P-450 complexes formed from nitroalkanes in reducing conditions, *Eur. J. Biochem.* **76**:607–615.

124. Pessayre, D., Konstantinova-Mitcheva, M., Descatoire, V., Cobert, B., Wandscheer, J.-C., Level, R., Feldmann, G., Mansuy, D., and Benhamou, J.-P., 1981, Hypoactivity of cytochrome P-450 after triacetyloleandomycin administration, *Biochem. Pharmacol.* **30**:559–564.

125. Wrighton, S. A., Maurel, P., Schuetz, E. G., Watkins, P. B., Young, B., and Guzelian, P. S., 1985, Identification of the cytochrome P-450 induced by macrolide antibiotics in rat liver as the glucocorticoid responsive cytochrome P-450$_p$, *Biochemistry* **24**:2171–2178.

126. Watkins, P. B., Wrighton, S. A., Schuetz, E. G., Maurel, P., and Guzelian, P. S., 1986, Macrolide antibiotics inhibit the degradation of the glucocorticoid-responsive cytochrome P-450p in rat hepatocytes in vivo and in primary monolayer culture, *J. Biol. Chem.* **261**:6264–6271.

127. Hines, R. N., and Prough, R. A., 1980, The characterization of an inhibitory complex formed with cytochrome P-450 and a metabolite of 1,1-disubstituted hydrazines, *J. Pharmacol. Ther.* **214**:80–86.

128. Muakkasah, S. F., Bidlack, W. R., and Yang, W. C. T., 1981, Mechanism of the inhibitory action of isoniazid on microsomal drug metabolism, *Biochem. Pharmacol.* **30**:1651–1658.

129. Moloney, S. J., Snider, B. J., and Prough, R. A., 1984, The interactions of hydrazine derivatives with rat-hepatic cytochrome P-450, *Xenobiotica* **14**:803–814.

130. Muakkassah, S. F., Bidlack, W. R., and Yang, W. C. T., 1982, Reversal of the effects of isoniazid on hepatic cytochrome P-450 by potassium ferricyanide, *Biochem. Pharmacol.* **31**:249–251.

131. Mahy, J.-P., Battioni, P., Mansuy, D., Fisher, J., Weiss, R., Mispelter, J., Morgenstern-Badarau, I., and Gans, P., 1984, Iron porphyrin–nitrene complexes: Preparation from 1,1-dialkylhydrazines: Electronic

structure from NMR, Mössbauer, and magnetic susceptibility studies and crystal structure of the [tetrakis(*p*-chlorophenyl) porphyrinato-(2,2,6,6-tetramethyl-1-piperidyl) nitrene]iron complex, *J. Am. Chem. Soc.* **106**:1699–1706.

132. Mansuy, D., Battioni, P., and Mahy, J.-P., 1982, Isolation of an iron–nitrene complex from the dioxygen and iron porphyrin dependent oxidation of a hydrazine, *J. Am. Chem. Soc.* **104**:4487–4489.

133. Ortiz de Montellano, P. R., 1985, Alkenes and alkynes, in: *Bioactivation of Foreign Compounds* (M. Anders, ed.), Academic Press, New York, pp. 121–155.

134. De Matteis, F., 1978, Loss of liver cytochrome P-450 caused by chemicals, in: *Heme and Hemoproteins, Handbook of Experimental Pharmacology*, Vol. 44 (F. De Matteis and W. N. Aldridge, eds.), Springer-Verlag, Berlin, pp. 95–127.

135. Ortiz de Montellano, P. R., and Correia, M. A., 1983, Suicidal destruction of cytochrome P-450 during oxidative drug metabolism, *Annu. Rev. Pharmacol. Toxicol.* **23**:481–503.

136. Ortiz de Montellano, P. R., and Mico, B. A., 1980, Destruction of cytochrome P-450 by ethylene and other olefins, *Mol. Pharmacol.* **18**:128–135.

137. Collman, J. P., Hampton, P. D., and Brauman, J. I., 1986, Stereochemical and mechanistic studies of the "suicide" event in biomimetic P-450 olefin epoxidation, *J. Am. Chem. Soc.* **108**:7861–7862.

138. Brady, J. F., Ishizaki, H., Fukuto, J. M., Lin, M. C., Fadel, A., Gapac, J. M., and Yang, C. S., 1991, Inhibition of cytochrome P-450 2E1 by diallyl sulfide and its metabolites, *Chem. Res. Toxicol.* **4**:642–647.

139. Ortiz de Montellano, P. R., Stearns, R. A., and Langry, K. C., 1984, The allylisopropylacetamide and novonal prosthetic heme adducts, *Mol. Pharmacol.* **25**:310–317.

140. Ortiz de Montellano, P. R., Mangold, B. L. K., Wheeler, C., Kunze, K. L., and Reich, N. O., 1983, Stereochemistry of cytochrome P-450-catalyzed epoxidation and prosthetic heme alkylation, *J. Biol. Chem.* **258**:4208–4213.

141. Ortiz de Montellano, P. R., Kunze, K. L., Beilan, H. S., and Wheeler, C., 1982, Destruction of cytochrome P-450 by vinyl fluoride, fluroxene, and acetylene: Evidence for a radical cation intermediate in olefin oxidation, *Biochemistry* **21**:1331–1339.

142. Kunze, K. L., Mangold, B. L. K., Wheeler, C., Beilan, H. S., and Ortiz de Montellano, P. R., 1983, The cytochrome P-450 active site, *J. Biol. Chem.* **258**:4202–4207.

143. Collman, J. P., Hampton, P. D., and Brauman, J. I., 1986, Stereochemical and mechanistic studies of the "suicide" event in biomimetic P-450 olefin epoxidation, *J. Am. Chem. Soc.* **108**:7861–7862.

144. Collman, J. P., Hampton, P. D., and Brauman, J. I., 1990, Suicide inactivation of cytochrome P-450 model compounds by terminal olefins. Part I: A mechanistic study of heme N-alkylation and epoxidation, *J. Am. Chem. Soc.* **112**:2977–2986.

145. Collman, J. P., Hampton, P. D., and Brauman, J. I., 1990, Suicide inactivation of cytochrome P-450 compounds by terminal olefins. Part II: Steric and electronic effects in heme N-alkylation and epoxidation, *J. Am. Chem. Soc.* **112**:2986–2998.

146. Mansuy, D., Devocelle, L., Artaud, I., and Battioni, J.-P., 1985, Alkene oxidations by iodosylbenzene catalyzed by iron-porphyrins: Fate of the catalyst and formation of N-alkyl-porphyrin green pigments from monosubstituted alkenes as in cytochrome P-450, *Nouv. J. Chim.* **9**:711–716.

147. Artaud, I., Devocelle, L., Battioni, J.-P., Girault, J.-P., and Mansuy, D., 1987, Suicidal inactivation of iron porphyrin catalysts during alk-1-ene oxidation: Isolation of a new type of N-alkylporphyrin, *J. Am. Chem. Soc.* **109**:3782–3783.

148. Traylor, T. G., Nakano, T., Mikztal, A. R., and Dunlap, B. E., 1987, Transient formation of N-alkylhemins during hemin-catalyzed epoxidation of norbornene. Evidence concerning the mechanisms of epoxidation, *J. Am. Chem. Soc.* **109**:3625–3632.

149. Traylor, T. G., and Mikztal, A. R., 1989, Alkene epoxidations catalyzed by iron(III), manganese(III), and chromium(III) porphyrins. Effects of metal and porphyrin substituents on selectivity and regiochemistry of epoxidation, *J. Am. Chem. Soc.* **111**:7443–7448.

150. Nakano, T., Traylor, T. G., and Dolphin, D., 1990, The formation of N-alkylporphyrins during epoxidation of ethylene catalyzed by iron(III) meso-tetrakis(2,6-dichlorophenyl)porphyrin, *Can. J. Chem.* **10**:1859–1866.

151. Tian, Z.-Q., Richards, J. L., and Traylor, T. G., 1995, Formation of both primary and secondary N-alkylhemins during hemin-catalyzed epoxidation of terminal alkenes, *J. Am. Chem. Soc.* **117**:21–29.

152. Dexter, A. F., and Hager, L. P., 1995, Transient heme N-alkylation of chloroperoxidase by terminal alkenes and alkynes, *J. Am. Chem. Soc.* **117**:817–818.

153. Ortiz de Montellano, P. R., and Kunze, K. L., 1980, Self-catalyzed inactivation of hepatic cytochrome P-450 by ethynyl substrates, *J. Biol. Chem.* **255**:5578–5585.

154. Guengerich, F. P., 1990, Mechanism-based inactivation of human liver microsomal cytochrome P-450 IIIA4 by gestodene, *Chem. Res. Toxicol.* **3**:363–371.

155. De Matteis, F., Abbritti, G., and Gibbs, A. H., 1973, Decreased liver activity of porphyrin-metal chelatase in hepatic porphyria caused by 3,5-diethoxycarbonyl-1,4-dihydrocollidine: Studies in rats and mice, *Biochem. J.* **134**:717–727.

156. De Matteis, F., and Gibbs, A., 1972, Stimulation of liver 5-aminolaevulinate synthetase by drugs and its relevance to drug-induced accumulation of cytochrome P-450, *Biochem. J.* **126**:1149–1160.

157. Gayarthri, A. K., and Padmanaban, G., 1974, Biochemical effects of 3,5-diethoxycarbonyl-1,4-dihydrocollidine in mouse liver, *Biochem. Pharmacol.* **23**:2713–2725.

158. Tephly, T. R., Gibbs, A. H., Ingall, G., and De Matteis, F., 1980, Studies on the mechanism of experimental porphyria and ferrochelatase inhibition produced by 3,5-diethoxycarbonyl-1,4-dihydrocollidine, *Int. J. Biochem.* **12**:993–998.

159. Cole, S. P. C. C., and Marks, G. S., 1984, Ferrochelatase and N-alkylated porphyrins, *Mol. Cell. Biochem.* **64**:127–137.

160. Augusto, O., Beilan, H. S., and Ortiz de Montellano, P. R., 1982, The catalytic mechanism of cytochrome P-450: Spin-trapping evidence for one-electron substrate oxidation, *J. Biol. Chem.* **257**:11288–11295.

161. De Matteis, F., Hollands, C., Gibbs, A. H., de Sa, N., and Rizzardini, M., 1982, Inactivation of cytochrome P-450 and production of N-alkylated porphyrins caused in isolated hepatocytes by substituted dihydropyridines: Structural requirements for loss of haem and alkylation of the pyrrole nitrogen atom, *FEBS Lett.* **145**:87–92.

162. McCluskey, S. A., Marks, G. S., Sutherland, E. P., Jacobsen, N., and Ortiz de Montellano, P. R., 1986, Ferrochelatase-inhibitory activity and N-alkylprotoporphyrin formation with analogues of 3,5-diethoxycarbonyl-1,4-dihydro-2,4,6-trimethylpyridine (DDC) containing extended 4-alkyl groups: Implications for the active site of ferrochelatase, *Mol. Pharmacol.* **30**:352–357.

163. McCluskey, S. A., Riddick, D. S., Mackie, J. E., Kimmett, R. A., Whitney, R. A., and Marks, G. S., 1992, Inactivation of cytochrome P450 and inhibition of ferrochelatase by analogues of 3,5-diethoxycarbonyl-1,4-dihydro-2,4,6-trimethylpyridine with 4-nonyl and 4-dodecyl substituents, *Can. J. Physiol. Pharmacol.* **70**:1069–1074.

164. Tephly, T. R., Coffman, B. L., Ingall, G., Abou Zeit-Har, M. S., Goff, H. M., Tabba, H. D., and Smith, K. M., 1981, Identification of N-methylprotoporphyrin IX in livers of untreated mice and mice treated with 3,5-diethoxycarbonyl-1,4-dihydrocollidine: Source of the methyl group, *Arch. Biochem. Biophys.* **212**:120–126.

165. De Matteis, F., Gibbs, A. H., Farmer, P. B., and Lamb, J. H., 1981, Liver production of N-alkylated porphyrins caused by treatment with substituted dihydropyridines, *FEBS Lett.* **129**:328–331.

166. De Matteis, F., Gibbs, A. H., and Hollands, C., 1983, N-Alkylation of the haem moiety of cytochrome P-450 caused by substituted dihydropyridines. Preferential attack of different pyrrole nitrogen atoms after induction of various cytochrome P-450 isoenzymes, *Biochem. J.* **211**:455–461.

167. Tephly, T. R., Black, K. A., Green, M. D., Coffman, B. L., Dannan, G. A., and Guengerich, F. P., 1986, Effect of the suicide substrate 3,5-diethoxycarbonyl-2,6-dimethyl-4-ethyl-1,4-dihydropyridine on the metabolism of xenobiotics and on cytochrome P-450 apoproteins, *Mol. Pharmacol.* **29**:81–87.

168. Tephly, T. R., Black, K. A., Green, M. D., Coffman, B. L., Dannan, G. A., and Guengerich, F. P., 1986, Effect of suicide substrate 3,5-diethoxycarbonyl-2,6-dimethyl-4-ethyl-1,4-dihydropyridine on the metabolism of xenobiotics and on cytochrome P-450 apoproteins, *Mol. Pharmacol.* **29**:81–87.

169. Riddick, D. S., Park, S. S., Gelboin, H. V., and Marks, G. S., 1990, Effects of 4-alkyl analogues of 3,5-diethoxycarbonyl-1,4-dihydro-2,4,6-trimethylpyridine on hepatic cytochrome P-450 heme, apoproteins, and catalytic activities following *in vivo* administration to rats, *Mol. Pharmacol.* **37**:130–136.

170. Lee, J. S., Jacobsen, N. E., and Ortiz de Montellano, P. R., 1988, 4-Alkyl radical extrusion in the cytochrome P-450-catalyzed oxidation of 4-alkyl-1,4-dihydropyridines, *Biochemistry* **27**:7703–7710.

171. Böcker, R. H., and Guengerich, F. P., 1986, Oxidation of 4-aryl- and 4-alkyl-substituted 2,6-dimethyl-3,5-bis(alkoxycarbonyl)-1,4-dihydropyridines by human liver microsomes and immunochemical evidence for the involvement of a form of cytochrome P-450, *J. Med. Chem.* **29**:1596–1603.

172. McCluskey, S. A., Whitney, R. A., and Marks, G. S., 1989, Evidence for the stereoselective inhibition of chick embryo hepatic ferrochelatase by N-alkylated porphyrins, *Mol. Pharmacol.* **36**:608–614.

173. Kennedy, C. H., and Mason, R. P., 1990, A reexamination of the cytochrome P-450-catalyzed free radical production from dihydropyridine: Evidence of trace transition metal catalysis, *J. Biol. Chem.* **265**:11425–11428.

174. Ortiz de Montellano, P. R., and Kerr, D. E., 1985, Inactivation of myoglobin by *ortho*-substituted aryl hydrazines: Formation of prosthetic heme aryl-iron but not N-aryl adducts, *Biochemistry* **24**:1147–1152.

175. Lukton, D., Mackie, J. E., Lee, J. S., Marks, G. S., and Ortiz de Montellano, P. R., 1988, 2,2-Dialkyl-1,2-dihydroquinolines: Cytochrome P-450 catalyzed N-alkylporphyrin formation, ferrochelatase inhibition, and induction of 5-aminolevulinic acid synthase activity, *Chem. Res. Toxicol.* **1**:208–215.

176. Muakkassah, W. F., and Yang, W. C. T., 1981, Mechanism of the inhibitory action of phenelzine on microsomal drug metabolism, *J. Pharmacol. Exp. Ther.* **219**:147–155.

177. Ortiz de Montellano, P. R., Augusto, O., Viola, F., and Kunze, K. L., 1983, Carbon radicals in the metabolism of alkyl hydrazines, *J. Biol. Chem.* **258**:8623–8629.

178. Ortiz de Montellano, P. R., and Watanabe, M. D., 1987, Free radical pathways in the *in vitro* hepatic metabolism of phenelzine, *Mol. Pharmacol.* **31**:213–219.

179. Rumyantseva, G. V., Kennedy, C. H., and Mason, R. P., 1991, Trace transition metal-catalyzed reactions in the microsomal metabolism of alkyl hydrazines to carbon-centered free radicals, *J. Biol. Chem.* **266**:21422–21427.

180. Jonen, H. G., Werringloer, J., Prough, R. A., and Estabrook, R. W., 1982, The reaction of phenylhydrazine with microsomal cytochrome P-450: Catalysis of heme modification, *J. Biol. Chem.* **257**:4404–4411.

181. Mansuy, D., Battioni, P., Bartoli, J.-F., and Mahy, J.-P., 1985, Suicidal inactivation of microsomal cytochrome P-450 by hydrazones, *Biochem. Pharmacol.* **34**:431–432.

182. Delaforge, M., Battioni, P., Mahy, J.-P., and Mansuy, D., 1986, *In vivo* formation of σ-methyl and σ-phenyl-ferric complexes of hemoglobin and liver cytochrome P-450 upon treatment of rats with methyl and phenylhydrazine, *Chem. Biol. Interact.* **60**:101–114.

183. Raag, R., Swanson, B. S., Poulos, T. L., and Ortiz de Montellano, P. R., 1990, Formation, crystal structure, and rearrangement of a cytochrome P450$_{cam}$ iron-phenyl complex, *Biochemistry* **29**:8119–8126.

184. Ortiz de Montellano, P. R., and Kunze, K. L., 1981, Formation of N-phenylheme in the hemolytic reaction of phenylhydrazine with hemoglobin, *J. Am. Chem. Soc.* **103**:581–586.

185. Saito, S., and Itano, H. A., 1981, Beta-meso-phenylbiliverdin IX-alpha and N-phenylprotoporphyrin IX, products of the reaction of phenylhydrazine with oxyhemoproteins, *Proc. Natl. Acad. Sci. USA* **78**:5508–5512.

186. Augusto, O., Kunze, K. L., and Ortiz de Montellano, P. R., 1982, N-Phenylprotoporphyrin IX formation in the hemoglobin–phenylhydrazine reaction: Evidence for a protein-stabilized iron-phenyl intermediate, *J. Biol. Chem.* **257**:6231–6241.

187. Kunze, K. L., and Ortiz de Montellano, P. R., 1983, Formation of a sigma-bonded aryl-iron complex in the reaction of arylhydrazines with hemoglobin and myoglobin, *J. Am. Chem. Soc.* **105:**1380–1381.

188. Ortiz de Montellano, P. R., and Kerr, D. E., 1983, Inactivation of catalase by phenylhydrazine: Formation of a stable aryl-iron heme complex, *J. Biol. Chem.* **258:**10558–10563.

189. Ringe, D., Petsko, G. A., Kerr, D. E., and Ortiz de Montellano, P. R., 1984, Reaction of myoglobin with phenylhydrazine: A molecular doorstop, *Biochemistry* **23:**2–4.

190. Swanson, B. A., and Ortiz de Montellano, P. R., 1991, Structure determination and absolute stereochemistry of the four N-phenylprotoporphyrin IX regioisomers, *J. Am. Chem. Soc.* **113:**8146–8153.

191. Tuck, S. F., Graham-Lorence, S., Peterson, J. A., and Ortiz de Montellano, P. R., 1993, Active sites of the cytochrome P450$_{cam}$ (CYP101) F87W and F87A mutants. Evidence for significant structural reorganization without alteration of catalytic regiospecificity, *J. Biol. Chem.* **268:**269–275.

192. Swanson, B. A., Dutton, D. R., Yang, C. S., and Ortiz de Montellano, P. R., 1991, The active sites of cytochromes P450 IA1, IIB1, IIB2, and IIE1. Topological analysis by *in situ* rearrangement of phenyl-iron complexes, *J. Biol. Chem.* **266:**19258–19264.

193. Swanson, B. A., Halpert, J. R., Bornheim, L. M., and Ortiz de Montellano, P. R., 1992, Topological analysis of the active sites of cytochromes P450IIB4 (rabbit), P450IIB10 (mouse) and P450IIB11 (dog) by *in situ* rearrangement of phenyl-iron complexes, *Arch. Biochem. Biophys.* **292:**42–46.

194. Tuck, S. F., Peterson, J. A., and Ortiz de Montellano, P. R., 1992, Active site topologies of bacterial cytochromes P450 101 (P450$_{cam}$), P450 108 (P450$_{terp}$), and P450 102 (P450$_{BM-3}$): *In situ* rearrangement of their phenyl-iron complexes, *J. Biol. Chem.* **267:**5614–5620.

195. Tuck, S. F., Aoyama, Y., Yoshida, Y., and Ortiz de Montellano, P. R., 1992, Active site topology of *Saccharomyces cerevisiae* lanosterol 14α-demethylase (CYP51) and its A310D mutant (cytochrome P450$_{SG1}$), *J. Biol. Chem.* **267:**13175–13179.

196. Tuck, S. F., and Ortiz de Montellano, P. R., 1992, Topological mapping of the active sites of cytochromes P4502B1 and P4502B2 by *in situ* rearrangement of their aryl-iron complexes, *Biochemistry* **31:**6911–6916.

197. Tuck, S. F., Hiroya, K., Shimizu, T., Hatano, M., and Ortiz de Montellano, P. R., 1993, The cytochrome P450 1A2 (CYP1A2) active site: Topology and perturbations caused by Glu-318 and Thr-319 mutations, *Biochemistry* **32:**2548–2553.

198. Battioni, P., Mahy, J.-P., Delaforge, M., and Mansuy, D., 1983, Reaction of monosubstituted hydrazines and diazenes with rat-liver cytochrome P-450: Formation of ferrous-diazene and ferric sigma-alkyl complexes, *Eur. J. Biochem.* **134:**241–248.

199. Battioni, P., Mahy, J.-P., Gillet, G., and Mansuy, D., 1983, Iron porphyrin dependent oxidation of methyl- and phenylhydrazine: Isolation of iron(II)-diazene and sigma-alkyliron (III) (or aryliron(III)) complexes. Relevance to the reactions of hemoproteins with hydrazines, *J. Am. Chem. Soc.* **105:**1399–1401.

200. Campbell, C. D., and Rees, C. W., 1969, Reactive intermediates. Part III. Oxidation of 1-aminobenzotriazole with oxidants other than lead tetra-acetate, *J. Chem. Soc. C* **1969:**752–756.

201. Ortiz de Montellano, P. R., and Mathews, J. M., 1981, Autocatalytic alkylation of the cytochrome P-450 prosthetic haem group by 1-aminobenzotriazole: Isolation of an N,N-bridged benzyne-protoporphyrin IX adduct, *Biochem. J.* **195:**761–764.

202. Ortiz de Montellano, P. R., Mathews. J. M., and Langry, K. C., 1984, Autocatalytic inactivation of cytochrome P-450 and chloroperoxidase by 1-aminobenzotriazole and other aryne precursors, *Tetrahedron* **40:**511–519.

203. Ortiz de Montellano, P. R., and Costa, A. K., 1985, Dissociation of cytochrome P450 inactivation and induction, *Arch. Biochem. Biophys.* **251:**514–524.

204. Mico, B. A., Federowicz, D. A., Ripple, M. G., and Kerns, W., 1988, *In vivo* inhibition of oxidative drug metabolism by, and acute toxicity of, 1-aminobenzotriazole (ABT), *Biochem. Pharmacol.* **37:**2515–2519.

205. Mugford, C. A., Mortillo, M., Mico, B. A., and Tarloff, J. B., 1992, 1-Aminobenzotriazole-induced destruction of hepatic and renal cytochromes P450 in male Sprague–Dawley rats, *Fundam. Appl. Toxicol.* **19**:43–49.

206. Mathews, J. M., and Bend, J. R., 1986, N-Alkylaminobenzotriazoles as isozyme-selective suicide inhibitors of rabbit pulmonary microsomal cytochrome P-450, *Mol. Pharmacol.* **30**:25–32.

207. Mathews, J. M., and Bend, J. R., 1993, N-Aralkyl derivatives of 1-aminobenzotriazole as potent isozyme-selective mechanism-based inhibitors of rabbit pulmonary cytochrome P450 *in vivo, J. Pharmacol. Exp. Ther.* **265**:281–285.

208. Woodcroft, K. J., Szczepan, E. W., Knickle, L. C., and Bend, J. R., 1990, Three N-aralkylated derivatives of 1-aminobenzotriazole as potent isozyme-selective mechanism-based inhibitors of guinea pig pulmonary cytochrome P450 *in vitro, Drug Metab. Dispos.* **18**:1031–1037.

209. Moreland, D. E., Corbin, F. T., and McFarland, J. E., 1993, Effects of safeners on the oxidation of multiple substrates by grain sorghum microsomes, *Pestic. Biochem. Physiol.* **45**:43–53.

210. Cabanne, F., Huby, D., Gaillardon, P., Scalla, R., and Durst, F., 1987, Effect of the cytochrome P-450 inactivator 1-aminobenzotriazole on the metabolism of chlortoluron and isoproturon in wheat, *Pestic. Biochem. Biophys.* **28**:371–380.

211. Feyereisen, R., Langry, K. C., and Ortiz de Montellano, P. R., 1984, Self-catalyzed destruction of insect cytochrome P-450, *Insect Biochem.* **14**:19–26.

212. Capello, S., Henderson, L., DeGrazia, F., Liberato, D., Garland, W., and Town, C., 1990, The effect of the cytochrome P-450 suicide inactivator, 1-aminobenzotriazole, on the *in vivo* metabolism and pharmacologic activity of flurazepam, *Drug Metab. Dispos.* **18**:190–196.

213. Kaikaus, R. M., Chan, W. K., Lysenko, N., Ray, R., Ortiz de Montellano, P. R., and Bass, N. M., 1993, Induction of peroxisomal fatty acid β-oxidation and liver fatty acid-binding protein by peroxisome proliferators: Mediation via the cytochrome P450IVA1 ω-hydroxylase pathway, *J. Biol. Chem.* **268**:9593–9603.

214. Nichols, W. K., Larson, D. N., and Yost, G. S., 1990, Bioactivation of 3-methylindole by isolated rabbit lung cells, *Toxicol. Appl. Pharmacol.* **105**:264–270.

215. Whitman, D. W., and Carpenter, B. K., 1980, Experimental evidence for nonsquare cyclobutadiene as a chemically significant intermediate in solution, *J. Am. Chem. Soc.* **102**:4272–4274.

216. Stearns, R. A., and Ortiz de Montellano, P. R., 1985, Inactivation of cytochrome P450 by a catalytically generated cyclobutadiene species, *J. Am. Chem. Soc.* **107**:234–240.

217. Stejskal, R., Itabashi, M., Stanek, J., and Hruban, Z., 1975, Experimental porphyria induced by 3-[2-(2,4,6-trimethylphenyl)-thioethyl]-4-methylsydnone, *Virchows Arch.* **18**:83–100.

218. Ortiz de Montellano, P. R., and Grab, L. A., 1986, Inactivation of cytochrome P-450 during catalytic oxidation of a 3-[(arylthio)ethyl]sydnone: N-vinyl heme formation via insertion into the Fe-N bond, *J. Am. Chem. Soc.* **108**:5584–5589.

219. White, E. H., and Egger, N., 1984, Reaction of sydnones with ozone as a method of deamination: On the mechanism of inhibition of monoamine oxidase by sydnones, *J. Am. Chem. Soc.* **106**:3701–3703.

220. Chevrier, B., Weiss, R., Lange, M. C., Chottard, J.-C., and Mansuy, D., 1981, An iron(III)-porphyrin complex with a vinylidene group inserted into an iron-nitrogen bond: Relevance of the structure of the active oxygen complex of catalase, *J. Am. Chem. Soc.* **103**:2899–2901.

221. Latos-Grazynski, L., Cheng, R.-J., La Mar, G. N., and Balch, A. L., 1981, Reversible migration of an axial carbene ligand into an iron-nitrogen bond of a porphyrin: Implications for high oxidation states of heme enzymes and heme catabolism, *J. Am. Chem. Soc.* **103**:4271–4273.

222. Grab, L. A., Swanson, B. A., and Ortiz de Montellano, P. R., 1988, Cytochrome P-450 inactivation by 3-alkylsydnones: Mechanistic implications of N-alkyl and N-alkenyl heme adduct formation, *Biochemistry* **27**:4805–4814.

223. White, I. N. H., Smith, A. G., and Farmer, P. B., 1983, Formation of N-alkylated protoporphyrin IX in the livers of mice after diethylnitrosamine treatment, *Biochem. J.* **212**:599–608.

224. Ding, X., and Coon, M. J., 1988, Cytochrome P-450-dependent formation of ethylene from N-nitrosoethylamines, *Drug Metab. Dispos.* **16**:265–269.

225. Frater, Y., Brady, A., Lock, E. A., and De Matteis, F., 1993, Formation of N-methyl protoporphyrin in chemically-induced protoporphyria. Studies with a novel porphyrogenic agent, *Arch. Toxicol.* **67:**179–185.

226. De Matteis, F., and Gibbs, A. H., 1980, Drug-induced conversion of liver haem into modified porphyrins, *Biochem. J.* **187:**285–288.

227. Holley, A. E., Frater, Y., Gibbs, A. H., De Matteis, F., Lamb, J. H., Farmer, P. B., and Naylor, S., 1991, Isolation of two N-monosubstituted protoporphyrins, bearing either the whole drug or a methyl group on the pyrrole nitrogen atom, from liver of mice given griseofulvin, *Biochem. J.* **274:**843–848.

228. Gibbs, A. H., Naylor, S., Lamb, J. H., Frater, Y., and De Matteis, F., 1990, Copper-induced dealkylation studies of biologically N-alkylated porphyrins by fast atom bombardment mass spectrometry, *Anal. Chim. Acta* **241:**233–239.

229. Holley, A., King, L. J., Gibbs, A. H., and De Matteis, F., 1990, Strain and sex differences in the response of mice to drugs that induce protoporphyria: Role of porphyrin biosynthesis and removal, *J. Biochem. Toxicol.* **5:**175–182.

230. De Matteis, F., Gibbs, A. H., Martin, S. R., and Milek, R. L. B., 1991, Labeling *in vivo* and chirality of griseofulvin-derived N-alkylated protoporphyrins, *Biochem. J.* **280:**813–816.

231. Kunze, K. L., and Trager, W. F., 1993, Isoform-selective mechanism-based inhibition of human cytochrome P450 1A2 by furafylline, *Chem. Res. Toxicol.* **6:**649–656.

232. Clarke, S. E., Ayrton, A. D., and Chenery, R. J., 1994, Characterization of the inhibition of P4501A2 by furafylline, *Xenobiotica* **24:**517–526.

233. Hoag, M. K. P., Trevor, A. J., Kalir, A., and Castagnoli, N., 1987, NADPH-dependent metabolism, covalent binding to macromolecules, and inactivation of cytochrome(s) P450, *Drug Metab. Dispos.* **15:**485–490.

234. Osawa, Y., and Coon, M. J., 1989, Selective mechanism-based inactivation of the major phenobarbital-inducible P-450 cytochrome from rabbit liver by phencyclidine and its oxidation product, the iminium compound, *Drug Metab. Dispos.* **17:**7–13.

235. Owens, S. M., Gunnell, M., Laurenzana, E. M., and Valentine, J. L., 1993, Dose- and time-dependent changes in phencyclidine metabolite covalent binding in rats and the possible role of CYP2D1, *J. Pharmacol. Exp. Ther.* **265:**1261–1266.

236. Brady, J. F., Dokko, J., Di Stefano, E. W., and Cho, A. K., 1987, Mechanism-based inhibition of cytochrome P-450 by heterocyclic analogues of phencyclidine, *Drug Metab. Dispos.* **15:**648–652.

237. Koop, D. R., 1990, Inhibition of ethanol-inducible cytochrome P450IIE1 by 3-amino-1,2,4-triazole, *Chem. Res. Toxicol.* **3:**377–383.

238. Correia, M. A., Decker, C., Sugiyama, K., Caldera, P., Bornheim, L., Wrighton, S. A., Rettie, A. E., and Trager, W. F., 1987, Degradation of rat hepatic cytochrome P-450 heme by 3,5-dicarbethoxy-2,6-dimethyl-4-ethyl-1,4-dihydropyridine to irreversibly bound protein adducts, *Arch. Biochem. Biophys.* **258:**436–451.

239. Tierney, D. J., Haas, A. L., and Koop, D. R., 1992, Degradation of cytochrome P450 2E1: Selective loss after labilization of the enzyme, *Arch. Biochem. Biophys.* **293:**9–16.

240. Osawa, Y., and Pohl, L. R., 1989, Covalent bonding of the prosthetic heme to protein: A potential mechanism for the suicide inactivation or activation of hemoproteins, *Chem. Res. Toxicol.* **2:**131–141.

241. Davies, H. S., Britt, S. G., and Pohl, L. R., 1986, Carbon tetrachloride and 2-isopropyl-4-pentenamide-induced inactivation of cytochrome P-450 leads to heme-derived protein adducts, *Arch. Biochem. Biophys.* **244:**387–352.

242. Guengerich, P., 1986, Covalent binding to apoprotein is a major fate of heme in a variety of reactions in which cytochrome P-450 is destroyed, *Biochem. Biophys. Res. Commun.* **138:**193–198.

243. Riddick, D. S., and Marks, G. S., 1990, Irreversible binding of heme to microsomal protein during inactivation of cytochrome P450 by alkyl analogues of 3,5-diethoxycarbonyl-1,4-dihydro-2,4,6-trimethylpyridine, *Biochem. Pharmacol.* **40:**1915–1921.

244. Sugiyama, K., Yao, K., Rettie, A. E., and Correia, M. A., 1989, Inactivation of rat hepatic cytochrome P-450 isozymes by 3,5-dicarbethoxy-2,6-dimethyl-4-ethyl-1,4-dihydropyridine, *Chem. Res. Toxicol.* **2:**400–410.

245. Schaefer, W. H., Harris, T. M., and Guengerich, F. P., 1985, Characterization of the enzymatic and non-enzymatic peroxidative degradation of iron porphyrins and cytochrome P-450 heme, *Biochemistry* **24:**3254–3263.

246. Nerland, D. E., Iba, M. M., and Mannering, G. J., 1981, Use of linoleic acid hydroperoxide in the determination of absolute spectra of membrane-bound cytochrome P450, *Mol. Pharmacol.* **19:**162–167.

247. Karuzina, I. I., and Archakov, A. I., 1994, The oxidative inactivation of cytochrome P450 in monooxygenase reactions, *Free Radical Biol. Med.* **16:**73–97.

248. Yao, K., Falick, A. M., Patel, N., and Correia, M. A., 1993, Cumene hydroperoxide-mediated inactivation of cytochrome P450 2B1: Identification of an active site heme-modified peptide, *J. Biol. Chem.* **268:**59–65.

249. Correia, M. A., 1993, Drug-mediated heme-modification of cytochrome P450 apoproteins: Structural characterization and physiological implications, *Toxicologist* **13:**15.

250. Catalano, C. E., Choe, Y. S., and Ortiz de Montellano, P. R., 1989, Reactions of the protein radical in peroxide-treated myoglobin: Formation of a heme-protein cross-link, *J. Biol. Chem.* **264:**10534–10541.

251. Choe, Y. S., and Ortiz de Montellano, P. R., 1991, Differential additions to the myoglobin prosthetic heme group. Oxidative γ-meso substitution by alkylhydrazines, *J. Biol. Chem.* **266:**8523–8530.

252. Osawa, Y., Martin, B. M., Griffin, P. R., Yates, J. R., III, Shabanowitz, J., Hunt, D. F., Murphy, A. C., Chen, L., Cotter, R. J., and Pohl, L. R., 1990, Metabolism-based covalent bonding of the heme prosthetic group to its apoprotein during the reductive debromination of BrCCl₃ by myoglobin, *J. Biol. Chem.* **265:**10340–10346.

253. Osawa, Y., Highet, R. J., Bax, A., and Pohl, L. R., 1991, Characterization by NMR of the heme–myoglobin adduct formed during the reductive metabolism of BrCCl₃. Covalent bonding of the proximal histidine to the ring 1 vinyl group, *J. Biol. Chem.* **266:**3208–3214.

254. Kindt, J. T., Woods, A., Martin, B. M., Cotter, R. J., and Osawa, Y., 1992, Covalent alteration of the prosthetic heme of human hemoglobin by BrCCl₃. Cross-linking of heme to cysteine residue 93, *J. Biol. Chem.* **267:**8739–8743.

255. Correia, M. A., Yao, K., Wrighton, S. A., Waxman, D. J., and Rettie, A., 1992, Differential apoprotein loss of rat liver cytochromes P450 after their inactivation by 3,5-dicarbethoxy-2,6-dimethyl-4-ethyl-1,4-dihydropyridine: A case for distinct proteolytic mechanisms? *Arch. Biochem. Biophys.* **294:**493–503.

256. Guzelian, P. S., and Swisher, R. W., 1979, Degradation of cytochrome P-450 haem by carbon tetrachloride and 2-allyl-2-isopropylacetamide in rat liver in vivo and in vitro: Involvement of non-carbon monoxide-forming mechanisms, *Biochem. J.* **184:**481–489.

257. Ortiz de Montellano, P. R., and Kunze, K. L., 1980, Inactivation of hepatic cytochrome P-450 by allenic substrates, *Biochem. Biophys. Res. Commun.* **94:**443–449.

258. Hanzlik, R. P., Kishore, V., and Tullman, R., 1979, Cyclopropylamines as suicide substrates for cytochromes P-450, *J. Med. Chem.* **22:**759–761.

259. Macdonald, T. L., Zirvi, K., Burka, L. T., Peyman, P., and Guengerich, F. P., 1982, Mechanism of cytochrome P-450 inhibition by cyclopropylamines, *J. Am. Chem. Soc.* **104:**2050–2052.

260. Ortiz de Montellano, P. R., and Mathews, J. M., 1981, Inactivation of hepatic cytochrome P-450 by a 1,2,3-benzothiadiazole insecticide synergist, *Biochem. Pharmacol.* **30:**1138–1141.

261. De Groot, H., and Haas, W., 1981, Self-catalyzed O₂-independent inactivation of NADPH- or dithionite-reduced microsomal cytochrome P-450 by carbon tetrachloride, *Biochem. Pharmacol.* **30:**2343–2347.

262. Poli, G., Cheeseman, K., Slater, T. F., and Danzani, M. U., 1981, The role of lipid peroxidation in CCl₄-induced damage to liver microsomal enzymes: Comparative studies in vitro using microsomes and isolated liver cells, *Chem. Biol. Interact.* **37**:13–24.

263. Fernandez, G., Villaruel, M. C., de Toranzo, E. G. D., and Castro, J. A., 1982, Covalent binding of carbon to the heme moiety of cytochrome P-450 and its degradation products, *Res. Commun. Chem. Pathol. Pharmacol.* **35**:283–290.

264. De Groot, H., Harnisch, U., and Noll, T., 1982, Suicidal inactivation of microsomal cytochrome P-450 by halothane under hypoxic conditions, *Biochem. Biophys. Res. Commun.* **107**:885–891.

265. Reiner, O., and Uehleke, H., 1971, Bindung von Tetrachlorkohlenstoff an reduziertes mikrosomales Cytochrom P-450 und an Häm, *Hoppe-Seylers Z. Physiol. Chem.* **352**:1048–1052.

266. Cox, P. J., King, L. J., and Parke, D. V., 1976, The binding of trichlorofluoromethane and other haloalkanes to cytochrome P-450 under aerobic and anaerobic conditions, *Xenobiotica* **6**:363–375.

267. Roland, W. C., Mansuy, D., Nastainczyk, W., Deutschmann, G., and Ullrich, V., 1977, The reduction of polyhalogenated methanes by liver microsomal cytochrome P450, *Mol. Pharmacol.* **13**:698–705.

268. Mansuy, D., and Fontecave, M., 1983, Reduction of benzyl halides by liver microsomes: Formation of 478 nm-absorbing sigma-alkyl-ferric cytochrome P-450 complexes, *Biochem. Pharmacol.* **32**:1871–1879.

269. Mansuy, D., Lange, M., Chottard, J. C., Bartoli, J. F., Chevrier, B., and Weiss, R., 1978, Dichlorocarbene complexes of iron(II)-porphyrins—Crystal and molecular structure of Fe(TPP)(CCl₂)(H₂O), *Angew. Chem. Int. Ed. Engl.* **17**:781–782.

270. Ahr, H. J., King, L. J., Nastainczyk, W., and Ullrich, V., 1980, The mechanism of chloroform and carbon monoxide formation from carbon tetrachloride by microsomal cytochrome P-450, *Biochem. Pharmacol.* **29**:2855–2861.

271. Mansuy, D., Lange, M., Chottard, J. C., and Bartoli, J. F., 1978, Reaction du complexe carbenique Fe(II)(tetraphenylporphyrine)(CCl₂) avec les amines primaires: Formation d'isonitriles, *Tetrahedron Lett.* **33**:3027–3030.

272. Mansuy, D., and Battioni, J.-P., 1982, Isolation of sigma-alkyl-iron(III) or carbene-iron(II) complexes from reduction of polyhalogenated compounds by iron(II)-porphyrins: The particular case of halothane CF₃CHClBr, *J. Chem. Soc. Chem. Commun.* **1982**:638–639.

273. Ruf, H. H., Aur, H., Nastainczyk, W., Ullrich, V., Mansuy, D., Battioni, J.-P., Montiel-Montoya, R., and Trautwein, A., 1984, Formation of a ferric carbanion complex from halothane and cytochrome P-450: Electron spin resonance, electronic spectra and model complexes, *Biochemistry* **23**:5300–5306.

274. Castro, C. E., Wade, R. S., and Belser, N. O., 1985, Biodehalogenation: Reactions of cytochrome P-450 with polyhalomethanes, *Biochemistry* **24**:204–210.

275. Callot, H. J., and Scheffer, E., 1980, Model for the in vitro transformation of cytochrome P-450 into "green pigments," *Tetrahedron Lett.* **21**:1335–1338.

276. Lange, M., and Mansuy, D., 1981, N-Substituted porphyrins formation from carbene iron-porphyrin complexes: A possible pathway for cytochrome P-450 heme destruction, *Tetrahedron Lett.* **22**:2561–2564.

277. Chevrier, B., Weiss, R., Lange, M., Chotard, J. C., and Mansuy, D., 1981, An iron(III)-porphyrin complex with a vinylidene group inserted into an iron-nitrogen bond: Relevance to the structure of the active oxygen complex of catalase, *J. Am. Chem. Soc.* **103**:2899–2901.

278. Olmstead, M. M., Cheng, R.-J., and Balch, A. L., 1982, X-ray crystallographic characterization of an iron porphyrin with a vinylidene carbene inserted into an iron-nitrogen bond, *Inorg. Chem.* **21**:4143–4148.

279. Falzon, M., Nielsch, A., and Burke, M. D., 1986, Denaturation of cytochrome P-450 by indomethacin and other non-steroidal anti-inflammatory drugs: Evidence for a surfactant mechanism and a selective effect of a *p*-chlorophenyl moiety, *Biochem. Pharmacol.* **35**:4019–4024.

280. Guengerich, F. P., Dannan, G. A., Wright, T. S., Martin, M. V., and Kaminsky, L. S., 1982, Purification and characterization of liver microsomal cytochromes P-450: Electrophoretic, spectral, catalytic, and

immunochemical properties and inducibility of eight isozymes isolated from rats treated with phenobarbital or beta-naphthoflavone, *Biochemistry* **21**:6019–6030.

281. Halpert, J. R., 1995, Structural basis of selective cytochrome P450 inhibition, *Annu. Rev. Pharmacol. Toxicol.* **35**:29–53.

282. Covey, D. F., 1988, Aromatase inhibitors: Specific inhibitors of oestrogen biosynthesis, in: *Sterol Biosynthesis Inhibitors* (M. Berg and M. Plempel, eds.), Ellis Horwood, Cambridge, pp. 534–571.

283. Henderson, D., Habenicht, U.-F., Nishino, Y., Kerb, U., and El Etreby, M. F., 1986, Aromatase inhibitors and benign prostatic hyperplasia, *J. Steroid Biochem.* **25**:867–876.

284. Van Wauwe, J. P., and Janssen, P. A. J., 1989, Is there a case for P-450 inhibitors in cancer treatment? *J. Med. Chem.* **32**:2231–2239.

285. Kellis, J. T., Sheets, J. J., and Vickery, L. E., 1984, Amino-steroids as inhibitors and probes of the active site of cytochrome P-450$_{scc}$. Effects on the enzyme from different sources, *J. Steroid Biochem.* **20**:671–676.

286. Sheets, J. J., and Vickery, L. E., 1983, Active site-directed inhibitors of cytochrome P-450$_{scc}$: Structural and mechanistic implications of a side chain-substituted series of amino-steroids, *J. Biol. Chem.* **258**:11446–11452.

287. Sheets, J. J., and Vickery, L. E., 1982, Proximity of the substrate binding site and the heme-iron catalytic site in cytochrome P-450$_{scc}$, *Proc. Natl. Acad. Sci. USA* **79**:5773–5777.

288. Nagahisa, A., Foo, T., Gut, M., and Orme-Johnson, W. H., 1985, Competitive inhibition of cytochrome P-450$_{scc}$ by (22R)- and (22S)-22-aminocholesterol: Side chain stereochemical requirements for C-22 amine coordination to the active-site heme, *J. Biol. Chem.* **260**:846–851.

289. Vickery, L. E., and Singh, J., 1988, 22-Thio-23,24-bisnor-5-cholen-3β-ol: An active site-directed inhibitor of cytochrome P450$_{scc}$, *J. Steroid Biochem.* **29**:539–543.

290. Nagahisa, A., Spencer, R. W., and Orme-Johnson, W. H., 1983, Acetylenic mechanism-based inhibitors of cholesterol side chain cleavage by cytochrome P-450$_{scc}$, *J. Biol. Chem.* **258**:6721–6723.

291. Olakanmi, O., and Seybert, D. W., 1990, Modified acetylenic steroids as potent mechanism-based inhibitors of cytochrome P-450$_{scc}$, *J. Steroid Biochem.* **36**:273–280.

292. Krueger, R. J., Nagahisa, A., Gut, M., Wilson, S. R., and Orme-Johnson, W. H., 1985, Effect of P-450$_{scc}$ inhibitors on corticosterone production by rat adrenal cells, *J. Biol. Chem.* **260**:852–859.

293. Nagahisa, A., Orme-Johnson, W. H., and Wilson, S. R., 1984, Silicon mediated suicide inhibition: An efficient mechanism-based inhibitor of cytochrome P-450$_{scc}$ oxidation of cholesterol, *J. Am. Chem. Soc.* **106**:1166–1167.

294. Trahanovsky, W. S., and Himstedt, A. L., 1974, Oxidation of organic compounds with cerium(IV). XX. Abnormally rapid rate of oxidative cleavage of (beta-trimethylsilylethyl)-phenylmethanol, *J. Am. Chem. Soc.* **96**:7974–7976.

295. Vickery, L. E., and Singh, J., 1988, 22-Thio-23,24-bisnor-5-cholen-3β-ol: An active site-directed inhibitor of cytochrome P450$_{scc}$, *J. Steroid Biochem.* **29**:539–543.

296. Miao, E., Zuo, C., Nagahisa, A., Taylor, B. J., Joardar, S., Byon, C., Wilson, S. R., and Orme-Johnson, W. H., 1990, Cytochrome P450$_{scc}$ mediated oxidation of (20S)-22-nor-22-thiacholesterol: Characterization of mechanism-based inhibition, *Biochemistry* **29**:2199.

297. Vanden Bossche, H., 1992, Inhibitors of P450-dependent steroid biosynthesis: From research to medical treatment, *J. Steroid Biochem. Mol. Biol.* **43**:1003–1021.

298. Brodie, A. M. H., Marsh, D., and Brodie, H. J., 1979, Aromatase inhibitors. IV. Regression of hormone-dependent, mammary tumors in the rat with 4-acetoxy-4-androstene-3,17-dione, *J. Steroid Biochem.* **10**:423–429.

299. Henderson, I. C., and Canellos, G. P., 1980, Cancer of the breast (The past decade), *N. Engl. J. Med.* **302**:78–90.

300. Santen, R. J., Worgul, T. J., Samojlik, E., Interrante, A., Boucher, A. E., Lipton, A., Harvey, H. A., White, D. S., Smart, E., Cox, C., and Wells, S. A., 1981, A randomized trial comparing surgical adrenalectomy with aminoglutethimide plus hydrocortisone in women with advanced breast cancer, *N. Engl. J. Med.* **305**:545–551.

301. Brodie, A. M. H., Dowsett, M., and Coombes, R. C., 1988, Aromatase inhibitors as new endocrine therapy for breast cancer, *Cancer Treat. Res.* **39**:51–65.

302. Brodie, A. M. H., Banks, P. K., Inkster, S. E., Dowsett, M., and Coombes, R. C., 1990, Aromatase inhibitors and hormone-dependent cancers, *J. Steroid Biochem. Mol. Biol.* **37**:327–333.

303. Henderson, D., Habenicht, U.-F., Nishino, Y., and El Etreby, M. F., 1987, Estrogens and benign prostatic hyperplasia: The basis for aromatase inhibitor therapy, *Steroids* **50**:219–233.

304. Schweikert, H.-U., and Tunn, U. W., 1987, Effects of the aromatase inhibitor testolactone on human benign prostatic hyperplasia, *Steroids* **50**:191–199.

305. Phillips, G. B., Castelli, W. P., Abbott, R. D., and McNamara, P. M., 1983, Association of hyperestrogenemia and coronary heart disease in men in the Framingham cohort, *Am. J. Med.* **74**:863–869.

306. Harris, A. L., Powles, T. J., Smith, I. E., Coombes, R. C., Ford, H. T., Gazet, J. C., Harmer, C. L., Morgan, M., White, H., Parsons, C. A., and McKinna, J. A., 1983, Aminoglutethimide for the treatment of advanced postmenopausal breast cancer, *Eur. J. Cancer Clin. Oncol.* **19**:11–17.

307. Foster, A. B., Jarman, M., Leung, C.-S., Rowlands, M. G., Taylor, G. N., Plevey, R. G., and Sampson, P., 1985, Analogues of aminoglutethimide: Selective inhibition of aromatase, *J. Med. Chem.* **28**:200–204.

308. Foster, A. B., Jarman, M., Leung, C.-S., Rowlands, M. G., and Taylor, G. N., 1983, Analogues of aminoglutethimide: Selective inhibition of cholesterol side-chain cleavage, *J. Med. Chem.* **26**:50–54.

309. Bhatnagar, A. S., Hausler, A., Schieweck, K., Browne, L. J., Bowman, R., and Steele, R. E., 1990, Novel aromatase inhibitors, *J. Steroid Biochem. Mol. Biol.* **37**:363–367.

310. Lipton, A., Harvey, H. A., Demers, L. M., Hanagan, J. R., Mulagha, M. T., Kochak, G. M., Fitzsimmons, S., Sanders, S. I., and Santen, R. J., 1990, A phase I trial of CGS 16949A: A new aromatase inhibitor, *Cancer* **65**:1279–1285.

311. Santen, R. J., Demers, L. M., Adlercreutz, H., Harvey, H., Santner, S., Sanders, S., and Lipton, A., 1989, Inhibition of aromatase with CGS 16949A in postmenopausal women, *J. Clin. Endocrinol. Metab.* **68**:99–106.

312. Stein, R. C., Dowsett, M., Davenport, J., Hedley, A., Ford, H. T., Gazet, J.-C., and Coombes, R. C., 1990, Preliminary study of the treatment of advanced breast cancer in postmenopausal women with the aromatase inhibitor CGS 16949A, *Cancer Res.* **50**:1381–1384.

313. Demers, L. M., Melby, J. C., Wilson, T. E., Lipton, A., Harvey, H. A., and Santen, R. J., 1990, The effects of CGS 16949A, an aromatase inhibitor on adrenal mineralocorticoid biosynthesis, *J. Clin. Endocrinol. Metab.* **70**:1162–1166.

314. Wouters, W., De Coster, R., Tuman, R. W., Bowden, C. R., Bruynseels, J., Vanderpas, H., Van Rooy, P., Amery, W. K., and Janssen, P. A. J., 1989, Aromatase inhibition by R 76713: Experimental and clinical pharmacology, *J. Steroid Biochem.* **34**:427–430.

315. Wouters, W., De Coster, R., Van Dun, J., Krekels, M. D. W. G., Dillen, A., Raeymaekers, A., Freyne, E., Van Gelder, J., Sanz, G., Venet, M., and Janssen, M., 1990, Comparative effects of the aromatase inhibitor R76713 and of its enantiomers R83839 and R83842 on steroid biosynthesis *in vitro* and *in vivo*, *J. Steroid Biochem. Mol. Biol.* **37**:1049–1054.

316. Vanden Bossche, H., Willemsens, G., Roels, I., Bellens, D., Moereels, H., Coene, M.-C., Le Jeune, L., Lauwers, W., and Janssen, P. A. J., 1990, R 76713 and enantiomers: Selective, nonsteroidal inhibitors of the cytochrome P450-dependent oestrogen synthesis, *Biochem. Pharmacol.* **40**:1707–1718.

317. Flynn, G. A., Johnston, J. O., Wright, C. L., and Metcalf, B. W., 1981, The time-dependent inactivation of aromatase by 17-β-hydroxy-10-methylthioestra-1,4-dien-3-one, *Biochem. Biophys. Res. Commun.* **103**:913–918.

318. Wright, J. N., van Leersum, P. T., Chamberlin, S. G., and Akhtar, M., 1989, Inhibition of aromatase by steroids substituted at C-19 with halogen, sulphur, and nitrogen, *J. Chem. Soc. Perkin Trans. I* **1989**:1647–1655.

319. Wright, J. N., Slatcher, G., and Akhtar, M., 1991, "Slow-binding" sixth-ligand inhibitors of cytochrome P-450 aromatase. Studies with 19-thiomethyl- and 19-azido-androstenedione, *Biochem. J.* **273**:533–539.

320. Delaisi, C., Coucet, B., Hartmann, C., Tric, B., Gourvest, J. F., and Lesuisse, D., 1992, RU54115, a tight-binding aromatase inhibitor potentially useful for the treatment of breast cancer, *J. Steroid Biochem. Mol. Biol.* **41**:773–777.

321. Bednarski, P. J., and Nelson, S. D., 1989, Interactions of thiol-containing androgens with human placental aromatase, *J. Med. Chem.* **32**:203–213.

322. Geelen, J. A. A., Deckers, G. H., Van Der Wardt, J. T. H., Loozen, H. J. J., Tax, L. J. W., and Kloosterboer, H. J., 1991, Selection of 19-(ethyldithio)-androst-4-ene-3,17-dione (ORG 30958): A potent aromatase inhibitor *in vivo*, *J. Steroid Biochem. Mol. Biol.* **38**:181–188.

323. Lovett, J. A., Darby, M. V., and Counsell, R. E., 1984, Synthesis and evaluation of 19-aza- and 19-aminoandrostenedione analogues as potential aromatase inhibitors, *J. Med. Chem.* **27**:734–740.

324. Shih, M.-J., Carrell, M. H., Carrell, H. L., Wright, C. L., Johnston, J. O., and Robinson, C. H., 1987, Stereoselective inhibition of aromatase by novel epoxysteroids, *J. Chem. Soc. Chem. Commun.* **1987**:213–214.

325. Childers, W. E., and Robinson, C. H., 1987, Novel 10β-thiiranyl steroids as aromatase inhibitors, *J. Chem. Soc. Chem. Commun.* 320–321.

326. Childers, W. E., Silverton, J. V., Kellis, J. T., Vickery, L. E., and Robinson, C. H., 1991, Inhibition of human placental aromatase by novel homologated 19-oxiranyl and 19-thiiranyl steroids, *J. Med. Chem.* **34**:1344–1349.

327. Kellis, J. T., Childers, W. E., Robinson, C. H., and Vickery, L. E., 1987, Inhibition of aromatase cytochrome P-450 by 10-oxirane and 10-thiirane substituted androgens. Implications for the structure of the active site, *J. Biol. Chem.* **262**:4421–4426.

328. Njar, V. C. O., Safi, E., Silverton, J. V., and Robinson, C. H., 1993, Novel 10β-aziridinyl steroids: Inhibitors of aromatase, *J. Chem. Soc. Perkin Trans. I* **N10**:1161–1168.

329. Metcalf, B. W., Wright, C. L., Burkhan, J. P., and Johnston, J. O., 1981, Substrate-induced inactivation of aromatase by allenic and acetylenic steroids, *J. Am. Chem. Soc.* **103**:3221–3222.

330. Johnston, J. O., 1987, Biological characterization of 10-(2-propynyl)estr-4-ene-3,17-dione (MDL 18,962), an enzyme-activated inhibitor of aromatase, *Steroids* **50**:105–120.

331. Johnston, J. O., Wright, C. L., and Metcalf, B. W., 1984, Time-dependent inhibition of aromatase in trophoblastic tumor cells in tissue culture, *J. Steroid Biochem.* **20**:1221–1226.

332. Covey, D. G., Hood, W. F., and Parikh, V. D., 1981, $10\text{-}\beta$-Propynyl-substituted steroids: Mechanism-based enzyme-activated irreversible inhibitors of estrogen biosynthesis, *J. Biol. Chem.* **256**:1076–1079.

333. Brandt, M. E., Puett, D., Covey, D. F., and Zimniski, S. J., 1988, Characterization of pregnant mare's serum gonadotropin-stimulated rat ovarian aromatase and its inhibition by 10-propargylestr-4-ene-3,17-dione, *J. Steroid Biochem.* **34**:317–324.

334. Marcotte, P. A., and Robinson, C. H., 1982, Synthesis and evaluation of 10-beta-substituted 4-estrene-3,17-diones as inhibitors of human placental microsomal aromatase, *Steroids* **39**:325–344.

335. Numazawa, M., Mutsumi, A., Asano, N., and Ito, Y., 1993, A time-dependent inactivation of aromatase by 19-substituted androst-4-ene-3,6,17-diones, *Steroids* **58**:40–46.

336. Marcotte, P. A., and Robinson, C. H., 1982, Design of mechanism-based inactivators of human placental aromatase, *Cancer Res.* **42**:3322–3325.

337. Marcotte, P. A., and Robinson, C. H., 1982, Inhibition and inactivation of estrogen synthetase (aromatase) by fluorinated substrate analogues, *Biochemistry* **21**:2773–2778.

338. Numazawa, M., Mutsumi, A., Hoshi, K., Oshibe, M., Ishikawa, E., and Kigawa, H., 1991, Synthesis and biochemical studies of 16- and 19-substituted androst-4-enes as aromatase inhibitors, *J. Med. Chem.* **34**:2496–2504.

339. Mann, J., and Pietrzak, B., 1987, Preparation of aromatase inhibitors. Synthesis of 19,19-difluoro-4-hydroxyandrost-4-ene-3,7-dione and related compounds, *J. Chem. Soc. Perkin Trans. I* **1987**:385–388.

340. Furth, P. S., and Robinson, C. H., 1989, Tritium release from [19-³H]19,19-difluoroandrost-4-ene-3,17-dione during inactivation of aromatase, *Biochemistry* **28**:1254–1259.

341. Covey, D. F., and Hood, W. F., 1982, Aromatase enzyme catalysis is involved in the potent inhibition of estrogen biosynthesis caused by 4-acetoxy- and 4-hydroxy-4-androstene-3,17-dione, *Mol. Pharmacol.* **21**:173–180.

342. Brodie, A. M. H., Garrett, W. M., Hendrickson, J. R., Tsai-Morris, C.-H., Marcotte, P. A., and Robinson, C. H., 1981, Inactivation of aromatase in vitro by 4-hydroxy-4-androstene-3,17-dione and 4-acetoxy-4-androstene-3,17-dione and sustained effects in vivo, *Steroids* **38**:693–702.

343. Brodie, A. M. H., 1994, Aromatase inhibitors in the treatment of breast cancer, *J. Steroid Biochem. Mol. Biol.* **49**:281–287.

344. Di Salle, E., Giudici, D., Briatico, G., and Ornati, G., 1990, Novel irreversible aromatase inhibitors, *Ann. N.Y. Acad. Sci.* **595**:357–367.

345. Di Salle, E., Giudici, D., Ornati, G., Briatico, G., D'Alessio, R., Villa, V., and Lombardi, P., 1990, 4-Aminoandrostenedione derivatives: A novel class of irreversible aromatase inhibitors. Comparison with FCE 24304 and 4-hydroxyandrostenedione, *J. Steroid Biochem. Mol. Biol.* **37**:369–374.

346. Di Salle, E., Briatico, G., Giudici, D., Ornati, G., and Zaccheo, T., 1989, Aromatase inhibition and experimental antitumor activity of FCE 24304, MDL 18962 and SH 489, *J. Steroid Biochem.* **34**:431–434.

347. Marsh, D. A., Brodie, H. J., Garrett, W., Tsai-Morris, C.-H., and Brodie, A. M., 1985, Aromatase inhibitors. Synthesis and biological activity of androstenedione derivatives, *J. Med. Chem.* **28**:788–795.

348. Brodie, A. M. H., Brodie, H. J., Garrett, W. M., Hendrickson, J. R., Marsh, D. H., and Tsai-Morris, C.-H., 1982, Effect of an aromatase inhibitor, 1,4,6-androstatriene-3,17-dione, on 7,12-dimethyl-[a]-anthracene-induced mammary tumors in the rat and its mechanism of action *in vivo*, *Biochem. Pharmacol.* **31**:2017–2023.

349. Henderson, D., Norbisrath, G., and Kerb, U., 1986, 1-Methyl-1,4-androstadiene-3,17-dione (SH 489): Characterization of an irreversible inhibitor of estrogen biosynthesis, *J. Steroid Biochem.* **24**:303–306.

350. Numazawa, M., Mutsumi, A., Hoshi, K., and Tanaka, Y., 1992, Androst-5-ene-7,17-dione: A novel class of suicide substrate of aromatase, *Biochem. Biophys. Res. Commun.* **186**:32–39.

351. Covey, D. F., and Hood, W. F., 1981, Enzyme-generated intermediates derived from 4-androstene-3,6,17-trione and 1,4,6-androstatriene-3,17-dione cause a time-dependent decrease in human placental aromatase activity, *Endocrinology* **108**:1597–1599.

352. Numazawa, M., Tsuji, M., and Mutsumi, A., 1987, Studies on aromatase inhibition with 4-androstene-3,6,17-trione: Its 3β-reduction and time-dependent irreversible binding to aromatase with human placental microsomes, *J. Steroid Biochem.* **28**:337–344.

353. Numazawa, M., Midzuhashi, K., and Nagaoka, M., 1994, Metabolic aspects of the 1β-proton and the 19-methyl group of androst-4-ene-3,6,17-trione during aromatization by placental microsomes and inactivation of aromatase, *Biochem. Pharmacol.* **47**:717–726.

354. Longcope, C., Femino, A., and Johnston, J. O., 1988, Inhibition of peripheral aromatization in baboons by an enzyme-activated aromatase inhibitor (MDL 18,962), *Endocrinology* **122**:2007–2011.

355. Johnston, J. O., 1990, Studies with the steroidal aromatase inhibitor, 19-acetylenic androstenedione (MDL 18,962), *J. Cancer Res. Clin. Oncol.* **116**:880.

356. Covey, D. F., Hood, W. F., Bensen, D. D., and Carrell, H. L., 1984, Hydroperoxides as inactivators of aromatase: 10-Beta-hydroperoxy-4-estrene-3,17-dione, crystal structure and inactivation characteristics, *Biochemistry* **23**:5398–5406.

357. Covey, D. F., Hood, W. F., and McMullan, P. C., 1986, Studies of the inactivation of human placental aromatase by 17α-ethynyl-substituted 10β-hydroperoxy and related 19-nor steroids, *Biochem. Pharmacol.* **35**:1671–1674.

358. Bednarski, P. J., Porubek, D. J., and Nelson, S. D., 1985, Thiol-containing androgens as suicide substrates of aromatase, *J. Med. Chem.* **28**:775–779.

359. Shih, M.-J., Carrell, M. H., Carrell, H. L., Wright, C. L., Johnston, J. O., and Robinson, C. H., 1987, Stereoselective inhibition of aromatase by novel epoxysteroids. *J. Chem Soc. Chem. Commun.* **1987:**213–214.

360. Burkhart, J. P., Peet, N. P., Wright, C. L., and Johnston, J. O., 1991, Novel time-dependent inhibitors of human placental aromatase, *J. Med. Chem.* **34:**1748–1750.

361. Vanden Bossche, H., Willemsens, G., Cools, W., Marichal, P., and Lauwers, W., 1983, Hypothesis on the molecular basis of the antifungal activity of N-substituted imidazoles and triazoles, *Biochem. Soc. Trans.* **11:**665–667.

362. Mercer, E. I., 1991, Sterol biosynthesis inhibitors: Their current status and modes of action, *Lipids* **26:**584–597.

363. Berg, M., and Plempel, M. (eds.), 1988, *Sterol Biosynthesis Inhibitors*, Ellis Horwood, Cambridge.

364. Vanden Bossche, H., Lauwers, W., Willemsens, G., Marichal, P., Cornelissen, F., and Cools, W., 1984, Molecular basis for the antimycotic and antibacterial activity of N-substituted imidazoles and triazoles: The inhibition of isoprenoid biosynthesis, *Pestic. Sci.* **15:**188–198.

365. Heeres, J., De Brabander, M., and Vanden Bossche, H., 1982, Ketoconazole: Chemistry and basis for selectivity, in: *Current Chemotherapy and Immunotherapy*, Vol. 2 (P. Periti and G. G. Grossi, eds.), American Society of Microbiology, Washington, DC, pp. 1007–1009.

366. Willemsens, G., Cools, W., and Vanden Bossche, H., 1980, Effects of miconazole and ketoconazole on sterol synthesis in a subcellular fraction of yeast and mammalian cells, in: *The Host–Invader Interplay* (H. Vanden Bossche, ed.), Elsevier/North-Holland, Amsterdam, pp. 691–694.

367. Murray, M., Ryan, A. J., and Little, P. J., 1982, Inhibition of rat hepatic microsomal aminopyrine N-demethylase activity by benzimidazole derivatives: Quantitative structure–activity relationships, *J. Med. Chem.* **25:**887–892.

368. Santen, R. J., Vanden Bossche, H., Symoens, J., Brugmans, J., and DeCoster, R., 1983, Site of action of low dose ketoconazole or androgen biosynthesis in men, *J. Clin. Endocrinol. Metab.* **57:**732–736.

369. Gahder, P., Mercer, E. I., Baldwin, B. C., and Wiggins, T. E., 1983, A comparison of the potency of some fungicides as inhibitors of sterol 14-demethylation, *Pestic. Biochem. Physiol.* **19:**1–10.

370. Nes, W. R., 1974, Role of sterols in membranes, *Lipids* **9:**596–612.

371. Yeagle, P. L., Martin, R. B., Lala, A. K., Lin, H.-K., and Block, K., 1977, Differential effects of cholesterol and lanosterol on artificial membranes, *Proc. Natl. Acad. Sci. USA* **74:**4924–4926.

372. Freter, C. E., Laderson, R. C., and Sibert, D. F., 1979, Membrane phospholipid alterations in response to sterol depletion of LM cells, *J. Biol. Chem.* **254:**6909–6916.

373. Dolle, R. E., Allaudeen, H. S., and Kruse, L. I., 1990, Design and synthesis of 14α-methyl-15-aza-D-homosterols as novel antimycotics, *J. Med. Chem.* **33:**877–880.

374. Frye, L. L., Cusack, K. P., Leonard, D. A., and Anderson, J. A., 1994, Oxolanosterol oximes: Dual-action inhibitors of cholesterol biosynthesis, *J. Lipid Res.* **35:**1333–1344.

375. Aoyama, Y., Yoshida, Y., Sonoda, Y., and Sato, Y., 1987, 7-Oxo-24,25-dihydrolanosterol: A novel lanosterol 14α-demethylase (*P-450* 14DM) inhibitor which blocks electron transfer to the oxyferro intermediate, *Biochim. Biophys. Acta* **922:**270–277.

376. Cooper, A. B., Wright, J. J., Ganguly, A. K., Desai, J., Loenberg, D., Parmegiani, R., Feingold, D. S., and Sud, I. J., 1989, Synthesis of 14-α-aminomethyl substituted lanosterol derivatives; inhibitors of fungal ergosterol biosynthesis, *J. Chem. Soc. Chem. Commun.* **1989:**898–900.

377. Frye, L. L., Cusack, K. P., and Leonard, D. A., 1993, 32-Methyl-32-oxylanosterols: Dual-action inhibitors of cholesterol biosynthesis, *J. Med. Chem.* **36:**410–416.

378. Frye, L. L., and Robinson, C. H., 1988, Novel inhibitors of lanosterol 14α-methyl demethylase, a critical enzyme in cholesterol biosynthesis, *J. Chem. Soc. Chem. Commun.* **1988:**129–131.

379. Mayer, R. J., Adams, J. L., Bossard, M. J., and Berkhout, T. A., 1991, Effects of a novel lanosterol 14α-demethylase inhibitor on the regulation of 3-hydroxy-3-methylglutaryl-coenzyme A reductase in Hep G2 cells, *J. Biol. Chem.* **266:**20070–20078.

380. Frye, L. L., and Robinson, C. H., 1990, Synthesis of potential mechanism-based inactivators of lanosterol 14α-demethylase, *J. Org. Chem.* **55:**1579–1584.

381. Tuck, S. F., Robinson, C. H., and Silverton, J. V., 1991, Assessment of the active-site requirements of lanosterol 14α-demethylase: Evaluation of novel substrate analogues as competitive inhibitors, *J. Org. Chem.* **56**:1260–1266.

382. Bossard, M. J., Tomaszek, T. A., Gallagher, T., Metcalf, B. W., and Adams, J. L., 1991, Steroidal acetylenes: Mechanism-based inactivators of lanosterol 14α-demethylase, *Bioorg. Chem.* **19**:418–432.

383. Trzaskos, J. M., Magolda, R. L., Favata, M. F., Fischer, R. T., Johnson, P. R., Chen, H. W., Ko, S. S., Leonard, D. A., and Gaylor, J. L., 1993, Modulation of 3-hydroxy-3-methylglutaryl-CoA reductase by 15α-fluorolanost-7-en-3β-ol. A mechanism-based inhibitor of cholesterol biosynthesis, *J. Biol. Chem.* **268**:22591–22599.

384. Angelastro, M. R., Laughlin, M. E., Schatzman, G. L., Bey, P., and Blohm, T. R., 1989, 17β-(Cyclo-propylamino)-androst-5-en-3β-ol, a selective mechanism-based inhibitor of cytochrome P450$_{17α}$ (steroid 17α-hydroxylase/C$_{17-20}$ lyase), *Biochem. Biophys. Res. Commun.* **162**:1571–1577.

385. Berg, A. M., Kickman, A. B., Miao, E., Cochran, A., Wilson, S. R., and Orme-Johnson, W. H., 1990, Effects of inhibitors of cytochrome P-45017α on steroid production in mouse Leydig cells and mouse and pig testes microsomes, *Biochemistry* **29**:2193.

386. Viger, A., Coustal, S., Perard, S., Chappe, B., and Marquet, A., 1988, Synthesis and activity of new inhibitors of aldosterone biosynthesis, *J. Steroid Biochem.* **30**:469–472.

387. Viger, A., Coustal, S., Perard, S., Piffeteau, A., and Marquet, A., 1989, 18-Substituted progesterone derivatives as inhibitors of aldosterone biosynthesis, *J. Steroid Biochem.* **33**:119–124.

388. Gomez-Sanchez, C. E., Chiou, S., and Yamakita, N., 1993, 18-Ethynyl-deoxycorticosterone inhibition of steroid production is different in freshly isolated compared to cultured calf zona glomerulosa cells, *J. Steroid Biochem. Mol. Biol.* **46**:805–810.

389. Johnston, J. O., Wright, C. L., Bohnke, R. A., and Kastner, P. R., 1991, Inhibition of aldosterone biosynthesis in primates by 18-acetylenic deoxycorticosterone, *Endocrinology* **128(Suppl.)**:Abstract 24.

390. Shak, S., and Goldstein, I., 1984, Omega-oxidation is the major pathway for the catabolism of leukotriene B$_4$ in human polymorphonuclear leukocytes, *J. Biol. Chem.* **259**:10181–10187.

391. Kupfer, D., 1982, Endogenous substrates of monooxygenases: Fatty acids and prostaglandins, in: *Hepatic Cytochrome P450 Monooxygenase System* (J. B. Sehenkman and D. Kupfer, eds.), Pergamon Press, Elmsford, NY, pp. 157–190.

392. Kupfer, D., 1980, Endogenous substrates of monooxygenases: Fatty acids and prostaglandins, *Pharmacol. Ther. A* **11**:469–496.

393. Fulco, A. J., 1991, P450$_{BM-3}$ and other inducible bacterial P450 cytochromes: Biochemistry and regulation, *Annu. Rev. Pharmacol. Toxicol.* **31**:177–203.

394. Ortiz de Montellano, P. R., and Reich, N. O., 1984, Specific inactivation of hepatic fatty acid hydroxylases by acetylenic fatty acids, *J. Biol. Chem.* **259**:4136–4141.

395. Salaun, J. P., Reichhart, D., Simon, A., Durst, F., Reich, N. O., and Ortiz de Montellano, P. R., 1984, Autocatalytic inactivation of plant cytochrome P-450 enzymes: Selective inactivation of the lauric acid in-chain hydroxylase from *Helianthus tuberosus* L. by unsaturated substrate analogs, *Arch. Biochem. Biophys.* **232**:1–7.

396. CaJacob, C. A., and Ortiz de Montellano, P. R., 1986, Mechanism-based *in vivo* inactivation of lauric acid hydroxylases, *Biochemistry* **25**:4705–4711.

397. Hirt, D. L., and Jacobson, H. R., 1991, Functional effects of cytochrome P450 arachidonate metabolites in the kidney, *Semin. Nephrol.* **11**:148–155.

398. McGiff, J. C., Quilley, C. P., and Carroll, M. A., 1993, The contribution of cytochrome P450-dependent arachidonate metabolites to integrated renal function, *Steroids* **58**:573–579.

399. Zou, A.-P., Ma, Y.-H., Sui, Z.-H., Ortiz de Montellano, P. R., Clark, J. E., Masters, B. S., and Roman, R. J., 1994, Effects of 17-octadecynoic acid, a suicide-substrate inhibitor of cytochrome P450 fatty acid ω-hydroxylase, on renal function in rats, *J. Pharmacol. Exp. Ther.* **268**:474–481.

400. Shak, S., Reich, N. O., Goldstein, I. M., and Ortiz de Montellano, P. R., 1985, Leukotriene B₄ ω-hydroxylase in human polymorphonuclear leukocytes: Suicidal inactivation by acetylenic fatty acids, *J. Biol. Chem.* **260:**13023–13028.

401. Williams, D. E., Muerhoff, A. S., Reich, N. O., CaJacob, C. A., Ortiz de Montellano, P. R., and Masters, B. S. S., 1989, Prostaglandin and fatty acid ω and (ω-1) oxidation in rabbit lung. Acetylenic fatty acid mechanism based inactivators as specific inhibitors, *J. Biol. Chem.* **264:**749–756.

402. Shirane, N., Sui, Z., Peterson, J. A., and Ortiz de Montellano, P. R., 1993, Cytochrome P450$_{BM-3}$ (CYP102): Regiospecificity of oxidation of ω-unsaturated fatty acids and mechanism-based inactivation, *Biochemistry* **32:**13732–13741.

403. Kikuta, Y., Kusunose, E., Endo, K., Yamamoto, S., Sogawa, K., Fujii-Kuriyama, Y., and Kusunose, M., 1993, A novel form of cytochrome P-450 family 4 in human polymorphonuclear leukocytes. cDNA cloning and expression of leukotriene B₄ ω-hydroxylase, *J. Biol. Chem.* **268:**9376–9380.

404. Clancy, R. M., Dahinden, C. A., and Hugli, T. E., 1984, Oxidation of leukotrienes at the ω end: Demonstration of a receptor for the 20-hydroxy derivative of leukotriene B₄ on human neutrophils and implications for the analysis of leukotriene receptors, *Proc. Natl. Acad. Sci. USA* **81:**5729–5733.

Part IV

Regulatory Mechanisms and Physiological Roles of Cytochrome P450

CHAPTER 10

Induction of Cytochrome P450 Enzymes That Metabolize Xenobiotics

JAMES P. WHITLOCK, JR. and MICHAEL S. DENISON

1. Introduction

One of the many interesting aspects of the cytochrome P450 enzymes is that some are inducible; that is, following exposure of the cell to an inducing chemical, enzyme activity increases, in some cases by orders of magnitude. The induction phenomena were first recognized because they produced alterations in pharmacological responses to drugs or other xenobiotics. For example, animals chronically exposed to barbiturates become "tolerant" to the hypnotic effects of these drugs, because they induce the cytochrome P450 enzymes responsible for their own metabolism.[1] Similarly, the induction of cytochrome P450 enzymes reduced the incidence of neoplasia in animals exposed to chemical carcinogens.[2] Such examples illustrate two interesting points. First, inducers are often substrates for the induced enzymes; thus, enzyme activity increases only as needed. Second, enzyme induction usually enhances detoxification, particularly when low to moderate concentrations of substrate are present; thus, under most conditions, induction is a protective mechanism, whereby the cell can detoxify lipophilic compounds that might otherwise accumulate. Both characteristics are likely to facilitate the survival of the cell in a potentially toxic chemical environment.

Induction of cytochrome P450 enzyme activity can be disadvantageous in some instances. For example, the induced enzymes often metabolize numerous substrates; therefore, enzyme induction by one chemical may lead to increased metabolism of other compounds; if one happens to be a drug with a low therapeutic index, the increase in

JAMES P. WHITLOCK, JR. • Department of Molecular Pharmacology, Stanford University School of Medicine, Stanford, California 94305. *MICHAEL S. DENISON* • Department of Environmental Toxicology, University of California, Davis, Davis, California 95616.

Cytochrome P450: Structure, Mechanism, and Biochemistry (Second Edition), edited by Paul R. Ortiz de Montellano. Plenum Press, New York, 1995.

metabolism may lead to the loss of efficacy.[3] Occasionally, enzyme induction may enhance chemical toxicity. For example, high concentrations of the analgesic acetaminophen (e.g., such as those attained following a suicide attempt) saturate detoxification pathways, leading to reactions via cytochrome P450 that generate reactive electrophiles, which bind to cellular macromolecules and produce hepatic necrosis. Cytochrome P450 induction increases the severity of acetaminophen toxicity by increasing the production of reactive metabolites and accentuating the imbalance between detoxification and activation pathways for the drug.[4]

From a more basic standpoint, the induction of cytochrome P450 enzyme activity constitutes an interesting biological response for analyzing the mechanisms by which small molecules (e.g., drugs, hormones, environmental contaminants) can alter the cellular phenotype, often by enhancing the transcription of a specific gene(s). The knowledge obtained from analyzing induction at the molecular level can provide insights into gene regulation that are of relatively broad interest.

Others have reviewed cytochrome P450 enzyme induction in detail.[5–8] Here, we briefly summarize the induction mechanisms for several cytochrome P450 enzymes that metabolize xenobiotics. We focus first on the induction of cytochrome P450 1A1, because the mechanistic information is most detailed for this system.

2. Aromatic Hydrocarbon-Inducible Cytochrome P450s

2.1. Background

One of the first examples of cytochrome P450 enzyme induction was the finding that ingestion of polycyclic aromatic hydrocarbons (PAHs), such as 3-methylcholanthrene (3-MC), increased the rate of carcinogen metabolism in rodent liver.[9] Subsequent experiments have linked the enhanced metabolism to a form of cytochrome P450 that is now designated as cytochrome P450 1A1. This enzyme metabolizes various PAHS; it is often studied using the aryl hydrocarbon hydroxylase (AHH) assay, which measures its ability to oxygenate the carcinogen benzo(a)pyrene.[10] AHH activity is inducible in numerous tissues and in many cells in culture; the latter property has facilitated the use of molecular genetic techniques to analyze the induction mechanism.

Initial studies revealed that induction required intact cells and that inhibitors of RNA and protein synthesis, such actinomycin D and cycloheximide, blocked induction; such findings suggested that induction might involve gene transcription. In addition, early studies of various inducers of AHH activity revealed structural similarities among them, a finding that suggested the existence of a specific cellular target molecule (i.e., "receptor") at which the inducers might act.

The development of techniques for purification of cytochrome P450 enzymes from microsomal membranes led to the generation of antibodies that were directed against specific forms of cytochrome P450. The antibodies were used in combination with *in vitro* transcription techniques to demonstrate that induction of cytochrome P450 1A1 activity was associated with an increase in its translatable mRNA. This information, in turn, led to the use of a differential hybridization strategy to clone P450 1A1 cDNA from mouse liver.[11] Subsequently, nuclear "run-on" experiments revealed that induction reflected an increase in the rate of transcription of the CYP1A1 gene.[12,13]Thus, the initial studies of cytochrome P450 1A1 induction revealed that it involved increased CYP1A1 gene transcription. At this relatively early stage of analysis, the induction mechanism appeared similar to that for steroid hormones, which also act at the transcriptional level. By analogy with steroid-responsive systems, a specific intracellular protein (possibly related to steroid receptors) was postulated to mediate the induction of CYP1A1 transcription by PAHs. However, such a receptor could not be detected, using PAHs as ligands.

2.2. The Ah Receptor

The observation that the environmental contaminant 2,3,7,8-tetrachlorodibenzo-*p*-dioxin (TCDD) was an unusually potent inducer of AHH activity was crucial to a better understanding of the induction mechanism. Poland *et al.*, using radiolabeled TCDD as a ligand, showed that mouse liver cytosol contained a protein that bound TCDD saturably (i.e., about 10^5 binding sites per cell), reversibly, and with high affinity (i.e., in the low nanomolar range), consistent with TCDD's potency as an inducer. Studies with congeners of TCDD revealed that ligand binding exhibited structural specificity. In addition, ligand binding appeared to obey the law of mass action. Thus, the TCDD-binding protein had the properties expected for a receptor. The protein has been designated as the "Ah receptor" because it binds and mediates the response to other *a*romatic *h*ydrocarbons (such as 3-MC) in addition to TCDD.[14,15]

Two types of evidence implicate the Ah receptor in the induction of CYP1A1 transcription. One is biochemical; studies of structure–activity relationships reveal that, within groups of structurally related compounds, there is a correlation between receptor binding affinity and potency as an inducer. The second is genetic; inbred mouse strains differ quantitatively in their responsiveness to aromatic hydrocarbons. For example, TCDD is about tenfold more potent in inducing cytochrome P450 1A1 in the more responsive strains (typified by C57BL/6) than in the less responsive strains (typified by DBA/2). The polymorphism is genetic in origin, and the more responsive phenotype is expressed as an autosomal dominant trait. The segregation pattern for cytochrome P450 1A1 induction is identical to that for the binding of TCDD to the Ah receptor; thus, the same genetic locus governs both TCDD binding and enzyme induction. These biochemical and genetic findings implicate the Ah receptor in the induction mechanism.[15]

Genetic experiments in mouse hepatoma cells provide additional insights into the mechanism of cytochrome P450 1A1 induction. The use of cell fusion to analyze variant cells that exhibit diminished responsiveness to TCDD reveals several complementation groups, a finding that implies that several genes contribute to the induction mechanism.[16,17] One type of variant is defective in TCDD binding; these cells apparently contain an altered Ah receptor and respond poorly to TCDD. In a second type of variant, TCDD binding is identical to wild type (implying that the receptor is normal), but the liganded receptor fails to accumulate in the nucleus, and the cells fail to respond to TCDD. These cells are defective in a protein termed Arnt, which is described in more detail below. These data imply that the induction of cytochrome P450 1A1 transcription requires both a ligand-binding protein (the Ah receptor) and a second protein (Arnt), which mediates the binding of the liganded receptor to DNA.[18]

Inability to purify the Ah receptor in quantities sufficient for detailed biochemical study hampered analysis of the CYP1A1 induction mechanism for a number of years. However, the recent cloning and expression of receptor cDNA has provided new insights into the induction process. Cloning of receptor cDNA was facilitated by the development of an ^{125}I-labeled photoaffinity ligand, which permitted the covalent tagging of the protein and the use of denaturing procedures for purification.[19] Bradfield *et al.* used two-dimensional polyacrylamide gel electrophoresis to isolate the covalently tagged receptor and determine its N-terminal amino acid sequence.[20] The sequence information, in turn, was used to generate antipeptide antibodies[21] and to design oligonucleotides for isolating receptor cDNA.[22,23] The deduced amino acid sequence of the C57BL/6 mouse receptor reveals that it has a mass of about 89 kDa, with the following features: (1) a basic helix–loop–helix (bHLH) domain, located toward the NH_2 end of the receptor. By analogy with other bHLH proteins, the basic region probably contributes to DNA binding and the HLH region to dimerization with other proteins, such as Arnt. (2) Two regions, designated as "PAS," because of their sequence homology with the regulatory proteins *Per*, *Arnt*, and *Sim*. The PAS regions may also participate in interactions between the receptor and other proteins. (3) A glutamine-rich region, toward the C-terminal end, which might contribute a transcriptional activation function to the receptor. The *in vitro* translated receptor does not bind strongly to DNA, even in the presence of an inducer; acquisition of DNA-binding capability requires that the receptor interact with the Arnt protein. Thus, the transcriptionally active form of the receptor is heteromeric. Its deduced sequence reveals that the Ah receptor has no structural similarities to the steroid/thyroid/retinoid family of receptors; thus, it is a novel type of ligand-activated transcription factor, distinct from those described previously. Immunohistochemical studies using antireceptor antibodies reveal that, in mouse hepatoma cells, the unliganded receptor is cytoplasmic; exposure to an inducer leads to the nuclear accumulation of the receptor.[24]

2.3. The Arnt Protein

Hankinson and his colleagues identified a human cDNA that complements the defect in the class of mouse hepatoma cells in which CYP1A1 induction fails to occur even though the Ah receptor is normal. The corresponding protein was designated as "Arnt" (for *Ah receptor nuclear translocator*), because of its assumed role in translocating the liganded Ah receptor from cytoplasm to nucleus.[25,26] Its deduced amino acid sequence reveals that Arnt has an organization that is similar to the Ah receptor: (1) a bHLH domain, located toward the N-terminus; (2) two PAS homologies; (3) a glutamine-rich region toward the C-terminus. The Arnt protein does not bind TCDD and does not bind to DNA in the absence of the liganded Ah receptor.[27–29] Immunohistochemical studies, using anti-Arnt antibodies, reveal that Arnt resides in the nucleus in uninduced mouse hepatoma cells and that exposure to inducer produces no detectable change in its intracellular distribution. Furthermore, the nuclear accumulation of the Ah receptor occurs in Arnt-defective cells.[24] These findings argue against a primary role for Arnt in receptor "translocation." Instead, it seems likely that Arnt confers DNA-binding capability on the liganded receptor by dimerizing with it. In support of this idea, immunoprecipitation experiments reveal that the liganded receptor and Arnt can interact in solution; furthermore, deletion of Arnt's bHLH domain is associated with loss of the receptor–Arnt interaction.[27]

2.4. Other Proteins

The unliganded Ah receptor associates with the 90-kDa heat shock protein (hsp90) *in vitro*.[30] The hsp90–receptor interaction might (1) stabilize the unliganded receptor in a configuration that facilitates ligand binding and/or (2) inhibit the unliganded receptor from binding inappropriately to DNA.[31] It is unknown whether hsp90 influences the induction of cytochrome P450 1A1 in intact cells.

Several groups have made observations *in vitro* suggesting that phosphorylation of the Ah receptor and/or Arnt (possibly by protein kinase C) influences their heterodimerization and/or the binding of the heteromer to DNA.[32–35] Other mammalian transcription factors also undergo cycles of phosphorylation and dephosphorylation; in many cases, the functional significance of the modification is unknown.[36] Likewise, the role of protein phosphorylation in the mechanism of cytochrome P450 1A1 induction remains to be defined more clearly.

The "superinduction" of cytochrome P450 1A1 by temporary inhibition of protein synthesis suggests the possible existence of an inhibitory protein(s) that modulates the induction process.[37] The existence of a dominant negative class of induction-defective mouse hepatoma cells is consistent with the existence of an inhibitory regulator of CYP1A1 gene expression.[38] Chemical cross-linking studies suggest that additional proteins, which remain to be characterized, may also participate in the induction mechanism.[39–42]

2.5. Activation of Transcription

Induction of CYP1A1 transcription by TCDD is rapid (i.e., it occurs within minutes) and is direct (i.e., it does not require ongoing protein synthesis). The latter finding implies that the protein components required for induction (such as the Ah receptor and Arnt) exist constitutively within the cell.

Recombinant DNA and transfection techniques were used to analyze the transcriptional response in detail. Potential regulatory DNA domains from the CYP1A1 gene were inserted into plasmids containing a "reporter" gene (chloramphenicol acetyltransferase); the chimeric genes were tested for inducibility by transfection into wild-type, Ah receptor-defective, and Arnt-defective cells. Such studies revealed the existence of an inducible, Ah receptor-dependent, and Arnt-dependent transcriptional enhancer located upstream of the CYP1A1 gene.[43] Additional transfection studies revealed the existence of a second control element with the properties of a transcriptional promoter. Neither the enhancer nor the promoter is active in the absence of the other, raising the issue as to the mechanism by which the two DNA control elements function in concert.[44] Chromatin structure may play a role in this process, as described below.

Analyses of enhancer DNA *in vitro*, using an electrophoretic mobility shift technique, revealed that it participated in inducible, Ah receptor-dependent, and Arnt-dependent protein–DNA interactions, whose characteristics were those expected for the binding of the receptor–Arnt heteromer to DNA.[45,46] More detailed analyses revealed that the heteromer recognizes a specific "core" nucleotide sequence, which is present in multiple copies within the enhancer.[47] Mutational studies have defined specific nucleotide pairs that are important for heteromer binding and for enhancer function.[47–49] Additional studies reveal that the receptor heteromer binds within the major DNA groove and contacts several guanine residues within the recognition sequence.[50] Experiments with an ^{125}I-labeled inducer imply that the heteromer binds in a 1:1 ratio to its recognition sequence.[45] Electrophoretic mobility shift experiments also reveal that the binding of the receptor heteromer to its recognition sequence bends the DNA *in vitro*. The bend occurs very near the site of the protein–DNA interaction.[51] The latter finding suggested that distortion of enhancer structure might contribute to the induction mechanism.

2.6. Chromatin Structure

A ligation-mediated polymerase chain reaction technique was used to analyze the CYP1A1 induction mechanism in intact cells. These studies reveal that few (if any) proteins bind to the inactive (i.e., uninduced) enhancer; thus, the inactive enhancer appears relatively inaccessible to DNA-binding proteins *in vivo*. From a mechanistic standpoint, this finding argues against the idea that a DNA-binding repressor maintains the enhancer in an inactive configuration. Exposure to an inducer leads to the rapid occupation of six binding sites for the receptor–Arnt heteromer, while few additional proteins bind to the enhancer. These data suggest that the heteromer activates transcription via a mechanism that does not require other enhancer-binding proteins.[52]

The CYP1A1 promoter contains no binding sites for the receptor–Arnt heteromer. Furthermore, the proteins that bind to the promoter are expressed constitutively (i.e., they are present in uninduced cells). However, in intact uninduced cells, these proteins fail to bind to the promoter; thus, like the uninduced enhancer, the uninduced promoter appears inaccessible to DNA-binding proteins. Exposure to inducer leads to the rapid occupation of protein binding sites on the promoter. Therefore, induction is associated with an increase in accessibility of both the enhancer and promoter regions of the CPY1A1 gene. These changes are primary responses, because they are insensitive to actinomycin D at a concentration that inhibits CYP1A1 transcription by > 95%. Thus, the binding of the receptor heteromer to the enhancer has the effect of increasing the accessibility of the promoter to constitutively expressed transcription factors.[53,54]

Studies of the chromatin structure of the CYP1A1 gene reveal that the inactive enhancer/promoter region assumes a nucleosomal structure and that nucleosomes are specifically positioned at the promoter.[55] Its nucleosomal organization provides an explanation for the inaccessibility of the regulatory region in uninduced cells. The loss of nucleosomes at the promoter constitutes a plausible mechanism for the increase in CYP1A1 transcription that occurs in induced cells.

On the enhancer, the six binding sites for the receptor heteromer are arranged in an irregular pattern. The absence of regular spacing between the sites suggests that enhancer activation does not require protein–protein interactions between adjacent, DNA-bound receptor heteromers. Instead, the irregular spacing of binding sites may reflect constraints imposed by chromatin structure, because the receptor heteromer must bind to nucleosomes. For example, as the DNA helix wraps around the histone core of the nucleosome, the major groove (which contains the binding sites for the receptor heteromer) is periodically accessible and inaccessible. Therefore, increasing the number of binding sites at irregular intervals along the enhancer increases the probability that at least one site will be accessible, even when the DNA is nucleosomal. In addition, the receptor heteromer contacts a relatively short (6 base pair) DNA segment, increasing the probability that the entire binding site will be accessible in the nucleosome. Thus, the multiplicity, irregular distribution, and small size of the binding sites may have evolved as a mechanism for overcoming the steric constraint imposed by the nucleosomal organization of the inactive enhancer *in vivo*.

2.7. Future Research

Cloning and expression of cDNAs encoding the Ah receptor and Arnt permit more detailed mechanistic studies of the cytochrome P450 1A1 induction mechanism. It is now feasible to use mutagenesis techniques to study the structure and function of the Ah receptor and Arnt, to identify their cognate genes and to study their regulation, to analyze the possible contributions of protein modification to induction, and to identify additional proteins that may participate in the induction response.

The induction of cytochrome P450 1A1 is an interesting model system for analyzing the mechanism by which a relatively small molecule, such as TCDD, can activate the expression of a specific gene. Chromatin structure and nucleosome positioning often have important effects on mammalian gene transcription.[56] Future analyses of CYP1A1 induction may provide new insights into the mechanisms by which the Ah receptor/Arnt heteromer triggers the alterations in chromatin structure and nucleosome positioning that lead to an increase in DNA accessibility. Such findings may be of relatively broad interest, and may increase our understanding of the function of bHLH transcriptional regulatory proteins, in general.

Interindividual variation in cytochrome P450 1A1 activity may reflect genetic differences that affect (1) the catalytic properties of the enzyme itself or (2) the regulation of the CYP1A1 gene. In principle, such genetic variation can predispose the individual to the adverse effects of xenobiotics. Identification of the proteins that regulate cytochrome P450 1A1 induction increases the feasibility of determining whether polymorphisms in the Ah receptor and/or Arnt are associated with altered susceptibility to some xenobiotics.

2.8. Conclusion

Several factors have facilitated the mechanistic analyses of cytochrome P450 1A1 induction: (1) uninduced activity (background) is low and induced activity (signal) is high; (2) induction occurs in cells in culture; (3) a high potency, nonmetabolized inducer (TCDD) is available; (4) induction exhibits a genetic polymorphism in mice; (5) methods are available to isolate induction-defective cells, which enables genetic analyses. As described in more detail in subsequent sections, analyses of the induction of other cytochrome P450 enzymes suffer from the lack of one or more of these factors. Therefore, for other cytochrome P450 enzymes, our understanding of the induction mechanism is less complete than it is for cytochrome P450 1A1.

3. Phenobarbital-Inducible Cytochrome P450s

3.1. Background

The induction of drug- and steroid-metabolizing enzyme activity by phenobarbital (PB) was one of the seminal observations leading to scientific interest in cytochrome P450.[5] PB induces proliferation of hepatic endoplasmic reticulum and increased expression of several phase I and II enzymes, including several forms of cytochrome P450.[1,57–59] In addition, numerous compounds that are structurally unrelated to PB (e.g., DDT, dieldrin, chlordane, *trans*-stilbene oxide, diortho-substituted polychlorinated biphenyls, allylisopropylacetamide) induce a similar pattern of enzyme activities, and are considered to be "PB-like."[8,59,60] PB induces multiple forms of cytochrome P450 in rat liver, including, CYP2A1, CYP2B1, CYP2B2, CYP2C6, CYP2C7, CYP2C11, CYP3A1, and CYP3A2.[59,60] Some of the induction responses are substantial (e.g., 50- to 100-fold with CYP2B1/2) whereas others are much smaller (e.g., 2- to 4-fold with CYP2A1 and CYP2C6).

Two of the major cytochrome P450s induced by PB in rat liver, CYP2B1 and CYP2B2, have been the focus of intensive research. CYP2B1 metabolizes a broad spectrum of lipophilic drugs and steroids, while the less metabolically active CYP2B2 exhibits a somewhat different spectrum of catalytic activity.[59,61–63] In uninduced animals, CYP2B2 levels are low but detectable; the level of 2B1 is at least five- to ten-fold lower and is often unmeasurable.[62,64,65] The CYP2B1/2 genes share 97% amino acid sequence identity, and they are coordinately regulated and induced in parallel by PB and PB-like chemicals.[57,59,66–68] PB rapidly (within 30–60 min) increases the rate of transcription of the CYP2B1/2 genes; increased transcription appears to account for the subsequent mRNA accumulation and increased CYP2B1/2 enzymatic activity.

Induction of cytochrome P450 enzymes by PB and PB-like chemicals occurs in numerous species. In chick embryo liver, PB rapidly induces (within 30 min) CYP2H1 and CYP2H2, which are closely related structurally to CYP2B1/2.[69,70] The high level of inducible expression (about 50-fold) appears to involve both transcriptional and posttranscriptional mechanisms and occurs in chick embryo hepatocytes *in vivo* and in primary cells in culture.[70–72] PB-inducible P450s have also been observed in *Bacillus megaterium*. Induction of these bacterial fatty acid monooxygenases (CYP102 and CYP106) by PB and other barbiturates occurs extremely rapidly (<5 min), with a maximal rate of synthesis occurring in less than 30 min (compared to 12–48 hr in rat and chicken primary hepatocytes); induction reflects an increase in the rate of transcription of these genes.[73] Taken together, these observations imply that PB acts primarily at the level of transcription to induce cytochrome P450 activity.

Because PB induces cytochrome P450 in bacteria, birds, and mammals, it might be anticipated that the induction mechanism would be highly conserved; however, this is not the case. For example, studies in rat hepatocytes using cycloheximide reveal that ongoing protein synthesis is required for induction of CYP2B1/2 mRNA by PB.[74,75] Such findings suggest that the activation of CYP2B1/2 transcription requires prior or concurrent synthesis of a PB-dependent stimulatory factor; therefore, PB may act indirectly to activate transcription in this system. In contrast, inhibition of protein synthesis by cycloheximide synergistically enhances ("superinduces") CYP3A mRNA accumulation following exposure of rat hepatocytes to PB.[75] It is unclear whether the synergistic effect of cycloheximide involves a transcriptional or posttranscriptional effect. Interestingly, cycloheximide alone induces CYP3A1 mRNA in female but not male rat liver, a phenomenon that may reflect the loss of a labile repressor protein.[75] Thus, in rat hepatocytes, PB induces CYP2A1/2 and CYP3A by different mechanisms, which are distinguishable by their sensitivity to cycloheximide. Similar differences occur in other systems. For example, cycloheximide blocks the induction of CYP102 by PB in *B. megaterium*; however, in chicken hepatocytes, cycloheximide induces and, in combination with PB, superinduces CYP2H1/2. These observations have been interpreted to mean that a labile repressor regulates expression of the CYP2H1/2 genes.[74,76,77] Together, the results of experiments using cycloheximide imply that PB can induce cytochrome P450 enzymes by more than one mechanism.

Hormones, growth factors, and cytokines also influence the response of cytochrome P450 genes to PB. For example, in primary rat hepatocytes, growth hormone blocks the induction of CYP2B1/2 and CYP3A gene transcription by PB.[65,78,79] Negative regulation by growth hormone may also account for the diminished responsiveness of CYP2B enzymes to PB in female rats, because females exhibit higher levels of growth hormone *in vivo*.[65] In addition, interleukin 6 inhibits the response of the CYP2B1/2 genes to PB.[80] Understanding of the molecular mechanism by which these factors modulate the response to PB requires more complete knowledge of the mechanism(s) of PB action.

3.2. A Phenobarbital Receptor?

The mechanism by which the cell recognizes PB and PB-like chemicals and the pathway(s) by which the PB signal activates the transcriptional machinery are unknown. Many have speculated about the presence of a cellular receptor for PB; however, the structural diversity among PB-like inducers is difficult to reconcile with the existence of a specific receptor analogous to other well-characterized receptors, which exhibit stereospecificity in ligand binding. Studies using radiolabeled PB have failed to detect a specific PB-binding protein[81]; however, the lack of binding could reflect the low affinity of PB for its hypothetical target protein (implied by the relatively high concentration of PB needed for induction).[68] A relatively potent PB-like inducer, 1,4-bis[2-(3,5-dichloropyridyloxy)] benzene (TCPOBOP), has been identified. Although TCPOBOP was 600-fold more potent than PB as an inducer of cytochrome P450 2B1 in mice, subsequent studies revealed that it did not induce cytochrome P450 in guinea pigs or rats.[82–84] Therefore, TCPOBOP probably does not act by a mechanism identical to that of PB. The species-dependent mechanism by which TCPOBOP induces cytochrome P450 2B1 is unknown. The question of how the cell recognizes and transduces the PB signal remains unanswered.

3.3. Activation of Transcription

Compared to our understanding of CYP1A1 induction by TCDD, we know relatively little about the mechanism by which PB activates cytochrome P450 gene transcription. One impasse has been the lack of continuous cell lines that respond to PB. Only recently have there become available cell culture systems that are suitable for the transfection experiments necessary to analyze the *cis-* and *trans-*acting factors involved in the induction response.

In *B. megaterium*, deletion analysis of the upstream regulatory region of the barbiturate-inducible CYP102 (BM-3) gene revealed the presence of a 17-bp regulatory element (designated as the "Barbie Box"), which appears to be involved in gene regulation by PB.[73,85-87] Comparison of the flanking regions of the bacterial CYP102 and CYP106 (BM-1) genes to those of the rat CYP2B1/2 genes revealed the presence of consensus Barbie Boxes upstream of all four genes, in approximately the same location (-200 to -300 relative to the transcriptional start site of the bacterial genes and -100 to -119 relative to that of the rat genes). Gel retardation studies revealed that synthetic oligonucleotides containing the consensus sequence bound proteins from both bacterial and rat liver nuclei in a PB-dependent manner.[85] Notably, PB treatment reduced the binding of the bacterial proteins to DNA but increased the binding of the rat liver nuclear proteins to DNA. These results suggest that the *B. megaterium* protein(s) may act as a repressor (which is released from the DNA in response to PB) and the rat liver nuclear protein(s) acts as an activator (which binds to the DNA in response to PB). Additional experiments, involving incubation of bacterial extracts or rat liver nuclear extracts with PB *in vitro*, imply that these proteins are constitutively expressed in their respective cells and that the response to PB involves an alteration in their DNA binding activity.[85] These results contrast with those of Rangarajan and Padmanaban, who identified an 85-kDa putative transcription factor in rat liver nuclei, whose binding to DNA was induced by exposure of the animals to PB.[88] Their findings imply that the putative transcription factor is synthesized *de novo* following PB administration, because cycloheximide blocked the PB-inducible protein–DNA interaction. It is noteworthy that their DNase I footprint spanned a 32-bp region that contained the 17-bp Barbie Box identified by Fulco and co-workers.[73,85] Taken together, these studies imply the existence of a PB-dependent protein(s) that binds to DNA near the promoter region of these PB-responsive cytochrome P450 genes. Whether the protein(s) is expressed constitutively or is induced by PB appears unresolved, and its function remains to be determined.

Shaw *et al.* have studied the induction of cytochrome P450 2C6 by PB in a rat hepatoma (FAO) cell line. Using recombinant DNA and transfection techniques, they localized a PB-responsive DNA region to a 1.4-kb fragment containing the promoter and upstream flanking regions of the CYP2B1 and CYP2B2 genes. In addition, they demonstrated that PB-responsiveness was inhibited by the glucocorticoid/progesterone antagonist RU486. These latter results suggested that PB might act indirectly to induce gene expression, by causing the accumulation of an endogenous steroid, which might act as the primary inducer of cytochrome P450 2B1/2 via a steroid receptor.[89] This hypothetical indirect induction mechanism might account for the ability of chemicals of widely divergent structure to induce a PB-like response.

May and colleagues also used molecular biological approaches to analyze the upstream regulatory region of the chicken CYP2H1 genes, using a series of reporter constructs transfected into primary chick embryo hepatocytes.[90] Their experiments localized a PB-responsive element(s) to a region between -5.9 and -1.1 kb upstream of the start site of transcription. These results indicate that PB-responsive elements can reside substantially farther upstream of the target gene than the Barbie Boxes described by Fulco and co-workers. Whether the PB-responsive DNA domain(s) upstream of the CYP2H1 gene has the properties typical of a transcriptional enhancer remains to be determined.

As an alternative approach to the identification of PB-responsive DNA elements, Omiecinski and colleagues generated transgenic mouse lines that contained the rat CYP2B2 gene with different lengths of 5′-flanking DNA.[91] One transgene contained 800 bp of flanking DNA and the other contained 19 kb of flanking DNA. Rat CYP2B2 activity was constitutively expressed by the shorter transgene; however, it was not PB-inducible. In contrast, the larger transgene exhibited little constitutive activity and was highly inducible by PB. These findings imply that a PB-responsive DNA regulatory element(s) for the rat CYP2B2 gene is contained within the -0.8 to -19 kb region upstream of the transcription start site. These results also imply that the CYP2B2 gene may contain more than one type of PB-responsive element. It is possible that the Barbie Boxes identified by Fulco and co-workers may be part of the core promoter for the CYP2B2 gene, and that the domain(s) farther upstream identified by Omiecinski *et al.* may function as a PB-responsive enhancer.

3.4. Future Research

Compared to the situation for aromatic hydrocarbon-inducible cytochrome P450s, we understand relatively little about the mechanism by which PB and PB-like chemicals induce P450 gene expression. The lack of convenient PB-responsive cell culture systems has been a major impediment to detailed molecular analysis of the induction mechanism. Additionally, the apparent lack of substantial genetic polymorphisms in PB responsiveness within animal species has hindered mechanistic analyses. The development of techniques for isolating cells that vary in responsiveness to PB would facilitate studies of the induction mechanism in the future.

Recently, investigators have described cell culture systems that are suitable for the transfection experiments required to analyze the *cis-* and *trans-*acting factors involved in the response to PB. These systems utilize either primary hepatocytes (rat or chicken) or a feeder layer or a laminin-rich matrix, which permits stable expression of hepatic phenotypes, including PB-inducible increases in P450 2B1/2 mRNA and protein. In addition, several continuous cell lines that respond to PB have been identified. For example, rat hepatoma (FAO) and human hepatoma (HepG2) cell lines respond to PB with induction of P450 2C6[89] or glutathione *S*-transferase activity,[92,93] respectively. The availability of these PB-responsive cell lines should facilitate mechanistic analyses of the response to PB; their utility in transient transfection experiments to study the promoter regions of PB-inducible P450 genes has been demonstrated.[89,90] Identification of a PB-responsive element(s) by transfection provides an avenue for purification and cloning of the cognate transactivating factor(s).

Recent studies in HepG2 cells reveal that PB stimulates the DNA binding of the transcription factor AP-1 and activates transcription (in an AP-1-dependent manner) of the glutathione *S*-transferase Ya and quinone reductase genes.[93] Because the DNA binding activity of AP-1 (fos/jun) appears to be regulated by the redox state of a cysteine residue in the DNA binding domains of the two proteins, these findings suggested that PB might activate AP-1 by increasing intracellular oxidant levels, thereby activating gene transcription in an indirect manner. Future studies may elucidate the mechanism by which PB activates AP-1 and may also assess the potential role of AP-1 in the induction of P450 2B1/2. However, the failure of P450 2B1/2 to respond to PB in HepG2 cells implies that additional factors are involved in the induction of these genes.

The lack of structural similarity of PB-like inducers, together with the species-specific action of the most potent inducer, TCPOBOP, is difficult to reconcile with the concept of a specific "PB receptor." It seems more likely that these compounds induce cytochrome P450 enzymes by an indirect mechanism, the details of which are unknown. The differential responsiveness of cytochrome P450s to PB, the different effects of cycloheximide on PB-responsiveness, and the differential loss of PB-inducible responses in cells in culture suggest that PB alters gene expression via more than one mechanism. These observations indicate that the question of PB-responsiveness remains an intriguing experimental issue for the future.

4. Peroxisome Proliferator-Inducible Cytochrome P450s

4.1. Background

A variety of structurally dissimilar compounds, including fibrate hypolipidemic drugs (such as clofibrate and nafenopin), phthalate ester plasticizers [such as di-(2-ethyl-hexyl)phthalate], and halogenated aromatic solvents (such as trichloroacetic acid) induce peroxisome proliferation in mammalian liver.[94] This response is characterized by increased activity of several peroxisomal enzymes involved in the β-oxidation of fatty acids. In addition, exposure to peroxisome proliferators leads to an increase in hepatic smooth endoplasmic reticulum and the induction of lauric acid ω-hydroxylase activity, which is catalyzed by cytochrome P450 4A1.[95,96] This enzyme, together with the peroxisomal β-oxidation enzymes, may comprise a pathway that is of particular importance in the metabolism of long-chain fatty acids.[94]

In rat liver, clofibrate rapidly (within 1 hr) increases the transcription rate of the CYP4A1 gene, which is followed by an increase in the corresponding mRNA, protein, and lauric acid ω-hydroxylase activity.[97] Similarly, peroxisome proliferators rapidly increase the transcription rates of the genes for fatty acyl-CoA oxidase (ACO) and enoyl-CoA hydratase/3-hydroxyacyl-CoA dehydrogenase, components of the peroxisomal fatty acid β-oxidation pathway.[98] Thus, enzyme induction by peroxisome proliferators involves increases in gene transcription.

4.2. Peroxisome Proliferator Activated Receptor

The transcriptional nature of the response to peroxisome proliferators led to the hypothesis that they might act via a receptor protein related to the steroid hormone nuclear receptor superfamily. Oligonucleotide probes, derived from a conserved amino acid sequence among the nuclear receptors, were used to clone new members of the receptor superfamily from a mouse liver cDNA library. Because the target genes for these so-called "orphan" receptors were unknown, the new receptors were tested for function by expressing chimeric proteins, which contained the DNA-binding domain of a steroid receptor linked to the putative ligand-binding domain of the orphan receptor. For such a chimera, transfection experiments revealed that peroxisome proliferators activated the expression of a steroid-responsive target gene. Furthermore, a compound's potency as a peroxisome proliferator correlated with its ability to activate the chimeric receptor. These observations implied that the orphan cDNA encoded a protein [designated as "peroxisome proliferator activated receptor" (PPAR)] that mediates the biological response to peroxisome proliferators.[99] An analogous approach, using chimeric receptors and transfection, has been used to show that certain fatty acids, as well as peroxisome proliferators, can activate the rat liver PPAR.[100] It is currently unclear whether peroxisome proliferators directly bind to and activate the PPAR, or whether they act indirectly (e.g., by generating a "second messenger" that acts at the PPAR). Direct binding of ligand to the PPAR has not been demonstrated; furthermore, the structural diversity of the peroxisome proliferators suggests that they might act indirectly to induce gene transcription.

The deduced amino acid sequences for the mouse,[99,101] rat,[100] human,[102] and frog[103] PPARs imply that they belong to the superfamily of steroid/thyroid/retinoid receptors (as would be expected, given the cloning approach used). The putative DNA-binding domain of the PPAR exhibits the greatest degree of homology with other nuclear hormone receptors; otherwise, the different PPARs are more related to each other than to other members of the receptor superfamily.

4.3. Activation of Transcription

Nuclear hormone receptors are ligand-activated, DNA-binding proteins that activate transcription by binding to specific enhancer DNA sequences, which are often located upstream of the target genes.[104] The homology between the PPAR and other nuclear hormone receptors implied that the PPAR might have a similar mechanism of action, an idea that was consistent with the observations that induction of peroxisomal enzymes and cytochrome P450 4A1 occurred at the level of transcription. Tugwood *et al.* examined the DNA upstream of the clofibrate-inducible rat liver ACO gene for the presence of a domain(s) that could mediate the transcriptional effects of peroxisomal proliferators. Using chimeric genes, which were tested for function by transfection, they identified a DNA domain that conferred responsiveness to the peroxisome proliferator Wy-14,643 on a chloramphenicol acetyltransferase reporter gene; furthermore, electrophoretic mobility shift experiments suggested that the PPAR binds to its cognate response element *in vitro*.[105] Others have reported similar observations using the frog[103] and human[102] PPARs.

Using an analogous molecular biological approach, Muerhoff *et al.* identified a peroxisome proliferator-inducible, PPAR-dependent DNA domain upstream of the rabbit cytochrome P450 4A6 gene.[106] These observations imply that the PPAR acts via a mechanism analogous to that described for the steroid/thyroid/retinoid family of nuclear receptors.

4.4. Future Research

The molecular biological studies reveal similarities between the cytochrome P450 4A genes and the ACO gene in their transcriptional activation by the PPAR; however, the induction mechanisms for the two genes may differ. For example, in primary cultures of rat hepatocytes exposed to peroxisome proliferators, accumulation of P450 4A1 mRNA precedes the accumulation of ACO mRNA; furthermore, cycloheximide inhibits the induced increase in ACO mRNA, but has a lesser effect on the accumulation of P450 4A1 mRNA.[107] These findings suggest that induction of CYP4A1 transcription is a primary response to peroxisome proliferators and precedes the induction of ACO transcription, which is a secondary response. Kaikaus *et al.* extended these findings by showing that, in primary rat hepatocytes, mechanism-based inactivators of cytochrome P450 4A1 inhibited the induction of ACO, but not the induction of P450 4A1 by peroxisome proliferators. Furthermore, hexadecanedioic acid, a long-chain fatty acid formed by cytochrome P450 4A1, induced ACO, but not P450 4A1.[108] These data suggest that, following exposure to peroxisome proliferators, induction of the cytochrome P450 4A1 enzyme, with the consequent generation of long-chain fatty acids, is required for the subsequent induction of peroxisomal β-oxidation enzymes. This hypothesis is consistent with the observation that tissues that contain microsomal ω-oxidation systems (i.e., cytochrome P450 4A enzymes) are the tissues that respond to peroxisomal proliferators. This sequential model for the response to peroxisome proliferators appears to be an interesting area for future research.

The lack of obvious structural similarity among compounds that activate transcription via the PPAR contrasts with the situation for other members of the nuclear receptor superfamily, where structure–activity relationships are much clearer. Thus, the question remains as to the identity of the ligand(s) that actually binds to the PPAR. In principle, it should be possible to identify such ligands (e.g., using binding techniques). Their discovery might provide new insights into the contributions of the PPAR, peroxisomal enzymes, and cytochrome P450 4A enzymes to lipid metabolism and cellular homeostasis. The conservation of the PPAR's primary structure across species suggests that it mediates signaling pathways that are of fundamental importance to the organism.

In some cases, PPAR-dependent pathways may converge with other signaling systems to generate novel patterns of gene regulation. For example, the PPAR may form heterodimers with another member(s) of the nuclear receptor family, permitting the target gene to respond to a more diverse set of chemical signals.[109,110] Future studies of potential dimerization partners for the PPAR may provide new insights into the mechanisms by which peroxisome proliferators produce their biological effects.

5. Steroid-Inducible Cytochrome P450s

5.1. Background

In the early 1970s, Selye analyzed a class of steroidal compounds that protected rats against certain toxic chemicals. The most potent such "catatoxic" steroid, pregnenolone-16α-carbonitrile (PCN), produced (1) proliferation of hepatic endoplasmic reticulum, (2) an increase in hepatic cytochrome P450 content, and (3) induction of hepatic xenobiotic-metabolizing enzyme activity.[111] Following these initial observations, Elshourbagy and Guzelian demonstrated that PCN induced a novel form of cytochrome P450, which had distinctive biochemical, catalytic, and immunological properties.[112] The enzymes of this class (now designated as the cytochrome P450 3A family) metabolize a variety of substrates, including the hormone testosterone, the antibiotics erythromycin and triacetyloleandomycin, and drugs such as lovastatin, nifedipine, midazolam, mephenytoin, and cyclosporine.[113–117] The ability of the enzyme to metabolize different types of drug may lead to clinically significant drug interactions.[114] In addition, human genetic polymorphisms appear to exist, which might cause interindividual differences in responses to drugs that are metabolized by cytochrome P450 3A enzymes.[115,118]

5.2. Mechanism of Induction

Exposure of rats either to PCN or to dexamethasone leads within several hours to an increased rate of hepatic cytochrome P450 3A gene transcription, which is followed by an increase in mRNA accumulation and enzyme content.[119–122] Thus, induction occurs at the transcriptional level. The cellular target molecule that mediates the transcriptional response is unknown. By analogy with other steroid-inducible systems, a nuclear receptor could be involved. However, studies with dexamethasone and other steroid hormones imply that cytochrome P450 3A induction does not occur by the usual glucocorticoid receptor pathway. For example, the rank order of potency of steroids that induce P450 3A differs from the rank order of steroids that induce the tyrosine aminotransferase (TAT) gene, which is known to be under the control of the glucocorticoid receptor; furthermore, some steroids induce P450 3A gene expression, but inhibit TAT expression.[123] These differences in structure–activity relationships imply that the classical glucocorticoid receptor does not control the CYP3A gene; whether a novel, as yet undiscovered, member of the nuclear receptor superfamily regulates CYP3A induction is unknown.

Triacetyloleandomycin (TAO), a macrolide antibiotic, induces the cytochrome P450 3A protein; however, the mechanism is not transcriptional. Studies in rat liver and in cultured hepatocytes reveal that TAO produces no increase in the rate of P450 3A protein synthesis; instead, TAO decreases the rate of P450 3A protein degradation to about one-quarter of normal.[124] The molecular mechanism by which TAO inhibits enzyme degradation is unknown; however, because TAO is a substrate for cytochrome P450 3A, direct binding of TAO (or a metabolite) to the enzyme protein is probably involved, in view of observations that the TAO-stabilized protein is catalytically inactive.

PB and "PB-like" compounds (e.g., certain polychlorinated biphenyls) induce cytochrome P450 3A. However, structure–activity studies reveal that the rank order of potency of PB-like compounds as inducers of P450 3A differs from their rank order of potency as inducers of P450 2B enzymes; therefore, the mechanism by which PB acts as an inducer probably differs for the two cytochrome P450 families.[125]

5.3. Future Research

The development of cell culture systems in which the CYP3A gene(s) respond to steroids, PB, and other inducers should facilitate molecular analyses of the induction mechanism in the future.[126,127] The discovery of more potent inducers of the PCN type and the development of a photoaffinity ligand might permit the identification of the receptor protein that presumably mediates the induction response. Isolation of induction-defective cells would allow the application of genetic techniques to the study of this interesting system.

6. Ethanol-Inducible Cytochrome P450s

Interest in ethanol and cytochrome P450 enzymes stems from observations of ethanol's toxicity to the liver and its effects on the activities of hepatic xenobiotic-metabolizing enzymes.[128,129] Koop *et al.* reported an ethanol-inducible form, cytochrome P450 2E1, in rabbit liver[130]; analogous forms exist in rats and humans. Cytochrome P450 2E1 metabolizes several xenobiotics, including ethanol, benzene, carbon tetrachloride, alkylnitrosamines, and acetaminophen[131]; the enzyme also metabolizes acetone, which has led to the suggestion that it contributes to gluconeogenesis during fasting.[132]

Regulation of cytochrome P450 2E1 expression occurs at several levels. In rats, the increase in enzyme activity that occurs shortly after birth coincides with increased transcription of the CYP2E1 gene.[133] The mechanism that governs the developmental regulation of the CYP2E1 gene is unknown; the increase in transcription is associated with decreased methylation of CYP2E1 DNA during the neonatal period.[133] Spontaneously diabetic rats and rats with chemically induced diabetes exhibit increased levels of cytochrome P450 2E1 that appear to reflect mRNA stabilization and not increased transcription.[134] The increased concentration of ketones in the diabetic state may reduce the rate of P450 2E1 mRNA turnover[135]; however, the mechanism remains to be determined.

In most instances, inducers of cytochrome P450 2E1 produce an increase in the amount of enzyme protein without a corresponding increase in the cognate mRNA. Several compounds, such as ethanol, propanol, acetone, benzene, trichloroethylene, isoniazid, and imidazole, act in this fashion, which probably reflects the stabilization of enzyme protein.[136] For example, exposure of rats to acetone is associated with a decrease in the turnover of hepatic cytochrome P450 2E1.[137] In addition, in cultured primary hepatocytes, the rate of loss of cytochrome P450 2E1 protein is decreased in the presence of enzyme substrates. Spectroscopic studies imply that binding of the substrate to the enzyme contributes to protein stabilization.[138] The mechanism by which substrates might selectively stabilize P450 2E1 is unknown. Eliasson *et al.* have speculated that cAMP-dependent phosphorylation of the protein might be involved.[139]

7. Conclusion

Induction of cytochrome P450 enzymes occurs by various mechanisms in organisms that have diverged extensively during evolution. Therefore, it is likely that the induction phenomenon is advantageous from an evolutionary standpoint, possibly as a mechanism by which terrestrial organisms confront potentially toxic compounds in the diet and in the environment.[140] Knowledge of the induction mechanisms may provide insights into the strategies that organisms use to adapt to a changing environment. Such information may also generate a better understanding of the biochemical pathways by which cells recognize and respond to chemical stimuli. These insights are likely to be of broad interest. Knowledge of induction mechanism and the factors that control enzyme levels may reveal genetic polymorphisms that affect an individual's susceptibility to the adverse effects of certain compounds. Such information may be useful from a public health standpoint in identifying individuals who are at particular risk from exposure to certain xenobiotics.

References

1. Remmer, H., and Merker, H. J., 1963, Drug-induced changes in the liver endoplasmic reticulum: Association with drug-metabolizing enzymes, *Science* **142**:1657–1658.

2. Conney, A. H., Miller, E. C., and Miller, J. A., 1956, The metabolism of methylated aminoazo dyes. 5. Evidence for induction of enzyme synthesis in the rat by 3-methylcholanthrene, *Cancer Res.* **16**:450–459.

3. Park, B. K., and Breckenridge, A. M., 1981, Clinical implications of enzyme induction and enzyme inhibition, *Clin. Pharmacokinet.* **6**:1–24.

4. Thomas, S. H. L., 1993, Paracetamol (acetaminophen) poisoning, *Pharmacol. Ther.* **60**:91–120.

5. Conney, A. H., 1967, Pharmacological implications of microsomal enzyme induction, *Pharmacol. Rev.* **19**:317–366.

6. Whitlock, J. P., Jr., 1986, The regulation of cytochrome P450 gene expression, *Annu. Rev. Pharmacol. Toxicol.* **26**:333–369.

7. Nebert, D. W., and Gonzalez, F. J., 1987, P450 genes: Structure, evolution, and regulation, *Annu. Rev. Biochem.* **56**:945–993.

8. Okey, A. B., 1990, Enzyme induction in the cytochrome P450 system, *Pharmacol. Ther.* **45**:241–298.

9. Conney, A. H., 1982, Induction of microsomal enzymes by foreign compounds and carcinogenesis by polycyclic aromatic hydrocarbons, *Cancer Res.* **42**:4875–4917.

10. Gelboin, H. V., 1980, Benzo(a)pyrene metabolism, activation, and carcinogenesis: Role and regulation of mixed-function oxidases and related enzymes, *Physiol. Rev.* **60**:1107–1166.

11. Negishi, M., Swan, D. C., Enquist, L. W., and Nebert, D. W., 1981, Isolation and characterization of a cloned DNA sequence associated with the murine Ah locus and 3-methylcholanthrene-induced form of cytochrome P450, *Proc. Natl. Acad. Sci. USA* **78**:800–804.

12. Gonzalez, F. J., Tukey, R. H., and Nebert, D. W., 1984, Structural gene products of the Ah locus. Transcriptional regulation of cytochrome P_1-450 and P_3-450 mRNA levels by 3-methylcholanthrene, *Mol. Pharmacol.* **26**:117–121.

13. Israel, D. I., and Whitlock, J. P., Jr., 1984, Regulation of cytochrome P_1-450 gene transcription by 2,3,7,8-tetrachlorodibenzo-p-dioxin in wild type and variant mouse hepatoma cells, *J. Biol. Chem.* **259**:5400–5402.

14. Poland, A. P., Glover, E., and Kende, A. S., 1976, Stereospecific, high affinity binding of 2,3,7,8-tetrachlorodibenzo-p-dioxin by hepatic cytosol: Evidence that the binding species is the receptor for the induction of aryl hydrocarbon hydroxylase, *J. Biol. Chem.* **251**:4936–4946.

15. Poland, A., and Knutson, J. C., 1982, 2,3,7,8-Tetrachlorodibenzo-p-dioxin and related aromatic hydrocarbons: Examination of the mechanism of toxicity, *Annu. Rev. Pharmacol. Toxicol.* **22**:517–554.

16. Hankinson, O., 1983, Dominant and recessive aryl hydrocarbon hydroxylase-deficient mutants of the mouse hepatoma line, Hepa 1, and assignment of the recessive mutants to three complementation groups, *Somatic Cell Genet.* **9**:497–514.

17. Miller, A. G., Israel, D. I., and Whitlock, J. P., Jr., 1983, Biochemical and genetic analysis of variant mouse hepatoma cells defective in the induction of benzo(a)pyrene-metabolizing enzyme activity, *J. Biol. Chem.* **258**:3523–3527.

18. Whitlock, J. P., Jr., 1993, Mechanistic aspects of dioxin action, *Chem. Res. Toxicol.* **6**:754–763.

19. Poland, A., Glover, E., Ebetino, F. H., and Kende, A. S., 1986, Photoaffinity labeling of the Ah receptor, *J. Biol. Chem.* **261**:6352–6365.

20. Bradfield, C. A., Glover, E., and Poland, A., 1991, Purification and N-terminal amino acid sequence of the Ah receptor from the C57BL/6J mouse, *Mol. Pharmacol.* **39**:13–19.

21. Poland, A., Glover, E., and Bradfield, C. A., 1991, Characterization of polyclonal antibodies to the Ah receptor prepared by immunization with a synthetic peptide hapten, *Mol. Pharmacol.* **39**:20–26.

22. Burbach, K. M., Poland, A., and Bradfield, C. A., 1992, Cloning of the Ah-receptor cDNA reveals a novel ligand-activated transcription factor, *Proc. Natl. Acad. Sci. USA* **89**:8185–8189.

23. Ema, M., Sugawa, K., Wanatabe, N., Chujoh, Y., Matsushita, N., Gotoh, O., Funae, Y., and Fujii-Kuriyama, Y., 1992, cDNA cloning and structure of mouse putative Ah receptor, *Biochem. Biophys. Res. Commun.* **184:**246–253.

24. Pollenz, R. C., Sattler, C. A., and Poland, A., 1994, The arylhydrocarbon receptor and aryl hydrocarbon receptor nuclear translocator protein show distinct subcellular localizations in Hepa 1c1c7 cells by immunofluorescence microscopy, *Mol. Pharmacol.* **45:**428–438.

25. Hoffman, E. C., Reyes, H., Chu, F.-F., Sander, F., Conley, L. H., Brooks, B. A., and Hankinson, O., 1991, Cloning of a factor required for activity of the Ah dioxin receptor, *Science* **252:**954–958.

26. Reyes, H., Reiz-Porszasz, S., and Hankinson, O., 1992, Identification of the Ah receptor nuclear translocator protein (Arnt) as a component of the DNA binding form of the Ah receptor, *Science* **256:**1193–1195.

27. Whitelaw, M., Pongratz, I., Wilhelmsson, A., Gustafsson, J. A., and Poellinger, L., 1993, Ligand-dependent recruitment of the Arnt coregulator determines DNA recognition by the dioxin receptor, *Mol. Cell. Biol.* **13:**2504–2514.

28. Matsushita, N., Sogawa, K., Ema, M., Yoshido, A., and Fujii-Kuriyama, Y., 1993, A factor binding to the xenobiotic responsive element (XRE) of P-4501A1 gene consists of at least two helix–loop–helix proteins, Ah receptor and Arnt, *J. Biol. Chem.* **268:**21002–21006.

29. Probst, M. R., Reisz-Porsasz, S., Agbunag, R. V., Ong, M. S., and Hankinson, O., 1993, Role of the aryl hydrocarbon (Ah) receptor nuclear translocator protein (ARNT) in aryl hydrocarbon (dioxin) receptor action, *Mol. Pharmacol.* **44:**511–518.

30. Perdew, G. H., 1988, Association of the Ah receptor with the 90 kDa heat shock protein, *J. Biol. Chem.* **263:**13802–13805.

31. Pongratz, I., Mason, G. G. F., and Poellinger, L., 1992, Dual roles of the 90 kDa heat shock protein hsp90 in modulating functional activities of the dioxin receptor, *J. Biol. Chem.* **267:**13728–13734.

32. Okino, S. T., Pendurthi, U. R., and Tukey, R. H., 1992, Phorbol esters inhibit the dioxin receptor-mediated transcriptional activation of the mouse Cyp1a1 and Cyp1a2 genes by 2,3,7,8 tetrachloro-dibenzo-p-dioxin, *J. Biol. Chem.* **267:**6991–6998.

33. Carrier, F., Owens, R. A., Nebert, D. W., and Puga, A., 1992, Dioxin-dependent activation of murine Cyp1A1 gene transcription requires protein kinase C-dependent phosphorylation, *Mol. Cell. Biol.* **12:**1856–1863.

34. Berghard, A., Gradin, K., Pongratz, I., Whitelaw, M., and Poellinger, L., 1993, Cross-coupling of signal transduction pathways: The dioxin receptor mediates induction of cytochrome P450A1 expression via a protein kinase C mechanism, *Mol. Cell. Biol.* **13:**677–689.

35. Schafer, M. W., Madhukar, B. V., Swanson, H. I., Tullis, K., and Denison, M. S., 1993, Protein kinase C is not involved in Ah receptor transformation and DNA binding, *Arch. Biochem. Biophys.* **307:**267–271.

36. Hunter, T., and Karin, M., 1992, The regulation of transcription by phosphorylation, *Cell* **70:**375–387.

37. Lusska, A., Wu, L., and Whitlock, J. P., Jr., 1992, Superinduction of CYP1A1 transcription by cycloheximide, *J. Biol. Chem.* **267:**15146–15151.

38. Watson, A. J., Weir-Brown, K. I., Bannister, R. M., Chu, F. F., Reisz-Porszasz, S., Fujii-Kuriyama, Y., Sogawa, K., and Hankinson, O., 1992, Mechanism of action of a repressor of dioxin-dependent induction of Cyp1A1 gene transcription, *Mol. Cell. Biol.* **12:**2115–2123.

39. Gasiewicz, T. A., Elferink, C. J., and Henry, E. C., 1991, Characterization of multiple forms of the Ah receptor: Recognition of a dioxin-responsive enhancer involves heteromer formation, *Biochemistry* **30:**2909–2916.

40. Perdew, G. H., 1992, Chemical cross-linking of the cytosolic and nuclear forms of the Ah receptor in hepatoma cell line 1c1c7, *Biochem. Biophys. Res. Commun.* **182:**55–62.

41. Elferink, C. J., Gasiewicz, T. A., and Whitlock, J. P., Jr., 1990, Protein–DNA interactions at a dioxin-responsive enhancer: Evidence that the transformed Ah receptor is heteromeric, *J. Biol. Chem.* **265:**20708–20712.

42. Swanson, H. I., Tullis, K., and Denison, M. S., 1993, Binding of transformed Ah receptor complex to a dioxin responsive transcriptional enhancer: Evidence for two distinct heteromeric DNA-binding forms, *Biochemistry* **32**:12841–12849.

43. Jones, P. B. C., Durrin, L. K., Galeazzi, D. R., and Whitlock, J. P., Jr., 1986, Control of cytochrome P_1-450 gene expression: Analysis of a dioxin-responsive enhancer system, *Proc. Natl. Acad. Sci. USA* **83**:2802–2806.

44. Jones, K. W., and Whitlock, J. P., Jr., 1990, Functional analysis of the transcriptional promoter for the CYP1A1 gene, *Mol. Cell. Biol.* **10**:5098–5105.

45. Denison, M. S., Fisher, J. M., and Whitlock, J. P., Jr., 1989, Protein–DNA interactions at recognition sites for the dioxin–Ah receptor complex, *J. Biol. Chem.* **264**:16478–16482.

46. Hapgood, J., Cuthill, S., Denis, M., Poellinger, L., and Gustafsson, J.-A., 1989, Specific protein–DNA interactions at a xenobiotic-responsive element: Copurification of dioxin receptor and DNA-binding activity, *Proc. Natl. Acad. Sci. USA* **86**:60–64.

47. Lusska, A., Shen, E., and Whitlock, J. P., Jr., 1993, Protein–DNA interactions at a dioxin-responsive enhancer: Analysis of six *bona fide* DNA-binding sites for the liganded Ah receptor, *J. Biol. Chem.* **268**:6575–6580.

48. Yao, E. F., and Denison, M. S., 1992, DNA sequence determinants for binding of transformed Ah receptor to a dioxin-responsive enhancer, *Biochemistry* **31**:5060–5067.

49. Shen, E. S., and Whitlock, J. P., Jr., 1992, Protein–DNA interactions at a dioxin-responsive enhancer: Mutational analysis of the DNA binding site for the liganded Ah receptor, *J. Biol. Chem.* **267**:6815–6819.

50. Shen, E. S., and Whitlock, J. P., Jr., 1989, The potential role of DNA methylation in the response to 2,3,7,8-tetrachlorodibenzo-p-dioxin, *J. Biol. Chem.* **264**:17754–17758.

51. Elferink, C. J., and Whitlock, J. P., Jr., 1990, 2,3,7,8-Tetrachlorodibenzo-p-dioxin inducible, Ah receptor-mediated bending of enhancer DNA, *J. Biol. Chem.* **265**:5718–5721.

52. Wu, L., and Whitlock, J. P., Jr., 1993, Mechanism of dioxin action: Receptor–enhancer interactions in intact cells, *Nucleic Acids Res.* **21**:119–125.

53. Durrin, L. K., and Whitlock, J. P., Jr., 1989, 2,3,7,8-Tetrachlorodibenzo-p-dioxin: Ah receptor-mediated change in cytochrome P_1-450 chromatin structure occurs independent of transcription, *Mol. Cell. Biol.* **9**:5733–5737.

54. Wu, L., and Whitlock, J. P., Jr., 1992, Mechanism of dioxin action: Ah receptor-mediated increase in promoter accessibility *in vivo*, *Proc. Natl. Acad. Sci. USA* **89**:4811–4815.

55. Morgan, J. E., and Whitlock, J. P., Jr., 1992, Transcription-dependent and transcription-independent nucleosome disruption induced by dioxin, *Proc. Natl. Acad. Sci. USA* **89**:11622–11626.

56. Kornberg, R. D., and Lorch, Y., 1992, Chromatin structure and transcription, *Annu. Rev. Cell Biol.* **8**:563–587.

57. Hardwick, J., Gonzalez, F. J., and Kasper, C. B., 1983, Transcriptional regulation of rat liver epoxide hydrolase, NADPH-cytochrome P-450 oxidoreductase and cytochrome P-450b genes by phenobarbital, *J. Biol. Chem.* **258**:8081–8085.

58. MacKenzie, P. I., 1986, Rat liver UDP-glucuronosyl transferase: Sequence and expression of a cDNA encoding a phenobarbital-inducible form, *J. Biol. Chem.* **261**:6119–6125.

59. Waxman, D. J., and Azaroff, L., 1992, Phenobarbital induction of cytochrome P-450 gene expression, *Biochem. J.* **281**:577–592.

60. Gonzalez, F. J., 1989, The molecular biology of cytochrome P-450s, *Pharmacol. Rev.* **40**:243–288.

61. Wilson, N. M., Christou, M., Turner, C. R., Wrighton, S. A., and Jefcoate, C. R., 1984, Binding and metabolism of benzo(a)pyrene and 7,12-dimethylbenz(a)anthracene by seven purified forms of P-450, *Carcinogenesis* **5**:1475–1483.

62. Christou, M., Wilson, N. M., and Jefcoate, C. R., 1987, Expression and function of three P-450 isozymes in rat hepatic tissues, *Arch. Biochem. Biophys.* **258**:519–534.

63. Wolf, C. R., Miles, J. S., Seiman, S., Burke, M. D., Rosendowski, B. N., Kelly, K., and Smith, W. E., 1988, Evidence that catalytic differences of two structurally homologous forms of cytochrome P450 are related to their heme environment, *Biochemistry* **27**:1597–1603.

64. Suwa, Y., Mizukami, Y., Sogawa, K., and Fuji-Kuriyama, Y., 1985, Gene structure of a major form of phenobarbital-inducible P-450 in rat liver, *J. Biol. Chem.* **260**:7980–7981.

65. Yamazoe, Y., Shimada, M., Murayama, N., and Kato, R., 1987, Suppression of levels of phenobarbital-inducible rat liver cytochrome P-450 by pituitary hormone, *J. Biol. Chem.* **262**:7423–7428.

66. Atchinson, M., and Adesnick, M., 1983, A cytochrome P450 multigene family: Characterization of a gene activated by phenobarbital administration, *J. Biol. Chem.* **258**:11285–11295.

67. Omiecinski, C. J., Walz, F. J., Jr., and Vlasuk, G. P., 1985, Phenobarbital induction of rat liver cytochromes P-450b and P-450c: Quantitation of specific RNAs by hybridization to synthetic oligodeoxyribonucleotide probes, *J. Biol. Chem.* **260**:3247–3250.

68. Kocarek, T. A., Schuetz, E. G., and Guzelian, P. S., 1990, Differentiated induction of cytochrome P450b/e and P450p mRNAs by dose of phenobarbital in primary cultures of adult rat hepatocytes, *Mol. Pharmacol.* **38**:440–444.

69. Mattschloss, L. A., Holbs, A. A., Steggles, A. W., May, B. K., and Elliott, W. H., 1986, Isolation and characterization of genomic clones for two chicken phenobarbital-inducible cytochrome P-450 genes, *J. Biol. Chem.* **261**:9438–9443.

70. Hansen, A. J., and May, B. K., 1989, Sequence of a chicken phenobarbital-inducible cytochrome P-450 cDNA: Regulation of two P-450 mRNAs transcribed from different genes, *DNA* **8**:179–191.

71. Hamilton, J. W., Bement, W. J., Sinclair, P. R., Sinclair, J. F., and Wetterhahn, K. E., 1988, Expression of 5-aminolevulinate synthase and cytochrome P-450 mRNAs in chicken embryo hepatocytes *in vivo* and in culture, *Biochem. J.* **255**:267–275.

72. Hamilton, J. W., and Wetterhahn, K. E., 1989, Differential effects of chromium (VI) on constitutive and inducible gene expression in chick embryo liver *in vivo* and correlation with chromium (VI)-induced DNA damage, *Mol. Carcinogen.* **2**:274–286.

73. Fulco, A. J., 1991, P450BM-3 and other inducible bacterial P-450 cytochromes: Biochemistry and regulation, *Annu. Rev. Pharmacol. Toxicol.* **31**:177–203.

74. Bhat, G. J., Rangarajan, P. N., and Padmanaban, G., 1987, Differential effects of cycloheximide on rat liver cytochrome P-450 gene transcription in the whole animal and hepatoma cell culture, *Biochem. Biophys. Res. Commun.* **148**:1118–1123.

75. Burger, H., Schuetz, E. G., Schuetz, J. D., and Guzelian, P. S., 1990, Divergent effects of cycloheximide on the induction of class II and class III cytochrome P450 mRNAs in cultures of adult rat hepatocytes, *Arch. Biochem. Biophys.* **281**:204–211.

76. Hamilton, J. W., Bement, W. J., Sinclair, P. R., Sinclair, J. F., Alcedo, J. A., and Wetterhahn, K. E., 1992, Inhibition of protein synthesis increases the transcription of the phenobarbital-inducible CYP2H1 and CYP2H2 genes in chick embryo hepatocytes, *Arch. Biochem. Biophys.* **298**:96–104.

77. Dogra, S. C., Hahn, C. N., and May, B. K., 1993, Superinduction by cycloheximide of cytochrome P4502H1 and 5-aminolevulinate synthase gene transcription in chick embryo liver, *Arch. Biochem. Biophys.* **300**:531–534.

78. Schuetz, E. G., Schuetz, J. D., May, B., and Guzelian, P. S., 1990, Regulation of cytochrome P-450b/e and P-450p gene expression by growth hormone in adult rat hepatocytes cultured on a reconstituted basement membrane, *J. Biol. Chem.* **265**:1188–1192.

79. Waxman, D. J., Morrissey, J. J., Naik, S., and Jauregui, H. O., 1990, Phenobarbital induction of cytochrome P-450. High level, long term responsiveness of primary rat hepatocyte cultures to drug induction and glucocorticoid dependence of phenobarbital response, *Biochem. J.* **271**:113–119.

80. Williams, J. F., Bement, W. J., Sinclair, J. F., and Sinclair, P. R., 1991, Effect of interleukin 6 on phenobarbital induction of cytochrome P-450IIB in cultured rat hepatocytes, *Biochem. Biophys. Res. Commun.* **178**:1049–1055.

81. Tieney, B., and Bresnick, E., 1981, Differences in the binding of 3-methylcholanthrene and phenobarbitone to rat liver cytosolic and nuclear protein fractions, *Arch. Biochem. Biophys.* **210**:729–739.

82. Poland, A., Mak, I., Glover, E., Boatman, R. J., Ebetino, F. H., and Kende, A. S., 1980, 1,4-Bis[2-(3,5-dichloropyridyloxy)]benzene, a potent phenobarbital-like inducer of microsomal monooxygenase activity, *Mol. Pharmacol.* **18:**571–580.

83. Poland, A., Mak, I., and Glover, E., 1981, Species differences in responsiveness to 1,4-bis[2-(3,5-dichloropyridyloxy)] benzene, a potent phenobarbital-like inducer of microsomal monooxygenase activity, *Mol. Pharmacol.* **20:**442–450.

84. Smith, G., Henderson, C. J., Parker, M. G., White, R., Bars, R. G., and Wolf, C. R., 1993, 1,4-Bis[2-(3,5-dichloropyridyloxy)]benzene, an extremely potent modulator of mouse hepatic cytochrome P-450 gene expression, *Biochem. J.* **289:**807–813.

85. He, J., and Fulco, A. J., 1991, A barbiturate-regulated protein binding to a common sequence in the cytochrome P450 genes of rodents and bacteria, *J. Biol. Chem.* **266:**7864–7869.

86. Shaw, G., and Fulco, A. J., 1993, Inhibition by barbiturates of the binding of Bm3R1 repressor to its operator site on the barbiturate-inducible cytochrome P450BM-3 gene of *Bacillus megaterium, J. Biol. Chem.* **268:**2997–3004.

87. Shaw, G., and Fulco, A. J., 1992, Barbiturate-mediated regulation of expression of the cytochrome P-450BM-3 gene of *Bacillus megaterium* by Bm3R1 protein, *J. Biol. Chem.* **267:**5515–5526.

88. Rangarajan, P. N., and Padmanaban, G., 1989, Regulation of cytochrome P-450b/e gene expression by a heme- and phenobarbitone-modulated transcription factor, *Proc. Natl. Acad. Sci. USA* **86:**3963–3967.

89. Shaw, P. M., Adesnik, M., Weiss, M. C., and Corcos, L., 1993, The phenobarbital-induced transcriptional activation of cytochrome P-450 genes is blocked by the glucocorticoid-progesterone antagonist RU486, *Mol. Pharmacol.* **44:**775–783.

90. Hahn, C. N., Hansen, A. J., and May, B. K., 1991, Transcriptional regulation of the chicken CYP2H1 gene: Localization of a phenobarbital-responsive enhancer domain, *J. Biol. Chem.* **266:**17031–17039.

91. Ramsden, R., Sommer, K. M., and Omiecinski, C. J., 1993, Phenobarbital induction and tissue-specific expression of the rat CYP2B2 gene in transgenic mice, *J. Biol. Chem.* **268:**21722–21726.

92. Doostdar, H., Grant, M. H., Melvin, W. T., Wolf, C. R., and Burke, M. D., 1993, The effects of inducing agents on cytochrome P450 and UDP-glucuronyl-transferase activities in human HEPG2 hepatoma cells, *Biochem. Pharmacol.* **46:**629–635.

93. Pinkus, R., Bergelson, S., and Daniel, V., 1993, Phenobarbital induction of AP-1 binding activity mediates activation of glutathione S-transferase and quinone reductase gene expression, *Biochem. J.* **290:**637–640.

94. Lock, E. A., Mitchell, A. M., and Elcombe, C. R., 1989, Biochemical mechanisms of induction of hepatic peroxisome proliferation, *Annu. Rev. Pharmacol. Toxicol.* **29:**145–163.

95. Orton, T. C., and Parker, G. L., 1982, The effect of hypolipidemic drugs on the hepatic microsomal drug metabolizing enzyme system of the rat: Induction of cytochrome(s) P-450 with specificity toward terminal hydroxylation of lauric acid, *Drug Metab. Dispos.* **10:**110–115.

96. Tamburini, P. P., Masson, H., Bains, S. K., Makowski, R. J., Morris, B., and Gibson, G. G., 1984, Multiple forms of hepatic cytochrome P450: Purification, characterization, and comparison of a novel clofibrate-induced isozyme with other forms of cytochrome P-450, *Eur. J. Biochem.* **139:**235–246.

97. Hardwick, J. P., Song, B.-J., Huberman, E., and Gonzalez, F. J., 1987, Isolation, complementary DNA sequence, and regulation of rat hepatic lauric acid ω-hydroxylase (cytochrome P-450$_{LA\omega}$), *J. Biol. Chem.* **262:**801–810.

98. Reddy, J. K., Goel, S. K., Nemali, M. R., Carrino, J. J., Laffler, T. G., Reddy, M. K., Sperbeck, S. J., Osumi, T., Hashimoto, T., Lalwani, N. D., and Rao, M. S., 1986, Transcriptional regulation of peroxisomal fatty acyl-CoA oxidase and enoyl-CoA hydratase/3-hydroxyacyl-CoA dehydrogenase in rat liver by peroxisome proliferators, *Proc. Natl. Acad. Sci. USA* **83:**1747–1751.

99. Issemann, I., and Green, S., 1990, Activation of a member of the steroid hormone receptor superfamily by peroxisome proliferators, *Nature* **347:**645–650.

100. Göttlicher, M., Widmark, E., Li, Q., and Gustafsson, J.-A., 1992, Fatty acids activate a chimera of the clofibric acid-activated receptor and the glucocorticoid receptor, *Proc. Natl. Acad. Sci. USA* **89:**4653–4657.

101. Zhu, Y., Alvares, K., Huang, Q., Rao, M. S., and Reddy, J. K., 1993, Cloning of a new member of the peroxisome proliferator-activated receptor gene family from mouse liver, *J. Biol. Chem.* **268:**26817–26820.

102. Sher, T., Yi, H.-F., McBride, O. W., and Gonzalez, F. J., 1993, cDNA cloning, chromosomal mapping, and functional characterization of the human peroxisome proliferator activated receptor, *Biochemistry* **32:**5598–5604.

103. Dreyer, C., Krey, G., Keller, H., Givel, F., Helftenbein, G., and Wahli, W., 1992, Control of the peroxisomal β-oxidation pathway by a novel family of nuclear hormone receptors, *Cell* **68:**879–887.

104. Evans, R. M., 1988, The steroid and thyroid hormone receptor superfamily, *Science* **240:**889–895.

105. Tugwood, J. D., Issemann, I., Anderson, R. G., Bundell, K. R., McPheat, W. L., and Green, S., 1992, The mouse peroxisome proliferator activated receptor recognizes a response element in the 5′ flanking sequence of the rat acyl CoA oxidase gene, *EMBO J.* **11:**433–439.

106. Muerhoff, A. S., Griffin, K. J., and Johnson, E. F., 1992, The peroxisome proliferator-activated receptor mediates the induction of CYP4A6, a cytochrome P450 fatty acid ω-hydroxylase, by clofibric acid, *J. Biol. Chem.* **267:**19051–19053.

107. Bell, D. R., and Elcombe, C. R., 1991, Induction of acyl-CoA oxidase and cytochrome P450IVA1 RNA in rat primary hepatocyte cultures by peroxisome proliferators, *Biochem. J.* **280:**249–253.

108. Kaikaus, R. M., Chan, W. K., Lysenko, N., Ray, R., Ortiz de Montellano, P. R., and Bass, N. M., 1993, Induction of peroxisomal fatty acid β-oxidation and liver fatty acid-binding protein by peroxisome proliferators, *J. Biol. Chem.* **268:**9593–9603.

109. Kliewar, S. A., Umesono, K., Noonan, D. J., Heyman, R. A., and Evans, R. A., 1992, Convergence of 9-*cis* retinoic acid and peroxisome proliferator signalling pathways through heterodimer formation of their receptors, *Nature* **358:**771–774.

110. Bogazzi, F., Hudson, L. D., and Nikodem, V. M., 1994, A novel heterodimerization partner for thyroid hormone receptor, *J. Biol. Chem.* **269:**11683–11686.

111. Selye, H., 1971, Hormones and resistance, *J. Pharmacol. Sci.* **60:**1–28.

112. Elshourbagy, N. A., and Guzelian, P. S., 1980, Separation, purification, and characterization of a novel form of hepatic cytochrome P-450 from rats treated with pregnenolone-16α-carbonitrile, *J. Biol. Chem.* **255:**1279–1285.

113. Wang, R. W., Kari, P. H., Lu, A. Y. N., Thomas, P. E., Guengerich, F. P., and Vyas, K. P., 1991, Biotransformation of lovastatin. IV. Identification of cytochrome P4503A proteins as the major enzymes responsible for the oxidative metabolism of lovastatin in rat and human liver microsomes, *Arch. Biochem. Biophys.* **290:**355–361.

114. Kronbach, T., Fischer, V., and Meyer, U. A., 1988, Cyclosporine metabolism in human liver: Identification of a cytochrome P-450 III gene family as the major cyclosporine-metabolizing enzyme explains interactions of cyclosporine with other drugs, *Clin. Pharmacol. Ther.* **43:**630–635.

115. Guengerich, F. P., Martin, M. V., Beaune, P. H., Kremers, P., Wolff, T., and Waxman, D. J., 1986, Characterization of rat and human liver microsomal cytochrome P-450 forms involved in nifedipine oxidation, a prototype for genetic polymorphisms in oxidative drug metabolism, *J. Biol. Chem.* **261:**5051–5060.

116. Watkins, P. B., Wrighton, S. A., Maurel, P., Schuetz, E. G., Mendez-Picon, G., Parker, G. A., and Guzelian, P. S., 1985, Identification of an inducible form of cytochrome P-450 in human liver, *Proc. Natl. Acad. Sci. USA* **82:**6310–6314.

117. Kronbach, T., Mathys, D., Umeno, M., Gonzalez, F. J., and Meyer, U. A., 1989, Oxidation of midazolam and triazolam by human liver cytochrome P450IIIA4, *Mol. Pharmacol.* **36:**89–96.

118. Aoyama, T., Yamano, S., Waxman, D. J., Lapenson, D. P., Meyer, U. A., Fischer, V., Tyndale, R., Inaba, T., Kalow, W., Gelbirn, H. V., and Gonzales, F. J., 1989, Cytochrome P-450 hPCN3, a novel

cytochrome P-450 IIIA gene product that is differentially expressed in adult human liver, *J. Biol. Chem.* **264**:10388–10395.

119. Hardwick, J. P., Gonzalez, F. J., and Kasper, C. B., 1983, Cloning of DNA complementary to cytochrome P450 induced by pregnenolone-16α-carbonitrile, *J. Biol. Chem.* **258**:10182–10186.

120. Wrighton, S. A., Schuetz, E. G., Watkins, P. B., Maurel, P., Barwick, J., Bailey, B. S., Hartle, H. T., Young, B., and Guzelian, P. S., 1985, Demonstration in multiple species of inducible hepatic cytochromes P-450 and their mRNAs related to the glucocorticoid-inducible cytochrome P-450 of the rat, *Mol. Pharmacol.* **28**:312–321.

121. Gonzalez, F. J., Song, B.-J., and Hardwick, J. P., 1986, Pregnenolone-16α-carbonitrile-inducible P-450 gene family: Gene conversion and differential regulation, *Mol. Cell. Biol.* **6**:2969–2976.

122. Simmons, D. L., McQuiddy, P., and Kasper, C. B., 1987, Induction of the hepatic mixed-function oxidase system by synthetic glucocorticoids, *J. Biol. Chem.* **262**:326–332.

123. Schuetz, E. G., and Guzelian, P. S., 1984, Induction of cytochrome P-450 by glucocorticoids in rat liver. II. Evidence that glucocorticoids regulate induction of cytochrome P-450 by a non-classical receptor mechanism, *J. Biol. Chem.* **259**:2007–2012.

124. Watkins, P. B., Wrighton, S. A., Schuetz, E. G., Maurel, P., and Guzelian, P. S., 1986, Macrolide antibiotics inhibit the degradation of the glucocorticoid-responsive cytochrome P-450p in rat hepatocytes *in vivo* and in primary monolayer culture, *J. Biol. Chem.* **261**:6264–6271.

125. Schuetz, E. G., Wrighton, S. A., Safe, S. H., and Guzelian, P. S., 1986, Regulation of cytochrome P-450p by phenobarbital and phenobarbital-like inducers in adult rat hepatocytes in primary monolayer culture and *in vivo*, *Biochemistry* **25**:1124–1133.

126. Schuetz, E. G., Omiecinski, C. J., Li, D., Muller-Eberhard, U., Kleinman, H. K., Elswick, B., and Guzelian, P. S., 1988, Regulation of gene expression in adult rat hepatocytes cultured on a basement membrane matrix, *J. Cell. Physiol.* **134**:309–323.

127. Schuetz, E. G., Schuetz, J. D., Strom, S. C., Thompson, M. T., Fisher, R. A., Molowa, D. T., Li, D., and Guzelian, P. S., 1993, Regulation of human liver cytochromes P-450 in family 3A in primary and continuous culture of human hepatocytes, *Hepatology* **18**:1254–1262.

128. Lieber, C. S., and DeCarli, L. M., 1968, Ethanol oxidation by hepatic microsomes: Adaptive increase after ethanol feeding, *Science* **162**:917–918.

129. Koop, D. R., and Coon, M. J., 1986, Ethanol oxidation and toxicity: Role of alcohol P-450 oxygenase, *Alcohol. Clin. Exp. Res.* **10**:445–49s.

130. Koop, D. R., Morgan, E. T., Tarr, G. E., and Coon, M. J., 1982, Purification and characterization of a unique isozyme of cytochrome P-450 from liver microsomes of ethanol-treated rabbits, *J. Biol. Chem.* **257**:8472–8480.

131. Koop, D. R., and Tierney, D. J., 1990, Multiple mechanisms in the regulation of ethanol-inducible cytochrome P450IIE1, *BioEssays* **12**:429–435.

132. Casazza, J. P., Felver, M. E., and Veech, R. L., 1984, The metabolism of acetone in rat, *J. Biol. Chem.* **259**:231–236.

133. Umeno, M., Song, B.-J., Kozak, C., Gelboin, H. V., and Gonzalez, F. J., 1988, The rat P450IIE1 gene: Complete intron and exon sequence, chromosome mapping, and correlation of developmental expression with specific 5′ cytosine demethylation, *J. Biol. Chem.* **263**:4956–4962.

134. Song, B.-J., Matsunaga, T., Hardwick, J. P., Park, S. S., Veech, R. L., Yang, C. S., Gelboin, H. V., and Gonzalez, F. J., 1987, Stabilization of cytochrome P450j messenger ribonucleic acid in the diabetic rat, *Mol. Endocrinol.* **1**:542–547.

135. Dong, Z., Hong, J., Ma, Q., Li, D., Bullock, J., Gonzalez, F. J., Park, S. S., Gelboin, H. V., and Yang, C. S., 1988, Mechanism of induction of cytochrome P-450 ac (P-450j) in chemically induced and spontaneously diabetic rats, *Arch. Biochem. Biophys.* **263**:29–35.

136. Koop, D. R., Crump, B. C., Nordblom, G. D., and Coon, M. J., 1985, Immunochemical evidence for induction of the alcohol-oxidizing cytochrome P-450 of rabbit liver microsomes by diverse agents: Ethanol, imidazole, trichloroethylene, acetone, pyrazole, and isoniazid, *Proc. Natl. Acad. Sci. USA* **82**:4065–4069.

137. Song, B.-J., Veech, R. L., Park, S. S., Gelboin, H. V., and Gonzalez, F. J., 1989, Induction of rat hepatic N-nitrosodimethylamine demethylase by acetone is due to protein stabilization, *J. Biol. Chem.* **264:**3568–3572.

138. Eliasson, E., Johansson, I., and Ingelman-Sundberg, M., 1988, Ligand-dependent maintenance of ethanol-inducible cytochrome P-450 in primary hepatocyte cell cultures, *Biochem. Biophys. Res. Commun.* **150:**436–443.

139. Eliasson, E., Johansson, I., and Ingelman-Sundberg, M., 1990, Substrate, hormone, and cAMP-regulated cytochrome P450 degradation, *Proc. Natl. Acad. Sci. USA* **87:**3225–3229.

140. Gonzalez, F. J., and Nebert, D. W., 1990, Evolution of the P450 gene superfamily, *Trends Genet.* **6:**182–186.

Hormonal Regulation of Liver Cytochrome P450 Enzymes

DAVID J. WAXMAN and THOMAS K. H. CHANG

1. Introduction

Endogenous steroids and other naturally occurring lipophilic substances serve as important substrates for cytochrome P450 (CYP) enzymes found in liver and other tissues,[1,2] including the primary steroidogenic tissues (Kagawa and Waterman, this volume). Steroid hormones are metabolized by liver P450 enzymes with a higher degree of regio- and stereoselectivity than many foreign compound substrates,[3] suggesting that these endogenous lipophiles serve as physiological P450 substrates. Eight of the twelve mammalian P450 gene families described as of 1993[4] encode enzymes that catalyze steroid hydroxylations. Two of these gene families (CYP2 and CYP3) encode liver P450 enzymes required for the hydroxylation of steroid hormones and bile acids, and two encode P450s that participate in the conversion of cholesterol to bile acids in the liver, where they contribute in a major way to cholesterol homeostasis (CYP7, CYP27).[5]

The physiological requirements with respect to steroid hormone hydroxylation differ between the sexes, and not surprisingly, several steroid hydroxylase liver P450s are expressed in a sex-dependent manner.[3,6] Rat P450 enzymes 2C11 and 2C12 are prototypic examples of sex-specific liver P450 enzymes, and they have been a major focus of studies of the underlying endocrine factors, as well as the cellular and molecular regulatory mechanisms that govern sex-specific liver gene expression. CYP2C11 is the major male-specific androgen 16α- and 2α-hydroxylase in adult rat liver, and is induced at puberty in males but not females[7,8] under the influence of neonatal androgenic imprinting

DAVID J. WAXMAN • Division of Cell and Molecular Biology, Department of Biology, Boston University, Boston, Massachusetts 02215. THOMAS K. H. CHANG • Division of Pharmacology and Toxicology, Faculty of Pharmaceutical Sciences, The University of British Columbia, Vancouver, British Columbia, V6T 1Z3, Canada.

Cytochrome P450: Structure, Mechanism, and Biochemistry (Second Edition), edited by Paul R. Ortiz de Montellano. Plenum Press, New York, 1995.

(programming).[9] By contrast, the steroid sulfate 15β-hydroxylase CYP2C12 is expressed in a female-specific manner in adult rats.[9,10] As discussed more fully later in this chapter, the imprinting effects of neonatal androgen on liver expression of these steroid hydroxylase P450 genes are mediated by the hypothalamus and its regulation of pituitary GH secretory patterns.[11] Thyroid hormone status is a second key component of the basic endocrine regulation of these liver steroid hydroxylase P450s.[12–14] Corresponding patterns of sex-dependent, GH regulation have been reported for several mouse liver steroid hydroxylase P450s[15–18] and also for several non-P450 liver enzymes.[e.g.,11,19–22] Human liver P450 enzyme levels and their associated drug metabolism activities may also be determined, in part, by age, sex, and/or hormone status.[23–25] Studies of the underlying mechanisms governing the endocrine regulation of rat liver P450 enzymes may thus be of general importance for our understanding of the hormonal regulation of liver-expressed genes both in rodent models and in man. Endocrine-regulated steroid hydroxylase P450s contribute in an important way to foreign compound metabolism in the liver,[26] and studies of their hormonal control may shed light on the underlying basis for the influence of hormone status on a broad range of P450-catalyzed drug metabolism and carcinogen activation reactions.[27–29]

This chapter reviews studies leading to the identification of the key endocrine regulatory factors and the underlying mechanisms through which these factors operate to control the expression of liver P450 enzymes. Primary emphasis is given to studies on the hormonally regulated P450s expressed in rat liver, the best studied model system. Also discussed are environmental and pathophysiologic factors that can perturb hormonal status and the impact of these factors on the expression of sex-dependent P450s.

2. Sex-Dependent Rat Liver P450s: Developmental Regulation

Several P450 enzymes are expressed in rat liver in a sex-specific manner, where they are subject to complex developmental regulation and endocrine control[3,6] (Table I). CYP2C11, the major male-specific androgen 2α- and 16α-hydroxylase of adult liver, is not expressed in immature rats and is induced dramatically at puberty (beginning at 4–5 weeks of age) in male but not female rats.[7,8] A similar developmental profile is found for two other male-specific rat liver P450s, CYP2A2[30,31] and 2C13.[32,33] By contrast, a fourth adult male-specific liver P450, the steroid 6β-hydroxylase CYP3A2, is expressed in prepubertal rat liver at similar levels in both sexes, but is markedly suppressed at puberty only in females.[9,34–36] The steroid sulfate 15β-hydroxylase CYP2C12 is expressed at a moderate level in both male and female rats at 3–4 weeks of age. Beginning at puberty, however, CYP2C12 levels are further increased in females while they are fully suppressed in males.[9,10] Two female-predominant liver enzymes are also induced at puberty in adult female rats. These are CYP2C7,[32,37] which catalyzes retinoic acid 4-hydroxylation,[38] and steroid 5α-reductase, which is a P450-independent enzyme that plays an important role in steroid metabolism in adult female rats.[9,27] Finally, CYP2A1 is a female-predominant steroid 7α-hydroxylase that is expressed in both sexes shortly after birth, but is suppressed at puberty to a greater extent in male than in female rat liver[9,39,40] (Table I). Each of these sex-dependent P450 enzymes is primarily expressed in liver, although low-level extrahepa-

TABLE I
Hormonal Regulation of Gender-Dependent Rat Liver P450 Enzymes

CYP enzyme[a]	Testosterone hydroxylase activities[b]	Hormonal regulation[c]		
		Androgenic imprinting[d]	Thyroid hormone[e]	
Male-specific				
2A2	15α	+ +	+/–	
2C11	2α, 16α	+ +	+/–	
2C13	6β,[f] 15α	+ +	ND	
3A2	6β, 2β	+ +	– –	
4A2	(see footnote g)	ND	–	
Female-specific				
2C12	15β[h]	– –	+/–	
Female-predominant[i]				
2A1	7α	ND	–	
2C7	16α	ND	+ +	
5α-reductase	—	– –	+ +	

[a]P450 gene designations are based on the systematic nomenclature of Nelson et al.[4] Table modified from Ref. 5.
[b]The major sites of testosterone hydroxylation catalyzed by the individual P450 proteins are shown. Testosterone metabolites specific to the P450's activity in rat liver microsomal incubations are underlined. Based on Refs. 3, 9, 170 and references therein.
[c]See Fig. 2 for a summary of the effects of GH secretory patterns on P450 enzyme expression. "++" indicates a positive effect on adult enzyme expression, while "– –" indicates a suppressive effect. "–" indicates a lesser degree of suppression, while "+/–" indicates no major effect. ND, not determined in a definitive manner.
[d]For further details see Refs. 31, 33, 45.
[e]Based on Refs. 12–14, 40, 72.
[f]Purified CYP2C13 exhibits high testosterone hydroxylase activity in a purified enzyme system, but this enzyme makes only marginal contributions to liver microsomal testosterone hydroxylation.[171]
[g]P450 4A2 catalyzes fatty acid ω-hydroxylation, but does not catalyze testosterone hydroxylation.
[h]15β-hydroxylation of steroid sulfates. CYP2C12 also catalyzes weak testosterone 15α- and 1α-hydroxylase activities.
[i]Liver expression of these enzymes is readily detectable in both male and female rats, but at a 3- to 10-fold higher level in females as compared to males.

tic expression may occur in some cases.[41,42] Studies of liver P450 expression during senescence have revealed a general loss of gender-dependent enzyme expression that reflects a decrease in male P450 levels and an increase in expression of female-predominant P450 enzymes in aging male rats.[43,44]

3. Hormonal Control of Liver P450 Expression

3.1. Regulation by Gonadal Hormones

3.1.1. Testosterone

3.1.1a. Distinct Effects of Neonatal Androgen and Adult Androgen. Gonadal hormones play an essential role in determining the expression of the major sex-specific rat liver P450 forms at adulthood. In the case of testosterone, there are two distinct postnatal developmental periods of hormone production, neonatal and postpubertal, and each period makes a distinct contribution to the expression of the sex-dependent liver P450s at adulthood.

Castration of male rats at birth eliminates both periods of androgen production, and this in turn abolishes the normal adult expression of each of the male-specific P450s: CYP2A2,[31] 2C11,[8,9,32,45,46]2C13,[33] and 3A2.[9,45,46]2C11[47] and 2C13 mRNA levels[33] are also abolished in birth-castrated rats, indicating that enzyme expression is regulated at a pretranslational step. Treatment of birth-castrated male rats with testosterone during the neonatal period leads to a partial restoration of the expression of these male-specific P450 forms at adulthood.[9,31,33] A brief period of neonatal androgen exposure is thus sufficient to "imprint" or irreversibly program the male rat to express these P450 enzymes later on in adult life. These effects of neonatal androgen on male-specific P450 enzymes are very similar to the androgenic imprinting effects observed in earlier studies of liver microsomal steroid hydroxylase activities,[48,49] several of which can now be associated with specific liver P450 forms.[3] However, neonatal testosterone given to birth-castrated male rats only partially restores 2C11[9,46] or 2C13[33] to normal adult male levels, indicating that neonatal androgen alone is insufficient for full adult expression of these male-specific P450s. Consistent with this observation, the combination of neonatal androgen treatment with adult androgen exposure results in complete restoration of normal adult male expression of the male-specific P450s.[9] Moreover, testosterone treatment of adult male rats that were castrated either neonatally or prepubertally, can substantially increase the expression of 2C11[45,46,50,51] and 2C13.[33] However, in contrast to the irreversible imprinting effects of neonatal androgen treatment, the effects of adult androgen exposure are likely to be reversible, as evidenced by the partial loss of 2C11 in male rats castrated at adulthood[8,9] and by the reversal of this loss by the synthetic androgen methyltrienolone.[47] Similarly, the continued presence of testosterone at adulthood is also required to maintain normal adult expression of 3A2, since castration at 90 days of age reduces hepatic 3A2 mRNA levels by > 80%, but this can be restored by subsequent administration of testosterone to the adult rat.[52] Thus, while neonatal testosterone imprints the rat for expression of the male-specific P450 enzymes beginning at puberty, a time when the demand for P450-dependent liver steroid metabolism is increased in adult rats, the additional presence of androgen during the pubertal and postpubertal periods is required to maintain full enzyme expression during adult life.

3.1.1b. Testosterone Suppression of Female Enzymes. In contrast to the positive regulation by testosterone of the male-specific enzymes, testosterone suppresses expression of the female-specific CYP2C12 as well as the female-predominant enzymes 2A1 and steroid 5α-reductase. Hepatic 2C12 content is reduced in intact, adult female rats exposed chronically to testosterone[45] or to the synthetic androgen methyltrienolone.[10] Similarly, treatment of neonatally ovariectomized rats with testosterone, either pubertally or neonatally, results in a major decrease in microsomal steroid 5α-reductase activity.[45] Birth castration of male rats increases the adult levels of hepatic 2A1, but testosterone administration to these animals remasculinizes (i.e., decreases) the levels of this P450.[53] Androgens thus exert a suppressive effect on liver 2A1 expression. Studies of the effect of testosterone on the expression of the female-predominant 2C7 are inconclusive.[50,54]

3.1.1c. Mechanisms of Testosterone Regulation. The cellular and hormonal mechanisms by which neonatal testosterone imprints the adult expression of liver P450 enzymes are only partially understood. Testosterone's primary effects on liver P450 profiles are mediated by the hypothalamic–pituitary axis[55] and its control of the sex-dependent pattern of pituitary growth hormone (GH) secretion.[56,57] Consistent with this conclusion, testosterone has only minor effects on liver enzyme profiles in hypophysectomized rats in most[11] but not all[58,59] instances. As will be discussed in Section 3.2, these GH secretory patterns play a key role in regulating the expression of the sex-dependent P450 forms.

3.1.2. Estrogen

Whereas testosterone has a major positive regulatory influence on the male-specific P450 forms, estrogen plays a somewhat lesser role in the expression of the female-specific and the female-predominant liver P450 enzymes. Ovariectomy at birth reduces, but does not abolish, expression of hepatic CYP2C7, 2C12, and steroid 5α-reductase in adult female rats[9,45,50] and normal adult enzyme levels can be restored by estrogen replacement. By contrast, estradiol suppresses hepatic 2C11 in both intact and castrated male rats.[45,50] However, the absence of 2C11 in adult female rats is not related to a negative effect of estrogen. Thus, ovariectomy alone does not lead to 2C11 expression in female rats.[9,45,50] In male rats, the suppression of 2C11 by estradiol may be irreversible, as demonstrated by the major loss of this P450 in adult male rats exposed to estradiol during the neonatal period or at puberty. However, this effect is not a consequence of a direct action of estradiol on the liver, since estradiol does not impact on hepatic 2C11 levels in hypophysectomized rats.[58] Rather, the effects of estradiol on liver P450 expression involve action via the hypothalamic–pituitary axis,[55,60] and appear to result from an estrogen-dependent increase in the interpeak baseline levels of plasma GH.[55,60] This effect of estradiol may be sufficient to alter the sex-specific effects of the GH secretory pattern since, as discussed in greater detail later in this chapter, recognition of a "masculine" GH pulse by hepatocytes requires an obligatory recovery period during which there is no plasma GH and hence no stimulation of hepatocyte GH receptors.[61] In addition, estrogen may antagonize the induction of 2C11 by testosterone as suggested by the absence of androgen imprinting of this P450 in intact female rats treated with neonatal or pubertal testosterone.[50,51] Indeed, the stimulatory effect of testosterone on pulsatile GH secretion can be blocked by the presence of intact ovaries in female rats.[57] However, the precise neuroendocrine mechanisms responsible for the antagonistic effects of estrogen on androgen imprinting remain to be elucidated.

3.2. Growth Hormone Regulation of Liver P450s

3.2.1. Sex-Specific GH Secretory Profiles

As discussed in Section 3.1, gonadal steroids do not act directly at the liver to regulate the sex-specific patterns of liver P450 expression. Rather, their effects on liver P450s are primarily mediated via the gonadal–hypothalamic–pituitary axis and its sex-dependent regulation of pituitary GH secretory patterns. Plasma GH profiles are sexually differentiated in many species, including humans,[62–64] although the differences between the sexes are most dramatic in rodents.[65–67] In the rat, GH is secreted by the pituitary gland in adult males in an intermittent, or pulsatile, manner that is characterized by high peaks of hormone in plasma (150–200 ng/ml) each 3.5–4 hr followed by a period of very low or undetectable circulating GH (<1–2 ng/ml) (Fig. 1A). By contrast, in the adult female rat, GH is secreted

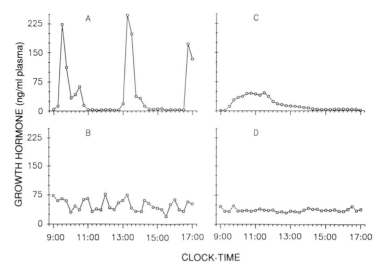

FIGURE 1. Sex-dependence of plasma GH profiles in adult rats. Shown are plasma GH profiles measured in unrestrained and unstressed male (panel A) and female rats (panel B), as well as hypophysectomized rats administered GH by subcutaneous injection (panel C), to mimic the male pattern of intermittent GH stimulation, or continuously, via an osmotic minipump (panel D), to mimic the female secretory pattern. Data shown are taken from Ref. 61.

more frequently (multiple pituitary secretory events per hour) and in a manner such that the plasma GH pulses overlap and the hormone is continually present in circulation at significant levels (~15–40 ng/ml) at all times[11] (Fig. 1B). Hypophysectomy and GH replacement experiments carried out in a number of laboratories have demonstrated that these sex-dependent plasma GH profiles are, in turn, responsible for establishing and for maintaining the sex-dependent patterns of liver P450 gene expression.[8,10,31,61,68]

Three distinct responses of liver P450s to plasma GH profiles can be discerned (Fig. 2). These are:

1. Continuous plasma GH, a characteristic of adult female rats, stimulates expression of female-specific and female-dominant liver enzymes, such as CYP2C12, 2A1, and 2C7 and steroid 5α-reductase.[10,13,69]

2. Intermittent plasma GH pulses, which are characteristic of adult male rats, induce expression of the male-specific liver enzyme CYP2C11 and its associated steroid 2α-hydroxylase activity[8,61,68] (Fig. 3). The effects of intermittent GH exposure on other male-specific liver P450s (2A2, 2C13, 3A2, 4A2) are less clear. Expression of this latter group of P450 enzymes is not obligatorily dependent on GH pulses, when judged by their high level of expression in hypophysectomized rats of both sexes.[31,33,36] On the other hand, expression of 2A2 and 3A2 in liver can be stimulated by intermittent GH pulses given to adult male rats that are depleted of circulating GH by neonatal monosodium glutamate treatment (Waxman, Ram, Pampori, and Shapiro, 1995, *Molec. Pharmacol.*, in press).

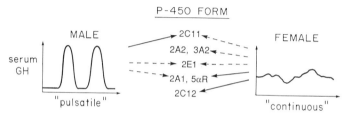

FIGURE 2. Role of GH secretory profiles in the expression of rat hepatic P450 enzymes and steroid 5α-reductase. Influence of pulsatile versus continuous plasma GH on the expression of hepatic enzymes whose expression in adult rats is male-specific (2C11, 2A2, 3A2), female-predominant (2A1, 5αR), or female-specific (2C12). Stimulation of enzyme expression is indicated by a solid line, and suppression of enzyme expression by a dashed line. Other pituitary-determined hormones (e.g., thyroid hormone) may be required for the full effects of GH on some of these hepatic enzymes (see text). Abbreviation: 5αR, steroid 5α-reductase. Figure modified from Ref. 53. In addition to those P450s shown, CYP4A2 and 2C13 are male-specific P450s that are regulated in liver in a manner similar to CYP2A2,[33,72] while CYP2C7 is a female-predominant P450 whose expression is stimulated by continuous GH in a manner similar to CYP2A1 and steroid 5α-reductase.[13,54]

3.   GH can also have negative effects on liver P450 enzyme expression, as revealed by the marked suppression of each of the male-specific P450s following continuous GH treatment (Figs. 2 and 3). This suppression occurs in both intact and hypophysectomized rats, demonstrating that GH suppression is not simply a consequence of the destruction of plasma GH pulses by continuous GH treatment. GH suppression is also a key determinant of the lower responsiveness of female rats to phenobarbital induction of CYP2B1,[70,71] and probably also the lower responsiveness of the females to the induction of 4A enzymes by peroxisome proliferators such as clofibrate.[72]

3.2.2. Molecular Mechanisms of GH Action

GH regulates liver P450 expression at the level of steady-state mRNA, all but ruling out potential translational and posttranslational mechanisms, such as GH regulation at the level of P450 protein turnover. Induction of CYP2C12 mRNA by continuous GH does require, however, ongoing protein synthesis,[73] suggesting either an indirect induction mechanism or a requirement for one or more protein components that may have a short half-life. An analysis of liver nuclear RNA pools has established that unprocessed, nuclear 2C11 and 2C12 RNAs (hnRNA) respond to circulating GH profiles in a manner that is indistinguishable from the corresponding mature, cytoplasmic mRNAs.[74] Consequently, transport of 2C11 and 2C12 mRNA to the cytoplasm, and cytoplasmic P450 mRNA stability are unlikely to be important GH-regulated control points for sex-specific P450 expression. Moreover, nuclear run-on transcription analyses have established that GH regulates the sex-specific expression of the 2C11 and 2C12 genes at the level of transcript initiation.[74,75] Transcription is also the major step for regulation of 2A2 and 2C13 mRNAs,[74,75] whose male-specific patterns of expression appear to be primarily a consequence of the suppressive effects of continuous GH exposure in adult female rats.[31] Thus,

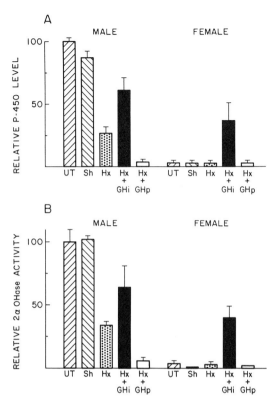

FIGURE 3. Intermittent GH stimulates expression of CYP2C11 in adult male and female rat liver. Adult male and female rats were either untreated (UT), sham-operated (Sh), or hypophysectomized (Hx) at 8 weeks of age. Hx rats were treated with GH administered for 7 days by either of two protocols: GHi—intermittent hormone injection, twice daily, given subcutaneously (cf. Fig. 1, panel C); or GHp—continuous GH infusion via an Alzet osmotic minipump implanted subcutaneously (cf. Fig. 1, panel D). Rats were sacrificed and isolated liver microsomes were then assayed for relative CYP2C11-dependent testosterone 2α-hydroxylase activity (panel B). Data based on studies by Waxman *et al.*[31] Whereas intermittent GH stimulates CYP2C11 protein and activity in hypophysectomized males and females (Hx + GHi), continuous GH infusion suppresses the basal CYP 2C11 expression observed in hypophysectomized males (Hx + GHp).

transcription initiation is the key step at which the three distinct effects of GH outlined in Section 3.2.1 are operative: stimulation of 2C11 expression by pulsatile GH, suppression of each of the male-specific P450s by continuous GH, and stimulation of 2C12 expression by continuous GH.[74]

Recent studies have established that the 5′-flanking DNA segments of the 2C11 gene[76] and the 2C12 gene[77] both contain specific DNA sequences that interact in a sex-dependent and GH-regulated manner with DNA-binding proteins (putative transcription factors) that are differentially expressed in male versus female rat liver nuclei.[74,78] These DNA sequences are hypothesized to include GH response elements which contribute to the sex-specific transcription of the 2C11 and 2C12 genes. In addition, two negative regulatory elements ("silencer elements") have recently been identified in the 2C11 promoter; however, their significance with respect to GH regulation and sex-specific P450 expression is unclear.[79] Further studies are required to establish the functional significance of these putative regulatory elements for the expression of the 2C11 and 2C12 genes.

3.2.3. Cellular Mechanisms: GH Signaling

The cellular mechanisms whereby pituitary GH secretory profiles differentially regulate expression of the sex-dependent liver P450s are only partially understood. GH can act directly on the hepatocyte to regulate liver P450 expression, as demonstrated by the responsiveness of primary rat hepatocyte cultures to continuous GH-stimulated expression of CYP2C12 mRNA; however, these effects do not appear to involve production of IGF-I, a mediator of several of GH's secondary effects on extrahepatic tissues.[15,73,80] Discrimination by the hepatocyte between male and female plasma GH profiles is likely to occur at the cell surface, where a higher level of GH receptors (see below) is found in female as compared to male rats.[81] This sex difference in the abundance of cell surface GH receptors may, at least in part, be related to differential effects of intermittent versus continuous GH stimulation.[82] Conceivably, these differences in receptor levels could play a role in the activation of distinct intracellular signaling pathways by chronic (female) as compared to intermittent (male) GH stimulation (also see below).

3.2.3a. Significance of GH Pulse Frequency. Studies have been carried out to determine which of the three descriptive features of a GH pulse—namely, hormone pulse duration, pulse height, or pulse frequency—is required for proper recognition of a GH pulse as "masculine." Direct measurement of the actual plasma GH profiles achieved when GH is administered to hypophysectomized rats by twice daily s.c. GH injection (i.e., the intermittent GH replacement protocol most commonly used to stimulate CYP2C11 expression) has revealed broad peaks of circulating GH, which last as long as 5–6 hr[61] (Fig. 1C). Since these sustained GH "pulses" do stimulate expression of the male-specific 2C11 (provided that they are not administered in close succession), it is apparent that GH pulse duration need not be tightly regulated to elicit this response from the hepatocyte. Similarly, studies on the requirements for GH pulse height carried out using GH-deficient rats (either dwarf rats or rats depleted of adult circulating GH by neonatal monosodium glutamate treatment) have established that GH pulse height is also not a critical factor for stimulation of 2C11 expression.[83,84] This finding can be understood in terms of the K_d of the GH–GH receptor complex, which at 10^{-10} M (~2 ng/ml)[85] is only 1% of the peak plasma hormone level in adult male rats. Rather, GH pulse frequency appears to be the most critical determinant for GH stimulation of a male pattern of liver P450 expression. This finding has been established in studies of hypophysectomized rats given physiologic replacement doses of GH by intermittent intravenous injections at frequencies of 2, 4, 6, or 7 times per day.[61] Analysis of liver 2C11 levels in these rats at the conclusion of a 7-day period of GH pulsation revealed that a normal male pattern of liver 2C11 gene expression can be induced in hypophysectomized rats by 6 pulses of GH per day (which approximates the normal male plasma GH pulse frequency), as well as by frequencies of only 2 or 4 pulses per day. However, hypophysectomized rats given 7 daily GH pulses do not respond. Therefore, the hepatocyte no longer recognizes the pulse as "masculine" if GH pulsation becomes too frequent. This indicates that hepatocytes require a minimum of GH off-time (~2.5 hr in the hypophysectomized rat model used in these studies), which implies a need for an obligatory recovery period to effectively stimulate 2C11 expression, a condition that is not met in the case of hepatocytes exposed to GH continuously (female hormone profile). This recovery period could serve to reset an intracellular signaling apparatus, or perhaps may provide time needed for replenishment of GH receptors at the cell surface.

3.2.3b. Role of GH Receptor. The effects of GH on hepatocytes and other responsive cells are transduced by GH receptor (GHR), a 620-amino-acid cell surface transmembrane protein.[85] *In vivo* studies carried out in intact male rats have demonstrated that the GHR internalizes to an intracellular compartment coincident with its stimulation by plasma GH pulses, and that it reappears at the cell surface at the time of the next hormone pulse.[86,87] It remains to be determined whether or not internalization of GHR, which may be ligand driven,[82] is required for transduction of the effects of GH on liver P450 expression.

GHR is a member of the cytokine receptor superfamily[88] and is comprised of a 246-amino-acid extracellular domain that binds GH, a single transmembrane segment, and a 350-amino-acid intracellular domain that participates in the intracellular signaling events stimulated by GH.[85,89] X-ray crystallographic and other studies have shown that a single molecule of GH binds in a stepwise manner to *two* molecules of GHR to yield receptor dimers: $GH + GHR \rightarrow GH–GHR \xrightarrow{+GHR} GH–(GHR)_2$.[90,91] The four-helix bundle protein GH initially binds to a single receptor molecule by contacts that involve GH's Site 1, and this is followed by the recruitment of a second molecule of GHR, which interacts with Site 2 on the GH molecule to give the heterotrimeric $GH–(GHR)_2$ complex. Receptor dimerization appears to be reversible, and the equilibrium may be shifted in favor of monomer formation in the continued presence of excess GH: $GH–(GHR)_2 \xrightarrow{+GH} 2\ GH–GHR$. Functional studies have demonstrated that formation of the dimeric receptor complex is necessary, and perhaps sufficient, for stimulation of at least some GH-induced intracellular signaling events.[92] This conclusion is based, however, on studies carried out with chimeric GHR–cytokine receptor molecules. Moreover, while GHR dimerization is clearly required for some GH responses, other GH responses might not proceed via receptor dimerization[93] and, indeed, could be dependent on the formation of monomeric GH–GHR complexes. Conceivably, the distinct patterns of liver gene expression induced by continuous plasma GH (female GH pattern; CYP2C12 expression) as compared to pulsatile GH (male GH pattern; 2C11 expression) might arise from distinct GH signaling pathways perhaps stimulated by monomeric (GH–GHR) as compared to dimeric [$GH–(GHR)_2$] hormone receptor complexes (Fig. 4). Several GH mutants and analogues that bind GHR without effecting receptor dimerization have been identified.[91,92] and could be useful in testing this hypothesis.

3.2.3c. GH Signaling Pathways. The mechanisms through which GH signals the hepatocyte to express sex-specific patterns of cytochrome P450 gene expression are unknown. Studies of the intracellular signaling pathways induced by GH in other cell types (primarily 3T3-F442A preadipocytes, Obl771 adipose cells, and IM-9 lymphocytes) indicate that a complex array of events may occur following GH binding to GHR. This idea is consistent with the large number of physiologic effects that GH can have on responsive tissues and

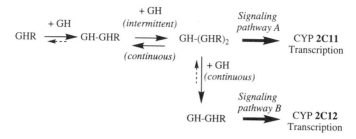

FIGURE 4. Hypothetical role of GH-induced receptor dimerization in discrimination between intermittent and continuous GH stimulation of hepatocytes. According to this model, the dimeric receptor complex GH–(GHR)$_2$ would be the predominant active complex in the case of the intermittent plasma GH profiles present in male rats, and would stimulate intracellular signaling pathway A. By contrast, the monomeric GH–GHR complex would predominate in the case of female rats exposed to GH continuously, and this would activate a distinct intracellular signaling pathway (pathway B). GHR, GH receptor.

cell types. Little is known about the importance of these pathways for the sex-dependent regulation of CYP genes in hepatocytes. It may be useful to consider, as a working hypothesis, the proposal that intermittent GH and continuous GH activate distinct intracellular signaling pathways. These could conceivably involve the activation of distinct pathways by monomeric GH–GHR versus dimeric GH–(GHR)$_2$ complexes (Fig. 4; Fig. 5, pathways A versus B, C). It is alternatively possible that intermittent and continuous GH both activate hepatocytes via dimeric receptor complexes, but that the frequency of GH stimulation determines which one of several alternative intracellular events are triggered in response to receptor binding (Fig. 5, pathways B versus C).

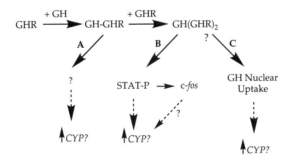

FIGURE 5. GH activates multiple intracellular signaling events. Intermittent and continuous GH are hypothesized to activate distinct intracellular pathways which in turn lead to activation of distinct subsets of GH-regulated CYP genes. Pathway A corresponds to a hypothetical pathway involving a monomeric GH–GHR complex (see also Fig. 3). Pathway B corresponds to a known pathway for activation of the c-*fos* gene by GH, and proceeds via STAT protein(s) which undergo tyrosine phosphorylation and nuclear translocation (see Fig. 6). Potentially, some of the effects of GH on CYP expression could proceed via the c-*fos* activation pathway. The possible contributions of this pathway to the effects of GH on CYP enzyme expression are not known, nor is the potential role of GH taken up into the nucleus, pathway C. It is unknown whether this latter nuclear translocation event proceeds via the dimeric GH–(GHR)$_2$ complex, as shown, or whether a novel alternative pathway leads to nuclear uptake of GH.

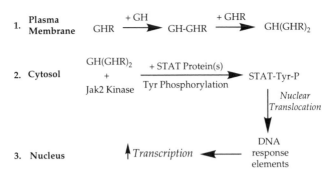

FIGURE 6. Proposed role of STAT proteins in GH signaling. GH binding to GHR at the plasma membrane stimulates receptor dimerization. This process is associated with recruitment (or activation) of the Janus-type kinase Jak2 at the inner surface of the plasma membrane, where Jak2 catalyzes the tyrosine phosphorylation of one or more STAT proteins (STAT, signal transducer/activator of transcription). The phosphorylated STAT(s), perhaps in association with other proteins, translocate to the nucleus, where they bind as homo- and heteromeric complexes to DNA response elements in target genes, leading to activation of gene transcription. This scheme is modeled on the one developed for the role of STATs in the activation of interferon responsive genes by interferon-α and -γ.[100]

At least three distinct, but probably interdependent, early responses to GH have been observed:

1. In several cell types GH stimulates tyrosine phosphorylation of multiple cellular polypeptides, including GHR itself.[94,95] Some of these phosphorylations are catalyzed by a GHR-associated kinase that has recently been identified as Jak2,[96] a member of the Janus family of tyrosine kinases,[97] while others may be catalyzed by downstream kinases.[95] One of these is MAP kinase, whose phosphorylation activates a cascade of kinases; this cascade ultimately leads to the activation of the ribosomal protein S6, thereby accounting for the stimulatory effects of GH on protein synthesis.[98,99] Recent studies suggest that GH can stimulate tyrosine phosphorylation of a cytoplasmic STAT protein (signal transducer/activator of transcription) that is related to the STAT proteins involved in the transcriptional responses which are stimulated by growth factors and cytokines such as interferon-γ, epidermal growth factor (EGF), and various interleukins.[100] Tyrosine phosphorylation of STAT proteins is associated with nuclear translocation and activation of a latent DNA binding activity, resulting in the transcriptional activation of target genes (Fig. 6). Further studies are required to ascertain whether STAT proteins participate in the transcriptional activation of CYP2C11, 2C12, and other GH-regulated liver P450s.

2. Another early event induced by GH involves phospholipase C-catalyzed diacyl-glycerol formation, via an inositol phosphate-*independent* pathway, which leads to the activation of protein kinase C.[101,102] Activated protein kinase C may, in turn, trigger multiple secondary events, including internalization of GHR from the cell surface to a Golgi fraction[86] and to the cell nucleus,[103,104] leading to downregulation of GHR at the cell surface.[105] These findings may help explain several key observations, including: (a) the stimulation of liver GHR internalization each 4 hr by plasma GH pulses in adult male rats[82,86]; (b) the finding that a well-defined minimum recovery period is required for hepatocytes to respond to a succeeding pulse of GH with the expression of CYP2C11[61];

and (c) the observed loss of GH responsiveness, including responsiveness to 2C12 induction, in cells chronically treated with activators of protein kinase C.[106,107]

3. Enhanced transcription of the c-*fos* gene is a third early response to GH.[107] c-*fos* transcription leads to activation of the lipoprotein lipase gene both in liver cells[108] and in adipocytes[109] and appears to account for the stimulation of lipase gene expression by GH. While the activation of c-*fos* by GH may be mediated by the GH-induced diacylglycerol formation and PKC activation pathway described above,[101] recent studies suggest that c-*fos* activation results from GH stimulation of Jak2-catalyzed tyrosine phosphorylation of STAT1α/p91, a component of a GH-inducible nuclear DNA-binding complex that binds to and directly activates the c-*fos* promoter.[110,111]

While significant advances in our understanding of the intracellular events stimulated by GH have thus been made in adipocytes and other non-hepatic cells, relatively little is known about the signaling pathways stimulated by GH in hepatocytes[63] or their possible role in mediating the sex-specific effects of plasma GH profiles on liver P450 gene expression. Recent studies carried out in our laboratory have examined the occurrence of GH signaling pathways in liver cells, with particular emphasis on any differential effects that intermittent versus continuous GH may have on intracellular signaling via tyrosine phosphorylation of nuclear factors that could contribute to sex-specific P450 gene transcription. To investigate the influence of plasma GH secretory patterns on intracellular hepatocyte signaling, the patterns of liver nuclear protein tyrosine phosphorylation were compared in male and female rats. An Mr ~93,000 polypeptide that is related to the prolactin-activated mammary Stat 5 was thus found to be tyrosine phosphorylated to a high level in male but not female rat liver.[111a] GH, but not prolactin, rapidly stimulated tyrosine phosphorylation of this protein, designated liver Stat 5, when given to hypophysectomized rats. Intermittent plasma GH pulses triggered repeated liver Stat 5 phosphorylation, while continuous GH exposure led to desensitization and a dramatic decline in nuclear levels of this GH-activated liver transcription factor. Intermittent GH pulsation also stimulated translocation of liver Stat 5 protein from the cytosol to the nucleus and activated its specific DNA-binding activity toward a mammary Stat 5-binding DNA sequence derived from the rat β-casein gene. Liver Stat 5 is thus a GH-activated DNA-binding protein that undergoes tyrosine phosphorylation and nuclear translocation in response to intermittent plasma GH stimulation, and may serve as an intracellular mediator of the stimulatory effects of GH pulses on male-specific liver P450 gene expression.[111a]

The broad range of physiological and metabolic effects that GH has on target tissues,[11] and in particular the occurrence of sex-dependent effects of GH secretory profiles on long bone growth and liver gene expression, suggest that GH is likely to activate several independent, or perhaps parallel signaling pathways, even within a single cell type. This idea is supported by the finding that GH can activate multiple Stat proteins in hepatocytes; these include liver Stat 5,[111a] as well as Stat 3 and Stat 1 (Ram P.A., Park S.-H., Choi H. K. and Waxman D. J., unpublished). It seems likely that these three GH-responsive Stat proteins will contribute to the activation of different subsets of GH-responsive liver genes, and conceivably may respond in a differential manner to the temporal pattern of circulating plasma GH levels. This would help explain the broad diversity of GH's effects, as well as the restriction of GH's sex and plasma hormone profile-dependent effects to a subset of

GH-activatable P450 genes. For instance, the rapid activation of the c-*fos* gene by GH[101,107] could be mediated in part by the binding of GH-activated Stat 3 and/or Stat 1 to the Stat-binding DNA sequence (SIE site) found within a regulatory element upstream of that gene, while activation of the liver cytochrome P450 gene CYP 2C11 by intermittent GH pulsation[74,75] could be mediated by the binding of GH-activated liver Stat 5 to cognate regulatory elements within or adjacent to 2C11. However, since the stimulatory effects of GH pulses on 2C11 gene expression may take a full 1 to 2 days to be manifest,[6] as compared to the activation of liver Stat 5 within 15 min of GH treatment,[111a] the induction of 2C11 gene transcription by GH pulsation may be an *indirect* response to liver Stat 5 activation. Further studies are required to determine the molecular mechanisms through which liver Stat 5 activates target genes in response to GH pulses, as well as the cellular mechanism through which hepatocytes exposed to continuous GH in female rats become desensitized with respect to liver Stat 5 activation.

3.3. Thyroid Hormone

3.3.1. Thyroid Regulation of Cytochrome P450

Although GH is the major regulator of specific liver P450s, thyroid hormone also plays a critical role. The major thyroid hormones, T3 and T4, positively regulate some[13,39] but not all[40] of the female-predominant liver P450 enzymes, while they negatively regulate several of the male-specific enzymes[12,14] (Table I). These effects of thyroid hormone are operative at the level of mRNA expression, and are independent of the indirect effects that thyroid hormone has on liver P450 levels as a consequence of its effects on liver GH receptors[112] and its stimulation of GH gene transcription and GH secretion by the pituitary.[113] Molecular studies of these effects of thyroid hormone have not been carried out.

3.3.2. Thyroid Regulation of NADPH-Cytochrome P450 Reductase

Thyroid hormone is also required for full expression of NADPH-cytochrome P450 reductase, a flavoenzyme that catalyzes electron transfer to all microsomal P450s. P450 reductase is an obligatory, and often rate-limiting electron-transfer protein that participates in all microsomal P450-catalyzed drug oxidation and steroid hydroxylase reactions.[114,115] This thyroid hormone dependence of P450 reductase enzyme expression is evidenced by the major decrease (>80% reduction) in liver microsomal P450 reductase activity and P450 reductase mRNA levels that occurs following hypophysectomy[69] or in response to methimazole-induced hypothyroidism.[116] It is further supported by the reversal of this activity loss when thyroxine (T4), but not GH or other pituitary-dependent factors, is given at a physiologic replacement dose.[69,116] Restoration of liver P450 reductase activity *in vivo* by T4 replacement also effects a substantial increase in liver microsomal P450 steroid hydroxylase activities. A similar effect can be achieved when liver microsomes isolated from hypophysectomized rats are supplemented with exogenous, purified P450 reductase, which preferentially stimulates steroid hydroxylation catalyzed by microsomes prepared from thyroid-deficient animals.[69] P450 reductase levels are also modulated by thyroid hormone status in several extrahepatic tissues.[116] Conceivably, interindividual differences in P450 reductase levels may occur in response to physiological or pathophysiological differences in circulating thyroid hormone levels and could be an important contributory factor to individual differences in cytochrome P450 reductase/cytochrome P450-catalyzed carcinogen metabolism and carcinogen activation reactions.

4. Alteration of Liver P450 Expression by Hormonal Perturbation

As described in this chapter, many, although not all, liver P450s are under hormone regulatory controls. An individual's circulating hormonal profile can, however, be altered under certain situations, including drug therapy, exposure to chemicals found in the environment, and disease states such as diabetes and liver cirrhosis. The resultant changes in circulating hormone levels or alterations in hormone secretory dynamics could, therefore, influence the expression of specific liver P450s. The following sections describe some of the factors that are known to cause hormonal perturbation and discuss the impact of these changes on liver P450 expression and on P450-dependent drug and xenobiotic metabolism and toxicity.

4.1. Modulation by Drugs and Other Foreign Chemicals

The anticancer drugs cisplatin,[117,118] cyclophosphamide,[119,120] and ifosphamide[120] alter the profile of P450 enzyme expression in liver and perhaps other tissues, at least in part related to the hormonal perturbations that these cytotoxic agents induce. Treatment of adult male rats with a single dose of cisplatin depletes serum androgen, and this effect is persistent for up to 28 days after drug administration.[117] Serum androgen depletion by cisplatin is associated with a feminization of hepatic liver enzyme expression. Thus, cisplatin-treated male rats have elevated levels of the female-predominant CYP2A1, 2C7, and steroid 5α-reductase, but have reduced levels of the male-specific 2A2, 2C11, and 3A2.[117,118] These effects of cisplatin may be related in part to the drug's action on the testes[121,122]; however, effects on the hypothalamus are also suggested to contribute to the observed depletion of circulating testosterone and the resultant alteration in liver P450 expression.[117] Consistent with this hypothesis, cisplatin treatment of adult female rats severely decreases circulating estradiol levels and significantly reduces the expression of the estrogen-dependent 2A21, 2C7, and 2C12.[118]

Serum testosterone depletion also occurs in cyclophosphamide- and ifosphamide-treated adult male rats and this depletion is associated with feminization of liver enzyme profiles[119,120] in a manner similar to that produced by cisplatin. While endogenous androgen secretion in cyclophosphamide-treated rats can be stimulated by the luteinizing hormone analogue chorionic gonadotropin, the resultant increase in serum testosterone does not reverse the loss of hepatic CYP2C11 expression.[119] This observation is analogous to the earlier finding that the suppression of 2C11 by 3,4,5,3',4',5'-hexachlorobiphenyl[123] is not causally related to the associated depletion of serum testosterone.[124] Consequently, modulation of liver enzyme expression by cyclophosphamide may involve action at the hypothalamic–pituitary axis, which establishes the sex-dependent plasma GH profile that in turn dictates the expression of 2C11 and other sex-dependent liver P450 enzymes, as discussed earlier in this chapter. That this suppression of 2C11 is not obligatorily dependent on serum androgen depletion, can be demonstrated by the finding that the anticancer drug

1-(2-chloroethyl)-3-cyclohexyl-1-nitrosourea (CCNU; lomustine) substantially suppresses 2C11 in the absence of a significant decrease in circulating testosterone levels.[125] The mechanism by which CCNU suppresses 2C11 is not known, but the drug may act directly at the hypothalamic–pituitary axis to influence the signaling elements in the ultradian rhythm of circulating GH.

Several drugs and other foreign chemicals that serve as prototypic inducers of liver P450 enzymes have been shown to suppress CYP2C11 expression in liver. Adult male rats administered the 2E1 inducer ethanol by a total enteral nutrition system have reduced hepatic 2C11 and 3A2 levels, whereas their 2A1 activity is unaltered.[126] The same ethanol treatment alters the dynamics of plasma GH secretion by decreasing the GH pulse amplitude and increasing the GH pulse frequency. The increased frequency of GH pulses thus explains the reduced expression of 2C11 after chronic ethanol intake because hepatocytes require a minimum "off time" in order to express the male pattern of GH secretion that stimulates 2C11 expression.[61] Other P450 inducers such as 3-methylcholanthrene,[123,127] 2,3,7,8-tetrachlorodibenzo-p-dioxin,[128] several polybrominated biphenyl congeners,[129] phenobarbital,[7,127,130] and dexamethasone[131] also decrease hepatic 2C11 protein or activity levels in adult male rats. The mechanism(s) by which these drugs and xenobiotics modulate 2C11 expression is not known, but 3-methylcholanthrene,[123] 2,3,7,8-tetrachlorodibenzo-p-dioxin,[132] and dexamethasone[133] are each known to decrease serum testosterone levels. Interestingly, the Ah receptor, which is a key mediator of CYP1A induction, may play a role in hepatic 2C11 suppression by polycyclic aromatic hydrocarbons as suggested by the observation that compounds that have higher affinities for the Ah receptor are more effective in decreasing 2C11 levels.[129] Further investigations will be required to elucidate the various mechanisms by which P450 inducers downregulate the expression of 2C11 and other sex-dependent rat hepatic P450 enzymes.

4.2. Pathophysiologic State

4.2.1. Diabetes

Uncontrolled insulin-dependent diabetes is not only accompanied by defective carbohydrate metabolism, which results in hyperglycemia, hyperlipidemia, and hyperketonemia, but is also associated with hormonal perturbation, including a reduction in circulating testosterone,[134–136] thyroid hormone, and plasma GH.[137,138] As described earlier in this chapter, these hormones regulate either directly or indirectly many liver P450 enzymes. Accordingly, the diabetic state is associated with profound changes in the levels of various hepatic P450 enzymes. Whereas diabetes leads to induction of several rat liver P450 forms, including CYP2A1,[138,139] 2B1,[140,141] 2C7,[139] 2E1,[142–145] 4A2,[139] and 4A3,[139] it suppresses 2A2,[138] 2C11,[138,140,141,145] and 2C13.[138] Changes in the levels of some of these liver P450s (e.g., 2C11 and 2E1) have also been shown at the mRNA level,[140,141,146,147] indicating that modulation occurs at a pretranslational step. These alterations in P450 protein or P450 mRNA levels in diabetes are reversible by insulin.

In the diabetic male rat, the profile of GH secretion is altered in such a manner that it resembles more closely the pattern found in the normal female rat.[137] The induction of CYP2A1 and 2C7 in diabetic male rats can therefore be explained, at least in part, as a response to the more continuous pattern of GH secretion, which stimulates these P450 forms.[13,31,54,148] In contrast, this pattern of GH secretion reduces 2A2 and 2C13 levels because continuous GH administration suppresses these two P450s.[31,33,149] 2C11 expression is obligatorily dependent on the intermittent male pattern of plasma GH secretion.[61] Therefore, the more continuous secretion of GH in diabetic male rats[137] would be expected to suppress this P450. In the case of 2B1, GH pulse height is the suppressive signal[70] and accordingly, the reduction in GH peak concentration in diabetic male rats[138] leads to increases in 2B1 levels.[140,141] However, a GH-independent mechanism is likely to contribute to the observed diabetes-induced alterations in specific liver P450 expression. GH, independent of its plasma profile, is suppressive toward hepatic 2E1,[53] but the levels of this P450 are substantially elevated in both diabetic male and female rats.[138,150,151] 2E1 induction in diabetes has been attributed to increased plasma concentrations of ketone bodies.[143,152]

4.2.2. Liver Cirrhosis

Gonadal abnormalities occur in liver cirrhosis. Adult male rats fed a chronic choline-deficient diet to induce cirrhosis have enhanced serum estradiol concentrations[153] and reduced testicular weight[154] and serum testosterone levels.[153] In association with the perturbation in hormonal status is a major decline in hepatic CYP2C11 content,[153] and this decline is not accompanied by induction of steroid 5α-reductase activity.[155] Whether the alteration in serum steroid hormone levels is directly or indirectly responsible for the apparent demasculinization of liver P450 remains to be determined.

4.3. Impact on Drug Metabolism and Procarcinogen Activation

As discussed in Section 4.1, the anticancer drug cisplatin provides an example of a foreign compound that depletes serum testosterone and consequently feminizes the expression of liver P450s in adult male rats. This type of alteration in the profile of liver P450 enzymes could have important pharmacological consequences, as suggested by the finding that cisplatin suppression of CYP2C11 decreases liver P450-catalyzed activation of anticancer prodrugs, such as cyclophosphamide[118,156,157] and ifosphamide[157] when assayed in isolated microsomal systems. Liver P450 activation of these latter two drugs is required for their anticancer drug activity,[158] and 2C11 contributes significantly to this metabolic pathway in male rat liver.[156,157] Clinical studies indicate that cisplatin may exert effects on circulating hormone levels in human cancer patients that are similar, though not identical, to those seen in rats.[159] If these hormone perturbations in turn alter P450 levels in human liver, this could have important implications regarding drug–drug interactions when cisplatin is administered in combination chemotherapy with agents such as cyclophosphamide. As discussed above, drug metabolism in diabetic rats can be altered as a consequence of perturbations of GH secretory patterns and the resultant changes in the expression of multiple liver P450 enzymes. Accordingly, diabetes is associated with a decrease in P450-mediated *in vitro* hepatic metabolism of imipramine,[136,160] lidocaine,[136] codeine,[161] and chlorpromazine.[161] In addition, alteration of liver P450 expression in diabetes is postulated to be responsible for the enhanced *in vitro* metabolic activation of certain chemical carcinogens, including Try-P-1 (3-amino-1,4-dimethyl-5*H*-pyrido[4,5-*b*]indole) and Try-P-2 (3-amino-1-methyl-5*H*-pyrido[4,3-*b*]indole).[162] These examples demonstrate the potential for alterations in liver P450 expression that potentially lead to reduced drug metabolism and enhanced procarcinogen activation. Further investigations will be necessary to determine the extent to which these events occur in humans and to evaluate their true pharmacological and toxicological significance.

5. Conclusion

Multiple liver P450 enzymes are subject to complex and distinct hormonal regulatory controls. Pituitary GH secretory patterns and circulating thyroid hormone levels are the most important endocrine regulators. GH regulates the sex-specific expression of liver P450s and their associated steroid hydroxylation and xenobiotic metabolism activities through transcriptional mechanisms, while thyroid hormone acts directly to influence the expression of individual P450 enzymes as well as indirectly via effects on pituitary GH secretion and NADPH-cytochrome P450 reductase gene expression. Gonadal hormones also modulate the expression of these P450s, but the effects are largely indirect and are mediated by the hypothalamic–pituitary axis and its control of plasma GH levels. Recent studies with human subjects have revealed sex differences in GH secretory dynamics,[64] which can be modulated by gonadal hormones.[163,164] This sexually dimorphic GH secretion could conceivably contribute to the apparent sex differences in the levels of human hepatic CYP1A2[165] and 3A activities[166] and in the extent of hepatic P450-catalyzed biotransformation of multiple therapeutic agents.[167–169] Factors such as drugs and other foreign chemicals as well as pathophysiologic states, including diabetes and liver cirrhosis, can affect hormone secretion, and this in turn may impact on the expression of individual liver P450s and their potential for drug metabolism and procarcinogen activation.

ACKNOWLEDGMENTS.   Studies carried out in D.J.W.'s laboratory were supported in part by the National Institutes of Health (Grant DK33765) and the American Cancer Society (Grant CN-14).

References

1. Zimniak, P., and Waxman, D. J., 1993, P450 metabolism of endogenous steroid hormone, bile acid and fatty acid substrates, in: *Handbook of Experimental Pharmacology, Cytochrome P450* (J. B. Schenkman and H. Greim, eds.), Springer-Verlag, Berlin, pp. 123–144.
2. Coon, M. J., and Koop, D. R., 1983, P450 oxygenases in lipid transformation, *Enzymes* **16**:645–677.
3. Waxman, D. J., 1988, Interactions of hepatic cytochromes P-450 with steroid hormones. Regioselectivity and stereospecificity of steroid metabolism and hormonal regulation of rat P-450 enzyme expression, *Biochem. Pharmacol.* **37**:71–84.
4. Nelson, D. R., Kamataki, T., Waxman, D. J., Guengerich, F. P., Estabrook, R. W., Feyereisen, R., Gonzalez, F. J., Coon, M. J., Gunsalus, I. C., Gotoh, O., Okuda, K., and Nebert, D. W., 1993, The P450 superfamily: Update on new sequences, gene mapping, accession numbers, early trivial names of enzymes, and nomenclature, *DNA Cell Biol.* **12**:1–51.
5. Waxman, D. J., 1992, Regulation of liver-specific steroid metabolizing cytochromes P450: Cholesterol 7α-hydroxylase, bile acid 6β-hydroxylase, and growth hormone-responsive steroid hormone hydroxylases, *J. Steroid Biochem. Mol. Biol.* **43**:1055–1072.

6. Zaphiropoulos, P. G., Mode, A., Norstedt, G., and Gustafsson, J. A., 1989, Regulation of sexual differentiation in drug and steroid metabolism, *Trends Pharmacol. Sci.* **10**:149–153.

7. Waxman, D. J., 1984, Rat hepatic cytochrome P-450 isoenzyme 2c. Identification as a male-specific, developmentally induced steroid 16 α-hydroxylase and comparison to a female-specific cytochrome P-450 isoenzyme, *J. Biol. Chem.* **259**:15481–15490.

8. Morgan, E. T., MacGeoch, C., and Gustafsson, J.-A., 1985, Hormonal and developmental regulation of expression of the hepatic microsomal steroid 16α-hydroxylase cytochrome P-450 apoprotein in the rat, *J. Biol. Chem.* **260**:11895–11898.

9. Waxman, D. J., Dannan, G. A., and Guengerich, F. P., 1985, Regulation of rat hepatic cytochrome P-450: Age-dependent expression, hormonal imprinting, and xenobiotic inducibility of sex-specific isoenzymes, *Biochemistry* **24**:4409–4417.

10. MacGeoch, C., Morgan, E. T., and Gustafsson, J. A., 1985, Hypothalamo-pituitary regulation of cytochrome P-450(15) β apoprotein levels in rat liver, *Endocrinology* **117**:2085–2092.

11. Jansson, J.-O., Ekberg, S., and Isaksson, O., 1985, Sexual dimorphism in the control of growth hormone secretion, *Endocrine Rev.* **6**:128–150.

12. Waxman, D. J., Ram, P. A., Notani, G., LeBlanc, G. A., Alberta, J. A., Morrissey, J. J., and Sundseth, S. S., 1990, Pituitary regulation of the male-specific steroid 6 β-hydroxylase P-450 2a (gene product IIIA2) in adult rat liver. Suppressive influence of growth hormone and thyroxine acting at a pretranslational level, *Mol. Endocrinol.* **4**:447–454.

13. Ram, P. A., and Waxman, D. J., 1990, Pretranslational control by thyroid hormone of rat liver steroid 5 α-reductase and comparison to the thyroid dependence of two growth hormone-regulated CYP2C mRNAs, *J. Biol. Chem.* **265**:19223–19229.

14. Ram, P. A., and Waxman, D. J., 1991, Hepatic P450 expression in hypothyroid rats: Differential responsiveness of male-specific P450 forms 2a (IIIA2), 2c (IIC11), and RLM2 (IIA2) to thyroid hormone, *Mol. Endocrinol.* **5**:13–20.

15. Noshiro, M., and Negishi, M., 1986, Pretranslational regulation of sex-dependent testosterone hydroxylases by growth hormone in mouse liver, *J. Biol. Chem.* **261**:15923–15927.

16. Squires, E. J., and Negishi, M., 1988, Reciprocal regulation of sex-dependent expression of testosterone 15 α-hydroxylase (P-450(15 α)) in liver and kidney of male mice by androgen. Evidence for a single gene, *J. Biol. Chem.* **263**:4166–4171.

17. Harada, N., and Negishi, M., 1988, Substrate specificities of cytochrome P-450, C-P-450(16)α and P-450(15)α, and contribution to steroid hydroxylase activities in mouse liver microsomes, *Biochem. Pharmacol.* **37**:4778–4780.

18. Aida, K., and Negishi, M., 1993, A trans-acting locus regulates transcriptional repression of the female-specific steroid 15 α-hydroxylase gene in male mice, *J. Mol. Endocrinol.* **11**:213–222.

19. Srivastava, P. K., and Waxman, D. J., 1993, Sex-dependent expression and growth hormone regulation of class α and class μ glutathione-S-transferases in adult rat liver, *Biochem. J.* **294**:159–165.

20. Jeffery, S., Carter, N. D., Clark, R. G., and Robinson, I. C. A., 1990, The episodic secretory pattern of growth hormone regulates liver carbonic anhydrase III, *Biochem. J.* **266**:69–74.

21. Robertson, J. A., Haldosen, L. A., Wood, T. J., Steed, M. K., and Gustafsson, J. A., 1990, Growth hormone pretranslationally regulates the sexually dimorphic expression of the prolactin receptor gene in rat liver, *Mol. Endocrinol.* **4**:1235–1239.

22. Rudling, M., Norstedt, G., Olivecrona, H., Reihner, E., Gustafsson, J. A., and Angelin, B., 1992, Importance of growth hormone for the induction of hepatic low density lipoprotein receptors, *Proc. Natl. Acad. Sci. USA* **89**:6983–6987.

23. Watkins, P. B., Murray, S. A., Winkelman, L. G., Heuman, D. M., Wrighton, S. A., and Guzelian, P. S., 1989, Erythromycin breath test as an assay of glucocorticoid-inducible liver cytochromes P-450. Studies in rats and patients, *J. Clin. Invest.* **83**:688–697.

24. Levitsky, L. L., Schoeller, D. A., Lambert, G. H., and Edidin, D. V., 1989, Effect of growth hormone therapy in growth hormone-deficient children on cytochrome P-450-dependent 3-N-demethylation of caffeine as measured by the caffeine $^{13}CO_2$ breath test, *Dev. Pharmacol. Ther.* **12**:90–95.

25. Redmond, G. P., Bell, J. J., Nichola, P. S., and Perel, J. M., 1980, Effect of growth hormone on human drug metabolism: Time course and substrate specificity, *Pediatr. Pharmacol.* **1**:63–70.

26. Guengerich, F. P., 1987, Enzymology of rat liver cytochromes P450, in: *Mammalian Cytochromes P450* (F. P. Guengerich, ed.), CRC Press, Boca Raton, FL, pp. 1–54.

27. Colby, H. D., 1980, Regulation of hepatic drug and steroid metabolism by androgens and estrogens, *Adv. Sex Horm. Res.* **4**:27–71.

28. Kato, R., 1974, Sex-related differences in drug metabolism, *Drug Metab. Rev.* **3**:1–32.

29. Skett, P., 1987, Hormonal regulation and sex differences of xenobiotic metabolism, *Prog. Drug Metab.* **10**:85–139.

30. Thummel, K. E., Favreau, L. V., Mole, J. E., and Schenkman, J. B., 1988, Further characterization of RLM2 and comparison with a related form of cytochrome P450, RLM2b, *Arch. Biochem. Biophys.* **266**:319–333.

31. Waxman, D. J., LeBlanc, G. A., Morrissey, J. J., Staunton, J., and Lapenson, D. P., 1988, Adult male-specific and neonatally programmed rat hepatic P-450 forms RLM2 and 2a are not dependent on pulsatile plasma growth hormone for expression, *J. Biol. Chem.* **263**:11396–11406.

32. Bandiera, S., Ryan, D. E., Levin, W., and Thomas, P. E., 1986, Age- and sex-related expression of cytochromes p450f and P450g in rat liver, *Arch. Biochem. Biophys.* **248**:658–676.

33. McClellan, G. P., Linko, P., Yeowell, H. N., and Goldstein, J. A., 1989, Hormonal regulation of male-specific rat hepatic cytochrome P-450g (P-450IIC13) by androgens and the pituitary, *J. Biol. Chem.* **264**:18960–18965.

34. Gonzalez, F. J., Song, B. J., and Hardwick, J. P., 1986, Pregnenolone 16 α-carbonitrile-inducible P-450 gene family: Gene conversion and differential regulation, *Mol. Cell. Biol.* **6**:2969–2976.

35. Sonderfan, A. J., Arlotto, M. P., Dutton, D. R., McMillen, S. K., and Parkinson, A., 1987, Regulation of testosterone hydroxylation by rat liver microsomal cytochrome P-450, *Arch Biochem. Biophys.* **255**:27–41.

36. Yamazoe, Y., Murayama, N., Shimada, M., Yamauchi, K., Nagata, K., Imaoka, S., Funae, Y., and Kato, R., 1988, A sex-specific form of cytochrome P-450 catalyzing propoxycoumarin O-depropylation and its identity with testosterone 6 β-hydroxylase in untreated rat livers: Reconstitution of the activity with microsomal lipids, *J. Biochem.* **104**:785–790.

37. Gonzalez, F. J., Kimura, S., Song, B. J., Pastewka, J., Gelboin, H. V., and Hardwick, J. P., 1986, Sequence of two related P-450 mRNAs transcriptionally increased during rat development. An R.dre.1 sequence occupies the complete 3' untranslated region of a liver mRNA, *J. Biol. Chem.* **261**:10667–10672.

38. Leo, M. A., Iida, S., and Lieber, C. S., 1984, Retinoic acid metabolism by a system reconstituted with cytochrome P-450, *Arch. Biochem. Biophys.* **234**:305–312.

39. Arlotto, M. P., and Parkinson, A., 1989, Identification of cytochrome P450a (P450IIA1) as the principal testosterone 7 α-hydroxylase in rat liver microsomes and its regulation by thyroid hormones, *Arch. Biochem. Biophys.* **270**:458–471.

40. Yamazoe, Y., Ling, X., Murayama, N., Gong, D., Nagata, K., and Kato, R., 1990, Modulation of hepatic level of microsomal testosterone 7 α-hydroxylase, P-450a (P450IIA), by thyroid hormone and growth hormone in rat liver, *J. Biochem.* **108**:599–603.

41. de Waziers, I., Cugnenc, P. H., Yang, C. S., Leroux, J. P., and Beaune, P. H., 1990, Cytochrome P 450 isoenzymes, epoxide hydrolase and glutathione transferases in rat and human hepatic and extrahepatic tissues, *J. Pharmacol. Exp. Ther.* **253**:387–394.

42. Friedberg, T., Siegert, P., Grassow, M. A., Bartlomowicz, B., and Oesch, F., 1990, Studies of the expression of the cytochrome P450IA, P450IIB, and P450IIC gene family in extrahepatic and hepatic tissues, *Environ. Health Perspect.* **88**:67–70.

43. Fujita, S., Morimoto, R., Chiba, M., Kitani, K., and Suzuki, T., 1989, Evaluation of the involvement of a male specific cytochrome P-450 isozyme in senescence-associated decline of hepatic drug metabolism in male rats, *Biochem. Pharmacol.* **38**:3925–3931.

44. Robinson, R. C., Nagata, K., Gelboin, H. V., Rifkind, J., Gonzalez, F. J., and Friedman, F. K., 1990, Developmental regulation of hepatic testosterone hydroxylases: Simultaneous activation and repression of constitutively expressed cytochromes P450 in senescent rats, *Arch. Biochem. Biophys.* **277:**42–46.

45. Dannan, G. A., Guengerich, F. P., and Waxman, D. J., 1986, Hormonal regulation of rat liver microsomal enzymes. Role of gonadal steroids in programming, maintenance, and suppression of Δ^4-steroid 5 α-reductase, flavin-containing monooxygenase, and sex-specific cytochromes P-450, *J. Biol. Chem.* **261:**10728–10735.

46. Shimada, M., Murayama, N., Yamazoe, Y., Kamataki, T., and Kato, R., 1987, Further studies on the persistence of neonatal androgen imprinting on sex-specific cytochrome P-450, testosterone and drug oxidations, *Jpn. J. Pharmacol.* **45:**467–478.

47. Janeczko, R., Waxman, D. J., LeBlanc, G. A., Morville, A., and Adesnik, M., 1990, Hormonal regulation of levels of the messenger RNA encoding hepatic P450 2c (IIC11), a constitutive male-specific form of cytochrome P450, *Mol. Endocrinol.* **4:**295–303.

48. Einarsson, K., Gustafsson, J. A., and Stenberg, A., 1973, Neonatal imprinting of liver microsomal hydroxylation and reduction of steroids, *J. Biol. Chem.* **248:**4987–4997.

49. Gustafsson, J. A., Mode, A., Norstedt, G., and Skett, P., 1983, Sex steroid induced changes in hepatic enzymes, *Annu. Rev. Physiol.* **45:**51–60.

50. Bandiera, S., and Dworschak, C., 1992, Effects of testosterone and estrogen on hepatic levels of cytochromes P450 2C7 and P450 2C11 in the rat, *Arch. Biochem. Biophys.* **296:**286–295.

51. Cadario, B. J., Bellward, G. D., Bandiera, S., Chang, T. K. H., Ko, W. W. W., Lemieux, E., and Pak, R. C. K., 1992, Imprinting of hepatic microsomal cytochrome P-450 enzyme activities and cytochrome P-450IIC11 by peripubertal administration of testosterone in female rats, *Mol. Pharmacol.* **41:**981–988.

52. Ribeiro, V., and Lechner, M. C., 1992, Cloning and characterization of a novel CYP3A1 allelic variant: Analysis of CYP3A1 and CYP3A2 sex-hormone-dependent expression reveals that the CYP3A2 gene is regulated by testosterone, *Arch. Biochem. Biophys.* **293:**147–152.

53. Waxman, D. J., Morrissey, J. J., and LeBlanc, G. A., 1989, Female-predominant rat hepatic P-450 forms j (IIE1) and 3 (IIA1) are under hormonal regulatory controls distinct from those of the sex-specific P-450 forms, *Endocrinology* **124:**2954–2966.

54. Sasamura, H., Nagata, K., Yamazoe, Y., Shimada, M., Saruta, T., and Kato, R., 1990, Effect of growth hormone on rat hepatic cytochrome P-450f mRNA: A new mode of regulation, *Mol. Cell. Endocrinol.* **68:**53–60.

55. Mode, A., and Norstedt, G., 1982, Effects of gonadal steroid hormones on the hypothalamo–pituitary–liver axis in the control of sex differences in hepatic steroid metabolism in the rat, *J. Endocrinol.* **95:**181–187.

56. Jansson, J. O., and Frohman, L. A., 1987, Differential effects of neonatal and adult androgen exposure on the growth hormone secretory pattern in male rats, *Endocrinology* **120:**1551–1557.

57. Jansson, J. O., and Frohman, L. A., 1987, Inhibitory effect of the ovaries on neonatal androgen imprinting of growth hormone secretion in female rats, *Endocrinology* **121:**1417–1423.

58. Kamataki, T., Shimada, M., Maeda, K., and Kato, R., 1985, Pituitary regulation of sex-specific forms of cytochrome P-450 in liver microsomes of rats, *Biochem. Biophys. Res. Commun.* **130:**1247–1253.

59. LeBlanc, G. A., and Waxman, D. J., 1990, Regulation and ligand-binding specificities of two sex-specific bile acid-binding proteins of rat liver cytosol, *J. Biol. Chem.* **265:**5654–5661.

60. Carlsson, L., Eriksson, E., Seeman, H., and Jansson, J. O., 1987, Oestradiol increases baseline growth hormone levels in the male rat: Possible direct action on the pituitary, *Acta Physiol. Scand.* **129:**393–399.

61. Waxman, D. J., Pampori, N. A., Ram, P. A., Agrawal, A. K., and Shapiro, B. H., 1991, Interpulse interval in circulating growth hormone patterns regulates sexually dimorphic expression of hepatic cytochrome P450, *Proc. Natl. Acad. Sci. USA* **88:**6868–6872.

62. Asplin, C. M., Faria, A. C. S., Carlsen, E. C., Vaccaro, V. A., Barr, R. E., Iranmanesh, A., Lee, M. M., Veldhuis, J. D., and Evans, W. S., 1989, Alterations in the pulsatile mode of growth hormone release in men and women with insulin-dependent diabetes mellitus, *J. Clin. Endocrinol. Metab.* **69:**239–245.

63. Johnson, R. M., Napier, M. A., Cronin, M. J., and King, K. L., 1990, Growth hormone stimulates the formation of sn-1,2-diacylglycerol in rat hepatocytes, *Endocrinology* **127:** 2099–2103.

64. Winer, L. M., Shaw, M. A., and Baumann, G., 1990, Basal plasma growth hormone levels in man: New evidence for rhythmicity of growth hormone secretion, *J. Clin. Endocrinol. Metab.* **70:**1678–1686.

65. Eden, S., 1979, Age- and sex-related differences in episodic growth hormone secretion in the rat, *Endocrinology* **105:**555–560.

66. Tannenbaum, G. S., Martin, J. B., and Colle, E., 1976, Ultradian growth hormone rhythm in the rat: Effects of feeding, hyperglycemia, and insulin-induced hypoglycemia, *Endocrinology* **99:**720–727.

67. Tannenbaum, G. S., and Martin, J. B., 1976, Evidence for an endogenous ultradian rhythm governing growth hormone secretion in the rat, *Endocrinology* **98:**562–570.

68. Kato, R., Yamazoe, Y., Shimada, M., Murayama, N., and Kamataki, T., 1986, Effect of growth hormone and ectopic transplantation of pituitary gland on sex-specific forms of cytochrome P450 and testosterone and drug oxidations in rat liver, *J. Biochem.* **100:**895–902.

69. Waxman, D. J., Morrissey, J. J., and LeBlanc, G. A., 1989, Hypophysectomy differentially alters P-450 protein levels and enzyme activities in rat liver: Pituitary control of hepatic NADPH cytochrome P-450 reductase, *Mol. Pharmacol.* **35:**519–525.

70. Shapiro, B. H., Pampori, N. A., Lapenson, D. P., and Waxman, D. J., 1994, Growth hormone-dependent and -independent sexually dimorphic regulation of phenobarbital-induced hepatic cytochromes P450 2B1 and 2B2, *Arch. Biochem. Biophys.* **312:**234–239.

71. Yamazoe, Y., Shimada, M., Murayama, N., and Kato, R., 1987, Suppression of levels of phenobarbital-inducible rat liver cytochrome P-450 by pituitary hormone, *J. Biol. Chem.* **262:**7423–7428.

72. Sundseth, S. S., and Waxman, D. J., 1992, Sex-dependent expression and clofibrate inducibility of cytochrome P450 4A fatty acid ω-hydroxylases. Male specificity of liver and kidney CYP4A2 mRNA and tissue-specific regulation by growth hormone and testosterone, *J. Biol. Chem.* **267:**3915–3921.

73. Tollet, P., Enberg, B., and Mode, A., 1990, Growth hormone (GH) regulation of cytochrome P-450IIC12, insulin-like growth factor-I (IGF-I), and GH receptor messenger RNA expression in primary rat hepatocytes: A hormonal interplay with insulin, IGF-I, and thyroid hormone, *Mol. Endocrinol.* **4:**1934–1942.

74. Sundseth, S. S., Alberta, J. A., and Waxman, D. J., 1992, Sex-specific, growth hormone-regulated transcription of the cytochrome P450 2C11 and 2C12 genes, *J. Biol. Chem.* **267:**3907–3914.

75. Legraverend, C., Mode, A., Westin, S., Strom, A., Eguchi, H., Zaphiropoulos, P. G., and Gustafsson, J.-A., 1992, Transcriptional regulation of rat P-450 2C gene subfamily members by the sexually dimorphic pattern of growth hormone secretion, *Mol. Endocrinol.* **6:**259–266.

76. Morishima, N., Yoshioka, H., Higashi, Y., Sogawa, K., and Fujii-Kuriyama, Y., 1987, Gene structure of cytochrome P450 (M-1) specifically expressed in male rat liver, *Biochemistry* **26:**8279–8285.

77. Zaphiropoulos, P. G., Westin, S., Strom, A., Mode, A., and Gustafsson, J. A., 1990, Structural and regulatory analysis of a cytochrome P450 gene (CYP2C12) expressed predominantly in female rat liver, *DNA Cell Biol.* **9:**49–56.

78. Zhao, S., and Waxman, D. J., 1994, Interaction of sex- and growth hormone (GH)-dependent liver nuclear factors with CYP2C12 promoter, *FASEB J.* **8:**A1250.

79. Strom, A., Eguchi, H., Mode, A., Legraverend, C., Tollet, P., Stromstedt, P. E., and Gustafsson, J., 1994, Characterization of the proximal promoter and two silencer elements in the CYP2C11 gene expressed in rat liver, *DNA Cell Biol.* **13:**805–819.

80. Guzelian, P. S., Li, D., Schuetz, E. G., Thomas, P., Levin, W., Mode, A., and Gustafsson, J. A., 1988, Sex change in cytochrome P-450 phenotype by growth hormone treatment of adult rat hepatocytes maintained in a culture system on matrigel, *Proc. Natl. Acad. Sci. USA* **85:**9783–9787.

81. Baxter, R. C., and Zaltsman, Z., 1984, Induction of hepatic receptors for growth hormone (GH) and prolactin by GH infusion is sex independent, *Endocrinology* **115**:2009–2014.

82. Bick, T., Hochberg, Z., Amit, T., Isaksson, O. G., and Jansson, J. O., 1992, Roles of pulsatility and continuity of growth hormone (GH) administration in the regulation of hepatic GH-receptors, and circulating GH-binding protein and insulin-like growth factor-I, *Endocrinology* **131**:423–429.

83. Shapiro, B. H., MacLeod, J. N., Pampori, N. A., Morrissey, J. J., Lapenson, D. P., and Waxman, D. J., 1989, Signaling elements in the ultradian rhythm of circulating growth hormone regulating expression of sex-dependent forms of hepatic cytochrome P450, *Endocrinology* **125**:2935–2944.

84. Legraverend, C., Mode, A., Wells, T., Robinson, I., and Gustafsson, J. A., 1992, Hepatic steroid hydroxylating enzymes are controlled by the sexually dimorphic pattern of growth hormone secretion in normal and dwarf rats, *FASEB J.* **6**:711–718.

85. Leung, D. W., Spencer, S. A., Cachianes, G., Hammonds, R. G., Collins, C., Henzel, W. J., Barnard, R., Waters, M. J., and Wood, W. I., 1987, Growth hormone receptor and serum binding protein: Purification, cloning and expression, *Nature* **330**:537–543.

86. Bick, T., Youdim, M. B. H., and Hochberg, Z., 1989, Adaptation of liver membrane somatogenic and lactogenic growth hormone (GH) binding to the spontaneous pulsation of GH secretion in the male rat, *Endocrinology* **125**:1711–1717.

87. Bick, T., Youdim, M. B. H., and Hochberg, Z., 1989, The dynamics of somatogenic and lactogenic growth hormone binding: Internalization to Golgi fractions in the male rat, *Endocrinology* **125**:1718–1722.

88. Kelly, P. A., Ali, S., Rozakis, M., Goujon, L., Nagano, M., Pellegrini, I., Gould, D., Djiane, J., Edery, M., Finidori, J., and Pustel-Vinay, M. C., 1993, The growth hormone/prolactin receptor family, *Recent Prog. Horm. Res.* **48**:123–164.

89. Colosi, P., Wong, K., Leong, S. R., and Wood, W. I., 1993, Mutational analysis of the intracellular domain of the human growth hormone receptor, *J. Biol. Chem.* **268**:12617–12623.

90. de Vos, A. M., Ultsch, M., and Kossiakoff, A. A., 1992, Human growth hormone and extracellular domain of its receptor: Crystal structure of the complex, *Science* **255**:306–312.

91. Cunningham, B. C., Ultsch, M., de Vos, A. M., Mulkerrin, M. G., Clauser, K. R., and Wells, J. A., 1991, Dimerization of the extracellular domain of the human growth hormone receptor by a single hormone molecule, *Science* **254**:821–825.

92. Fuh, G., Cunningham, B. C., Fukunaga, R., Nagata, S., Goeddel, D. V., and Wells, J. A., 1992, Rational design of potent antagonists to the human growth hormone receptor, *Science* **256**:1677–1680.

93. Staten, N. R., Byatt, J. C., and Krivi, G. G., 1993, Ligand-specific dimerization of the extracellular domain of the bovine growth hormone receptor, *J. Biol. Chem.* **268**:18467–18473.

94. Wang, X., Moller, C., Norstedt, G., and Carter-Su, C., 1993, Growth hormone-promoted tyrosyl phosphorylation of a 121-kDa growth hormone receptor-associated protein, *J. Biol. Chem.* **268**:3573–3579.

95. Campbell, G. S., Christian, L. J., and Carter-Su, C., 1993, Evidence for involvement of the growth hormone receptor-associated tyrosine kinase in actions of growth hormone, *J. Biol. Chem.* **268**:7427–7434.

96. Argetsinger, L. S., Campbell, G. S., Yang, X., Witthuhn, B. A., Silvennoinen, O., Ihle, J. N., and Carter-Su, C., 1993, Identification of JAK2 as a growth hormone receptor-associated tyrosine kinase, *Cell* **74**:237–244.

97. Silvennoinen, O., Witthuhn, B. A., Quelle, F. W., Cleveland, J. L., Yi, T., and Ihle, J. N., 1993, Structure of the murine Jak2 protein-tyrosine kinase and its role in interleukin 3 signal transduction, *Proc. Natl. Acad. Sci. USA* **90**:8429–8433.

98. Moller, C., Hansson, A., Enberg, B., Lobie, P. E., and Norstedt, G., 1992, Growth hormone (GH) induction of tyrosine phosphorylation and activation of mitogen-activated protein kinases in cells transfected with rat GH receptor cDNA, *J. Biol. Chem.* **267**:23403–23408.

99. Campbell, G. S., Pang, L., Miyasaka, T., Saltiel, A. R., and Carter-Su, C., 1992, Stimulation by growth hormone of MAP kinase activity in 3T3-F442A fibroblasts, *J. Biol. Chem.* **267**:6074–6080.

100. Darnell, J. E., Jr., Kerr, I. M., and Stark, G. R., 1994, Jak-STAT pathways and transcriptional activation in response to IFNs and other extracellular signaling proteins, *Science* **264:**1415–1421.

101. Doglio, A., Dani, C., Grimaldi, P., and Ailhaud, G., 1989, Growth hormone stimulates c-fos gene expression by means of protein kinase C without increasing inositol lipid turnover, *Proc. Natl. Acad. Sci. USA* **86:**1148–1152.

102. Catalioto, R. M., Ailhaud, G., and Negrel, R., 1990, Diacylglycerol production induced by growth hormone in Ob1771 preadipocytes arises from phosphatidylcholine breakdown, *Biochem. Biophys. Res. Commun.* **173:**840–848.

103. Lobie, P. E., Barnard, R., and Waters, M. J., 1991, The nuclear growth hormone receptor binding protein. Antigenic and physicochemical characterization, *J. Biol. Chem.* **266:**22645–22652.

104. Lobie, P. E., Mertani, H., Morel, G., Morales-Bustos, O., Norstedt, G., and Waters, M. J., 1994, Receptor-mediated nuclear translocation of growth hormone, *J. Biol. Chem.* **269:**21330–21339.

105. Suzuki, K., Suzuki, S., Saito, Y., Ikebuchi, H., and Terao, T., 1990, Human growth hormone-stimulated growth of human cultured lymphocytes (IM-9) and its inhibition by phorbol diesters through down-regulation of the hormone receptors, *J. Biol. Chem.* **265:**11320–11327.

106. Tollet, P., Legraverend, C., Gustafsson, J. A., and Mode, A., 1991, A role for protein kinases in the growth hormone regulation of cytochrome P4502C12 and insulin-like growth factor-I messenger RNA expression in primary adult rat hepatocytes, *Mol. Endocrinol.* **5:**1351–1358.

107. Gurland, G., Ashcom, G., Cochran, B. H., and Schartz, J., 1990, Rapid events in growth hormone action. Induction of c-fos and c-jun transcription in 3T3-F442A preadipocytes, *Endocrinology* **127:**3187–3195.

108. Francis, S. M., Enerback, S., Moller, C., Enberg, B., and Norstedt, G., 1993, A novel in vitro model for studying signal transduction and gene regulation via the growth hormone receptor, *Mol. Endocrinol.* **7:**972–978.

109. Barcellini-Couget, S., Pradines-Figueres, A., Roux, P., Dani, C., and Ailhaud, G., 1993, The regulation by growth hormone of lipoprotein lipase gene expression is mediated by c-fos protooncogene, *Endocrinology* **132:**53–60.

110. Meyer, D. J., Campbell, G. S., Cochran, B. H., Argetsinger, L. S., Larner, A. C., Finbloom, D. S., Carter-Su, C., and Schwartz, J., 1994, Growth hormone induces a DNA binding factor related to the interferon-stimulated 91-kDa transcription factor, *J. Biol. Chem.* **269:**4701–4704.

111. Gronowski, A. M., and Rotwein, P., 1994, Rapid changes in nuclear protein tyrosine phosphorylation after growth hormone treatment in vivo, *J. Biol. Chem.* **269:**7874–7878.

111a. Waxman, D. J., Ram, P. A., Park S-H, and Choi, H. K., 1995, Intermittent plasma growth hormone triggers tyrosine phosphorylation and nuclear translocation of a liver-expressed, Stat 5-related DNA-binding protein: Proposed role as an intracellular regulator of male-specific liver gene transcription, *J. Biol. Chem.* **270:**13262–13270.

112. Hochberg, Z., Bick, T., and Harel, Z., 1990, Alterations of human growth hormone binding by rat liver membranes during hypo- and hyperthyroidism, *Endocrinology* **126:**325–329.

113. Samuels, H. H., Forman, B. M., Horowitz, Z. D., and Ye, Z. S., 1988, Regulation of gene expression by thyroid hormone, *J. Clin. Invest.* **81:**957–967.

114. Kaminsky, L. S., and Guengerich, F. P., 1985, Cytochrome P450 isozyme/isozyme functional interactions and NADPH-cytochrome P450 reductase concentrations as factors in microsomal metabolism of warfarin., *Eur. J. Biochem.* **149:**479–489.

115. Miwa, G. T., West, S. B., and Lu, A. Y. H., 1978, Studies on the rate-limiting enzyme component in the microsomal monooxygenase system. Incorporation of purified NADPH cytochrome c-reductase and cytochrome P450 into rat liver microsomes, *J. Biol. Chem.* **253:**1921–1929.

116. Ram, P. A., and Waxman, D. J., 1992, Thyroid hormone stimulation of NADPH P450 reductase expression in liver and extrahepatic tissues. Regulation by multiple mechanisms, *J. Biol. Chem.* **267:**3294–3301.

117. LeBlanc, G. A., and Waxman, D. J., 1988, Feminization of rat hepatic P-450 expression by cisplatin. Evidence for perturbations in the hormonal regulation of steroid-metabolizing enzymes, *J. Biol. Chem.* **263:**15732–15739.

118. LeBlanc, G. A., Sundseth, S. S., Weber, G. F., and Waxman, D. J., 1992, Platinum anticancer drugs modulate P-450 mRNA levels and differentially alter hepatic drug and steroid hormone metabolism in male and female rats, *Cancer Res.* **52:**540–547.

119. LeBlanc, G. A., and Waxman, D. J., 1990, Mechanisms of cyclophosphamide action on hepatic P-450 expression, *Cancer Res.* **50:**5720–5726.

120. Chang, T. K. H., and Waxman, D. J., 1993, Cyclophosphamide modulates rat hepatic cytochrome P450 2C11 and steroid 5α-reductase activity and messenger RNA levels through the combined action of acrolein and phosphoramide mustard, *Cancer Res.* **53:**2490–2497.

121. Maines, M. D., and Mayer, R. D., 1985, Inhibition of testicular cytochrome P-450-dependent steroid biosynthesis by cis-platinum, *J. Biol. Chem.* **260:**6063–6068.

122. Maines, M. D., Sluss, P. M., and Iscan, M., 1990, cis-Platinum-mediated decrease in serum testosterone is associated with depression of luteinizing hormone receptors and cytochrome P-450scc in rat testis, *Endocrinology* **126:**2398–2406.

123. Yeowell, H. N., Waxman, D. J., Wadhera, A., and Goldstein, J. A., 1987, Suppression of the constitutive, male-specific rat hepatic cytochrome P-450 2c and its mRNA by 3,4,5,3′,4′,5′-hexachlorobiphenyl and 3-methylcholanthrene, *Mol. Pharmacol.* **32:**340–347.

124. Yeowell, H. N., Waxman, D. J., LeBlanc, G. A., Linko, P., and Goldstein, J. A., 1989, Suppression of male-specific cytochrome P450 2c and its mRNA by 3,4,5,3′,4′,5′-hexachlorobiphenyl in rat liver is not causally related to changes in serum testosterone, *Arch. Biochem. Biophys.* **271:**508–514.

125. Chang, T. K. H., Chen, H., and Waxman, D. J., 1994, 1-(2-chloroethyl)-3-cyclohexyl-1-nitrosourea (CCNU) modulates rat liver microsomal cyclophosphamide and ifosphamide activation by suppressing cytochrome P450 2C11 messenger RNA levels, *Drug Metab. Dispos.* **22:**673–679.

126. Badger, T. M., Ronis, M. J. J., Lumpkin, C. K., Valentine, C. R., Shahare, M., Irby, D., Huang, J., Mercado, C., Thomas, P., Ingelman-Sundberg, M., and Crouch, J., 1993, Effects of chronic ethanol on growth hormone secretion and hepatic cytochrome P450 isozymes of the rat, *J. Pharmacol. Exp. Ther.* **264:**438–447.

127. Shimada, M., Murayama, N., Yamauchi, K., Yamazoe, Y., and Kato, R., 1989, Suppression in the expression of a male-specific cytochrome P450, P450-male: Difference in the effect of chemical inducers on P450-male mRNA and protein in rat livers, *Arch. Biochem. Biophys.* **270:**578–587.

128. Gustafsson, J. A., and Ingelman-Sundberg, M., 1979, Changes in steroid hormone metabolism in rat liver microsomes following administration of 2,3,7,8-tetrachlorodibenzo-p-dioxin (TCDD), *Biochem. Pharmacol.* **28:**497–499.

129. Dannan, G. A., Guengerich, F. P., Kaminsky, L. S., and Aust, S. D., 1983, Regulation of cytochrome P-450. Immunochemical quantitation of eight isozymes in liver microsomes of rats treated with polybrominated biphenyl congeners, *J. Biol. Chem.* **258:**1282–1288.

130. Guengerich, F. P., Dannan, G. A., Wright, S. T., Martin, M. V., and Kaminsky, L. S., 1982, Purification and characterization of liver microsomal cytochromes p-450: Electrophoretic, spectral, catalytic, and immunochemical properties and inducibility of eight isozymes isolated from rats treated with phenobarbital or β-naphthoflavone, *Biochemistry* **21:**6019–6030.

131. Levin, W., Thomas, P. E., Ryan, D. E., and Wood, A. W., 1987, Isozyme specificity of testosterone 7 α-hydroxylation in rat hepatic microsomes: Is cytochrome P-450a the sole catalyst? *Arch. Biochem. Biophys.* **258:**630–635.

132. Moore, R. W., Potter, C. L., Theobald, H. M., Robinson, J. A., and Peterson, R. E., 1985, Androgenic deficiency in male rats treated with 2,3,7,8-tetrachlorodibenzo-p-dioxin, *Toxicol. Appl. Pharmacol.* **79:**99–111.

133. Stahl, F., Gotz, F., and Dorner, G., 1984, Plasma testosterone levels in rats under various conditions, *Exp. Clin. Endocrinol.* **84:**277–284.

134. Murray, F. T., Orth, J., Gunsalus, G., Weisz, J., Li, J. B., Jefferson, L. S., Musto, N. A., and Bardin, C. W., 1981, The pituitary–testicular axis in the streptozotocin diabetic male rat: Evidence for gonadotroph, Sertoli cell and Leydig cell dysfunction, *Int. J. Androl.* **4:**265–280.

135. Warren, B. L., Pak, R., Finlayson, M., Gontovnick, L., Sunahara, G., and Bellward, G. D., 1983, Differential effects of diabetes on microsomal metabolism of various substrates. Comparison of streptozotocin and spontaneously diabetic Wistar rats, *Biochem. Pharmacol.* **32:**327–335.

136. Skett, P., Cochrane, R. A., and Joels, L. A., 1984, The role of androgens in the effect of diabetes mellitus on hepatic drug metabolism in the male rat, *Acta Endocrinol.* **107:**506–512.

137. Tannenbaum, G. S., 1981, Growth hormone secretory dynamics in streptozotocin diabetes: Evidence of a role for endogenous circulating somatostatin, *Endocrinology* **108:**76–82.

138. Thummel, K. E., and Schenkman, J. B., 1990, Effects of testosterone and growth hormone treatment on hepatic microsomal P450 expression in the diabetic rat, *Mol. Pharmacol.* **37:**119–129.

139. Shimojo, N., Ishizaki, T., Imaoka, S., Funae, Y., Fujii, S., and Okuda, K., 1993, Changes in amounts of cytochrome P450 isozymes and levels of catalytic activities in hepatic and renal microsomes of rats with streptozotocin-induced diabetes, *Biochem. Pharmacol.* **46:**621–627.

140. Yamazoe, Y., Murayama, N., Shimada, M., Yamauchi, K., and Kato, R., 1989, Cytochrome P450 in livers of diabetic rats: Regulation by growth hormone and insulin, *Arch. Biochem. Biophys.* **268:**567–575.

141. Donahue, B. S., and Morgan, E. T., 1990, Effects of vanadate on hepatic cytochrome P-450 expression in streptozotocin-diabetic rats, *Drug Metab. Dispos.* **18:**519–526.

142. Favreau, L. V., Malchoff, D. M., Mole, J. E., and Schenkman, J. B., 1987, Responses to insulin by two forms of rat hepatic microsomal cytochrome P-450 that undergo major (RLM6) and minor (RLM5b) elevations in diabetes, *J. Biol. Chem.* **262:**14319–14326.

143. Bellward, G. D., Chang, T., Rodrigues, B., McNeill, J. H., Maines, S., Ryan, D. E., Levin, W., and Thomas, P. E., 1988, Hepatic cytochrome P-450j induction in the spontaneously diabetic BB rat, *Mol. Pharmacol.* **33:**140–143.

144. Dong, Z. G., Hong, J. Y., Ma, Q. A., Li, D. C., Bullock, J., Gonzalez, F. J., Park, S. S., Gelboin, H. V., and Yang, C. S., 1988, Mechanism of induction of cytochrome P-450ac (P-450j) in chemically induced and spontaneously diabetic rats, *Arch. Biochem. Biophys.* **263:**29–35.

145. Ma, Q., Dannan, G. A., Guengerich, F. P., and Yang, C. S., 1989, Similarities and differences in the regulation of hepatic cytochrome P-450 enzymes by diabetes and fasting in male rats, *Biochem. Pharmacol.* **38:**3179–3184.

146. Song, B. J., Matsunaga, T., Hardwick, J. P., Park, S. S., Veech, R. L., Yang, C. S., Gelboin, H. V., and Gonzalez, F. J., 1987, Stabilization of cytochrome P450j messenger ribonucleic acid in the diabetic rat, *Mol. Endocrinol.* **1:**542–547.

147. Yamazoe, Y., Murayama, N., Shimada, M., Imaoka, S., Funae, Y., and Kato, R., 1989, Suppression of hepatic levels of an ethanol-inducible P-450DM/j by growth hormone: Relationship between the increased level of P-450DM/j and depletion of growth hormone in diabetes, *Mol. Pharmacol.* **36:**716–722.

148. Westin, S., Strom, A., Gustafsson, J. A., and Zaphiropoulos, P. G., 1990, Growth hormone regulation of the cytochrome P-450IIC subfamily in the rat: Inductive, repressive, and transcriptional effects on P-450f (IIC7) and P-450PB1 (IIC6) gene expression, *Mol. Pharmacol.* **38:**192–197.

149. Zaphiropoulos, P. G., Strom, A., Robertson, J. A., and Gustafsson, J. A., 1990, Structural and regulatory analysis of the male-specific rat liver cytochrome P-450 g: Repression by continuous growth hormone administration, *Mol. Endocrinol.* **4:**53–58.

150. Donahue, B. S., Skottner, L. A., and Morgan, E. T., 1991, Growth hormone-dependent and -independent regulation of cytochrome P-450 isozyme expression in streptozotocin-diabetic rats, *Endocrinology* **128:**2065–2076.

151. Barnett, C. R., Rudd, S., Flatt, P. R., and Ioannides, C., 1993, Sex differences in the diabetes-induced modulation of rat hepatic cytochrome P450 proteins, *Biochem. Pharmacol.* **45:**313–319.

152. Barnett, C. R., Petrides, L., Wilson, J., Flatt, P. R., and Ioannides, C., 1992, Induction of rat hepatic mixed-function oxidases by acetone and other physiological ketones: Their role in diabetes-induced changes in cytochrome P450 proteins, *Xenobiotica* **22**:1441–1450.

153. Murray, M., Cantrill, E., Mehta, I., and Farrell, G. C., 1992, Impaired expression of microsomal cytochrome P450 2C11 in choline-deficient rat liver during the development of cirrhosis, *J. Pharmacol. Exp. Ther.* **261**:373–380.

154. Murray, M., Zaluzny, L., and Farrell, G. C., 1986, Drug metabolism in cirrhosis. Selective changes in cytochrome P-450 isozymes in the choline-deficient rat model, *Biochem. Pharmacol.* **35**:1817–1824.

155. Murray, M., Zaluzny, L., and Farrell, G. C., 1987, Impaired androgen 16 α-hydroxylation in hepatic microsomes from carbon tetrachloride-cirrhotic male rats, *Gastroenterology* **93**:141–147.

156. Clarke, L., and Waxman, D. J., 1989, Oxidative metabolism of cyclophosphamide: identification of the hepatic monooxygenase catalysts of drug activation, *Cancer Res.* **49**:2344–2350.

157. Weber, G. F., and Waxman, D. J., 1993, Activation of the anti-cancer drug ifosphamide by rat liver microsomal P450 enzymes, *Biochem. Pharmacol.* **45**:1685–1694.

158. Sladek, N. E., 1988, Metabolism of oxazaphosphorines, *Pharmacol. Ther.* **37**:301–355.

159. LeBlanc, G. A., Kantoff, P. W., Ng, S. F., Frei, E., III, and Waxman, D. J., 1992, Hormonal perturbations in patients with testicular cancer treated with cisplatin, *Cancer* **69**:2306–2310.

160. Rouer, E., Lemoine, A., Cresteil, T., Rouet, P., and Leroux, J. P., 1987, Effects of genetically or chemically induced diabetes on imipramine metabolism. Respective involvement of flavin monooxygenase and cytochrome P450-dependent monooxygenases, *Drug Metab. Dispos.* **15**:524–528.

161. Dixon, R. L., Hart, L. G., and Fouts, J. R., 1961, The metabolism of drugs by liver microsomes from alloxan-diabetic rats, *J. Pharmacol. Exp. Ther.* **133**:7–11.

162. Ioannides, C., Bass, S. L., Ayrton, A. D., Trinick, J., Walker, R., and Flatt, P. R., 1988, Streptozotocin-induced diabetes modulates the metabolic activation of chemical carcinogens, *Chem. Biol. Interact.* **68**:189–202.

163. Ho, K. Y., Evans, W. S., Blizzard, R. M., Veldhuis, J. D., Merriam, G. R., Samojlik, E., Furlanetto, R., Rogol, A. D., Kaiser, D. L., and Thorner, M. O., 1987, Effects of sex and age on the 24-hour profile of growth hormone secretion in man: Importance of endogenous estradiol concentrations, *J. Clin. Endocrinol. Metab.* **64**:51–58.

164. Mansfield, M. J., Rudlin, C. R., Crigler, J., Jr., Karol, K. A., Crawford, J. D., Boepple, P. A., and Crowley, W., Jr., 1988, Changes in growth and serum growth hormone and plasma somatomedin-C levels during suppression of gonadal sex steroid secretion in girls with central precocious puberty, *J. Clin. Endocrinol. Metab.* **66**:3–9.

165. Relling, M. V., Lin, J. S., Ayers, G. D., and Evans, W. E., 1992, Racial and gender differences in N-acetyltransferase, xanthine oxidase, and CYP1A2 activities, *Clin. Pharmacol. Ther.* **52**:643–658.

166. Hunt, C. M., Westerkam, W. R., and Stave, G. M., 1992, Effect of age and gender on the activity of human hepatic CYP3A, *Biochem. Pharmacol.* **44**:275–283.

167. Wilson, K., 1984, Sex-related differences in drug disposition in man, *Clin. Pharmacokinet.* **9**:189–202.

168. Lew, K. H., Ludwig, E. A., Milad, M. A., Donovan, K., Middleton, E., Jr., Ferry, J. J., and Jusko, W. J., 1993, Gender-based effects on methylprednisolone pharmacokinetics and pharmacodynamics, *Clin. Pharmacol. Ther.* **54**:402–414.

169. Walle, T., Walle, U. K., Mathur, R. S., Palesch, Y. Y., and Conradi, E. C., 1994, Propranolol metabolism in normal subjects: Association with sex steroid hormones, *Clin. Pharmacol. Ther.* **56**:127–132.

170. Ryan, D. E., and Levin, W., 1990, Purification and characterization of hepatic microsomal cytochrome P-450, *Pharmacol. Ther.* **45**:153–239.

171. McClellan, G. P., Waxman, D. J., Caveness, M., and Goldstein, J. A., 1987, Phenotypic differences in expression of cytochrome P-450g but not its mRNA in outbred male Sprague–Dawley rats, *Arch. Biochem. Biophys.* **253**:13–25.

Regulation of Steroidogenic and Related P450s

NORIO KAGAWA and MICHAEL R. WATERMAN

One broad classification of P450s is as two groups, one containing forms that metabolize endogenous substrates and the other, forms that metabolize exogenous substrates (xenobiotics). Those forms that metabolize endogenous substrates generally convert less active compounds into more active ones. This is particularly true of the P450s that participate in steroid hormone biosynthesis, where the less active steroid cholesterol is converted into more active mineralocorticoids, glucocorticoids, progestins, and sex hormones. Interestingly, hepatic P450s of the xenobiotic-metabolizing class may play important roles in inactivating these steroid hormones. In the case of vitamin D we also find that distinct P450s play roles in producing the most active forms of this hormone and other P450s are important in their inactivation. This chapter will summarize our understanding of the function and regulation of P450 systems involved in production of active steroids and related compounds.

1. Steroidogenic Pathways and Reactions

Conversion of cholesterol to steroid hormones occurs primarily in adrenal cortex, gonads, and placenta. It is very likely that the brain also catalyzes many or all of the reactions found in the better known steroidogenic tissues. Estrogen production also occurs in adipose tissue and certain members of the steroidogenic pathways are found in other tissues (e.g., retina and stomach) suggesting that additional sites of steroid hormone

---

NORIO KAGAWA and MICHAEL R. WATERMAN • Department of Biochemistry, Vanderbilt University School of Medicine, Nashville, Tennessee 37232.

Cytochrome P450: Structure, Mechanism, and Biochemistry (Second Edition), edited by Paul R. Ortiz de Montellano. Plenum Press, New York, 1995.

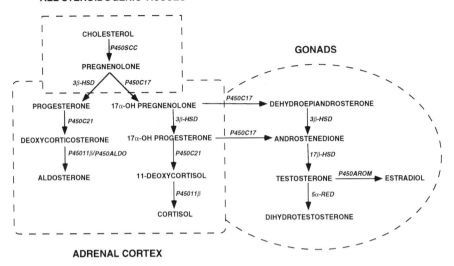

FIGURE 1. Generic steroidogenic pathways. The enzymes are: $P450_{scc}$, cholesterol side chain cleavage cytochrome P450; $P450_{c17}$, 17α-hydroxylase cytochrome P450; $P450_{c21}$, 21-hydroxylase cytochrome P450; $P450_{11\beta}$, 11β-hydroxylase cytochrome P450; $P450_{aldo}$, aldosterone synthase cytochrome P450; $P450_{arom}$, aromatase cytochrome P450; 3β-HSD, 3β-hydroxysteroid dehydrogenase; 17β-HSD, 17β-hydroxysteroid dehydrogenase; 5α-RED, 5α-reductase.

production have yet to be uncovered. Herein, however, attention will be focused on activities in the traditional steroidogenic tissues as schematized in Fig. 1.

The conversion of cholesterol to pregnenolone is the initial step in all steroidogenic pathways and is considered to be the rate-limiting step in steroidogenesis. This reaction takes place in the mitochondrion and is catalyzed by an integral membrane protein of the inner mitochondrial membrane, cholesterol side chain cleavage cytochrome P450 ($P450_{scc}$).[1] This steroidogenic P450, like the others indicated in Fig. 1 and like other mammalian P450s, has a modest turnover number. However, the rate-limiting step in steroidogenesis is not the conversion of cholesterol to pregnenolone by $P450_{scc}$ but rather is the process involved in making the substrate, cholesterol, available for $P450_{scc}$. The inner mitochondrial membrane, like all membranes, contains cholesterol, which is essential in maintaining the proper fluidity. Beyond this pool of structural cholesterol, steroidogenic mitochondria also contain in their membranes a steroidogenic pool of cholesterol which provides substrate for steroid hormone biosynthesis.[2] While the rate-limiting step in steroidogenesis is measured as the production of pregnenolone, the rate-limiting event is, in fact, the mobilization of cholesterol to the vicinity of $P450_{scc}$ in the inner mitochondrial membrane from lipid stores outside the mitochondrion.

$P450_{scc}$, like other mitochondrial P450s, faces the matrix side of the inner mitochondrial membrane and receives reducing equivalents from NADPH via a mini-electron transport chain in the matrix. An FAD-containing flavoprotein, ferredoxin reductase (adrenodoxin reductase in the adrenal), is reduced by NADPH and transfers electrons to a two-iron two-sulfur protein, ferredoxin (adrenodoxin in the adrenal), which in turn reduces $P450_{scc}$:

$$NADPH \rightarrow Ferredoxin\ reductase \rightarrow Ferredoxin \rightarrow P450_{scc}$$
$$(FAD) \qquad\qquad (2Fe-2S) \qquad (protoheme)$$

There are other mitochondrial P450s involved in steroidogenesis, 11β-hydroxylase P450 ($P450_{11\beta}$) and aldosterone synthase ($P450_{aldo}$), and these enzymes capture reducing equivalents in exactly the same way.

The microsomal steroidogenic P450s—17α-hydroxylase ($P450_{c17}$), 21-hydroxylase ($P450_{c21}$), and aromatase ($P450_{arom}$)—are integral proteins in the endoplasmic reticulum and are reduced by the ubiquitous microsomal flavoprotein, NADPH-cytochrome P450 reductase (P450 reductase):

$$NADPH \rightarrow P450\ reductase \rightarrow P450$$
$$(FAD\ \&\ FMN) \qquad (protoheme)$$

Both mitochondrial and microsomal P450s are integral proteins of their respective membranes. There is little understanding, however, of how mitochondrial P450s, including the steroid hydroxylases, are anchored in the inner mitochondrial membrane. They are synthesized in the cytosol as higher-molecular-weight precursors[3] and their amino-terminal precursor extensions are removed on mitochondrial uptake.[4] How $P450_{scc}$ finds its way to the inner mitochondrial membrane, binds heme, and becomes firmly anchored in this membrane remains unknown. Microsomal P450s including steroid hydroxylases are synthesized and become associated with the endoplasmic reticulum through the signal recognition particle pathway.[5] They are anchored in the membrane by a hydrophobic amino-terminal sequence although other regions of the tertiary structure of microsomal P450s also participate in membrane association.[6] The requirement for a functional microsomal P450 is a generic signal anchor sequence, the precise sequence having little effect on the activity of the P450.[7] $P450_{arom}$ is an unusual microsomal P450 because its amino-terminal sequence extends into the lumen of the endoplasmic reticulum and it is glycosylated.[8] The glycosylation seems to have no effect on $P450_{arom}$ activity.[9]

Inspection of Fig. 1 indicates that the complement of steroidogenic P450s varies from one steroidogenic tissue to another. The adrenal cortex contains the largest number of steroidogenic P450s including three forms that are found exclusively or predominantly in the adrenal cortex: $P450_{c21}$, $P450_{11\beta}$, and $P450_{aldo}$. Generally the adrenal cortex can be considered to contain two different types of steroidogenic cells: those that contain $P450_{c17}$ and $P450_{11\beta}$ produce glucocorticoids (fasciculata/reticularis) and those that do not contain $P450_{c17}$ but do contain $P450_{aldo}$ and produce the major mineralocorticoid, aldosterone (glomerulosa). The fasciculata/reticularis also produces dehydroepiandrosterone and androstenedione, which are C_{19} androgen precursors of the sex hormones. The steroidogenic cell type in the testis (Leydig cell) contains $P450_{scc}$ and $P450_{c17}$ which participate in the synthesis of the male sex hormones testosterone and dihydrotestosterone from cholesterol. In the ovarian follicle two types of steroidogenic cells are found, theca and granulosa. The theca expresses $P450_{scc}$ and $P450_{c17}$ while the granulosa expresses $P450_{scc}$ and $P450_{arom}$. Thus, estrogen production occurs in the granulosa cells while the immediate precursor, androstenedione, is produced in the theca. Following ovulation, the theca and granulosa cells differentiate into luteal cells of the corpus luteum. The bovine corpus luteum produces large quantities of progesterone because $P450_{c17}$ is not present in these cells and $P450_{scc}$ levels increase over those found in the follicular theca and granulosa cells.[10] The placenta also contains steroidogenic enzymes including $P450_{scc}$, $P450_{c17}$, and $P450_{arom}$ permitting synthesis of progesterone and estrogen.

The primary activities catalyzed by the steroid hydroxylases are summarized in Fig. 2. The side chain cleavage of cholesterol is a complicated three-step reaction. First is hydroxylation at position 22 on the cholesterol side chain followed by hydroxylation at position 20. The side chain is then cleaved by an O_2 + NADPH-dependent step, the details of which are not yet known. Convincing evidence indicates that the intermediates, 22-hydroxycholesterol and 20,22-dihydroxycholesterol, do not leave the active site.[11] Only following side chain cleavage to produce isocaproic acid and pregnenolone is a product from this concerted reaction released. As far as is known, all forms of $P450_{scc}$ in all species and all tissues follow this same reaction path.

$P450_{c17}$ represents a cross road in steroidogenic pathways because it can catalyze two distinct reactions. First is 17α-hydroxylation of either pregnenolone or progesterone. This

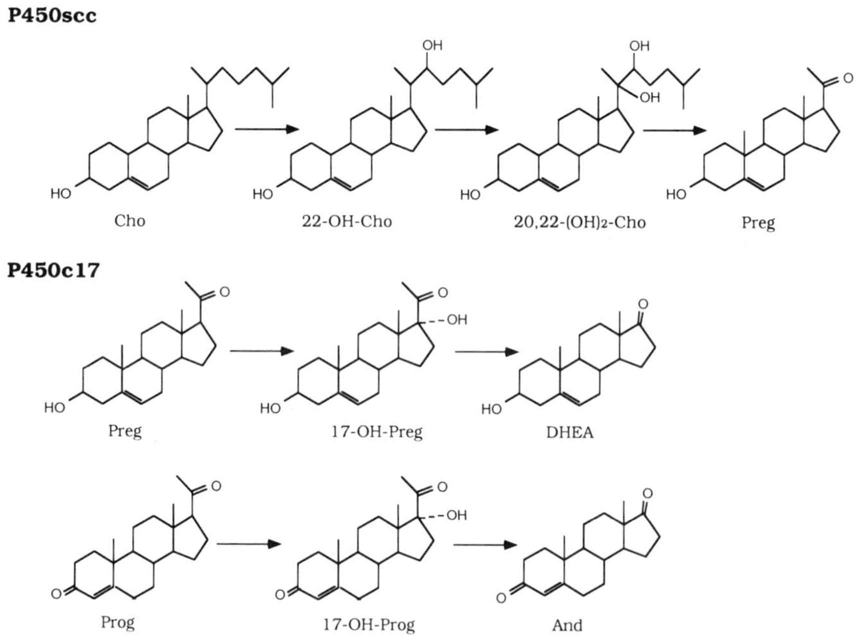

P450scc

| Cho | 22-OH-Cho | 20,22-(OH)₂-Cho | Preg |

P450c17

| Preg | 17-OH-Preg | DHEA |

| Prog | 17-OH-Prog | And |

FIGURE 2. Enzymatic activities associated with various steroid hydroxylases. Different substrates and intermediates are designated as: Cho, cholesterol; Preg, pregnenolone; 17-OH-Preg, 17α-hydroxypregnenolone; DHEA, dehydroepiandrosterone; Prog, progesterone; 17-OH-Prog, 17α-hydroxyprogesterone; And, androstenedione; DOC, deoxycorticosterone; DOS, 11-deoxycortisol; Cor, corticosterone; 18-OH-Cor, 18-hydroxycorticosterone; Aldo, aldosterone; Csol, cortisol; 19-OH-And, 19-hydroxyandrostenedione; 19-oxo-And, 19-oxo-androstenedione; Est, estrogen.

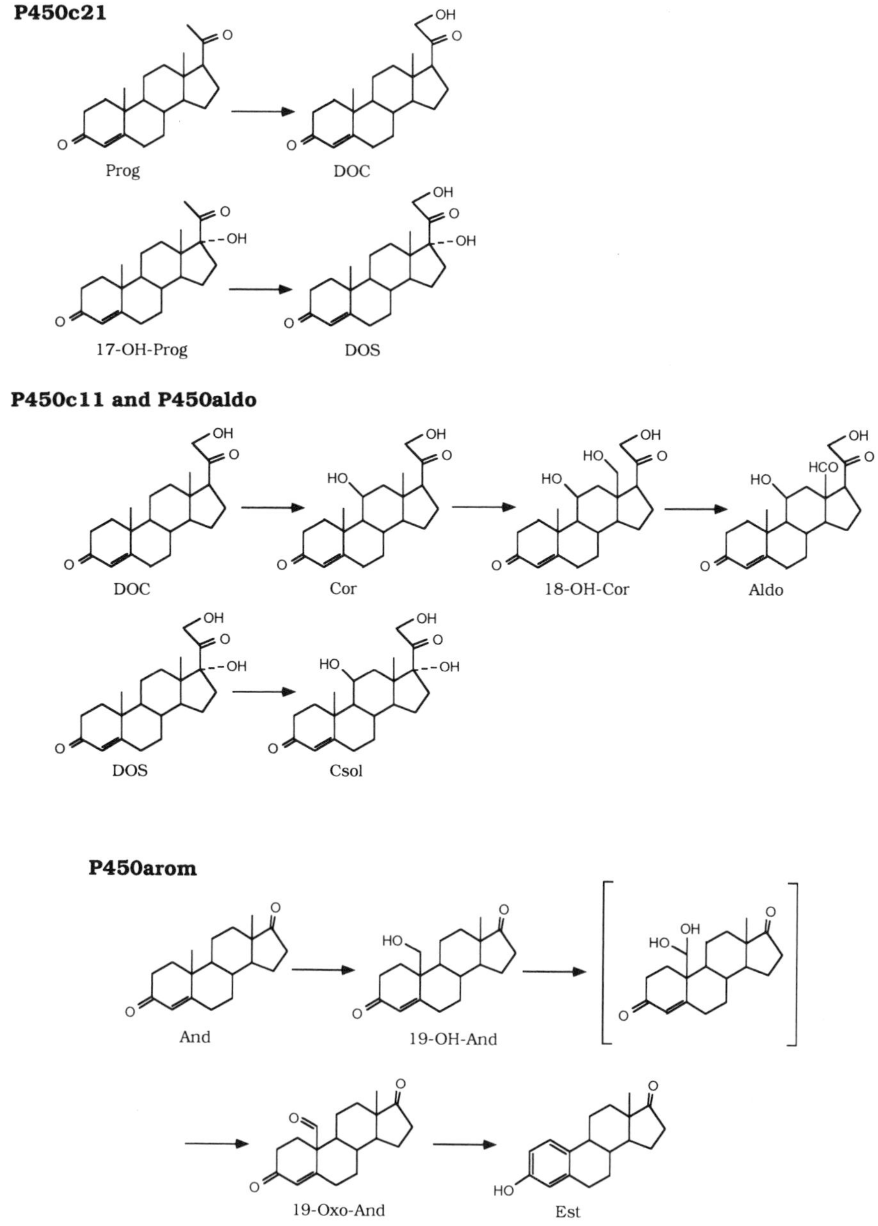

P450c21

Prog → DOC

17-OH-Prog → DOS

P450c11 and P450aldo

DOC → Cor → 18-OH-Cor → Aldo

DOS → Csol

P450arom

And → 19-OH-And →

19-Oxo-And → Est

FIGURE 2. (*continued*)

is a standard P450-dependent mixed-function oxidation reaction and is required for the production of the potent glucocorticoid, cortisol. The second reaction catalyzed by P450$_{c17}$ known as 17,20-lyase converts the C$_{21}$ steroids (17α-hydroxypregnenolone and 17α-hydroxyprogesterone) to C$_{19}$ steroids (dehydroepiandrosterone and androstenedione). Thus, the production of C$_{19}$ androgens is also a three-step process involving production of 17α-hydroxy intermediates followed by two additional steps, each requiring O$_2$ and NADPH, which are not yet fully characterized mechanistically. It is evident from studies with purified[12] and recombinant forms[13] of P450$_{c17}$ that both activities arise from a common active site. In the adrenal, a portion of the 17α-hydroxylated products of the first step must be released from the active site to be used in cortisol biosynthesis. Perhaps all are released and then a fraction rebinds for the 17,20-lyase reaction leading to androgens. In the gonads it makes little difference since these tissues do not contain pathways that would siphon off potential androgens as cortisol biosynthesis does in the adrenal. It is apparent that, from a physiological standpoint, cortisol biosynthesis in the adrenal and sex hormone biosynthesis in the gonads are the most important roles of P450$_{c17}$. Thus, it would make sense that P450$_{c17}$ activities not be concerted in the adrenal but be concerted in the gonads. In fact, 17α-hydroxylated steroids are readily detected in the testis, indicating that P450$_{c17}$ activity is not concerted in this tissue.[14] P450$_{c17}$ from some species (e.g., mouse, rat, chicken, fish) readily catalyzes the production of both dehydroepiandrosterone and androstenedione, while that from others (bovine, human) does not efficiently produce androstenedione and virtually all of the lyase product is a dehydroepiandrosterone.[15] Thus, P450$_{c17}$ in some species catalyzes androgen production via both Δ4 and Δ5 pathways while in others via only the Δ5 pathway. Because of the presence of 3β-hydroxysteroid dehydrogenase, which converts pregnenolone and 17α-hydroxypregnenolone to progesterone and 17α-hydroxyprogesterone, sex steroid biosynthesis, in testis and ovaries of cows and humans, would seem to be diminished. In these species conversion of the intermediates from the Δ5 pathway to the Δ4 pathway leads to 17α-hydroxyprogesterone, which is not a lyase substrate. However, cytochrome b_5, which is in relatively high concentration in human testis,[16] greatly enhances the conversion of 17α-hydroxyprogesterone to androstenedione by human P450$_{c17}$, preventing 17α-hydroxyprogesterone from being a dead-end product.[17] Why there is such species variation in P450$_{c17}$ activities is unknown. It seems that evolutionarily more ancient forms of P450$_{c17}$ catalyze both Δ4 and Δ5 lyase reactions (bird, fish) while more recent forms on the evolutionary tree (human) do not without the assistance of cytochrome b_5. The physiological reason for this evolutionary switch is interesting but not apparent.

P450$_{arom}$ catalyzes a very complicated series of reactions leading to the production of estrogen. This seems to be a concerted reaction and P450$_{arom}$, like P450$_{scc}$, appears to be the same in all species. This enzyme is well known to be expressed in adipose tissue and the brain allowing for the production of estrogen from these sites. In songbirds, brain P450$_{arom}$ is essential for the development of the song center in males. Testosterone produced by the testis crosses the blood–brain barrier and is converted to estrogen which triggers developmental events in the brain leading to formation of the song center.[18]

P450$_{c21}$ is the simplest of the steroid hydroxylases, being located solely in the two adrenal cortical steroidogenic regions (fasciculata/reticularis and glomerulosa) and having only two substrates, 17α-hydroxyprogesterone, an intermediate in cortisol biosynthesis, and progesterone, an intermediate in corticosterone and aldosterone biosynthesis. $P450_{c21}$ in all species seems to have the same catalytic properties. Extraadrenal 21-hydroxylase activity is known, but this is not catalyzed by the steroidogenic $P450_{c21}$ but rather by members of other P450 gene families.[19]

Adrenal cortex in several species including rat, mouse, and human contains both $P450_{11\beta}$ and $P450_{aldo}$.[20] Cattle, on the other hand, apparently utilize a single $P450_{11\beta}$ for production of both cortisol and aldosterone.[21] The biosynthesis of aldosterone requires 18-hydroxylation followed by production of the 18-oxo product. Thus, $P450_{aldo}$ is apparently able to catalyze 11β-hydroxylation, 18-hydroxylation, and the formation of the 18-oxo product. $P450_{11\beta}$ in most species catalyzes predominantly the 11β-hydroxylation reaction.

In summary, we find that in all species $P450_{scc}$, $P450_{arom}$, and $P450_{c21}$ have the same activities and presumably each functions by a single mechanism in all species. Most species seem to contain distinct enzymes, $P450_{11\beta}$ and $P450_{aldo}$ for glucocorticoid and mineralo-corticoid syntheses, respectively, although at least one species (bovine) utilizes a single enzyme for both. The mechanisms of 11β-hydroxylation and aldosterone synthase activities are apparently the same in all species. $P450_{c17}$, on the other hand, is quite variable between species and utilizes somewhat different mechanisms which are more or less dependent on cytochrome b_5 in different species. This variation associated with $P450_{c17}$ is not easily understood although it is noteworthy that this enzyme commits steroids into the pathways leading to sex hormone production. Perhaps physiological variations among different species associated with sex hormones and reproduction are the basis for variation in $P450_{c17}$ activities.

2. Regulation of Steroidogenesis

The regulation of steroidogenesis has a developmental component, a tissue-specific component, and a maintenance component. Developmental regulation leads to the timely appearance of steroidogenic activities during embryogenesis and cell differentiation. Tissue-specific regulation ensures the production of steroid hormones in steroidogenic tissues and not in other tissues. The maintenance component ensures that optimal steroidogenic capacity exists in steroidogenic tissues throughout life so that steroid hormones can be produced on demand.

2.1. Development

In the mouse, expression of genes encoding steroidogenic enzymes is detected shortly after appearance of steroidogenic organs as recognizable entities.[22] The only clearly established use for steroid hydroxylases in fetal development is in the generation of the male phenotype (e.g., male secondary sex characteristics).[23] Without production of testosterone by the fetal testis, male secondary sex characteristics do not appear and the phenotype of the 46XY karyotype is female. Beyond this well-established role for $P450_{scc}$ and $P450_{c17}$, it is not clear that fetal production of steroid hormones is important. In fact, in three examples of mutations in steroidogenic pathways, absence of steroid hormone production leads to no readily detectable phenotype at birth other than the absence of maleness. The three examples are a mutant rabbit having a deletion of the gene encoding $P450_{scc}$,[24] a rare human genetic disease with no detectable $P450_{scc}$ activity (congenital lipoid adrenal hyperplasia),[25] and the knockout mouse for the orphan nuclear receptor steroidogenic factor 1 (SF-1) which has no adrenals or gonads[26]; newborns which produce no detectable steroid hormones have no discernible phenotypes beyond absence of maleness in all three and absence of steroidogenic organs in the latter. In each case, the newborns die within a few days of birth without steroid hormone replacement therapy primarily because of the absence of aldosterone leading to defects in the maintenance of salt balance. Thus, even though all mammalian fetuses produce steroid hydroxylases quite early in their development, it is not possible to pinpoint a key role for these hormones (beyond testosterone) in development. Perhaps maternal steroids are sufficient to meet fetal demands in these mutant cases. Having said this, there is one example of expression of adrenal $P450_{c17}$ which affects cortisol production and suggests an important role for this steroid hormone in development. In the middle trimester of gestation, $P450_{c17}$ disappears from the bovine fetal adrenal.[27] This protein is present before and after this period, and during the absence of $P450_{c17}$ in the adrenal no cortisol is detected in the fetal circulation. $P450_{c17}$ is produced in the fetal bovine gonads (particularly testis) as expected. The physiological reasons for this peculiarity in developmental expression of $P450_{c17}$ in the adrenal are unknown but suggest that the absence of cortisol at specific times in fetal life is important. Thus, the question remains open as to the importance of fetal production of steroid hormones beyond testosterone.

Our understanding of the biochemical basis of developmental expression of steroid hydroxylase genes in the fetus is still quite primitive. However, as will be outlined below in the discussion of tissue-specific regulation, the discovery of a zinc-finger orphan receptor alternatively named steroidogenic factor 1 (SF-1)[28] or Ad4-binding protein (Ad4BP)[29] has provided initial insight into this process.

In the adult female we find another example of developmental regulation of steroid hydroxylases. As noted previously, thecal cells of the ovarian follicle express both $P450_{scc}$ and $P450_{c17}$ while the granulosa cells express $P450_{scc}$. At the time of ovulation there is a surge of the peptide hormone LH and the follicle undergoes differentiation into the corpus luteum. Included in this transition is the differentiation of thecal and granulosa cells into luteal cells. In the bovine ovary, coupled with the LH surge there is greatly enhanced expression of $P450_{scc}$ which is presumed to be mediated by cAMP.[30] At the same time the level of $P450_{scc}$ is increasing, expression of $P450_{c17}$ is shut off and this enzyme disappears.[30] As will be seen later, it is clearly established that both genes (CYP11A and CYP17) are activated transcriptionally by cAMP in the adrenal and this same mechanism (under the control of peptide hormones) is thought to function in maintenance of steroidogenic capacity in testis and ovary. It is surprising that in the presence of the LH surge and thus elevated cAMP levels, $P450_{c17}$ disappears during differentiation of thecal cells to luteal cells. The physiological basis of this event is clear, however. The corpus luteum produces large amounts of progesterone which are important in implantation of the blastocyst and the maintenance of early pregnancy. Absence of $P450_{c17}$ prevents the formation of 17α-hydroxyprogesterone. The biochemical basis of expression of $P450_{scc}$ and $P450_{c17}$ in opposite directions during differentiation of the bovine ovarian follicle into the corpus luteum, when in other cases levels of these proteins are regulated coordinately in the same direction, is unknown.

2.2. Tissue-Specific Regulation

Steroidogenesis occurs in relatively few tissues and presumably cell-specific factors are important in regulating the location of expression of these activities. In addition to the adrenal and gonads, placenta and brain are other important sites for steroidogenesis. The physiology of different species, of course, differs and thus it may not be surprising to find that different species have different steroidogenic pathways. The pathways shown in Fig. 1 are generic, and there are well-known species variations in these pathways. Rodents do not produce cortisol and thus utilize corticosterone as their major glucocorticoid. While the physiological reason for this is unknown, its biochemical basis is the absence of a $P450_{c17}$ in the adrenal. In another example, cited above, humans and rats utilize both $P450_{11\beta}$ and $P450_{aldo}$ for the biosynthesis of glucocorticoids (fasciculata/reticularis) and mineralocorticoids (glomerulosa), respectively. Cows, on the other hand, utilize a single form of $P450_{11\beta}$ for both biosynthetic pathways. The basis of variations such as these in steroidogenesis, from both physiological and evolutionary views, are unknown and will be interesting to decipher as will the mechanism by which regulation of expression of steroid hydroxylase genes is limited to a few tissues. The discovery that all steroid hydroxylase genes contain one or more binding sites for the orphan receptor SF-1 was made simultaneously by Morohashi, Omura, and colleagues[29] and by Parker and colleagues.[28] In adrenal, ovary, and testis, SF-1 transcripts appear early in fetal development slightly before the appearance of transcripts encoding the steroid hydroxylases.[22] Thus, by inference (the presence of SF-1 binding sites in each gene and the appearance of SF-1 transcripts slightly before steroid hydroxylase transcripts in the proper cells), SF-1 is thought to be important in regulating developmental/tissue-specific expression of all steroid hydroxylase genes. An interesting aside is that SF-1 levels remain high in the fetal mouse testis presumably associated with the requirement for testosterone for development of maleness. In the fetal mouse ovary, however, SF-1 levels are very low and no detectable steroid hydroxylase transcripts are seen, apparently because there is no requirement for the female sex hormone, estrogen, during fetal development.

It is now evident that SF-1 plays a more profound role in development than simply turning on steroid hydroxylase expression in the appropriate tissues. The SF-1 knockout mouse has no adrenals or gonads indicating a role for this orphan receptor in organogenesis.[26] Also, SF-1 expression has been found in the pituitary, a tissue devoid of steroid hydroxylases.[31] Thus, SF-1 probably has multiple roles in developmental/tissue-specific gene expression, one being associated with steroid hydroxylase genes.

To look at the other side of the coin, we can ask whether SF-1 is expressed at all sites where we find steroid hydroxylases. Apparently not. SF-1 expression in the brain is localized to the hippocampus and hypothalmus[26] while steroid hydroxylase expression is much broader.[32,33] In addition, $P450_{arom}$ is expressed in adipose tissue and SF-1 has not yet been found in this tissue. Certainly tissue-specific expression of steroid hydroxylase genes is complex. SF-1 is very important in this process in traditional steroidogenic cells.

However, expression of steroid hydroxylases in tissues not normally considered to be steroidogenic may not require SF-1. Also, since SF-1 is an orphan receptor of the steroid hormone/thyroid hormone nuclear receptor family, the nature of the putative ligand binding to this receptor is intriguing but unknown.

2.3. Maintenance of Steroidogenic Capacity

Throughout life, it is important that steroidogenesis be regulated such that optimal capacity to produce steroid hormones is maintained. In this way, when demands for glucocorticoids and mineralocorticoids beyond the normal circulating levels occur, for example, those resulting from inflammation, the hormone synthetic pathways can respond. It is apparent that maintenance of steroidogenic capacity (levels of steroidogenic enzymes) is complex, involving growth factors, cytokines, and peptide hormones.

From a temporal view, the maintenance of steroidogenesis is regulated acutely and chronically. In addition to maintaining optimal levels of steroidogenic enzymes and necessary proteins, a poorly understood but essential system is in place in traditional steroidogenic cells in the adrenal cortex, ovary, and testis whereby the availability of substrate (cholesterol) for steroidogenesis is acutely regulated. The primary trigger for the mobilization of cholesterol from lipid stores to the inner mitochondrial membrane in the vicinity of $P450_{scc}$ is a peptide hormone derived from the interior pituitary, ACTH serving this role in the adrenal cortex.[34] ACTH binds to its cell surface receptor which activates adenylate cyclase leading to elevated levels of intracellular cAMP. Through the action of protein kinase A (PKA), cholesterol is transported via an unknown mechanism to the vicinity of $P450_{scc}$. This process can be divided into two steps, transport of cholesterol to the mitochondrion and movement of cholesterol across the outer mitochondrial membrane to the inner membrane. Very recently a protein called steroidogenic acute regulatory (STAR) protein has been identified which appears to be important in this latter step.[35] However, we know little of the details of this cholesterol transport system. We understand in much greater detail the basis of the chronic regulation of steroidogenesis which is required for the maintenance of normal levels of steroidogenic enzymes.

2.3.1. cAMP-Independent Regulation of Steroid Hydroxylases

Primary cultures of bovine adrenocortical cells maintained for several days in the absence of ACTH lose approximately 50% of their normal levels of steroid hydroxylases, except for $P450_{c17}$ which essentially disappears.[36] This demonstrates that in bovine adrenal, steroid hydroxylase levels, with the exception of $P450_{c17}$, can be regulated in part by cAMP-independent processes. These regulators presumably include growth factors and cytokines, and while the biochemical mechanisms are not at all clear, considerable species variation exists at this level of regulation. The unique physiology associated with different species and steroidogenic tissues may be an important contributor to this variation. There are numerous reports of regulation of steroid hydroxylase protein and mRNA levels by factors other than cAMP; examples include insulinlike growth factor, epidermal growth factor, interferons, calcium, angiotensin II, phorbol esters, salt, androgens, and transforming growth factor-β. The biochemical details of how these effectors control the expression of steroid hydroxylase genes are not known. An example of our level of understanding cAMP-independent regulation is seen in a comparison of the human and bovine CYP17 genes. Analysis of human CYP17 suggests that the cAMP-responsive sequence (CRS) lies between -184/-104 bp[37] and is unrelated to the CRS elements (CRSI and CRSII) in bovine CYP17 which are described in detail below. Human CYP17 also contains a negative regulatory sequence element responsive to phorbol esters which has been localized at a site (-310/-184 bp) which is distinct from the cAMP response element. In bovine CYP17, the negative regulatory element for phorbol esters is located within CRSI which is one of the two cAMP-responsive elements in this gene.[38] Interestingly, the sequence in the human gene corresponding to bovine CRSI lies between -310/-184 bp. Thus, it appears that CYP17 in these two species may share a common negative response element for phorbol esters, but they do not share common cAMP-responsive elements. The DNA-binding proteins involved in phorbol ester repression of CYP17 expression or any of the regulatory functions of growth factors and cytokines on steroidogenesis remain unknown.

Evidence from different experimental approaches now suggests that ACTH can also influence the levels of steroid hydroxylases independently of cAMP.[39,40] It can be expected that cAMP-independent regulation provides a basal level of transcription of steroid hydroxylase genes in steroidogenic tissues on which cAMP-responsive transcriptional regulation is superimposed. The unique physiology associated with steroidogenesis in different species and different steroidogenic tissues may be an important contributor to the variation in both cAMP-independent regulation and cAMP-dependent regulation

2.3.2. cAMP-Dependent Regulation of Steroid Hydroxylase Levels

In experiments carried out as early as 1969 it was shown that hypophysectomy of rats leads to a substantial decline in the activities of steroid hydroxylases, and that treatment of hypophysectomized animals with pharmacological doses of peptide hormones derived from the anterior pituitary (ACTH in the adrenal) can restore these enzymatic activities in both the adrenal cortex and testis.[41-43] We now know that these peptide hormones activate adenylate cyclase leading to elevated levels of intracellular cAMP. In 1985 we showed that the action of ACTH leading to elevated levels of steroid hydroxylase activities (via cAMP) is at the transcriptional level.[44] Nuclear run-on experiments, which established the transcriptional activation of steroid hydroxylase genes by ACTH via cAMP, also demonstrated that it took several hours for enhanced transcription to be observed. This suggested that the CRE/CREB system might not be involved since it responds much more rapidly to cAMP.[45] Also, the enhanced transcription of the different steroid hydroxylase genes in the bovine adrenal cortex (CYP11A, CYP11B, CYP17, CYP21) was observed at approximately the same time indicating that these genes were turned on in a coordinate fashion.[44] Thus, as schematized below, the hypothesis at the outset of experiments designed to locate cAMP-responsive sequences (CRS) associated with the steroid hydroxylase genes was that all of the genes would share a common CRS recognized by common transcription factors, and that this system might be a novel cAMP-responsive system, at least distinct from the CRE/CREB system:

Hypothesis

| ACTH | adenylate → cyclase | cAMP | protein → kinase A | X | → → | Steroid hydroxylase CYP transcription |

In the normal animal, this is not an induction process as observed for regulation of expression of many other CYP genes by xenobiotics. Rather the natural pattern of peptide hormone release from the anterior pituitary exerts continued transcriptional pressure on steroid hydroxylase genes leading to maintenance of optimal steroidogenic capacity.

In addition to studies on the expression of the genes encoding $P450_{scc}$, $P450_{11\beta}$, $P450_{aldo}$, $P450_{c21}$, $P450_{c17}$, and $P450_{arom}$, several such studies have been carried out on the gene encoding the iron–sulfur protein adrenodoxin which is required for function of all mitochondrial P450s including $P450_{scc}$, $P450_{11\beta}$, and $P450_{aldo}$ as well as the 1α-hydroxylase of 25-hydroxyvitamin D. ACTH-dependent transcription of the adrenodoxin gene was found to be coordinate with that of the steroid hydroxylase genes in nuclear run-on assays[44] and it was predicted that the adrenodoxin gene would contain the same CRS as the steroid hydroxylase genes as defined by the above illustrated hypothesis.

Analysis of the 5′-flanking regions of the bovine adrenocortical steroid hydroxylase and adrenodoxin genes coupled to reporter genes has led to the surprising conclusion that each gene contains its own distinct cAMP-responsive elements. These sequences are presented in Table I. This is surprising, not only because of the coordinated transcription of these genes in response to cAMP but also since SF-1 (Ad4BP) binding sites important

Table I
cAMP-Responsive Sequences in Bovine Steroid Hydroxylase Genes

Bovine CYP11A	CRSI	−118 −100 ACTGAGTCTGGGAGGAGCT
	CRSII	−70 −50 GACCGCCCTGTCAGCTTCTCA
Bovine CYP11B	CRE(Ad1)	−67 −60 TGACGTGA
Bovine CYP21 CRS	CRS	−129 −115 CTGTTTTGTGGGCGG
Bovine CYP17	CRSI	−243 −225 TTGATGGACAGTGAGCAAG
	CRSII	−76 TGAGCATTAACATAAAGTCAAGGAGAAGGT CAGGGG −40
Bovine adrenodoxin	CRS	+698 GCCAGGGGGCGGGGCGGAGCGCCGGCAGGG GGCGGGGCTGCGC +745

in developmental and/or tissue-specific expression are found in each of the steroid hydroxylase genes. Why should the coordinated cAMP-dependent transcription of these genes not utilize a common transcriptional system? Figure 3 illustrates that the steroid hydroxylase genes have distinct evolutionary patterns which might provide an explanation for the origin of distinct CRS elements in these genes. However, the presence of SF-1 binding sequences in each of these genes makes the evolutionary distance between them an unlikely explanation for the diversity among CRS elements.

Investigation of the biochemistry of cAMP-dependent transcriptional regulation of steroid hydroxylase pathways has been carried out primarily in three species: bovine, human, and mouse. The bovine adrenocortical microsomal steroid hydroxylases are the sole known members of two different gene families, CYP21 encoding $P450_{c21}$ and CYP17 encoding $P450_{c17}$. The CRS in bovine CYP21 contains overlapping binding sites for two nuclear proteins, one being Sp1 and the other being a 78-kDa protein.[46–48] Binding of this latter protein, designated ASP, is required for cAMP-responsiveness of bovine CYP21, not binding of Sp1. An ASP binding site is located next to, but not overlapping with, the upstream CRS of the bovine CYP11A gene encoding the bovine mitochondrial steroid

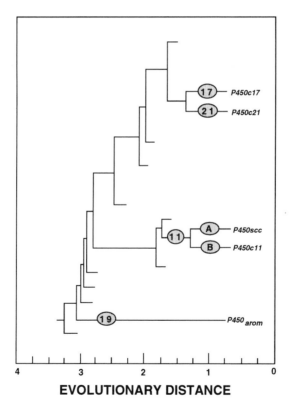

FIGURE 3. Evolutionary relationship of genes encoding steroidogenic P450s. Figure adapted from an evolutionary scheme of the CYP superfamily by Nebert *et al.*[94]

hydroxylase, $P450_{scc}$.[49] However, bovine CYP21 requires ASP binding for cAMP-dependent transcription while bovine CYP11A requires the transcription factor Sp1 binding (see below) for this purpose.

The bovine CYP17 gene contains two distinct CRS elements,[50] each binding its own group of nuclear proteins. This gene is novel among the bovine steroid hydroxylase genes in that it seems to be strictly regulated by cAMP. In primary cultures of bovine adrenocortical cells maintained in serum but in the absence of ACTH, $P450_{c17}$ disappears.[51] This indicates that the $P450_{c17}$ level is strictly dependent on cAMP. The CYP17 CRS closest to the promoter (CRSII) binds SF-1 and other proteins including COUP-TF.[52] The upstream bovine CYP17 CRS (CRSI) is found to bind four nuclear proteins.[53] Two of these proteins contain peptide sequences not found in the protein data base. The other two proteins are homeodomain proteins (helix–turn–helix DNA-binding motif) of the PBX family of genes. These PBX proteins contain a PKA phosphorylation site but have not yet been shown to be phosphorylated in response to ACTH. CRSI of bovine CYP17 is the first cellular target for PBX to be discovered, although this DNA-binding protein is expressed in most tissues.[54]

The most closely related steroid hydroxylase genes are those encoding mitochondrial P450s, the CYP11 gene family. Of the three members of this family, $P450_{scc}$, $P450_{11\beta}$, and $P450_{aldo}$, bovine adrenals are found to contain only $P450_{scc}$ and $P450_{11\beta}$. Even though these genes are closely related, they seem to have quite distinct cAMP-responsive elements. Bovine $P450_{scc}$ contains two CRSs near the promoter region which play an important role in cAMP responsiveness of this gene.[49,55] Each of these is a binding site for Sp1. Sp1 is a ubiquitous, zinc-finger-containing transcription factor which is not generally thought to be involved in cAMP-dependent transcription. Nevertheless, these Sp1-binding elements[56] have been located by deletion analysis of the 5'-flanking region of bovine CYP11A by virtue of their ability to enhance transcription of reporter genes in response to elevation of cAMP levels in steroidogenic cell cultures. This role has been further substantiated by the observation that overexpression of the catalytic subunit of PKA enhances transcription of reporter genes driven by these CRS elements.[57] Since Sp1 does not contain a PKA phosphorylation site, it might be that an unidentified nuclear protein interacts with Sp1, perhaps in response to phosphorylation, in order to activate CYP11A transcription in response to the binding of ACTH to its adrenal cell surface receptor. Perhaps this unidentified protein will prove to be localized predominantly in steroidogenic cells.

The other bovine mitochondrial steroid hydroxylase, $P450_{11\beta}$, is regulated by cAMP through a near-consensus CRE sequence which presumably binds the leucine zipper-containing transcription factor, CREB.[57] Transcription mediated by this near-perfect CRE is strongly enhanced by cooperation of an upstream Ad4BP(SF-1) binding site. Thus, the two bovine mitochondrial steroid hydroxylases, which are members of the same gene family (CYP11), utilize quite different cAMP-responsive systems for their coordinate expression in response to ACTH.

The bovine adrenodoxin gene contains two CRS elements the most important of which is located within the first intron.[58] This CRS contains two potential binding sites for Sp1, suggesting that it may be regulated in a manner similar to that observed for CYP11A. However, this CRS has also been found to bind another protein which has yet to be characterized (P.-Y. Cheng, unpublished). Perhaps unlike the bovine $P450_{scc}$ gene, additional DNA-binding proteins as well as Sp1 are required for cAMP-dependent transcription of the adrenodoxin gene.

As a summary, it is evident from this discussion and Table I that each of the genes encoding bovine adrenocortical steroid hydroxylases utilizes a different cAMP-responsive system in maintenance of optimal steroid hydroxylase levels. Characterization of the cAMP-responsive mechanisms for adrenocortical steroid hydroxylases in other species shows both similarities and differences to results obtained with the bovine genes. The human and mouse contain two members of the CYP11B subfamily, CYP11B1 encoding $P450_{11\beta}$ and CYP11B2 encoding $P450_{aldo}$. The mouse gene encoding $P450_{11\beta}$ does not contain a CRE-like sequence and responds slowly to cAMP,[59] while that encoding $P450_{aldo}$ does contain such a sequence and responds more rapidly to cAMP.[60] The human adrenodoxin gene contains an Sp1-binding site,[61] but a comparison of proteins that bind to this CRS element with those binding to the bovine adrenodoxin CRS has not yet been made. The two Sp1-binding regions near the TATA box in the bovine CYP11A gene are also present in the human, mouse, and rat genes and are located in regions that have been reported to participate in cAMP-responsiveness.[62–64] The human CYP11A gene also contains a CRS far upstream from the Sp1-binding sites that participates in cAMP-responsiveness.[62,65,66] This region contains a CRE, and CREB binding to this site may function in combination with the Sp1-binding sites near the TATA box in achieving the full cAMP-responsiveness of this gene. Deletion analysis of the promoter region of human CYP17 suggests that a functional CRS lies in a region that has no sequence homology to either bovine CRSI or CRSII.[37] The mouse $P450_{c17}$ gene contains a CRS that has no sequence homology to CRS elements in either the human or bovine CYP17 genes.[67] The mouse $P450_{c21}$ gene, like the mouse $P450_{scc}$ gene, contains multiple elements involved in cAMP-responsiveness.[68] One such element binds SF-1 and the transcription factor NGF1-B,[69] although the biochemical roles of these factors in cAMP-responsiveness have yet to be elucidated. Human CYP21 contains an ASP-binding site that is nearly identical to that in bovine CYP21[46] and participates in cAMP-dependent transcription, in addition to a CRE-like sequence far upstream of the promoter.[70]

2.4. Summary and Future Directions

The steroidogenic P450s produce ligands for the zinc-finger-containing nuclear receptors. They, themselves, are regulated by peptide hormones derived from the anterior pituitary. Consequently, steroid hormone biosynthesis is a link between these peptide hormones and function of the nuclear receptor superfamily.[71] Goals of future research in this area include further characterization of the transcription factors involved in both cAMP-independent and cAMP-dependent transcription of these genes and the biochemical basis of the coupling of these DNA-binding proteins to the basal transcription machinery. From the results presented above we can anticipate that this coupling will be quite complex and that study of the details of transcription of each steroid hydroxylase in more than one species will probably be necessary to understand in detail the interaction of peptide hormones, growth factors, and cytokines to the activation of nuclear receptors.

In addition, the picture will not become clear without careful analysis of how the different levels of regulation of transcription of steroid hydroxylase genes interact. As noted above, for example, fetal bovine $P450_{c17}$ is not expressed in the middle trimester because of an absence of fetal production of ACTH. This suggests an important interaction of developmental requirements for steroid hormones with the mechanism required for long-term maintenance of steroidogenesis. Obviously, investigation of one level of transcriptional regulation of steroid hydroxylase genes cannot be carried out in a vacuum. Rather, investigation of each level must consider the biochemistry of the other levels as well. Also, since the physiology of steroidogenesis is very species dependent, it will be practical to study this process in at least a limited number of species. Mouse is an important system because of the potential for genetic applications and human is important for obvious reasons. Other species such as bovine, ovine, and porcine have commercial importance. Therefore, it is difficult to restrict study of the regulation of steroidogenesis to only one or two species. However, the complexity within any one species presents a considerable experimental challenge and that presented by multiple species is truly daunting.

3. Regulation of P450arom (CYP19) Expression

CYP19 provides the most complex pattern of gene expression yet found in the P450 superfamily of genes. In part, this is probably related to the expression of $P450_{arom}$ in a variety of different tissues including ovary, placenta, adipose, and brain. The human CYP19 gene contains nine coding exons (numbered II through X). The upstream region of this gene, however, is very complex containing multiple promoters as indicated in Fig. 4. The promoter designated II resides just upstream from the site of initiation of translation while promoters I.1, I.2, I.3, and I.4 lie farther upstream.[72-77] Transcription initiated by these different promoters produces at least seven alternatively spliced transcripts, each encoding the same $P450_{arom}$.[78]

In human placenta, two transcripts predominate, the one starting from promoter I.1 which is located at least 35 kb upstream from exon II being the major mRNA.[72] In the human corpus luteum, on the other hand, the major transcript is derived from the proximal promoter II, the promoter nearest exon II. A minor transcript in this tissue derived from

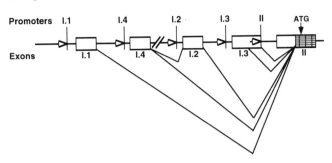

FIGURE 4. Schematic diagram of the upstream region of the human CYP19 gene. Positions of promoters relative to the four upstream untranslated exons (I.1, I.2, I.3, I.4) are indicated as is promoter II. In exon II the translated region is indicated in the hatched region. Figure adapted from Simpson *et al.*.[78]

promoter I.3 has been detected by the rapid amplification of cDNA ends (RACE) procedure.[75] The transcript arising from promoter II is also the primary one in RNA in human ovarian follicles as well as granulosa cells in culture. Thus, the same CYP19 promoter is used in both the follicular and luteal phases of the human ovarian cycle and the same promoter has been found to function in both rat[79] and chicken[80] ovary. In adipose tissue, the adipose stromal cells which surround the adipocytes contain most of the aromatase activity. In these cells, transcripts arising from promoters I.3 and I.4 represent the major species of $P450_{arom}$ mRNA. When adipose stromal cells in culture are treated with dexamethasone in serum or with dibutyryl cAMP and phorbol ester in the absence of serum, more than 20- and 150-fold increases in CYP19 expression are observed, respectively.[81,82] Interestingly, accompanying this increase in expression is a switch in promoter usage, promoter II specific-transcripts becoming the major mRNA form. Thus, both gene expression and promoter activity are regulated by these stimulators.

The complexity of tissue-specific expression of human CYP19 has made study of cellular mechanisms of CYP19 gene expression complicated. Toda et al.[83] have found using a number of reporter gene constructs in transfection studies that a cell-specific enhancer element is localized between -242/-186 bp upstream of the promoter I.1 start site of transcription. A sequence near the proximal (II) promoter has been found to convey cAMP responsiveness in primary cultures of rat granulosa cells.[84] This sequence includes an SF-1 binding site (AGGTCA).

CYP19 is a particularly important P450 gene for many reasons including its being quite ancient, its required for female sexual development, and its potential role in breast cancer. However, nature has thrown investigators of this gene a curve because of the large and complex nature of the regulatory region of this gene. Nevertheless, because of the importance of this gene, we can expect that considerable understanding of the regulation of this gene will be forthcoming in the near future.

4. P450s and Vitamin D Metabolism

At least three distinct forms of cytochrome P450 participate in metabolism of vitamin D_3 as outlined in Fig. 5. The vitamin D 25-hydroxylase has been localized to both the endoplasmic reticulum and the mitochondria of the liver. The mitochondrial $P450_{c25}$ has been cloned and contains an expected mitochondrial, precursor sequence.[85] $P450_{c25}$ is a member of the CYP27 gene family[86] and there has been confusion about the biological function of this P450. It seems clear that the hepatic CYP27 gene product participates in both vitamin D metabolism and bile acid biosynthesis. In addition to being the hepatic vitamin D 25-hydroxylase, this mitochondrial P450 also serves the role as the cholesterol 27-hydroxylase in bile acid biosynthesis in the liver.[87] It appears that there is little regulation of hepatic vitamin D 25-hydroxylation. Certainly the details of the important dual activities of this unique form of hepatic P450 and their regulation are a fascinating topic for investigation .

As a result of application of the techniques of molecular biology to the study of P450s, virtually all of the known forms of cytochrome P450 in mammals have been cloned. A glaring exception to this statement is the 1α-hydroxylase of 25-hydroxy vitamin D_3.

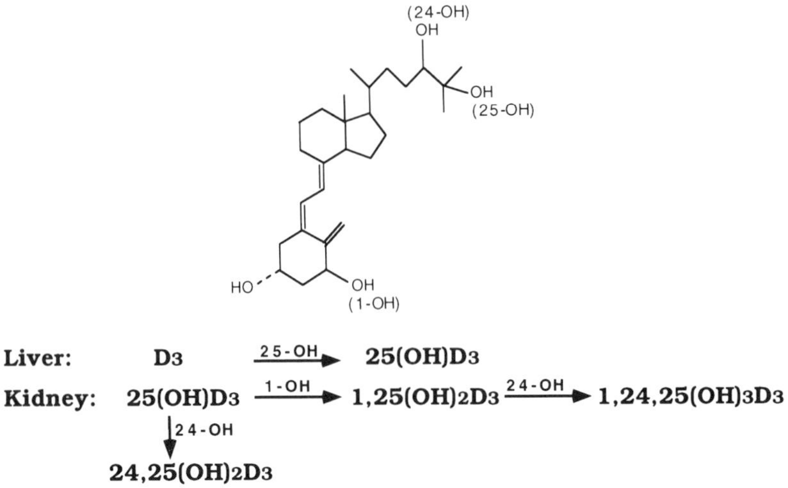

Liver: D3 $\xrightarrow{\text{2 5 - OH}}$ 25(OH)D3

Kidney: 25(OH)D3 $\xrightarrow{\text{1 - OH}}$ 1,25(OH)2D3 $\xrightarrow{\text{2 4 - OH}}$ 1,24,25(OH)3D3

$\Big\downarrow$ 2 4 - OH

24,25(OH)2D3

FIGURE 5. (a) The biological hydroxylation sites on vitamin D_3 and (b) the hepatic and renal distribution of the vitamin D hydroxylases. The enzymes indicated are 25-OH, vitamin D-25 hydroxylase; 1-OH, 25-hydroxy vitamin D-1α-hydroxylase; 24-OH, vitamin D-24-hydroxylase.

1α,25-Dihydroxy vitamin D_3 is the metabolite of vitamin D having the greatest biological activity. In particular, this hormone is key in the absorption of dietary Ca^{2+} in the intestine and the resorption of Ca^{2+} in bone. The 1α-hydroxylase activity is localized in kidney mitochondria. Thus, 25-hydroxy vitamin D_3 produced in the liver is transported to the kidney where the most active form of vitamin D is produced by 1α-hydroxylation. Several laboratories have made strong efforts to clone the 1 α-hydroxylase, without success. The enzyme has been purified from chick,[88] cow,[89] and pig[90] but cloning has remained elusive. The 1α-hydroxylase activity in kidney is highly regulated by both parathyroid hormone via a cAMP-dependent pathway and by Ca^{2+} concentration.[91] The cloning of this gene and characterization of regulatory elements, as has been done for the steroid hydroxylases, is of considerable importance.

While 1α,25-dihydroxy vitamin D_3 is the most active vitamin D metabolite, 25-hydroxy vitamin D_3 has the same activities to a lesser extent. Thus, in the regulation of calcium homeostasis it is important that both 1α,25-dihydroxy and 25-hydroxy vitamin D_3 can be inactivated as necessary. The third important form of P450 involved in vitamin D metabolism serves that role. $P450_{c24}$ is localized in kidney mitochondria, apparently in different locations than the 1α-hydroxylase. The cDNA encoding $P450_{c24}$ has been cloned,[92] and this protein is the product of the CYP24 gene.[86] $P450_{c24}$ is distributed quite widely, not found solely in the kidney, with the same gene product apparently being found in all tissues. Renal 24-hydroxylase is regulated in a similar fashion to the 1α-hydroxylase and extrarenal 24-hydroxylase is also regulated by 1α,25-dihydroxy vitamin D_3.[91] Thus, as levels of the most potent vitamin D metabolite increase, the capacity for its inactivation also increases.

In addition to cloning the 1α-hydroxylase, characterization of the biochemical details of transcription of these three very important forms of cytochrome P450 are major challenges for the future. These enzymes are not particularly abundant but play essential roles in development, growth, and homeostasis. Thus, detailed study of their localization, activities, and regulation are necessary to complete our understanding of calcium homeostasis.

5. Conclusions

This chapter began by suggesting that P450s can be classified into two groups, those metabolizing endogenous substrates and those metabolizing xenobiotics (drugs and carcinogens). From the viewpoint of the forms that metabolize endogenous substrates, this classification probably is valid. There is little evidence that steroid hydroxylases and vitamin D hydroxylases have roles in xenobiotic metabolism. Of course, there are additional members of this class of enzymes such as forms involved in fatty acid, prostaglandin, and arachidonic acid metabolism described elsewhere in this volume, as well as additional members yet to be discovered. The brain would seem to be a rich source for these enzymes although tissues such as kidney and intestine can be expected to also contain novel new forms of P450 involved in endogenous substrate metabolism. Another consideration of endogenous P450-mediated substrate metabolism is the increasing body of evidence indicating that already known forms such as the steroid hydroxylases will be discovered in nontraditional sites such as localization of $P450_{c17}$ in stomach.[93] In short, there is a great deal more work to be done to round out our understanding of endogenous substrate metabolism by the P450 superfamily; new enzymes to be found, new mechanisms of regulation for already known enzymes to be uncovered, and novel functions for both new and unknown P450s.

However, it can be predicted that the classification into two groups will break down because of the discovery of roles for so-called xenobiotic-metabolizing P450s in endogenous metabolism. Several drug-metabolizing enzymes are found expressed early in development, prior to challenge by our chemical environment. Perhaps these enzymes play important roles in metabolism of endogenous compounds involved in development (such as retinoids) and in fact can also be classed as P450s involved in endogenous metabolism. There can be no doubt that in the next few years the distinction between P450s involved in endogenous substrate metabolism and P450s involved in xenobiotic metabolism will become more blurred.

ACKNOWLEDGMENTS.   The authors thank their many colleagues for contributions to understanding the regulation of steroidogenesis. Support of the USPHS (grant DK28350) is also acknowledged. The editorial assistance of Marlene Jayne was invaluable.

References

1. Simpson, E. R., 1979, Cholesterol side-chain cleavage cytochrome P450 and the control of steroidogenesis, *Mol. Cell. Endocrinol.* **13**:213–227.

2. Jefcoate, C. R., McNamara, B. C., Artemenko, I., and Yamazaki, T., 1992, Regulation of cholesterol movement to mitochondrial cytochrome P450scc in steroid hormone synthesis, *J. Steroid Biochem. Mol. Biol.* **43**:751–767.

3. DuBois, R. N., Simpson, E. R., Tuckey, J., Lambeth, J. D., and Waterman, M. R., 1981, Evidence for a higher molecular weight precursor of cholesterol side chain cleavage cytochrome P450 and induction of mitochondrial and cytosolic proteins by ACTH in adult bovine adrenal cells, *Proc. Natl. Acad. Sci. USA* **78**:1028–1032.

4. Matocha, M. F., and Waterman, M. R., 1985, Synthesis and processing of mitochondrial steroid hydroxylases. In vivo maturation of the precursor forms of cytochrome P450scc, P45011β and adrenodoxin, *J. Biol. Chem.* **260**:12259–12265.

5. Sakaguchi, M., Mihara, K., and Sato, R., 1987, A short amino-terminal segment of microsomal cytochrome P450 functions both as an insertion signal and as a stop-transfer sequence, *EMBO J.* **6**:2425–2431.

6. Clark, B. J., and Waterman, M. R., 1991, The folding of bovine 17α-hydroxylase into a functional hemoprotein in COS 1 cells requires the hydrophobic amino terminal sequence, *J. Biol. Chem.* **266**:5898–5904.

7. Clark, B. J., and Waterman, M. R., 1992, Functional expression of bovine 17α-hydroxylase in COS 1 cells is dependent upon the presence of an amino-terminal signal anchor sequence, *J. Biol. Chem.* **267**:24568–24574.

8. Shimozawa, O., Salcaguchi, M., Ogawa, H., Harada, N., Mihara, K., and Omura, T., 1993, Core glycosylation of cytochrome P450(arom), *J. Biol. Chem.* **268**:21399–21402.

9. Amarneh, B., Corbin, C. J., Peterson, J. A., Simpson, E. R., and Graham-Lorence, S., 1993, Functional domains of human aromatase cytochrome P450 characterized by molecular modelling and site-directed mutagenesis, *Mol. Endocrinol.* **7**:1617–1624.

10. Rodgers, R. J., Waterman, M. R., and Simpson, E. R., 1986, Cytochromes P450scc, P45017α, adrenodoxin and reduced nicotinamide adenine dinucleotide phosphate-cytochrome P450 reductase in bovine follicles and corpora lutea. Changes in specific contents during the ovarian cycle, *Endocrinology* **118**:1366–1374.

11. Lambeth, J. D., Kitchen, S. E., and Farooqui, A. A., 1982, Cytochrome P450scc–substrate interactions, *J. Biol. Chem.* **257**:1876–1884.

12. Nakajin, S., Shinoda, M., Haniu, M., Shively, J. E., and Hall, P. F., 1984, C21 steroid side chain cleavage enzyme from porcine adrenal microsomes, *J. Biol. Chem.* **259**:3971–3976.

13. Zuber, M. X., Simpson, E. R., and Waterman, M. R., 1986, Expression of bovine 17α-hydroxylase cytochrome P450 cDNA in non-steroidogenic (COS 1) cells, *Science* **234**:1258–1261.

14. Inano, H., and Tamaoki, B.-i., 1978, *In vitro* effect of 16α-hydroxyprogesterone on the enzyme activities related to androgen production in human testes, *Acta Endocrinol.* **88**:768–777.

15. Fevold, H. R., Lorence, M. C., McCarthy, J. L., Trant, J. M., Kagimoto, M., Waterman, M. R., and Mason, J. I., 1989, Rat testis P45017α: Characterization of a full-length cDNA encoding a unique steroid hydroxylase capable of catalyzing both P4 and P5- 17,20-lyase reactions, *Mol. Endocrinol.* **3**:968–976.

16. Mason, J. I., and Estabrook, R. W., 1973, Testicular cytochrome P450 and iron–sulfur protein as related to steroid metabolism, *Ann. N.Y. Acad. Sci.* **212**:406–419.

17. Katagiri, M., Kagawa, N., and Waterman, M. R., 1994, The role of cytochrome b5 in the biosynthesis of androgens by human P450c17, *Arch. Biochem. Biophys.* **317**:343–347

18. Schlinger, B. A., and Arnold, A. P., 1991, Brain is the major site of estrogen synthesis in a male songbird, *Proc. Natl. Acad. Sci. USA* **88**:4191–4194.

19. Johnson, E. F., Kronback, T., and Hsu, M.-H., 1992, Analysis of the catalytic specificity of cytochrome P450 enzymes through site-directed mutagenesis, *FASEB J.* **6**:700–705.

20. Ogishima, T., Mitani, F., and Ishimura, Y., 1989, Isolation of aldosterone synthase cytochrome P450 from zona glomerulosa mitochondria of rat adrenal cortex, *J. Biol. Chem.* **264**:10935–10938.

21. Ogishima, T., Mitani, F., and Ishimura, Y., 1989, Isolation of two distinct cytochromes P45011β with aldosterone synthase activity from bovine adrenocortical mitochondria, *J. Biochem.* **105:**497–499.
22. Ikeda, Y., Shen, W.-H., Ingraham, H. A., and Parker, K. L., 1994, Developmental expression of mouse steroidogenic factor-1, an essential regulator of the steroid hydroxylases, *Mol. Endocrinol.* **8:**654–662.
23. Waterman, M. R., and Keeney, D. S., 1992, Genes involved in androgen biosynthesis and the male phenotype, *Horm. Res.* **38:**217–221.
24. Yang, X., Iwamoto, K., Wang, M., Artwohl, J., Mason, J. I., and Pang, S., 1993, Inherited congenital adrenal hyperplasia in the rabbit is caused by a deletion in the gene encoding cytochrome P450 cholesterol side-chain cleavage enzyme, *Endocrinology* **132:**1977–1982.
25. Sakai, Y., Yanase, T., Okabe, Y., Hara, T., Waterman, M. R., Takayanagi, R., Haji, J., and Nawata, N., 1994, No mutation in cytochrome P450scc in a patient with congenital lipoid adrenal hyperplasia, *J. Clin. Endocrinol. Metab.* **79:**1198–1201.
26. Luo, X., Ikeda, Y., and Parker, K. L., 1994, A cell-specific nuclear receptor is essential for adrenal and gonadal development and sexual differentiation, *Cell* **77:**481–490.
27. Lund, J., Faucher, D. J., Ford, S. P., Porter, J. C., Waterman, M. R., and Mason, J. I., 1988, Developmental expression of bovine adrenocortical steroid hydroxylases: Regulation of P45017α expression leads to episodic fetal cortisol production, *J. Biol. Chem.* **263:**16195–16201.
28. Ikeda, Y., Lala, D. S., Luo, X., Kim, E., Moisan, M.-P., and Parker, K. L., 1993, Characterization of the mouse FTZ-F1 gene, which encodes a key regulator of steroid hydroxylase gene expression, *Mol. Endocrinol.* **7:**852–860.
29. Honda, S., Morohashi, K., Nomura, M., Takeya, M., Kitajimi, M., and Omura, T., 1993, Ad4BP regulating steroidogenic P450 genes is a member of steroid hormone receptor superfamily, *J. Biol. Chem.* **268:**7479–7502.
30. Rodgers, R. J., Waterman, M. R., and Simpson, E. R., 1987, Levels of messenger ribonucleic acid encoding cholesterol side chain cleavage cytochrome P450, 17α-hydroxylase cytochrome P450, adrenodoxin, and low density lipoprotein receptor in bovine follicles and corpora lutea throughout the ovarian cycle, *Mol. Endocrinol.* **1:**274–279.
31. Barnhart, K. M., and Mellon, P. L., 1994, The orphan nuclear receptor, steroidogenic factor-1, regulates the glycoprotein hormone α-subunit gene in pituitary gonadotropes, *Mol. Endocrinol.* **8:**878–885.
32. Mellon, S. H., and Deschepper, C. F., 1993, Neurosteroid biosynthesis: Genes for adrenal steroidogenic enzymes are expressed in the brain, *Brain Res.* **629:**283–292.
33. Lauber, M. E., and Lichtensteiger, W., 1994, Pre- and postnatal ontogeny of aromatase cytochrome P450 messenger ribonucleic acid expression in the male rat brain studied by *in situ* hybridization, *Endocrinology* **135:**1661–1668.
34. Simpson, E. R., and Waterman, M. R., 1988, Action of ACTH to regulate the synthesis of steroidogenic enzymes in adrenal cortical cells, *Annu. Rev. Physiol.* **50:**427–440.
35. Clark, B. J., Wells, J., King, S. R., and Stocco, D. M., 1994, The purification, cloning, and expression of a novel luteinizing hormone-induced mitochondrial protein in MA-10 mouse Leydig tumor cells, *J. Biol. Chem.* **269:**28318–28322.
36. Waterman, M. R., and Simpson, E. R., 1989, Regulation of steroid hydroxylase gene expression is multifactorial in nature, *Recent Prog. Horm. Res.* **45:**533–566.
37. Brentano, S. T., Picado-Leonard, J., Mellon, S. H., Moore, C. C. D., and Miller, W. L., 1990, Tissue-specific, cyclic adenosine 3′,5′-monophosphate-induced, and phorbol ester-repressed transcription from the human P450c17 promoter in mouse cells, *Mol. Endocrinol.* **4:**1972–1979.
38. Bakke, M., and Lund, J., 1992, A novel 3′,5′-cyclic adenosine monophosphate-responsive sequence in the bovine CYP17 gene is a target of negative regulation by protein kinase C, *Mol. Endocrinol.* **6:**1323–1331.
39. Hanukoglu, I., Feuchtwanger, R., and Hanukoglu, A., 1990, Mechanism of corticotropin and cAMP induction of mitochondrial cytochrome P450 system enzymes in adrenal cortex cells, *J. Biol. Chem.* **265:**20602–20608.

40. Enyeart, J. J., Mlinar, B., and Enyeart, J. A., 1993, T-Type Ca^{2+} channels are required for adrenocorticotropin-stimulated cortisol production by bovine adrenal zona fasciculata cells, *Mol. Endocrinol.* **7:**1031–1040.

41. Kimura, T., 1969, Effects of hypophysectomy and ACTH administration on the level of adrenal cholesterol side-chain desmolase, *Endocrinology* **85:**492–499.

42. Purvis, J. L., Canick, J. A., Latif, S. A., Rosenbaum, J. H., Hologgitar, J., and Menard, R. H., 1973, Lifetime of microsomal cytochrome P450 and steroidogenic enzymes in rat testis as influenced by human chorionic gonadotropin, *Arch. Biochem. Biophys.* **159:**39–49.

43. Purvis, J. L., Canick, J. A., Mason, J. I., Estabrook, R. W., and McCarthy, J. L., 1973, Lifetime of adrenal cytochrome P450 as influenced by ACTH, *Ann. N.Y. Acad. Sci.* **212:**319–342.

44. John, M. E., John, M. C., Boggaram, V., Simpson, E. R., and Waterman, M. R., 1986, Transcriptional regulation of steroid hydroxylase genes by corticotropin, *Proc. Natl. Acad. Sci. USA* **83:**4715–4719.

45. Roesler, W. J., Vandenbar, G. R., and Hanson, R. W., 1988, Cyclic AMP and the induction of eucaryotic gene transcription, *J. Biol. Chem.* **263:**9063–9066.

46. Kagawa, N., and Waterman, M. R., 1991, Evidence that an adrenal-specific nuclear protein regulates cAMP responsiveness of the human CYP21B (P450c21) gene, *J. Biol. Chem.* **266:**11199–11204.

47. Kagawa, N., and Waterman, M. R., 1992, Purification and characterization of a transcription factor which appears to regulate cAMP responsiveness of the human CYP21B gene, *J. Biol. Chem.* **267:**21213–21219.

48. Zanger, U. M., Kagawa, N., Lund, J., and Waterman, M. R., 1992, Distinct biochemical mechanisms for cAMP-dependent transcription of CYP17 and CYP21, *FASEB J.* **6:**713–719.

49. Momoi, K., Waterman, M. R., Simpson, E. R., and Zanger, U. M., 1992, 3′,5′-Cyclic adenosine monophosphate-dependent transcription of the CYP11A (cholesterol side chain cleavage cytochrome P450) gene involves a DNA response element containing a putative binding site for transcription factor Sp1, *Mol. Endocrinol.* **6:**1682–1690.

50. Lund, J., Ahlgren, R., Wu, D., Kagimoto, M., Simpson, E. R., and Waterman, M. R., 1990, Transcriptional regulation of the bovine CYP17 (P45017α) gene, *J. Biol. Chem.* **265:**3304–3312.

51. Zuber, M. X., John, M. E., Olcamura, T., Simpson, E. R., and Waterman, M. R., 1986, Bovine adrenocortical cytochrome P45017α, regulation of gene expression by ACTH and elucidation of primary sequence, *J. Biol. Chem.* **261:** 2475–2482.

52. Bakke, M., and Lund, J., 1994, Mutually exclusive interactions of two nuclear orphan receptors determine activity of a cAMP-responsive sequence in the bovine CYP17 gene, *Mol. Endocrinol.* **9:**327–339.

53. Kagawa, N., Ogo, A., Takahashi, Y., Iwamatsu, A., and Waterman, M. R., 1994, A cAMP-responsive sequence (CRS1) of CYP17 is a cellular target for the homeodomain protein Pbx1, *J. Biol. Chem.* **269:**18716–18719.

54. Monica, K., Galili, N., Nourse, J., Saltman, D., and Cleary, M. L., 1991, PBX2 and PBX3, new homeobox genes with extensive homology to the human proto-oncogene PBX1, *Mol. Cell. Biol.* **11:**6149–6157.

55. Ahlgren, R., Simpson, E. R., Waterman, M. R., and Lund, J., 1990, Characterization of the promoter/regulatory region of the bovine CYP11A (P450scc) gene: Basal and cAMP-dependent expression, *J. Biol. Chem.* **265:**3313–3319.

56. Venepally, P., and Waterman, M. R., 1994, Two Sp1-binding sites mediate cAMP-induced transcription of the bovine CYP11A gene through the protein kinase A pathway, *J. Biol. Chem.* submitted for publication.

57. Honda, S., Morohashi, K., and Omura, T., 1990, Novel cAMP regulatory elements in the promoter region of bovine P45011β gene, *J. Biochem.* **108:**1042–1049.

58. Chen, J.-Y., and Waterman, M. R., 1992, Two promoters in the bovine adrenodoxin gene and the role of associated, unique cAMP-responsive sequences, *Biochemistry* **31:**2400–2407.

59. Rice, D. A., Aitken, L. D., Vandenbark, G. R., Mouw, A. R., Franklin, A., Schimmer, B. P., and Parker, K. L., 1989, A cAMP-responsive element regulates expression of the mouse steroid 11β-hydroxylase gene, *J. Biol. Chem.* **264:**14011–14015.

60. Domalik, L. J., Chaplin, D. D., Kirkman, M. S., Wu, R. C., Liu, W. W., Howard, T. A., Seldin, M. F., and Parker, K. L., 1991, Different isozymes of mouse 11β-hydroxylase produce mineralocorticoids and glucocorticoids, *Mol. Endocrinol.* **5:**1853–1861.

61. Chang, C.-Y., Huang, C., Guo, I.-C., Tsai, H.-M., Wu, D.-A., and Chung, B.-C., 1992, Transcription of the human ferredoxin gene through a single promoter which contains the 3′,5′-cyclic adenosine monophosphate-responsive sequence and SP1-binding site, *Mol. Endocrinol.* **6:**1362–1370.

62. Guo, I.-C., Tsai, H.-M., and Chung, B.-C., 1994, Actions of two different cAMP-responsive sequence and an enhancer of the human CYP11A1 (P450scc) gene in adrenal Y1 and placenta JEG-3 cells, *J. Biol. Chem.* **269:**6362–6369.

63. Rice, D. A., Kirkman, M. S., Aitken, L. D., Mouw, A. R., Schimmer, B. P., and Parker, K. L., 1990, Analysis of the promoter region of the gene encoding mouse cholesterol side-chain cleavage enzyme, *J. Biol. Chem.* **265:**11713–11720.

64. Oonk, R. B., Parker, K. L., Gibson, J. L., and Richards, J. S., 1990, Rat cholesterol side-chain cleavage cytochrome P450 (P450scc) gene. Structure and regulation by cAMP *in vitro, J. Biol. Chem.* **265:**22392–22401.

65. Watanabe, N., Inoue, H., and Fujii-Kuriyama, Y., 1994, Regulatory mechanisms of cAMP-dependent and cell-specific expression of human steroidogenic cytochrome P450scc (CYP11A1) gene, *Eur. J. Biochem.* **222:**825–834.

66. Moore, C. C. D., Brentano, S. T., and Miller, W. L., 1990, Human P450scc gene transcription is induced by cyclic AMP and repressed by 12-O-tetradecanoylphorbol-13-acetate and A23187 through independent cis elements, *Mol. Cell. Biol.* **10:**6013–6023.

67. Youngblood, G. L., and Payne, A. H., 1992, Isolation and characterization of the mouse P450 17α-hydroxylase/C$_{17-20-}$ lyase gene (CYP17): Transcriptional regulation of the gene by cyclic adenosine 3′,5′-monophosphate in MA-10 Leydig cells, *Mol. Endocrinol.* **6:**927–934.

68. Parissenti, A., Parker, K. L., and Schimmer, B. P., 1993, Identification of promoter elements in the mouse 21-hydroxylase (Cyp21) gene that require a functional cAMP-dependent protein kinase, *Mol. Endocrinol.* **7:**283–290.

69. Wilson, T., Mouw, A. R., Weaver, C. A., Millbrandt, J., and Parker, K. L., 1993, The orphan nuclear receptor NGF1-B regulates steroid 21-hydroxylase gene expression, *Mol. Cell. Biol.* **13:**861–868.

70. Watanabe, N., Kitazume, M., Fujisawa, J., Yoshida, M., and Fujii-Kuriyama, Y., 1993, A novel cAMP-dependent regulatory region including a sequence like the cAMP-responsive element, far upstream of the human CYP21A2 gene, *Eur. J. Biochem.* **214:**521–531.

71. Waterman, M. R., 1994, Biochemical diversity of cAMP-dependent transcription of steroid hydroxylase genes in the adrenal cortex, *J. Biol. Chem.* **269:**27783–27786.

72. Means, G. D., Mahendroo, J., Corbin, C. J., Mathis, J. M., Powell, F. E., Mendelson, C. R., and Simpson, E. R., 1989, Structural analysis of the gene encoding human aromatase cytochrome P450, the enzyme responsible for estrogen biosynthesis, *J. Biol. Chem.* **264:**19385–19391.

73. Harada, N., Yamada, K., Saito, K., Kibe, N., Dohmae, S., and Takagi, Y., 1990, Structural characterization of the human estrogen synthetase (aromatase) gene, *Biochem. Biophys. Res. Commun.* **166:**365–372.

74. Toda, K., Terashima, M., Kamamoto, T., Sumimoto, H., Yamamoto, Y., Sagara, Y., Ikeda, H., and Shizuta, Y., 1990, Structural and functional characterization of human aromatase P450 gene, *Eur. J. Biochem.* **193:**559–565.

75. Means, G. D., Kilgore, M. W., Mahendroo, M. S., Mendelson, C. R., and Simpson, E. R., 1991, Tissue-specific promoters regulate aromatase cytochrome P450 gene expression in human ovary and fetal tissues, *Mol. Endocrinol.* **5:**2005–2013.

76. Kilgore, M. W., Means, G. D., Mendelson, C. R., and Simpson, E. R., 1992, Alternative promotion of aromatase cytochrome P450 expression in human fetal tissues, *Mol. Cell. Endocrinol.* **83:**R9–R16.

77. Harada, N., Utsumi, T., and Talcagi, Y., 1993, Tissue-specific expression of the human aromatase cytochrome P450 gene by alternative use of multiple exons I and promoters, and switching of tissue-specific exons I in carcinogenesis, *Proc. Natl. Acad. Sci. USA* **90:**11312–11316.

78. Simpson, E. R., Mahendroo, M. S., Means, G. D., Kilgore, M. W., Hinshelwood, M. M., Graham-Lorence, S., Amarneh, B., Ito, Y., Fisher, C. R., Michael, M. D., Mendelson, C. R., and Bulun, S. E., 1994, Aromatase cytochrome P450, the enzyme responsible for estrogen biosynthesis, *Endocrine Rev.* **15:**342–355.

79. Hickey, G. T., Krasnow, J. S., Beattie, W. G., and Richards, J. S., 1990, Aromatase cytochrome P450 in rat ovarian granulosa cells before and after luteinization: Adenosine 3,5-monophosphate-dependent and independent regulation. Cloning and sequencing of rat aromatase cDNA and 5 genomic DNA, *Mol. Endocrinol.* **4:**3–12.

80. Matsumine, H., Herbst, M. A., Ou, S.-H. I., Wilson, J. D., and McPhaul, M. J., 1991, Aromatase mRNA in the extragonadal tissues of chickens with the Henny-feathering trait is derived from a distinct promoter structure that contains a segment of a retroviral long terminal repeat, *J. Biol. Chem.* **266:**19900–19907.

81. Mendelson, C. R., Corbin, C. J., Smith, M. E., Smith, J., and Simpson, E. R., 1986, Growth factors suppress, and phorbol esters potentiate the action of dibutyryl cyclic AMP to stimulate aromatase activity of human adipose stromal cells, *Endocrinology* **118:**968–973.

82. Evans, C. T., Corbin, C. J., Saunders, C. T., Merrill, J. C., Simpson, E. R., and Mendelson, C. R., 1987, Regulation of estrogen biosynthesis in human adipose stromal cells: Effects of dibutyryl cyclic AMP, epidermal growth factor, and phorbol esters on the synthesis of aromatase cytochrome P450, *J. Biol. Chem.* **262:**6914–6920.

83. Toda, K., Miyahara, K., Kawamoto, T., Ikeda, H., Sagara, Y., and Shizuta, Y., 1992, Characterization of a cis-acting regulatory element involved in human aromatase P450 gene expression, *Eur. J. Biochem.* **205:**303–309.

84. Fitzpatrick, S. L., and Richards, J. S., 1993, Cis-acting elements of the rat aromatase promoter required for cAMP induction in ovarian granulosa cells and constitutive expression in R2C Leydig cells, *Mol. Endocrinol.* **7:**341–354.

85. Usui, E., Noshiro, M., and Okuda, K., 1990, Molecular cloning of cDNA for vitamin D_3 25-hydroxylase from rat liver mitochondria, *FEBS Lett.* **262:**135–138.

86. Nelson, D. R., Kamataki, T., Waxman, D. J., Guengerich, F. P., Estabrook, R. W., Feyereisen, R., Gonzalez, F. J., Coon, M. J., Gunsalus, I. C., Gotoh, O., Okuda, K., and Nebert, D. W., 1993, The P450 superfamily: Update on new sequences, gene mapping, accession numbers, early trivial names of enzymes, and nomenclature, *DNA Cell Biol.* **12:**1–51.

87. Okuda, K.-I., 1994, Liver mitochondrial P450 involved in cholesterol catabolism and vitamin D activation, *J. Lipid Res.* **35:**361–372.

88. Mandel, M. L., Swartz, S. J., and Ghazarian, J. G., 1990, Avian kidney mitochondrial hemeprotein P4501α: Isolation, characterization and NADPH-ferredoxin reductase-dependent activity, *Biochim. Biophys. Acta* **1034:**239–246.

89. Hiwatashi, A., Nishii, Y., and Ichikawa, Y., 1982, Purification of cytochrome P4501a (25-hydroxyvitamin D_3-1α-hydroxylase) of bovine.kidney mitochondria, *Biochem. Biophys. Res. Commun.* **105:** 320–327.

90. Gray, R. W., and Ghazarian, J. G., 1989, Solubilization and reconstitution of kidney 25-hydroxyvitamin D_3 1α- and 24-hydroxylases from vitamin D-replete pigs, *Biochem. J.* **259:**561–568.

91. Armbrecht, H. J., Nemani, R. K., and Wongsurawat, N., 1993, Regulation of calcium metabolism by the vitamin D hydroxylases, in: *Advances in Molecular and Cell Biology* (C. R. Jefcoate, ed.) in press.

92. Ohyama, Y., Noshiro, M., and Okuda, K., 1991, Cloning and expression of cDNA encoding 24-hydroxyvitamin D_3 25-hydroxylase, *FEBS Lett.* **278:**195–198.

93. Valle, L. D., Belvedere, P., Simontacchi, C., and Colombo, L., 1992, Extraglandular hormonal steroidogenesis in aged rats, *J. Steroid Biochem. Mol. Biol.* **43:**1095–1098.

94. Nebert, D. W., Nelson, D. R., Coon, M. J., Estabrook, R. W., Feyereisen, R., Fujii-Kuriyama, Y., Gonzalez, F. J., Guengerich, F. P., Gunsalus, I. C., Johnson, E. F., Loper, J. C., Sato, R., Waterman, M. R., and Waxman, D. J., 1991, The P450 superfamily: Update on new sequences, gene mapping, and recommended nomenclature, *DNA Cell Biol.* **10:**1–14.

Cytochrome P450 and the Metabolism of Arachidonic Acid and Oxygenated Eicosanoids

JORGE H. CAPDEVILA, DARRYL ZELDIN, KEIKO MAKITA, ARMANDO KARARA, and JOHN R. FALCK

1. Introduction

Eukaryotic cells contain substantial amounts of arachidonic acid (AA; 5,8,11,14-ei-cosatetraenoic acid) esterified predominantly to the *sn*-2 position of cellular glycerophospholipids. As with many lipid-derived mediators, e.g., cholesterol, phosphoinositides, diglycerides, AA serves a structural role, as a component of cellular membranes, and an important functional role, as a participant in a variety of receptor/agonist-mediated signaling cascades.[1-4] In the absence of stimuli, the intracellular levels of nonesterified AA are nearly undetectable. However, most organ cells possess an elaborate enzymatic machinery that, in response to a variety of stimuli, catalyzes: (1) the hydrolytic cleavage of the AA molecule form selected, hormonally sensitive phospholipid pools, (2) the transduction of chemical information into the fatty acid molecular template by means of regio- and stereospecific oxygenation reactions, and (3) the decoding of that chemical information either by receptor-mediated processes or, alternatively, by the direct effects of these oxygenated metabolites on metabolic pathways[1-4] (Fig. 1). As a net result, these processes provide cells with a rapid and versatile on/off molecular switch for the intra- or intercellular transduction and/or amplification of functionally meaningful information.

JORGE H. CAPDEVILA • Departments of Medicine and Biochemistry, Vanderbilt University Medical School, Nashville, Tennessee 37232. *DARRYL ZELDIN, KEIKO MAKITA, and ARMANDO KARARA* • Department of Medicine, Vanderbilt University Medical School, Nashville, Tennessee 37232. *JOHN R. FALCK* • Department of Molecular Genetics, Southwestern Medical Center, Dallas, Texas 75235.

Cytochrome P450: Structure, Mechanism, and Biochemistry (Second Edition), edited by Paul R. Ortiz de Montellano. Plenum Press, New York, 1995.

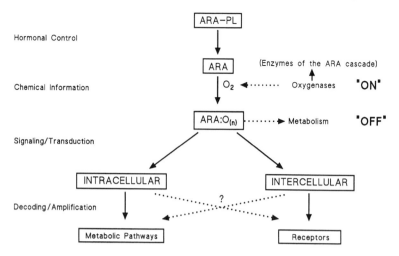

FIGURE 1. Metabolism, bioactivation, and signaling properties of the arachidonic acid molecular template. ARA, arachidonic acid; ARA-PL, arachidonoyl-glycerophospholipids.

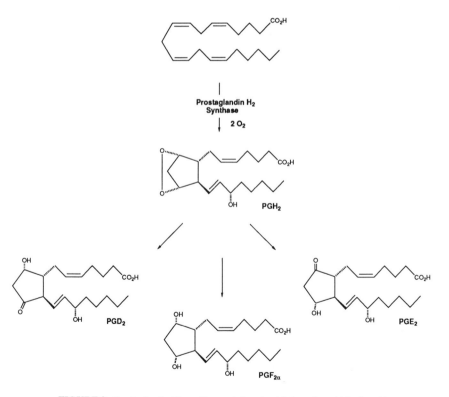

FIGURE 2. Prostaglandin H_2 synthase-catalyzed oxidation of arachidonic acid.

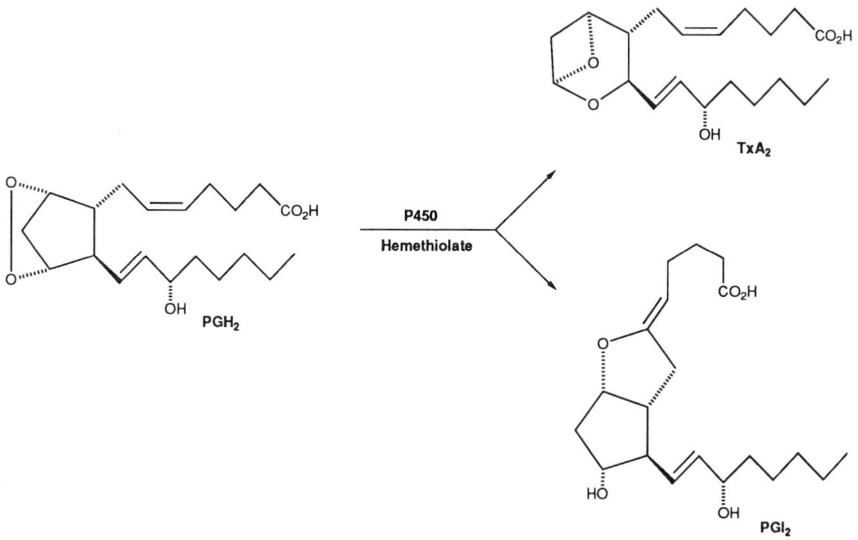

FIGURE 3. Isomerization of prostaglandin H_2 by thromboxane and prostacyclin synthase.

The AA cascade, consisting of prostaglandin H_2 synthase, lipoxygenases, and, more recently, cytochrome P450, is a premier example of the role that lipid-derived mediators play in cell and organ function.[1-8] Metabolism by prostaglandin H_2 synthase generates an unstable cyclic endoperoxide [prostaglandin H_2 (PGH$_2$)] that rearranges enzymatically or chemically to generate several prostaglandins (PGs), prostacyclin (PGI$_2$), or thromboxane A_2 (TXA$_2$) (Figs. 2 and 3). Metabolism by the lipoxygenases generates several regioisomeric allylic hydroperoxides containing a *cis,trans* conjugated-diene functionality. One of these, 5-hydroperoxyeicosatetranoic acid (5-HPETE), is the precursor in the biosynthesis of leukotrienes (Fig. 4). The physiological and biomedical significance of prostanoids and leukotrienes has been extensively documented.[1-4] Among these are their critical roles in pulmonary, vascular, and renal physiology as well as in the pathophysiology of inflammation, asthma, and, more recently, cancer.[9] The reactions catalyzed by PGH$_2$ synthase and lipoxygenases are mechanistically similar to those of free radical-mediated autoxidation of polyunsaturated fatty acids. Reactions are initiated by regioselective hydrogen atom abstraction from a bis-allylic methylene carbon, followed by regio- and enantioselective coupling of the resulting carbon radical to ground-state molecular oxygen. The kinetics, regiochemistry, and chirality of these reactions are all under strict enzymatic control. Thus, in contradistinction to the cytochrome P450-catalyzed, redox-coupled, activation of molecular oxygen and delivery to ground-state carbon, PGH$_2$ synthase, and lipoxygenases are typical dioxygenases that catalyze substrate instead of oxygen activation.

The participation of microsomal cytochrome P450 in the hydroxylation of the ultimate and penultimate carbon atoms of short- and mid-chain saturated fatty acids (ω, ω-1 oxidations) is well established. The physiological significance of these reactions, catalyzed

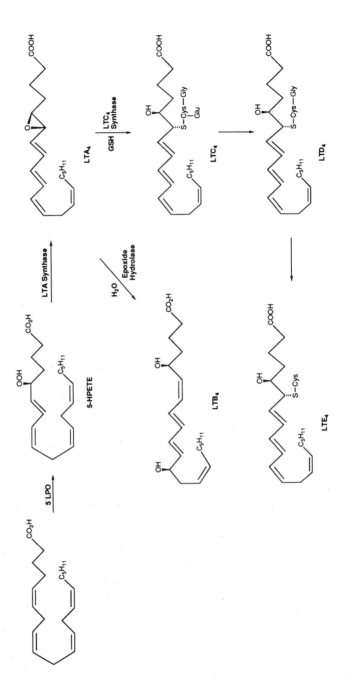

FIGURE 4. Enzymology of leukotriene biosynthesis.

predominantly by members of the P450 4A gene subfamily, is yet to be determined, although they may be of importance for fatty acid catabolism. The studies of the involvement of microsomal P450 in the metabolism of AA were initiated in 1969 by the demonstration that polyunsaturated fatty acids, including AA, interacted with the P450 heme moiety and inhibited the microsomal metabolism of several drugs.[10] These studies were confirmed and expanded, nearly 10 years later, by Pessayre *et al.*[11] The then growing pharmacological and toxicological importance of P450 focused the attention of most researchers on the study of its role in the transformation of xenobiotics and, thus, little further work was done in this area. In 1976, Cinti and Feinstein demonstrated the presence of P450 in human platelets and, importantly, that AA-induced platelet aggregation could be blocked by known P450 inhibitors.[12] It was not until 1981 that the role of cytochrome P450 in the oxidative metabolism of AA was unequivocally demonstrated. Thus, microsomal fractions and purified preparations of P450, reconstituted with purified NADPH-P450 reductase and cytochrome b_5, were shown to actively catalyze the oxidative metabolism of AA to a series of polar products, chromatographically distinct from prostanoids.[13,14] Soon after, the structural characterization of most of the products generated from AA by incubates containing liver or kidney microsomal fractions was completed.[15–19] It was evident from the outset that the physiological importance of the AA substrate made these observations unique and likely to be functionally significant. Furthermore, interest in these novel reactions of P450 was stimulated by (1) the initial demonstration that some of its products displayed potent biological activities[20,21] and, (2) the documentation of its participation in the *in vivo* metabolism of endogenous AA pools.[22] These earlier studies established the metabolism of AA by P450 as a formal metabolic pathway, P450 as an endogenous member of the AA metabolic cascade, and, more importantly, suggested functional roles for this enzyme in the bioactivation of the fatty acid and, thus, in cell and organ physiology. In recent years, the study of the biochemistry and biological significance of these reactions has developed into an area of intense research. Although the physiological and/or pathophysiological implications of this pathway of AA metabolism remain to be fully understood, work from several laboratories is beginning to establish biochemical and functional correlations that can be interpreted as suggestive of a physiological function.[5–8] Here we will focus primarily on recent advances in the biochemistry of cytochrome P450-catalyzed eicosanoid metabolism. Functional aspects that, in our view, hold the greatest promise for the future will be briefly highlighted. The potential physiologic importance of these reactions and their metabolites has been reviewed.[5–8]

The participation of several P450 isoforms in the metabolism of prostanoids was well established prior to the demonstration of its role in arachidonate oxygenation. Soon after the structural characterization of the leukotrienes, several groups demonstrated their oxidation by microsomal P450.[4] As with AA, the potential functional significance of these reactions has stimulated considerable interest in their enzymology and regulation.[5–8] For clarity, we will discuss first those reactions associated with the metabolism of oxygenated eicosanoids by P450 and then concentrate on the studies of its role(s) in AA metabolism.

2. Metabolism of Oxygenated Eicosanoids

In addition to the catalysis of AA metabolism, microsomal cytochrome P450 also participates in the enzymatic transformation of several oxygenated eicosanoids by mediating both NADPH-independent and -dependent reactions. This differential requirement for NADPH and redox changes in the heme iron illustrate the marked differences in oxygen chemistry for these reactions, i.e., the activation and delivery of atmospheric oxygen to the substrate versus the isomerization of AA peroxides.

2.1. NADPH-Independent Reactions; "Peroxide Isomerase"

The peroxidase activity of P450 was initially described in 1974 by O'Brien and collaborators.[23] Since then, the hemoprotein has been shown to catalyze the metabolism of a wide variety of organic hydroperoxides including fatty acid peroxides.[24–27] This activity of P450 is associated with the enzyme's ferric state (Fe^{3+}) and does not require electron transfer from NADPH. Compared to the hemoprotein monooxygenase activity, these reactions exhibit unusually high catalytic rates, reflecting perhaps, (1) a more straightforward and/or simpler oxygen chemistry, (2) lack of enzymatic electron transfer from NADPH to the protein heme-iron center through the flavoprotein cytochrome P450 reductase, and (3) the inherent simplicity of a bimolecular process as compared with the multistep, multicomponent interactions required for the monooxygenase reactions. The mechanism by which the hemoprotein cleaves the peroxide O–O bond, i.e., homolytic or heterolytic scission, plays a decisive role in determining the catalytic outcome of these reactions (see Chapter 15).

Studies of the P450 peroxidase activity, initially thought to be an artifact, provided important insights into the reaction mechanism(s) and the nature of the "active oxygen" involved in the monooxygenase activity of P450.[27] However, the recent characterization of several heme-thiolate coordinated hemoproteins with varying degrees of sequence homology to the P450 gene family, and reports of their ability to isomerize arachidonate peroxides to physiologically important products have evoked interest in these reactions and their functional relevance. A homolytic pathway was proposed for the formation of 11- and 13-hydroxy-14,15-epoxyeicosatrienoic acids from 15-HPETE by rat liver microsomes.[24] Spectral studies initially suggested that human platelet thromboxane synthase was a P450 type of hemoprotein.[25,26] The heterolytic cleavage of the PGH_2 endoperoxide and subsequent oxygen atom or oxenoid transfer[28] was proposed by Ullrich and collaborators to account for the enzymology of PGI_2 and TXA_2 formation from PGH_2[25,26] (Fig. 3). The cDNAs coding for the human platelet and lung thromboxane synthases were cloned, sequenced, and shown to code for cysteine-heme coordinated hemoproteins with ≤ 35% overall amino acid identity to members of the P450 3 gene family.[29,30] Similarly, a cDNA coding for bovine endothelial prostacyclin synthase was recently cloned, sequenced, and expressed as a catalytically functional polypeptide.[31] The enzyme's deduced amino acid sequence showed the presence of the conserved cysteine, typical of all P450s, and a 32% identity to human cholesterol 7α-hydroxylase, a member of the P450 7 gene family.[31] The significance of these pathways and their products to human vascular homeostasis represents one of the best examples of the relevance of AA and its metabolic cascade to clinical medicine.

In plants, the allene oxide derived from the 13-hydroperoxide of linoleic acid is the key precursor in the biogenesis of jasmonic acid, a plant growth hormone.[32] A flaxseed peroxidase responsible for the heterolytic cleavage of 13-hydroperoxy linoleic acid and the formation of the corresponding allene oxide was purified, characterized, and its cDNA cloned and expressed by Brash and collaborators.[32] Sequence analysis indicated that the flaxseed allene oxide synthase had ≤ 25% overall identity to other P450s and that the protein contains the conserved cysteine residue involved in heme coordination.[32]

The discovery and recent association of the peroxidase function of P450 as it relates to the biosynthesis of autacoids of animal or vegetal origin has opened novel and exciting research areas. The significance of these P450-supported pathways to cell physiology and/or pathophysiology has only begun to be explored. The identification of these enzymes, many of which are members of the arachidonate cascade with well-established physiological roles, as homologues of the P450 gene family should facilitate studies of their molecular properties as well as current efforts to delineate the role of these hemoproteins in the control of cell, organ, and body physiology. It is expected that the recent cloning and characterization of these enzymes will stimulate future research into areas such as the potential linkage between cardiovascular dysfunction and genetic and/or functional polymorphisms in the structure or the regulation of the relevant enzymes.

2.2. NADPH-Dependent Reactions

As mentioned, P450 plays a well-established role in the NADPH-dependent metabolism of several bioactive oxygenated eicosanoids. These reactions are of importance in that they (1) increase eicosanoid structural diversity and, hence, informational content, (2) may alter the pharmacological profile of the substrate, and (3) may participate in the regulation of steady-state and/or stimulated levels of physiologically relevant molecules. While it has been generally accepted that these reactions, for the most part, attenuate the biological activity of their substrate and that they are involved in catabolic processes, recent studies have indicated that some ω oxidized prostanoids show unique and potent biological properties.[6] However, in most cases the sequence of steps leading to ω/ω-1 oxidized prostanoids from endogenous AA pools remains to be clarified. Further evaluation should be helpful in establishing the regulation and functional significance of these reactions *in vivo.*

Microsomal P450 oxidizes a variety of eicosanoids that, in addition to AA, includes prostanoids, leukotrienes, HETEs, and EETs. For the most part, these reactions result in the hydroxylation of the eicosanoid at the ultimate (C-20 or ω carbon) or penultimate carbon atoms (C-19 or ω-1 carbon). More recently, Wong and collaborators described the epoxidation of infused PGI_2 by a perfused kidney preparation.[33] Additionally, 5,6- and 8,9-EET are also metabolized by PGH_2 synthase.[34,35] The former leads to a variety of 5,6-oxygenated prostanoids.[34] Oxidation of the latter was stereo-dependent, i.e., 8(S),9(R)-EET formed 11(R)-hydroxy-8(S),9(R)-epoxyeicosatrienoic acid exclusively, whereas the 8(R),9(S)-enantiomer formed both C-11 and C-15 hydroxylated metabolites.[35] A detailed study of secondary metabolism of 12(R)-HETE and 14,15-EET by P450 has been reported.[36,37] To date, however, none of the transformations described in this section, with the exception of ω/ω-1 hydroxylation, have been shown to occur *in vivo* and from endogenous precursors.

2.2.1. ω/ω-1 Oxidation of Prostanoids

The presence of C_{19} hydroxylated prostanoids in human semen was reported by Hamberg and Samuelsson in 1966.[38] Since then, C-19 and C-20 hydroxylations have become recognized routes for the metabolism of several prostanoids. More recently, the hydroxylation at the C-18 (ω-3) carbon of PGE_2 was reported.[39] Early studies demonstrated that prostanoid ω and ω-1 oxidation was NADPH-dependent, localized to the endoplasmic reticulum,[40] and catalyzed by microsomal P450.[41] The recognized biological significance of prostanoids has continued to stimulate interest in these reactions and their physiological significance.

Reconstitution studies using purified enzymes or recombinant P450s showed that most of these reactions were catalyzed by members of the P450 4A gene subfamily. At present, approximately ten 4A isoforms have been cloned and/or isolated and purified from rats (4A1, 4A2, 4A3, and 4A8), rabbits (4A4, 4A5, 4A6, 4A7), or humans (4A9 and 4A11).[42] Many of these enzymes have been characterized enzymatically.[42–58] A summary of their relative substrate specificities and regioselectivities for fatty acid or prostanoid hydroxylation is shown in Table I.

Although the catalysis of NADPH-dependent prostanoid ω-1 and ω-2 oxidation by microsomal preparations is well established, the isoform(s) responsible for these reactions remains to be unequivocally identified. Thus, while several recombinant and/or purified 4A isoforms catalyze both ω and ω-1 oxidation of saturated fatty acids such as laurate and palmitate, none catalyze preferentially AA or prostanoid ω-1 oxidation at significant rates (Table I).

2.2.2. ω/ω-1 Oxidation of Leukotrienes and Other Eicosanoids

The ω-oxidation of leukotriene B_4 (LTB_4) has been documented in whole animals, isolated cells, and subcellular fractions.[4,59–62] Studies from several laboratories demon-

TABLE I
Metabolism of Fatty Acids and Prostanoids by P450 4A Subfamily Isoforms

4A isoform	Species	Enzymatic activities[42–58]
4A1	Rat	ω-oxidation of laurate and arachidonate
4A2	Rat	ω/ω-1 oxidation of laurate
4A3	Rat	ω/ω-1 oxidation of laurate
4A8	Rat	unknown
4A4	Rabbit	ω-oxidation of palmitate, arachidonate, and of prostaglandins A, E, D, and $F_{2\alpha}$
4A5	Rabbit	ω/ω-1 oxidation of laurate and palmitate; some ω-oxidation of PGA_1 and arachidonate
4A6	Rabbit	ω-oxidation of laurate, palmitate, and arachidonate; low PGA_1 ω-oxidation
4A7	Rabbit	ω-oxidation of laurate, palmitate, arachidonate, and PGA_1, inactive toward PGE_1
4A9	Human	ω-oxidation of laurate
4A11	Human	ω-oxidation of laurate

strated that the ω-oxidation of LTB_4 was NADPH-dependent, localized in the endoplasmic reticulum, and catalyzed by P450.[4,59–62] Although not as extensively studied as fatty acid or prostanoid hydroxylations, these reactions are also catalyzed by members of the P450 4A gene subfamily and appear to also serve catabolic roles. Recently, a novel LTB_4 ω-hydroxylase was cloned from a human polymorphonuclear leukocyte cDNA library and shown to code for a novel P450 isoform, P450 4F3.[63] Recombinant P450 4F3 actively catalyzed the ω-oxidation of LTB_4 with a K_m of 0.71 µM.[63] However, the activities of P450 4F3 toward prostanoids or fatty acids were not reported.[63] It is of interest that, early on, biochemical studies indicated that leukocyte LTB_4 ω-oxidation was catalyzed by a unique P450 isoform, distinct from those involved in fatty acid and prostanoid metabolism.[64,65]

The ω-oxidation of 12(S)-HETE by polymorphonuclear leukocytes was demonstrated in 1984 by Wong et al.[66] and Marcus et al.[67] Moreover, the last authors also showed that endogenous AA pools could be converted to 12,20-dihydroxyeicosatetraenoic acid by a coincubated mixture of human platelets and polymorphonuclear leukocytes, thus providing one of the first examples of intercellular eicosanoid metabolism. Both 5- and 15-HETE are known to undergo ω-oxidation by P450.[68,69] Finally, a P450 similar to that responsible for LTB_4 metabolism has been implicated in the ω-oxidation of lipoxin A_4 and B_4 by human neutrophils and polymorphonuclear leukocytes.[70,71]

3. The Arachidonic Acid Monooxygenase

Studies of the enzymology of the AA cascade were initially dominated by those reactions associated with PGH_2 synthase or with PGH_2 isomerases such as prostacyclin and thromboxane synthase.[1–3] In 1979, with the discovery that leukotriene C_4 and the slow-reacting substance of anaphylaxis (SRSA) were the same molecule,[4] emphasis shifted toward the characterization of mammalian lipoxygenases. In both instances, the functional significance of these metabolites preceded the biochemical and molecular characterization of the relevant enzymes. Indeed, the physiological significance of prostanoids and leukotrienes has been the driving force responsible for the intensive research in this area. On the other hand, and with the exception of those mitochondrial P450 isoforms involved in cholesterol and steroid metabolism, studies of the enzymology, biochemistry, and molecular biology of microsomal P450s were stimulated by their role in transforming xenobiotics and their relevance to toxicology, pharmacology, and chemical carcinogenesis. The last 10 years has witnessed an increased interest in the physiological significance of microsomal P450 in the metabolism of several endogenous substrates. AA, as the physiological precursor for numerous important lipidic mediators of cell function, has served as a focal point for many of these studies.[5–8]

As mentioned, the catalysis of AA oxidation by liver and kidney microsomes was first reported in 1981.[13] In those early studies it was demonstrated that the reactions were NADPH-dependent, required a functional hemoprotein, and were unrelated to NADPH-dependent microsomal lipid peroxidation.[13,14] As with most enzymes of the arachidonate cascade, P450 did not metabolize phospholipid-bound AA nor its methyl ester.[14] For example, incubation mixtures containing $[1-^{14}C]=AA$, rat liver or kidney microsomal fractions, and NADPH catalyzed the rapid oxidation of the fatty acid (3–6 and 0.2–0.5

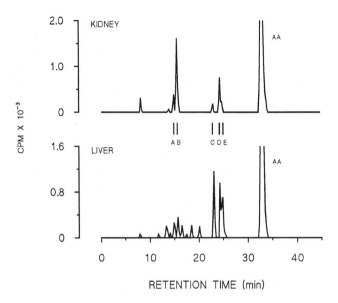

FIGURE 5. Regioselectivity of arachidonic acid metabolism by microsomal fractions isolated from rat liver and kidney. Rat liver and kidney microsomes (0.5 and 1.0 mg protein/ml, respectively) were incubated with [1-^{14}C]arachidonic acid (1–4 µCi/µmole, 100 µM, final concentration) and NADPH (1 mM, final concentration). After 5 min (liver) or 10 min (kidney) at 30°C, organic soluble products were extracted and resolved by reversed-phase high-pressure chromatography as described.[14] Shown are radiochromatograms derived from incubates containing 0.2 mg of liver (bottom) or 0.6 mg of kidney microsomes (top). A, ω-1 alcohol; B, ω-alcohol; C, 14, 15 EET; D, 11, 12-EET; E, 8, 9-EET.

nmole product/min per mg microsomal protein for liver and kidney microsomes, respectively, at 30°C) to several polar products. As shown in Fig. 5, radioactive metabolites could be extracted into organic solvents and resolved by reversed-phase HPLC. The detailed structural characterization of metabolites isolated from incubates containing rat liver or kidney microsomal fractions[7,8,15–19]demonstrated that, under conditions favoring primary metabolism, i.e., low protein concentrations and short incubation times, the microsomal enzymes oxidized AA by one or more of the following types of reactions:

1. *Allylic oxidation* (*lipoxygenase-like reaction*) to generate six regioisomeric hydroxyeicosatetraenoic acids containing a *cis,trans*-conjugated dienol functionality (HETEs; 5-, 8-, 9-, 11-, 12-, and 15-HETE)

2. *Hydroxylations at or near the terminal sp³ carbon* (*ω/ω-1 oxygenase reaction*) affording 16-, 17-, 18-, 19-, and 20-hydroxyeicosatetraenoic acids (16-, 17-, 18-, 19-, and 20-OH-AA) (ω, ω-1, ω-2, ω-3, and ω-4 alcohols)

3. *Olefin epoxidation* (*epoxygenase reaction*) furnishing four regioisomeric epoxyeicosatrienoic acids (EETs; 5,6-, 8,9-, 11,12-, and 14,15-EET) (Fig. 6).

As with most P450-catalyzed reactions, the type of products generated from AA during metabolism are highly dependent on the tissue source of microsomal enzymes,

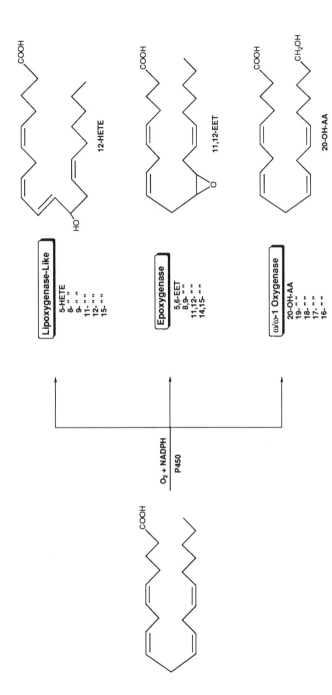

FIGURE 6. Reactions catalyzed by microsomal P450 during the metabolism of arachidonic acid.

animal species, sex, age, hormonal status, diet, and exposure to xenobiotics.[5–8] For example, metabolism of AA by microsomal fractions isolated from rat liver generated four regioisomeric EETs as the major reaction products ($\geq 85\%$ of total products) (Fig. 5). On the other hand, under similar conditions, the products of the reaction catalyzed by rat kidney microsomes corresponded to a mixture of ω/ω-1 alcohols and EETs (65 and 35% of the total products, respectively) (Fig. 5). Studies utilizing inducers of microsomal P450 or, alternatively, reconstituted systems containing solubilized and purified rat liver P450 isoforms demonstrated that the hemoprotein controls, in an isoform-specific fashion, the regioselectivity of oxygen insertion into the fatty acid template at two different levels: (1) the *type of reaction catalyzed,* i.e., olefin epoxidation (EETs), allylic oxidation (HETEs), or hydroxylations at the C-16–C-20 sp^3 carbons (ω, ω-1, ω-2, ω-3, and ω-4 alcohols), (2) to a lesser extent, the *regioselectivity of oxygen insertion,* i.e., epoxidation at either of the four olefins, allylic oxidation initiated at any of the three bis-allylic methylenes, or hydroxylation at C-16–C-20, and (3) stereochemical control of oxygen insertion.

The classification of the P450-derived arachidonate metabolites proposed in Fig. 6, and above, was based on the chemistry of the reactions catalyzed by the enzyme system. Furthermore, this classification has provided a rational and useful framework for most of the subsequent studies of the biochemistry, molecular biology, and functional significance of this pathway for arachidonate metabolism. Consequently, our discussion of the involvement of microsomal P450 in the oxygenated, NADPH-dependent, metabolism of AA will be based on that classification.

3.1. Allylic Oxidations (Lipoxygenase-Like Reactions)

This activity of microsomal cytochrome P450 results in the formation of six regioisomeric allylic alcohols containing a characteristic *cis,trans*-conjugated dienol functionality (HETEs) (Fig. 6). These HETEs are structurally similar to those generated by several plant and mammalian lipoxygenases.[72] However, for the P450-catalyzed reaction, no evidence has been found for the formation of an intermediate hydroperoxide.[17] The stereochemical characterization of the HETEs generated by rat liver microsomal P450 demonstrated that while 5-,8-,9-,11-, and 15-HETE were formed as nearly racemic mixtures, 12(R)-HETE was generated enantioselectively (81% optical purity).[73] In mammals, the catalysis of 12(R)-HETE formation is unique to the P450 enzyme system. All of the mammalian 12-lipoxygenases characterized to date are highly selective for the S-enantiomer.[72] The presence of 12(R)-HETE in human skin and its increased generation during the inflammatory conditions associated with psoriasis has been reported.[74] It has been proposed that 12(R)-HETE is the product generated from AA by bovine cornea epithelium by a P450-mediated reaction.[75] However, the chirality of this product and the participation of P450 in its formation remains controversial.[76] Importantly, *in vitro* studies have demonstrated that 12(R)-HETE is a powerful and enantioselective inhibitor of Na^+/K^+-ATPase activity.[75] Finally, the enzymatic formation of 12(R)-hydroxy-5,8,14-eicosatrienoic acid, a potent ocular proinflammatory and angiogenic substance in rabbits, has been elucidated.[77]

The mechanism of HETE formation by microsomal P450 has yet to be defined. The possibility that their formation may be preceded by P450-catalyzed hydroxylation at C-6,

C-10, or C-13, followed by chemical and/or enzymatic isomerization of the resulting bis-allylic hydroxyeicosatetraenoic acids to chiral and/or racemic HETEs has been explored.[78] Microsomal fractions from human, monkey, and rat liver catalyze the bis-allylic hydroxylation of AA to 7-, 10-, and 13-hydroxyeicosatetraenoic acids.[78] Under mildly acidic conditions, these bis-allylic alcohols readily rearranged to HETEs; however, a precursor–product relationship between them and the P450-derived HETEs has not been demonstrated.[8,78] The contribution of P450 to the *in vivo* formation of HETEs, as well as the molecular characterization of P450 isoforms responsible for these reactions, remain to be determined. Nevertheless, the unique chirality of these products and their associated biological activities continue to stimulate interest in their study. Areas to be clarified include: (1) the role of P450 in the biosynthesis of endogenous HETE pools, (2) the identification and molecular characterization of the individual P450 isoforms responsible for this reaction, (3) the uniqueness of 12(*R*)-HETE biosynthesis by the P450 enzymes, and (4) the mechanism of HETE formation by the P450 enzymes and its relationship to that of lipoxygenases.

3.2. Hydroxylations at C-16–C-20 (ω/ω-1 Oxygenase Reaction)

3.2.1. Introduction

The ω and/or ω-1 hydroxylation of saturated, midchain fatty acids is the oldest and best characterized role of microsomal P450 in fatty acid metabolism. For instance, lauric acid ω-oxidation was the reaction utilized to demonstrate the first functional reconstitution of a purified liver microsomal P450.[79] It has been the general consensus that these reactions participate in the catabolism of several midchain fatty acids prior to degradation by β-oxidation and/or urinary excretion. In 1981, AA joined the list of substrates for these P450-mediated reactions when 19- and 20-OH-AA were isolated from incubates containing AA, NADPH, and rabbit kidney cortex microsomes.[13,15,16] It is of interest that the microsomal ω and ω-1 hydroxylation of saturated fatty acids, in particular of lauric acid, proceeds at rates that are substantially higher than those obtained with AA. However, regardless of the structure of the fatty acid substrate, i.e., saturated (e.g., laurate) or polyunsaturated (e.g., arachidonate), ω/ω-1 oxidation entails the delivery by P450 of a reactive oxygen to unactivated sp^3 carbons. Therefore, it is likely that for these reactions, the oxygen chemistries and the reaction mechanism(s) are similar and independent of the degree of saturation in the fatty acid. Nevertheless, the AA molecular template imposes additional steric requirements on the P450 protein catalyst. Hydroxylation at the thermodynamically less reactive C-16 through C-20 and not at the chemically comparable C-2 through C-4 suggest a rigid and highly structured binding site for the AA molecule. This AA binding site must be capable of positioning the acceptor carbon atoms not only in optimal proximity to the heme-bound active oxygen but also with complete segregation of the reactive olefins and bis-allylic methylene carbons (Fig. 6).[80]

3.2.2. Enzymology, Isoform Specificity

AA ω/ω-1 oxygenation has been demonstrated in microsomal fractions isolated from organs such as liver, kidney, lung, intestine, olfactory epithelium, and anterior pituitary.[6–8] It is, however, in renal tissues that these reactions are the most prevalent and thus better

TABLE II
Regioselectivity of Purified P450 ω/ω-1AA Oxygenase Isoforms[a]

Regioisomer	1A1 (%)	1A2 (%)	4A1 (%)
16-OH-AA	8	47	ND
17-OH-AA	19	20	ND
18-OH-AA	19	20	ND
19-OH-AA	46	ND	ND
20-OH-AA	8	ND	100
% of total metabolites	87	45	100

[a]Values are averages calculated from three different experiments with S.E. < 15% of the mean.

characterized.[6–8] Fatty acid ω/ω-1 oxidation is regulated *in vivo* by a wide variety of factors including animal age, diet, starvation, administration of fatty acids, hypolipidemic drugs, dioxins, flavonoids, aspirin, steroids, mineralocorticoids, and diabetes.[6–8,81] More recently, 16-, 17-, and 18-OH-AA (ω-2, ω-3, and ω-4 alcohols) were added to the list of metabolites generated by the P450 AA ω/ω-1 oxygenase reaction.[39,80] Thus, while monkey seminal vesicle microsomes metabolized AA to 18(R)-OH-AA,[39] liver microsomes isolated from β-naphthoflavone-treated rats generated a mixture of the corresponding 16-, 17-, 18-, and 19-alcohols.[80]

Several rat, rabbit, and human members of the P450 4A gene subfamily have been purified and/or cloned and expressed. P450 4A isoforms appear to be highly specialized for fatty acid and/or prostanoid metabolism.[45–58] All enzymatically characterized P450 4A proteins (either purified or recombinant) catalyze saturated fatty acid ω-oxidation and most also hydroxylate AA at the C-20 carbon (Table I). To date, no member of the P450 4A gene subfamily has been shown to be selective only for ω-1 oxidation of fatty acids, including AA (Table I). Studies with inducers of liver microsomal P450 indicated that AA hydroxylation at C-16, C-17, C-18, and C-19, but not at C-20, was induced by β-naphthoflavone and dioxin.[80,82] Reconstitution experiments using purified liver P450 1A1 and 1A2, the two major liver P450 isoforms induced by animal treatment with β-naphthoflavone or dioxins, demonstrated that these isoforms were more or less regioselective for arachidonate oxidation at the C-16 – C-19 carbons (87 and 45% of total products for P450 1A1 and 1A2, respectively)[80] (Table II). Furthermore, while P450 1A1 oxidized AA preferentially at C-19, oxygenation by P450 1A2 occurred predominantly at C-16.[80] Whether these P450 1A1 and 1A2 regioselectivities are unique to AA or common to all fatty acids remains to be determined. Furthermore, it is of interest that despite the limited sequence homology that exists between 1A and 4A isoforms, these enzymes show a distinct regioselectivity for the adjacent C-19 and C-20 carbons.[80] Finally, purified P450 2E1, an isoform induced in rat liver by diabetes, fasting, and alcohol, converted AA to 19(S)- and 18(R)-OH-AA (with 72 and nearly 100% optical purity, respectively) as its major reaction products.[83]

3.2.3. Functional Significance, Prohypertensive Role

The relevance of kidney microsomal P450, and in particular of products of the AA ω/ω-1 oxygenase reaction, to the pathophysiology of experimental hypertension was initially proposed by McGiff and collaborators.[6] In early studies, it was shown that deoxycorticosterone acetate (DOCA) was a powerful inducer of the rabbit kidney AA ω/ω-1 oxygenase activity.[21] Significantly, the time course of P450 induction by DOCA closely paralleled the time-dependent changes in renal salt and water transport induced by the adrenal corticoid.[84] Studies with the spontaneous hypertensive rat model (SHR/WKY model) indicated that, in these animals, the developmental phase of hypertension was linked to attendant increases in the activities of the renal AA ω/ω-1 oxygenase reaction.[6] Moreover, hypertensive SHR animals could be made normotensive after heme oxygenase induction by SnCl$_2$ administration.[6] The normotensive effects of SnCl$_2$ were attributed to a selective depletion of the rat kidney P450 AA ω/ω-1 oxygenases.[6] The preferential expression of the P450 4A2 gene in hypertensive SHR animals has been reported.[85] Additionally, several metabolites of this P450-catalyzed pathway of AA metabolism exhibit potent *in vivo* and *in vitro* effects on renal function, including natriuresis, alterations in ion transport, and vasoconstriction.[6,86] The modulation of renal Na$^+$/K$^+$-ATPases by both 19- and 20-OH-AA[87] as well as the prostaglandin synthase-dependent vasoconstrictor activity of 20-OH-AA and its urinary excretion have been reported.[88,89] However, the majority of the 20-OH-AA present in rat urine was found conjugated to glucuronic acid, an established route for the excretion of hydroxylated compounds.[90]

3.3. Olefin Epoxidation (Epoxygenase Reaction): The Cytochrome P450 AA Epoxygenase

3.3.1. Introduction

In the last few years, this P450-catalyzed pathway of AA metabolism has attracted considerable attention based, in part, on (1) the potent biological activities of its metabolites[5–7] and (2) the early demonstration of the EETs as endogenous constituents of several organ tissues, human urine, and plasma.[7] In mammals, the epoxidation of polyunsaturated fatty acids to bis-allylic, *cis*-epoxides is unique to the P450 enzyme system and, at difference with the fatty acid ω/ω-1 oxygenase, more or less selective for AA.[7,8,91] Thus, while the enzymatic or nonenzymatic reduction and/or isomerization of polyunsaturated fatty acid hydroperoxides can yield epoxides or epoxy-alcohol derivatives, these products are structurally different from those generated by the P450 enzymes.[7,8] Only *cis*-epoxides are generated by the P450 system (Fig. 7).[18] This is most consistent with olefin epoxidation via a concerted process or, alternatively, that the protein catalyst restricts the degrees of freedom for C–C rotation in the transition state.

The catalysis of AA epoxidation by P450 was initially suggested by the demonstration that 11,12- and 14,15-dihydroxyeicosatrienoic acids were formed by incubates containing kidney cortex microsomes, AA, and NADPH.[15] Soon thereafter, Chacos *et al.* proved, for the first time, that rat liver microsomal fractions catalyzed the NADPH-dependent formation of 5,6-,8,9-,11,12-, and 14,15-EET[18] (Fig. 7). These studies unambiguously established microsomal P450 as an active AA epoxygenase and focused attention on its biochemical and physiological implications.[6–8] The discovery of several EET-associated biological activities suggested, early on, a potential functional significance for these reactions.[20,21] To date, the catalysis of EET formation by purified P450s, microsomal fractions, or isolated cell preparations has been demonstrated in numerous tissues, includ-

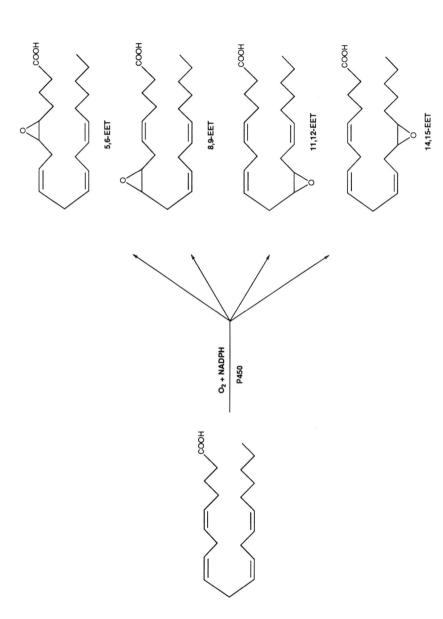

FIGURE 7. The cytochrome P450 arachidonic acid epoxygenase.

ing liver, kidney, pituitary, brain, adrenal, endothelium, and ovary.[6–8] Finally, the establishment of EETs as endogenous constituents of rat liver, human urine, and rabbit kidney moved the epoxygenase from the domain of *in vitro* biochemical reactions to that of an endogenous metabolic pathway[22] and instituted microsomal P450 as a participant in the metabolism of endogenous fatty acids such as AA.

3.3.2. Enzymology, Isoform Specificity

The oxygenated metabolism of AA is characterized by the increasing complexity of its metabolic pathways and by the diversity of functional properties attributed to many of its metabolites. While this metabolic versatility allows for the transfer and storage of a high degree of functionally meaningful chemical information, the integration of these multiple reactions, their products and associated biological activities to cell, tissue, organ, and body physiology is an important and difficult task. It would be greatly facilitated by an understanding of the biochemistry of the reactions involved, the molecular properties of the relevant enzymes, and the chemistry of their metabolites. Additionally, the well-established catalytic, structural, and genetic heterogeneity of the P450 enzyme system further complicates the assignment of a metabolic transformation to distinct P450 isoforms.[42,81] The multiplicity of P450 isoforms involved in AA epoxidation by liver microsomes was initially suggested by observed changes in EET chirality resulting from animal treatment with P450 inducers.[92] For example, phenobarbital treatment increased, in a time-dependent fashion, the regio- and enantiofacial selectivity of the rat liver microsomal epoxygenase(s).[92] The phenobarbital-induced increases in stereoselectivity resulted in a remarkable inversion in absolute configuration of the EETs produced by the microsomal enzymes (Table III).[92] These studies were important in that they demonstrated, in contrast to cyclooxygenase and lipoxygenases enzymes, that the enantioselectivity of the epoxygenase was variable. Among the enzymes of the arachidonate cascade, the P450 epoxygenase is unique in that its regio- and stereochemical selectivity is under regulatory control and can be experimentally altered, *in vivo*, by animal manipulation (Table III).[92] It was subsequently demonstrated that arachidonate epoxidation is highly enantioselective and that P450 controls, in an isoform-specific fashion, the regio- and enantioselectivity of the epoxygenase reaction.[82,92–95] Reconstitution of the P450 AA monooxygenase activity using purified P450 isoforms and/or recombinant proteins demonstrated that members of the P450 2 gene family were the only isoforms that metabolized the fatty acid exclusively by enantioselective epoxidation.[82,93–95] Isoforms of the 2B and 2C subfamily so far identified as epoxygenases include rat 2B1, 2B2, 2C11, and 2C23,[92,94] rabbit 2C1 and 2C2,[95] and human 2C8 and 2C9/10.[96] While P450s 1A1, 1A2, and 2E1 are active AA ω/ω-1 oxygenases, they also produce low and variable amounts of EETs ($\leq 20\%$ of total products).[82,83,92] A unique case is that of a P450 purified from the livers of dioxin-treated chick embryos.[82] This protein has several structural features typical of proteins of the 1A gene subfamily, but metabolizes AA to EETs as the major reaction products (75% of total products).[82]

The structural characterization of the EETs generated by incubates containing purified 2B and 2C P450 epoxygenases showed that, among these proteins, none was selective for the epoxidation of a single fatty acid olefin. Thus, although highly enantioselective, AA

TABLE III
Effect of Phenobarbital Treatment on the Regio- and Stereoselectivity of the Rat Liver
Microsomal AA Epoxygenase[a]

EET regioisomer	Control	Phenobarbital
8,9-EET (rate of formation)[b]	0.38	0.58
% S,R	32	78
% R,S	68	22
11,12-EET (nmole/min/mg protein)	0.59	0.75
% S,R	19	83
% R,S	81	17
14,15-EET (nmole/min/mg protein)	0.42	0.98
% S,R	67	25
% R,S	33	75

[a]Values are averages calculated from at least four different experiments with S.E. ≤ 15% of the mean. The rates of total metabolism for microsomal fractions from control and phenobarbital-treated rats were 2.2±0.3 and 4.2±0.1 nmole/min/per mg protein, respectively.
[b]Reaction rates are in nmole/min/per mg microsomal protein at 30°C.

epoxidation shows a more limited regioselectivity.[92–96] Cross-contamination, a common problem with purified forms of proteins that share extensive sequence identity, complicates the interpretation of results obtained with purified P450 isoforms.[92] The ability of a single 2C P450 protein to catalyze the enantioselective epoxidation of more than one fatty acid olefin was clearly demonstrated using recombinant proteins.[94–96] For example, the first recombinant epoxygenase characterized, P450 2C23, catalyzed the enantioselective epoxidation of AA to 8,9-, 11,12-, and 14,15-EET (27, 54, and 19% of total products, respectively).[94] The recombinant protein generated 8(R),9(S)-, 11(R),12(S)-, and 14(S), 15(R)-EETs with optical purities of 95, 85, and 75%, respectively.[94] The regio- and enantioselective formation of 11,12- and 14,15-EET by recombinant P450 2C2 and a P450 2CAA, purified from rabbit kidney cortex, was reported recently.[95] It is of significance that the degrees of stereochemical selectivity displayed by AA epoxygenase isoforms are unusually high for P450-catalyzed oxidations of unbiased, noncyclic molecules such as AA.[92–96]

Antibody inhibition and product chirality experiments indicate that majority of the constitutive arachidonate epoxygenases in rat liver and kidney microsomes belong to the P450 2C gene subfamily (Fig. 8).[93] By analogy, it was concluded that the predominant human epoxygenase(s) are also members of the P450 2C gene subfamily. Recently, cDNAs coding for human P450 2C8 and 2C9/10 were cloned, expressed, and shown to catalyze the regio- and enantioselective epoxidation of AA.[96] In rat, rabbit, and human, the P450 2C gene subfamily codes for a highly homologous group of proteins, many of which are expressed constitutively.[81,97] Additionally, some 2C isoforms are sex-specific and/or expressed under developmental and hormonal control.[81,97] The extensive sequence identity between the members of this gene subfamily has complicated the interpretation of catalytic assignments done using purified 2C isoforms. Furthermore, it has been demonstrated that P450 protein structural homology can often accompanied by significant catalytic heterogeneity.[98,99] For example, P450s 2C8 and 2C9/10 share extensive sequence homology, but recombinant 2C8 and 2C9/10 epoxidize AA with distinct regio- and stereochemical selectivities.[96] To overcome these limitations, recombinant DNA techniques, with their unique selectivity and amplification powers, are being utilized for the enzymatic characterization of human and rat 2C P450 isoforms. As more recombinant 2C proteins become available, reconstitution studies will be useful in defining the extent to which these P450 isoforms contribute to the epoxidation of endogenous AA pools.

3.3.3. The P450 AA Epoxygenase: A Member of the "AA Metabolic Cascade"

Inasmuch as *in vitro* studies are an indispensable tool for the biochemical, enzymatic, and molecular characterization of metabolic pathways, they provide only limited information concerning the *in vivo* significance of the products and enzymes involved. Additionally, in view of its well-known catalytic versatility, the participation of microsomal P450 in the *in vitro* epoxidation of AA was not completely unexpected. It was therefore apparent that the uniqueness and the significance of the P450 AA epoxygenase reaction were going to be defined by whether or not (1) the enzyme system participated in the *in vivo* metabolism of the fatty acid and (2) its products played significant roles in cell and organ physiology. To address these important questions, and since asymmetric synthesis is an accepted requirement for the biosynthetic origin of most eicosanoids, a method for the quantification and chiral characterization of the EET pools present in biological samples was developed.[100] Chiral analysis of the EETs present endogenously in the rat liver (approximately

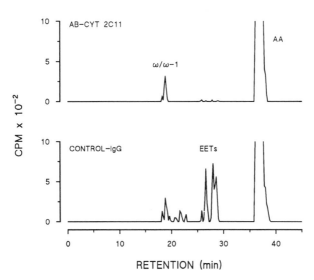

FIGURE 8. Selective inhibition of the rat liver arachidonic acid epoxygenase by a polyclonal antibody raised against purified P450 2C11. Rat liver microsomal fractions (0.5 μM P450, final concentration) were incubated with [1-^{14}C]arachidonic acid (10 μCi/μmole, 100 μM, final concentration) and NADPH (1 mM, final concentration) in the presence of nonimmune (bottom) or anti-P450 2C11 IgG (top) (20 mg of protein, each). After 5 min at 30°C the reaction products were extracted, resolved, and analyzed as described.[14] Shown are the radiochromatograms derived from control, nonimmune (bottom), or anti-P450 2C11 (top)-containing incubates.

TABLE IV
Effect of Phenobarbital Treatment on EET Biosynthesis by Rat Liver[a]

EET regioisomer	Control	Phenobarbital
8,9-EET (concentration)[b]	0.28	1.90
% S,R	79	99
% RS	21	1
11,12-EET (concentration)	0.14	0.17
% S,R	63	83
% R,S	37	17
14,15-EET (concentration)	0.36	1.20
% S,R	23	6
% R,S	77	94

[a]Values are averages calculated from at least three different experiments with S.E. $\leq$ 17% of the mean. The concentration of EET in the livers of control and phenobarbital-treated rats were 0.81±0.1 and 3.3±0.2 µg of total EETs/g liver, respectively.
[b]Concentrations in µg total EETs/g liver.

1 µg total EETs/g wet tissue) showed the biosynthesis of 8,9-, 11, 12-, and 14,15-EET in a 4:1, 2:1, and 3:1 ratio of antipodes, respectively (Table IV).[100] To probe the role of P450 in the *in vivo* catalysis of AA epoxidation by P450, the effects of inducers on the size and the chirality of the rat liver EET pools were determined.[100] Animal treatment with phenobarbital resulted in a 3.7-fold increase in liver microsomal P450 and it induced, selectively, the biosynthesis of 8(S),9(R)- and 14(R),15(S)-EET which were then obtained as nearly optically pure enantiomers (Table IV).[100] These results (1) demonstrated the enzymatic origin of the EETs present in rat liver, (2) documented a new metabolic function for P450 in the epoxidation of endogenous AA pools, and (3) established P450 as a member of the *in vivo* "arachidonic acid metabolic cascade."[7,100] A comparison of the data in Tables III and IV shows that only after phenobarbital pretreatment do the absolute configurations of the endogenous EETs match that of EETs formed *in vitro* by the microsomal enzymes.[92,100] This apparent discrepancy between the chirality of the EETs present, endogenously, in control livers and those generated, *in vitro,* by the microsomal proteins suggests that (1) factors other than rates of biosynthesis control the *in vivo* steady-state concentrations of liver EETs or (2) epoxygenase isoforms responsible for endogenous EET biosynthesis are lost during isolation and/or analysis of the microsomal enzymes. At present, the existence of endogenous chiral EETs has been demonstrated in rat liver, kidney, brain, plasma, and urine, and in human kidney, plasma, and urine.[7]

A distinctive feature of the endogenous EET pools in rat liver and kidney was their presence esterified to the *sn*-2 position of several cellular glycerophospholipids ($\geq$ 92% of the total liver EETs).[101] Chiral analysis of the fatty acids at *sn*-2 revealed an enantioselective preference for 8(S),9(R)-, 11(S),12(R)-, and 14(R),15(S)-epoxyeicosatrienoates in all three classes of phospholipids, with 55% of the total liver EETs in phosphatidylcholine, 32% in phosphatidylethanolamine, and 12% in phosphatidylinositols.[101] EET-phospholipid formation required a multistep process, initiated by the P450 enantioselective epoxidation of AA, ATP-dependent activation to the corresponding EET-CoA derivatives, and EET enantiomer-selective lysolipid acylation.[101] The asymmetric nature of the esterified EET cogently demonstrated that rat liver biosynthesized these lipids from endogenous precursors, enzymatically and under normal physiological conditions.[101] The observed *in vivo* EET esterification process appears to be unique since most endogenously formed eicosanoids are either secreted, excreted, or undergo oxidative metabolism and excretion. There are, however, reports of esterification by isolated cells of exogenously added HETEs and EETs.[102–104]

The biosynthesis of endogenous pools of phospholipids containing esterified EET moieties in rat liver, kidney, brain, and plasma and in human kidney and plasma[7,105] indicates new and potentially important functional roles for P450. As a participant in the arachidonate cascade, microsomal P450 may play a central role(s) in the biosynthesis of unique cellular glycerolipids and, thus, in the control of membrane physicochemical properties and/or the generation of novel lipid-derived mediators. Furthermore, these studies also show, in contrast to most eicosanoids, the potential for the cellular generation of preformed bioactive EETs via hydrolytic reactions, thus obviating the need for AA oxidative metabolism.

3.3.4. Functional Significance of the Epoxygenase Metabolites

As noted previously, work from several laboratories has identified a variety of potent biological activities displayed by the EETs (reviewed in Ref. 5–7) Among these activities, we emphasize: (1) *Vasoactive properties:* The 5,6- and 14,15-EETs have been reported to act as vasodilators at the systemic level and in tissue microcirculation.[6] In contrast, 8(*S*), 9(*R*)-EET, the major circulating enantiomer in plasma,[105] appears to be a powerful and stereoselective renal vasoconstrictor.[6] The reported requirement for prior prostaglandin H_2 synthase metabolism may be responsible for the apparent organ-specific activities of 5,6-EET.[106] (2) *Ion transport:* The 5,6- and 14,15-EETs have been shown to increase cytosolic Ca^{2+} concentrations in several cell preparations including pituitary, hepatocytes, and kidney mesangial and proximal tubule cells.[6] Additionally, the EETs show proximal and distal effects on renal Na^+ transport.[6] (3) *Peptide hormone release:* The EETs, at nanomolar concentrations, serve as potent, regioselective *in vitro* secretagogues for several brain, pituitary, and pancreatic hormones such as somatostatin, luteinizing hormone, growth hormone, vasopressin, prolactin, insulin, and glucagon.[6] Finally, and in light of their significance to cell and body physiology, two areas of current interest in our laboratory will be briefly discussed.

3.3.4a. Membrane Biology.

The biosynthesis of EET-containing cellular phospholipids may provide the molecular basis for some of the biological properties attributed to the EETs.[7] Many of these biological activities can be interpreted in terms of the ability of the EETs to become incorporated into cellular lipids[7] and, consequently, to alter cell membrane permeability and/or fusogenic properties and cause changes in ion fluxes or peptide hormone secretion.[6,7] Exogenously added EETs have been shown to modify the intracellular concentrations of ions such as Ca^{2+}, Na^+, K^+, and H^{+5-7} and the permeability of cell membranes to water or peptide hormones.[5–7] Studies of lipid peroxidation demonstrated that the oxidation of membrane phospholipids profoundly changes the physicochemical properties and, therefore, the structural and functional properties of biological membranes.[107–111] These effects included changes in the permeability of the membrane to ions, and its fusogenic properties as well as in the activities of several membrane-bound enzymes.[107-111] The lipid bilayer provides the matrix in which structural and functional membrane proteins carry out their cellular functions. Alterations in the structure and the chemical composition of these lipids are known to affect the activities and function of membrane proteins.[111] The documented, P450-dependent, enzymatic machinery for the epoxidation of bilayer phospholipids could thus serve as a tool for the structural control of distinct membrane domains. Based on these studies, as well as the capacity of synthetic 8,9-epoxyeicosatrienoylphosphocholine to alter the Ca^{2+} permeability of synthetic liposomes, we proposed a functional role for microsomal P450 in the control of the cell membrane microenvironment structure and, hence, functional properties.[110-112] We envision a process in which enzyme-controlled EET formation and acylation induces localized changes in the fluidity of selected membrane microenvironments. The process could then be reversed by lipase-catalyzed hydrolysis of the acylated EET, followed by enzymatic hydration to the corresponding DHET. Of interest, under conditions that are optimal for EET esterification, no DHET acylation could be detected.[101]

3.3.4b.   Renal Physiology and Hypertension.   The first report of a renal effect by products of the P450 pathway of AA metabolism, published in 1984,[21] revealed the inhibition of Na^+ and K^+ transport in the isolated rabbit cortical collecting tubule by 5,6-EET.[21] Since then, the list of renal effects attributed to the products of the P450 arachidonate monooxygenase has expanded continuously (reviewed in Ref. 6). Among these are: (1) modulation of the renal Na^+/K^+-ATPases, (2) changes in renal cell Ca^{2+} concentrations, (3) changes in proximal tubule Na^+ transport, (4) inhibition of water transport, (5) renal vasodilation, (6) natriuresis, and (7) EET enantioselective renal vasoconstriction.[5-7] As previously noted, the proposal of a role for the P450 monooxygenase in the pathophysiology of hypertension[6] focused interest on the potential significance of this enzyme system to renal and body physiology. More recently, the relevance of the epoxygenase reaction to renal function has been highlighted by the demonstration of (1) the EETs as endogenous constituents of rat and human kidney and urine,[93,113,114] (2) the marked increase in the urinary excretion of DHETs observed during pregnancy-induced hypertension,[114] and (3) the regulation of the rat kidney epoxygenase by excess dietary salt.[93]

Increased salt intake results in increased renal salt excretion. This adaptive process prevents excessive salt retention, volume expansion, and, one of its detrimental sequelae, hypertension. In rats, excess dietary salt markedly increases the urinary excretion of epoxygenase metabolites.[93] Metabolic, immunological, and nucleic acid hybridization studies suggested that high-salt diets induced a P450 isoform that is either absent or present at very low concentrations in the kidneys of untreated animals.[93] These data, in conjunction with the documented capacity of the EETs to inhibit proximal and distal nephron Na^+ absorption,[6] suggested that the salt-inducible P450 AA epoxygenase(s) may be involved in the kidney's adaptive response to an increased salt intake.[93] Two recent observations have provided strong support for this hypothesis: (1) In salt-loaded Sprague–Dawley rats, the administration of clotrimazole resulted in the inhibition of the renal epoxygenase, a reduction in the urinary excretion of EETs, DHETs, and Na^+, and in the development of salt-sensitive and clotrimazole-dependent systemic hypertension.[115] The salt-sensitive and clotrimazole-dependent hypertension was fully abrogated when either the clotrimazole treatment or the salt loading was discontinued.[115] (2) In the Dahl rat model of genetic salt-sensitive hypertension,[116] the salt-sensitive phenotype has been associated with an inability of the kidneys to induce their AA epoxygenase in response to excess dietary salt.[115] Importantly, in hypertensive Dahl salt-sensitive rats, salt sensitivity has recently been associated with a mutation in the gene coding for a P450 2C isoform.[117] While salt-sensitive hypertension in humans and in the Dahl rat is most likely a polygenic trait,[116] the data summarized indicated that a genetically controlled or experimental impairment of the renal epoxygenase(s) leads to the development of systemic high blood pressure.[115,116] The indication that alterations in the function and/or regulation of individual P450 protein(s) predispose animals to salt-sensitive hypertension is not only of paramount biochemical and physiological significance but also, and more importantly, of high significance for clinical medicine.

In conclusion, results from several groups have demonstrated the central role that the cytochrome P450 system plays in the metabolism and the bioactivation of AA. These studies have not only expanded the list of biologically significant eicosanoids, but also documented new and functionally significant endogenous roles for microsomal P450. These observations may prove vital to our understanding of lipid metabolism, membrane biology, and the physiological significance of P450 in the formation of lipid-derived mediators.

ACKNOWLEDGMENT. The authors thank the USPHS–NIH for supporting the research carried out in their laboratories and cited herein.

References

1. Needleman, P., Turk, J., Jakschik, B. A., Morrison, A. R., and Lefkowith, J. B., 1986, Arachidonic acid metabolism, *Annu. Rev. Biochem.* **55**:62–102 and references therein.
2. Smith, W. L., Marnett, L. J., and DeWitt, D. L., 1991, Prostaglandin and thromboxane biosynthesis, *Pharmacol. Ther.* **49**:153–179, and references therein.
3. Smith, W. L., 1992, Prostanoid biosynthesis and mechanism of action, *Am. J. Physiol.* **263**:F181–F191, and references therein.
4. Ford-Hutchinson, A. W., Gresser, M., and Young, R. N., 1994, 5-Lipoxygenase, *Annu. Rev. Biochem.* **63**:383–417, and references therein.
5. Fitzpatrick, F. A., and Murphy, R. C., 1989, Cytochrome P-450 metabolism of arachidonic acid: Formation and biological actions of "epoxygenase"-derived eicosanoids, *Pharmacol. Rev.* **40**:229–241, and references therein.
6. McGiff, J. C., 1991, Cytochrome P-450 metabolism of arachidonic acid, *Annu. Rev. Pharmacol. Toxicol.* **31**:339–369, and references therein.
7. Capdevila, J. H., Falck, J. R., and Estabrook, R. W., 1992, Cytochrome P-450 and the arachidonate cascade, *FASEB J.* **6**:731–736, and references therein.
8. Oliw, E. H., 1994, Oxygenation of polyunsaturated fatty acids by cytochrome P450 monooxygenases, *Prog. Lipid Res.* **33**:329–354, and references therein.
9. Marnett, L. J., 1992, Aspirin and the potential role of prostaglandins in colon cancer, *Cancer Res.* **52**:5575–5589, and references therein.
10. DiAgustine, R. P., and Fouts, J. R., 1969, The effects of unsaturated fatty acids on hepatic microsomal drug metabolism and cytochrome P-450, *Biochem. J.* **115**:547–554.

11. Pessayre, D., Mazel, P., Descatoire, V., Rogier, E., Feldmann, G., and Benhamou, J. P., 1979, Inhibition of hepatic drug-metabolizing enzymes by arachidonic acid, *Xenobiotica* **9**:301–310.

12. Cinti, D. L., and Feinstein, M. B., 1976, Platelet cytochrome P-450: A possible role in arachidonate-induced aggregation, *Biochem. Biophys. Res. Commun.* **73**:171–179.

13. Capdevila, J. H., Parkhill, L., Chacos, N., Okita, R., Masters, B. S., and Estabrook R. W., 1981, The oxidative metabolism of arachidonic acid by purified cytochromes P-450, *Biochem. Biophys. Res. Commun.* **101**:1357–1363.

14. Capdevila, J. H., Chacos, N. Werringloer, J., Prough, R. A., and Estabrook, R. W., 1981, Liver microsomal cytochrome P-450 and the oxidative metabolism of arachidonic acid, *Proc. Natl. Acad. Sci. USA* **78**:5362–5366.

15. Oliw, E. H., and Oates, J. A., 1981, Oxygenation of arachidonic acid by hepatic microsomes of the rabbit. Mechanism of biosynthesis of two vicinal diols, *Biochim. Biophys. Acta* **666**:327–340.

16. Morrison, A. R., and Pascoe, N., 1981, Metabolism of arachidonic acid through NADPH-dependent oxygenase of renal cortex, *Proc. Natl. Acad. Sci. USA* **78**:7375–7378.

17. Capdevila, J., Marnett, L. J., Chacos, N., Prough, R. A., and Estabrook, R. W. 1982, Cytochrome P-450-dependent oxygenation of arachidonic acid to hydroxyeicosatetraenoic acids, *Proc. Natl. Acad. Sci. USA* **79**:767–770.

18. Chacos, N., Falck, J. R., Wixtrom, C., and Capdevila, J, 1982, Novel epoxides formed during the liver cytochrome P-450 oxidation of arachidonic acid, *Biochem. Biophys. Res. Commun.* **104**:916–922.

19. Oliw, E. H., Guengerich, F. P., and Oates J. A., 1982, Oxygenation of arachidonic acid by hepatic monooxygenases, *J. Biol. Chem.* **257**:3771–3781.

20. Capdevila, J. H., Chacos, N., Falck, J. R., Manna, S., Negro-Vilar, A., and Ojeda, S. R., 1983, Novel hypothalamic arachidonate products stimulate somatostatin release from the median eminence, *Endocrinology* **113**:421–423.

21. Jacobson, H. R., Corona, S., Capdevila, J. H., Chacos, N., Manna, S., Womack, A., and Falk, J. R., 1984, in: *Prostaglandins and Membrane Ion Transport* (P. Braquet, J. C. Frolich, S. Nicosia, and R. Garay, eds.), Raven Press, New York.

22. Capdevila, J. H., Pramanik, B., Napoli, J. L., Manna, S., and Falck, J. R., 1984, Arachidonic acid epoxidation: Epoxyeicosatrienoic acids are endogenous constituents of rat liver, *Arch. Biochem. Biophys.* **231**:511–517.

23. Rahimtula, A. D., and O'Brien, P. J., 1974, Hydroperoxide catalyzed liver microsomal aromatic hydroxylation reactions involving cytochrome P-450, *Biochem. Biophys. Res. Commun.* **60**:440–447.

24. Weiss, R. H., Arnold, J. L., and Estabrook, R. W., 1987, Transformation of an arachidonic acid hydroperoxide into epoxyhydroxy and trihydroxy fatty acids by liver microsomal cytochrome P-450, *Arch. Biochem. Biophys.* **252**:334–338.

25. Ullrich, V., Castle, L., and Haurand, M., 1982, Cytochrome P-450 as an oxene transferase, in: *Oxygenases and Oxygen Metabolism* (M. Nozaki, S. Yamamoto, Y. Ishimura, M. J. Coon, L. Ernster, and R. W. Estabrook, eds.), Academic Press, New York.

26. Hecker, M., and Ullrich, V., 1989, On the mechanism of prostacyclin and thromboxane A_2 biosynthesis, *J. Biol. Chem.* **264**:141–150.

27. White, R. E., and Coon, M. G., 1980, Oxygen activation by cytochrome P-450, *Annu. Rev. Biochem.* **49**:315–356, and references therein.

28. Hamilton, G. A., 1964, Oxidation by molecular oxygen. II. The oxygen atom transfer mechanism for mixed-function oxidases and the model mixed-function oxidases, *J. Am. Chem. Soc.* **86**:3391–3396.

29. Yokoyama, C., Miyata, A., Ihara, H., Ullrich, V., and Tanabe, T., 1991, Molecular cloning of human platelet thromboxane A synthase, *Biochem. Biophys. Res. Commun.* **178**:1479–1484.

30. Ohashi, K., Ruan, K. H., Kulmacz, R. J., Wu, K. K., and Wang, L. H., 1992, Primary structure of human thromboxane synthase determined from the cDNA sequence, *J. Biol. Chem.* **267**:789–793.

31. Hara, S., Miyata, A., Yokoyama, C., Inoue, H., Burgger, R., Lottspeich, F., Ullrich, V., and Tanabe, T., 1994, Isolation and molecular cloning of prostacyclin synthase from bovine endothelial cells, *J. Biol. Chem.* **269**:19897–19903.

32. Song, W. C., Funk, C. D., and Brash, A. R., 1993, Molecular cloning of an allene oxide synthase: A cytochrome P450 specialized for the metabolism of fatty acid hydroperoxides, *Proc. Natl. Acad. Sci. USA* **90**:8519–8523.

33. Wong, P. Y. K., Malik, K. U., Taylor, B. M., Schneider, W. P., McGiff, J. C., and Sun, F. F., 1985, Epoxidation of prostacyclin in the rabbit kidney, *J. Biol. Chem.* **260**:9150–9153.

34. Oliw, E. H., 1984, Metabolism of 5(6)-epoxyeicosatrienoic acid by ram seminal vesicles formation of novel prostaglandin E_1 metabolites, *Biochim. Biophys. Acta* **793**:408–415.

35. Zhang, J. Y., Prakash, C., Yamashita, K., and Blair, I. A., 1992, Regiospecific and enantioselective metabolism of 8,9-epoxyeicosatrienoic acids by cyclooxygenase, *Biochem. Biophys. Res. Commun.* **183**:138–143.

36. Jajjo, H. K., Capdevila, J. H., Falck, J. R., Bhatt, R. K., and Blair, I. A., 1992, Metabolism of 12(R)-hydroxyeicosatetraenoic acid by rat liver microsomes, *Biochim. Biophys. Acta* **1123**:110–116.

37. Capdevila, J., Mosset, P., Yadagiri, P., Sun Lumin, and Falck, J. R., 1988, NADPH-dependent microsomal metabolism of 14,15-epoxyeicosatrienoic acid to diepoxides and epoxyalcohols, *Arch. Biochem. Biophys.* **261**:122–132.

38. Hamberg, M., and Samuelsson, B., 1966, Prostaglandins in human seminal plasma. Prostaglandins and related factors, *J. Biol. Chem.* **241**:257–263.

39. Oliw, E. H., 1989, Biosynthesis of 18(R)-hydroxyeicosatetraenoic acid from arachidonic acid by microsomes of monkey seminal vesicles, *J. Biol. Chem.* **264**:17845–17853.

40. Israelson, U., Hamberg, M., and Samuelsson, B., 1969, Biosynthesis of 19-hydroxy-prostaglandin A_1, *Eur. J. Biochem.* **11**:390–394.

41. Kupfer, D., 1982, Endogenous substrates of monooxygenases: Fatty acids and prostaglandins, in: *Hepatic Cytochrome P-450 Monooxygenase System*, (J. B. Schenkman and D. Kupfer, eds.), Pergamon Press, Elmsford, NY, pp. 157–182.

42. Nelson, D. R., Kamataki, T., Waxman, D. J., Guengerich, F. P., Estabrook, R. W., Feyereisen, R., Gonzalez, F. J., Coon, M. J., Gunsalus, I. C., Gotoh, O., Okuda, K., and Nebert, D. W., 1993, The P450 superfamily: Update on new sequences, gene mapping, accession numbers, early trivial names of enzymes and nomenclature, *DNA Cell Biol.* **12**:1–51, and references therein.

43. Hardwick, J. P., Song, B. J., Huberman, E., and Gonzalez, F. J., 1987, Isolation, complementary DNA sequence, and regulation of rat hepatic lauric acid ω-hydroxylase (cytochrome $P450_{LA\omega}$), *J. Biol. Chem.* **262**:801–810.

44. Sharma, R. K., Doig, M. V., Lewis, D. F. V., and Gibson, G., 1989, Role of hepatic and renal cytochrome P450 IVA1 in the metabolism of lipid substrates, *Biochem. Pharmacol.* **38**:3621–3629.

45. Imaoka, S., Tanaka, S., and Funae, Y., 1989, ω and (ω-1)-hydroxylation of lauric acid and arachidonic acid by rat renal cytochrome P-450, *Biochem. Int.* **18**:731–740.

46. Imaoka, S., Nagashima, K., and Funae, Y., 1990, Characterization of three cytochrome P450s purified from renal microsomes of untreated male rats and comparison with human renal cytochrome P450, *Arch. Biochem. Biophys.* **276**:473–480.

47. Kimura, S., Hanioka, N., Matsunaga, E., and Gonzalez, F. J., 1989, The rat clofibrate-inducible CYP4A gene subfamily I. Complete intron and exon sequence of the CYP4A1 and CYP4A2 genes, unique exon organization and identification of a conserved 19-bp upstream element, DNA **8**:503–516.

48. Kimura, S., Hardwick, J. P., Kozac, C. A., and Gonzalez, F. J., 1989, The rat clofibrate inducible CYP4A subfamily II. cDNA sequence of IVA3, mapping of the *Cyp4a* locus to mouse chromosome 4, and coordinated and tissue specific regulation of the CYP4A genes, *DNA* **8**:517–525.

49. Stromsteadt, M., Hayashi, S., Zaphiropoulos, P. G., and Gustafsson, J.A., 1990, Cloning and characterization of a novel member of the cytochrome P450 subfamily IVA in rat prostate, *DNA Cell Biol.* **8**:569–577.

50. Matsubara, S., Yamamoto, S., Sogawa, K., Yokotani, N., Fujii-Kuriyama, Y., Haniu, M., Shively, J. E., Gotoh, O., Kusunose E., and Kusunose, M., 1987, cDNA cloning and inducible expression during pregnancy of the mRNA for rabbit pulmonary prostaglandin ω-hydroxylase (cytochrome $P-450_p$-2), *J. Biol. Chem.* **262**:13366–13371.

51. Johnson, E. F., Walker, D. L., Griffin, K. J., Clark, J. E., Okita, R. T., Muerhoff, S., and Masters, B. S., 1990, Cloning and expression of three rabbit kidney cDNAs encoding lauric acid ω-hydroxylases, *Biochemistry* **29**:873–879.

52. Roman, L. J., Palmer, C. N. A., Clark, J. E., Muerhoff, S. A., Griffin, K. L., Johnson, E. F., and Masters, B. S. S., 1993, Expression of rabbit cytochrome P450A which catalyzes the ω-hydroxylation of arachidonic acid, fatty acids, and prostaglandins, *Arch. Biochem. Biophys.* **307**:57–65.

53. Yokotani, N., Kusunose, E., Sogawa, K., Kawashima, H., Kinosaki, M., Kusunose, M., and Fujii-Kuriyama, Y., 1991, cDNA cloning and expression of the mRNA for cytochrome P-450kd which shows a fatty acid ω-hydroxylating activity. *Eur. J. Biochem.* **196**:531–536.

54. Yokotani, N., Bernhardt, R., Sogawa, K., Kusunose, E., Gotoh, O., Kusunose, M., and Fujii-Kuriyama, Y., 1989, Two forms of ω-hydroxylase toward prostaglandin A and laurate. cDNA cloning and their expression, *J. Biol. Chem.* **264**:21665–21669.

55. Sawamura, A., Kusunose, E., Satouchi, K., and Kusunose, M., 1993, Catalytic properties of rabbit kidney fatty acid ω-hydroxylase cytochrome P-450ka2 (CYP4A7), *Biochim. Biophys. Acta* **1168**:30–36.

56. Kawashima, H., Kusunose, E., Kubota, I., Maekawa, M., and Kusunose, M., 1992, Purification and NH2-terminal amino acid sequences of human and rat kidney fatty acid ω-hydroxylases, *Biochim. Biophys. Acta* **1123**:156–162.

57. Palmer, C. N. A., Richardson, T. H., Griffin, K. J., Hsu, M., Muerhoff, A. S., Clark, J. E., and Johnson, E. F., 1993, Characterization of a cDNA encoding a human kidney cytochrome P-450 4A fatty acid ω-hydroxylase and the cognate enzyme expressed in *Escherichia coli*, *Biochim. Biophys. Acta* **1172**:161–166.

58. Imaoka, S., Ogawa, H., Kimura, S., and Gonzalez, F. J., 1993, Complete cDNA sequence and cDNA-directed expression of CYP4A11, a fatty acid omega-hydroxylase expressed in human kidney, *DNA Cell Biol.* **12**:893–899.

59. Jubiz, W., Radmark, O., Malmsten, C., Hansson, G., Lindgren, J. A., Palmblad, J., Uden, A. M., and Samuelsson, B., 1981, A novel leukotriene produced by stimulation of leukocytes with formyl-methionylleucylphenylalanine, *J. Biol. Chem.* **257**:6106–6110.

60. Powell, W. S., 1984, Properties of leukotriene B4 20-hydroxylase from polymorphonuclear leukocytes, *J. Biol. Chem.* **259**:3082–3089.

61. Shak, S., and Goldstein, I. M., 1984, ω-Oxidation is the major pathway for the catabolism of leukotriene B4 in human polymorphonuclear leukocytes, *J. Biol. Chem.* **259**:10181–10187.

62. Romano, M. C., Eckardt, R. D., Bender, P. E., Leonard, T. B., Straub, K. M., and Newton, J. F., 1987, Biochemical characterization of hepatic microsomal leukotriene B4 hydroxylases, *J. Biol. Chem.* **262**:1590–1595.

63. Kikuta, Y., Kusunose, E., Endo, K., Yamamoto, S., Sogawa, K., Fujii-Kuriyama, Y., and Kusunose, M., 1993, A novel form of cytochrome P-450 family 4 in human polymorphonuclear leukocytes, *J. Biol. Chem.* **268**:9376–9380.

64. Soberman, R. J., Okita, R. T., Fitzsimmons, B., Rokach, J., Spur, B., and Austen, K. F., 1987, Stereochemical requirements for substrate specificity of LTB4 20-hydroxylase, *J. Biol. Chem.* **262**:12421–12427.

65. Sumimoto, J., Takeshige, K., and Minakami, S., 1988, Characterization of human neutrophil leukotriene B4 omega-hydroxylase, a system involving a unique cytochrome P-450 and NADPH-cytochrome P-450 reductase, *Eur. J. Biochem.* **172**:315–324.

66. Wong, P. Y. K., Westlund, P., Hamberg, M., Granstrom, E., Chao, P. W. H., and Samuelsson, B., 1984, ω-Hydroxylation of 12-L-hydroxy-5,8,10,14-eicosatetraenoic acid in human polymorphonuclear leukocytes, *J. Biol. Chem.* **259**:2683–2686.

67. Marcus, A. J., Safier, L. B., Ullman, H. L., Broekman, M. J., Islam, N., Oglesby, T. D., and Gorman, R., 1984, 12S,20-Dihydroxyeicosatetraenoic acid: A new eicosanoid synthesized by neutrophils from 12S-hydroxyeicosatetraenoic acid produced by thrombin- or collagen-stimulated platelets, *Proc. Natl. Acad. Sci. USA* **81**:903–907.

68. Flaherty, J. T., and Nishihira, J., 1987, 5-Hydroxyeicosatetraenoate promotes Ca^{2+} and protein kinase C mobilization in neutrophils, *Biochem. Biophys. Res. Commun.* **148:**575–581.

69. Okita, R. T., Soberman, R. J., Bergholte, J. M., Masters, B. S. S., Hayes, R., and Murphy, R. C., 1987, ω-Hydroxylation of 15-hydroxyeicosatetraenoic acid by lung microsomes from pregnant rabbits, *Mol. Pharmacol.* **32:**706–709.

70. Boucher, J. L, Delaforge, M., and Mansuy, D., 1991, Metabolism of lipoxins A_4 and B_4 and of their all-trans isomers by human leukocytes and rat liver microsomes, *Biochem. Biophys. Res. Commun.* **177:**134–139.

71. Mizukami, Y., Sumimoto, H, and Minakami, S., 1993, Omega hydroxylation of lipoxin B_4 by human neutrophil microsomes: Identification of omega-hydroxy metabolite of lipoxin B_4 and catalysis by leukotriene B_4 omega-hydroxylase (cytochrome P-450LTB omega), *Biochim. Biophys. Acta* **1168:**87–93.

72. Samuelsson, B., Dahlen, S. E., Lindgren, J. A., Rouzer, C. A., and Serhan, C. H., 1987, Leukotrienes and lipoxins: Structures, biosynthesis and biological effects, *Science* **237:**1171–1176, and references therein.

73. Capdevila, J. H., Yadagiri, P., Manna, S., and Falck, J. R., 1986, Absolute configuration of the hydroxyeicosatetraenoic acids (HETEs) formed during catalytic oxygenation of arachidonic acid by microsomal cytochrome P-450, *Biochem. Biophys. Res. Commun.* **141:**1007–1011.

74. Schwartzman, M. L., Balazy, M., Masferrer, J., Abraham, N. G., McGiff, J. C., and Murphy R. C., 1987, 12(*R*)-Hydroxyeicosatetraenoic acid: A cytochrome P450-dependent arachidonate metabolite that inhibits Na^+,K^+-ATPase in the cornea, *Proc. Natl. Acad. Sci. USA* **84:**8125–8129.

75. Oliw, E. H., 1993, Biosynthesis of 12(*S*)-hydroxyeicosatetraenoic acid by bovine corneal epithelium, *Acta Physiol. Scand.* **147:**117–121.

76. Woollard, P. M., 1986, Stereochemical differences between 12-hydroxy-5,8,10,14-eicosatetraenoic acid in platelets and psoriatic lesions, *Biochem. Biophys. Res. Commun.* **141:**1007–1011.

77. Murphy, R. C., Falck, J. R., Lumin, S., Yadagiri, P., Zirrolli, J. A., Balazy, M., Masferrer, J., Abraham, N. G., and Schwartzman, M. L., 1988, 12(*R*)-Hydroxyeicosatrienoic acid: A vasodilator cytochrome P-450-dependent arachidonate metabolite from bovine corneal epithelium, *J. Biol. Chem.* **263:**17197–17202.

78. Oliw, E. H., 1993, *bis*-Allylic hydroxylation of linoleic acid and arachidonic acid by human hepatic monooxygenases, *Biochim. Biophys. Acta* **1166:**258–263.

79. Lu, A. Y. H., Junk, K., and Coon, M. J., 1969, Resolution of the cytochrome P-450-containing ω-hydroxylation system of liver microsomes into three components, *J. Biol. Chem.* **244:**3714–3721.

80. Falck, J. R., Lumin, S., Blair, I., Dishman, E., Martin, M. V., Waxman, D. J., Guengerich, F. P., and Capdevila, J. H., 1990, Cytochrome P-450-dependent oxidation of arachidonic acid to 16-, 17-, and 18-hydroxyeicosatetraenoic acids, *J. Biol. Chem.* **265:**10244–10249.

81. Gonzalez, F. J., 1989, The molecular biology of cytochromes P450s, *Pharmacol. Rev.* **40:**243–288, and references therein.

82. Rifkind, A. B., Kanetoshi, A., Orlinick, J., Capdevila, J. H., and Lee, C., 1994, Purification and biochemical characterization of two major cytochrome P-450 isoforms induced by 2,3,7,8-tetrachlorodibenzo-*p*-dioxin in chick embryo liver, *J. Biol. Chem.* **269:**3387–3396.

83. Laethem, R. M., Balazy, M., Falck, J. R., Laethem, C. L., and Koop, D. R., 1993, Formation of 19(S)-, 19(R)-, and 18(R)-hydroxyeicosatetraenoic acids by alcohol-inducible cytochrome P450 2E1, *J. Biol. Chem.* **268:**12912–12918.

84. Lapuerta, L., Chacos, N., Falck, J. R., Jacobson, H., and Capdevila, J. H., 1988, Renal microsomal cytochrome P-450 and the oxidative metabolism of arachidonic acid, *Am. J. Med. Sci.* **31:**275–279.

85. Iwai, N., and Inagami, T., 1990, Isolation of preferentially expressed genes in the kidneys of hypertensive rats, *Hypertension* **17:**161–169.

86. Gebremedhin, D., Ma, Y. H., Imig, J. D., Harder, D. R., and Roman, R. J., 1993, Role of cytochrome P-450 in elevating renal vascular tone in spontaneously hypertensive rats, *J. Vasc. Res.* **30:**53–60.

87. Escalante, B., Sessa, W. C., Falck, J. R., Yadagin, P., and Schwartzman, M. L., 1990, Cytochrome P450-dependent arachidonic acid metabolites, 19- and 20-hydroxyeicosatetraenoic acids, enhance sodium-potassium ATPase activity in vascular smooth muscle, *J. Cardiovasc. Pharmacol.* **16**:438–443.

88. Schwartzman, M. L., Falck, J. R., Yadagiri, P., and Escalante, B., 1989, Metabolism of 20-hydroxyeicosatetraenoic acid by cyclooxygenase. Formation and identification of novel endothelium-dependent vasoconstrictor metabolites, *J. Biol. Chem.* **264**:11658–11662.

89. Schwartzman, M. L., Omata, K., Lin, F., Bhatt, R. K., Falck, J. R., and Abraham, N. G., 1991, Detection of 20-hydroxyeicosatetraenoic acid in rat urine, *Biochem. Biophys. Res. Commun.* **180**:445–449.

90. Prakash, C., Zhang, J. Y., Falck, J. R., Chauhan, K., and Blair, I. A., 1992, 20-Hydroxyeicosatetraenoic acid is excreted as a glucuronide conjugate in human urine, *Biochem. Biophys. Res. Commun.* **185**:728–733.

91. Capdevila, J. H., Kim, Y. R., Martin-Wixtrom, C., Falck, J. R., Manna, S., and Estabrook, R. W., 1985, Influence of a fibric acid type of hypolipidemic agent on the oxidative metabolism of arachidonic acid by liver microsomal cytochrome P-450, *Arch. Biochem. Biophys.* **243**:8–19.

92. Capdevila, J. H., Karara, A., Waxman, D. J., Martin, M. V., Falck, J. R., and Guengerich, F. P., 1990, Cytochrome P-450 enzyme-specific control of the regio- and enantiofacial selectivity of the microsomal arachidonic acid epoxygenase, *J. Biol. Chem.* **265**:10865–10871.

93. Capdevila, J. H., Wei, S., Yan, Y., Karara, A., Jacobson, H. R., Falck, J. R., Guengerich F. P., and DuBois, R. N., 1992, Cytochrome P-450 arachidonic acid epoxygenase. Regulatory control of the renal epoxygenase by dietary salt loading, *J. Biol. Chem.* **267**:21720–21726.

94. Karara, A., MaKita, K., Jacobson, J. R., Falck, J. R., Guengerich, F. P., DuBois, R. N., and Capdevila, J. H., 1993, Molecular cloning, expression, and enzymatic characterization of the rat kidney cytochrome P-450 arachidonic acid epoxygenase, *J. Biol. Chem.* **268**:13565–13570.

95. Daikh, B. E., Laethem, R. M., and Koop, D., 1994, Stereoselective epoxidation of arachidonic acid by cytochrome P-450s 2CAA and 2C2, *J. Pharmacol. Exp. Ther.* **269**:1130–1135.

96. Capdevila, J. H., Jacobson, H., and Zeldin, D., 1994, Molecular cloning, expression and enzymatic characterization of a human kidney cytochrome P450 arachidonic acid epoxygenase, *J. Am. Soc. Nephrol.* **5**:677.

97. Lund, J., Zaphiropoulos, P. G., Mode, A., Warner, M., and Gustafsson, J. A., 1991, Hormonal regulation of cytochrome P450 gene expression, *Adv. Pharmacol.* **22**:325–354, and references therein.

98. Lindberg, R. L., and Negishi, M., 1989, Alterations of mouse cytochrome P450coh substrate specificity by mutation of a single amino-acid residue, *Nature* **339**:632–634.

99. Johnson, E. F., Kronbach, T. K., and Hsu, M. H., 1992, Analysis of the catalytic specificity of cytochrome P450 enzymes through site-directed mutagenesis, *FASEB J.* **6**:700–705, and references therein.

100. Karara, A., Dishman, E., Blair, I., Falck, J. R., and Capdevila, J. H., 1989, Endogenous epoxyeicosatrienoic acids. Cytochrome P-450 controlled stereoselectivity of the hepatic arachidonic acid epoxygenase, *J. Biol. Chem.* **264**: 19822–19827.

101. Karara, A., Dishman, E., Falck, J. R., and Capdevila, J. H., 1991, Endogenous epoxyeicosatrienoyl-phospholipids. A novel class of cellular glycerolipids containing epoxidized arachidonate moieties, *J. Biol. Chem.* **266**:7561–7569.

102. Brezinski, M., and Serhan, C. N., 1990, Selective incorporation of 15(S)-hydroxyeicosatetraenoic acid in phosphatidylinositol of human neutrophils: Agonist-induced deacylation and transformation of stored hydroxyeicosanoids, *Proc. Natl. Acad. Sci. USA* **87**:6248–6252.

103. Legrand, A. B., Lawson, J. A., Meyrick, B. O., Blair, I. A., and Oates, J. A., 1991, Substitution of 15-hydroxyeicosatetraenoic acid in the phosphoinositide signaling pathway, *J. Biol. Chem.* **266**:7570–7577.

104. Bertstrom, K., Kayganich, K., Murphy, R. C., and Fitzpatrick, F. A., 1992, Incorporation and distribution of epoxyeicosatrienoic acids into cellular phospholipids, *J. Biol. Chem.* **267**:3686–3690.

105. Karara, A., Wei, S., Spady, D., Swift, L., Capdevila, J. H., and Falck, J. R., 1992, Arachidonic acid epoxygenase: Structural characterization and quantification of epoxyeicosatrienoates in plasma, *Biochem. Biophys. Res. Commun.* **182**:1320–1325.

106. Carrol, M. A., Garcia, M. P., Falck, J. R., and McGiff, J. C., 1990, 5,6-Epoxyeicosatrienoic acid, a novel arachidonate metabolite. Mechanism of vasoactivity in the rat, *Circ. Res.* **67**:1082–1088.

107. Frei, B., Winterhalter, K. H., and Richter, C., 1985, Quantitative and mechanistic aspects of the hydroperoxide-induced release of Ca^{+2} from rat liver mitochondria, *Eur. J. Biochem.* **149**:633–639.

108. Sevanian, A., and Hochstein, P., 1985, Mechanisms and consequences of lipid peroxidation in biological systems, *Annu. Rev. Nutr.* **5**:365–390, and references therein.

109. Gast, K., Zirwer, D., Ladhoff, M., Schreiber, J., Koelsch, R., and Kretschmer, K., 1982, Auto-oxidation-induced fusion of lipid vesicle, *Biochim. Biophys. Acta* **686**:99–109.

110. Hauser, H., and Poupart, G., 1992, Lipid structure, in: *The Structure of Biological Membranes* (P. Yeagle, ed.), CRC Press, Ann Arbor, MI., pp. 3–71 and references therein.

111. Yeagle, P., 1992, The dynamics of membrane lipids, in: *The Structure of Biological Membranes* (P. Yeagle, ed.), CRC Press, Ann Arbor, MI, pp. 157–174, and references therein.

112. Capdevila, J. H., Jin, Y., Karara, A., and Falck, J. R., 1993, Cytochrome P450 epoxygenase dependent formation of novel endogenous epoxyeicosatrienoyl-phospholipids, in: *Eicosanoids and Other Bioactive Lipids in Cancer, Inflammation and Radiation Injury* (S. Nigam, L. J. Marnett, K. V. Honn, and T. L. Walden, Jr., eds.), Kluwer Academic, Boston, pp. 11–15.

113. Karara, A., Dishman, E., Jacobson, H., Falck, J. R., and Capdevila, J. H., 1990, Arachidonic acid epoxygenase. Stereochemical analysis of the endogenous epoxyeicosatrienoic acids of human kidney cortex, *FEBS Lett.* **268**:227–230.

114. Catella, F., Lawson, J. A., Fitzgerald, D. J., and Fitzgerald, G. A., 1990, Endogenous biosynthesis of arachidonic acid epoxides in humans: Increased formation in pregnancy-induced hypertension, *Proc. Natl. Acad Sci. USA* **87**:5893–5897.

115. Makita, K., Takahashi, K., Karara, A., Jacobson, H. R., Falck, J. R., and Capdevila, J. H., 1994, Experimental and/or genetically controlled alterations of the renal microsomal cytochrome P450 epoxygenase induce hypertension in rats fed a high salt diet, *J. Clin. Invest.* **94**:2414–2420.

116. Rapp, J. P., 1982, Dahl salt-susceptible and Dahl salt-resistant rats. A review, *Hypertension* **4**:753–763, and references therein.

117. Manuscript in preparation.

Human Cytochrome P450 Enzymes

F. PETER GUENGERICH

1. Background and General Approaches

In the past decade there has been considerable progress in the characterization of individual human P450 enzymes. These advances were initiated by early work on the purification of individual enzymes from human liver and other sources. The early work in the area was guided by a focus on the most abundant and easily purified enzymes.[1–4] It is now apparent, in retrospect, that these proteins were in the 2C and 3A families. Efforts were shifted to attempts to purify individual P450s on the basis of catalytic activities with the evidence that in some cases a single P450 could be identified in this way; e.g., the P450 now known as P450 2D6 was found to be under monogenic control.[5] The approach is technically demanding because of the need to do separations in the presence of detergents and then remove them before analysis of catalytic activity. Nevertheless, human P450s 1A1,[6] 1A2,[7] 2A6,[8] 2C8,[9] 2C9,[9,10] 2D6,[7,11,12] 2E1,[13,14] 3A4,[15] 3A5,[16] 4A11,[17] and lanosterol 14α-demethylase[18] were isolated in this general manner. Another approach that has been used, sometimes in a mode complementary to catalytic specificity, is purification on the basis of immunochemical similarity to animal P450s.[7,11,19,20]

With the development of methods of recombinant DNA technology it became possible to isolate cDNAs corresponding to individual P450s. Screening of expression libraries with antibodies and oligonucleotide probes derived from partial amino acid sequences has been done to isolate clones, which have been subjected to sequence analysis.[21] More recently it has become possible to isolate known cDNAs directly from human or other mRNA samples by reverse transcription/polymerase chain reaction amplification. This principle can be applied with degenerate primers to isolate new P450 cDNAs, although it does not appear that such an approach has yet been applied to isolate new human P450s. Along with the capability to isolate cDNAs came techniques for the isolation of genes and assignment

F. PETER GUENGERICH • Department of Biochemistry and Center in Molecular Toxicology, Vanderbilt University School of Medicine, Nashville, Tennessee 37232.

Cytochrome P450: Structure, Mechanism, and Biochemistry (Second Edition), edited by Paul R. Ortiz de Montellano. Plenum Press, New York, 1995.

TABLE I
Characteristics of Human P450 Enzymes

P450	Chromosome location [22]	Known inducers	Approx. % total hepatic P450	Extent of variability in level, fold	Poly-morphism	Noninvasive markers
1A1	15q22-qter	TCDD	<1	~100	+	
1A2	15q22-qter	Smoking, charred food	12	40	(+)	Caffeine
1B	2	TCDD	<1			
2A6	19q13.1-13.2		4	30	+	Coumarin
2A7	19q13.1-13.2		?	?		
2B6	19q12-13.2		<1	50		
2C8	10q24.1-24.3					
2C9	10q24.1-24.3	Barbiturates, rifampicin	20 (total 2C)	25 (total 2C)	(+)	Hexobarbital, tolbutamide, warfarin
2C10[a]	10q24.1-24.3					
2C17[a]	10q24.1-24.3					
2C18	10q24.1-24.3					
2C19	10q24.1-24.3	Barbiturates, rifampicin		~100	+	(S)-Mephenytoin
2D6	22q13.1		4	>1000	+	Debrisoquine, dextromethorphan
2E1	10	Ethanol, isoniazid	6	20	(+)	Chlorzoxazone caffeine
2F1	19		?	?		
3A4	7q22.1	Barbiturates, rifampicin, dexamethasone	28	20		Nifedipine, lidocaine, erythromycin, midazolam, dapsone, 6β-hydroxy-cortisol
3A5	7q22.1		0–8	>100	+	
3A7	7q22.1		<1	?	?	
4A9			?	?	?	
4A11	1	Clofibrate, etc.?	?	?		
4F2			?	?		
4F3			?	?		
5			?	?b		
7	8q11-q12		?	?		
11A1	15q23-q24			b		
11B1	8q21-q22			b		
11B2	8q21-q22			b		
17	10q24.3			b		
19	15q21			b		
21A2	6p			b		
27	2q33-qter		?	b		

[a]These are considered potentially allelic variants of other subfamily members.[22]

[b]The levels of these P450s are relatively constant; when the activity is deficient (because of inactive enzyme or enzyme deficiency), the result is usually a disease state.

of chromosomal locations (Table I). Another aspect of recombinant technology that has proven to be extremely useful is protein expression in heterologous systems, which has facilitated the assignment of catalytic specificity. The specific details of these approaches will generally not be presented, only the conclusions reached with the methods.

Although the overall levels of P450 vary only a few fold among most humans, there is considerably greater variation in the levels of many of the individual forms of P450 (Fig. 1). This is seen in different races, with similar patterns (Fig. 2, Table II), although some polymorphisms show strong racial links.[26–28] As discussed later, these variations in levels of P450s can have considerable consequences.

At this point, mention should be made of the *in vitro* approaches that can be used to define the catalytic specificity of individual P450 enzymes.[29]

1. One approach often used initially in this laboratory is an attempt to measure levels of the catalytic activity under consideration in a number of different liver microsomal samples (usually ≥10) and then correlate these with accepted marker catalytic activities of other P450s in the same set of samples (Fig. 3). Some of the commonly accepted reactions for each P450 are mentioned in Table III. Alternatively, levels of marker enzymes can be estimated by immunochemical techniques, usually immunoblotting, to augment specificity.[34] If the reaction is catalyzed by a particular P450, then there should be a high correlation.[35] A simplistic way of looking at the results is that r^2, the correlation coefficient (determined with absolute or rank values), is an estimate of the fraction of the variance accounted for by the relationship between the two variables. If two P450s are under common regulation, it is obvious that this approach will not distinguish which one is involved in the reaction.

2. Another approach involves the addition of inhibitors to crude preparations such as microsomes. Such experiments can be done with several kinds of inhibitors, including (a) substrates, (b) competitive inhibitors (other than substrates), (c) noncompetitive inhibitors (including mechanism-based), and (d) inhibitory antibodies. Substrates can be used as inhibitors, although they are usually not as potent as the other inhibitors, and sometimes there is not effective competition.[36,37] Inhibitors are usually used to titrate catalytic activity[38] with the maximum (or extrapolated) inhibition corresponding in principle to the fraction of the reaction catalyzed by the particular P450 (or other P450s also inhibited by the compound or inhibitor). The successful application of the approach depends on the selectivity of the inhibitors, of course. Antibodies are not commercially available at this time (in the amounts needed for such studies), but these can be prepared against (a) the purified human p450,[10] (b) an animal ortholog,[39] (c) a purified recombinant P450,[40] or (d) peptides synthesized based on portions of primary sequence.[41] Monoclonal antibodies have also been used,[42,43] although many are not inhibitory.[44,45] Antibody specificity cannot be assumed, as cross-reactivity across P450 families has been documented even with monoclonal antibodies[46,47] and (polyclonal) antibodies raised against recombinant P450s.[48] A list of chemical inhibitors generally used to selectively inhibit individual human P450s is presented in Table III.

3. The P450 responsible for a particular catalytic activity can be purified from tissues on the basis of monitoring this activity through the fractionation. The purified P450, in principle, should be at least as active as in the microsomal preparation from which it was

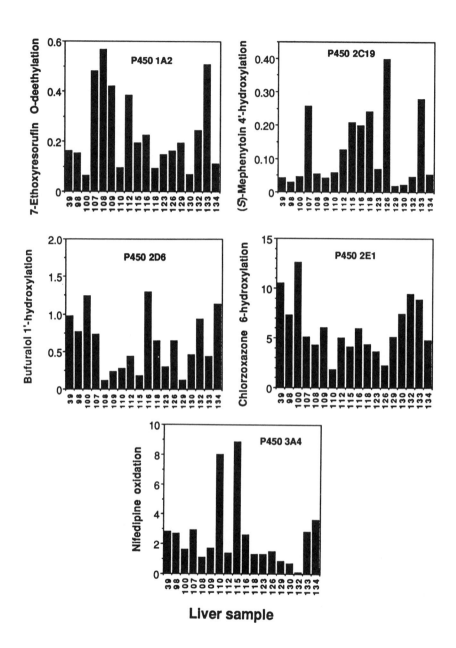

FIGURE 1. Variation in levels of five human P450s in 18 randomly selected liver samples. Individual P450s and catalytic activities are indicated on each chart.[23] Sample numbers refer to codes from this laboratory.

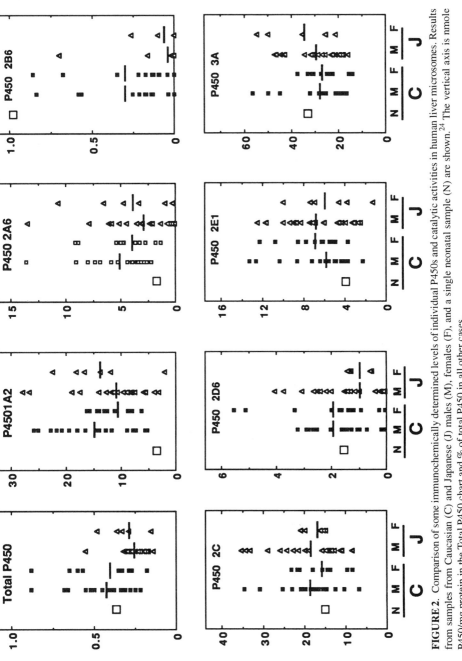

FIGURE 2. Comparison of some immunochemically determined levels of individual P450s and catalytic activities in human liver microsomes. Results from samples from Caucasian (C) and Japanese (J) males (M), females (F), and a single neonatal sample (N) are shown.[24] The vertical axis is nmole P450/mg protein in the Total P450 chart and % of total P450 in all other cases.

TABLE II
Contents of Liver Microsomal P450 Enzymes[a] in Japanese and Caucasian Populations[24]

	n	Total P450 (spectral)	P450							Total P450 (immunochemical sum)
			1A2	2A6	2B6	2C	2D6	2E1	3A4	
			pmole P450/mg protein (% of total P450)							
Total	72	309±175 (100±57)	37±24 (13±7)	13±13 (4.0±4.1)	0.68±1.4 (0.15±0.26)	55±28 (20±8)	4.5±2.9 (1.7±1.2)	20±13 (6.6±3.1)	87±53 (29±10)	217±107 (73±17)
Japanese	40	233±102 (100±44)	26±20 (12±7)	6.5±7.3 (2.8±3.2)	0.14±0.62 (0.03±0.12)	46±23 (21±9)	3.0±1.9 (1.4±1.2)	15±9 (6.4±3.1)	72±44 (30±11)	168±80 (74±18)
Caucasian	32	406±199 (100±49)	50±22 (14±6)	21±14 (5.6±4.7)	1.4±1.8 (0.29±0.32)	68±29 (18±7)	6.4±2.8 (1.9±1.2)	26±14 (6.9±3.1)	106±58 (27±10)	277±106 (73±16)

[a]Total P450 contents in liver microsomes were determined spectrally[25] and individual forms of P450 were assayed immunochemically. All values are the means and standard deviations. Shown in parentheses are relative contents (% of total P450) of individual P450 forms.

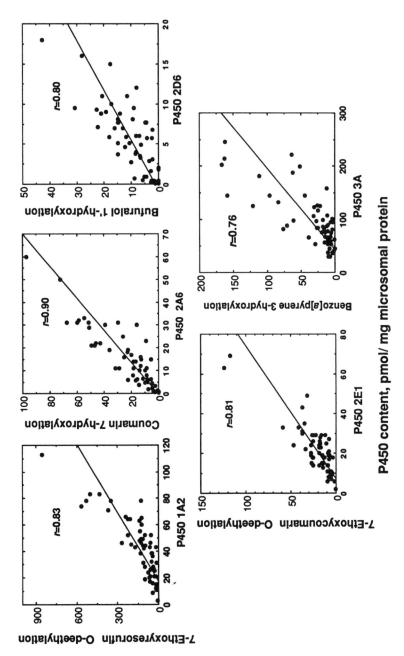

FIGURE 3. Correlation of catalytic activities with immunochemically determined levels of five P450s in human liver microsomes. The correlation coefficients (*r*) were determined using linear regression analysis.[24]

TABLE III
Useful Markers and Inhibitors of Major Human Liver P450 Enzymes

P450	Marker substrate	Inhibitors
1A2	Phenacetin O-deethylation	7,8-Benzoflavone
	7-Ethoxyresorufin O-deethylation	Fluvoxamine
		Furafylline[a]
2A6	Coumarin 7-hydroxylation	Diethyldithiocarbamate
2C9	Tolbutamide (methyl) hydroxylation	Sulfaphenazole
2C19	(S)-Mephenytoin 4-hydroxylation	
2D6	Bufuralol 1-hydroxylation	Quinidine
	Debrisoquine 4-hydroxylation	Ajmalicine
2E1	Chlorzoxazone 6-hydroxylation	4-Methylpyrazole
		Diethyldithiocarbamate
3A4	Nifedipine oxidation	Gestodene
		Troleandomycin
4A11	Lauric acid 12-hydroxylation	
7	Cholesterol 7α-hydroxylation	

[a]This has been reported to be a very selective inhibitor of human P450 1A2[30–32] but work in this laboratory indicates considerably less inhibition or specificity.[33]

isolated. This approach is generally technically demanding and, even when a catalytically active protein is purified, it may be difficult to evaluate the rate in terms of the contribution of the particular P450 to the overall catalytic activity. The approach has been used to gain information about the major p450s[6–10,13–15,17,18,49] but has been largely supplanted by the option of expressing recombinant proteins when cDNA sequences are available (*vide infra*). Obtaining suitable human tissues, especially extrahepatic, may be a problem. However, there are probably still some human P450s that have not been characterized, and if other approaches do not provide a clear answer regarding the role of a known P450 then this may be the only approach.

4. Demonstration of catalytic activity with recombinant P450 proteins. This approach has been very popular since the first successful expression of mammalian P450s in vector-based systems.[50,51] Since then, human and other P450s have been expressed in several cell lines of mammalian and insect origin, yeast, and bacteria.[52–58] Some of these P450s have been purified and can be used in reconstituted enzyme systems in the same way that P450s purified from tissues have been.[40,59–63] The catalytic activities of these recombinant P450s can also be examined in microsomes or in intact cells.[51,64,65] Also, it has been possible to express human P450 fusion proteins with the NADPH-P450 reductase attached.[36,66] Analysis of the abilities of the different proteins to catalyze a particular reaction provides at least qualitative insight into the contributions of individual P450s. As in the case of P450s purified from tissues, it is difficult to use these results alone to quantitatively address the roles of individual P450s in a particular activity.

The above approaches all have their own advantages and disadvantages, and the most reliable conclusions are reached when results from one line of evidence corroborate others. Variations of the first two approaches can also be used in *in vivo* settings. For instance, if two different reactions are both catalyzed by the same enzyme, then there should be a positive correlation when they are compared to each other in a set of human volunteers. Some of the chemical inhibitors can also be used *in vivo* (Table III). Another approach involves induction studies in humans. For instance, P450 3A4 and at least some P450 2C enzymes are induced by barbiturates and rifampicin. If a reaction is postulated to be catalyzed by one of these enzymes, then pretreatment of people with these compounds should accelerate the oxidation of the drugs (*in vivo*). Induction studies can also be done in isolated hepatocytes.[67]

2. *In Vivo* Significance of Human P450 Enzymes

The P450s are the main enzymes involved in the biotransformation of drugs. Much of our knowledge of human P450s is, indeed, derived from studies with drugs, since these chemicals can be administered to humans in defined amounts and rates of formation of products can be readily ascertained. Even when P450-catalyzed oxidation of a drug is not a main step in metabolism and conjugation reactions are dominant, there may be minor oxidation reactions that influence toxicity (e.g., acetaminophen[68]). The P450s are also the major enzymes involved in the oxidation of other xenobiotics such as carcinogens and pesticides and endobiotics such as steroids, fat-soluble vitamins, and eicosanoids. Deficiencies in some of the P450s involved in some of the steroid hydroxylations are known to be debilitating.[28,69–71]

Although the total level of liver microsomal P450 does not vary considerably among individual humans (generally ~3-fold[72]), there is considerably more variation in the levels of individual P450s (Tables I, II; Figs. 1, 2). These can generally vary in concentration by 1–3 orders of magnitude, and these differences are often reflected in parameters of drug metabolism. The variation in human hepatic P450 levels and activities is related to several factors. The first is genetic polymorphisms where heritable DNA changes lead to lack of production of the P450, lack of inducibility of a P450, or synthesis of a form of the P450 with altered catalytic activity. A listing of recognized P450 genetic polymorphisms is presented in Table I. (A polymorphism is usually defined as a genetically determined difference affecting ≥ 2% of the population under consideration. Those base changes that occur less frequently are often referred to as "genetic deficiencies" or "inborn errors of metabolism" if they are clearly associated with disease, or "rare alleles" if they are not so serious.) These will be mentioned under the heading of each P450 family.

Another reason for variation in levels of P450 enzymes is enzyme induction. Several of the P450s are known to be induced by drugs, foodstuffs, or societal habits such as ingestion of tobacco smoke and alcoholic beverages. A third mechanism of modulation of P450 catalytic activities in humans is enzyme inhibition. The level of a P450 may actually be lowered in some cases of irreversible inhibition, or the catalytic activity may be attenuated even in the absence of a decrease in the level of the protein. Both drugs[73,74] and foodstuffs[75] have been shown to act as inhibitors.

A number of unanticipated adverse reactions have been observed with drugs because of unusually high or low levels of individual P450 enzymes. In most cases a physician prescribes a set dose of a drug for most patients in the absence of a reason to do otherwise, and historically this dose regimen has been developed in the pharmaceutical company on

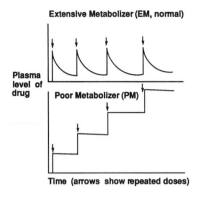

FIGURE 4. Significance of unusually low metabolism of a drug by P450s or other enzymes. The typical pattern is seen in the upper panel, where the plasma level of the drug is maintained in a certain range when a particular repetitive dose is prescribed. Unusually slow metabolism (lower panel) results in an elevated plasma level of the drug.

the basis of a rate of biotransformation and clearance worked out in a few healthy individuals. However, if the patient who takes this "typical" dose is very deficient in the major P450(s) that oxidizes this drug, then an unusually high plasma level of the drug may build up, particularly with multiple dosing (Fig. 4). If the drug has inherent pharmacological activity, then exaggerated effects may be seen if the therapeutic window is not very wide. On the other hand, some drugs are administered as "prodrugs" that require enzymatic oxidation to generate their active forms from precursors. If an individual has unusually low enzymatic activity, then such a drug would not be likely to be effective if administered at the recommended dose. The situation would be reversed in individuals who have unusually high levels of the particular P450 enzyme involved in the oxidation. That is, they will tend to oxidize drugs with direct pharmacological activity very rapidly and render them ineffective. Prodrugs would be activated too quickly and might yield dangerously high levels of active drug.

There are also some other possible ramifications of unusually high or low levels of a particular P450. Some disease states are associated with these. Deficiencies in steroidogenic P450s can be lethal or debilitating, as pointed out earlier. Many endobiotics are substrates for the liver microsomal P450s, although there is not clear evidence that alterations in levels of these P450s are predisposing for disease. It has been postulated that individuals deficient in P450 2D6 ("poor metabolizers") are overrepresented for Parkinson's disease,[76–78] conceivably because of lack of oxidation of an unknown natural product.[79] It has also been reported that such P450 2D6-deficient individuals are at *decreased* smoking-associated risk of lung cancer,[80–85] although searches for involved carcinogens have been rather negative[86,87] and the epidemiological conclusions are not without controversy.[88–93]

3. Human Extrahepatic P450s

P450 was originally discovered in the liver endoplasmic reticulum and most of the attention has been given to these enzymes, since the highest concentration of P450 is here. The amount of microsomal P450 in a human liver is considerable (e.g., ~7500 nmoles in a 1.5-kg liver[38]) and is certainly capable of oxidizing many substrates in a "first-pass" effect.[94]

Obviously the P450s localized in steroidogenic tissues have specific roles in steroid anabolism (see also Chapter 12). Some extrahepatic human P450s undoubtedly have key roles in maintenance of homeostasis and signal transduction, although the suggestive evidence comes essentially completely from work with experimental animals and will not be discussed here.

Extrahepatic P450s have been shown to have roles in the metabolism of drugs. The lungs and nasal tissues contain appreciable amounts of P450, and these sites may be important in oxidations that occur when compounds enter the pulmonary system. Renal P450s may also be of significance in processing compounds generated in the kidneys or transported there from the liver. One of the extrahepatic tissues in which there has been particular attention is the small intestine.[95] P450 3A4 is abundant here,[96] and the enzyme appears to play a major role in the metabolism of many orally administered drugs.[97,98] In this regard it may considerably reduce the oral bioavailability through inactivation. Distinguishing between the roles of human hepatic and intestinal P450 enzymes in metabolism is not always easy unless careful comparisons are made of the pharmacokinetics when the drug is administered orally and intravenously. Inducers often act on both the hepatic and extrahepatic enzymes. Further, there are other considerations regarding the activation of toxins and carcinogens. For instance, very low levels of the carcinogen aflatoxin B_1 are consumed and essentially only through the diet. The compound is activated (8,9-epoxidation) by P450 3A4, which is abundant in liver and intestine. Activation in the intestine would lead to DNA adduction there, but the cells are sloughed off in a few days and there is not really an opportunity for tumor development. Activation in the liver appears to initiate hepatocarcinogenesis, which is the malady associated with aflatoxin B_1. The amount of P450 3A4 in the small intestine is probably in excess of the daily consumption of aflatoxin B_1 and the question can be raised as to how it traverses the intestinal wall to get to the liver.

4. Individual P450 Subfamilies

4.1. P450 1A

The human P450 1A subfamily contains two members, P450 1A1 and P450 1A2, which are ~70% identical in their sequences. Both have been studied extensively because of their roles in the activation of carcinogens.

P450 1A1 is expressed at only very low levels in human liver and is essentially an extrahepatic enzyme.[6,99,100] The level can be induced by smoking, although the response varies considerably. Some of the tissues in which the enzyme has been studied are lung, placenta, and lymphocytes. Attempts to correlate the inducibility of the enzyme with the incidence of cigarette smoking-induced lung cancer incidence have been nonconclusive[101–105] although there does appear to be a correlation between cancer and the "absolute" level of enzyme.[106,107] More recently an association between an *Msp*I polymorphism and lung cancer incidence has been observed in a Japanese study[108] but not in any work with Caucasians.[109,110] The *Msp*I polymorphism has been considered to be linked to a Val/Ile polymorphism in the protein coding region,[111,112] but linkage disequilibrium has been observed.[110,113] A distinct polymorphism has been reported in African-Americans[114] but its relation to cancer is unknown.

As in the case of P450 1A1 enzymes from experimental animal species, human P450 1A1 is efficient in catalyzing a number of oxidations of polycyclic aromatic hydrocarbons.[40,115] The trace of P450 1A1 in human liver may be responsible for the low level of benzo[a]pyrene 7,8-epoxidation.[116,117] Human P450 1A1 has been expressed in mammalian cells, yeast, and *Escherichia coli* [40,118–121] and has also been shown to oxidize some drugs.[122,123] Such oxidation may be important in the disposition of some drugs, although the lack of the enzyme in liver seems to preclude a major role in the biotransformation of these materials when they are systemically available.

Human P450 1A2, in contrast, is expressed essentially only in the liver and not in extrahepatic tissues,[116] although there are occasional reports of low extrahepatic expression.[95,124] Both human P450s 1A1 and 1A2 are under the regulation of the A*h* locus, involving the interaction of Ah receptor/ARNT heterodimeric complexes (activated by binding of polycyclic hydrocarbons, etc.) with upstream enhancer elements.[125] However, the P450 1A2 gene also contains putative AP-1 sites that may interact with Ah responsive elements ("XRE" or "DRE" sites) to explain the tissue-selective induction.[126] One of the noninvasive assays used to screen individuals for levels of P450 1A2 expression involves the rate of caffeine N^3-demethylation.[127,128] Levels of P450 1A2 and the rate of caffeine oxidation both vary ~ 40-fold among the population. A trimodal distribution of the activity has been interpreted as evidence for genetic polymorphism.[128] This view is consonant with the conclusions of Vesell[129] from twin studies with antipyrine, which P450 1A2 also contributes to the clearance of.[130] However, no differences in the coding or upstream regions of the genes of individuals with high and low P450 1A2 levels have been found to date.[131] A possibility is that there is polymorphism in a *trans*-acting element. It should be pointed out, however, that another laboratory has used a different parameter (involving secondary metabolites of caffeine) to reach a different conclusion, that the distribution of P450 1A2 is monomorphic.[132] Theophylline may also be an acceptable noninvasive probe.[133,134]

A list of drug substrates appears in Table IV. The significance of P450 1A2 in drug oxidation is not great, although dietary alterations leading to theophylline ineffectiveness can now be interpreted in terms of this enzyme.[141] The antiulcer drug omeprazole can induce P450 1A2 although not particularly effectively.[142–144] P450 1A2 has not been reported to catalyze many steroid hydroxylations except 17β-estradiol 2-hydroxylation.[62,145]

There is considerable interest in the ability of P450 1A2 to catalyze the *N*-hydroxylation of carcinogenic aryl amines and heterocyclic amines.[118,127] The oxidized products can modify DNA, either directly or following conjugation with acyl or sulfate groups.[146] The heterocyclic amines are of considerable interest in that they are found in charbroiled food and cigarette smoke.[147] There is some epidemiological evidence that elevated levels of P450 1A2 can be a predisposing factor to colon cancer, although the risk is marginal unless there is a high level of *N*-acetyltransferase and high consumption of charbroiled meat.[131]

TABLE IV
Drug Substrates for Human P450 1A2

Drug	References
Acetaminophen (ring)[a]	135
Antipyrine (4, 3-methyl)	130, 136
Bufuralol (1 and 4 others)	137
Ondansetron (7, 8)	123
Phenacetin	7
Tacrine	138
Tamoxifen (N-demethylation)	122
Theophylline (1, 3, 8)	133, 139
Warfarin	140

[a]Sites of oxidation are shown in parentheses.

Both human P450 1A1 and 1A2 have been expressed in several vector systems.[40,62,118–120,145,148,149] Those isolated from E. coli are almost completely high-spin iron proteins, although this property does not seem directly related to their catalytic activities.[40,62]

4.2. P450 1B

Recently, cDNA clones of mouse, rat, and human P450 1B1 have been reported.[150,151] The human cDNA clone was isolated from a keratinocyte cell line on the basis of its inducibility with TCDD; the P450 1B proteins are also expressed constitutively at low levels (mRNA) in several tissues, including heart, brain, placenta, lung, liver, skeletal muscle, kidney, spleen, thymus, prostate, testis, ovary, small intestine, colon, and peripheral blood leukocytes.[151] Indirect evidence indicates a role for rat P450 1B1 in the oxidation of 7,12-dimethylbenz[a]anthracene.[150] The human 1B1 protein has not yet been expressed, however, and its catalytic specificity is unknown.

4.3. P450 2A

P450 2A6 appears to be the only member of the subfamily expressed in human liver (this was formerly termed P450 2A3).[8,152,153] The level of expression is rather low (Tables I, II)[8] and the inducibility has not been examined directly.

A genetic polymorphism has been reported; the incidence of poor metabolizers in a Caucasian population is only ~ 2%.[154–157] This polymorphism seems to be associated with a Leu-to-His change in the coding region.

The purified[8] and recombinant[152,153] P450 2A6 enzymes have been shown to catalyze a number of reactions including 7-ethoxycoumarin O-deethylation and the activation of the carcinogens 6-aminochrysene and aflatoxin B_1. However, because of the low level of P450 2A6 expressed in most human liver samples, the contribution of P450 2A6 to these activities seems to be minimal.[8,158] Several lines of evidence indicate that most of the coumarin 7-hydroxylation activity in human liver microsomes can be attributed to P450 2A6, and this activity appears to be a useful marker for the enzyme.[8,152] Coumarin 7-hydroxylation has also been used as an in vivo diagnostic assay.[154,155,157]

P450 2A6 seems to have overlapping catalytic specificity with P450 2E1 in the activation of nitrosamines.[159,160] In human liver microsomes, P450 2E1 preferentially catalyzes the oxidation of N-nitrosodimethylamine but P450 2A6 preferentially oxidizes N-nitrosodiethylamine.[159,160] Both proteins contribute to the oxidation of tobacco-specific nitrosamines.[160–162] There is precedent for the extrahepatic expression of P450 2A subfamily enzymes in experimental animals,[163] and it may be possible that in human extrahepatic tissues such as nasal mucosa (and other tissues related to head and neck tumors) the expression of P450s 2A6 and 2E1 is of significance to the oxidation of these chemicals.

P450 2A7 has a sequence 94% identical to P450 2A6.[152] An attempt to express the protein did not yield a protein with the expected spectral properties,[152] so little is known about the catalytic selectivity of this P450. Little information is available regarding its tissue-specific expression.

4.4. P450 2B

Historically the P450 2B family enzymes played major roles in the development of the rat and rabbit P450 models. These were among the first microsomal P450s purified[164–167] and show the most dramatic induction by barbiturates. In early work with humans it was generally assumed that similar enzymes would be important in human liver.[1]

The only P450 2B enzyme known to be expressed in human liver is P450 2B6.[20,168] The level of expression of P450 2B6 in human liver is very low; in the work of Mimura et al.[20] the highest level found in any of 60 livers examined only accounted for ~ 1% of the total P450. Immunochemical cross-reactivity with rat P50 2B1 is poor and studies with antibodies (raised against rat P450 2B1) are not very definitive.[1,20] We have found considerably stronger cross-reactivity with anti-monkey P450 2B preparations.[20]

Purified and recombinant P450 2B6 enzymes have been shown to catalyze some reactions,[20] including nicotine,[169] but it is not clear that P450 2B6 dominates in any particular reaction in microsomes. Thus, no diagnostic markers are currently available for in vivo use. It is difficult to know if barbiturates induce this enzyme in human liver. Among the 60 liver samples examined by Mimura et al.,[20] only 4 contained appreciable levels of P450 2B6 and these same samples were relatively high in P450 3A4, which is known to be inducible by barbiturates and other substances (vide infra). However, the overall correlation of levels of P450s 2B6 and 3A4 was not particularly strong and it cannot be concluded that there is coordinate regulation.

P450 2B7 appears to be a pseudogene and is not expressed.[22]

4.5. P450 2C

P450 2C subfamily proteins were among the first to be purified from human liver.[1,2] This subfamily is the most complex of the P450s found in humans. DNA blotting experiments indicate the presence of ~ 7 genes/genelike sequences.[9] Six cDNAs have been reported, although questions exist about the separate identity of some (Table I).[22] The P450s in this subfamily are involved in the oxidation of some important drugs. These P450s are under a variety of modes of regulation, including inducibility and genetic polymorphism. Since the P450 2C subfamily members are all >80% identical, evaluation of distinct roles has not been trivial because of lack of discrimination of probes and difficulties in protein isolation.

P450 2C8 has been purified[9,170,171] and cloned (cDNA)[9,172,173] in a number of laboratories. The level of this protein does not appear to be as high as that of P450 2C9(10) (*vide infra*). Since most antibody and nucleic acid probes have not discriminated among members of this subfamily and little information is available concerning substrates that are preferentially substrates for P450 2C8, the regulation of this gene is not well understood. Recently some reactions have been identified as being preferentially catalyzed by P450 2C8. One reaction is the 6α-hydroxylation of taxol, the major path of oxidation of this oncological drug (J. W. Harris, personal communication).[174] A protein that appears to correspond to P450 2C8 was originally suggested as being involved in the 4-hydroxylation of retinol and retinoic acid,[170] and the report has been confirmed using assays with recombinant proteins.[175]

Several lines of evidence suggest that P450 2C9 may be generally the most abundantly expressed of the human P450 2C subfamily proteins, at least in liver.[9,10,176] P450 2C9 seems to have the highest catalytic activity in several reactions, including tolbutamide (methyl) hydroxylation,[177,178] tienilic acid hydroxylation,[179] phenytoin 4'-hydroxylation,[180] and (S)-warfarin 7-hydroxylation.[140] There is *in vivo* evidence that treatment of an individual with barbiturates or rifampicin induces oxidation of tolbutamide, which may be interpreted as evidence of inducibility of P450 2C9.[181] Genetic polymorphisms in phenytoin 4'-hydroxylation[182] and tolbutamide hydroxylation[183] have been suggested but not confirmed.

The protein coding sequence of P450 2C10 differs from that of P450 2C9 in only two codons.[9,184] It may be an allelic variant of P450 2C9.[22] The cDNA clone was derived from the same (single liver) cDNA library as a P450 2C9 clone.[9] Although the protein coding regions of these two cDNAs were very similar, the 3' tails diverged considerably and selective probes recognized both sequences in human liver mRNA.[9] It is conceivable that the two cDNAs are the result of a library artifact or that other cDNAs contain similar 3' tails. Little difference has been seen to date in the catalytic activities of expressed P450 2C9 and 2C10.[185] For the purposes of this review the actual sequences used in expression studies are utilized in tables of catalytic specificity (Table V).

P450 2C18 appears to be a minor member of the P450 2C subfamily, as judged by the frequency of isolation of clones from cDNA libraries.[176,198] P450 2C18 protein has been expressed in yeast and mammalian cell systems[176,198] and shows (low) catalytic activity toward tolbutamide and tienilic acid (P. H. Beaune, personal communication).[198] To date no reactions have been shown to be preferentially catalyzed by this enzyme.

A P450 2C17 cDNA has been reported by Romkes *et al.*[176] but the protein has not been expressed. There is some question as to its existence as a distinct gene.[22]

A P450 2C19 cDNA has been reported by Romkes *et al.*[176] Although nearly all of the P450 2C subfamily members have been considered at one time or another to be the polymorphic (S)-mephenytoin 4'-hydroxylase,[9,10,176,199] the evidence now appears to indicate that this is P450 2C19.[47,200] The recombinant enzyme has a relatively high catalytic activity[200] and studies in progress in several laboratories are directed toward the identification of the defect(s). The genetic polymorphism, first reported by Wedlund *et*

TABLE V
Drug Substrates for Human P450 2C Enzymes

Drug[a]	P450 2C enzymes reported to be catalysts	References
Diazepam (N-demethylation)	2C19?	186, 187
Diclofenac (4)	2C9	188
Hexobarbital (3)	2C9	61, 177, 189
Ibuprofen (methyl, 2)	2C9	190
Losartan (aldehyde, acid)	?	191
Mefanamic acid (3-methyl)	2C9	190
(S)-Mephenytoin (4)	2C19	10
Mephobarbital	?	192, 193
Nirvanol (4)	2C19	10
Omeprazole	2C19	194, 195
Phenytoin (4)	2C9	180
Piroxicam (5)	2C9	190
Retinoic acid	2C8	170, 175
Retinol	2C8	170
Tenoxicam (5)	2C9	190
Tienilic acid	2C8, 2C9, 2C18	179
Tolbutamide (methyl)	2C8, 2C9, 2C18	177, 196
Trimethadone (N-demethylation)	2C9	197
(S)-Warfarin (7)	2C9	140

[a]Sites of oxidation are shown in parentheses.

al.[201] and Küpfer and Preisig,[202] is of interest in that it is seen at a frequency of ~ 3% in Caucasians but nearly 20% in Asians (Orientals).[26] Thus, drugs whose metabolism is primarily dependent on this enzyme are of particular concern when administered to races where metabolism may be considerably slower, particularly if the therapeutic window is narrow. A number of other reactions have been suggested to be catalyzed by the same enzyme, on the basis of cosegregation of *in vivo* disposition with (S)-mephenytoin 4'-hydroxylation (Table V). It will be of interest to learn if recombinant P450 2C19 has high catalytic activities in these instances.

Sulfaphenazole appears to be a strong and selective inhibitor of P450 2C9[177,203], it does not inhibit P450 2C19 as well. However, selectivity with other P450 2C enzymes has not been examined. It is possible to prepare antipeptide (2C19) antibodies,[200] but their abilities to inhibit catalytic activities have not been examined.

4.6. P450 2D

The only human P450 2D enzyme expressed is P450 2D6. CYP2D7 and CYP2D8 are pseudogenes.[22,204,205]This P450 has a particularly interesting history in that it was the first one shown to be under monogenic regulation, i.e., show a genetic polymorphism related to functional activity in drug metabolism.[5] Smith first studied the basis for attenuated debrisoquine clearance after a personal hypotensive episode in a drug metabolism study.[206] Independently, Eichelbaum and Dengler found a genetic polymorphism in the oxidation of sparteine.[207] Eventually this work and that on some other drugs[208,209] culminated in the purification of the rat[210–212] and human[7,11,213] P450 2D enzymes. Antibodies produced against rat P450 2D1 recognized (human) P450 2D6[39,210] and were used to clone the human cDNA and gene.[204,205]

The P450 2D6 polymorphism has been extensively characterized. Most phenotypically deficient Caucasians (~ 7% of the total population) can be characterized in terms of a small number of genetic changes.[154,204,214–217] Among the most common are a G → A transition at an intron/exon boundary that leads to abnormal mRNA splicing, a frameshift in another exon, and a complete gene deletion.[154,218,219] These and other genotypic differences are now readily analyzed by polymerase chain reaction and restriction fragment length polymorphism methods.[154,220,221] Considerable racial differences are seen in P450 2D6. Few Asians (Orientals) or Africans are bona fide "poor metabolizers" (< 1%).[26,27] However, the dominant Chinese allele differs from the major Caucasian one and seems to have somewhat less catalytic activity (Ingelman-Sundberg, personal communication), at least toward some of the substrates. The genetic basis for a lower prevalence of P450 2D6 deficiency in African-Americans has also been described.[222] Some individuals have unusually high levels of P450 2D6 activity. In some of these cases this phenomenon has been attributed to stable gene duplication.[223] The possibility also exists that some alleles may code for a form of the protein with higher activity. However, it should be pointed out that studies with rat P450 2D1 show that some point mutations do not affect all catalytic activities of the enzyme in the same way.[224]

P450 2D6 has been implicated in at least 30 different drug oxidation reactions (Table VI). When the contributions of different enzymes to drug metabolism are considered, P450 2D6 has a major role but the value may be biased to the high side because of the ease of analysis of this enzyme and the search for the involvement of P450 2D6 and genetic polymorphism in drug oxidations. In general there is very limited metabolism of these drugs in the individuals who are phenotypic poor metabolizers, unless there are alternate pathways involving other enzymes. In situations where the therapeutic window is narrow, adverse reactions can occur.

In addition to the long list of substrates (Table VI), many P450 2D6 inhibitors have also been identified (see Chapter 9).[271–273] Many of these have very low K_i values and are effective both *in vitro* and *in vivo*. For instance, quinidine can be given to people to change their effective phenotype from extensive to poor.[274,275] Such inhibition can have clinical significance in terms of adverse drug interactions. For instance, administration of a potent inhibitor along with a P450 2D6 substrate could have the same effect as administering the drug unknowingly to a poor metabolizer. Some substrates may be encountered in foods or even synthesized *in vivo*. Although exogenous codeine has been reported to be a substrate for P450 2D6,[234] the codeine formed endogenously is not.[276]

The most specific substrates and inhibitors of P450 2D6 have the common property of a basic nitrogen, either an aliphatic amine or a guanidino moiety.[86] Crude pharmacophore models were first constructed with a putative point anionic charge in the protein set 5–7 Å away from the $(FeO)^{3+}$ site of the heme.[86] More sophisticated pharmacophore models have been developed based on more information from both substrates and inhibitors.[273,277,278] In quinidine, the inhibition appears to be competitive but there is no

TABLE VI
Drug Substrates for P450 2D6

Drug	References
Ajmaline	225
Amiflamine	226
Amitriptyline	227
Apridine	228
Bufuralol (1)[a]	213, 229
Bupranolol	230
CGP 15210 G (bis[cis-3-hydroxy-4-(2,3-dimethyl-phenoxy)piperidine]sulfate)	231
Clomipramine	232
Clozapine	233
Codeine	234
Debrisoquine (4)	211, 235–238
Deprenyl	239
Desipramine	12, 240
Dextromethorphan (O-demethylation)	241–243
Encainide	7, 39, 244
Flecainide	245
Guanoxan	246
Haloperidol	247
Hydrocodone	248
Imipramine	249
Indoramin	250
MDL 73005 (8-[[2-(2,3-dihydro-1,4-benzo-dioxin-2-yl)methylamino]ethyl]-8-azaspiro[4,5] decane-7,9-dione) (M$_2$, M$_3$, M$_4$)	251
4-Methoxyamphetamine	252
Methoxyphenamine	253
Metropolol	208, 254, 255
Mexiletine	256
Minaprine	257
Nortriptyline	258
Paroxetine	259
Perhexiline	260
Perphenazine	261
Phenformin	262
Propafenone	263, 264
N-Propylajmaline	265
Propranolol	7, 39, 266
Sparteine (Δ^2, Δ^5)	39, 267
Thioridazine	268
Timolol	269, 270

[a]Sites of oxidation are shown in parentheses.

FIGURE 5. Pharmacophore model for the active site of P450 2D6. The two panels of the figure show two views (90° difference) between superpositions of low-energy conformers of all inhibitors examined in the study.[273]

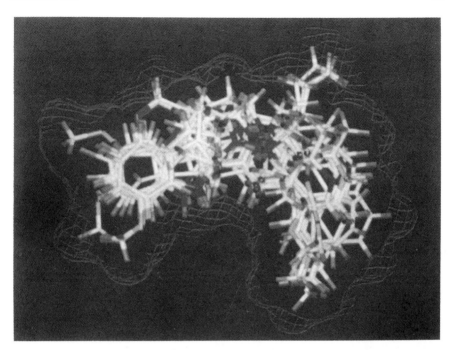

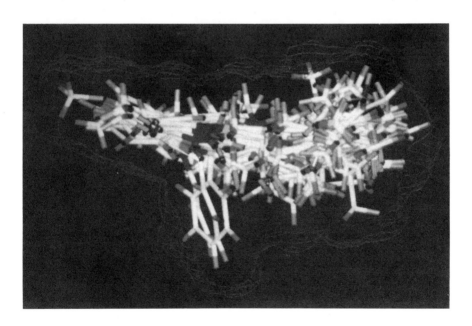

oxidation of quinidine by P450 2D6 (P450 3A4 catalyzes two oxidations[279]). This information was useful in development of one of the pharmacophore models (Fig. 5).[273] Recently, however, P450 2D6 has been clearly shown to oxidize a drug to cleave a C–N bond.[239] Presumably the catalytic mechanism involves single electron transfer from the nitrogen atom, a process that is probably operative over a range of at least 5 Å.[280] Thus, every model will have some limitations.

Low levels of P450 2D expression have been detected in some extrahepatic rat tissues, including brain.[281] However, the poor metabolizer phenotypic individuals appear to function normally, at least until they encounter drugs with low therapeutic indices that are not being oxidized by P450 2D6 at sufficient rates.[282] There has been considerable interest in the possible association of P450 2D6 phenotypes with various diseases, particularly parkinsonism and various cancers.[76–78,80,83,85,90–93,283–286] In most cases there is no known compound involved that serves as a substrate, and many of the epidemiological studies are still controversial. However, the general hypothesis that P450 levels may be linked to diseases of environmental etiology is appealing and deserves further consideration.

4.7. P450 2E

P450 2E1 is the only gene in this subfamily in most species, including humans (rabbits are an exception[22,287]). The enzyme has been of interest in humans and experimental animal models because of the possible relevance to alcoholism, chemical carcinogenesis, diabetes, and other maladies.[14,288,289] The enzyme appears to be expressed in liver and a number of other tissues. P450 2E1 may be of particular importance in the etiology of disease in certain extrahepatic tissues, e.g., in tobacco- and alcohol-associated cancers of the head and neck area.[29,289,290] Levels of human hepatic P450 2E1 vary by an order of magnitude.[14,291,292] In experimental animals the regulation of P450 2E1 has been found to be very complex, involving aspects of developmental and hormonal regulation, transcriptional and posttranscriptional control, and protein stabilization.[289,293–298] It is not clear exactly which of these aspects are involved in humans, although it has been shown that the level of hepatic P450 2E1 is elevated in alcoholics.[299] Some polymorphisms have been identified in humans, including one in a 5′-upstream region where a transcription factor (HNF-1) putatively binds.[300] However, as yet none of these have been associated with levels of the enzyme or catalytic activity.

P450 2E1 has been purified from human liver,[13,14] has been expressed in recombinant systems,[63,301–303] and has been purified from a recombinant source.[63] Cytochrome b_5 appears to be necessary for optimal activity of the enzyme, at least with regard to several catalytic activities.[13,63] The effect appears to be on both V_{max} and K_m.[13,63] This observation is of interest in that kinetic hydrogen isotope effects are seen with human and animal P450 2E1 enzymes, for which an isotope effect is seen on K_m but not V_{max}.[304,305] The increased K_m parameter probably is not a consequence of altered substrate affinity but probably reflects a combination of individual rate constants and is more likely a ratio of a rate-limiting step divided by the rate of hydrogen abstraction.[306] More remains to be done regarding this hypothesis.

TABLE VII
Drug Substrates for Human P450 2E1

Drug[a]	Reference
Acetaminophen (ring)	135
Caffeine (8?)	133
Chlorzoxazone (6)	291
Enflurane	307

[a]Site of oxidation is shown in parentheses.

Few drugs are oxidized by P450 2E1 (Table VII). The enzyme contributes to acetaminophen oxidation, but clearly the most relevant reaction is chlorzoxazone 6-hydroxylation.[291] This is probably the best marker activity for human (and animal) P450 2E1 *in vitro*.[291] Human P450 1A1 has some activity,[123] although the K_m is higher and the contribution in liver is probably negligible because of the low level of hepatic P450 1A1 (*vide supra*). The *in vivo* assay seems to reflect the hepatic enzyme level, even if the extent of variation appears to be somewhat less than that seen in *in vitro* assays.[308]

Although relatively few drugs are oxidized by P450 2E1, the list of carcinogens is quite extensive.[14,29] Most of these are relatively small compounds, suggesting a small access channel to the active site. As mentioned above, one possible explanation for the deuterium isotope effects on K_m may be that the substrate only binds *after* the activated oxygen complex has been formed,[306] in a manner resembling the flavin-containing monooxygenase[309] and bacterial methane monooxygenase.[310] There appears to be some overlap in catalytic specificity between P450 2A6 and 2E1.[160,311] Also, both are inhibited by diethyldithiocarbamate (the reduced form of disulfiram).[160] 4-Methylpyrazole appears to be a more selective inhibitor of P450 2E1 than diethyldithiocarbamate (Table III).[160]

4.8. P450 2F

A P450 2F1 cDNA clone was isolated from a human lung cDNA library by screening with a P450 2C9 probe.[312] Expression is seen in lung and at low levels in liver.[312,313] The protein has not been isolated but has been expressed in a vaccinia-based system. It shows low *O*-dealkylation activity toward 7-ethoxycoumarin, 7-propoxycoumarin, 7-pentoxyresorufin, and 7-benzyloxyresorufin.[312] However, no activity toward endogenous substrates or carcinogens has been reported.

DNA blotting work suggests the presence of a closely related gene or pseudogene.[312] Further information about this aspect has not been published.

4.9. P450 3A

The P450 3A subfamily enzymes play very prominent roles in human metabolism of xenobiotics.[314] These were among the first human P450s to be isolated.[3,4] P450 3A4 appears to be the most abundant human P450 (Tables I, II) and has very broad substrate specificity (Table VIII).

A protein termed P450$_{NF}$ was isolated from human liver microsomes on the basis of its ability to oxidize the drug nifedipine.[15] This protein is now termed P450 3A4. Antibodies raised against the protein were used to screen a cDNA library (prepared from

TABLE VIII
Drug Substrates for P450 3A4 (Including Steroids)

Drug[a]	References
Acetaminophen (quinone formation)	135
Alfentanil (noralfentanil)	315
Alpidem (propyl α,β)	316
Amiodarone (N-deethylation)	317
Bayer R4407 [(+)K8644)] (dihydropyridine)	318
Bayer R5417 [(−)K8644)] (dihydropyridine)	318
Benzphetamine (N-demethylation)	15
Budesonide (6β)	319, 320
Codeine (N-demethylation)	321
Cortisol (6β)	322, 323
Cyclophosphamide (high K_m)	324
Cyclosporin A (AM9, AM1, AM4N; nomenclature formerly M1, M17, M21)	325, 326
Cyclosporin G	(P. Koch, personal communication)
Dapsone (N)	327
Dehydroepiandrosterone 3-sulfate (16α)	323
Dextromethorphan (N-demethylation)	241
Diazepam (3)	186, 187
Diltiazem	328
Ebastine (alcohol)	329
Ergot CQA 206–291	330
Erythromycin (N-demethylation)	19, 323
17β-Estradiol (2, 4)	15, 331, 332
17α-Ethynylestradiol (2)	333
Felodipine (dihydropyridine)	318
FK506	334, 335
Gestodene	38
Imipramine (N-demethylation)	249
Lidocaine (N-deethylation)	336, 337
Losortan (aldehyde, acid)	191
Lovastatin (6β, 6 -exo-methylene, 3,3,5-dihydrodiol)	37, 338
MDL 73005 (8-[[2-(2,3-dihydro-1,4-benzo-dioxin-2-yl)methy-lamino]ethyl]8-azaspiro[4,5]decane-7,9-dione) (M_1, M_5)	251
Midazolam (1, 4)	339
Nifedipine (dihydropyridine)	15
Niludipine (dihydropyridine)	340
Nimodipine (dihydropyridine)	340
Nisoldipine (dihydropyridine)	340
Nitrendipine (dihydropyridine)	340
Omeprazole	195, 341
Progesterone (6β, some 16α)	332, 342
Quinidine (3, N)	279
Rapamycin (41, others)	334
Sertindole (N-dealkylation)	343
Sulfamethoxazole (N)	344
Sulfentanil	23
Tamoxifen (N-demethylation)	122, 345

TABLE VIII
(Continued)

Drug[a]	References
Tamoxifen (*N*-demethylation)	122, 345
Taxol (2-phenyl)	174, 346
Terfenadine (*t*-butyl, *N*-dealkylation)	347
Testosterone (6β, trace 15β, 2β)	15, 342, 348
Triazolam	339
Trimethadone (*N*-demethylation)	197
Troleandomycin (N)	349
Verapamil	350
Warfarin (*R*-10, *S*-dehydro)	323
Zatosetron (N)	351
Zonisamide	352

[a]Sites of oxidation are shown in parentheses, if identified.

a single liver sample) and isolate a full-length cDNA ("NF25").[353] Watkins *et al.*[19] isolated a protein "HLp" from human liver microsomes on the basis of its immunochemical cross-reactivity with a rat P450 3A enzyme.[354] Three cDNA clones were overlapped to obtain a sequence now known as P450 3A3,[355] which differs from P450 3A4 in 14 nucleotide positions, including an apparent 3-base frameshift. Later, Aoyama *et al.*[348] isolated a cDNA (pCN1) corresponding to P450 3A4 (NF25) except for one nucleotide difference. Genomic blotting analysis with long, nonoverlapping cDNA probes is consistent with (a minimum of) three genes or genelike sequences in this gene family.[353] Studies on the hybridization of specific probes have clearly shown that P450 3A4 is the dominant P450 3A subfamily mRNA expressed in all human liver samples examined.[356,357] The identity of P450 3A3 as a separate entity may be questioned. Another cDNA clone, termed "NF10," was isolated from the same library as P450 3A4 ("NF25") and found to differ only in a 3-base frameshift (different than "HLp").[356] Specific probes recognizing the frameshift did not detect mRNA and this frameshift may be considered a library artifact. However, the NF10 clone was found to contain two separate polyadenylation signals, and these were shown to give rise to the two size classes of mRNA.[356] The relative use of the two signals seems unrelated to levels of mRNA or protein or catalytic activity. Protein and mRNA are clearly inducible by barbiturates, rifampicin, and dexamethasone.[19,67,358]

P450 3A5 has been isolated from human liver microsomes[16] and the cDNA has been reported by two groups.[348,359] The coding sequence is 88% identical to that of P450 3A4. P450 3A5 is polymorphically expressed, appearing in only about one-fourth of humans.[360,361] When it is expressed, the level in liver is usually about one-third that of P450 3A4.[360,361]

A P450 termed "HLFa" was isolated from human fetal liver microsomes by Kamataki and Kitada.[3,362] This P450 is now termed 3A7; the cDNA clone was obtained from a fetal liver library. mRNA blotting studies indicate that P450 3A7 is expressed in fetal but not in adult liver.[363] A more recent study indicated expression of P450 3A7 in adult endometrium and placenta.[364] The possibility exists that expression might occur in tumors, as is the case for many "fetal" enzymes,[365] but such a study has apparently not been done.

Kamataki and his associates have isolated 25-kb genes for both P450 3A4 and 3A7.[366] The structures of the two genes are very similar, with only (two) major stretches that differ. However, studies to date have not implicated these variable regions in binding fetal versus adult tissue factors. P450 3A4 is inducible by barbiturates, glucocorticoids (e.g., dexamethasone), and some macrolide antibiotics (e.g., troleandomycin), both in cultured cells[67,358] and in human liver.[19] The level of P450 3A4 in human liver can account for as much as 60% of the total.[38] Although some individuals have very little of the enzyme,[333] the mean level is still high (Table I). P450 3A4 is also expressed in other tissues, including the lung[116,367] and especially the small intestine.[95] The presence of relatively high levels of P450 3A4 in the small intestine is of considerable significance in the metabolism of orally administered drugs, where extensive transformation may occur during absorption across the wall.[368]

The P450 3A subfamily enzymes have a broad range of substrates and figure very prominently in the metabolism of drugs, particularly P450 3A4. One of the historical difficulties with the P450 3A subfamily enzymes (from humans and experimental animals) has been the reconstitution of optimal conditions for catalytic activity.[15,354,369] The usual conditions involving addition of L-α-1,2-dilauroyl-sn-glycero-3-phosphocholine (dilauroyl lecithin) and an equimolar amount of NADPH-P450 reductase are generally not very effective.[15,370] Purified P450 3A4 has been reported to be more active in a mixture of dilauroyl and dioleoyl lecithins, phosphatidylserine, and cholate.[370] The presence of cytochrome b_5 is usually required. This laboratory has also found that such a system, used with bacterial recombinant or liver-derived P450 3A4, is also enhanced by the inclusion of reduced glutathione or N-acetylcysteine.[60] The mechanistic basis is unknown. Shet et $al.$[36] expressed and purified a fused P450 3A4/rat NADPH-P450 reductase protein and found that the reconstitution conditions varied, with regard to the need for certain lipids and cytochrome b_5, depending on the substrate. The enzyme was active toward some substrates in Tris buffer and not phosphate,[36] and no glutathione enhancement was seen. Recent work with the P450 3A4 protein in this laboratory has confirmed the glutathione stimulation[60] in several different buffers. A high concentration of $MgCl_2$ (30 mM) is very stimulatory; this can be partially replaced by high NaCl concentrations (0.4 M). One way to avoid potential problems in interpretation of catalytic activity studies with this enzyme is to measure rates in recombinant systems in which P450 3A4 is incorporated into the endoplasmic reticulum of a cell. This approach seems to yield relatively high rates of oxidation with mammalian cells. However, the expression levels are relatively low. In yeast, the endogenous cytochrome b_5 and NADPH-P450 reductase do not couple well with P450 3A4 and low activity is seen in yeast microsomes.[323,349] Pompon and his associates developed a system in which human cytochrome b_5 and NADPH-P450 reductase are inserted into the $Saccharomyces$ $cerevisiae$ genome and found considerably higher catalytic activities of P450 3A4.[371] With any system, care must be taken in the interpretation of rate data, with regard to their relevance in the normal hepatic system.

Despite the difficulties in reconstituting P450 3A4 activities mentioned above, considerable information is now available to support the view that P450 3A4 is a major contributor to the oxidation of a large number of drugs (Table VIII). Part of this evidence comes from studies involving immunoinhibition, correlation, and selective inhibition with chemicals such as troleandomycin and gestodene.[314] Some of the evidence has been corroborated *in vivo* (e.g., in studies involving barbiturate induction, inhibition, etc.). A number of measurements of rates of catalysis with P450s 3A5 and 3A7 have been published.[348,361] In general, the same compounds are substrates for the different P450 3A subfamily members, but some major differences have been noted.[348,361] Few direct comparisons between the purified or recombinant enzymes have been made[348,361] and even then there has been no systematic effort to determine how relevant the membrane environment conditions are or how V_{max} and K_m values compare. Currently, the available information on catalytic activities and knowledge of levels of expression of the different enzymes in adult human liver and intestine argues that P450 3A4 has the dominant role for most compounds.

The list of substrates for P450 3A4 contains a variety of different structures, and the question arises as to how one enzyme can have such a broad catalytic specificity.[314,372] Closer inspection of the exact transformations catalyzed indicates that with each substrate there is considerable regio- and stereoselectivity in the oxidations. For instance, testosterone is oxidized almost exclusively at the 6β-position.[15,342] Thus, the catalytic site cannot simply be a very large and loose pocket in which chemicals can tumble and sites of oxidation are governed largely by thermodynamic considerations. The largest known P450 substrate is cyclosporin A (M_r 1201), which is oxidized primarily at three positions.[325,339,348] One possibility is that the active site has a flat mouth to accommodate large substrates and still displays high regioselectivity. However, several observations suggest that the active site topology may be more complex.

Conney and his associates noted that certain flavonoids could directly stimulate certain catalytic activities in human liver microsomes, including benzo[*a*]pyrene 3-hydroxylation and aflatoxin B_1 8,9-epoxidation.[373,374] These activities have been shown to be catalyzed largely by P450 3A4 in human liver microsomes, and both are stimulated by 7,8-benzoflavone (α-naphthoflavone) as in microsomes.[375,376] Interestingly, 7,8-benzoflavone also *inhibits* the 3α-hydroxylation of aflatoxin B_1 by P450 3A4,[377] shifting the balance of oxidation from detoxication (aflatoxin Q_1 formation) to *exo*-epoxidation. Another point is that aflatoxin B_2, which cannot be converted to an epoxide because of saturation at the 8,9 bond, inhibits the epoxidation reaction but not 3α-hydroxylation.[377] The opposite pattern is seen with 2,6-dimethyl-4-(2-nitrophenyl)-3,5-pyridinedicarboxylic acid dimethyl ester (oxidation product of nifedipine).[378] Sometimes, known P450 3A4 substrates compete for oxidation but in other cases they do not.[36,37] These and other findings suggest that an allosteric model might be applicable. Such an explanation has been postulated for the kinetic behavior of rabbit P450 3A6.[369] Other suggestions have also been offered to explain related observations with other P450s, such as enhanced P450/NADPH-P450 reductase interaction[379] and enhanced substrate affinity.[380] Some support for the allosteric model has been seen when the aflatoxin B_1 concentration has been varied, with either human liver microsomes or a system containing bacterial recombinant P450 3A4. The curves seen when either 3α-hydroxylation or 8,9-epoxidation is plotted (versus aflatoxin B_1 concentration) are sigmoidal.[378] When 7,8-benzoflavone is present, the expected increase in epoxidation and decrease in 3α-hydroxylation are seen, and the curves become hyperbolic.[378] These observations may be interpreted in terms of a model in which the substrate aflatoxin B_1 occupies only the catalytic site at low concentrations. 7,8-Benzoflavone occupies the putative allosteric site. At high aflatoxin B_1 concentrations it can also occupy the allosteric site. When the allosteric site is occupied, it changes the conformation to the catalytic site to move the 8,9-double bond of aflatoxin B_1 closer to the (FeO) complex (or alternatively, to stabilize the transition state leading to epoxidation, relative to 3α-hydroxylation). The proximity of the allosteric and catalytic sites is unknown, even if this paradigm is valid. However, a better understanding of the active site of P450 3A4 is clearly needed, in light of the large number of drugs oxidized by this enzyme and the potential for drug interactions.[49,314]

4.10. P450 4A

P450 4A11 has been purified from human kidney[17] and the cDNA has been cloned.[17,22] The enzyme is a fatty acid ω- and ω-1 hydroxylase, as commonly is the case for other 4A subfamily enzymes. In experimental animals this gene family is complex,[22] and more human P450 4A subfamily genes may be expected. (P450 4A9 has been reported anecdotally[22]). Further, expression of the protein in liver is also likely. In rodents the P450 4A genes are induced by peroxisomal proliferators.[381] It is not clear if such a response will be seen in humans, since evidence from cell culture suggests a weak response.[382] Although most drugs, steroids, and carcinogens would not be expected to be substrates, human P450 4A11 and any related enzymes might well be expected to catalyze ω- and ω-1 hydroxylations (and possibly ω-2 hydroxylations and epoxidations) of prostaglandins, leukotrienes, and related eicosanoids.[383]

4.11. P450 4B

A P450 4B1 cDNA was isolated from a human lung cDNA expression library using a (rat) P450 4A1 probe.[384] Expression is seen in human lung but not liver. The protein has been expressed in a vaccinia virus system and appears to be functional as judged by its ferrous-CO spectrum. However, no catalytic activities have been described to date, despite searches with mutagens that are known to be activated by animal P450s with structural similarity.[384]

4.12. P450 4F

Anecdotal reports of two human P450s in the 4F subfamily, 4F2 and 4F3, have appeared.[22] Further details are not currently available, but P450 4F3 is described as LTB_ω, presumably indicating ω-hydroxylation activity toward leukotriene B.

4.13. P450 5

P450 5 is the thromboxane synthase, which converts prostaglandin H_2 to thromboxane A_2.[385,386] The action of this enzyme leads to platelet aggregation and is counteracted by another P450, prostacyclin synthase, which converts prostaglandin H_2 to prostacyclin (prostaglandin I_2)[387] (the human P450 prostacyclin synthase has not been characterized at the level of its primary sequence).

This is a rather unusual P450 enzyme in that it catalyzes a rearrangement that does not involve the net gain or loss of electrons.[388] Accordingly the enzyme, even though microsomal, does not accept electrons from NADPH-P450 reductase.[386] A mechanism has been proposed to explain the catalytic cycle and involves heterolytic cleavage of the peroxide bond by the P450 (ferric) iron atom.[389]

Because of the adverse effects of thromboxanes, there is considerable interest in the development of inhibitors of this enzyme as drugs (as well as inhibitors of thromboxane receptors). Many compounds have been examined and used to help characterize aspects of the active site of P450 5.[390]

4.14. P450 7

P450 7 is the cholesterol 7α-hydroxylase, the enzyme catalyzing the first step in bile acid formation.[391] Extensive studies on this enzyme and its role in bile acid synthesis have been carried out in experimental animals, particularly rats.[392,393] The human cDNA and gene have been cloned.[394] Insofar as is known, the enzyme is highly specific in catalyzing this reaction and other substrates are not expected. The availability of the sequences may facilitate searches for polymorphisms related to metabolic alterations, and expression of the protein may also be useful in the development of inhibitors.

4.15. P450 11A

At this point we will begin to consider a group of P450s that have important roles in steroidogenesis. This set appears to be rather dissimilar from most of those that have been considered thus far. Their catalytic specificity seems to be highly restricted to steps known to be associated with well-recognized physiological processes (P450 7 is an exception in this regard). These P450s are not expressed in liver but rather localized to tissues in which the individual reaction products are utilized. Since the products are thought to be important for normal homeostasis, the levels of these P450s seem less prone to the wide interindividual variations seen with others (e.g., P450 families 1, 2, 3) and deficiencies in activity can lead to serious disease states. P450s in this group include 11A1, 11B1, 11B2, 17, 19, and 21. This part of the chapter will consider some pertinent aspects of the human P450s of this group. Many details of regulation involve studies from experimental animal models in *in vivo*, cell culture, and molecular systems and the reader is referred to Chapter 12, which covers the general function and regulation of the P450 11A1, 11B1, 11B2, 17, and 21 proteins. For reference in this chapter, an overview of the major functions of these steroidogenic P450s is presented here (Fig.6). [28]

P450 11A1 is the classic P450$_{SCC}$, or cholesterol side chain cleavage enzyme, that converts cholesterol to pregnenolone. The human gene is $\geq$ 20 kb long, consisting of nine exons and eight introns.[395] This is a mitochondrial enzyme (and utilizes electrons from adrenodoxin); a "precursor" sequence is present for import and subsequent cleavage. Interestingly, the "mature" (processed) form of P450 11A1 has only a single Cys, which defines the location of the heme-binding segment.[395] A *cis*-acting DNA element is located in the region between -1697 and -1523, considerably upstream from that in the bovine homologue.[28] Basal expression may also involve the sequence near -150.

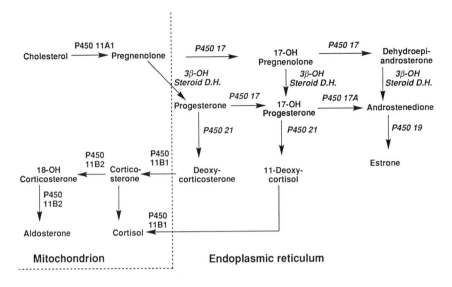

FIGURE 6. Steroid oxidations catalyzed by P450 enzymes.[28]

Deficiency in P450 11A1 activity can lead to hypertension, feminization, and gluco-corticoid insufficiency. Three deficient individuals were examined but no mutations were found in the protein coding regions of the P450 11A1 genes[396]; presumably the defect(s) is at the expression level.

4.16. P450 11B

In some species, including humans, there are two genes, P450 11B1 and P450 11B2, with distinct $P450_{11\beta}$ and $P450_{aldo}$ activities (Fig. 6); i.e., $P450_{aldo}$ is an 18-hydroxylase. These are also mitochondrial enzymes, and at least in some species there is tissue-specific localization.[28] The finding of separate genes is relatively recent and, since species diversity exists, details about the human genes and their function are incomplete. As with the P450 11A, 11B, 17, and 21 genes, there is cAMP regulation, although by distinct mechanisms.

In animal models salt is an important regulator of $P450_{aldo}$ activity,[397] although this aspect has not been carefully examined in humans. In one case of 11β-hydroxylase deficiency (leading to hypertension and virilization), amino acid substitution has been identified.[71] Further information about the two human genes in this subfamily will be of considerable interest.

4.17. P450 17

The human P450 17 gene contains eight exons and is the only one in this family.[28] In contrast to the mitochondrial P450 11 family enzymes, P450 17 is microsomal and receives electrons from the same NADPH-P450 reductase protein as the other microsomal P450s. The gene is regulated by cAMP through the action of three interactive elements at sites -313, -550, and -1500 upstream of the start site (N. Kagawa and M. R. Waterman, personal communication).

As pointed out earlier (Fig. 6), the P450 17 proteins can catalyze both the 17α-hydroxylation of progesterone *and* pregnenolone and the 17,20-lyase cleavage of the 17α-hydroxy products. The ratio of these activities seems to vary considerably among the P450 17 proteins isolated from different species. Recombinant human P450 17 appears to be primarily a 17α-hydroxylase under typical reconstitution conditions and has little lyase activity. However, the addition of cytochrome b_5 markedly stimulates the lyase activity (and has little effect on the 17α-hydroxylation activity) (M. Katagiri and M. R. Waterman, personal communication). The pattern seen in the presence of cytochrome b_5 resembles that seen in some tissue microsomes, and the situation is thought to resemble the *in vivo* situation. The structural similarity of the P450 17 enzymes of different sources is not clear, since chimeric proteins made between bovine and human P450 17 proteins generally have less catalytic activity than either native protein.[398] Recombinant human P450 17 has also been reported to catalyze some 16α-hydroxylation of progesterone.[399]

Deficiencies in P450 17 activity are known and lead to absence of cortisol and androgen production and hypertension.[28] Several mutants have been characterized and attributed to base pair mutations, a four-base insert, and a mutation that introduces a termination codon early.[400–403] To date no defects in the regulatory region have been reported but may be expected in the future.[28,403]

4.18. P450 19

P450 19 is classically called *aromatase*. Ryan[404] first demonstrated the P450-catalyzed conversion of androgens to estrogens in human placental microsomes. This reaction has been of great interest. Placental microsomes have long been a traditional source of material for assays and material for purification, primarily because of availability as compared to other human tissues. The enzyme is important in reproductive tissues and other sites as diverse as testis and adipose tissue. In experimental animals aromatase inhibitors have been shown to modulate male sexual differentiation in the brain.[405] A key area of interest involves attenuation of the enzyme in breast cancer, since the role of estrogens in promoting the growth of hormone-dependent mammary tumors is well established. Approximately 60% of premenopausal and 75% of postmenopausal patients have tumors responsive to estrogens.[405]

The overall reaction scheme is shown in Fig. 7. Testosterone, 16α-hydroxytestosterone, and androstenedione all serve as substrates, yielding 17β-estradiol, estratriol, and estrone as the respective products.[406,407] The scheme includes three reactions, known to occur with rates in the order $k_3 >> k_2 >> k_1$.[408] Thus, the first reaction is rate-limiting, and intermediates do not accumulate. The possibility existed that the reactions are catalyzed by distinct enzymes. Early efforts to purify the enzymes were fraught with many of the usual difficulties, but progress was made.[409–412] Kellis and Vickery[406] purified the human placental enzyme by conventional chromatography and demonstrated the complete aromatization of androstenedione and other androgens. Later, Yoshida and Osawa[413] utilized a monoclonal antibody-coupled column method to purify active homogeneous protein. One of the earlier, cruder preparations of P450 19[412] was used to prepare a monoclonal antibody, which was used to isolate a cDNA from a library.[414] The cDNA was expressed

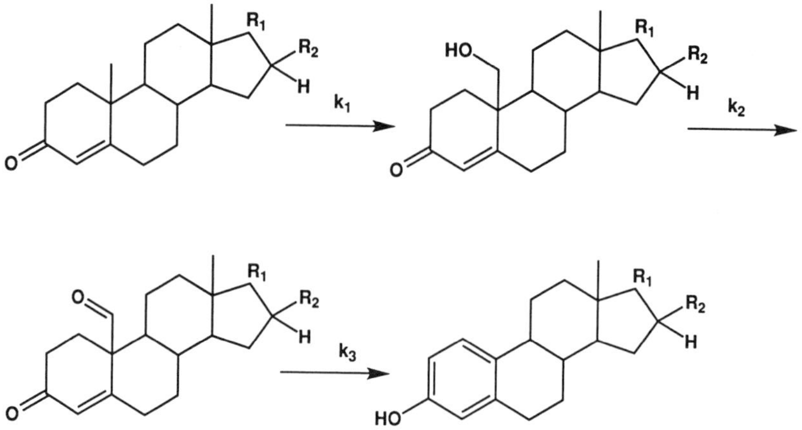

FIGURE 7. Aromatization reactions catalyzed by P450 19. The three distinct steps are shown, with the three possible substrates. R_1: -OH, R_2: -H, testosterone; R_1: = O, R_2: bond-H, androstenedione; R_1, R_2: bond-OH, 16-hydroxytestosterone.

in COS-1 monkey kidney tumor cells and also shown to catalyze the complete aromatization sequence.

The P450 19 gene is the only one in this family in humans or animals. It is quite complex, $\phi70$ kb in length, with 10 exons.[415] The promoter region contains the consensus TATA and CAAT boxes and two AP-1 binding sites. A *cis* regulatory element is present in the -500 to -243 region, flanked by an element between -242 and -183 needed for transcriptional regulation.[415] Another element (hATRE-1) between -2238 and -2214 is involved in the promoter-driven response.[416] mRNA expression in cell culture is induced by the phorbol ester 12-*O*-tetradecanoylphorbol 13-acetate. The region between -2141 and -2098 (hATRE-2) is involved in this response.[416] Alternative splicing of the aromatase gene has been documented.[417–419] A 109-bp segment between exons 1 and 2 is a new exon.[417] P450 C19 mRNA is formed with 10 exons in most tissues but from 9 exons in human prostate and from 10 *or* 11 exons in placenta.[419] Changes in the splicing have been observed in liver during development and in adipose tissue during breast cancer development.

The catalytic mechanism of the aromatization reaction has also been a matter of interest. As indicated earlier, purification and expression work demonstrated that a single enzyme can catalyze all three steps. The first two steps in the sequence are readily interpreted in terms of classical hydrogen abstraction/oxygen rebound P450 mechanisms[420] (Fig. 8), but the last step is not.[421] A number of possibilities have been raised[422] but the most likely seems to be the one originally proposed by Akhtar *et al.*[423] and developed by Robinson and Cole.[408,422,424,425] This paradigm involves a ferric peroxide attack to remove the aldehyde and aromatize the A ring (Fig. 9). Although this type of

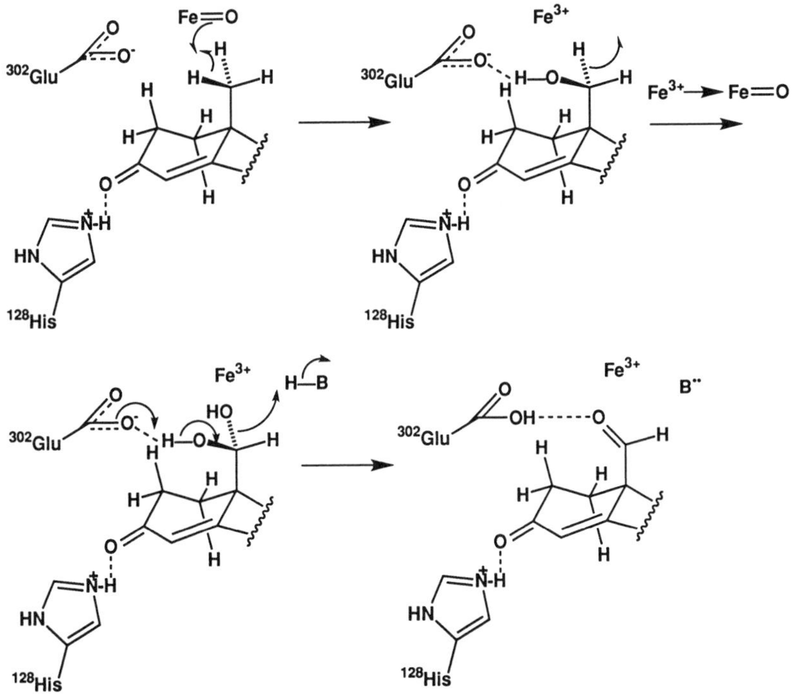

FIGURE 8. Postulated details of first and second oxidation steps of aromatization by P450 19.[408]

reaction was originally considered unsatisfying, there is now supporting evidence in model reactions catalyzed by other, nonsteroidogenic P450s.[426]

P450 19 inhibitors have been studied extensively because of their potential for use in estrogen suppression with breast and other estrogen-sensitive tumors.[405] 4-Hydroxyandrostanedione and aminoglutethimide have been evaluated.[405,427,428] 7,8-Benzoflavone, the inhibitor of P450 1A enzymes and allosteric modulator of P450 3A4 (*vide supra*), is a potent competitive inhibitor of P450 19 (K_i 70 nM)[429] and other flavones are also inhibitory.[430] However, natural flavonoids have not been examined in detail. There is considerable interest in the development of more selective P450 19 inhibitors, including both competitive[424,428] and mechanism-based.[424] In this latter category are molecules with C-19-propargyl and difluoromethyl groups and mercapto and dienone moieties placed elsewhere. With increased enzyme specificity or the use of nonsteroid compounds, it may be possible to avoid effects on other enzymes involved in steroid metabolism.

The advances in the characterization of P450 19 and the literature base have led to a number of studies on structure–function relationships. Affinity and photoaffinity labels have been developed[407,424] although apparently not used to directly map the substrate binding site. Site-directed mutagenesis studies have been done by several groups, although the ability to do large-scale expression work has been a hindrance to date. The difficulty may be related to the glycosylation near the N-terminus, apparently unusual, that results

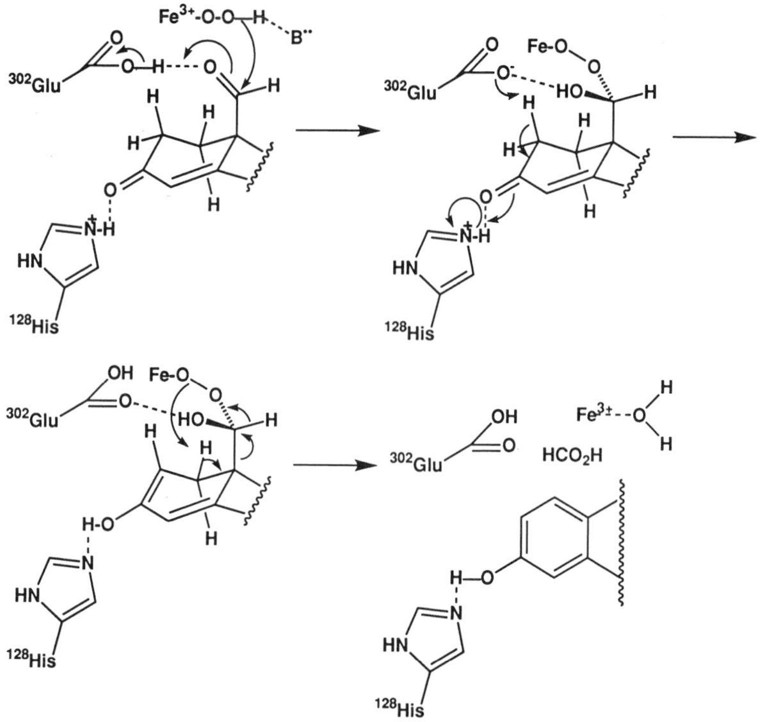

FIGURE 9. Postulated final oxidation step in aromatization catalyzed by P450 19.[408]

in placement of the N-terminus in the lumen of the endoplasmic reticulum.[431,432] The F406R and P308F mutants had major consequences and Zhou *et al*.[433] suggested interaction of the steroid D ring with Pro308. Mutations at Pro308 and especially Glu302 inhibited function, and these results are the basis for the inclusion of Glu302 in Figs. 8 and 9. The entire region in the area of residues 298–313 appears to be important.[432,434,435] This corresponds, at least in models, to the I-helix of bacterial P450 101.[435] Chen and Zhou[435] have suggested which exons and subdomains of P450 19 have particular functions. Graham-Lorence *et al*.[434] have used the P450 101 structure to derive a model for the active site of P450 19. Such models have also been considered to be in agreement with the current information from site-directed mutagenesis experiments and considerations of known substrates and inhibitors.[436] The further usefulness of such models in prediction remains to be demonstrated.

4.19. P450 21

The human P450 21A2 gene codes for the adrenal-specific steroid 21-hydroxylase, which is responsible for the 21-hydroxylation of both progesterone and 17-hydroxypro-gesterone (Fig. 6). CYP21A1 is a pseudogene. It is not transcribed but is a source of mutations related to recombinations and gene duplication because of its close similarity.[69]

(The nomenclature convention in mice is the opposite: 21a1 is the active gene and 21a2 is the pseudogene.) The functional human gene (21A2) contains a single cAMP-responsive site in the -126 to -113 bp region.[28,437] The ubiquitous transcription factor Sp1 and a protein termed ASP both bind here. The ASP protein seems to be required for cAMP-responsiveness. ASP is found only in adrenal and a few other cells.[437]

More than 90% of known cases of congenital adrenal hyperplasia have been attributed to deficiencies in P450 21A2 activity.[28,70] This is a relatively frequent inborn error of metabolism, occurring with a frequency of between 1/5000 and 1/20,000 individuals, depending on the population.[70] The most severe form of the disease involves a salt-wasting syndrome. Aldosterone is absent, sodium is lost, and potassium and protons are retained by the kidney.[28] Death can result. In less severe cases the accumulation of non-(21)hydroxylated steroids can lead to excess androgen production and virilization of the external genitalia.[28,70]

As pointed out earlier, most of the mutations in P450 21B can be linked to the similarity of the pseudogene CYP21A1, which has been postulated to have arisen by gene duplication.[438,439] CYP21A1 has an 8-bp deletion in exon 3, a single bp insertion in exon 7, and a single base change leading to a stop codon in exon 8.[28] The activity of P450 21A2 appears to be highly sensitive to minor alterations in its structure, including single base changes.[70] A number of mutations have been identified.[69,70,440-443] The severity of the disease (congenital adrenal hyperplasia) appears to be linked in an adverse fashion to the residual catalytic activity in these enzymes.[69]

4.20. P450 27

P450 27 catalyzes the first step (27-hydroxylation) in the oxidation of the side chain of sterol intermediates in bile acid synthesis. This is another mitochondrial P450. Cali and Russell[444] cloned a cDNA from a human fibroblast library, although the gene is expected to be expressed in liver and possibly other tissues. The disease cerebrotendinous xanthomatosis is caused by mutations in the gene.[445] P450 27 hydroxylates a number of steroids, as well as vitamin D_3.[444]

4.21. Lanosterol 14α-Demethylase

Sonoda et al.[18] recently purified a human liver lanosterol 14α-demethylase. A systematic P450 number has not yet been assigned, in the absence of a complete primary sequence. This protein may be a homologue of yeast P450 51.[446,447] Its catalytic specificity beyond lanosterol has not been explored.

5. Future Needs and Directions

It should be apparent from this chapter that there are still major scientific gaps in our knowledge of human P450 research. More P450s appear to be present in several of the families and subfamilies and have not as yet been characterized. Another point is that relatively few genes have been isolated, compared to the number of cDNAs. A challenging task with all of these genes is the characterization of regulation in the face of an increasing number of regulatory elements, which may interact with each other. Moreover, a serious limitation is the availability of appropriate human cells to utilize for transfections, etc. Another problem in the basic science is the better definition of active site structures. As pointed out, pharmacophore models hold promise with some of the P450s whose sites are fairly well defined (Fig. 5).[273,277,278] However, this approach will probably not be of much utility with the enzymes having relatively broad catalytic specificity (e.g., P450 3A4). It is also unlikely that modeling of complete sequences into the three known bacterial P450 structures[448–451] will provide significant new insight into the fine details of substrate docking.[452] Ultimately, crystal structures of some of the mammalian/human P450s will be necessary to answer many questions. Further, there is the matter of P450 2E1 to consider, where it has been postulated that the substrate does not interact until the (FeO) complex is formed (*vide supra*).

There are also research needs in more applied aspects of human P450 work. Numerous expression systems have been developed and are already in practical and commercial use. However, all of these systems can probably be improved. The mammalian systems present advantages for the measurement of endpoints of toxicity and mutagenicity, but expression levels are relatively low[453,454] or else, in the case of vaccinia, there are technical problems limiting the widespread application and cost constraints on the application to high-volume work. Only relatively recently have high-level expression systems for human (and other) P450s become available.[52,121,455,456] Deficiencies still exist. Baculovirus[52] is expensive, yeast require supplementation with cDNAs for accessory proteins,[121] and bacteria often require tailoring of 5′ sequences.[456]

Another need is better *in vivo* validation of assays. These assays have proven to be extremely valuable in relating the results of *in vitro* investigations to actual clinical situations. Nevertheless, there are a number of potential reasons why a particular drug may not be appropriate as an *in vivo* screen, even if it seems to be oxidized by a single P450 *in vitro*. Urinary, single time-point assays are desirable but not always possible. When there is extensive extrahepatic metabolism or blood flow is limiting, the assay may not appropriately reflect the hepatic concentration of the enzyme. In the case of P450 3A4, there is still controversy as to which assay gives the most meaningful results, since the parameters from different assays are not corroborating.[457] Even with caffeine as a marker of P450 1A2[127] there are questions regarding which metabolite is most useful.[128,132] As pointed out earlier, different parameters yield different suggestions regarding genetic polymorphism. Noninvasive assays can be extremely useful in the process of drug development and in epidemiological studies involving cancer and other diseases. Again, the point should be emphasized that *in vitro* studies are useful in their own right and for postulating roles of certain P450s *in vivo*, but pharmacokinetic studies are needed for validation of hypotheses concerning roles of P450s *in vivo*.

The availability of new information about human P450 enzymes and new resources and reagents has led to a number of changes in the process of drug development. A problem with animal models has been that metabolism turned out to be very different in humans and led to dramatic changes in efficacy and safety when a lead drug reached clinical trials. Today many pharmaceutical companies have implemented strategies in which drug metabolism is a more integral part of the entire process. Human liver samples or recombinant enzymes can be used to screen a set of different chemicals under development in order to predict which might be expected to be retained longer and have better bioavailability. When a lead candidate is selected, the approaches described in this chapter can be used to determine which P450s are most likely to be involved in its metabolism. This information is useful in predicting possible drug interactions resulting from competition and induction. It is also possible to move quickly to *in vivo* studies to evaluate possible scenarios raised by *in vitro* work and to avoid doing unnecessary interaction studies. In the early stages of drug development, it is possible to compare patterns of *in vitro* metabolism in human samples with those of animals in order to better select an animal model for certain tests, including safety, since similar metabolites will be produced in that system. If the major P450s involved in the oxidation of a particular drug can be identified, it is now possible to use recombinant P450 enzymes to quickly produce the metabolites for purification and characterization. In a recent example from this laboratory, 35 nmole of P450 1A2 was used to generate metabolites of bufuralol, which were characterized by their mass, UV, and NMR spectra.[137] It is also possible to use information about rates of metabolism early in the development process to decide what chemical modifications to make to a drug to retard its metabolism.

Ever since the discovery of the P450 enzymes there has been an appreciation of their roles in the metabolism of carcinogens.[458] Even in the 1950s it was recognized that modifications of levels of P450s could influence the risk of cancer from particular chemicals.[459–462] This view was reinforced in genetic studies with animals.[463,464] Humans are known to vary dramatically with regard to levels of individual P450s (e.g., Tables I, II and Figs. 1, 2) and the hypothesis that these differences influence cancer susceptibility has long been attractive.[465] This research area has attracted further interest with the demonstration of catalytic selectivity of human P450s in the activation and detoxication of carcinogens, particularly those to which humans are routinely exposed (Table IX). *In vitro* assays have been interesting but the epidemiology associating variations in individual P450s with cancer risk has been difficult. The complexity may be related to the fact that cancer is a multistep process and that there are usually numerous contributors among the genes. The relationship of P450 1A1 and its inducibility to lung cancer has been studied for some time, with various results,[101,105] but there does seem to be a relationship between basal levels of the enzyme and cancer risk in smokers.[105,107] Some evidence has also been obtained that increased P450 1A2 levels are a contributing factor in the development of colon cancer associated with ingestion of heterocyclic amines, although the relationship is not strong in the absence of high *N*-acetyltransferase activity.[131] P450 2D6 has been studied for an association with various cancers but, as mentioned before, the results are still conflicting.[76,77,80,83,85,90–93,283–286,470] Other P450s are now being studied for their relationship to cancer risk (e.g., 2E1,[471] 3A4[472]) but definitive results are not available in any case. The epidemiology is certainly challenging, in part because the levels of many of the P450s are inducible and assessing the level of enzyme activity at a certain time point may not be indicative of the situation when tumor initiation occurred. In this regard, genotypic assays suggest an advantage but this does not seem to have been realized yet (*vide supra*).

TABLE IX

Human P450 Enzymes Involved in the Activation of Carcinogens[14,29,116,160,375,376,466–469]

P450 1A1	P450 1A2	P450 2A6	P450 2E1	P450 3A4
Benzo[a]pyrene and other polycyclic hydrocarbons	PhIP	N,N-Dimethylnitrosamine (DMN)	Benzene	Aflatoxin B$_1$
2-Amino-1-methyl-6-phenylimidazo-[4,5-b]pyridine(PhIP)	2-Amino-6-methyl-dipyrido[1,2-a:3,2'-d]imidazole (Glu P-1)	N,N-Diethylnitrosamine (DEN)	Styrene	Aflatoxin G$_1$
	2-Aminodipyrido[1,2-a:3,2'-d]imidazole (Glu P-2)	NNK	Acrylonitrile	Sterigmatocystin
	2-Amino-3-methylimidazo[4,5-f]quinoline (IQ)	4-(Methylnitrosamino)-1-(3-pyridyl)-1-butanol (NNAL)	Vinyl carbamate	7,8-Dihydroxy-7,8-dihydrobenzo(a)pyrene and some other polycyclic hydrocarbons
	2-Amino-3,5-dimethylimidazo[4,5-f]quinoline (MeIQ)	Nornitrosonicotine (NNN)	Vinyl chloride	17β-Estradiol
	2-Amino-3,8-dimethylimidazo[4,5-f]quinoline (MeIQx)		Vinyl bromide	6-Aminochrysene
	3-Amino-1-methyl-5H-pyrido[4,3-b]indole(Trp P-2)		Ethyl carbamate	Senecionine
	4-Aminobiphenyl		Trichloroethylene	4,4'-Methylene-bis(2-chloroaniline) (MOCA)
	2-Naphthylamine		Carbon tetrachloride	tris(2,3-Dibromopropyl) phosphate
	2-Aminofluorene		Chloroform	
	2-Acetylaminofluorene		DMN	
	4-(Methylnitrosamino)-1-(3-pyridyl)-1-butanone (NNK)		DEN	
			NNK	
			NNAL	
			NNN	
			Butadiene	

6. Conclusions

In the past decade there has been considerable development of knowledge about human P450 enzymes. As expected, most of the P450s involved in steroid anabolism have the same roles as their orthologs in experimental animals. The P450s involved in the metabolism of xenobiotics have some similarities to animal P450s but there are also many differences. Most of the oxidations of xenobiotics can be accounted for by the action of a relatively small number of P450s (5–10, depending on one's perspective). There are some considerable differences between humans and experimental animals within the P450 subfamilies, both in catalytic specificity and regulation. There are certainly many needs remaining in both basic and applied research, but it has already been possible to make considerable use of the knowledge of human P450s in the overall process of drug discovery and development. Further developments may be expected in other areas in which P450s have roles, including chemical carcinogenesis, steroidogenesis, and various diseases where the etiology is poor. Indeed, the entire field of xenobiotic metabolism has changed considerably within a short period of time because of the characterization of the human P450s.

ACKNOWLEDGMENTS. Research in the author's laboratory has been supported by USPHS grants CA44353 and ES00267. Thanks are extended to individuals who provided unpublished results, to Drs. M. R. Waterman and F. F. Kadlubar for comments, and to the members of the author's laboratory for their involvement in some of the original research.

References

1. Wang, P., Mason, P. S., and Guengerich, F. P., 1980, Purification of human liver cytochrome P-450 and comparison to the enzyme isolated from rat liver, *Arch. Biochem. Biophys.* **199**:206–219.
2. Beaune, P., Dansette, P., Flinois, J. P., Columelli, S., Mansuy, D., and Leroux, J. P., 1979, Partial purification of human liver cytochrome P-450, *Biochem. Biophys. Res. Commun.* **88**:826–832.
3. Kitada, M., and Kamataki, T., 1979, Partial purification and properties of cytochrome P450 from homogenates of human fetal livers, *Biochem. Pharmacol.* **28**:793–797.
4. Wang, P. P., Beaune, P., Kaminsky, L. S., Dannan, G. A., Kadlubar, F. F., Larrey, D., and Guengerich, F. P., 1983, Purification and characterization of six cytochrome P-450 isozymes from human liver microsomes, *Biochemistry* **22**:5375–5383.
5. Smith, R. L., Idle, J. R., Mahgoub, A. A., Sloan, T. P., and Lancaster, R., 1978, Genetically determined defects of oxidation at carbon centres of drugs, *Lancet* **i**:943–944.
6. Shimada, T., Yun, C.-H., Yamazaki, H., Gautier, J.-C., Beaune, P. H., and Guengerich, F. P., 1992, Characterization of human lung microsomal cytochrome P-450 1A1 and its role in the oxidation of chemical carcinogens, *Mol. Pharmacol.* **41**:856–864.
7. Distlerath, L. M., Reilly, P. E. B., Martin, M. V., Davis, G. G., Wilkinson, G. R., and Guengerich, F. P., 1985, Purification and characterization of the human liver cytochromes P-450 involved in debrisoquine 4-hydroxylation and phenacetin *O*-deethylation, two prototypes for genetic polymorphism in oxidative drug metabolism, *J. Biol. Chem.* **260**:9057–9067.
8. Yun, C.-H., Shimada, T., and Guengerich, F. P., 1991, Purification and characterization of human liver microsomal cytochrome P-450 2A6, *Mol. Pharmacol.* **40**:679–685.
9. Ged, C., Umbenhauer, D. R., Bellew, T. M., Bork, R. W., Srivastava, P. K., Shinriki, N., Lloyd, R. S., and Guengerich, F. P., 1988, Characterization of cDNAs, mRNAs, and proteins related to human liver microsomal cytochrome P-450 (S)-mephenytoin 4′-hydroxylase, *Biochemistry* **27**:6929–6940.

10. Shimada, T., Misono, K. S., and Guengerich, F. P., 1986, Human liver microsomal cytochrome P-450 mephenytoin 4-hydroxylase, a prototype of genetic polymorphism in oxidative drug metabolism. Purification and characterization of two similar forms involved in the reaction, *J. Biol. Chem.* **261**:909–921.

11. Gut, J., Catin, T., Dayer, P., Kronbach, T., Zanger, U., and Meyer, U.A., 1986, Debrisoquine/sparteine-type polymorphism of drug oxidation: Purification and characterization of two functionally different human liver cytochrome P-450 isozymes involved in impaired hydroxylation of the prototype substrate bufuralol, *J. Biol. Chem.* **261**:11734–11743.

12. Birgersson, C., Morgan, E. T., Jörnvall, H., and von Bahr, C., 1986, Purification of a desmethylimipramine and debrisoquine hydroxylating cytochrome P-450 from human liver, *Biochem. Pharmacol.* **35**:3165–3166.

13. Wrighton, S. A., Thomas, P. E., Ryan, D. E., and Levin, W., 1987, Purification and characterization of ethanol-inducible human hepatic cytochrome P-450HLj, *Arch. Biochem. Biophys.* **258**:292–297.

14. Guengerich, F. P., Kim, D.-H., and Iwasaki, M., 1991, Role of human cytochrome P-450 IIE1 in the oxidation of several low molecular weight cancer suspects, *Chem. Res. Toxicol.* **4**:168–179.

15. Guengerich, F. P., Martin, M. V., Beaune, P. H., Kremers, P., Wolff, T., and Waxman, D. J., 1986, Characterization of rat and human liver microsomal cytochrome P-450 forms involved in nifedipine oxidation, a prototype for genetic polymorphism in oxidative drug metabolism, *J. Biol. Chem.* **261**:5051–5060.

16. Wrighton, S.A., and VandenBranden, M., 1989, Isolation and characterization of human fetal liver cytochrome P450HLp2: A third member of the P450III gene family, *Arch. Biochem. Biophys.* **268**:144–151.

17. Kawashima, H., Kusunose, E., Kubota, I., Maekawa, M., and Kusunose, M., 1992, Purification and NH2-terminal amino acid sequences of human and rat kidney fatty acid ω-hydroxylases, *Biochim. Biophys. Acta* **1123**:156–162.

18. Sonoda, Y., Endo, M., Ishida, K., Sato, Y., Fukusen, N., and Fukuhara, M., 1993, Purification of a human P-450 isozyme catalyzing lanosterol 14α-demethylation, *Biochim. Biophys. Acta* **1170**:92–97.

19. Watkins, P. B., Wrighton, S. A., Maurel, P., Schuetz, E. G., Mendez-Picon, G., Parker, G. A., and Guzelian, P. S., 1985, Identification of an inducible form of cytochrome P-450 in human liver, *Proc. Natl. Acad. Sci. USA* **82**:6310–6314.

20. Mimura, M., Baba, T., Yamazaki, Y., Ohmori, S., Inui, Y., Gonzalez, F. J., Guengerich, F. P., and Shimada, T., 1993, Characterization of cytochrome P450 2B6 in human liver microsomes, *Drug Metab. Dispos* **21**:1048–1056.

21. Gonzalez, F. J., 1989, The molecular biology of cytochrome P450s, *Pharmacol. Rev.* **40**:243–288.

22. Nelson, D. R., Kamataki, T., Waxman, D. J., Guengerich, F. P., Estabrook, R. W., Feyereisen, R., Gonzalez, F. J., Coon, M. J., Gunsalus, I. C., Gotoh, O., Okuda, K., and Nebert, D. W., 1993, The P450 superfamily: Update on new sequences, gene mapping, accession numbers, early trivial names of enzymes, and nomenclature, *DNA Cell Biol.* **12**:1–51.

23. Tateishi, T., Krivoruk, Y., Wood, A. J. J., Guengerich, F. P., and Wood, M., 1994, Identification of human liver P450 3A4 as the enzyme responsible for sulfentanil N-dealkylation, in: *Abstracts, 12th Int. Congr. Pharmacol.* (July 24–29, Montreal).

24. Shimada, T., Yamazaki, H., Mimura, M., Inui, Y., and Guengerich, F. P., 1994, Interindividual variations in human liver cytochrome P450 enzymes involved in the oxidation of drugs, carcinogens, and toxic chemicals: Studies with liver microsomes of 30 Japanese and 30 Caucasians, *J. Pharmacol. Exp. Ther.* **270**:414–423.

25. Omura, T., and Sato, R., 1964, The carbon monoxide-binding pigment of liver microsomes. I. Evidence for its hemoprotein nature, *J. Biol. Chem.* **239**:2370–2378.

26. Nakamura, K., Goto, F., Ray, W. A., McAllister, C. B., Jacqz, E., Wilkinson, G. R., and Branch, R. A., 1985, Interethnic differences in genetic polymorphism of debrisoquin and mephenytoin hydroxylation between Japanese and Caucasian populations, *Clin. Pharmacol. Ther.* **38**:402–408.

27. Relling, M. V., Cherrie, J., Schell, M. J., Petros, W. P., Meyer, W. H., and Evans, W. E., 1991, Lower prevalence of the debrisoquin oxidative poor metabolizer phenotype in American black versus white subjects, *Clin. Pharmacol. Ther.* **50**:308–313.

28. Keeney, D.S., and Waterman, M. R., 1993, Regulation of steroid hydroxylase gene expression: Importance to physiology and disease, *Pharmacol. Ther.* **58**:301–317.

29. Guengerich, F. P., and Shimada, T., 1991, Oxidation of toxic and carcinogenic chemicals by human cytochrome P-450 enzymes, *Chem. Res. Toxicol.* **4**:391–407.

30. Sesardic, D., Boobis, A., Murray, B., Murray, S., Segura, J., De La Torre, R., and Davies, D., 1990, Furafylline is a potent and selective inhibitor of cytochrome P450 1A2 in man, *Br. J. Clin. Pharmacol.* **29**:651–663.

31. Kunze, K. L., and Trager, W. F., 1993, Isoform-selective mechanism-based inhibition of human cytochrome-P450 1A2 by furafylline, *Chem. Res. Toxicol.* **6**:649–656.

32. Gallagher, E. P., Wienkers, L. C., Stapleton, P. L., Kunze, K. L., and Eaton, D. L., 1994, Role of human microsomal and human complementary DNA-expressed cytochrome-P4501A2 and cytochrome-P4503A4 in the bioactivation of aflatoxin B_1, *Cancer Res.* **54**:101–108.

33. Hammons, G. J., Milton, D., Guengerich, F. P., Tukey, R. H., and Kadlubar, F. F., 1994, Human cytochrome P450 1A2-catalyzed *N*-hydroxylation of carcinogenic heterocyclic amines, *FASEB J.* **8**:A1381.

34. Guengerich, F. P., Wang, P., and Davidson, N. K., 1982, Estimation of isozymes of microsomal cytochrome P-450 in rats, rabbits, and humans using immunochemical staining coupled with sodium dodecyl sulfate-polyacrylamide gel electrophoresis, *Biochemistry* **21**:1698–1706.

35. Beaune, P., Kremers, P. G., Kaminsky, L. S., de Graeve, J., and Guengerich, F. P., 1986, Comparison of monooxygenase activities and cytochrome P-450 isozyme concentrations in human liver microsomes, *Drug Metab. Dispos.* **14**:437–442.

36. Shet, M. S., Fisher, C. W., Holmans, P. L., and Estabrook, R. W., 1993, Human cytochrome P450 3A4: Enzymatic properties of a purified recombinant fusion protein containing NADPH-P450 reductase, *Proc. Natl. Acad. Sci. USA* **90**:11748–11752.

37. Wang, R. W., Kari, P. H., Lu, A.Y.H., Thomas, P. E., Guengerich, F. P., and Vyas, K. P., 1991, Biotransformation of lovastatin. IV. Identification of cytochrome P-450 3A proteins as the major enzymes responsible for the oxidative metabolism of lovastatin in rat and human liver microsomes, *Arch. Biochem. Biophys.* **290**:355–361.

38. Guengerich, F. P., 1990, Mechanism-based inactivation of human liver cytochrome P-450 IIIA4 by gestodene, *Chem. Res. Toxicol.* **3**:363–371.

39. Distlerath, L. M., and Guengerich, F. P., 1984, Characterization of a human liver cytochrome P-450 involved in the oxidation of debrisoquine and other drugs using antibodies raised to the analogous rat enzyme, *Proc. Natl. Acad. Sci. USA* **81**:7348–7352.

40. Guo, Z., Gillam, E.M.J., Ohmori, S., Tukey, R. H., and Guengerich, F. P., 1995, Expression of modified human cytochrome P450 1A1 in *Escherichia coli*. Effects of 5′ substitution, purification, spectral characterization, reconstitution conditions and catalytic properties, *Arch. Biochem. Biophys.* **317**:374–384.

41. Edwards, R. J., Singleton, A. M., Sesardic, D., Boobis, A. R., and Davies, D. S., 1988, Antibodies to a synthetic peptide that react specifically with a common surface region on two hydrocarbon-inducible isoenzymes of cytochrome P-450 in the rat, *Biochem. Pharmacol.* **37**:3735–3741.

42. Fujino, T., Park, S. S., West, D., and Gelboin, H. V., 1982, Phenotyping of cytochromes P-450 in human tissues with monoclonal antibodies, *Proc. Natl. Acad. Sci. USA* **79**:3682–3686.

43. Reik, L. M., Levin, W., Ryan, D. E., Maines, S. L., and Thomas, P. E., 1985, Monoclonal antibodies distinguish among isozymes of the cytochrome P-450b subfamily, *Arch. Biochem. Biophys.* **242**:365–382.

44. Gelboin, H. V., and Friedman, F. K., 1985, Monoclonal antibodies for studies on xenobiotic and endobiotic metabolism, *Biochem. Pharmacol.* **34**:2225–2234.

45. Goldfarb, I., Korzekwa, K., Krausz, K. W., Gonzalez, F., and Gelboin, H. V., 1993, Cross-reactivity of thirteen monoclonal antibodies with ten vaccinia cDNA expressed rat, mouse and human cytochrome P450s, *Biochem. Pharmacol.* **46:**787–790.

46. Wrighton, S. A., VandenBranden, M., Becker, G. W., Black, S. D., and Thomas, P. E., 1992, Two monoclonal antibodies recognizing different epitomes on rat cytochrome P450 IIB1 react with human IIE1, *Mol. Pharmacol.* **41:**76–82.

47. Wrighton, S. A., Stevens, J. C., Becker, G. W., and VandenBranden, M., 1993, Isolation and characterization of human liver cytochrome P450 2C19: Correlation between 2C19 and S-mephenytoin 4′-hydroxylation, *Arch. Biochem. Biophys.* **306:**240–245.

48. Soucek, P., Guo, Z., Sandhu, P., Martin, M. V., and Guengerich, F. P., 1994, Immunochemical cross-reactivity of human cytochrome P450 enzymes analyzed with recombinant proteins, *FASEB J.* **8:**A1248.

49. Wrighton, S. A., and Stevens, J. C., 1992, The human hepatic cytochromes P450 involved in drug metabolism, *Crit. Rev. Toxicol.* **22:**1–21.

50. Oeda, K., Sakaki, T., and Ohkawa, H., 1985, Expression of rat liver cytochrome P-450MC cDNA in *Saccharomyces cerevisiae, DNA* **4:**203–210.

51. Zuber, M. X., Simpson, E. R., and Waterman, M. R., 1986, Expression of bovine 17α-hydroxylase cytochrome P-450 cDNA in nonsteroidogenic (COS 1) cells, *Science* **234:**1258–1261.

52. Gonzalez, F. J., Kimura, S., Tamura, S., and Gelboin, H. V., 1991, Expression of mammalian cytochrome P450 using baculovirus, *Methods Enzymol.* **206:**93–99.

53. Guengerich, F. P., Brian, W. R., Sari, M.-A., and Ross, J.-T., 1991, Expression of mammalian cytochrome P450 enzymes using yeast-based vectors, *Methods Enzymol.* **206:**130–145.

54. Gonzalez, F. J., Aoyama, T., and Gelboin, H. V., 1991, Expression of mammalian cytochrome P450 using vaccinia virus, *Methods Enzymol.* **206:**85–92.

55. Clark, B. J., and Waterman, M. R., 1991, Heterologous expression of mammalian P450 in COS cells, *Methods Enzymol.* **206:**100–108.

56. Porter, T. D., and Larson, J. R., 1991, Expression of mammalian P450s in *Escherichia coli, Methods Enzymol.* **206:**108–116.

57. Doehmer, J., and Oesch, F., 1991, V79 Chinese hamster cells genetically engineered for stable expression of cytochromes P450, *Methods Enzymol.* **206:**117–123.

58. Crespi, C. L., 1991, Expression of cytochrome P450 cDNAs in human B lymphoblastoid cells: Applications to toxicology and metabolite analysis, *Methods Enzymol.* **206:**123–129.

59. Imai, Y., and Nakamura, M., 1989, Point mutations at threonine-301 modify substrate specificity of rabbit liver microsomal cytochromes P-450 (laurate (ω-1)-hydroxylase and testosterone 16α-hydroxylase), *Biochem. Biophys. Res. Commun.* **158:**717–722.

60. Gillam, E. M. J., Baba, T., Kim, B.-R., Ohmori, S., and Guengerich, F. P., 1993, Expression of modified human cytochrome P450 3A4 in *Escherichia coli* and purification and reconstitution of the enzyme, *Arch. Biochem. Biophys.* **305:**123–131.

61. Sandhu, P., Baba, T., and Guengerich, F. P., 1993, Expression of modified cytochrome P450 2C10 in *Escherichia coli,* purification, and reconstitution of catalytic activity, *Arch. Biochem. Biophys.* **306:**443–450.

62. Sandhu, P., Guo, Z., Baba, T., Martin, M. V., Tukey, R. H., and Guengerich, F. P., 1994, Expression of modified human cytochrome P450 1A2 in *Escherichia coli.* Stabilization, purification, characterization, and catalytic activities of the enzyme, *Arch. Biochem. Biophys.* **309:**168–177.

63. Gillam, E.M.J., Guo, Z., and Guengerich, F. P., 1994, Expression of modified human cytochrome P450 2E1 in *Escherichia coli,* purification, and spectral and catalytic properties, *Arch. Biochem. Biophys.* **312:** 59–66.

64. Sakaki, T., Oeda, K., Yabusaki, Y., and Ohkawa, H., 1986, Monooxygenase activity of *Saccharomyces cerevisiae* cells transformed with expression plasmids carrying rat cytochrome P-450MC cDNA, *J. Biochem.* **99:**741–749.

65. Dogra, S., Doehmer, J., Glatt, H., Mölders, H., Siegert, P., Friedberg, T., Seidel, A., and Oesch, F., 1990, Stable expression of rat cytochrome P-450IA1 cDNA in V79 Chinese hamster cells and their use in mutagenicity testing, *Mol. Pharmacol.* **37**:608–613.

66. Fisher, C. W., Shet, M. S., Caudle, D. L., Martin-Wixtrom, C.A., and Estabrook, R. W., 1992, High-level expression in *Escherichia coli* of enzymatically active fusion proteins containing the domains of mammalian cytochromes P450 and NADPH-P450 reductase flavoprotein, *Proc. Natl. Acad. Sci. USA* **89**:10817–10821.

67. Morel, F., Beaune, P. H., Ratanasavanh, D., Flinois, J.-P., Yang, C.-S., Guengerich, F. P., and Guillouzo, A., 1990, Expression of cytochrome P-450 enzymes in cultured human hepatocytes, *Eur. J. Biochem.* **191**:437–444.

68. Dahlin, D. C., Miwa, G. T., Lu, A.Y. H., and Nelson, S. D., 1984, *N*-Acetyl-*p*-benzoquinone imine: A cytochrome P-450-mediated oxidation product of acetaminophen, *Proc. Natl. Acad. Sci. USA* **81**:1327–1331.

69. Higashi, Y., Hiromasa, T., Tanae, A., Miki, T., Nakura, J., Kondo, T., Ohura, T., Ogawa, E., Nakayama, K., and Fujii-Kuriyama, Y., 1991, Effects of individual mutations in the P-450n (C21) pseudogene on the P-450 (C21) activity and their distribution in the patient genomes of congenital steroid 21-hydroxylase deficiency, *J. Biochem.* **109**:638–644.

70. Miller, W.L., and Morel, Y., 1989, The molecular genetics of 21-hydroxylase deficiency, *Annu. Rev. Genet.* **23**:371–393.

71. White, P. C., Dupont, J., New, M. I., Leiberman, E., Hochberg, Z., and Rösler, A., 1991, A mutation in CYP11Ba (Arg 448→His) associated with steroid 11β-hydroxylase deficiency in Jews of Moroccan origin, *J. Clin. Invest.* **87**:1664–1667.

72. Distlerath, L. M., and Guengerich, F. P., 1987, Enzymology of human liver cytochromes P-450, in: *Mammalian Cytochromes P-450, Vol. 1* (F. P. Guengerich, ed.), CRC Press, Boca Raton, FL, pp. 133–198.

73. Tinel, M., Belghiti, J., Descatoire, V., Amouyal, G., Letteron, P., Geneve, J., Larrey, D., and Pessayre, D., 1987, Inactivation of human liver cytochrome P-450 by the drug methoxsalen and other psoralen derivatives, *Biochem. Pharmacol.* **36**:951–955.

74. Ortiz de Montellano, P. R., and Correia, M. A., 1983, Suicidal destruction of cytochrome P-450 during oxidative drug metabolism, *Annu. Rev. Pharmacol. Toxicol.* **23**:481–503.

75. Bailey, D. G., Spence, J. D., Munoz, C., and Arnold, J. M. O., 1991, Interaction of citrus juices with felodipine and nifedipine, *Lancet* **337**:268–269.

76. Barbeau, A., Roy, M., Bernier, G., Campanella, G., and Paris, S., 1987, Ecogenetics of Parkinson's disease: Prevalence and environmental aspects in rural areas, *Can. J. Neurol. Sci.* **14**:36–41.

77. Barbeau, A., Roy, M., Paris, S., Cloutier, T., Plasse, L., and Poirier, J., 1985, Ecogenetics of Parkinson's disease: 4-Hydroxylation of debrisoquine, *Lancet* **ii**:1213–1215.

78. Armstrong, M., Daly, A. K., Cholerton, S., Bateman, D. N., and Idle, J. R., 1992, Mutant debrisoquine hydroxylation genes in Parkinson's disease, *Lancet* **339**:1017–1018.

79. Fonne-Pfister, R., Bargetzi, M. J., and Meyer, U. A., 1987, MPTP, the neurotoxin inducing Parkinson's disease, is a potent competitive inhibitor of human and rat cytochrome P450 isozymes (P450bufI, P450dbl) catalyzing debrisoquine 4-hydroxylation, *Biochem. Biophys. Res. Commun.* **148**:1144–1150.

80. Ayesh, R., Idle, J. R., Ritchie, J. C., Crothers, M. J., and Hetzel, M. R., 1984, Metabolic oxidation phenotypes as markers for susceptibility to lung cancer, *Nature* **312**:169–170.

81. Idle, J. R., Armstrong, M., Boddy, A. V., Boustead, C., Cholerton, S., Cooper, J., Daly, A. K., Ellis, J., Gregory, W., Hadidi, H., Höfer, C., Holt, J., Leathart, J., McCracken, N., Monkman, S. C., Painter, J. E., Taber, H., Walker, D., and Yule, M., 1992, The pharmacogenetics of chemical carcinogenesis, *Pharmacogenetics* **2**:246–258.

82. Sugimura, H., Caporaso, N. E., Shaw, G. L., Modali, R. V., Gonzalez, F. J., Hoover, R. N., Resau, J. H., Trump, B. F., Weston, A., and Harris, C. C., 1990, Human debrisoquine hydroxylase gene polymorphisms in cancer patients and controls, *Carcinogenesis* **11**:1527–1530.

83. Kaisary, A., Smith, P., Jaczq, E., McAllister, C. B., Wilkinson, G. R, Ray, W. A., and Branch, R. A., 1987, Genetic predisposition to bladder cancer: Ability to hydroxylate debrisoquine and mephenytoin as risk factors, *Cancer Res.* **47**:5488–5493.

84. Law, M. R., Hetzel, M. R., and Idle, J. R., 1989, Debrisoquine metabolism and genetic predisposition to lung cancer, *Br. J. Cancer* **54**:686–687.

85. Caporaso, N. E., Tucker, M. A., Hoover, R. N., Hayes, R. B., Pickle, L. W., Issaq, H. J., Muschik, G. M., Green-Gallo, L., Buivys, D., Aisner, S., Resau, J. H., Trump, B. F., Tollerud, D., Weston, A., and Harris, C. C., 1990, Lung cancer and the debrisoquine metabolic phenotype, *J. Natl. Cancer Inst.* **82**:1264–1271.

86. Wolff, T., Distlerath, L. M., Worthington, M. T., Groopman, J. D., Hammons, G. J., Kadlubar, F. F., Prough, R. A., Martin, M. V., and Guengerich, F. P., 1985, Substrate specificity of human liver cytochrome P-450 debrisoquine 4-hydroxylase probed using immunochemical inhibition and chemical modeling, *Cancer Res.* **45**:2116–2122.

87. Shimada, T., and Guengerich, F. P., 1991, Activation of amino-α-carboline, 2-amino-1-methyl-6-phenylimidazo[4,5-*b*]pyridine, and a copper phthalocyanine cellulose extract of cigarette smoke condensate by cytochrome P450 enzymes in rat and human liver microsomes, *Cancer Res.* **51**:5284–5291.

88. Roots, I., Drakoulis, N., and Brockmöller, J., 1993, Still an open question: Does active CYP2D6 predispose to lung cancer? in: *8th International Conference on Cytochrome P450: Biochemistry, Biophysics, and Molecular Biology* (October 24–28, Lisbon), p. 159.

89. Ritter, J., Somasundaram, R., Heinemeyer, G., and Roots, I., 1986, The debrisoquine hydroxylation phenotype and the acetylator phenotype as genetic risk factors for the occurrence of larynx and pharynx carcinoma, *Acta Pharmacol. Toxicol.* **59(Suppl. 5)**:221.

90. Drakoulis, N., Minks, T., Ploch, M., Otte, F., Heinemeyer, G., Kampf, D., Loddenkemper, R., and Roots, I., 1986, Questionable association of debrisoquine hydroxylator phenotype and risk for bronchial carcinoma, *Acta Pharmacol. Toxicol.* **59(Suppl. 5)**:220.

91. Speirs, C. J., Murray, S., Davies, D. S., Mabadeje, A. F. B., and Boobis, A. R., 1990, Debrisoquine oxidation phenotype and susceptibility to lung cancer, *Br. J. Clin. Pharmacol.* **29**:101–109.

92. Ladero, J. M., Benítez, J., González, J. F., Vargas, E., and Díaz-Rubio, M., 1991, Oxidative polymorphism of debrisoquine is not related to human colo-rectal cancer, *Eur. J. Clin. Pharmacol.* **40**:525–527.

93. Caporaso, N. E., Shields, P. G., Landi, M. T., Shaw, G. L., Tucker, M. A., Hoover, M., Sugimura, H., Weston, A., and Harris, C. C., 1992, The debrisoquine metabolic phenotype and DNA-based assays: Implications of misclassification for the association of lung cancer and the debrisoquine metabolic phenotype, *Environ. Health Perspect.* **98**:101–105.

94. Waller, D. G., Renwick, A. G., Gruchy, B. S., and George, C. F., 1984, The first pass metabolism of nifedipine in man, *Br. J. Clin. Pharmacol.* **18**:951–954.

95. Kaminsky, L.S., and Fasco, M. J., 1992, Small intestinal cytochromes P450, *Crit. Rev. Toxicol.* **21**:407–422.

96. Kolars, J. C., Schmiedlin-Ren, P., Schuetz, J. D., Fang, C ., and Watkins, P. B., 1992, Identification of rifampin-inducible P450IIIA4 (CYP3A4) in human small bowel enterocytes, *J. Clin. Invest.* **90**:1871–1878.

97. Meese, C. O., Fischer, C., Küpfer, A., Wisser, H., and Eichelbaum, M., 1991, Identification of the "major" polymorphic carbocysteine metabolite as *S*-(carboxymethylthio)-L-cysteine, *Biochem. Pharmacol.* **42**:R13–R16.

98. Turgeon, D. K., Normolle, D. P., Leichtman, A. B., Annesley, T. M., Smith, D. E., and Watkins, P. B., 1992, Erythromycin breath test predicts oral clearance of cyclosporine in kidney transplant recipients, *Clin. Pharmacol. Ther.* **52**:471–478.

99. Cresteil, T., and Eisen, H.J., 1988, Regulation of human cytochrome P₁-450 gene, in: *Liver Cells and Drugs, Colloque INSERM* (A. Guillouzo, ed.), INSERM/John Libbey Erotext Ltd., London, pp. 51–58.

100. Schweikl, H., Taylor, J. A., Kitareewan, S., Linko, P., Nagorney, D., and Goldstein, J. A., 1993, Expression of *CYP1A1* and *CYP1A2* genes in human liver, *Pharmacogenetics* **3**:239–249.

101. Kellerman, G., Luyten-Kellerman, M., and Shaw, C. R., 1973, Genetic variation of aryl hydrocarbon hydroxylase in human lymphocytes, *Am. J. Hum. Genet.* **25**:327–331.

102. Kellerman, G., Shaw, C. R., and Luyten-Kellerman, M., 1973, Aryl hydrocarbon hydroxylase inducibility and bronchogenic carcinoma, *N. Engl. J. Med.* **298**:934–937.

103. Paigen, B., Ward, E., Reilly, A., Houten, L., Gurtoo, H. L., Minowada, J., Steenland, K., Havens, M. B., and Sartori, P., 1981, Seasonal variation of aryl hydrocarbon hydroxylase activity in human lymphocytes, *Cancer Res.* **41**:2757–2761.

104. Leboeuf, R., Havens, M., Tabron, D., and Paigen, B., 1981, Arylhydrocarbon hydroxylase activity and cytochrome *P*-450 in human tissues, *Biochim. Biophys. Acta* **658**:348–355.

105. Kouri, R. E., McKinney, C. E., Levine, A. S., Edwards, B. K., Vesell, E. S., Nebert, D. W., and McLemore, T. L., 1984, Variations in aryl hydrocarbon hydroxylase activities in mitogen-activated human and nonhuman primate lymphocytes, *Toxicol. Pathol.* **12**:44–48.

106. Kouri, R. E., McKinney, C. E., Slomiany, D. J., Snodgrass, D. R., Wray, N. P., and McLemore, T. L., 1982, Positive correlation between high aryl hydrocarbon hydroxylase activity and primary lung cancer as analyzed in cryopreserved lymphocytes, *Cancer Res.* **42**:5030–5037.

107. McLemore, T. L., Adelberg, S., Liu, M. C., McMahon, N. A., Yu, S. J., Hubbard, W.C., Czerwinski, M., Wood, T. G., Storeng, R., Lubet, R. A., Eggleston, J. C., Boyd, M.R., and Hines, R. N., 1990, Expression of CYP1A1 gene in patients with lung cancer: Evidence for cigarette smoke-induced gene expression in normal lung tissue and for altered gene regulation in primary pulmonary carcinomas, *J. Natl. Cancer Inst.* **82**:1333–1339.

108. Hayashi, S., Watanabe, J., Nakachi, K., and Kawajiri, K., 1991, Genetic linkage of lung cancer-associated *Msp*I polymorphisms with amino acid replacement in the heme binding region of the human cytochrome P450IA1 gene, *J. Biochem.* **110**:407–411.

109. Tefre, T., Ryberg, D., Haugen, A., Nebert, D. W., Skaug, V., Brogger, A., and Borresen, A. L., 1991, Human *CYP1A1* (cytochrome $P_1$450) gene: Lack of association between the *Msp* I restriction fragment length polymorphism and incidence of lung cancer in a Norwegian population, *Pharmacogenetics* **1**:20–25.

110. Hirvonen, A., Husgafvel-Pursiainen, K., Karjalainen, A., Anttila, S., and Vainio, H., 1992, Point-mutational *Msp*1 and Ile-Val polymorphisms closely linked in the CYP1A1 gene: Lack of association with susceptibility to lung cancer in a Finnish study population, *Cancer Epidemiol. Biomarkers Prev.* **1**:485–489.

111. Hayashi, S. I., Watanabe, J., Nakachi, K., and Kawajiri, K., 1991, PCR detection of an A/G polymorphism within exon 7 of the CYP1A1 gene, *Nucl. Acids Res.* **19**:4797.

112. Drakoulis, N., Cascorbi, I., Brockmöller, J., Gross, C. R., and Roots, I., 1993, Exon-7 point mutation (M2; 4889A>G) in human CYP1A1 gene as susceptibility factor for lung cancer, in: *8th International Conference on Cytochrome P450: Biochemistry, Biophysics, and Molecular Biology* (October 24–28, Lisbon, Portugal), p. 204.

113. Wedlund, P. J., Kimura, S., Gonzalez, F. J., and Nebert, D. W., 1994, I462V mutation in the human CYP1A1 gene: Lack of correlation with either the Msp I 1.9 kb (M2) allele or CYP1A1 inducibility in a three-generation family of East Mediterranean descent, *Pharmacogenetics* **4**:21–26.

114. Crofts, F., Cosma, G. N., Currie, D., Taioli, E., Toniolo, P., and Garte, S. J., 1993, A novel CYP1A1 gene polymorphism in African-Americans, *Carcinogenesis* **14**:1729–1731.

115. Roberts-Thomson, S. J., McManus, M. E., Tukey, R. H., Gonzalez, F. J., and Holder, G. M., 1993, The catalytic activity of four expressed human cytochrome P450s towards benzo[a]pyrene and the isomers of its proximate carcinogen, *Biochem. Biophys. Res. Commun.* **192**:1373–1379.

116. Shimada, T., Martin, M. V., Pruess-Schwartz, D., Marnett, L. J., and Guengerich, F. P., 1989, Roles of individual human cytochrome P450 enzymes in the bioactivation of benzo(*a*)pyrene, 7,8-dihydroxy-7,8-dihydrobenzo(*a*)pyrene, and other dihydrodiol derivatives of polycydic aromatic hydrocarbons, *Cancer Res.* **49**:6304–6312.

117. Yun, C.-H., Shimada, T., and Guengerich, F. P., 1992, Roles of human liver cytochrome P-4502C and 3A enzymes in the 3-hydroxylation of benzo(*a*)pyrene, *Cancer Res.* **52**:1868–1874.

118. McManus, M. E., Burgess, W. M., Veronese, M. E., Huggett, A., Quattrochi, L. C., and Tukey, R. H., 1990, Metabolism of 2-acetylaminofluorene and benzo(*a*)pyrene and activation of food-derived heterocyclic amine mutagens by human cytochromes P-450, *Cancer Res.* **50**:3367–3376.

119. Eugster, H. P., Sengstag, C., Meyer, U. A., Hinnen, A., and Würgler, F. E., 1990, Constitutive and inducible expression of human cytochrome P450IA1 in yeast *Saccharomyces cerevisiae*: An alternative enzyme source for *in vitro* studies, *Biochem. Biophys. Res. Commun.* **172**:737–744.

120. Ching, M. S., Lennard, M. S., Tucker, G. T., Woods, H. F., Kelly, D. E., and Kelly, S. L., 1991, The expression of human cytochrome P450IA1 in the yeast *Saccharomyces cerevisiae, Biochem. Pharmacol.* **42**:753–758.

121. Renaud, J. P., Peyronneau, M. A., Urban, P., Truan, G., Cullin, C., Pompon, D., Beaune, P., and Mansuy, D., 1993, Recombinant yeast in drug metabolism, *Toxicology* **182**:39–52.

122. Simon, I., Berthou, F., Riche, C., Beaune, P., and Ratanasavanh, D., 1993, Both cytochrome P4501A and 3A4 are involved in the *N*-demethylation of tamoxifen, in: *Abstracts, 5th Europenn ISSX Meeting* (September 26–29, Tours), Vol. 3, p. 44.

123. Berthou, F., Carriere, V., Ratanasavanh, D., Goasduff, T., Morel, F., Gautier, J. C., Guillouzo, A., and Beaune, P., 1993, On the specificity of chlorzoxazone as drug probe of cytochrome P4502E1, in: *Abstracts, 5th European ISSX Meeting* (September 26–29, Tours), Vol. 3, p. 116.

124. Farin, F. M., and Omiecinski, C. J., 1993, Regiospecific expression of cytochrome P-450s and microsomal epoxide hydrolase in human brain tissue, *J. Toxicol. Environ. Health* **40**:317–335.

125. Whitlock, J. P., Jr., 1993, Mechanistic aspects of dioxin action, *Chem. Res. Toxicol.* **6**:754–763.

126. Quattrochi, L. C., Vu, T., and Tukey, R. H., 1994, The human CYP1A2 gene and induction by 3-methylcholanthrene. A region of DNA that supports Ah-receptor binding and promoter-specific induction, *J. Biol. Chem.* **269**:6949–6954.

127. Butler, M. A., Iwasaki, M., Guengerich, F. P., and Kadlubar, F. F., 1989, Human cytochrome P-450PA (P-450IA2), the phenacetin O-deethylase, is primarily responsible for the hepatic 3-demethylation of caffeine and N-oxidation of carcinogenic arylamines, *Proc. Natl. Acad. Sci. USA* **86**:7696–7700.

128. Butler, M. A., Lang, N. P., Young, J. F., Caporaso, N. E., Vineis, P., Hayes, R. B., Teitel, C. H., Massengill, J. P., Lawsen, M. F., and Kadlubar, F. F., 1992, Determination of CYP1A2 and NAT2 phenotypes in human populations by analysis of caffeine urinary metabolites, *Pharmacogenetics* **2**:116–127.

129. Vesell, E .S., and Penno, M. B., 1984, A new polymorphism of hepatic drug oxidation in humans: Family studies of antipyrine metabolites, *Fed. Proc.* **43**:2342–2347.

130. Engel, G., Knebel, N. G., Hofmann, U., and Eichelbaum, M., 1992, *In vitro* characterization of human cytochrome P450-enzymes involved in antipyrine metabolism, in: *Abstracts, 23rd European Workshop on Drug Metabolism* (September 21–25, Bergamo).

131. Lang, N. P., Butler, M. A., Massengill, J., Lawson, M., Stotts, R. C., Hauer-Jensen, M., and Kadlubar, F. F., 1994, Rapid metabolic phenotypes for acetyltransferase and cytochrome P4501A2 and putative exposure to food-borne heterocyclic amines increases the risk for colorectal cancer or polyps, *Cancer Epidemiol. Biomarkers Prev.* **3**:675–682.

132. Kalow, W., and Tang, B. K., 1993, The use of caffeine for enzyme assays: A critical appraisal, *Clin. Pharmacol. Ther.* **53**:503–514.

133. Gu, L., Gonzalez, F. J., Kalow, W., and Tang, B. K., 1992, Biotransformation of caffeine, paraxanthine, theobromine and theophylline by cDNA-expressed human CYP1A2 and CYP2E1, *Pharmacogenetics* **2**:73–77.

134. Robson, R. A., Matthews, A. P., Miners, J. O., McManus, M. E., Meyer, U. A., Hall, P. de la M., and Birkett, D. J., 1987, Characterisation of theophylline metabolism in human liver microsomes, *Br. J. Clin. Pharmacol.* **24**:293–300.

135. Patten, C., Thomas, P. E., Guy, R., Lee, M., Gonzalez, F. J., Guengerich, F. P., and Yang, C. S., 1993, Cytochrome P450 enzymes involved in acetaminophen activation by rat and human liver microsomes and their kinetics, *Chem. Res. Toxicol.* **6:**511–518.

136. Dahlqvist, R., Bertilsson, L., Birkett, D. J., Eichelbaum, M., Säwe, J., and Sjöqvist, P., 1984, Theophylline metabolism in relation to antipyrine, debrisoquine, and sparteine metabolism, *Clin. Pharmacol. Ther.* **35:**815–821.

137. Yamazaki, H., Guo, Z., Mimura, M., Gonzalez, F. J., Sugahara, C., Guengerich, F. P., and Shimada, T., 1994, Bufuralol hydroxylation by cytochrome P450 2D6, 1A1, and 1A2 enzymes in human liver microsomes, *Mol. Pharmacol.* **46:**568–577.

138. Woolf, T. F., Pool, W. F., Kukan, M., Bezek, S., Kunze, K., and Trager, W. F., 1993, Characterization of tacrine metabolism and bioactivation using heterologous expression systems and inhibition studies: Evidence for CYP1A2 involvement, in: *Abstracts, 5th North American ISSX Meeting* (October 17–21, Tucson), Vol. 4, p. 139.

139. Kunze, K. L., Wienkers, L. C., Thummel, K. E., and Trager, W. F., 1993, The use of furafylline together with liver screening techniques and enzyme kinetics to evaluate the role of P450 1A2 in the metabolism of theophylline at therapeutic concentrations in human liver, in: *Abstracts, 5th North American ISSX Meeting* (October 17–21, Tucson), Vol. 4, p. 138.

140. Rettie, A. E., Korzekwa, K. R., Kunze, K. L., Lawrence, R. F., Eddy, A. C., Aoyama, T., Gelboin, H. V., Gonzalez, F. J., and Trager, W. F., 1992, Hydroxylation of warfarin by human cDNA-expressed cytochrome P-450: A role for P-4502C9 in the etiology of (*S*)-warfarin drug interactions, *Chem. Res. Toxicol.* **5:**54–59.

141. Feldman, C. H., Hutchinson, V. E., Pippenger, C. E., Blemenfeld, T. A., Feldman, B. R., and Davis, W. J., 1980, Effect of dietary protein and carbohydrate on theophylline metabolism in children, *Pediatrics* **66:**956–962.

142. Diaz, D., Fabre, I., Daujat, M., Saintaubert, B., Bories, P., Michel, H., and Maurel, P., 1990, Omeprazole is an aryl hydrocarbon-like inducer of human hepatic cytochrome-P450, *Gastroenterology* **99:**737–747.

143. Quattrochi, L. C., and Tukey, R. H., 1993, Nuclear uptake of the Ah (dionan) receptor in response to omeprazole: Transcriptional activation of the human *CYP1A1* gene, *Mol. Pharmacol.* **43:**504–508.

144. Rost, K. L., and Roots, I., 1993, Accelerated caffeine metabolism after omeprazole treatment in breath, plasma, and urine, in: *8th International Conference on Cytochrome P450: Biochemistry, Biophysics, and Molecular Biology* (October 24–28, Lisbon), p. 203.

145. Fisher, C. W., Caudle, D. L., Martin-Wixtrom, C., Quattrochi, L. C., Tukey, R. H., Waterman, M. R., and Estabrook, R. W., 1992, High-level expression of functional cytochrome P450 1A2 in *Escherichia coli, FASEB J.* **6:**759–764.

146. Kadlubar, P. F., and Hammons, G. J., 1987, The role of cytochrome P-450 in the metabolism of chemical carcinogens, in: *Mammalian Cytochromes P-450*, Vol. 2 (F. P. Guengerich, ed.), CRC Press, Boca Raton, FL, pp. 81–130.

147. Sugimura, T., 1992, Multistep carcinogenesis: A 1992 perspective, *Science* **258:**603–607.

148. Crespi, C. L., Steimel, D. T., Aoyama, T., Gelboin, H. V., and Gonzalez, F. J., 1990, Stable expression of human cytochrome P450IA2 cDNA in a human lymphoblastoid cell line: Role of the enzyme in the metabolic activation of aflatoxin B_1, *Mol. Carcinog.* **3:**5–8.

149. Glatt, H. R., Pauly, K., Wölfel, C., Dogra, S., Seidel, A., Lee, H., Harvey, R. G., Oesch, F., and Doehmer, J., 1993, Stable expression of heterologous cytochromes P450 in V79 cells: Mutagenicity studies with polycyclic aromatic hydrocarbons, in: *Polycyclic Aromatic Compounds: Synthesis, Properties, Analytical Measurements, Occurrence and Biological Effects. (Proceedings of the Thirteenth International Symposium on Polynuclear Aromatic Hydrocarbons, October 1–4, 1991, Bordeaux, France)* (P. Garrigues and M. Lamotte, eds.) Gordon & Breach, New York pp. 1167–1174.

150. Savas, U., Bhattacharyya, K. K., Christou, M., Alexander, D. L., and Jefcoate, C. R., 1994, Mouse cytochrome P450EF, representative of a new 1B subfamily of cytochrome P450s. Cloning, sequence determination, and tissue expression, *J. Biol. Chem.* **264:** 14905–14911.

151. Sutter, T. R, Tang, Y. M., Hayes, C. L., Wo, Y.-Y. P., Jabs, E. W., Li, X., Yin, H., Cody, C. S., and Greenlee, W. F., 1994, Complete cDNA sequence of a human dioxin-inducible mRNA identifies a new gene subfamily of cytochrome P450 that maps to chromosome 2, *J. Biol. Chem.* **269:**13092–13099.

152. Yamano, S., Tatsuno, J., and Gonzalez, F. J., 1990, The CYP2A3 gene product catalyzes coumarin 7-hydroxylation in human liver microsomes, *Biochemistry* **29:**1322–1329.

153. Miles, J. S., McLaren, A. W., Forrester, L. M., Glancey, M. J., Lang, M. A., and Wolf, C. R., 1990, Identification of the human liver cytochrome P450 responsible for coumarin 7-hydroxylase activity, *Biochem. J.* **267:**365–371.

154. Daly, A. K., Cholerton, S., Gregory, W., and Idle, J. R., 1993, Metabolic polymorphisms, *Pharmacol. Ther.* **57:**129–160.

155. Cholerton, S., Idle, M. E., Vas, A., Gonzalez, F. J., and Idle, J. R., 1992, Comparison of a novel thin-layer chromatographic–fluorescence detection method with a spectrofluorometric method for the determination of 7-hydroxycoumarin in human urine, *J. Chromatogr.* **575:**325–330.

156. Daly, A. K., Cholerton, S., Armstrong, M., and Idle, J. R., 1994, Genotyping for polymorphisms in xenobiotic metabolism as a predictor of disease susceptibility, *Environ. Health Perspect.* **102:**55–61.

157. Rautio, A., Kraul, H., Kojo, A., Salmela, E., and Pelkonen, O., 1992, Interindividual variability of coumarin 7-hydroxylation in healthy volunteers, *Pharmacogenetics* **2:**227–233.

158. Crespi, C. L., Penman, B. W., Steimel, D. T., Gelboin, H. V., and Gonzalez, F. J., 1991, The development of a human cell line stably expressing human CYP3A4: Role in the metabolic activation of aflatoxin B_1 and comparison to CYP1A2 and CYP2A3, *Carcinogenesis* **12:**355–359.

159. Crespi, C. L., Penman, B. W., Leakey, J.A.E., Arlotto, M. P., Stark, A., Parkinson, A., Turner, T., Steimel, D. T., Rudo, K., Davies, R. L., and Langenbach, R., 1990, Human cytochrome P450IIA3: cDNA sequence, role of the enzyme in the metabolic activation of promutagens, comparison to nitrosamine activation by human cytochrome P450IIE1, *Carcinogenesis* **11:**1293–1300.

160. Yamazaki, H., Inui, Y., Yun, C. H., Mimura, M., Guengerich, F. P., and Shimada, T., 1992, Cytochrome P450 2E1 and 2A6 enzymes as major catalysts for metabolic activation of N-nitrosodialkylamines and tobacco-related nitrosamines in human liver microsomes, *Carcinogenesis* **13:**1789–1794.

161. Smith, T. J., Guo, Z., Gonzalez, P. J., Guengerich, F. P., Stoner, G. D., and Yang, C. S., 1992, Metabolism of 4-(methylnitrosamino)-1-(3-pyridyl)-1-butanone (NNK) in human lung and liver microsomes and cytochromes P-450 expressed in hepatoma cells, *Cancer Res.* **52:**1757–1763.

162. Crespi, C. L., Penman, B. W., Gelboin, H. V., and Gonzalez, F. J., 1991, A tobacco smoke-derived nitrosamine, 4-(methylnitrosamino)-1-(3-pyridyl)-1-butanone, is activated by multiple human cytochrome P450s including the polymorphic human cytochrome P4502D6, *Carcinogenesis* **12:**1197–1201.

163. Hong, J. Y., Ding, X., Smith, T. J., Coon, M. J., and Yang, C. S., 1992, Metabolism of 4-(methylnitrosamino)-1-(3-pyridyl)-1-butanone (NNK), a tobacco-specific carcinogen, by rabbit nasal microsomes and cytochrome P405s NMa and NMb, *Carcinogenesis* **13:**2141–2144.

164. Haugen, D. A., van der Hoeven, T. A., and Coon, M. J., 1975, Purified liver microsomal cytochrome P-450: Separation and characterization of multiple forms, *J. Biol. Chem.* **250:**3567–3570.

165. Imai, Y., and Sato, R., 1974, A gel-electrophoretically homogeneous preparation of cytochrome P-450 from liver microsomes of phenobarbital-pretreated rabbits, *Biochem. Biophys. Res. Commun.* **60:**8–14.

166. Guengerich, F. P., 1978, Separation and purification of multiple forms of microsomal cytochrome P-450. Partial characterization of three apparently homogeneous cytochromes P-450 prepared from livers of phenobarbital- and 3-methylcholanthrene-treated rats, *J. Biol. Chem.* **253:**7931–7939.

167. Ryan, D. E., Thomas, P. E., Korzeniowski, D., and Levin, W., 1979, Separation and characterization of highly purified forms of liver microsomal cytochrome P-450 from rats treated with polychlorinated biphenyls, phenobarbital, and 3-methylcholanthrene, *J. Biol. Chem.* **254:**1365–1374.

168. Yamano, S., Nhamburo, P. T., Aoyama, T., Meyer, U. A., Inaba, T., Kalow, W., Gelboin, H. V., McBride, O. W., and Gonzalez, F. J., 1989, cDNA cloning and sequence and cDNA-directed expression of human

P450 IIB1: Identification of a normal and two variant cDNAs derived from the CYP2B locus on chromosome 19 and differential expression of the IIB mRNAs in human liver, *Biochemistry* **28**:7340–7348.

169. Flammang, A. M., Gelboin, H. V., Aoyama, T., Gonzalez, F. J., and McCoy, G. D., 1992, Nicotine metabolism by cDNA-expressed human cytochrome P-450s, *Biochem. Arch.* **8**:1–8.

170. Leo, M. A., Lasker, J. M., Raucy, J. L., Kim, C. I., Black, M., and Lieber, C. S., 1989, Metabolism of retinol and retinoic acid by human liver cytochrome P450IIC8, *Arch. Biochem. Biophys.* **269**:305–312.

171. Wrighton, S. A., Thomas, P. E., Willis, P., Maines, S. L., Watkins, P. B., Levin, W., and Guzelian, P. S., 1987, Purification of a human liver cytochrome P-450 immunochemically related to several cytochromes P-450 purified from untreated rats, *J. Clin. Invest.* **80**:1017–1022.

172. Okino, S. T., Quattrochi, L. C., Pendurthi, U. R. , McBride, O. W., and Tukey, R. H., 1987, Characterization of multiple human cytochrome P450 1 cDNAs, *J. Biol. Chem.* **262**:16072–16079.

173. Kimura, S., Pastewka, J., Gelboin, H. V., and Gonzalez, F. J., 1987, cDNA and amino acid sequences of two members of the human P450IIC gene subfamily, *Nucl. Acids Res.* **15**:10053.

174. Harris, J. W., Rahman, A., Kim, B.-R., Guengerich, F. P., and Collins, J. M., 1994, Metabolism of taxol by human hepatic microsomes and liver slices: Participation of cytochrome P450 3A4 and of an unknown P450 enzyme, *Cancer Res.* **54**: 4026–4035.

175. Cosme, J., Maurel, P., Soucek, P., Guengerich, F. P., and Beaune, P. H., 1994, Cloning, expression and characterization of human cytochrome P-450 2C8 in heterologous systems, *Abstracts, 14th European Drug Metabolism Workshop* (July 3–8, Paris).

176. Romkes, M., Faletto, M. B., Blaisdell, J. A., Raucy, J. L., and Goldstein, J. A., 1991, Cloning and expression of complementary DNAs for multiple members of the human cytochrome P450IIC subfamily, *Biochemistry* **30**:3247–3255.

177. Brian, W. R., Srivastava, P. K., Umbenhauer, D. R., Lloyd, R. S., and Guengerich, F. P., 1989, Expression of a human liver cytochrome P-450 protein with tolbutamide hydroxylase activity in *Saccharomyces cerevisiae, Biochemistry* **28**:4993–4999.

178. Relling, M. V., Aoyama, T., Gonzalez, F. J., and Meyer, U. A., 1990, Tolbutamide and mephenytoin hydroxylation by human cytochrome P450s in the CYP2C subfamily, *J. Pharmacol. Exp. Ther.* **252**:442–447.

179. Lopez Garcia, M. P., Dansette, P. M., Valadon, P., Amar, C., Beaune, P. H., Guengerich, F. P., and Mansuy, D., 1993, Human liver P450s expressed in yeast as tools for reactive metabolite formation studies: Oxidative activation of tienilic acid by P450 2C9 and P450 2C10, *Eur. J. Biochem.* **213**:223–232.

180. Veronese, M. E., Mackenzie, P. I., Doecke, C. J., McManus, M. E., Miners, J. O., and Birkett, D. J., 1991, Tolbutamide and phenytoin hydroxylations by cDNA-expressed human liver cytochrome P4502C9, *Biochem. Biophys. Res. Commun.* **175**:1112–1118.

181. Zilly, W., Breimer, D. D., and Richter, E., 1977, Stimulation of drug metabolism by rifampicin in patients with cirrhosis or cholestasis measured by increased hexobarbital and tolbutamide clearance, *Eur. J. Clin. Pharmacol.* **11**:287–293.

182. Vasko, M. R., Bell, R. D., Daly, D. D., and Pippenger, C. E., 1980, Inheritance of phenytoin hypometabolism: A kinetic study of one family, *Clin. Pharmacol. Ther.* **27**:96–103.

183. Scott, J., and Poffenbarger, P. L., 1978, Pharmacogenetics of tolbutamide metabolism in humans, *Diabetes* **28**:41–51.

184. Umbenhauer, D. R., Martin, M. V., Lloyd, R. S., and Guengerich, F. P., 1987, Cloning and sequence determination of a complementary DNA related to human liver microsomal cytochrome P-450 S-mephenytoin 4-hydroxylase, *Biochemistry* **26**:1094–1099.

185. Srivastava, P. K., Yun, C.-H., Beaune, P. H., Ged, C., and Guengerich, F. P., 1991, Separation of human liver tolbutamine hydroxylase and (S)-mephenytoin 4'-hydroxylase cytochrome P-450 enzymes, *Mol. Pharmacol.* **40**:69–79.

186. Kato, R., Yasumori, T., Nagata, K., Yang, S. K., Chen, L. S., Murayama, N., and Yamazoe, Y., 1993, Oxidative metabolism of diazepam in human liver: Differential role of CYP2C and CYP3A depending

on substrate concentration, in: *8th International Conference on Cytochrome P450: Biochemistry, Biophysics, and Molecular Biology* (October 24–28, Lisbon), p. 241.

187. Yasumori, T., Nagata, K., Yang, S. K., Chen, L.-S., Murayama, N., Yamazoe, Y., and Kato, R., 1993, Cytochrome P450 mediated metabolism of diazepam in human and rat: Involvement of human CYP2C in *N*-demethylation in the substrate concentration-dependent manner, *Pharmacogenetics* **3**:291–301.

188. Leeman, T., Transon, C., and Dayer, P., 1993, Cytochrome P450TB (CYP2C): A major monooxygenase catalyzing diclofenac 4′-hydroxylation in human liver, *Life Sci.* **52**:29–34.

189. Knodell, R. G., Dubey, R. K., Wilkinson, G. R., and Guengerich, F. P., 1988, Oxidative metabolism of hexobarbital in human liver: Relationship to polymorphic S-mephenytoin 4-hydroxylation, *J. Pharmacol. Exp. Ther.* **245**:845–849.

190. Leeman, T. D., Transon, C., Bonnabry, P., and Dayer, P., 1993, A major role for cytochrome P450TB (CYP2C subfamily) in the actions of non-steroidal antiinflammatory drugs, *Drugs Exp. Clin. Res.* **19**:189–195.

191. Stearns, R. A., Chakravarty, P. K., Chen, R., and Chiu, S.H.L., 1993, Investigations into the mechanism of oxidation of losartan, an alcohol, to its active metabolite, a carboxylic acid derivative, in: *Abstracts, 5th North American ISSX Meeting* (October 17–21, Tucson), Vol. 4, p. 238.

192. Küpfer, A., and Branch, R. A., 1985, Stereoselective mephobarbital hydroxylation cosegregates with mephenytoin hydroxylation, *Clin. Pharmacol. Ther.* **38**:414–418.

193. Hall, S. D., Guengerich, F. P., Branch, R. A., and Wilkinson, G. R., 1987, Characterization and inhibition of mephenytoin 4-hydroxylase activity in human liver microsomes, *J. Pharmacol. Exp. Ther.* **240**:216–222.

194. Andersson, T., Regårdh, C. G., Dahl-Puustinen, M. L., and Bertilsson, L., 1990, Slow omeprazole metabolizers are also poor S-mephenytoin hydroxylators, *Ther. Drug Monitoring* **12**:415–416.

195. Curi-Pedrosa, R., Pichard, L., Bonfils, C., Jacqz-Aigrain, E., Guengerich, F. P., and Maurel, P., 1993, Major implication of cytochrome P450 3A4 in the oxidative metabolism of antisecretory drugs omeprazole and lansoprazole in human liver microsomes and hepatocytes, in: *Abstracts, 5th European ISSX Meeting (September 26–29, Tours), Vol. 3, p. 46.*

196. Knodell, R. G., Hall, S. D., Wilkinson, G. R., and Guengerich, F. P., 1987, Hepatic metabolism of tolbutamide: Characterization of the form of cytochrome P-450 involved in methyl hydroxylation and relationship to in vivo disposition, *J. Pharmacol. Exp. Ther.* **241**:1112–1119.

197. Nakamura, M., Tanaka, E., Misawa, S., Shimada, T., Imaoka, S., and Funae, Y., 1994, Trimethadione metabolism, a useful indicator for assessing hepatic drug-oxidizing capacity, *Biochem. Pharmacol.* **47**: 247–251.

198. Furuya, H., Meyer, U. A., Gelboin, H. V., and Gonzalez, F. J., 1991, Polymerase chain reaction-directed identification, cloning, and quantification of human CYP2C18 mRNA, *Mol. Pharmacol.* **40**:375–382.

199. Yasumori, T., Murayama, N., Yamazoe, Y., Nogi, Y., Fukasawa, T., and Kato, R., 1989, Expression of a human P-450IIC gene in yeast cells using galactose-inducible expression system, *Mol. Pharmacol.* **35**:443–449.

200. Goldstein, J. A., Faletto, M. B., Romkessparks, M., Sullivan, T., Kitareewan, S., Raucy, J. L., Lasker, J. M., and Ghanayem, B. I., 1994, Evidence that CYP2C19 is the major (*S*)-mephenytoin 4′-hydroxylase in humans, *Biochemistry* **33**:1743–1752.

201. Wedlund, P. J., Aslanian, W. S., McAllister, C. B., Wilkinson, G. R., and Branch, R. A., 1984, Mephenytoin hydroxylation deficiency in Caucasians: Frequency of a new oxidative drug metabolism polymorphism, *Clin. Pharmacol. Ther.* **36**:773–780.

202. Küpfer, A., and Preisig, R., 1984, Pharmacogenetics of mephenytoin: A new drug hydroxylation polymorphism in man, *Eur. J. Clin. Pharmacol.* **26**:753–759.

203. Miners, J. O., Smith, K. J., Robson, R. A., McManus, M. E., Veronese, M. E., and Birkett, D. J., 1988, Tolbutamide hydroxylation by human liver microsomes: Kinetic characterisation and relationship to other cytochrome P-450 dependent xenobiotic oxidations, *Biochem. Pharmacol.* **37**:1137–1144.

204. Gonzalez, F. J., Skoda, R. C., Kimura, S., Umeno, M., Zanger, U. M., Nebert, D. W., Gelboin, H. V., Hardwick, J. P., and Meyer, U. A., 1988, Characterization of the common genetic defect in humans deficient in debrisoquine metabolism, *Nature* **331**:442–446.

205. Kimura, S., Umeno, M., Skoda, R. C., Meyer, U. A., and Gonzalez, F. J., 1989, The human debrisoquine 4-hydroxylase (CYP2D) locus: Sequence and identification of the polymorphic CYP2D6 gene, a related gene, and a pseudogene, *Am. J. Hum. Genet.* **45**:889–904.

206. Idle, J. R., Mahgoub, A., Lancaster, R., and Smith, R. L., 1978, Hypotensive response to debrisoquine and hydroxylation phenotype, *Life Sci.* **22**:979–984.

207. Eichelbaum, M., Spannbrucker, N., Steincke, B., and Dengler, H. J., 1979, Defective N-oxidation of sparteine in man: A new pharmacogenetic defect, *Eur. J. Clin. Pharmacol.* **16**:183–187.

208. Lennard, M. S., Silas, J. H., Freestone, S., Tucker, G. T., Ramsay, L. E., and Woods, H. F., 1982, Impaired metabolism of metoprolol in poor hydroxylators of debrisoquine, *Br. J. Pharmscol.* **16**:572P–573P.

209. Bertilsson, L., Eichelbaum, M., Mellström, B., Säwe, J., Schulz, H. U., and Sjöqvist, F., 1980, Nortriptyline and antipyrine clearance in relation to debrisoquine hydroxylation in man, *Life Sci.* **27**:1673–1677.

210. Larrey, D., Distlerath, L. M., Dannan, G. A., Wilkinson, G. R., and Guengerich, F. P., 1984, Purification and characterization of the rat liver microsomal cytochrome P-450 involved in the 4-hydroxylation of debrisoquine, a prototype for genetic variation in oxidative drug metabolism, *Biochemistry* **23**:2787–2795.

211. Gonzalez, F. J., Matsunaga, T., Nagata, K., Meyer, U. A., Nebert, D. W., Pastewka, J., Kozak, C. A., Gillette, J., Gelboin, H. V., and Hardwick, J. P., 1987, Debrisoquine 4-hydroxylase: Characterization of a new P450 gene subfamily, regulation, chromosomal mapping, and molecular analysis of the DA rat polymorphism, *DNA* **6**:149–161.

212. Matsunaga, E., Zanger, U. M., Hardwick, J. P., Gelboin, H. V., Meyer, U. A., and Gonzalez, F. J., 1989, The CYP2D gene subfamily: Analysis of the molecular basis of the debrisoquine 4-hydroxylase deficiency in DA rats, *Biochemistry* **28**:7349–7355.

213. Gut, J., Gasser, R., Dayer, P., Kronbach, T., Catin, T., and Meyer, U. A., 1984, Debrisoquine-type polymorphism of drug oxidation: Purification from human liver of a cytochrome P450 isozyme with high activity for bufuralol hydroxylation, *FEBS Lett.* **173**:287–290.

214. Skoda, R. C., Gonzalez, F. J., Demierre, A., and Meyer, U. A., 1988, Two mutant alleles of the human cytochrome P-450dbl gene (P450C2D1) associated with genetically deficient metabolism of debrisoquine and other drugs, *Proc. Natl. Acad. Sci. USA* **85**:5240–5243.

215. Tyndale, R., Aoyama, T., Broly, F., Matsunaga, T., Inaba, T., Kalow, W., Gelboin, H. V., Meyer, U. A., and Gonzalez, F. J., 1991, Identification of a new variant CYP2D6 allele lacking the codon encoding Lys-281: Possible association with the poor metabolizer phenotype, *Pharmacogenetics* **1**:26–32.

216. Armstrong, M., Fairbrother, K., Idle, J. R., and Daly, A. K., 1994, The cytochrome P450 CYP2D6 allelic variant CYPD6J and related polymorphisms in a European population, *Pharmacogenetics* **4** 73–81.

217. Armstrong, M., Idle, J. R., and Daly, A. K., 1993, A polymorphic *Cfo*I site in exon 6 of the human cytochrome P50 CYP2D6 gene detected by the polymerase chain reaction, *Hum. Genet.* **91**:616–617.

218. Gonzalez, F. J., and Meyer, U. A., 1991, Molecular genetics of the debrisoquin-sparteine polymorphism, *Clin. Pharmacol. Ther.* **50**:233–238.

219. Gough, A. C., Miles, J. S., Spurr, N. K., Moss, J. E., Gaedigk, A., Eichelbaum, M., and Wolf, C. R., 1990, Identification of the primary gene defect at the cytochrome P450 CYP2D locus, *Nature* **347**:773–776.

220. Heim, M., and Meyer, U. A., 1990, Genotyping of poor metabolizers of debrisoquine by allele-specific PCR amplification, *Lancet* **336**:529–532.

221. Daly, A. K., Armstrong, M., Monkman, S. C., Idle, M. E., and Idle, J. R., 1991, Genetic and metabolic criteria for the assignment of debrisoquine 4-hydroxylation (cytochrome P4502D6) phenotypes, *Pharmacogenetics* **1**:33–41.

222. Evans, W. E., Relling, M. V., Rahman, A., McLeod, H. L., Scott, E. P., and Lin, J. S., 1993, Genetic basis for a lower prevalence of deficient CYP2D6 oxidative drug metabolism phenotypes in black Americans, *J. Clin. Invest.* **91**:2150–2154.

223. Johansson, I., Lundqvist, E., Bertilsson, L., Dahl, M. L., Sjoqvist, F., and Ingelman-Sundberg, M., 1993, Inherited amplification of an active gene in the cytochrome-P450 CYP2D locus as a cause of ultrarapid metabolism of debrisoquine, *Proc. Natl. Acad. Sci. USA.* **90**:11825–11829.

224. Matsunaga, E., Zeugin, T., Zanger, U. M., Aoyama, T., Meyer, U. A., and Gonzalez, F. J., 1990, Sequence requirements for cytochrome P-450IID1 catalytic activity: A single amino acid change (Ile^{380} Phe) specifically decreases V_{max} of the enzyme for bufuralol but not debrisoquine hydroxylation, *J. Biol. Chem.* **265**:17197–17201.

225. Köppel, C., Tenczer, J., and Arndt, I., 1989, Metabolic disposition of ajmaline, *Eur. J. Drug Metab. Pharmacokin.* **14**:309–316.

226. Alvan, G., Grind, M., Graffner, C., and Sjöqvist, F., 1984, Relationship of N-demethylation of amiflamine and its metabolite to debrisoquine hydroxylation polymorphism, *Clin. Pharmacol. Ther.* **36**:515–519.

227. Mellström, B., Säwe, J., Bertilsson, L., and Sjöqvist, F., 1986, Amitriptyline metabolism: Association with debrisoquin hydroxylation in nonsmokers, *Clin. Pharmacol. Ther.* **39**:369–371.

228. Ebner, T., and Eichelbaum, M., 1993, The metabolism of apridine in relation to the sparteine/debrisoquine polymorphism, *Br. J. Clin. Pharmacol.* **35**:426–430.

229. Dayer, P., Balant, L., Kupfer, A., Striberni, R., and Leemann, T., 1985, Effect of oxidative polymorphism (debrisoquine/sparteine type) on hepatic first-pass metabolism of bufuralol, *Eur. J. Clin. Pharmacol.* **28**:317–320.

230. Pressacco, J., Muller, R., and Kalow, W., 1993, Interactions of bupranolol with the polymorphic debrisoquine/spearteine monooxygenase (CYP2D6), *Eur. J., Clin. Pharmacol.* **45**:261–264.

231. Gleiter, C. H., Aichele, G., Nilsson, E., Hengen, N., Antonin, K. H., and Bieck, P. R., 1985, Discovery of altered pharmacokinetics of CGP 15 210 G in poor hydroxylators of debrisoquine during early drug development, *Br. J. Clin. Pharmacol.* **20**:81–84.

232. Balant-Gorgia, A. E., Balant, L. P., Genet, C., Dayer, P., Aeschlimann, J. M., and Garrone, G., 1986, Importance of oxidative polymorphism and levomepromazine treatment on the steady-state blood concentrations of clomipramine and its major metabolites, *Eur. J. Clin. Pharmacol.* **31**:449–455.

233. Fischer, V., Vogels, B., Maurer, G., and Tynes, R. E., 1992, The antipsychotic clozapine is metabolized by the polymorphic human microsomal and recombinant cytochrome P450 2D6, *J. Pharmacol. Exp. Ther.* **260**:1355–1360.

234. Dayer, P., Desmeules, J., Leemann, T., and Striberni, R., 1988, Bioactivation of the narcotic drug codeine in human liver is mediated by the polymorphic monooxygenase catalyzing debrisoquine 4-hydroxylation, *Biochem. Biophys. Res. Commun.* **152**:411–416.

235. Tucker, G. T., Silas, J. H., Iyun, A. O., Lennard, M. S., and Smith, A. J., 1977, Polymorphic hydroxylation of debrisoquine, *Lancet.* **ii**:718.

236. Evans, D.A.P., Harmer, D., Downham, D. Y., Whibley, E. J., Idle, J. R, Ritchie, J., and Smith, R. L., 1983, The genetic control of sparteine and debrisoquine metabolism in man with new methods of analysing bimodal distributions, *J. Med. Genet.* **20**:321–329.

237. Mahgoub, A., Idle, J. R., Dring, L. G., Lancaster, R., and Smith, R. L., 1977, Polymorphic hydroxylation of debrisoquine in man, *Lancet* **ii**:584–586.

238. Broly, F., Gaedigk, A., Heim, M., Eichelbaum, M., Morike, K., and Meyer, U. A., 1991, Debrisoquine/sparteine hydroxylation genotype and phenotype: Analysis of common mutations and alleles of CYP2D6 in a European population, *DNA Cell Biol.* **10**:545–558.

239. Grace, J. M., Kinter, M. T., and Macdonald, T.L., 1994, Atypical metabolism of deprenyl and its enantiomer, (*S*)-(+)*N*,α-dimethyl-*N*-propynylphenethylamine, by cytochrome P450 2D6, *Chem. Res. Toxicol.* **7**:286–290.

240. Spina, E., Steiner, E., Ericsson, Ö., and Sjöqvist, F., 1987, Hydroxylation of desmethylimipramine: Dependence on the debrisoquin hydroxylation phenotype, *Clin. Pharmacol. Ther.* **41**:314–319.

241. Coulter, C., Sanzgiri, U., and Parkinson, A., 1993, Evidence for the involvement of CYP3A enzymes in the *N*-demethylation of dextromethorphan, in: *Abstracts, 5th North American ISSX Meeting* (October 17–21, Tucson), Vol. 4, p. 183.

242. Mortimer, Ö., Lindström, B., Laurell, H., Bergman, U., and Rane, A., 1989, Dextromethorphan: Polymorphic serum pattern of the O-demethylated and didemethylated metabolites in man, *Br. J. Clin. Pharmacol.* **27:**223–227.

243. Küpfer, A., Schmid, B., Preisig, R., and Pfaff, G., 1984, Dextromethorphan as a safe probe for debrisoquine hydroxylation polymorphism, *Lancet* **ii:**517–518.

244. Woosley, R. L., Roden, D. M., Dai, G., Wang, T., Altenbern, D., Oates, J., and Wilkinson, G. R., 1986, Co-inheritance of the polymorphic metabolism of encainide and debrisoquin, *Clin. Pharmacol. Ther.* **39:**282–287.

245. Gross, A. S., Mikus, G., Fischer, C., Hertrampf, R, Gundert-Remy, U., and Eichelbaum, M., 1989, Stereoselective disposition of flecainide in relation to the sparteine debrisoquine metaboliser phenotype, *Br. J. Clin. Pharmacol.* **28:**555–566.

246. Sloan, P., Mahgoub, A., Lancaster, R., Idle, J. R., and Smith, R. L., 1978, Polymorphism of carbon oxidation of drugs and clinical implications, *Br. Med. J.* **2:**655–657.

247. Subramanyam, B., Woolf, T., and Castagnoli, N., Jr., 1991, Studies on the in vitro conversion of haloperidol to a potentially neurotoxic pyridinium metabolite, *Chem. Res. Toxicol.* **4:**123–128.

248. Otton, S. V., Schadel, M., Cheung, S. W., Kaplan, H. L., Busto, U. E., and Sellers, E. M., 1993, CYP2D6 phenotype determines the metabolic conversion of hydrocodone to hydromorphone, *Clin. Pharmacol. Ther.* **54:**463–472.

249. Lemoine, A., Gautier, J. C., Azoulay, D., Guengerich, F. P., Beaune, P., Maurel, P., and Leroux, J. P., 1993, The major pathway of imipramine metabolism is catalyzed by cytochrome P-450 1A2 and P-450 3A4 in human liver, *Mol. Pharmacol.* **43:**827–832.

250. Pierce, D. M., 1990, A review of the clinical pharmacokinetics and metabolism of the alpha 1-adrenoceptor antagonist indoramin, *Xenobiotica* **20:**1357–1367.

251. Dow, J. D., Laucher-Harsany, V., and Haegele, K. D., 1993, *In vitro* studies on the metabolism of the putative anxiolytic MDL 73005 using human liver microsomes and microsomes prepared from cells expressing a single human cytochrome P-450 isozyme, in: *Abstracts, 5th European ISSX Meeting* (September 26–29, Tours), Vol. 3, p. 48.

252. Kitchen, I., Tremblay, J., Andre, J., Dring, L. G., Idle, J. R., Smith, R. L., and Williams, R. T., 1979, Interindividual and interspecies variation in the metabolism of the hallucinogen 4-methoxyamphetamine, *Xenobiotica* **9:**397–404.

253. Roy, S. D., Hawes, E. M., McKay, G., Korchinski, E. D., and Midha, K. K., 1985, Metabolism of methoxyphenamine in extensive and poor metabolizers of debrisoquin, *Clin. Pharmacol. Ther.* **38:**128–133.

254. Ellis, S. W., Ching, M. S., Watson, P. F., Henderson, C. J., Simula, A. P., Lennard, M. S., Tucker, G. T., and Woods, H. F., 1992, Catalytic activities of human debrisoquine 4-hydroxylase cytochrome P450 (CYP2D6) expressed in yeast, *Biochem. Pharmacol.* **44:**617–620.

255. Lennard, M. S., Silas, J. H., Freestone, S., Ramsay, L. E., Tucker, G. T., and Woods, H. F., 1982, Oxidation phenotype—A major determinant of metoprolol metabolism and response, *N. Engl. J. Med.* **307:**1558–1560.

256. Turgeon, J., Fiset, C., Giguère, R., Gilbert, M., Moerike, K., Rouleau, J. R., Kroemer, H. K., Eichelbaum, M., Grech-Bélanger, O., and Bélanger, P. M., 1991, Influence of debrisoquine phenotype and of quinidine on mexiletine disposition in man, *J. Pharmacol. Exp. Ther.* **259:**789–798.

257. Marre, T., Fabre, G., Lacarelle, B., Bourrie, M., Catalin, J., Berger, Y., Rahmani, R., and Cano, J. P., 1992, Involvement of the cytochrome P-450IID subfamily in minaprine 4-hydroxylation by human microsomes, *Drug Metab. Dispos.* **20:**316–321.

258. Bertilsson, L., Mellström, B., Sjöqvist, F., Mårtensson, B., and Asberg, M., 1981, Slow hydroxylation of nortriptyline and concomitant poor debrisoquine hydroxylation: Clinical implications, *Lancet* **ii:**560–561.

259. Bloomer, J. C., Woods, P. R., Haddock, R. E., Lennard, M. S., and Tucker, G. T., 1992, The role of cytochrome P4502D6 in the metabolism of paroxetine by human liver microsomes, *Br. J. Clin. Pharmacol.* **33:**521–523.

260. Shah, R. R., Oates, N. S., Idle, J. R., Smith, R. L., and Lockhart, J.D.F., 1982, Impaired oxidation of debrisoquine in patients with perhexiline neuropathy, *Br. Med. J.* **284:**295–299.

261. Dahl-Puustinen, M. L., Lidén, A., Alm, C., Nordin, C., and Bertilsson, L., 1989, Disposition of perphenazine is related to polymorphic debrisoquin hydroxylation in human beings, *Clin. Pharmacol. Ther.* **46:**78–81.

262. Oates, N. S., Shah, R. R., Idle, J. R., and Smith, R. L., 1981, Phenformin-induced lactic acidosis associated with impaired debrisoquine hydroxylation, *Lancet* **i:**837–838.

263. Kroemer, H. K., Fischer, C., Meese, C. O., and Eichelbaum, M., 1991, Enantiomer–enantiomer interaction of (S)- and (R)-propafenone for cytochrome P450IID6-catalyzed 5-hydroxylation: *In vitro* evaluation of the mechanism, *Mol. Pharmacol.* **40:**135–142.

264. Lee, J. T., Kroemer, H. K., Silberstein, D. J., Funck-Brentano, C., Lineberry, M. D., Wood, A. J. J., Roden, D. M., and Woosley, R. L., 1990, The role of genetically determined polymorphic drug metabolism in the beta-blockade produced by propafenone, *N. Engl. J. Med.* **322:**1764–1768.

265. Zekorn, C., Achtert, G., Hausleiter, H. J., Moon, C. H., and Eichelbaum, M., 1985, Pharmacokinetics of N-propylajmaline in relation to polymorphic sparteine oxidation, *Klin. Wochenschr.* **63:**1180–1186.

266. Shaw, L., Lennard, M. S., Tucker, G. T., Bax, N.D.S., and Woods, H. F., 1987, Irreversible binding and metabolism of propranolol by human liver microsomes—relationship to polymorphic oxidation, *Biochem. Pharmacol.* **36:**2283–2288.

267. Osikowska-Evers, B., Dayer, P., Meyer, U. A., Robertz, G. E, and Eichelbaum, M., 1987, Evidence for altered catalytic properties of the cytochrome P-450 involved in sparteine oxidation in poor metabolizers, *Clin. Pharmacol. Ther.* **41:**320–325.

268. Baumann, P., Meyer, J. W., Amey, M., Baettig, D., Bryois, C., Jonzier-Perey, M., Koeb, L., Monney, C., and Woggon, B., 1992, Dextromethorphan and mephenytoin phenotyping of patients treated with thioridazine or amitriptyline, *Ther. Drug Monitor.* **14:**1–8.

269. McGourty, J. C., Silas, J. H., Fleming, J. J., McBurney, A., and Ward, J. W., 1985, Pharmacokinetics and beta-blocking effects of timolol in poor and extensive metabolizers of debrisoquin, *Clin. Pharmacol. Ther.* **38:**409–413.

270. Lewis, R. V., Lennard, M. S., Jackson, P. R., Tucker, G. T., Ramsay, L. E., and Woods, H. F., 1985, Timolol and atenolol: Relationships between oxidation phenotype, pharmacokinetics and pharmacodynamics, *Br. J. Clin. Pharmacol.* **19:**329–333.

271. Otton, S. V., Inaba, T., and Kalow, W., 1984, Competitive inhibition of sparteine oxidation in human liver by β-adrenoceptor antagonists and other cardiovascular drugs, *Life Sci.* **34:**73–80.

272. Fonne-Pfister, R., and Meyer, U. A., 1988, Xenobiotic and endobiotic inhibitors of cytochrome P-450dbl function, the target of the debrisoquine/sparteine type polymorphism, *Biochem. Pharmacol.* **37:**3829–3835.

273. Strobl, G. R., von Kruedener, S., Stöckigt, J., Guengerich, F. P., and Wolff, T., 1993, Development of a pharmacophore for inhibition of human liver cytochrome P-450 2D6: Molecular modeling and inhibition studies, *J. Med. Chem.* **36:**1136–1145.

274. Inaba, T., Tyndale, R. E., and Mahon, W. A., 1986, Quinidine: Potent inhibition of sparteine and debrisoquine oxidation in vivo, *Br. J. Clin. Pharmacol.* **22:**199–200.

275. Leemann, T., Dayer, P., and Meyer, U. A., 1986, Single-dose quinidine treatment inhibits metoprolol oxidation in extensive metabolizers, *Eur. J. Clin. Pharmacol.* **29:**739–741.

276. Mikus, G., Bochner, F., Eichelbaum, M., Horak, P., Somogyi, A. A., and Spector, S., 1994, Endogenous codeine and morphine in poor and extensive metabolisers of the YP2D6 (debrisoquine/sparteine) polymorphism, *J. Pharmacol. Exp. Ther.* **268:**546–551.

277. Islam, S. A., Wolf, C. R., Lennard, M. S., and Sternberg, M.J.E., 1991, A three-dimensional molecular template for substrates of human cytochrome P450 involved in debrisoquine 4-hydroxylation, *Carcinogenesis* **12:**2211–2219.

278. Koymans, L., Vermeulen, N.P.E., van Acker, S.A.B.E., te Koppele, J. M., Heykants, J.J.P., Lavrijsen, K., Meuldermans, W., and Donné-Op den Kelder, G. M., 1992, A predictive model for substrates of cytochrome P450-debrisoquine (2D6), *Chem. Res. Toxicol.* **5**:211–219.

279. Guengerich, F. P., Müller-Enoch, D., and Blair, I. A., 1986, Oxidation of quinidine by human liver cytochrome P-450, *Mol. Pharmacol.* **30**:287–295.

280. Macdonald, T. L., Gutheim, W.G., Martin, R. B., and Guengerich, F. P., 1989, Oxidation of substituted *N,N*-dimethylanilines by cytochrome P-450: Estimation of the effective oxidation–reduction potential of cytochrome P-450, *Biochemistry* **28**:2071–2077.

281. Niznik, H. B., Tyndale, R. F., Sallee, F. R., Gonzalez, F. J., Hardwick, J. P., Inaba, T., and Kalow, W., 1990, The dopamine transporter and cytochrome P450IID1 (debrisoquine 4-hydroxylase) in brain: Resolution and identification of two distinct [³H]GBR-12935 binding proteins, *Arch. Biochem. Biophys.* **276**:424–432.

282. Gonzalez, F. J., and Nebert, D. W., 1990, Evolution of the P450 gene superfamily: Animal–plant 'warfare,' molecular drive and human genetic differences in drug oxidation, *Trends Genet.* **66**:164–168.

283. Idle, J. R., Mahgoub, A., Sloan, T. P., Smith, R. L., Mbanefo, C. O., and Bababunmi, E. A., 1981, Some observations on the oxidation phenotype status of Nigerian patients presenting with *cancer, Cancer Lett.* **11**:331–338.

284. Barbeau, A., Roy, M., Paris, S., Cloutier, T., Plasse, L., and Poirier, J., 1985, Ecogenetics of Parkinson's disease: 4-Hydroxylation of debrisoquine, *Lancet* **ii**:1213–1215.

285. Boobis, A. R., and Davies, D. S., 1990, Debrisoquine oxidation phenotype and susceptibility to lung cancer, *Br. J. Clin. Pharmacol.* **30**:653–656.

286. Vallada, H., Collier, D., Dawson, E., Owen, M., Nanko, S., Murray, R, and Gill, M., 1992, Debrisoquine 4-hydroxylase (CYP2D) locus and possible susceptibility to schizophrenia, *Lancet* **340**:181–182.

287. Ding, X., Pernecky, S. J., and Coon, M. J., 1991, Purification and characterization of cytochrome P450 2E2 from hepatic microsomes of neonatal rabbits, *Arch. Biochem. Biophys.* **291**:270–276.

288. Koop, D. R., 1992, Oxidative and reductive metabolism by cytochrome P450 2E1, *FASEB J.* **6**:724–730.

289. Yang, C. S., Yoo, J.S.H., Ishizaki, H., and Hong, J., 1990, Cytochrome P450IIE1: Roles in nitrosamine metabolism and mechanisms of regulation, *Drug Metab. Rev.* **22**:147–159.

290. Uematsu, F., Kikuchi, H., Motomiya, M., Abe, T., Ishioka, C., Kanamaru, R, Sagami, I., and Watanabe, M., 1992, Human cytochrome P450IIE1 gene: *Dra*I polymorphism and susceptibility to cancer, *Tohoku J. Exp. Med.* **168**:113–117.

291. Peter, R., Böcker, R. G., Beaune, P. H., Iwasaki, M., Guengerich, F. P., and Yang, C.-S., 1990, Hydroxylation of chlorzoxazone as a specific probe for human liver cytochrome P-450 IIE1, *Chem. Res. Toxicol.* **3**:566–573.

292. Guengerich, F. P., and Turvy, C. G., 1991, Comparison of levels of several human microsomal cytochrome P-450 enzymes and epoxide hydrolase in normal and disease states using immunochemical analysis of surgical liver samples, *J. Pharmacol. Exp. Ther.* **256**:1189–1194.

293. Hong, J. Y., Ning, S. M., Ma, B. L., Lee, M. J., Pan, J., and Yang, C. S., 1990, Roles of pituitary hormones in the regulation of hepatic cytochrome P450IIE1 in rats and mice, *Arch. Biochem. Biophys.* **281**:132–138.

294. Pan, J., Hong, J. Y., and Yang, C. S., 1992, Post-transcriptional regulation of mouse renal cytochrome P450 2E1 by testosterone, *Arch. Biochem. Biophys.* **299**:110–115.

295. Ronis, M.J.J., Johansson, I., Hultenby, K., Lagercrantz, J., Glaumann, H., and Ingelman-Sundberg, M., 1991, Acetone-regulated synthesis and degradation of cytochrome P4502E and cytochrome P4502B1 in rat liver, *Eur. J. Biochem.* **198**:383–389.

296. Ingelman-Sundberg, M., and Jörnvall, H., 1984, Induction of the ethanol-inducible form of rabbit liver microsomal cytochrome P450 by inhibitors of alcohol dehydrogenase, *Biochem. Biophys. Res. Commun.* **124**:375–382.

297. Gonzalez, F. J., Kimura, S., Song, B. J., Pastewka, J., Gelboin, H. V., and Hardwick, J. P., 1986, Sequence of two related P450 mRNAs transcriptionally increased during rat development, *J. Biol. Chem.* **261**:10667–10672.

298. Song, B. J., Veech, R. L., Park, S. S., Gelboin, H. V., and Gonzalez, F. J., 1989, Induction of rat hepatic N-nitrosodimethylamine demethylase by acetone is due to protein stabilization, *J. Biol. Chem.* **264**:3568–3572.

299. Perrot, N., Nalpas, B., Yang, C. S., and Beaune, P. H., 1989, Modulation of cytochrome P450 isozymes in human liver, by ethanol and drug intake, *Eur. J. Clin. Invest.* **19**:549–555.

300. Hayashi, S., Watanabe, J., and Kawajiri, K., 1991, Genetic polymorphisms in the 5′-flanking region change transcriptional regulation of the human P450IIE1 gene, *J. Biochem.* **110**:559–565.

301. Patten, C. J., Ishizaki, H., Aoyama, T., Lee, M., Ning, S. M., Huang, W., Gonzalez, F. J., and Yang, C. S., 1992, Catalytic properties of the human cytochrome P450 2E1 produced by cDNA expression in mammalian cells, *Arch. Biochem. Biophys.* **299**:163–171.

302. Mapoles, J., Berthou, F., Alexander, A., Simon, F., and Ménez, J.-F., 1993, Mammalian PC-12 cell genetically engineered for human cytochrome P450 2E1 expression, *Eur. J. Biochem.* **214**:735–745.

303. Winters, D. K., and Cederbaum, A. I., 1992, Expression of a catalytically active human cytochrome P-4502E1 in *Escherichia coli*, *Biochim. Biophys. Acta* **1156**:43–49.

304. Wade, D., Yang, C. S., Metral, C. J., Roman, J. M., Hrabie, J. A., Riggs, C. W., Anjo, T., Keefer, L. K., and Mico, B. A., 1987, Deuterium isotope effect on denitrosation and demethylation of N-nitroso-dimethylamine by rat liver microsomes, *Cancer Res.* **47**:3373–3377.

305. Yang, C. S., Ishizaki, H., Lee, M., Wade, D., and Fadel, A., 1991, Deuterium isotope effect in the interaction of N-nitrosodimethylamine, ethanol, and related compounds with cytochrome P-450IIE1, *Chem. Res. Toxicol.* **4**:408–413.

306. Andersen, M. E., Clewell, H. J., III, Mahle, D. A., and Gearhart, J. M., 1993, Deuterium isotope effects of dichloromethane in vivo and the mechanism of substrate oxidation by cytochrome P450 2E1, *Toxicol. Appl. Pharmacol.* **128**: 158–165.

307. Kharasch, E. D., Thummel, K. E., and Mautz, D., 1993, Human enflurane metabolism *in vivo* by cytochrome P450 2E1, in: *Abstracts, 5th North American ISSX Meeting* (October 17–21, Tucson), Vol. 4, p. 143.

308. O'Shea, D., Davis, S. N., Kim, R. B., and Wilkinson, G. R., 1994, Effect of fasting and obesity in humans on the 6-hydroxylation of chloroxazone—A putative probe of CYP2E1 activity, *Clin. Pharmacol. Ther.* **56**: 359–367.

309. Ziegler, D. M., 1988, Flavin-containing monooxygenases: Catalytic mechanism and substrate specificities, *Drug Metab. Rev.* **19**:1–32.

310. Lee, S. K., Nesheim, J. C., and Lipscomb, J. D., 1993, Transient intermediates of the methane monooxygenase catalytic cycle, *J. Biol. Chem.* **268**:21569–21577.

311. Crespi, C. L., Penman, B. W., Leakey, J. A., Arlotto, M. P., Stark, A., Parkinson, A., Turner, T., Steimel, D. T., Rudo, K., Davies, R. L., and Langenbach, R., 1990, Human cytochrome P450IIA3: cDNA sequence, role of the enzyme in the metabolic activation of promutagens, comparison to nitrosamine activation by human cytochrome P450IIE1, *Carcinogenesis* **11**:1293–1300.

312. Nhamburo, P. T., Kimura, S., McBride, O. W., Kozak, C. A., Gelboin, H. V., and Gonzalez, F. J., 1990, The human CYP2F gene subfamily: Identification of a cDNA encoding a new cytochrome P450, cDNA-directed expression, and chromosome mapping, *Biochemistry* **29**:5491–5499.

313. Gonzalez, F. J., and Gelboin, H. V., 1993, Role of human cytochrome P-450s in risk assessment and susceptibility to environmentally based disease, *J. Toxicol. Environ. Health* **40**:289–308.

314. Guengerich, F. P., Gillam, E. M. J., Martin, M. V., Baba, T., Kim, B. R., Shimada, T., Raney, K. D., and Yun, C. H., 1994, The importance of cytochrome P450 3A enzymes in drug metabolism, in: *Schering Foundation Workshop, Assessment of the Use of Single Cytochrome P450 Enzymes in Drug Research* (March 23–25, Springer-Verlag Berlin), pp 161–186.

315. Yun, C.-H., Wood, M., Wood, A. J. J., and Guengerich, F. P., 1992, Identification of the pharmacogenetic determinants of alfentanil metabolism: Cytochrome P-450 3A4. An explanation of the variable elimination clearance, *Anesthesiology* **77**:467–474.

316. Gillet, G., Pichard, L., Filali-Ansary, A., Thénot, J. P., and Maurel, P., 1993, Identification of the major cytochromes P450 involved in the formation of plasma metabolites of alpidem in man, in: *Abstracts, 5th European ISSX Meeting* (September 26–29, Tours) Vol. 3, p. 47.

317. Fabre, G., Julian, B., Saint-Aubert, B., Joyeux, H., and Berger, Y., 1993, Evidence for CYP3A-mediated *N*-deethylation of aminodarone in human liver microsomal fractions, *Drug Metab. Dispos.* **21**:978–985.

318. Guengerich, F. P., Brian, W. R., Iwasaki, M., Sari, M.-A., Bäärnhielm, C., and Berntsson, P., 1991, Oxidation of dihydropyridine calcium channel blockers and analogues by human liver cytochrome P-450 IIIA4, *J. Med. Chem.* **34**:1838–1844.

319. Andersson, P., and Jönsson, G., 1993, The metabolism of budesonide in human liver is catalysed by cytochrome P450 3A isoenzymes, in: *Abstracts, 5th North American ISSX Meeting* (October 17–21, Tucson), Vol. 4, p. 236.

320. Jönsson, G., Andersson, P., and Äström, A., 1993, Budesonide is metabolised by cytochrome P450 3A isoenzymes in human liver, in: *8th International Conference on Cytochrome P450: Biochemistry, Biophysics, and Molecular Biology* (October 24–28, Lisbon), p. 242.

321. Pellinen, P., Honkakoski, P., Stenbäck, F., Niemitz, M., Alhava, E., Pelkonen, O., Lang, M., and Pasanen, M., 1994, Cocaine *N*-demethylation and the metabolism-related hepatotoxicity can be prevented by cytochrome P450 3A inhibitors, *Eur. J. Pharmacol. Environ. Toxicol. Pharmacol. Sect.* **270**:35–43.

322. Ged, C., Rouillon, J. M., Pichard, L., Combalbert, J., Bressot, N., Bories, P., Michel, H., Beaune, P., and Maurel, P., 1989, The increase in urinary excretion of 6β-hydroxycortisol as a marker of human hepatic cytochrome P450IIIA induction, *Br. J. Clin. Pharmacol.* **28**:373–387.

323. Brian, W. R., Sari, M.-A., Iwasaki, M., Shimada, T., Kaminsky, L. S., and Guengerich, F. P., 1990, Catalytic activities of human liver cytochrome P-450 IIIA4 expressed in *Saccharomyces cerevisiae, Biochemistry* **29**:11280–11292.

324. Waxman, D. J., Chang, T.K.H., and Chen, G., 1993, Role of individual human liver P450s and other enzymes in anti-cancer drug metabolism: Drug activation and drug resistance mechanisms, in: *8th International Conference on Cytochrome P450: Biochemistry, Biophysics, and Molecular Biology* (October 24–28, Lisbon), p. 71.

325. Combalbert, J., Fabre, I., Fabre, G., Dalet, I., Derancourt, J., Cano, J. P., and Maurel, P., 1989, Metabolism of cyclosporin A. IV. Purification and identification of the rifampicin-inducible human liver cytochrome P450 (cyclosporin A oxidase) as a product of P450IIIA gene subfamily, *Drug Metab. Dispos.* **17**:197–207.

326. Kronbach, T., Fischer, V., and Meyer, U. A., 1988, Cyclosporine metabolism in human liver: Identification of a cytochrome P-450III gene family as the major cyclosporine-metabolizing enzyme explains interactions of cyclosporine with other drugs, *Clin. Pharmacol. Ther.* **43**:630–635.

327. Fleming, C. M., Branch, R. A., Wilkinson, G. R., and Guengerich, F. P., 1992, Human liver microsomal *N*-hydroxylation of dapsone by cytochrome P450 3A4, *Mol. Pharmacol.* **41**:975–980.

328. Chauncey, M. A., and Ninomiya, S., 1990, Metabolic studies with model cytochrome P-450 systems, *Tetrahedron Lett.* **31**:5901–5904.

329. Stevens, J. C., Berger, P. L., and Bordeaux, K. G., 1993, Metabolism of ebastine by human liver microsomes and cDNA-expressed cytochrome P450 forms, in: *Abstracts, 5th European ISSX Meeting* (September 26–29, Tours), Vol. 3, p. 66.

330. Ball, S. E., Maurer, G., Zollinger, M., Ladona, M., and Vickers, A. E. M., 1992, Characterization of the cytochrome P-450 gene family responsible for the *N*-dealkylation of the ergot alkaloid CQA 206–291 in humans, *Drug Metab. Dispos.* **20**:56–63.

331. Kerlan, V., Dreano, Y., Bercovici, J. P., Beaune, P. H., Foch, H. H., and Berthou, F., 1992, Nature of cytochromes P450 involved in the 2-/4-hydroxylations of estradiol in human liver microsomes, *Biochem. Pharmacol.* **44:**1745–1756.

332. Waxman, D. J., Lapenson, D. P., Aoyama, T., Gelboin, H. V., Gonzalez, F. J., and Korzekwa, K., 1991, Steroid hormone hydroxylase specificities of eleven cDNA-expressed human cytochrome P450s, *Arch. Biochem. Biophys.* **290:**160–166.

333. Guengerich, F. P., 1988, Oxidation of 17α-ethynylestradiol by human liver cytochrome P-450, *Mol. Pharmscol.* **33:**500–508.

334. Sattler, M., Guengerich, F. P., Yun, C.-H., Christians, U., and Sewing, K. F., 1992, Human and rat liver microsomal cytochrome P450 3A enzymes are involved in biotransformation of FK506 and rapamycin, *Drug Metab. Dispos.* **20:**753–761.

335. Vincent, S. H., Karanam, B. V., Painter, S. K., and Chiu, S.H.L., 1992, In vitro metabolism of FK-506 in rat, rabbit, and human liver microsomes: Identification of a major metabolite and of cytochrome P450 3A as the major enzymes responsible for its metabolism, *Arch. Biochem. Biophys.* **294:**454–460.

336. Bargetzi, M. J., Aoyama, T., Gonzalez, F. J., and Meyer, U. A., 1989, Lidocaine metabolism in human liver microsomes by cytochrome P450IIIA4, *Clin. Pharmacol. Ther.* **46:**421–427.

337. Imaoka, S., Enomoto, K., Oda, Y., Asada, A., Fujimori, M., Shimada, T., Fujita, S., Guengerich, F. P., and Funae, Y., 1990, Lidocaine metabolism by human cytochrome P-450s purified from hepatic microsomes: Comparison of those with rat hepatic cytochrome P-450s, *J. Pharmacol. Exp. Ther.* **255:**1385–1391.

338. Vyas, K. P., Kari, P. H., Pitzenberger, S. M., Wang, R. W., and Lu, A. Y. H., 1993, Identification of 3′,5′-dihydro-3′,5′-diol-Δ⁴-lovastatin as a new cytochrome P450 3A catalyzed metabolite of lovastatin in rat and human liver microsomes, in: *Abstracts, 5th North American ISSX Meeting* (October 17–21, Tucson), Vol. 4, p. 35.

339. Kronbach, T., Mathys, D., Umeno, M., Gonzalez, F. J., and Meyer, U. A., 1989, Oxidation of midazolam and triazolam by human liver cytochrome P450IIIA4, *Mol. Pharmacol.* **36:**89–96.

340. Böcker, R. H., and Guengerich, F. P., 1986, Oxidation of 4-aryl- and 4-alkyl-substituted 2,6-dimethyl-3,5-bis(alkoxycarbonyl)-1,4-dihydropyridines by human liver microsomes and immunochemical evidence for the involvement of a form of cytochrome P-450, *J. Med. Chem.* **29:**1596–1603.

341. Andersson, T., Miners, J. O., Veronese, M. E., and Birkett, D. J., 1993, Primary and secondary metabolism of omeprazole in human liver microsomes, in: *Abstracts, 5th European ISSX Meeting* (September 26–29, Tours), Vol. 3, p. 45.

342. Waxman, D. J., Attisano, C., Guengerich, F. P., and Lapenson, D. P., 1988, Cytochrome P-450 steroid hormone metabolism catalyzed by human liver microsomes, *Arch. Biochem. Biophys.* **263:**424–436.

343. Mulford, D. J., Rodrigues, A. D., and Bopp, B. A., 1993, Identification of the human cytochrome P450 enzymes involved in the N-dealkylation of [¹⁴C]sertindole, in: *Abstracts, 5th North American ISSX Meeting* (October 17–21, Tucson), Vol. 4, p. 147.

344. Mitra, A. K., Kalhorn, T. F., Thummel, K. E., Unadkat, J. D., and Slattery, J. T., 1993, Metabolism of arylamines by human liver microsomal cytochrome P450(s), in: *Abstracts, 5th North American ISSX Meeting* (October 17–21, Tucson), Vol. 4, p. 148.

345. Jacolot, F., Simon, I., Dreano, Y., Beaune, P., Riche, C., and Berthou, F., 1991, Identification of the cytochrome P450IIIA family as the enzymes involved in the N-demethylation of tamoxifen in human liver microsomes, *Biochem. Pharmacol.* **41:**1911–1919.

346. Kumar, G. N., Thornburg, K. R., Walle, U. K., and Walle, T., 1993, Evidence for human liver microsomal CYP 3A mediated hydroxylation of taxol, in: *Abstracts, 5th North American ISSX Meeting* (October, Tucson), Vol. 4, p 145.

347. Yun, C. H., Okerholm, R. A., and Guengerich, F. P., 1993, Oxidation of the antihistaminic drug terfenadine in human liver microsomes: Role of cytochrome P450 3A(4) in N-dealkylation and C-hydroxylation, *Drug Metab. Dispos.* **21:**403–409.

348. Aoyama, T., Yamano, S., Waxman, D. J., Lapenson, D. P., Meyer, U. A., Fischer, V., Tyndale, R, Inaba, T., Kalow, W., Gelboin, H. V., and Gonzalez, F. J., 1989, Cytochrome P-450 hPCN3, a novel

cytochrome P450 IIIA gene product that is differentially expressed in adult human liver, *J. Biol. Chem.* **264:**10388–10395.

349. Renaud, J. P., Cullin, C., Pompon, D., Beaune, P., and Mansuy, D., 1990, Expression of human liver cytochrome P450 IIIA4 in yeast: A functional model for the hepatic enzyme, *Eur. J. Biochem.* **194:**889–896.

350. Kroemer, H. K., Gautier, J.-C., Beaune, P., Henderson, C., and Wolf, C. R., 1993, Identification of P450 enzymes involved in metabolism of verapamil in humans, *Naunyn-Schmiedebergs Arch. Pharmacol.* **348:**332–337.

351. Ring, B. J., Parli, C. J., George, M. C., and Wrighton, S. A., 1993, Interspecies comparison and role of CYP3A in zatosetron metabolism, in: *Abstracts, 5th North American ISSX Meeting* (October 17–21, Tucson), Vol. 4, p. 36.

352. Nakasa, H., Komiya, M., Ohmori, S., Rikihisa, T., and Kitada, M., 1993, Rat liver microsomal cytochrome P-450 responsible for reductive metabolism of zonisamide, *Drug Metab. Dispos.* **21:**777–781.

353. Beaune, P. H., Umbenhauer, D. R., Bork, R. W., Lloyd, R. S., and Guengerich, F. P., 1986, Isolation and sequence determination of a cDNA clone related to human cytochrome P-450 nifedipine oxidase, *Proc. Natl. Acad. Sci. USA* **83:**8064–8068.

354. Elshourbagy, N. A., and Guzelian, P. S., 1980, Separation, purification, and characterization of a novel form of hepatic cytochrome P450 from rats treated with pregnenolone-16α-carbonitrile, *J. Biol. Chem.* **255:**1279–1285.

355. Molowa, D. T., Schuetz, E. G., Wrighton, S. A., Watkins, P. B., Kremers, P., Mendez-Picon, G., Parker, G. A., and Guzelian, P. S., 1986, Complete cDNA sequence of a cytochrome P-450 inducible by glucocorticoids in human liver, *Proc. Natl. Acad. Sci. USA* **83:**5311–5315.

356. Bork, R. W., Muto, T., Beaune, P. H., Srivastava, P. K., Lloyd, R. S., and Guengerich, F. P., 1989, Characterization of mRNA species related to human liver cytochrome P-450 nifedipine oxidase and the regulation of catalytic activity, *J. Biol. Chem.* **264:**910–919.

357. Kolars, J., Schmiedelin-Ren, P., Dobbins, W., Merion, R., Wrighton, S., and Watkins, P., 1990, Heterogeneity of P-450 IIIA expression in human gut epithelia, *FASEB J.* **4:**A2242.

358. Daujat, M., Pichard, L., Fabre, I., Diaz, D., Maurice, M., Pineau, T., Blanc, P., Fabre, G., Fabre, J. M., Saint Aubert, B., and Maurel, P., 1990, Human P450IA and IIIA subfamilies: Regulation of expression and inducibility in primary cultures of human hepatocytes, in: *Drug Metabolizing Enzymes: Genetics, Regulation and Toxicology, Proceedings of the Eighth International Symposium on Microsomes and Drug Oxidations (Stockholm, June 25–29)* (M. Ingelman-Sundberg, J.-Å. Gustafsson, and S. Orrenius, eds.), p. 16.

359. Schuetz, J. D., Molowa, D. T., and Guzelian, P. S., 1989, Characterization of a cDNA encoding a new member of the glucocorticoid-responsive cytochromes P450 in human liver, *Arch. Biochem. Biophys.* **274:**355–365.

360. Wrighton, S. A., Ring, B. J., Watkins, P. B., and VandenBranden, M., 1989, Identification of a polymorphically expressed member of the human cytochrome P-450III family, *Mol. Pharmacol.* **86:**97–105.

361. Wrighton, S. A., Brian, W. R., Sari, M. A., Iwasaki, M., Guengerich, F. P., Raucy, J. L., Molowa, D. T., and VandenBranden, M., 1990, Studies on the expression and metabolic capabilities of human liver cytochrome P450IIIA5 (HLp3), *Mol. Pharmacol.* **38:**207–213.

362. Kitada, M., Kamataki, T., Itahashi, K., Rikihisa, T., Kato, R., and Kanakubo, Y., 1985, Purification and properties of cytochrome P-450 from homogenates of human fetal livers, *Arch. Biochem. Biophys.* **241:**275–280.

363. Komori, M., Nishio, K., Kitada, M., Shiramatsu, K., Muroya, K., Soma, M., Nagashima, K., and Kamataki, T., 1990, Fetus-specific expression of a form of cytochrome P-450 in human livers, *Biochemistry* **29:**4430–4433.

364. Schuetz, J. D., Kauma, S., and Guzelian, P. S., 1993, Identification of the fetal liver cytochrome CYP3A7 in human endometrium and placenta, *J. Clin. Invest.* **92:**1018–1024.

365. Hancock, R. L., 1992, Theoretical mechanisms for synthesis of carcinogen-induced embryonic proteins: XXVII. Intermediate generalizations (Part B), *Med. Hypoth.* **37**:6–11.

366. Hashimoto, H., Toide, K., Kitamura, R., Fujita, M., Tagawa, S., Itoh, S., and Kamataki, T., 1993, Gene structure of CYP3A4, an adult-specific form of cytochrome-P450 in human livers, and its transcriptional control, *Eur. J. Biochem.* **218**:585–595.

367. Kelly, J. D., Eaton, D. L., Guengerich, F. P., and Coulombe, R. A., Jr., 1995, Aflatoxin B1 activation in human lung: Role of cytochrome P450 3A, *Carcinogenesis,* in press.

368. Kolars, J. C., Awni, W. M., Merion, R. M., and Watkins, P. B., 1991, First-pass metabolism of cyclosporin by the gut, *Lancet* **338**:1488–1490.

369. Schwab, G. E., Raucy, J. L., and Johnson, E. F., 1988, Modulation of rabbit and human hepatic cytochrome P-450-catalyzed steroid hydroxylations by α-naphthoflavone, *Mol. Pharmacol.* **33**:493–499.

370. Imaoka, S., Imai, Y., Shimada, T., and Funae, Y., 1992, Role of phospholipids in reconstituted cytochrome P450 3A forms and mechanism of their activation of catalytic activity, *Biochemistry* **31**:6063–6069.

371. Peyronneau, M. A., Renaud, J. P., Truan, G., Urban, P., Pompon, D., and Mansuy, D., 1992, Optimization of yeast-expressed human liver cytochrome-P450 3A4 catalytic activities by coexpressing NADPH-cytochrome P450 reductase and cytochrome b_5, *Eur. J. Biochem.* **207**:109–116.

372. Guengerich, F. P., 1995, Cytochrome P450s of human liver. Classification and activity profile of the major enzymes, in: *Advances in Drug Metabolism in Man* (G. M. Pacifici and G. N. Fracchia, eds.), Medical Research Commission, European Communities, Brussels, pp. 179–231.

373. Buening, M. K., Fortner, J. G., Kappas, A., and Conney, A. H., 1978, 7,8-Benzoflavone stimulates the metabolic activation of aflatoxin B1 to mutagens by human liver, *Biochem. Biophys. Res. Commun.* **82**:348–355.

374. Buening, M. K., Chang, R. L., Huang, M. R., Fortner, J. G., Wood, A. W., and Conney, A. H., 1981, Activation and inhibition of benzo(*a*)pyrene and aflatoxin B1 metabolism in human liver microsomes by naturally occurring flavonoids, *Cancer Res.* **41**:67–72.

375. Shimada, T., and Guengerich, F. P., 1989, Evidence for cytochrome P-450NF, the nifedipine oxidase, being the principal enzyme involved in the bioactivation of aflatoxins in human liver, *Proc. Natl. Acad. Sci. USA* **86**:462–465.

376. Shimada, T., Iwasaki, M., Martin, M. V., and Guengerich, F. P., 1989, Human liver microsomal cytochrome P-450 enzymes involved in the bioactivation of procarcinogens detected by *umu* gene response in *Salmonella typhimurium* TA1535/pSK1002, *Cancer Res.* **49**:3218–3228.

377. Raney, K. D., Shimada, T., Kim, D.-H., Groopman, J. D., Harris, T. M., and Guengerich, F. P., 1992, Oxidation of aflatoxin B1 and related dihydrofurans by human liver microsomes: Significance of aflatoxin Q1 as a detoxication product, *Chem. Res. Toxicol.* **5**:202–210.

378. Guengerich, F. P., Kim, B.-R., Gillam, E. M. J., and Shimada, T., 1994, Mechanisms of enhancement and inhibition of cytochrome P450 catalytic activity, in: *Proceedings, 8th Int. Conf. on Cytochrome P450: Biochemistry, Biophysics, and Molecular Biology,* (M. C. Lechner, ed.) John Libbey Eurotext, Chichester, U.K., pp. 97–101.

379. Huang, M. T., Johnson, E. F., Muller-Eberhard, U., Koop, D. R., Coon, M. J., and Conney, A. H., 1981, Specificity in the activation and inhibition by flavonoids of benzo[*a*]pyrene hydroxylation by cytochrome P-450 isozymes from rabbit liver microsomes, *J. Biol. Chem.* **256**:10897–10901.

380. Johnson, E. F., Schwab, G. E., and Vickery, L. E., 1988, Positive effectors of the binding of an active site-directed amino steroid to rabbit cytochrome P-450 3c, *J. Biol. Chem.* **263**:17672–17677.

381. Rao, M. S., and Reddy, J. K., 1991, An overview of peroxisome proliferator-induced hepatocarcinogenesis, *Environ. Health Perspect.* **93**:205–209.

382. Gibson, G. G., 1993, Peroxisome proliferators: Paradigms and prospects, *Toxicol. Lett.* **68**:193–201.

383. Roman, L. J., Palmer, C.N.A., Clark, J. E., Muerhoff, A. S., Griffin, K. J., Johnson, E. F., and Masters, B.S.S., 1993, Expression of rabbit cytochromes P4504A which catalyze the ω-hydroxylation of arachidonic acid, fatty acids, and prostaglandins, *Arch. Biochem. Biophys.* **307**:57–65.

384. Nhamburo, P. T., Gonzalez, F. J., McBride, O. W., Gelboin, H. V., and Kimura, S., 1989, Identification of a new P450 expressed in human lung: Complete cDNA sequence, cDNA-directed expression, and chromosome mapping, *Biochemistry* **28**:8060–8066.

385. Haurand, M., and Ullrich, V., 1985, Isolation and characterization of thromboxane synthase from human platelets as a cytochrome P450 enzyme, *J. Biol. Chem.* **260**:15059–15067.

386. Nüsing, R., Schneider-Voss, S., and Ullrich, V., 1990, Immunoaffinity purification of human thromboxane synthase, *Arch. Biochem. Biophys.* **280**:325–330.

387. Graf, H., Ruf, H. H., and Ullrich, V., 1983, Prostacyclin synthase, a cytochrome P450 enzyme, *Angew. Chem. Int. Ed. Engl.* **22**:487–488.

388. Ullrich, V., and Graf, H., 1984, Prostacyclin and thromboxane synthase as P-450 enzymes, *Trends Pharmacol. Sci.* **5**:352–355.

389. Hecker, M., and Ullrich, V., 1989, On the mechanism of prostacyclin and thromboxane A_2 biosynthesis, *J. Biol. Chem.* **264**:141–150.

390. Hecker, M., Haurand, M., Ullrich, V., and Terao, S., 1986, Spectral studies on structure–activity relationships of thromboxane synthase inhibitors, *Eur. J. Biochem.* **157**:217–223.

391. Danielsson, H., and Sjövall, J., 1975, Bile acid metabolism, *Annu. Rev. Biochem.* **44**:233–253.

392. Li, Y. C., Wang, D. P., and Chiang, J. Y. L., 1990, Regulation of cholesterol 7α-hydroxylase in the liver: Cloning, sequencing, and regulation of cholesterol 7α-hydroxylase mRNA, *J. Biol. Chem.* **265**:12012–12019.

393. Pendak, W. M., Li, Y. C., Chiang, J. Y. L., Studer, E. J., Gurley, E. C., Heuman, D. M., Vlahcevic, Z. R., and Hylemon, P. B., 1991, Regulation of cholesterol 7α-hydroxylase mRNA and transcriptional activity by taurocholate and cholesterol in the chronic biliary diverted rat, *J. Biol. Chem.* **266**:3416–3421.

394. Nishimoto, M., Noshiro, M., and Okuda, K., 1993, Structure of the gene encoding human liver cholesterol 7α-hydroxylase, *Biochim. Biophys. Acta* **1172**:147–150.

395. Morohashi, K., Sogawa, K., Omura, T., and Fujii-Kuriyama, Y., 1987, Gene structure of human cytochrome P-450(SCC), cholesterol desmolase, *J. Biochem.* **101**:879–887.

396. Lin, D., Gitelman, S. E., Saenger, P., and Miller, W. L., 1991, Normal genes for the cholesterol side chain cleavage enzyme, P450scc, in congenital lipoid adrenal hyperplasia, *J. Clin. Invest.* **88**:1955–1962.

397. Lauber, M., Sugano, S., Ohnishi, T., Okamoto, M., and Müller, J., 1987, Aldosterone biosynthesis and cytochrome P450$_{11\beta}$: Evidence for two different forms of the enzyme in rats, *J. Steroid Biochem.* **26**:693–698.

398. Lorence, M. C., Trant, J. M., Clark, B. J., Khyatt, B., Mason, J. I., Estabrook, R. W., and Waterman, M. R., 1990, Construction and expression of human/bovine P45017α chimeric proteins: Evidence for distinct tertiary structures in the same P450 from two different species, *Biochemistry* **29**:9819–9824.

399. Swart, P., Swart, A. C., Waterman, M. R., Estabrook, R. W., and Mason, J. I., 1993, Progesterone 16α-hydroxylase activity is catalyzed by human cytochrome P450 17α-hydroxylase, *J. Clin. Endocrinol. Metab.* **77**:98–102.

400. Yanase, T., Kagimoto, M., Matsui, N., Simpson, E. R., and Waterman, M. R., 1988, Combined 17α-hydroxylase/17,20-lyase deficiency due to a stop codon in the N-terminal region of 17α-hydroxylase cytochrome P-450, *Mol. Cell. Endocrinol.* **59**:249–253.

401. Kagimoto, M., Winter, J.S.D., Kagimoto, K., Simpson, E. R., and Waterman, M. R., 1988, Structural characterization of normal and mutant human steroid 17α-hydroxylase genes: Molecular basis of one example of combined 17α-hydroxylase/17,20 lyase deficiency, *Mol. Endocrinol.* **2**:564–570.

402. Yanase, T., Kagimoto, M., Suzuki, S., Hashiba, K., Simpson, E. R., and Waterman, M. R., 1989, Deletion of a phenylalanine in the N-terminal region of human cytochrome P-45017α results in partial combined 17α-hydroxylase/17,20-lyase deficiency, *J. Biol. Chem.* **264**:18076–18082.

403. Yanase, T., Imai, T., Simpson, E. R., and Waterman, M. R., 1992, Molecular basis of 17α-hydroxylase/17,20-lyase deficiency, *J. Steroid Biochem. Mol. Biol.* **43**:973–979.

404. Ryan, K. J., 1958, Conversion of androstenedione to estrone by placental microsomes, *Biochim. Biophys. Acta* **27**:658–662.

405. Brodie, A. M. H., 1987, Aromatase inhibitors: Applications of inhibitors of estrogen biosynthesis, *ISI Atlas of Science*: *Pharmacology* 266–269.

406. Kellis, J. T., Jr.., and Vickery, L. E., 1987, Purification and characterization of human placental aromatase cytochrome P-450, *J. Biol. Chem.* **262**:4413–4420.

407. Chen, S., Besman, M. J., Shively, J. E., Yanagibashi, K., and Hall, P. F., 1989, Human aromatase, *Drug Metab. Rev.* **20**:511–517.

408. Oh, S.S., and Robinson, C. H., 1993, Mechanism of human placental aromatase: A new active site model, *J. Steroid Biochem. Mol. Biol.* **44**:389–397.

409. Tan, L., and Muto, N., 1986, Purification and reconstitution properties of human placental aromatase: A cytochrome P450-type monooxygenase, *Eur. J. Biochem.* **157**:243–250.

410. Hall, P. F., Chen, S., Nakajin, S., Shinoda, M., and Shively, J. E., 1987, Purification and characterization of aromatase from human placenta, *Steroids* **50**:37–50.

411. Chen, S., Shively, J. E., Nakajin, S., Shinoda, M., and Hall, P. F., 1986, Amino terminal sequence analysis of human placenta aromatase, *Biochem. Biophys. Res. Commun.* **135**:713–179.

412. Mendelson, C. R., Wright, E. E., Evans, C. T., Porter, J. C., and Simpson, E. R., 1985, Preparation and characterization of polyclonal and monoclonal antibodies against human aromatase cytochrome P-450 (P-450arom), and their use in its purification, *Arch. Biochem. Biophys.* **243**:480–491.

413. Yoshida, N., and Osawa, Y., 1991, Purification of human placental aromatase cytochrome P-450 with monoclonal antibody and its characterization, *Biochemistry* **30**:3003–3010.

414. Corbin, C. J., Graham-Lorence, S., McPhaul, M., Mason, J. I., Mendelson, C. R., and Simpson, E. R., 1988, Isolation of a full-length cDNA insert encoding human aromatase system cytochrome P-450 and its expression in nonsteroidogenic cells, *Proc. Natl. Acad. Sci. USA* **85**:8948–8952.

415. Toda, K., Terashima, M., Kawamoto, T., Sumimoto, H., Yokoyama, Y., Kuribayashi, I., Mitsuuchi, Y., Maeda, T., Yamamoto, Y., Sagara, Y., Ikeda, H., and Shizuta, Y., 1990, Structural and functional characterization of human aromatase P-450 gene, *Eur. J. Biochem.* **193**:559–565.

416. Toda, K., and Shizuta, Y., 1994, Identification and characterization of *cis*-acting regulatory elements for the expression of the human aromatase cytochrome P-450 gene, *J. Biol. Chem.* **269**:8099–8107.

417. Toda, K., and Shizuta, Y., 1993, Molecular cloning of a cDNA showing alternative splicing of the 5′-untranslated sequence of mRNA for human aromatase P-450, *Eur. J. Biochem.* **213**:383–389.

418. Harada, N., 1992, A unique aromatase (P-450$_{AROM}$) mRNA formed by alternative use of tissue-specific exons 1 in human skin fibroblasts, *Biochem. Biophys. Res. Commun.* **189**:1001–1007.

419. Harada, N., Utsumi, T., and Takagi, Y., 1993, Tissue-specific expression of the human aromatase cytochrome P-450 gene by alternative use of multiple exons 1 and promoters, and switching of tissue-specific exons 1 in carcinogenesis, *Proc. Natl. Acad. Sci. USA* **90**:11312–11316.

420. Guengerich, F. P., and Macdonald, T. L., 1993, Sequential electron transfer oxidation reactions catalyzed by cytochrome P-450 enzymes, in: *Advances in Electron Transfer Chemistry*, Vol. 3 (P. S. Mariano, ed.), JAI Press, Greenwich, CT, pp. 191–241.

421. Ortiz de Montellano, P. R., 1986, Oxygen activation and transfer, in: *Cytochrome P-450* (P. R. Ortiz de Montellano, ed.), Plenum Press, New York, pp. 217–271.

422. Cole, P. A., and Robinson, C. H., 1988, A peroxide model reaction for placental aromatase, *J. Am. Chem. Soc.* **110**:1284–1285.

423. Akhtar, M., Calder, M. R., Corina, D. L., and Wright, J. N., 982, Mechanistic studies on C-19 demethylation in oestrogen biosynthesis, *Biochem. J.* **201**:569–580.

424. Cole, P. A., and Robinson, C. H., 1990, Mechanism and inhibition of cytochrome P-450 aromatase, *J. Med. Chem.* **33**:2933–2942.

425. Cole, P. A., and Robinson, C. H., 1991, Mechanistic studies on a placental aromatase model reaction, *J. Am. Chem. Soc.* **113**:8130–8137.

426. Vaz, A. D. N., Roberts, E. S., and Coon, M. J., 1991, Olefin formation in the oxidative deformylation of aldehydes by cytochrome P-450. Mechanistic implications for catalysis by oxygen-derived peroxide, *J. Am. Chem. Soc.* **113:**5886–5887.

427. Brodie, A. M. H., Banks, P. K., Inskster, S. E., Dowsett, M., and Coombes, R. C., 1990, Aromatase inhibitors and hormone-dependent cancers, *J. Steroid Biochem. Mol. Biol.* **37:**327–333.

428. Bhatnagar, A. S., Häusler, A., Schieweck, K., Browne, L. J., Bowman, R., and Steele, R. E., 990, Novel aromatase inhibitors, *J. Steroid Biochem. Mol. Biol.* **37:**363–367.

429. Kellis, J. T., Jr., and Vickery, L. E., 1984, Inhibition of human estrogen synthetase (aromatase) by flavones, *Science* **225:**1032–1034.

430. Kellis, J. T., Jr., Nesnow, S., and Vickery, L. E., 1986, Inhibition of aromatase cytochrome P-450 (estrogen synthetase) by derivatives of α-naphthoflavone, *Biochem. Pharmacol.* **35:**2887–2891.

431. Shimozawa, O., Sakaguchi, M., Ogawa, H., Harada, N., Mihara, K., and Omura, T., 1993, Core glycosylation of cytochrome P-405(arom): Evidence for localization of N terminus of microsomal cytochrome P-450 in the lumen, *J. Biol. Chem.* **268:**21399–21402.

432. Chen, S., Zhou, D., Swiderek, K. M., Kadohama, N., Osawa, Y., and Hall, P. F., 1993, Structure–function studies of human aromatase, *J. Steroid Biochem. Mol. Biol.* **44:**347–356.

433. Zhou, D., Pompon, D., and Chen, S., 1991, Structure–function studies of human aromatase by site-directed mutagenesis: Kinetic properties of mutants Pro-308 > Phe, Tyr-361 > Phe, Tyr-361 > Leu, and Phe-406 > Arg, *Proc. Natl. Acad. Sci. USA* **88:**410–414.

434. Graham-Lorence, S., Khalil, M. W., Lorence, M. C., Mendelson, C. R., and Simpson, E. R., 1991, Structure–function relationships of human aromatase cytochrome P-450 using molecular modeling and site-directed mutagenesis, *J. Biol. Chem.* **266:**11939–11946.

435. Chen, S., and Zhou, D., 1992, Functional domains for aromatase cytochrome P450 inferred from comparative analyses of amino acid sequences and substantiated by site-directed mutagenesis experiments, *J. Biol. Chem.* **267:**22587–22594.

436. Laughton, C. A., 1993, A detailed molecular model for human aromatase, *J. Steroid Biochem. Mol. Biol.* **44:**399–407.

437. Kagawa, N., and Waterman, M. R., 1991, Evidence that an adrenal-specific nuclear protein regulates the cAMP responsiveness of the human *CYP21B* (P450C21) gene, *J. Biol. Chem.* **266:**11199–11204.

438. Higashi, Y., Yoshioka, H., Yamane, M., Gotoh, O., and Fujii-Kuriyama, Y., 1986, Complete nucleotide sequence of two steroid 21-hydroxylase genes tandemly arranged in human chromosome: A pseudogene and a genuine gene, *Proc. Natl. Acad. Sci. USA* **83:**2841–2845.

439. Gitelman, S. E., Bristow, J., and Miller, W. L., 1992, Mechanism and consequences of the duplication of the human C4/P450c21/gene X locus, *Mol. Cell. Biol.* **12:**2124–2134.

440. Higashi, Y., Tanae, A., Inoue, H., and Fujii-Kuriyama, Y., 1988, Evidence for frequent gene conversion in the steroid 21-hydroxylase P-450(C21) gene: Implications for steroid 21-hydroxylase deficiency, *Am. J. Hum. Genet.* **42:**17–25.

441. Helmberg, A., Tabarelli, M., Fuchs, M. A., Keller, E., Dobler, G., Schnegg, I., Knorr, D., Albert, E., and Kofler, R., 1992, Identification of molecular defects causing congenital adrenal hyperplasia by cloning and differential hybridization of polymerase chain reaction-amplified 21-hydroxylase (CYP21) genes, *DNA Cell Biol.* **11:**359–368.

442. Chiou, S. H., Hu, M. C., and Chung, B., 1990, A missense mutation at Ile172→Asn or Arg356→Trp causes steroid 21-hydroxylase deficiency, *J. Biol. Chem.* **265:**3549–3552.

443. New, M. I., 1994, 21-Hydroxylase deficiency congenital adrenal hyperplasia, *J. Steroid Biochem. Mol. Biol.* **48:**15–22.

444. Cali, J. J., and Russell, D. W., 1991, Characterization of human sterol 27-hydroxylase: A mitochondrial cytochrome P-450 that catalyzes multiple oxidation reactions in bile acid biosynthesis, *J. Biol. Chem.* **266:**7774–7778.

445. Cali, J. J., Hsieh, C. L., Francke, U., and Russell, D. W., 1991, Mutations in the bile acid biosynthetic enzyme sterol 27-hydroxylase underlie cerebrotendinous xanthomatosis, *J. Biol. Chem.* **266:**7779–7783.

446. Ishida, N., Aoyama, Y., Hatanaka, R., Oyama, Y., Imajo, S., Ishiguro, M., Oshima, T., Nakazato, H., Noguchi, T., Maitra, U.S., Mohan, V. P., Sprinson, D. B., and Yoshida, Y., 1988, A single amino acid substitution converts cytochrome P45014DM to an inactive form, cytochrome P450SG1: Complete primary structures deduced from cloned DNAs, *Biochem. Biophys. Res. Commun.* **155**:317–323.

447. Aoyama, Y., Yoshida, Y., Nishino, T., Katsuki, H., Maitra, U. S., Mohan, V. P., and Sprinson, D. B., 1987, Isolation and characterization of an altered cytochrome P-450 from a yeast mutant defective in lanosterol 14α-demethylation, *J. Biol. Chem.* **262**:14260–14264.

448. Coles, B. F., Welch, A. M., Hertzog, P. J., Lindsay Smith, J. R., and Garner, R. C., 1980, Biological and chemical studies on 8,9-dihydroxy-8,9-dihydro-aflatoxin B1 and some of its esters, *Carcinogenesis* **1**:79–90.

449. Poulos, T. L., Finzel, B. C., and Howard, A. J., 1987, High-resolution crystal structure of cytochrome P450cam, *J. Mol. Biol.* **195**:687–700.

450. Ravichandran, K. G., Boddupalli, S. S., Hasemann, C. A., Peterson, J. A., and Deisenhofer, J., 1993, Crystal structure of hemoprotein domain of P450 BM-3, a prototype for microsomal P450's, *Science* **261**:731–736.

451. Hasemann, C. A., Ravichandran, K. G., Peterson, J. A., and Deisenhofer, J., 1994, Crystal structure and refinement of cytochrome P450(terp) at 2.3 angstrom resolution, *J. Mol. Biol.* **236**:1169–1185.

452. Poulos, T. L., 1991, Modeling of mammalian P450s on basis of P450cam X-ray structure, *Methods Enzymol.* **206**:11–30.

453. Crespi, C. L., Langenbach, R., and Penman, B. W., 1993, Human cell lines, derived from AHH-1 TK+/– human lymphoblasts, genetically engineered for expression of cytochromes P450, *Toxicology* **82**:89–104.

454. Schmalix, W. A., Mäser, H., Kiefer, F., Reen, R., Wiebel, F. J., Gonzalez, F., Seidel, A., Glatt, H., Greim, H., and Doehmer, J., 1994, Stable expression of human cytochrome P450 1A1 cDNA in V79 Chinese hamster cells and metabolic activation of benzo[a]pyrene, *Eur. J. Pharmacol.* **248**:251–261.

455. Larson, J. R., Coon, M. J., and Porter, T. D., 1991, Alcohol-inducible cytochrome P-450IIE1 lacking the hydrophobic NH2-terminal segment retains catalytic activity and is membrane-bound when expressed in *Escherichia coli*, *J. Biol. Chem.* **266**:7321–7324.

456. Barnes, H. J., Arlotto, M. P., and Waterman, M. R., 1991, Expression and enzymatic activity of recombinant cytochrome P450 17α-hydroxylase in *Escherichia coli*, *Proc. Natl. Acad. Sci. USA* **88**:5597–5601.

457. Kinirons, M. T., Oshea, D., Downing, T. E., Fitzwilliam, A. T., Joellenbeck, L., Groopman, J. D., Wilkinson, G. R., and Wood, A. J. J., 1993, Absence of correlations among 3 putative in vivo probes of human cytochrome-P4503A activity in young healthy men, *Clin. Pharmacol. Ther.* **54**:621–629.

458. Mueller, G. C., and Miller, J. A., 1953, The metabolism of methylated aminoazo dyes. II. Oxidative demethylation by rat liver homogenates, *J. Biol. Chem.* **202**:579–587.

459. Richardson, H. L., Stier, A. R., and Borsos-Nachtnebel, E., 1952, Liver tumor inhibition and adrenal histologic responses in rats to which 3'-methyl-4-dimethylaminoazobenzene and 20-methylcholanthrene were simultaneously administered, *Cancer Res.* **12**:356–361.

460. Miller, E. C., Miller, J. A., and Conney, A. H., 1954, On the mechanism of the methylcholanthrene inhibition of carcinogenesis by 3'-methyl-4-dimethylaminoazobenzene, *Proc. Am. Assoc. Cancer Res.* **1**:32.

461. Miller, E. C., Miller, J. A., Brown, R. R., and MacDonald, J. C., 1958, On the protective action of certain polycyclic aromatic hydrocarbons against carcinogenesis by aminoazo dyes and 2-acetylaminofluorene, *Cancer Res.* **18**:469–477.

462. Conney, A. H., Miller, E. C., and Miller, J. A., 1956, The metabolism of methylated aminoazo dyes. V. Evidence for induction of enzyme synthesis in the rat by 3-methylcholanthrene, *Cancer Res.* **16**:450–459.

463. Nebert, D. W., 1978, Genetic control of carcinogen metabolism leading to individual differences in cancer risk, *Biochimie* **60**:1019–1029.

464. Nebert, D. W., 1989, The *Ah* locus: Genetic differences in toxicity, cancer, mutation, and birth defects, *Crit. Rev. Toxicol.* **20**:153–174.

465. Guengerich, F. P., 1988, Roles of cytochrome P-450 enzymes in chemical carcinogenesis and cancer chemotherapy, *Cancer Res.* **48**:2946–2954.

466. Guengerich, F. P., 1990, Characterization of roles of human cytochrome P450 enzymes in carcinogen metabolism, *Asia Pac. J. Pharmacol.* **5**:327–345.

467. Guengerich, F. P., 1992, Metabolic activation of carcinogens, *Pharmacol. Ther.* **54**:17–61.

468. Guengerich, F. P., 1993, The 1992 Bernard B. Brodie Award Lecture. Bioactivation and detoxication of toxic and carcinogenic chemicals, *Drug Metab. Dispos.* **21**:1–6.

469. Yamazaki, H., Mimura, M., Oda, Y., Inui, Y., Shiraga, T., Iwasaki, K., Guengerich, P. P., and Shimada, T., 1993, Roles of different forms of cytochrome P450 in the activation of the promutagen 6-amino-chrysene to genotoxic metabolites in human liver microsomes, *Carcinogenesis* **14**:1271–1278.

470. Caporaso, N. E., Shields, P. G., Landi, M. T., Shaw, G. L., Tucker, M. A., Hoover, M., Sugimura, H., Weston, A., and Harris, C. C., 1992, The debrisoquine metabolic phenotype and DNA-based assays: Implications of misclassification for the association of lung cancer and the debrisoquine metabolic phenotype, *Environ. Health Perspect.* **98**:101–105.

471. Uematsu, F., Kikuchi, H., Motomiya, M., Abe, T., Sagami, I., Ohmachi, T., Wakui, A., Kanamaru, R., and Watanabe, M., 1991, Association between restriction fragment length polymorphism of the human cytochrome P450IIE1 gene and susceptibility to lung cancer, *Jpn. J. Cancer Res.* **82**:254–256.

472. Kirby, G. M., Wolf, C. R, Neal, G. E., Judah, D. J., Henderson, C. J., Srivatanakul, P., and Wild, C. P., 1993, In vitro metabolism of aflatoxin B1 by normal and tumorous liver tissue from Thailand, *Carcinogenesis* **14**:2613–2620.

Heme–Thiolate Proteins Different from Cytochromes P450 Catalyzing Monooxygenations

DANIEL MANSUY and JEAN-PAUL RENAUD

1. Introduction

Many hemoproteins involving a histidine proximal iron ligand, like hemoglobins, myoglobins, peroxidases, cytochromes, and prostaglandin synthases, have been found in living organisms.[1] On the contrary, only a limited number of hemoprotein families where the heme iron is bound to a cysteinate proximal ligand have been discovered so far. The Fe(II)–CO complex of these hemoproteins is most often characterized by a visible spectrum exhibiting a redshifted Soret peak around 450 nm, which is related to the presence of the very electron-rich cysteinate ligand in *trans* position to CO.[2] This distinctive feature of heme–thiolate proteins has been found so far for only three classes of hemoproteins, the cytochromes P450, nitric oxide synthases, and chloroperoxidase. Another heme–thiolate protein, called protein H450, has been reported, but it remains much less known than the other three. Among these heme–thiolate proteins, the cytochromes P450 responsible for monooxygenation reactions (which will be called "classical P450s" in this chapter) have been extensively studied[3] and their mechanisms of dioxygen activation and substrate hydroxylation are now well established.[3,4] During these last years, some cytochromes P450 that catalyze reactions very different from monooxygenations have been discovered.[5] Moreover, a new class of heme–thiolate proteins, the NO synthases, which are closely related to P450s,[6–8] but which would not belong to the P450 superfamily,[9] has been discovered. This chapter will focus on these "nonclassical P450s" and on NO synthases,

DANIEL MANSUY and JEAN-PAUL RENAUD • Laboratoire de Chimie et Biochimie Pharmacologiques et Toxicologiques, URA 400 CNRS, Université Paris V, F-75270 Paris Cedex 06, France. *Present address of J.-P.R.:* UPR 9004 CNRS, Institut de Génétique et Biologie Moléculaire et Cellulaire, F-67400 Illkirch-Graffenstaden, France.

Cytochrome P450: Structure, Mechanism, and Biochemistry (Second Edition), edited by Paul R. Ortiz de Montellano. Plenum Press, New York, 1995.

with a special emphasis on a comparison of their biochemical and mechanistic properties with those of the two more classical heme–thiolate proteins, the "classical P450s" and the chloroperoxidase from *Caldariomyces fumago*.

2. Common Properties of "Classical P450s" Catalyzing Monooxygenations

More than 200 "classical P450s" have been cloned and sequenced so far[9] but many more P450s of this type are present in living organisms. They all catalyze monooxygenation reactions [Eq. (1)] and belong to the multigene superfamily of cytochromes P450 according to protein sequence homology. Although they greatly differ in their substrate specificity and biological functions, they exhibit some common properties which must be recalled before comparing them to other heme–thiolate proteins.

$$RH + O_2 + 2e^- + 2H^+ \xrightarrow{\text{P450}} ROH + H_2O \tag{1}$$

2.1. The Heme Binding Sequence of "Classical P450s"

The proximal cysteine ligand of the P450 iron is part of a heme binding decapeptide sequence which contains five highly conserved amino acid residues. The structure of this heme binding sequence was established for P450$_{cam}$,[10] P450 BM3,[11] and P450$_{terp}$.[12] It is a decapeptide loop with the cysteine proximal ligand at position 8 within the decapeptide (Cys^{357} in P450$_{cam}$) (Table I). The first amino acid residue is a phenylalanine whose phenyl group seems to protect the reactive cysteinate ligand. The fourth amino acid, a glycine, appears to be responsible for the initiation of the hairpin turn of the loop, whereas the sixth

TABLE I

The Proximal Cysteine-Containing Decapeptide Responsible for Heme Binding in "Classical P450s"[a]

	Cysteinate protection			Hairpin turn initiation		Interaction with heme propionate		Proximal cysteinate		Close to contact to the heme
P450 101 (cam)	F	G	H	G	S	H	L	C^{357}	L	G
P450 102 (BM3)	F	G	N	G	Q	R	A	C	I	G
P450 1A1 (rat)	F	G	L	G	K	R	K	C	I	G
P450 2B1 (rat)	F	S	T	G	K	R	I	C	L	G
P450 2E1 (rat)	F	S	A	G	K	R	V	C	V	G
P450 3A1 (rat)	F	G	N	G	P	R	N	C	I	G
P450 3A4 (human)	F	G	S	G	P	R	N	C	I	G
P450 4A1 (rat)	F	S	G	G	A	R	N	C	I	G
P450 10 (pond snail)	W	G	H	G	A	R	M	C	L	G
P450 (consensus)	F	G/S	x	G	x	R/H	x	C	hy	G

[a]The sequences of all of these P450s are given in Ref. 9. hy refers to hydrophobic amino acid (L, I, or V).

one, a positively charged amino acid residue, a histidine or an arginine, is involved in an ionic (or hydrogen) bond with one of the heme carboxylates. The tenth amino acid, a glycine, is small enough to come in close contact to the heme. Interestingly, three amino acids of this sequence (different from the proximal cysteine), the phenylalanine at position 1 and the glycines at positions 4 and 10, are highly conserved in "classical P450s" reported so far (Table I).[9,13] Moreover, a positively charged residue, from Arg or His, is almost always present at position 6, and a hydrophobic amino acid (most often Leu, Ile, or Val) is always located between the proximal cysteine and the invariant glycine. Finally, position 2 is occupied either by a glycine or by a serine. The role of some of these amino acid residues for heme binding in eukaryotic P450s has been investigated and confirmed by site-directed mutagenesis experiments.[13] All of these data suggest a remarkable conservation of the structure around the fifth axial ligand of iron in "classical P450s."

2.2. The Protein Binding Site for O_2

Another highly conserved domain in "classical P450s" has been found in the central region of helix I that forms an O_2 binding pocket in P450$_{cam}$. In this P450 a localized distortion and widening of helix I between Gly^{248} and Thr^{252} provides a binding pocket for molecular oxygen.[10] Thr^{252} seems to play an important role in stabilizing the P450$_{cam}$ Fe(II)–O_2 complex and in facilitating the heterolytic cleavage of its O–O bond after its one-electron reduction. A high sequence homology has been found between the region surrounding Thr^{252} in P450$_{cam}$ and a region surrounding a corresponding threonine in other "classical P450s" (Table II).[13] In addition to this threonine, glycine and alanine (or glycine) residues are well conserved at positions n-3 and n-4 relative to the threonine. Moreover, hydrophobic residues at positions n-5, n-6, and n-7 are always present, whereas a D or E amino acid is most often present at position n–1. Site-directed mutagenesis experiments performed not only with P450$_{cam}$ but also with eukaryotic P450s like P450 2C2 and P450 2C14 indicate that an OH-containing amino acid residue (Thr or Ser) at position 252 in P450$_{cam}$ or at a corresponding position in the other P450s is often essential for the catalytic activity of those P450s.[13]

TABLE II
The Distal Threonine-Containing Sequence Involved in O_2 Binding in P450s[a]

P450 (consensus)	x	x	x	x	G/A	G	x	D/E	T	x	x	x	x
P450 101 (P. putida)	L	L	L	V	G	G	L	D	T	V	V	N	F
P450 102 (B. megat.)	T	F	L	I	A	G	H	E	T	T	S	G	L
P450 1A1 (human)	D	L	F	G	A	G	F	D	T	V	T	T	A
P450 2A1 (rat)	S	L	F	F	A	G	S	E	T	V	S	S	T
P450 4A1 (rat)	T	F	M	F	E	G	H	D	T	T	A	S	G
P450 51 (S. cerevisiae)	G	V	L	M	G	G	Q	H	T	S	A	A	T
P450 71 (avocado)	D	M	F	S	G	G	T	D	T	T	A	V	T
P450 55 (F. oxysporum)	L	L	L	V	A	G	N	A	T	M	V	N	M
P450 5 (TXAS, human)	I	F	L	I	A	G	Y	E	I	I	T	N	T
P450 5 (TXAS, mouse)	L	F	L	I	A	G	H	E	V	I	T	N	T
P450 74 (AOS)	F	N	S	W	G	G	F	K	I	L	F	P	S

[a]Sequences of all P450s in Ref. 9 except for that of P450 74 in Ref. 42.

2.3. The Catalytic Cycle of P450-Dependent Monooxygenations

All P450-dependent classical monooxygenations seem to involve a common catalytic cycle of dioxygen activation (Fig. 1).[1–5,13,14] The first four intermediates of this cycle have been very well characterized. The resting state of P450 exists as a mixture of a hexacoordinate low-spin Fe(III) state with a water molecule *trans* to the proximal cysteinate ligand

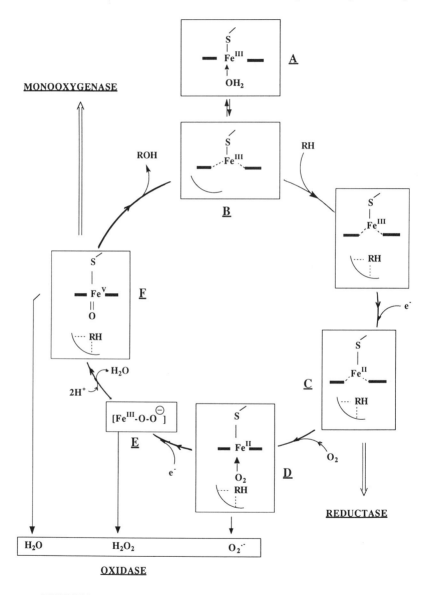

FIGURE 1. Catalytic cycle of cytochrome P450-dependent monooxygenases.

(*A*) and a pentacoordinate high-spin Fe(III) state (*B*). Binding of substrates to a protein site close to the heme generally shifts this equilibrium toward the pentacoordinate state which is more easily reduced to the pentacoordinate P450 Fe(II) high-spin state (*C*). This ferrous complex is able to bind several ligands including O_2 (*D*). The complexes derived from a one-electron reduction of P450 Fe(II)–O_2 are much less known, as they could not be characterized by any spectroscopic technique because of their very short lifetime. The first intermediate derived from this one-electron reduction should be a ferric peroxide complex (*E*), either Fe(III)–O–O⁻ or Fe(III)–OOH if the electron transfer is accompanied by a protonation of the reduced dioxygen molecule. The O–O bond of this intermediate may be cleaved in a homolytic or heterolytic manner. It is now generally accepted that the P450 active species responsible for most monooxygenation reactions is a high-valent iron–oxo complex derived from heterolytic cleavage of the O–O bond of the P450 Fe(III)–OOH intermediate. This high-valent iron–oxo species (*F*) is formally equivalent to the intermediate involved in the catalytic cycle of horseradish peroxidase (HRP), called compound I, which is a (porphyrin radical cation) Fe(IV)=O complex.[3] The iron–oxo nature of the P450 active species is suggested by many data coming from mechanistic studies of P450-dependent monooxygenations of various substrates[1-5] as well as from studies performed on iron porphyrin model systems.[1,3,14] It has been written as a (porphyrin radical cation) Fe(IV)=O species or an Fe(V)=O or Fe(IV)–O· species, this last electronic structure well expressing the free radical reactivity of this reactive intermediate.

The transfer of one oxygen atom from O_2 to substrates catalyzed by all "classical P450s" appears to be performed by this P450 Fe(V)=O species. It is a highly electrophilic compound that reacts with electron-rich centers of substrates like double bonds of alkenes or arenes, lone pairs of heteroatoms, or electrons involved in C–H bonds. This results in the epoxidation of alkenes, hydroxylation of aromatic rings, *N*-oxidation of amines, *S*-oxidation of thioethers, and hydroxylation of alkanes. The P450 iron–oxo species appears reactive enough to oxidize even the sulfur atom of thiophene compounds which is much less reactive than the sulfur atom of thioethers.[15-17] The derived thiophene sulfoxides are primary reactive metabolites that may further react with glutathione and protein nucleophiles.

More recently, "classical P450s" have been found to catalyze reactions different from simple monooxygenations, like dehydrogenations, oxidative deformylations, and dehydrations of aldoximes.

2.4. Catalytic Activities of "Classical P450s" Different from Simple Monooxygenations

2.4.1. P450s as Dehydrogenases

Alkyl groups of a few substrates have been found to undergo a P450-dependent dehydrogenation. Mechanistic studies have been performed in the case of valproic acid and testosterone [Eqs. (2) and (3)].[18-20] In fact, the P450-dependent oxidation of these substrates simultaneously leads to a hydroxylated metabolite and a dehydrogenated metabolite. The first step is common for the hydroxylation and dehydrogenation reactions and involves a free radical abstraction of a hydrogen atom from the substrate by P450 Fe(V)=O. The carbon-centered free radical derived from the substrate has two possible

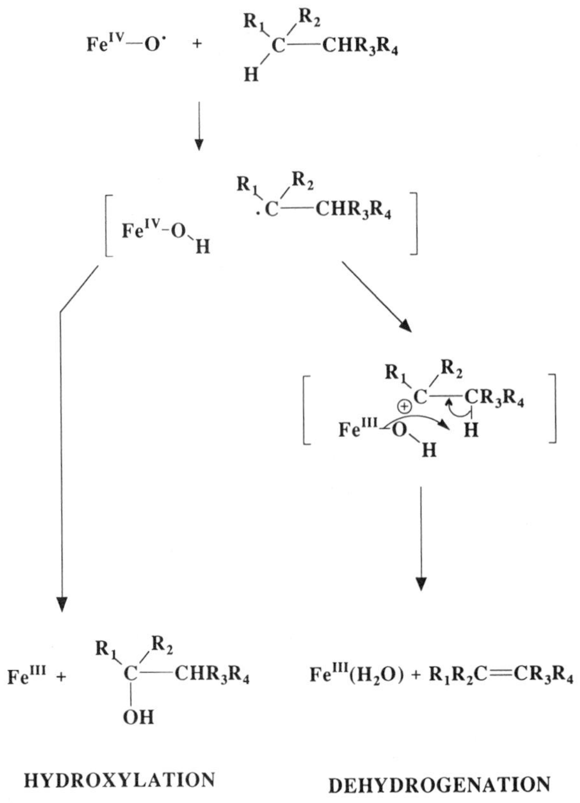

FIGURE 2. Common mechanism for P450-catalyzed hydroxylations and dehydrogenations.

fates in the presence of P450 Fe(IV)–OH also formed in the first step. The first one is an oxidative transfer of the OH ligand from P450 Fe(IV) to the free radical leading to a hydroxylated metabolite, as in classical monooxygenations. The second one is the elimination of a second hydrogen atom leading to an alkene metabolite. This reaction could be viewed as an electron transfer from the free radical to P450 Fe(IV)–OH followed by the loss of a β-proton from the resulting carbocation intermediate (Fig. 2). It has also been proposed that this second hydrogen atom removed in the dehydrogenation could be abstracted by Fe(IV)–OH acting as an ·OH radical bound to P450 Fe(III).[18–20]

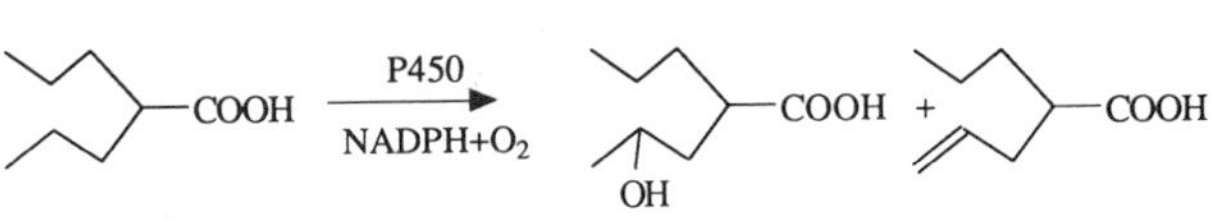

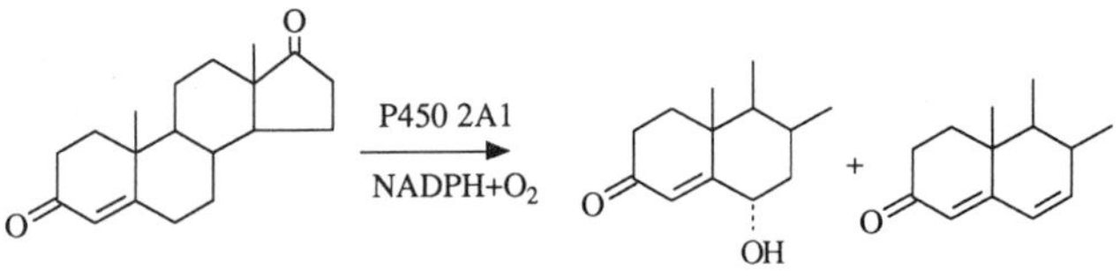

2.4.2. P450-Catalyzed Oxidative Deformylations

Several oxidative demethylations involved in the biosynthesis of steroids have long been known to be catalyzed by P450s.[21] The last step of these demethylations is an oxidative deformylation of an intermediate aldehyde leading to the formation of a double bond adjacent to the carbon bearing the CHO moiety and to the release of formate. More recently, such a C–C bond cleavage was found to occur on simpler substrates like cyclohexane carboxaldehyde with the formation of cyclohexene and formate [Eq. (4)].[22] Several mechanisms have been proposed for this deformylation reaction. However, most recent proposals favor the involvement of the P450 Fe(III)–O–O⁻ [or P450 Fe(III)–OOH] intermediate of the P450 catalytic cycle (intermediate E in Fig. 1) as the active species in such reactions.[21–24] Contrary to P450 Fe(V)=O, P450 Fe(III)–O–O⁻ should be a nucleophilic entity able to add to the carbonyl group of aldehydes with formation of an intermediate iron(III)–alkylperoxo complex (Fig. 3). Fragmentation of this complex may occur either by a concerted electrocyclic pathway or by a homolytic pathway starting with a homolytic O–O bond cleavage. Both mechanisms lead to the release of formate, the formation of an alkene, and regeneration of P450 in its resting state [Fe(III)–OH or Fe(III)(H₂O) after protonation]. The arguments which favor the involvement of P450 Fe(III)–OOH over P450 Fe(V)=O as the active species in this kind of reaction have been discussed in a recent review article.[21]

$$\text{cyclohexane-CHO} \quad \xrightarrow[\text{NADPH+O}_2]{\text{P450 LM}_2} \quad \text{cyclohexene} \quad + \ HCOOH \tag{4}$$

2.4.3. "Classical P450s" as Dehydrases: P450-Catalyzed Dehydration of Aldoximes

It has been recently reported that rat liver P450 Fe(II) catalyzed the dehydration of n-butyraldoxime to butyronitrile.[25] This reaction is not catalyzed by P450 Fe(III); it requires P450 in its reduced state though it does not consume any reducing agent. More recently, it has been shown that this reaction is quite general and occurs with several alkylaldoximes (heptanaldoxime, butyraldoxime, and phenylacetaldoxime) and arylaldoximes (benzaldoxime and 4-hexyloxy-benzaldoxime) [Eq. (5)].[26]

$$\overset{R}{\underset{H}{\diagdown}}C=N-OH \quad \xrightarrow{\text{rat liver P450 Fe(II)}} \quad R-C{\equiv}N \ + H_2O \tag{5}$$

R = n-propyl, n-hexyl, PhCH₂, Ph, 4-hexyloxy-Ph

Such P450-dependent dehydrations have been found to occur with Z-aldoximes but not with the corresponding E-aldoximes [Eq. (6)], as demonstrated in the case of benzaldoxime and 4-hexyloxy-benzaldoxime.[26] Quite interestingly, Z-aldoximes strongly interact with P450 Fe(II) with formation of a 442-nm-absorbing complex, and consumption of the aldoximes, disappearance of the P450 complexes, and formation of the corresponding nitriles are concomitant. On the contrary, E-aldoximes are not dehydrated under identical conditions and do not form 442-nm-absorbing complexes. These 442-nm-absorbing complexes thus appear as intermediates in the dehydration of Z-aldoximes. They presumably derive from the binding of Z-aldoximes to P450 Fe(II) via their nitrogen atom, the position of their Soret peak being similar to that of P450 Fe(II) complexes with nitrogen-containing

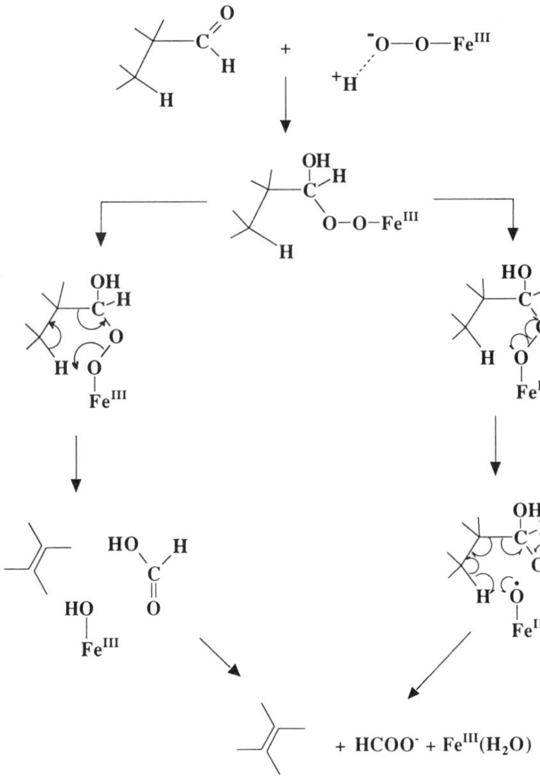

FIGURE 3. Possible mechanisms for P450-dependent oxidative deformylations of aldehydes (from Ref. 22).

ligands such as imidazoles or pyridines. Such a structure explains why only the oximes bearing a hydrogen substituent in position *trans* to the OH group, namely the Z-aldoximes, lead to 442-nm-absorbing complexes. The corresponding E-aldoximes and all ketoximes bear an alkyl or aryl substituent in that position that prevents their approach to the porphyrin ring.[26]

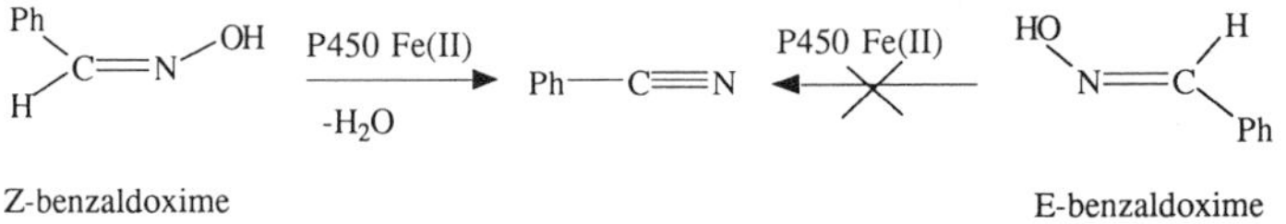

$$\text{(6)}$$

Z-benzaldoxime   E-benzaldoxime

Spectral studies have also shown that both Z- and E-aldoximes interact with P450 Fe(III) with formation of classical "reverse type I" spectra which should correspond to the binding of both kinds of aldoximes to P450 Fe(III) by their oxygen atom. After reduction of P450 Fe(III), only Z-aldoximes can bind to P450 Fe(II) by their nitrogen atom. This binding appears to be essential to start the dehydration process. Two mechanisms may be proposed to explain the role of P450 Fe(II) in the dehydration of Z-aldoximes (Fig. 4).[26] In the first one, coordination of the aldoxime nitrogen to P450 Fe(II) could lead to a charge

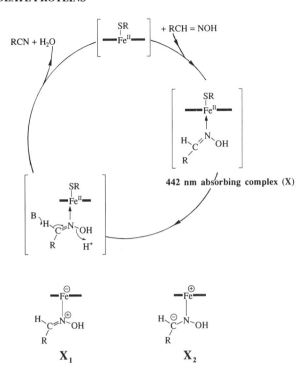

FIGURE 4. Catalytic cycle of the P450 Fe(II)-dependent dehydration of Z-aldoximes (from Ref. 26). X_1 and X_2 are possible mesomeric forms of the 442-nm-absorbing complex and starting points for the two possible dehydration mechanisms given in the text.

transfer from the C=N moiety to the iron that would make the CHR hydrogen more acidic and would facilitate its abstraction by a basic amino acid residue of the active site. This step would be followed by a loss of the oxime OH group presumably assisted by its protonation. The second mechanism is based, on the contrary, on charge transfer from the iron to the C=N bond, a direction of electron flow that appears more likely because of the high electron richness of P450 Fe(II) related to its ferrous state and the presence of an axial cysteinate ligand. Such a charge transfer should lead to the appearance of a negative charge on the aldoxime carbon and subsequent elimination of the aldoxime OH group, presumably after protonation. This would result in the intermediate formation of a complex that should be written, at least formally, as a P450 Fe(IV)–N=CHR complex involving a very acidic hydrogen atom β to the electrophilic iron (IV) species. Its loss as a proton would give the nitrile and would regenerate P450 Fe(III). Such a transient formation of a P450 Fe(IV) complex is not unlikely in P450 chemistry as the formation of P450 Fe(IV) complexes is well accepted in many P450 reactions.[3] Although more data are necessary to differentiate between these two mechanisms, the faster dehydration observed for alkylaldoximes than for arylaldoximes,[26] which can be explained by a faster cleavage of the N–O bond of alkylaldoximes because of the greater instability of their carbanionic intermediate, would seem to favor the second mechanism. In any case, whatever the mechanism may be, dehydration of aldoximes involves P450 Fe(II) as a key active species and not P450 Fe(II)O$_2$ or any subsequent intermediate of the catalytic cycle of P450-dependent monooxygenations.

The possible biological roles of this new P450-catalyzed reaction, the dehydration of aldoximes, remain to be established. However, 4-hydroxyphenyl acetaldoxime is an intermediate in the biosynthesis of cyanogenic glucosides from tyrosine in higher plants. One step in this biosynthetic pathway is the dehydration of Z-4-hydroxy-phenyl acetaldoxime to the corresponding nitrile.[27] Transformation of tyrosine into cyanogenic glucosides is catalyzed by microsomes of sorghum seedlings in the presence of NADPH, and the involvement of at least one cytochrome P450 has been proposed.[27] Further studies are necessary to determine the importance of the P450-catalyzed dehydration of aldoximes in plants or other living organisms.

3. Cytochromes P450 Catalyzing Reactions Different from Monooxygenations

The heme–thiolate proteins described in this section belong to the P450 superfamily because of the homology of their amino acid sequence with those of "classical P450s." However, they catalyze reactions very different from monooxygenations, dehydrogenations, or oxidative C–C bond cleavages. In fact, they do not even catalyze oxidation reactions.

3.1. P450s as Isomerases: Prostacyclin and Thromboxane Synthases[28]

Some years ago, it was reported that prostacyclin and thromboxane synthases (Fig. 5), two enzymes involved in the isomerization of prostaglandin H$_2$ (PGH$_2$) into either prostacyclin (PGI$_2$), a potent inhibitor of platelet aggregation and vasodilator of smooth muscle cells, or thromboxane (TXA$_2$), a potent proaggregatory and vasoconstricting agent, are P450 proteins.[29,30] Prostacyclin synthase (PGIS) was first purified from pig aorta microsomes.[29] It is a 49.2-kDa ferric hemoprotein exhibiting visible (417, 532, and 568 nm) and EPR (g = 1.9, 2.25, and 2.46) spectra similar to those reported for low-spin P450 Fe(III). In fact, these g values are characteristic for heme–thiolate proteins.[3] Moreover, reduction by dithionite in the presence of CO leads to a complex exhibiting a Soret peak at 451 nm, which confirms the presence of a cysteinate iron ligand. A 52-kDa hemoprotein with very similar characteristics was purified from bovine aorta.[31] More recently, the cDNAs for bovine and human endothelial PGIS have been cloned.[32,33] The protein sequence of human PGIS exhibits an 88.2% identity to the bovine enzyme, 33.8% to human cholesterol-7α-hydroxylase, and only 15.4% to human thromboxane synthase.[33] PGIS is widely expressed in human tissues and is particularly abundant in ovary, heart, skeletal muscle, lung, and prostate.[33]

Thromboxane synthase (TXAS) was first purified from human platelets.[30] It is also a ferric hemoprotein with an apparent molecular mass of 59 kDa and a visible spectrum characterized by peaks at 418, 537, and 570 nm. It is a heme–thiolate protein as shown by the EPR spectrum of its ferric state, with typical g = 1.9, 2.25, and 2.4 values, and by the

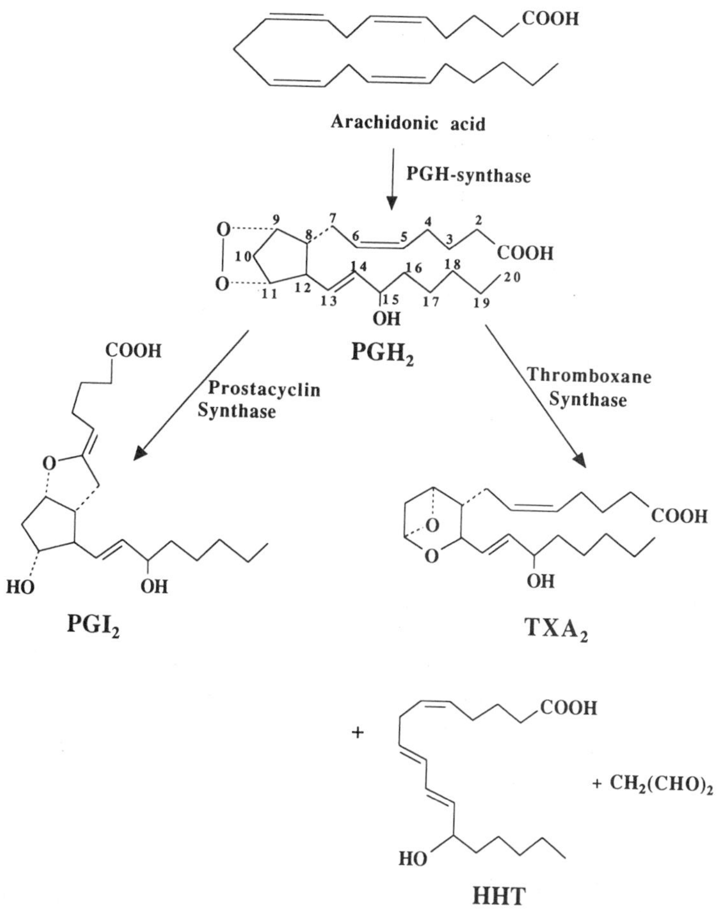

FIGURE 5. Biosynthesis of prostacyclin (PGI_2) and thromboxane A_2 (TXA_2) from arachidonic acid.

Soret peak of its Fe(II)–CO complex at 450 nm. The purified enzyme converts PGH_2 to thromboxane A_2 (or its hydrolyzed product thromboxane B_2) and to malondialdehyde and 12(S)-hydroxy-5(Z),8(E),10(E)-heptadecatrienoic acid (HHT) (Figs. 5, 8) in a 1:1:1 ratio, with an overall molecular activity of about 50 turnovers per second.[28] Polymerase chain reaction techniques have been used to isolate a cDNA clone containing the entire protein coding region of TXAS from a human lung[35] or platelet[34] cDNA library. A protein structure with 533 amino acids (60,487 Da) was deduced from the sequence of the platelet cDNA. It has a maximal 36% sequence homology with classical P450s.[34]

Several compounds containing a pyridine or imidazole moiety strongly bind to TXAS Fe(III).[36] Moreover, visible spectroscopic studies of the interaction of two PGH_2 analogues, 9,11-epoxymethano-PGH_2 and 11,9-epoxymethano-PGH_2, with TXAS and PGIS strongly suggest that PGH_2 is bound to TXAS Fe(III) by its peroxidic oxygen atom at C-9.

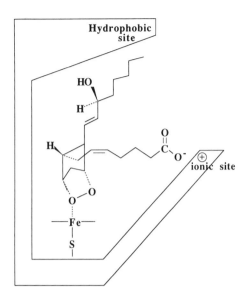

FIGURE 6. Active site topology proposed for thromboxane synthase (TXAS) (from Ref. 36).

On the contrary, PGH_2 appears to bind to PGIS Fe(III) by its peroxidic oxygen atom at C-11.[28] The geometry of the active site of TXAS was deduced from its interactions with a series of pyridine and imidazole inhibitors and with the aforementioned carbon analogues of PGH_2. It appears to involve both an ionic binding site for the carboxylate of PGH_2 and a hydrophobic site able to accommodate the PGH_2 alkyl chain (Fig. 6).[28,36]

3.1.1. Catalytic Cycle of PGIS and TXAS[28,37]

Contrary to P450s that catalyze monooxygenations, TXAS and PGIS catalyze the isomerization of PGH_2 (Fig. 5) without consumption of NADPH or NADH and without any need for electron transfer proteins such as P450 reductase. Moreover, they are not reducible by NADPH or NADH.[28] Thus, under physiological conditions, they should only exist in their ferric state (*A* or *B* of Fig. 1). In fact, this ferric state appears as the active species responsible for the isomerization reaction. Binding of the PGH_2 peroxidic oxygen atom at C-11 to PGIS Fe(III) is the first step of the PGIS catalytic cycle. The second step is an electron transfer from PGIS Fe(III) to the O–O peroxide moiety with formation of P450 Fe(IV) and an alkoxy radical at C-9 (Fig. 7). This radical is well located to add to the C-5–C-6 double bond leading to the bicyclic structure of PGI_2. The final step is a one-electron oxidation of the carbon radical at C-5 by P450 Fe(IV) and loss of the C-6 proton which is in the β-position to the intermediate carbocation. This last step is very similar to the last step of the P450-dependent dehydrogenation of some substrates (Fig. 2) as both reactions involve a one-electron oxidation of a substrate-derived free radical by P450 Fe(IV) followed by the loss of a proton β to the carbocationic center.

The first step of the catalytic cycle proposed for TXAS also involves a one-electron transfer from TXAS Fe(III) to the O–O bond of PGH_2 (Fig. 8). However, as PGH_2 is bound

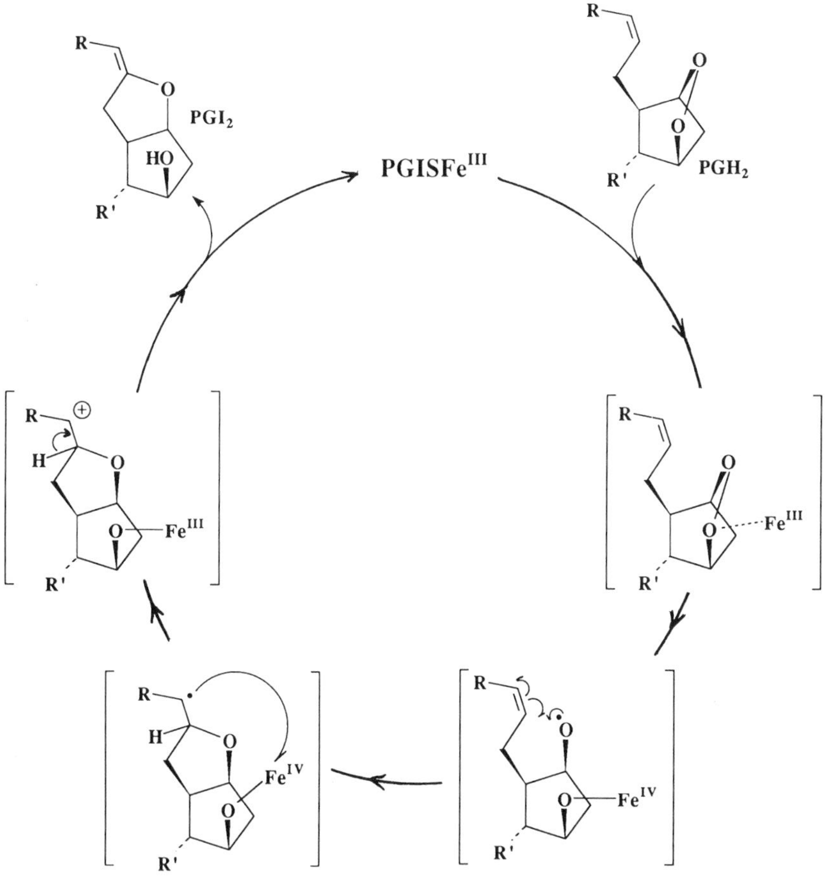

FIGURE 7. Proposed catalytic cycle for prostacyclin synthase (PGIS) (from Ref. 28).

to TXAS Fe(III) via its peroxidic oxygen atom at C-9,[37] this electron transfer leads to P450 Fe(IV) and an alkoxy radical at C-11. This radical undergoes a β-scission with formation of an aldehyde function and an allylic radical at C-12. This allylic radical has two possible fates. In the first one, it is oxidized by P450 Fe(IV) to the corresponding cation whose reaction with the aldehyde oxygen atom at C-11 leads to a cationic heterocycle that gives TXA$_2$ by participation of the oxygen atom at C-9 (Fig. 8). The second reaction that may occur at the level of the allylic radical at C-12 is a new β-scission leading to HHT and malondialdehyde, the two products observed in addition to TXA$_2$ during reaction of PGH$_2$ with TXAS (Fig. 8). In fact, this fragmentation of the intermediate C-12 radical leading to malondialdehyde is the only reaction observed with substrates slightly different from PGH$_2$, like PGH$_1$ which lacks the C-5–C-6 double bond, or 8-iso-PGH$_2$. The cyclization leading to TXA$_2$ only competes with this fragmentation because of a fast one-electron oxidation of the C-12 radical by P450 Fe(IV). If one disturbs the interaction between the

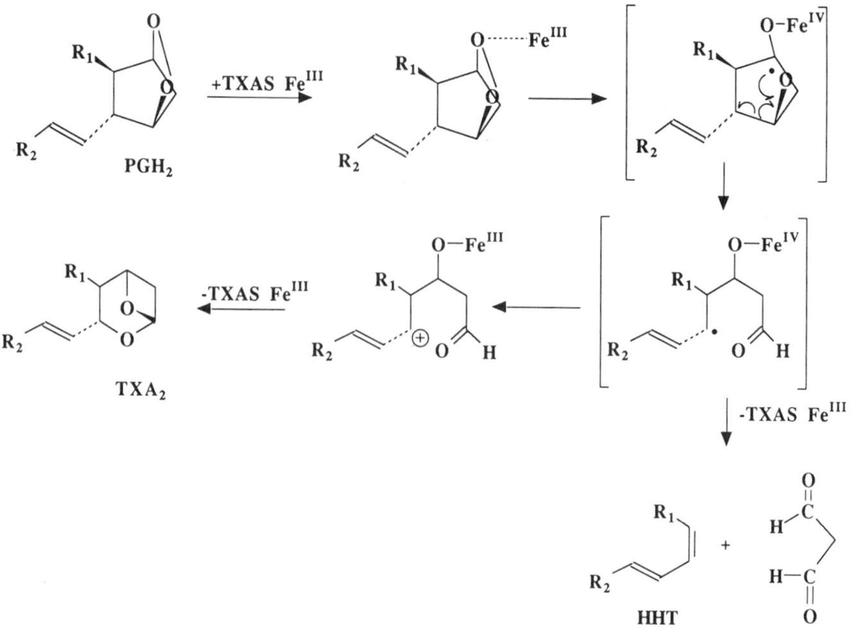

FIGURE 8. Possible mechanism of thromboxane synthase (TXAS) (from Ref. 28).

C-12 radical and P450 Fe(IV), as occurs with substrates different from PGH_2, only the fragmentation of the C-12 radical is observed. Interestingly, in reactions of PGIS with substrates like PGH_1 or 8-iso-PGH_2 for which the alkoxy radical at C-9 cannot undergo fast intramolecular cyclization, as in the case of PGH_2, fragmentation of the radical occurs with the eventual formation of malondialdehyde and a C_{17} fatty acid.[28]

These catalytic cycles of PGHS and TXAS, in contrast to "classical P450s," do not involve iron–O_2 or iron–oxo complexes. They only use Fe(III) as an active species and take profit of the ability of P450 Fe(III) to transfer an electron to peroxidic O–O bonds and the propensity of P450 to exist transiently in the Fe(IV) state.

Based on protein sequence homology, PGIS and TXAS appear to be two new families of P450s (called P450 5 for TXAS[9]). They exhibit a decapeptide sequence similar to the P450 heme binding sequence (Table III), with a glycine at positions n-6, n-4, and n+2 relative to the putative proximal cysteine, and the well-conserved phenylalanine at position n-7 for TXAS (human and mouse). However, it is noteworthy that this phenylalanine residue is replaced by a tryptophan in the case of PGIS (human and bovine).[32,33]

3.2. Allene Oxide Synthase: P450 as a Dehydrase

The conversion of specific lipid hydroperoxides to allene oxides that act as signaling molecules has been found in many plants as well as in algae, corals, and starfish.[38] The biosynthesis of jasmonic acid, a growth hormone and signaling molecule in plant defense, is performed in flaxseed, as in many plants, in several steps (Fig. 9).[39] The first one is a

TABLE III
Proximal Cysteine-Containing Peptide Involved in Heme Binding in P450s: Comparison with Other Heme–Thiolate Proteins[a]

	P450 (consensus)	F	G/S	×	G	×	R/H	×	C	hy	G
"Classical"	P450 101 (P. putida)	F	G	H	G	S	H	L	C^{357}	L	G
450s	P450 102 (B. megat.)	F	G	N	G	Q	R	A	C	I	G
	P450 1A1 (human)	F	G	M	G	K	R	K	C	I	G
	P450 1A1 (trout)	F	G	M	D	K	R	R	C	I	G
	P450 2A1 (rat)	F	S	T	G	K	R	F	C	L	G
	P450 3A1 (rat)	F	G	N	G	P	R	N	C	I	G
	P450 4A1 (rat)	F	S	G	G	A	R	N	C	I	G
	P450 10 (pond snail)	W	G	H	G	A	R	M	C	L	G
	P450 51 (S. cerevisiae)	F	G	G	G	R	H	R	C	I	G
	P450 71 (avocado)	F	G	A	G	R	R	G	C	P	G
Other	P450 5 (TXAS, human)	F	G	A	G	P	R	S	C	L	G
P450s	PGIS (human)	W	G	A	G	H	N	H	C	L	G
	PGIS (bovine)	W	G	A	G	H	N	Q	C	L	G
	P450 55 (F. oxysporum)	F	G	F	G	D	H	R	C	I	A
	P450 74 (AOS)	P	S	V	A	N	K	Q	C	A	G
NOSs	NOS (mm)	W	R	N	A	P	R	—	C^{194}	I	G
	NOS (rvsmc)	W	R	N	A	P	R	—	C^{197}	I	G
	NOS (rh)	W	R	N	A	P	R	—	C^{197}	I	G
	NOS (hh)	W	R	N	A	P	R	—	C^{200}	I	G
	NOS (hc)	W	R	N	A	P	R	—	C^{200}	I	G
	NOS (huvec)	W	R	N	A	P	R	—	C^{184}	V	G
	NOS (baec)	W	R	N	A	P	R	—	C^{186}	V	G
	NOS (rb)	W	R	N	A	S	R	—	C^{415}	V	G
	NOS (mb)	W	R	N	A	S	R	—	C^{415}	V	G
	NOS (hb)	W	R	N	A	S	R	—	C^{419}	V	G
Chloroperoxidase (CPO)		P	T	D	S	R	A	P	C	P	A

[a]The amino acid sequences of all of the "classical P450s" are derived from Ref. 9, those of P450 5 (human), PGIS (bovine), PGIS (human), P450 55A1, and P450 74 from Refs. 32, 33, 34, 42, and 48. The sequences of NOS from murine macrophages (mm), rat vascular smooth muscle cells (rvsmc), rat hepatocytes (rh), human hepatocytes (hh), human chondrocytes (hc), human umbilical vein endothelial cells (huvec), bovine aortic endothelial cells (baec), rat brain (rb), mouse brain (mb), and human brain (hb) are derived from Refs. 53, 55, 57, 58, 59, 60, 62, 63, 64, and 65, respectively. The sequence of chloroperoxidase is derived from Ref. 80. hy refers to hydrophobic amino acid (L, V, or I in general).

lipoxygenase-dependent oxidation of linolenic acid, and the second one, the dehydration of the derived hydroperoxide to an intermediate allene oxide. Cyclization of this intermediate leads to a cyclopentenone which is then transformed into jasmonic acid. Flaxseed allene oxide synthase (AOS) has been purified recently;[40] it is a 55-kDa hemoprotein that catalyzes the dehydration of 13-(S)-hydroperoxy linoleic acid to the corresponding allene oxide with a very high rate (around 1200 turnovers per second). The UV-visible spectrum of purified AOS exhibits a Soret peak at 392 nm as expected for a high-spin P450 Fe(III) (intermediate B of Fig. 1). Reduction of AOS by dithionite leads to a shift of the Soret peak at 407 nm and binding of CO to the appearance of a peak at 450 nm as expected for a heme–thiolate protein. As found for PGIS and TXAS, AOS is not reduced by NADPH or

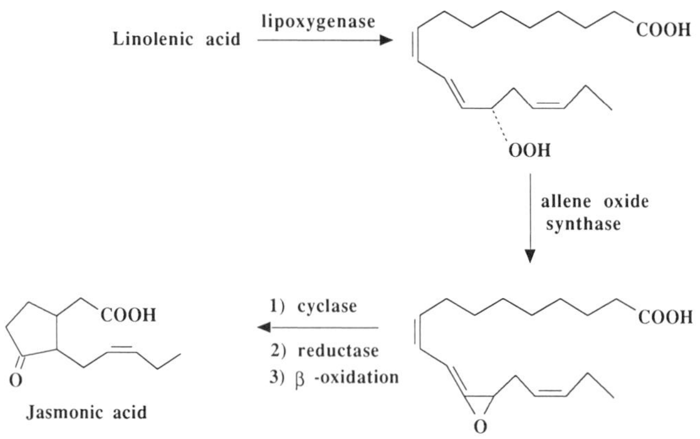

FIGURE 9. Biosynthesis of jasmonic acid from linolenic acid in plants.

NADH and does not appear to be associated with electron transfer proteins.[40] An unusual feature of flaxseed AOS is its low affinity (K~3 × 10^3 M^{-1}) for CO when compared to classical P450s (K~10^5 to 10^6 M^{-1}).[41] The primary structure of flaxseed AOS has been recently deduced from its cDNA.[42] The encoded protein of 536 amino acids has segments at the C-terminus that match certain well-conserved regions in classical P450s. Overall, it has less than 25% sequence similarity to other proteins and is most closely related to P450s with members of the CYP 3A family and thromboxane synthase at the top of the list. Flax AOS represents the first member of a P450 gene family designated as CYP74. Flax AOS cDNA encodes an N-terminal 58-amino-acid signal sequence characteristic of a mitochondrial or chloroplast transit peptide. Therefore, flax AOS is a type I P450 most likely located in chloroplasts.[42]

3.2.1. Catalytic Cycle of Allene Oxide Synthase (Fig.10)

As AOS Fe(III) is not reducible under physiological conditions, the species that interacts with the hydroperoxide substrate should be the resting ferric state itself, as in the case of PGIS and TXAS. The first step of the catalytic cycle should be a one-electron transfer from AOS Fe(III) to the O–O bond of 13-(S)-hydroperoxy linoleic acid resulting in the formation of AOS Fe(IV)–OH and a substrate alkoxy radical that may undergo an intramolecular cyclization to give an allylic free radical (Fig. 10). The second step is an oxidation of this allylic free radical by AOS Fe(IV) leading, at least formally, to an allylic cation which gives the allene oxide product by loss of a β-proton. This last step is remarkably similar to the last step of P450-catalyzed dehydrogenations (Fig. 2) and of the PGIS catalytic cycle (Fig. 7). In the three reactions, a P450 Fe(IV)–OR intermediate oxidizes a carbon-centered free radical with a final loss of a proton β to the carbon center. This mechanism has been recently confirmed by a study of the reaction of flaxseed AOS with another hydroperoxide, 8R-hydroxy-15S-hydroperoxyeicosa-5,9,11,13,17-pentaenoic acid.[43] This hydroperoxide is converted not only to an allene oxide but also to epoxydiols, the main product being 8R,13R-dihydroxy-14R,15S-epoxyeicosa-

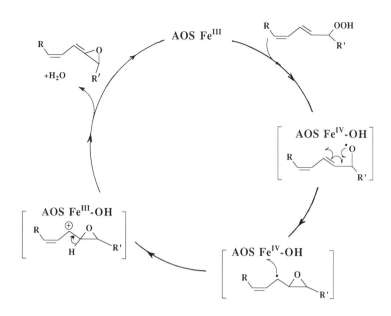

FIGURE 10. Catalytic cycle of allene oxide synthase (AOS) from flaxseed (from Ref. 43).

5Z,9E,11Z,17Z-tetraenoic acid. Formation of this diol occurs with complete retention of the two oxygen atoms of the hydroperoxide group of the starting substrate. These data strongly suggest a first step of this reaction identical to that found for the 13S-hydroperoxide of linoleic acid (Fig. 10), involving the formation of an allylic carbon-centered radical and AOS Fe(IV)–OH (Fig. 11). Now, the products derived from AOS-catalyzed transformation of the pentaenoic acid hydroperoxide come from two possible reactions of the intermediate free radical: (1) a one-electron oxidation by AOS Fe(IV)–OH followed by a β-proton loss leading to an allene oxide, and (2) an oxidative OH ligand transfer from AOS Fe(IV)–OH to the radical (oxygen rebound mechanism) which gives the isolated diol. This mechanism is very similar to that already discussed for the formation of dehydrogenated and hydroxylated products on oxidation of alkyl groups by "classical P450s" (Fig. 2).

Cytochromes P450 similar to flaxseed AOS which are not reducible by NADPH and exhibit a low affinity for CO appear to constitute a major portion of microsomal P450s in a variety of plant tissues, particularly from monocot species.[41] Such AOSs are probably involved in the biosynthesis of signaling molecules important for plant growth regulation and defense. In the animal kingdom, allene oxides derived from eicosatetraenoic acid may be involved in the biosynthesis of prostaglandins and clavulones,[44,45] and in other functions unknown so far.[38]

3.3. A P450 from *Fusarium oxysporum* Acts as a Nitric Oxide Reductase

A fungus, *Fusarium oxysporum*, is able to perform the anaerobic reduction of nitrate or nitrite to N_2O. The dissimilatory reduction of nitrate to nitrite is an energetically

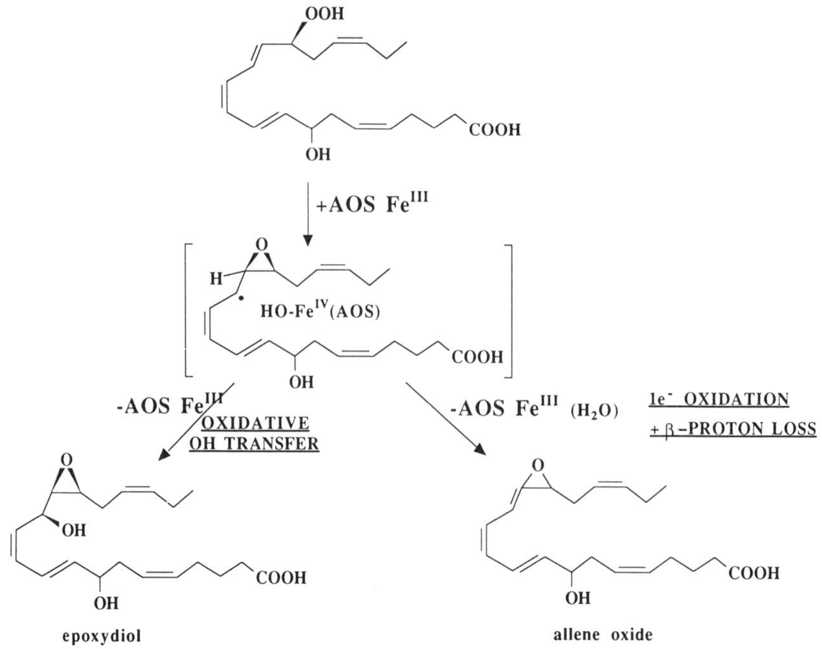

FIGURE 11. Formation of an allene oxide and an epoxydiol in the AOS-catalyzed transformation of 8R-hydroxy-15S-hydroperoxy eicosa-5,9,11,13,17-pentaenoic acid (from Ref. 43).

favorable process in this fungus, whereas the reduction of nitrite to N_2O might be energy consuming and function as a detoxification process.[46] A unique nitrate (nitrite)-inducible cytochrome P450, previously called P450 dNIR, is involved in this reduction of NO_2^- to N_2O.[46] It has been purified to homogeneity and found to be a very potent catalyst for the reduction of NO to N_2O, with maximum rates as high as 525 turnovers/sec.[47] For that reason, it was recently called P450$_{nor}$ ("nor" for $nitric$ $oxide$ $reductase$).[47] NADH, but not NADPH, is utilized as the electron donor and the stoichiometry of the reactants and product is 2:1:1 for NO: NADH:N_2O [Eq. (7)].[47]

$$2NO + NADH + H^+ \xrightarrow{\text{P450 nor}} N_2O + NAD^+ + H_2O \qquad (7)$$

A very intriguing feature of P450$_{nor}$ is its ability to catalyze the reduction of NO by NADH without the help of any electron-carrying mediator or P450 reductase.[47] A cDNA clone for this P450 was isolated recently.[48] Sequence determination revealed a polypeptide of 403 amino acids exhibiting higher homologies toward soluble bacterial P450s than toward eukaryotic P450s including those from yeasts and fungi. For instance, it exhibits about 40% sequence identity with a $Streptomyces$ P450 (P450 SU$_2$) but less than 20% identity with P450 14DM from $Saccharomyces$ $cerevisiae$, P450 bphA from $Aspergillus$ $niger$, and most P450s from mammals. Moreover, its amino-terminus region contains neither the signal-like, hydrophobic domain that is commonly observed in microsomal P450s nor the tagging prosequence that is essential for addressing to mitochondria. P450$_{nor}$

thus appears to be the first soluble P450 derived from eukaryotes. It has been classified into a new, fungal P450 gene family and called P450 55A1, the corresponding gene being *CYP 55*.[47]

3.3.1. Catalytic Cycle of NO Reduction by P450$_{nor}$

Native P450$_{nor}$ shows a visible spectrum which is characteristic of a mixture of high-spin and low-spin P450 Fe(III).[46] Addition of NO leads to a new spectrum similar to those previously reported for P450 Fe(III)–NO, indicating that NO does bind to the ferric state of P450$_{nor}$. Even under anaerobic conditions, P450$_{nor}$Fe(III) is not reduced by NADH. This is consistent with the lack of NADH oxidase activity of this P450. However, NADH reduces the Fe(III)–NO complex very rapidly.[47] All of these results suggest that the first two steps of the catalytic cycle of NO reduction by P450$_{nor}$ are (1) the binding of NO to P450 Fe(III), and (2) the one-electron reduction of P450 Fe(III)–NO by NADH (Fig. 12). The P450$_{nor}$-dependent reduction of NO by NADH is strongly inhibited by O_2 as well as by other iron ligands like CN^- and N_3^-, but not by CO.[47] This unexpected lack of inhibitory effect of CO is in agreement with P450 Fe(II) not being an intermediate in the catalytic cycle.

The mechanism of formation of N_2O from the P450 Fe(II)–NO complex is unclear. Dissociation of the ferrous P450–NO complex to yield P450 Fe(III) and NO^- and spontaneous dimerization of NO^- has been proposed to explain the formation of N_2O.[47] However, clear explanations for the reductive dimerization of NO to N_2O by an enzyme containing only a single heme center, as well as for the direct reduction of P450 Fe(III)–NO by NADH without any redox cofactor, remain to be found. Another possible mechanism could involve the binding of two NO molecules at the same P450 iron and a direct reduction of bound NO by NADH acting as a hydride donor. Studies are in progress in our laboratory to test this possible mechanism. Purified P450$_{nor}$ has been crystallized recently.[49] Its X-ray structure should shed some light on the unique features of this nonclassical P450.

The unique reaction catalyzed by P450$_{nor}$ could be involved as a detoxification pathway in denitrifying organisms.[47] Its very high turnover rate for a P450 would be important in protecting cells from toxic effects caused by NO.

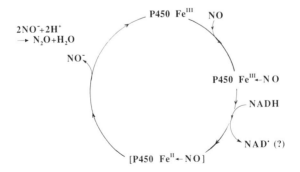

FIGURE 12. Catalytic cycle proposed for the reduction of NO to N_2O by P450 55A1 from *Fusarium oxysporum* (from Ref. 47).

4. Nitric Oxide Synthases (NOSs)

In the past few years nitric oxide has been discovered to be a biochemical of major importance.[50] It plays a central role in a great range of physiological processes, like neurotransmission, blood pressure control, and immune defense against tumor cells, bacteria, and parasites. In fact, it may be the first representative of a new class of biological messengers that exhibit a very simple chemical structure and can diffuse very quickly without the help of any transporters or channels.

4.1. Biosynthesis of NO in Mammals

Although NO can be generated in microbes from reduction of nitrite or oxidation of ammonia, its only biosynthetic pathway discovered so far in mammals is oxidation of L-arginine to citrulline and NO. This reaction is catalyzed by a particular class of monooxygenases called NO synthases. These enzymes catalyze the formation of NO from L-arginine in two steps [Eq. (8)].[51,52] The first step is the N-hydroxylation of one of the two equivalent guanidino nitrogens of L-arginine. This reaction exhibits all of the characteristics of classical P450-dependent monooxygenations, with the consumption of 1 mole of NADPH and O_2 and incorporation of one oxygen atom from O_2 into arginine. The second step is a three-electron oxidation of N^{ω}-hydroxy-L-arginine and involves an oxidative cleavage of the C=N–OH bond of N-hydroxyarginine with formation of citrulline and NO. This step differs from a classical monooxygenation as it consumes 1 mole of O_2 and *only 0.5 mole of NADPH*. NOSs are rather efficient enzymes with catalytic activities around 3 turnovers/sec.

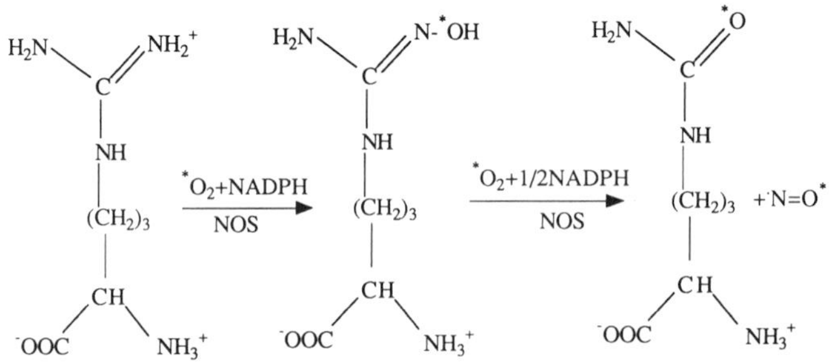

$$(8)$$

4.2. Characteristics of NOSs[51,52]

Several mammalian NOSs have been cloned and characterized during the last few years.[53–65] They belong to two main classes, the constitutive NOS (cNOS), such as brain NOS or NOS from endothelial cells, and the inducible NOS (iNOS), such as the enzymes that appear in murine macrophages or human liver after treatment with certain cytokines. Both iNOS and cNOS are dimeric enzymes containing two identical subunits consisting of a single polypeptide chain with a molecular mass between 130 and 150 kDa. All NOSs studied so far contain four prosthetic groups: flavin-adenine dinucleotide (FAD), flavin mononucleotide (FMN), tetrahydrobiopterin (BH_4) and iron protoporphyrin IX.[51,52] Each NOS subunit contains 1 mole of each of the flavins and of heme. The amount of BH_4 contained in each subunit of purified NOS varies from 0.1 to 1 mole, but the active enzyme *in vivo* is believed to contain 1 mole of BH_4 per subunit. Each NOS subunit contains two

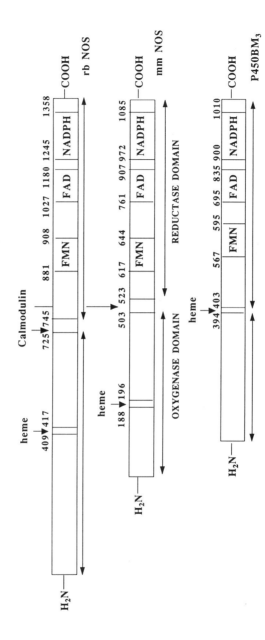

FIGURE 13. Domains and prosthetic groups of NO synthases: comparison with P450BM3 (rb and mm NOS represent rat brain and mouse macrophage NOS; see Table III). Possible location of the heme binding site for NOSs is proposed on the basis of the existence of the WRNAP(or S)RCI(or V)G sequence well conserved in NOSs and highly similar to the heme binding sequence in P450s (see Ref. 67 and text).

domains, a reductase and an oxygenase domain, which are separated by a sequence acting as a strong binding site for calmodulin (Fig. 13). This site allows regulation of the activity of cNOS by calmodulin and intracellular Ca^{2+} concentrations. The amino acid sequence of the reductase domain, which contains the binding sites for NADPH, FAD, and FMN, is very similar to that of NADPH cytochrome P450 reductase. It is likely that the function of this NOS domain is to store electrons from NADPH and to transfer them to a catalytic center located in the oxygenase domain. The oxygenase domain contains a 320-amino-acid sequence highly conserved among the NOS isoforms described so far. It is believed to contain binding sites for heme, BH_4, and L-arginine. In fact, it has been discovered recently that NOSs are heme–thiolate proteins, as indicated by the fact that the Soret peak of their Fe(II)–CO complex is around 450 nm.[6–8] Accordingly, purified NOSs exhibit spectral properties very similar to those of classical P450s: (1) they bind their substrate, L-arginine, with a shift of their Soret peak from 418 nm [low-spin Fe(III)] to 390 nm [high-spin Fe(III)], and (2) the EPR spectrum of native, ferric NOS shows g values characteristic of ferric heme–thiolate proteins.[7] The various roles of BH_4 are not clear even though it has been shown recently that BH_4 is required for assembly of the dimeric enzyme from its subunits *in vitro*,[66] which suggests that one of the possible roles of BH_4 may be to promote the assembly of the active enzyme.

4.3. Comparison between NOSs and P450s

The primary structure of NOSs, with a reductase and an oxygenase domain, is similar to that previously described for P450 BM3 from *Bacillus megaterium*,[11] which also contains a heme domain responsible for O_2 activation and a reductase domain with FAD, FMN, and NADPH binding sites within a single polypeptide chain (Fig. 13). It is thus tempting to believe that NOSs are closely related to P450s. However, the oxygenase domain of the NOSs cloned so far does not show significant overall amino acid sequence homology with P450s. Nevertheless, the following two results strongly suggest that, despite this very low overall sequence homology, NOSs exhibit great resemblance with P450s.

The Possible Heme Binding Sequence in NOS

NOSs do not contain the decapeptide sequence involving the proximal cysteine that is responsible for heme binding in P450s. However, if one looks for such a cysteine-containing sequence in the various NOSs cloned so far, one finds a nonapeptide sequence highly conserved in all NOSs: WRNAP(or S)RCV(or I)G (Table III).[67] Obviously the last part of this sequence is identical to that of P450s, as at position +1 relative to cysteine one always finds a hydrophobic amino acid (V or I) and at position +2 always a glycine. If one admits a one-amino-acid deletion (relative to P450s) at position −1, one finds in the NOS sequence at least three amino acids able to play roles identical to those of the P450 sequence: (1) a positively charged amino acid able to interact with a heme carboxylate at position −2, which is always an arginine in NOSs instead of an arginine or a histidine as in P450s, (2) an alanine residue at position −4 instead of a glycine in P450s responsible for initiation of the hairpin turn of the loop, and (3) a tryptophan residue at position −7 which obviously can play the role of the phenylalanine in P450s in protecting the proximal

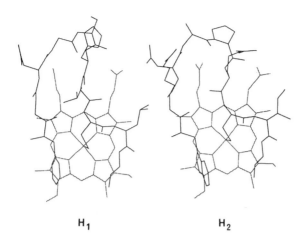

H_1 H_2

FIGURE 14. Comparison of the conformations of the heme binding sequences containing the proximal cysteine of P450 101 (from the X-ray structure described in Ref. 10) (H_1) and of NOS (H_2) (from molecular mechanics calculations; J.-P. Renaud and D. Mansuy, unpublished results) (see Ref. 67). In both cases, the heme with the bound cysteinate and the "heme binding loop" (FGHGSHLCLG for P450 101 and WRNAPRCIG for mm NOS) are shown.

cysteinate ligand. Furthermore, it is noteworthy that the phenylalanine found in most P450s at position –7 relative to the proximal cysteine is replaced by a tryptophan in two P450s, P450 10 from a pond snail and prostacyclin synthase (Table III). This suggests that the indole group of tryptophan can play a protective role similar to that of the phenyl group of phenylalanine. In the X-ray structure of P450$_{cam}$[10] the heme binding sequence appears as a decapeptide loop. Since the corresponding potential heme binding sequence of NOSs only contains nine amino acids, it was of interest to know whether this deletion of one amino acid, in passing from P450 to NOS, would greatly change the preferred conformation of the peptide. Molecular mechanics calculations give a conformation of the NOS nonapeptide (Fig. 14) very similar to that found for the decapeptide of P450$_{cam}$.[67] The CV(L or I)G parts of both loops are almost superimposed while the arginine close to the proximal cysteine in NOS interacts well with a heme carboxylate, as does the histidine in P450$_{cam}$. The indole group of the NOS tryptophan is located in a position similar to the phenyl group of the P450 phenylalanine and may protect the cysteinate ligand. Interestingly, the second arginine of the NOS sequence is well located to establish an ionic interaction with the second heme carboxylate. In a general manner, the highly conserved nonapeptide sequence in NOSs would exhibit characteristics for heme binding very similar to those of the heme binding sequence of P450s. This strongly suggests that the NOS proximal cysteine is that found in the WRNAP(or S)RCV(or I)G sequence. Very recent results of site-directed mutagenesis experiments performed either on human endothelial or on rat brain NOS support this proposition, as mutation of Cys184 in the former[68] and of Cys415 in the latter[99] (the cysteine of the nonapeptide sequence, see Table III) completely abolishes the NOS activity. Interestingly, the three main classes of NOSs appear to be characterized by specific heme binding sequences. Inducible NOSs from macrophages, hepatocytes, or chondrocytes all exhibit a WRNAPRCIG sequence whereas endothelial NOSs seem to be characterized by a slightly modified sequence WRNAPRCVG with a valine instead of an isoleucine, and constitutive brain NOSs all exhibit a WRNASRCVG sequence with a serine instead of a proline (Table III).

Catalysis of Oxidation of N^{ω}-Hydroxyarginine to Citrulline and NO by P450s

Because of the analogy between NOSs and P450s, it was interesting to know whether the two steps of the oxidation of L-arginine to citrulline and NO could be performed by P450s. The first step, the N^{ω}-hydroxylation of arginine, has not been found so far to be catalyzed by a P450. However, such N-hydroxylations are classical P450 reactions, which have been described for amidines like benzamidine[69] and pentamidine,[70] and guanidines like debrisoquine.[71]

The second step of NOS, which is a nonclassical monooxygenation reaction, has also been found recently to be catalyzed by some classical P450-dependent monooxygenases.[72] Rat liver P450s of the 3A subfamily appear especially active as catalysts for the oxidation of N^{ω}-hydroxyarginine to citrulline and nitrogen oxides, including NO[67,72] [Eq. (9)].

$$\text{L-Arg} + O_2 + \text{NADPH} \xrightarrow[\text{-NADP}^+]{\text{P450 3A}} N^{\omega}\text{-OH-Arg} + NO_2^- + NO \qquad (9)$$

In fact, these liver microsomal P450s are also able to catalyze the oxidative cleavage of arylamidoximes by O_2 and NADPH with formation of the corresponding arylamides and nitrogen oxides, including NO.[73] Even more recently, classical liver P450s have been found to be able to catalyze the oxidative cleavage of the C=N–OH bond of many N-hydroxyguanidines,[100] including N-hydroxydebrisoquine.[71] In a more general manner, they also catalyze the oxidative cleavage of the C=N–OH bond of various amidoximes, ketoximes, and aldoximes.[100] These data suggest the existence of a new P450 reaction, the oxidative cleavage of C=N–OH bonds, with formation of the corresponding C=O bonds and nitrogen oxides (Fig. 15).[101]

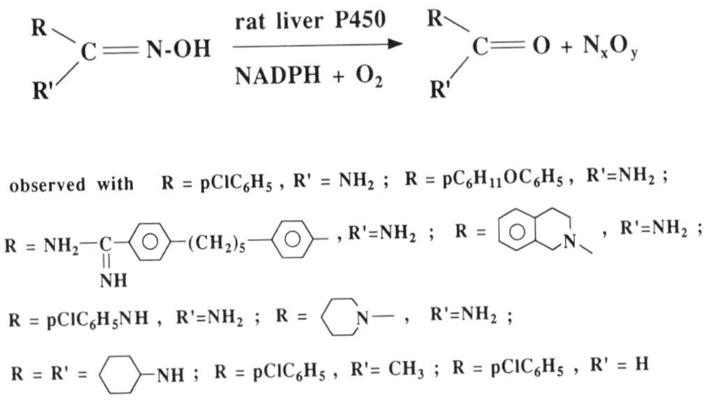

FIGURE 15. Oxidative cleavage of C=N–OH bonds of various compounds by "classical P450s" (from Refs. 70, 71, 73, and 100).

4.4. Mechanism of NOSs

The NOSs exhibit two other properties in common with "classical P450s." First, besides their monooxygenase function, they act as oxidases able to catalyze the reduction of O_2 by NADPH to O_2^{-}, H_2O_2, and H_2O, especially in the absence of L-arginine or after partial depletion of BH_4.[51,52] Second, both steps of the NOS-catalyzed oxidation of L-arginine to citrulline and NO are inhibited by carbon monoxide.[51] This indicates that the heme iron plays a key role in the activation of dioxygen and the oxidation of the substrate by NOS. The detailed mechanism of the NOS-catalyzed oxidation of L-arginine is not yet established. However, it is likely that the first step of the reaction, the N-hydroxylation of arginine, which exhibits characteristics very similar to those of classical P450-dependent N-hydroxylations, is performed by an Fe(V)=O intermediate according to a "classical" P450 catalytic cycle (Fig. 16) and by the "classical P450" mechanism for C–H or N–H bond hydroxylation. The mechanism of the second step is less clear, but it has been proposed that it involves two classical intermediates of the P450 catalytic cycle, the Fe(II)–O_2 [or Fe(III)–OO·] and the Fe(III)–OO^{-} species[51,52] (intermediates D and E of Fig. 1). This proposition is primarily based on the known high reactivity of the C=N–OH function of N-hydroxyarginine and other N-hydroxyguanidines toward nucleophiles and peracids.[52] This mechanism involves electron transfer from N-hydroxyarginine to the NOS Fe(II)–O_2 intermediate followed by nucleophilic addition of Fe(III)–O–O^{-} [or Fe(III)–OOH] to the C=N–OH carbon atom of the N-hydroxyarginine radical cation formed in the first step. This mechanism (Fig. 16) is very similar to the one previously proposed for P450-catalyzed deformylation of aldehydes (Fig. 3). Both reactions involve nucleophilic attack of Fe(III)–O–O^{-} on an electrophilic carbon (–HC=O or its nitrogen equivalent C=N–OH). As in P450-catalyzed deformylation, the last step of the NOS reaction is decomposition of the peroxide adduct by a six-center electrocyclic reaction which regenerates NOS Fe(III) and leads to citrulline and NO (instead of an alkene and HCOOH as in P450-dependent deformylations). In fact, a distinctive feature of the substrates of NOS and P450-dependent deformylases (N-hydroxyarginine and aldehydes), when compared to most substrates of P450-dependent monooxygenations, is the presence of a strong electrophilic center that should be very much reactive toward Fe(III)–O–O^{-}, which is nucleophilic in nature. Quite recently, a modification of this mechanism has been proposed.[74] It is based on electrochemical data that do not favor the electron transfer step between N-hydroxyarginine and the Fe(III)–OO· complex. It involves a free radical abstraction of the hydrogen atom of the OH group of N-hydroxyarginine by the Fe(III)–OO· intermediate, and a nucleophilic addition of Fe(III)–OOH to the N-hydroxyarginine iminoxyl radical formed in the first step.[74] Another possible mechanism for the second step of NOS could be a direct addition of the Fe(II)–O_2 <—> Fe(III)–O–O· intermediate to the C=N–OH bond, without electron transfer from N–OH-arginine to Fe(II)–O_2 (see below and Fig. 17).[101] The adduct resulting from this reaction is identical to that of the previous mechanism.

The mechanism of the oxidations of N^{ω}-hydroxyarginine and of other guanidoximes and arylamidoximes by microsomal P450s remains to be determined. However, preliminary results indicate that superoxide dismutase inhibits the P450-dependent oxidation of oximes, amidoximes, and guanidoximes[101] including N-hydroxypentamidine[70] and N-hydroxydebrisoquine.[71] This suggests that O_2^{-} formed by P450-dependent reduction of O_2

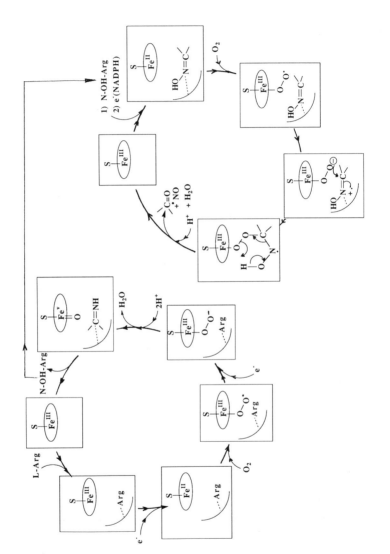

FIGURE 16. Catalytic cycles proposed for the two steps performed by NO synthases (from Refs. 51 and 52).

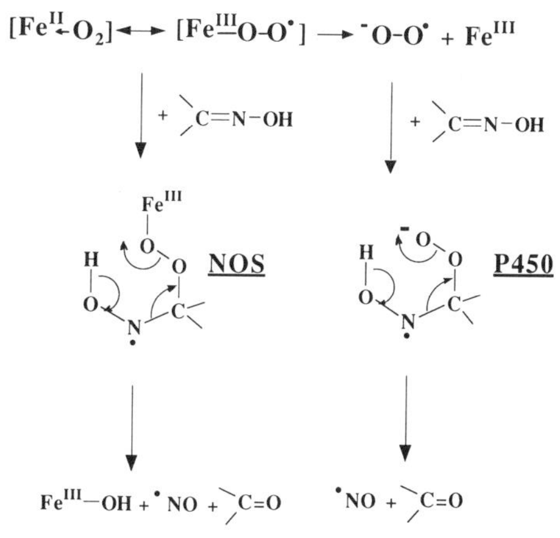

$$[Fe^{II}_{\cdot}O_2] \longleftrightarrow [Fe^{III}O\text{-}O^{\cdot}] \longrightarrow {}^{-}O\text{-}O^{\cdot} + Fe^{III}$$

FIGURE 17. Comparison between P450- and NOS-catalyzed oxidative cleavages of C=N–OH bonds.[101]

may react with substrates containing a C=N–OH function with formation of the corresponding C=O function and nitrogen oxides including NO. Quite recently, we found that $O_2{\cdot}^{-}$ rapidly reacts with compounds containing such C=N–OH functions to give the same products.[101] It is thus likely that either ${}^{-}O\text{-}O\cdot$ or its Fe(III) complex, Fe(III)–O–O·, is responsible for the oxidative cleavage of C=N–OH bonds with formation of nitrogen oxides. In the case of the observed P450-dependent oxidations of amidoximes and guanidoximes, $O_2{\cdot}^{-}$ could be the main oxidant responsible for these reactions, whereas, in the case of the NOS-dependent oxidation of N-hydroxyarginine, Fe(III)–O–O· would be the actual oxidant (Fig. 17).[101] This difference could be related to the more or less appropriate positioning of the substrate to be oxidized by Fe(III)–O–O· in the enzymatic active site. It is likely that this positioning is perfect in NOS, while it is much less adequate in microsomal P450s in the case of exogenous substrates containing a C=N–OH bond. In the latter case, the reaction of Fe(III)–O–O· would be much slower and dissociation to P450 Fe(III) and $O_2{\cdot}^{-}$ could occur before reaction with the substrate.[101] A detailed mechanism of these reactions remains to be established, as well as the role of BH_4 in the NOS reaction.

5. Hemoprotein H450

Hemoprotein H450 was first reported in 1976 as a new cytosolic hemoprotein in pig liver.[75] It exhibits a unique spectral property: its ferrous state displays a redshifted Soret peak in the absence of CO at nearly the same position (448 nm) as ferrous P450 does in the presence of CO. UV-visible, EPR, and magnetic circular dichroism studies showed that ferrous H450 was a hexacoordinate iron complex with an axial thiolate ligand, presumably from cysteine. The data are most consistent with the ligand *trans* to thiolate being either histidine or methionine.[76,77] The protein displays reversible pH-dependent spectral changes in both its ferric and ferrous states. The reversible pH effects on the spectral properties of the ferrous protein appear to involve protonation or displacement of the thiolate. Curiously, when CO is bound to ferrous H450, the Soret peak shifts to 420 nm[76] indicating that either CO displaces the thiolate endogenous ligand or its addition is accompanied by protonation of the thiolate.[77]

A cDNA clone coding for H450 was isolated from a rat liver cDNA library using anti-H450 antibody.[78] The deduced amino acid sequence contains 547 amino acids and corresponds to a molecular mass of 60,085 Da. H450 does not show any homology with cytochromes P450 or chloroperoxidase. An internal domain of H450, from residue 71 to 395, is highly similar to O-acetylserine (thiol)-lyases from E. coli and Salmonella typhimurium. This suggests that H450 and bacterial O-acetylserine (thiol)-lyase are derived from a common ancestor. However, H450 fails to exhibit O-acetylserine (thiol)-lyase activity.[78] In fact, the catalytic and biological functions of H450, which has been detected not only in rat liver but also in kidney and brain, remain to be determined.

6. Chloroperoxidase

Like the other hemoproteins described in this chapter, chloroperoxidase, isolated from the mold Caldariomyces fumago,[79] contains a ferric protoporphyrin IX bound to a cysteinate protein ligand.[80] Evidence for thiolate ligation comes from various spectroscopic studies[81,82] which show that chloroperoxidase and P450s have very similar iron coordination environments.

6.1. Comparison of the Heme Environments in Chloroperoxidase and P450s

Although the coordination chemistry and spectroscopic properties of chloroperoxidase and P450s are very similar because of the existence of an Fe–S bond in both hemoproteins, the heme environments of these proteins are very different. First, the sequence which contains the proximal cysteine of chloroperoxidase[80] is very different from the heme binding sequence of P450s (Table III). It contains none of the amino acids highly conserved in P450s, F, G, R(or H), L(or V or I), and G at positions $n-7$, $n-4$, $n-2$, $n+1$, and $n+2$, being replaced by P, S, A, P, and A in chloroperoxidase.

The distal side of the chloroperoxidase heme is somewhat similar to the distal environment of peroxidases, with the presence of a histidine and an arginine residue as shown by ^{1}H NMR of the chloroperoxidase Fe(III)–CN complex.[83] In most peroxidases, these residues are involved in catalysis of the dehydration of H_2O_2 and formation of peroxidase Compound I, the intermediate (porphyrin radical cation) Fe(IV)=O complex, that is equivalent to the P450 Fe(V)=O active species.[3] These differences in the distal environment of the heme in chloroperoxidase and P450 are easily understandable as the catalytic function of the former is to form a high-valent iron–oxo species by dehydration of H_2O_2 whereas that of the latter is to form an analogous species by reduction of O_2 bound to the iron.

The distal environment of the heme and the active site topology of chloroperoxidase were investigated by studying its reaction with phenylhydrazine and sodium azide.[84] These reactions lead to the formation of an Fe(III)–Ph complex and a δ-meso-azido adduct of the heme, respectively. Thus, both the heme iron and the δ-meso edge of the heme are readily accessible in this hemoprotein. Chloroperoxidase thus combines an accessible iron from the distal pocket of the heme, as in P450s, and a possible access to the porphyrin edge, as in peroxidases.

6.2. The Versatile Enzymatic Activity of Chloroperoxidase (Fig. 18)

Because its particular heme environment makes it a compromise between peroxidases and P450s, chloroperoxidase exhibits a versatile enzymatic activity. First, it acts as a *peroxidase* with a classical peroxidase catalytic cycle involving intermediate [porphyrin radical cation] Fe(IV)=O and [porphyrin] Fe(IV)=O complexes.[79] Thus, it is able to catalyze the oxidation of guaiacol or *p*-cresol by H_2O_2 in a manner similar to horseradish peroxidase (HRP). However, it is in general much poorer as a peroxidase than HRP.[85] It is also able to catalyze H_2O_2 decomposition to O_2 and H_2O.[86] In fact, it is the second most efficient catalytic hemoprotein after catalase.

Chloroperoxidase is also able to catalyze the chlorination of substrates either by H_2O_2 and Cl⁻, or by ClO⁻. Reaction of its resting Fe(III) state with ClO⁻ leads to a stable intermediate, called compound X, which seems to be an Fe(III)–OCl complex.[87] This intermediate may also be formed on reaction of Cl⁻ with chloroperoxidase Compound I (Fig. 19). Compounds containing acidic C–H bonds such as β-diketones and β-ketoacids are good substrates for the chlorination reaction (Fig. 18), which exhibits characteristics very similar to those of chlorinations by HOCl.[88] It is thus likely that Cl_2 or HOCl produced by compound X may diffuse from the active site and be the actual oxidant in chloroper-

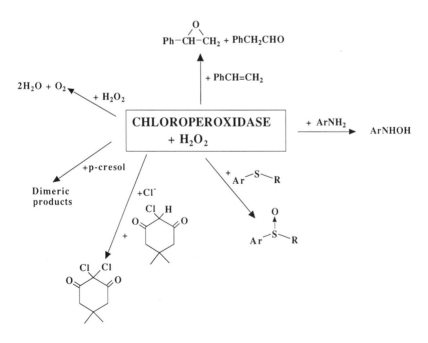

FIGURE 18. Various reactions catalyzed by chloroperoxidase.

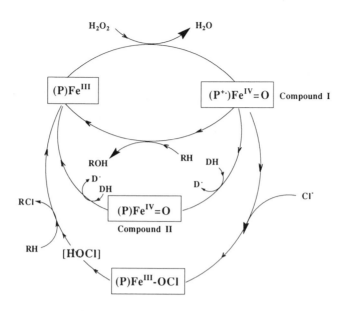

FIGURE 19. Catalytic cycles for the chloroperoxidase-dependent chlorinations, peroxidations, and monooxygenations of substrates.

oxidase-dependent chlorinations, in agreement with the nonstereoselective nature of these reactions.[87,88]

Finally, chloroperoxidase may catalyze the *monooxygenation of some substrates by H_2O_2*. For instance, alkenes are epoxidized,[89–91,93] thioethers are selectively oxidized to the corresponding sulfoxides,[92,94,95] and arylamines are *N*-hydroxylated.[96] It is noteworthy that in most of these reactions, an oxygen atom from H_2O_2 is incorporated into the product.[91,94,96] Moreover, the sulfoxidation of several alkylarylthioethers occurs with a very high enantiomeric selectivity.[92,95] These monooxygenation reactions are similar to those catalyzed by P450s and are likely to proceed either through a direct oxygen atom transfer from chloroperoxidase Compound I to the substrate, or through an electron transfer from the substrate to Compound I followed by a fast oxygen transfer from Compound II to the substrate radical cation formed in the first step. They are not possible in peroxidases in general because their iron is not accessible to substrates, but occur in chloroperoxidase because its iron is much more accessible.

7. Conclusion: How to Explain the Great Diversity of Reactions Catalyzed by Heme–Thiolate Proteins

Heme–thiolate proteins constitute a recently discovered family of hemoproteins, since the first classical P450s were found about 30 years ago, with most of the nonclassical P450s and NOSs only being discovered between 1990 and 1992. They exhibit only one common property: the *existence of an Fe–S bond between iron protoporphyrin IX and a cysteinate from the protein.* In general, this property is revealed by the appearance of a Soret peak around 450 nm for their Fe(II)–CO complex. However, the formation of a 450-nm-absorbing cysteinate–Fe(II)–CO complex is not an obligatory property for a heme–thiolate protein as (1) H450 is a heme–thiolate protein even though it binds CO with the appearance of a 420-nm-absorbing iron–CO complex, presumably because of a displacement of the thiolate ligand on CO binding,[77] and (2) some heme–thiolate proteins may not be able to form an Fe(II)–CO complex because, for instance, their ferric state is not reducible (PGIS and TXAS are difficult to reduce[28]) and/or their iron cannot bind CO (AOS exhibits a low affinity for CO[41]).

The nature of the reactions catalyzed by a newly discovered hemoprotein cannot be used as a criterion for characterizing it as a heme–thiolate protein since these proteins are able to catalyze a wide variety of reactions as shown in this chapter. In fact, the great diversity of reactions catalyzed by heme–thiolate proteins, most of which belong to the superfamily of P450s or, like NOSs, are closely related to them, can be easily explained if one takes into account the various oxidation states under which the P450 iron can exist under physiological conditions. These reactions are easily explained by considering the intrinsic reactivity of the iron intermediates involved in the P450-dependent monooxygenase catalytic cycle (Fig. 1). For the simplest reactions from a chemical point of view, such as isomerizations or certain dehydrations, *P450 Fe(III)* is the active species. It is sufficiently electron-rich, because of the presence of the cysteinate ligand, to transfer an electron to the O–O bond of an alkylperoxide or an alkylhydroperoxide. This reaction is the starting point of the isomerization of PGH_2 to prostacyclin or thromboxane, and of the dehydration of some fatty acid hydroperoxides by allene oxide synthases. For such simple reactions, a cysteinate–Fe(III) moiety surrounded by a protein able to bind the substrate in a proper conformation is sufficient for catalysis. In PGIS, TXAS, and AOS, the P450 iron plays two roles: (1) it promotes the one-electron cleavage of the substrate O–O bond and (2) it efficiently controls the intermediate carbon-centered radicals of the substrate (Figs. 7, 8, and 10).

P450 Fe(II) is more electron-rich and, consequently, it is able to react with a wider range of electron-accepting substrates. This is why "classical P450s" associated with electron transfer proteins are able to catalyze the reduction of various xenobiotics, including nitroarenes and polyhalogenated compounds.[3] The reductase activity of P450 55A1 exhibits a more peculiar mechanism as P450 55A1 Fe(III) is not linked to any electron transfer protein and is reduced directly by NADH only when it is bound to its substrate NO (Fig. 12). P450 Fe(II) may also act as a dehydration catalyst for Z-aldoximes presumably because the binding of these substrates to iron(II) leads to a charge transfer which facilitates the loss of H_2O.[26]

The oxidizing intermediates of the P450 catalytic cycle, Fe(III)–OO·, Fe(III)–OO⁻, and Fe(V)=O, may be formed either in systems able to activate dioxygen by reduction— i.e., in classical P450-dependent monooxygenases and in NOSs—or in systems able to activate H_2O_2 such as chloroperoxidase [in the case of Fe(III)–OO⁻ and Fe(V)=O]. The *Fe(II)–O₂* (or *Fe(III)–OO·*) species is the least reactive entity. It only seems to react with an especially reactive, electron-donating substrate, N^ω-hydroxyarginine, leading to the cleavage of its C=N–OH bond and formation of NO by NOS. This reaction of $O_2^{·-}$ bound to Fe(III) is easily understandable if one knows that $O_2^{\cdot-}$ itself oxidizes several compounds containing a C=N–OH function, like amidoximes and guanidoximes, leading to a cleavage of their C=N bond with formation of nitrogen oxides (see above). The *Fe(III)–O–O⁻* [or *Fe(III)–OOH*] *intermediate* is slightly more reactive; it is nucleophilic in nature and seems able to react with substrates containing an electrophilic center like aldehydes. This is the basis of oxidative deformylations catalyzed by P450-dependent monooxygenases, such as those involved in steroid demethylations. Finally, the *Fe(V)=O intermediate* is the most reactive state under which a heme–thiolate protein may exist. Contrary to the Fe(III)–O–O⁻ intermediate, it is strongly electrophilic in nature and can react with many substrates; it is responsible for all of the oxygen atom transfers (monooxygenations) catalyzed by "classical P450s," NOSs (first hydroxylation step) and chloroperoxidase. In P450s and NOSs this high-valent iron–oxo species is formed after reductive activation of dioxygen, whereas in chloroperoxidase it derives from dehydration of H_2O_2 in the active site.

In addition to their cysteinate–iron moiety, are there other structures common to heme–thiolate proteins? The proximal cysteine-containing decapeptide sequence, FG(or S)xGxR(orH)xCL(I or V)G, responsible for heme binding is well conserved in "classical P450s," in spite of exceptions such as P450 10 (from pond snail) where F is replaced by W, P450 71 (from avocado) where L is replaced by P, and trout P450 1A1 where G is replaced by D (Table III). As far as nonclassical P450s are concerned, this sequence is identical in thromboxane synthase, only slightly modified in P450 55A1 (A instead of terminal G) and in prostacyclin synthase (W and N instead of F and R), and markedly modified in allene oxide synthase [P, A, K, and A instead of F, G, R, and L(or V or I), respectively] (Table III). As discussed above, a similar sequence, WRNAP(or S)RCI(or V)G, is present and very well conserved in NOSs. This suggests that this sequence is crucial for proper coordination of the cysteinate sulfur atom to iron and proper protection of the reactive cysteinate ligand. However, the corresponding sequence in chloroperoxidase is completely different (Table III).

A second protein sequence that contains a threonine residue and appears to be responsible for O_2 binding and activation in monooxygenases is well conserved in "classical P450s" (Table II). Interestingly, protein sequence alignments suggest that the threonine residue, which appears to be crucial for O_2 activation in "classical P450s," is lacking in the corresponding sequence of allene oxide synthase (P450 74) and of thromboxane synthase (P450 5 from mouse and man). This is not unexpected for these two P450s act as dehydrases and isomerases that do not use O_2 in their catalytic cycle. On the contrary, P450 55A1 from *Fusarium oxysporum* exhibits a similar threonine-containing sequence that only differs from the P450 consensus sequence by replacement of the D (or E) residue before the threonine by an alanine. The role of such a threonine-containing sequence in a P450 that is responsible for NO reduction and does not use O_2 is unclear. It could facilitate NO binding and reduction by P450 55A1, much as it favors O_2 binding and reduction in "classical P450s."

The great diversity of reactions catalyzed by P450 enzymes and related heme–thiolate proteins is mainly due to the great diversity of the oxidation states that their iron can reach, and, in general, to the easy access of substrates to iron. Because of these two properties of P450s it is likely that new kinds of reactions will be discovered in the near future.

TABLE IV
Catalytic Activities of the Various Heme–Thiolate Proteins

Enzyme	Reaction	Activity (turnovers/sec)	Ref.
P450-dependent monooxygenase	$RH + O_2 + 2e^- + 2H^+ \rightarrow ROH + H_2O$	~0.1 $(10^{-3}-1)^a$	3
P450 BM3 (P450 102)	Fatty acid hydroxylation	27	97
PGIS	$PGH_2 \rightarrow PGI_2$	4	28
TXAS (P450 5)	$PGH_2 \rightarrow TXA_2$	24	28
	$PGH_2 \rightarrow HHT$	24	
	$PGH_2 \rightarrow TXA_2 + HHT$	48	
AOS (P450 74)	13-(S)-hydroperoxy-linoleic acid$\rightarrow$allene oxide + H_2O	1200	40
$P450_{nor}$ (P450 55A1)	$NO \xrightarrow{NADH} N_2O$	525	47
NOS	$Arg \xrightarrow[NADPH]{O_2} Cit + NO$	3	51

^{a}Certain P450s involved in the oxidation of endogenous compounds sometimes exhibit higher activities (between 1 and 7 turnovers/sec; for a recent example of such a high activity see Ref. 98).

It is often said that P450-dependent monooxygenases are relatively slow enzymes with rates between 10^{-3} and 10 turnovers/sec. This is due in general to relatively slow electron transfers within the multienzyme monooxygenase system, and these rates are higher (27 turnovers/sec) in self-sufficient P450 BM3 where the reductase is fused to the heme domain (Table IV). Furthermore, in P450s for which no electron transfer is required the rates are much higher (20 to 1200 turnovers/sec for allene oxide synthase, Table IV). In NO reductase (P450 55A1) which does not need any assistance of an electron transfer protein, the rate is also high (525 turnovers/sec) and approaches those of classical enzymes involved, for instance, in hydrolytic reactions.

A final difficult question concerns the role of the cysteinate ligand in heme–thiolate proteins. Although there is presently no clear answer to this question, it seems that this good electron donor ligand confers to iron a particular aptitude in electron transfer either to substrates (see Figs. 7, 8, and 10) or to intermediate free radicals derived from those substrates. Due to its cysteinate ligand, iron has a propensity to exist in high oxidation states and to control intermediate substrate-derived free radicals in a very efficient manner. In that regard, the last step of three different P450 reactions—i.e., the isomerization of prostacyclin (Fig. 7), the formation of allene oxides (Fig. 10), and the dehydrogenation of substrates (Fig. 2)—involves one-electron oxidation by Fe(IV) of a carbon-centered radical derived from the substrate, followed by loss of a β-proton. This one-electron oxidation which may be at the origin of the efficient control of intermediate radicals during the enzymatic process, may be favored by the presence of the thiolate ligand of the iron.

References

1. Mansuy, D., and Battioni, P., 1993, Dioxygen activation at heme centers in enzymes and synthetic analogs, in: *Bioinorganic Catalysis* (J. Reedijk, ed.), Dekker, New York, pp. 395–424.

2. Dawson, J. H., and Sono, M., 1987, Cytochrome P-450 and chloroperoxidase-thiolate-ligated heme enzymes: Spectroscopic determination of their active-site structures and mechanistic implications of thiolate ligation, *Chem. Rev.* **87:**1255–1276.

3. Ortiz de Montellano, P. R., 1986, *Cytochrome P-450: Structure, Mechanism, and Biochemistry,* Plenum Press, New York.

4. Guengerich, F. P., and MacDonald, T. L., 1984, Chemical mechanisms of catalysis by cytochromes P450: A unified view, *Acc. Chem. Res.* **17:**9–16.

5. For a short preliminary review on this subject, see: Mansuy, D., 1994, Cytochromes P450 and model systems: Great diversity of catalyzed reactions, *Pure Appl. Chem.* **66:**737–744.

6. White, K. A., and Marletta, M. A., 1992, Nitric oxide synthase is a cytochrome P450 type hemoprotein, *Biochemistry* **31:**6627–6630.

7. Stuehr, D. J., and Saito, M. I., 1992, Spectral characterization of brain and macrophage nitric oxide synthases, *J. Biol. Chem.* **267:**20547–20550.

8. McMillan, K., Bredt, D. S., Hirsh, D. J., Snyder, S. H., Clark, J. E., and Masters, B. S. S., 1992, Cloned, expressed rat cerebellar nitric oxide synthase contains stoichiometric amounts of heme which binds carbon monoxide, *Proc. Natl. Acad. Sci. USA* **89:**11141–11145.

9. Nelson, D. R., Kamataki, T., Waxman, D. J., Guengerich, F. P., Estabrook, R. W., Feyereisen, R., Gonzalez, F. J., Coon, M. J., Gunsalus, I. C., Gotoh, O., Okuda, K., and Nebert, D. W., 1993, The P450 superfamily: Update on new sequences, gene mapping, accession numbers and nomenclature, *DNA Cell. Biol.* **12:**1–51.

10. Poulos, T. L., Finzel, B. C., and Howard, A. J., 1987, High-resolution crystal structure of P450cam, *J. Mol. Biol.* **195:**687–700.

11. Ravichandran, K. G., Boddupalli, S. S., Hasemann, C. A., Peterson, J. A., and Deisenhofer, J., 1993, Crystal structure of hemoprotein domain of P450 BM-3, a prototype for microsomal P450's, *Science* **261:**731–736.

12. Hasemann, C. A., Ravichandran, K. G., Peterson, J. A., and Deisenhofer, J., 1994, Crystal structure and refinement of cytochrome P450 terp at 2.3 Å resolution, *J. Mol. Biol.* **236:**1169–1185.

13. Koymans, L., Donne-Op Den Kelder, G. M., Koppele Te, J. M., and Vermeulen, N. P. E., 1993, Cytochromes P450: Their active site structure and mechanism of oxidation, *Drug Metab. Rev.* **25:**325–387.

14. Mansuy, D., Battioni, P., and Battioni, J. P., 1989, Chemical model systems for drug-metabolizing cytochrome P450-dependent monooxygenases, *Eur. J. Biochem.* **184:**267–285.

15. Mansuy, D., Valadon, P., Erdelmeir, I., Lopez-Garcia, P., Amar, C., Girault, J. P., and Dansette, P., 1991, Thiophene S-oxides as new reactive metabolites: Formation by cytochrome P-450 dependent oxidation and reaction with nucleophiles, *J. Am. Chem. Soc.* **113:**7825–7826.

16. Dansette, P. M., Thang, D. C., El Amri, H., and Mansuy D., 1992, Evidence for thiophene-S-oxide as a primary reactive metabolite of thiophene in vivo: Formation of a dihydrothiophene sulfoxide mercapturic acid, *Biochem. Biophys. Res. Commun.* **186:**1624–1630.

17. Lopez-Garcia, P., Dansette, P., and Mansuy, D., 1994, Thiophene derivatives as new mechanism-based inhibitors of cytochromes P450: Inactivation of yeast-expressed human liver P450 2C9 by tienilic acid, *Biochemistry* **33:**166–175.

18. Rettie, A. E., Boberg, M., Rettenmeier, A. W., and Baillie, T. A., 1988, Cytochrome P450-catalyzed desaturation of valproic acid in vitro. Species differences, induction effects and mechanistic studies, *J. Biol. Chem.* **263:**13753–13758.

19. Nagata, K., Liberato, D. J., Gillette, J. R., and Sasame, M. A., 1986, An unusual metabolite of testosterone: 17α-Hydroxy-4,6-androsta-diene-3-one, *Drug Metab. Dispos.* **14:**559–565.

20. Korzekwa, K. R., Trager, W. F., Nagata, K., Parkinson, A., and Gillette, J. R., 1990, Isotope effect studies on the mechanism of the cytochrome P450IIA1-catalyzed formation of Δ^6-testosterone from testosterone, *Drug Metab. Dispos.* **18:**974–979.

21. Akhtar, M., and Wright, J. N., 1991, A unified mechanistic view of oxidative reactions catalyzed by P450 and related Fe-containing enzymes, *Nat. Prod. Rep.* **1991:**527–551.

22. Vaz, A. D. N., Roberts, E. S., and Coon, M. J., 1991, Olefin formation in the oxidative deformylation of aldehydes by cytochrome P450. Mechanistic implications for catalysis by oxygen-derived peroxide, *J. Am. Chem. Soc.* **113**:5886–5887.

23. Watanabe, Y., and Ishimura, Y. J., 1989, Aromatization of tetralone derivatives by 5,10,15,20-tetrakis (pentafluorophenyl)porphyrinatoiron(III) chloride/iodosobenzene and cytochrome P-450cam: A model study on aromatase cytochrome P-450 reaction, *J. Am. Chem. Soc.* **111**:410–411.

24. Cole, P. A., Bean, J. M., and Robinson, C. H., 1990, Conversion of a 3-desoxysteroid to 3-desoxyestrogen by human placental aromatase, *Proc. Natl. Acad. Sci. USA* **87**:2999–3003.

25. De Master, E. G., Shirota, F. N., and Nagasawa, H. T., 1992, A Beckmann-type dehydration of n-butyraldoxime catalyzed by cytochrome P450, *J. Org. Chem.* **57**:5074–5075.

26. Boucher, J. L., Delaforge, M., and Mansuy, D., 1994, Dehydration of alkyl and arylaldoximes as a new cytochrome P450-catalyzed reaction: Mechanism and stereochemical characteristics, *Biochemistry* **33**:7811–7818.

27. Halkier, B. A., Olsen, C. E., and Moeller, B. L., 1989, The biosynthesis of cyanogenic glucosides in higher plants, *J. Biol. Chem.* **264**:19487–19494.

28. For a recent review, see: Ullrich, V., and Ruf, H. H., 1994, Heme proteins in prostaglandin biosynthesis, in: *Metalloporphyrins in Catalytic Oxidations* (R. A. Sheldon, ed.), Dekker, New York, pp. 157–186.

29. Ullrich, V., Castle, L., and Weber, P., 1981, Spectral evidence for the cytochrome P450 nature of prostacyclin synthase, *Biochem. Pharmacol.* **30**:2033–2040.

30. Haurand, M., and Ullrich, V., 1985, Isolation and characterization of thromboxane synthase from human platelets as a cytochrome P450 enzyme, *J. Biol. Chem.* **260**:15059–15067.

31. De Witt, D. L., and Smith, W. L., 1983, Purification of prostacyclin synthase from bovine aorta by immunoaffinity chromatography, *J. Biol. Chem.* **258**:3285–3293.

32. Pereira, B., Wu, K. K., and Wang, L. H., 1994, Molecular cloning and characterization of bovine prostacyclin synthase, *Biochem. Biophys. Res. Commun.* **203**:59–66.

33. Miyata, A., Hara, S., Yokoyama, C., Inoue, H., Ullrich, V., and Tanabe, T., 1994, Molecular cloning and expression of human prostacyclin synthase, *Biochem. Biophys. Res. Commun.* **200**:1728–1734.

34. Yokoyama, C., Miyata, A., Ihara, H., Ullrich, V., and Tanabe, T., 1991, Molecular cloning of human platelet thromboxane A synthase, *Biochem. Biophys. Res. Commun.* **178**:1479–1484.

35. Ohashi, K., Ruan, K., Kulmacz, R. J., Wu, K. K., and Wang, L., 1992, Primary structure of human thromboxane synthase determined from the cDNA sequence, *J. Biol. Chem.* **267**:789–793.

36. Hecker, M., Haurand, M., Ullrich, V., and Terao, S., 1986, Spectral studies on structure activity relationships of thromboxane synthase inhibitors, *Eur. J. Biochem.* **157**:217–223.

37. Hecker, M., and Ullrich, V., 1989, On the mechanism of prostacyclin and thromboxane A2 biosynthesis, *J. Biol. Chem.* **264**:141–150.

38. Brash, A. R., Hughes, M. A., Hawkins, D. J., Boeglin, W. E., Song, W. C., and Meijer, L., 1991, Allene oxide and aldehyde biosynthesis in starfish oocytes, *J. Biol. Chem.* **266**:22926–22931.

39. Hamberg, M., and Fahlstadius, P., 1990, Allene oxide cyclase: A new enzyme in plant lipid metabolism, *Arch. Biochem. Biophys.* **276**:518–526.

40. Song, W. C., and Brash, A. R., 1991, Purification of an allene oxide synthase and identification of the enzyme as a cytochrome P450, *Science* **253**:781–784.

41. Lau, S. M. C., Harder, P. A., and O'Keefe, D. P., 1993, Low carbon monoxide affinity allene oxide synthase is the predominant cytochrome P450 in many plant tissues, *Biochemistry* **32**:1945–1950.

42. Song, W. C., Funk, C. D., and Brash, A. R., 1993, Molecular cloning of an allene oxide synthase: A cytochrome P450 specialized for the metabolism of fatty acid hydroperoxides, *Proc. Natl. Acad. Sci. USA* **90**:8519–8523.

43. Song, W. C., Baertschi, S. W., Boeglin, W. E., Harris, T. M., and Brash, A. R., 1993, Formation of epoxyalcohols by a purified allene oxide synthase, *J. Biol. Chem.* **268**:6293–6298.

44. Corey, E. J., d'Alarcoa, M., Matsuda, S. P. T., Lansbury, P. T., and Yamada, Y., 1987, Intermediacy of 8-(R)-HPETE in the conversion of arachidonic acid to pre-clavulone prostanoids, *J. Am. Chem. Soc.* **109**:289–290.

45. Brash, A. R., Baertschi, S. W., Ingram, C. D., and Harris, T. M., 1987, On non-cyclooxygenase prostaglandin synthesis in the sea whip coral, Plexaura homomalla: An 8(R)-lipoxygenase pathway leads to formation of an alpha-ketol and a racemic prostanoid, *J. Biol. Chem.* **262**:15829–15839.

46. Shoun, H., and Tanimoto, T., 1991, Denitrification by the fungus *Fusarium oxysporum* and involvement of cytochrome P450 in the respiratory nitrite reduction, *J. Biol. Chem.* **266**:11078–11082.

47. Nakahara, K., Tanimoto, T., Hatano, K., Usuda, K., and Shoun, H., 1993, Cytochrome P450 55A1 (P450 dNIR) acts as nitric oxide reductase employing NADH as the direct electron donor, *J. Biol. Chem.* **268**:8350–8355.

48. Kizawa, H., Tomura, D., Oda, M., Fukamizu, A., Hoshino, T., Gotoh, O., Yasui, T., and Shoun, H., 1991, Nucleotide sequence of the unique nitrate/nitrite-inducible cytochrome P450 cDNA from *Fusarium oxysporum*, *J. Biol. Chem.* **266**:10632–10637.

49. Nakahara, K., Shoun, H., Adachi, S., Tizuka, T., and Shiro, Y., 1994, Crystallisation and preliminary X-ray diffraction studies of nitric oxide reductase, cytochrome P450nor from *Fusarium oxysporum*, *J. Mol. Biol.* **239**:158–159.

50. Moncada, S., Palmer, R. M. J., and Higgs, E. A., 1991, Nitric oxide: Physiology, pathophysiology and pharmacology, *Pharmacol. Rev.* **43**:109–142.

51. Marletta, M. A., 1993, Nitric oxide synthase structure and mechanism, *J. Biol. Chem.* **268**:12231–12234.

52. Feldman, P. L., Griffith, O. W., and Stuehr, D. J., 1993, The surprising life of nitric oxide, *Chem. Eng. News* **71**:26–38.

53. Lyons, C. R., Orloff, G. J., and Cunningham, J. M., 1992, Molecular cloning and functional expression of an inducible nitric oxide synthase from a murine macrophage cell line, *J. Biol. Chem.* **267**:6370–6374.

54. Xie, Q. W., Cho, H. J., Calaycay, J., Mumford, R. A., Swiderek, K. M., Lee, T. D., Ding, A., Troso, T., and Nathan, C., 1992, Cloning and characterization of inducible nitric oxide synthase from mouse macrophages, *Science* **256**:225–228.

55. Nunokawa, Y., Ishida, N., and Tanaka, S., 1994, Cloning of inducible nitric oxide synthase in rat vascular smooth muscle cells, *Biochem. Biophys. Res. Commun.* **191**:89–94.

56. Adachi, H., Iida, S., Oguchi, S., Ohshima, H., Suzuki, H., Nagasaki, K., Kawasaki, H., Sugimura, T., and Esumi, H., 1993, Molecular cloning of a cDNA encoding an inducible calmodulin-dependent nitric oxide synthase from rat liver and its expression in COS1 cells, *Eur. J. Biochem.* **217**:37–43.

57. Wood, E. R., Berger, H., Jr., Sherman, P. A., and Lapetina, E. G., 1993, Hepatocytes and macrophages express an identical cytokine inducible nitric oxide synthase gene, *Biochem. Biophys. Res. Commun.* **191**:767–774.

58. Geller, D. A., Lowenstein, C. J., Shapiro, R. A., Nussler, A. K., Di Silvio, M., Wang, S. C., Nakayama, D. K., Simmons, R. L., Snyder, S. H., and Billiar, T. R., 1993, Molecular cloning and expression of inducible nitric oxide synthase from human hepatocytes, *Proc. Natl. Acad. Sci. USA* **90**:3491–3495.

59. Charles, I. G., Palmar, R. M. J., Hickery, M. S., Bayliss, M. T., Chubb, A. P., Hall, V. S., Moss, D. W., and Moncada, S., 1993, Cloning, characterization and expression of a cDNA encoding an inducible nitric oxide synthase from the human chondrocyte, *Proc. Natl. Acad. Sci. USA* **90**:11419–11423.

60. Marsden, P. A., Schappert, K. T., Chen, H. S., Flowers, M., Sundell, C. L., Wilcox, J. N., Lamas, S., and Michel, T., 1992, Molecular cloning and characterization of human endothelial nitric oxide synthase, *FEBS Lett.* **307**:287–293.

61. Lamas, S., Marsden, P. A., Li, G. K., Tempest, P., and Michel, T., 1992, Endothelial nitric oxide synthase: Molecular cloning and characterization of a distinct constitutive enzyme isoform, *Proc. Natl. Acad. Sci. USA* **89**:6348–6352.

62. Sessa, W. C., Harrison, J. K., Barber, C. M., Zeng, D., Durieux, M. E., D'Angelo, D. D., Lynch, K. R., and Peach, M. J., 1992, Molecular cloning and expression of a cDNA encoding endothelial cell nitric oxide synthase, *J. Biol. Chem.* **267**:15274–15276.

63. Bredt, D. S., Hwang, P. M., Glatt, C. E., Lowenstein, C., Reed, R. R., and Snyder, S. H., 1991, Cloned and expressed nitric oxide synthase structurally resembles cytochrome P-450 reductase, *Nature* **351**:714–718.

64. Ogura, T., Yokoyama, T., Fujisawa, H., Kurashima, Y., and Esumi, H., 1993, Structural diversity of neuronal nitric oxide synthase mRNA in the nervous system, *Biochem. Biophys. Res. Commun.* **193:**1014–1022.

65. Nakane, M., Schmidt, H. H. W., Pollock, J. S., Förstermann, J. S., and Murad, F., 1993, Cloned human brain nitric oxide synthase is highly expressed in skeletal muscle, *FEBS Lett.* **316:**175–180.

66. Baek, K. J., Thiel, B. A., Lucas, S., and Stuehr, D., 1993, Macrophage nitric oxide synthase subunits, *J. Biol. Chem.* **268:**21120–21129.

67. Renaud, J. P., Boucher, J. L., Vadon, S., Delaforge, M., and Mansuy, D., 1993, Particular ability of liver P450s3A to catalyze the oxidation of N^{ω}-hydroxyarginine to citrulline and nitrogen oxides and occurrence in NO synthases of a sequence very similar to the heme-binding sequence in P450s, *Biochem. Biophys. Res. Commun.* **192:**53–60.

68. Chen, P. F., Tsai, A. L., and Wu, K. K., 1994, Cysteine 184 of endothelial nitric oxide synthase is involved in heme coordination and catalytic activity, *J. Biol. Chem.* **269:**25062–25066.

69. Clement, B., Immel, M., Pfunder, H., Schmitt, S., and Zimmerman, M., 1991, in: *N-oxidation of Drugs* (P. Hlavica and L. A. Damani, eds.), Chapman & Hall, London, pp. 185–204.

70. Clement, B., and Jung, F., 1994, N-hydroxylation of the antiprotozoal drug pentamidine catalyzed by rabbit liver cytochrome P450 2C3 or human liver microsomes, microsomal retroreduction, and further oxidative transformation of the formed amidoximes, *Drug Metab. Dispos.* **22:**486–497.

71. Clement, B., Schultze-Mosgau, M. H., and Wohlers, H., 1993, Cytochrome P450-dependent N-hydroxylation of a guanidine (debrisoquine), microsomal catalyzed reduction and further oxidation of the N-hydroxy-guanidine metabolite to the urea derivative, *Biochem. Pharmacol.* **46:**2249–2267.

72. Boucher, J. L., Genet, A., Vadon, S., Delaforge, M., Henry, Y., and Mansuy, D., 1992, Cytochrome P450 catalyzes the oxidation of N^{ω}-hydroxy-L-arginine by NADPH and O_2 to nitric oxide and citrulline, *Biochem. Biophys. Res. Commun.* **187:**880–886.

73. Andronik-Lion, V., Boucher, J. L., Delaforge, M., Henry, Y., and Mansuy, D., 1992, Formation of nitric oxide by cytochrome P450-catalyzed oxidation of aromatic amidoximes, *Biochem. Biophys. Res. Commun.* **185:**452–458.

74. Korth, H. G., Sustmann, R., Tahter, C., Burther, A. R., and Ingold, K. U., 1994, On the mechanism of the nitric oxide synthase catalyzed conversion of N-hydroxy-L-arginine to citrulline and nitric oxide, *J. Biol. Chem.* **269:**17776–17779.

75. Kim, I. C., and Deal, W. C., 1976, Isolation and properties of a new, soluble, hemoprotein (H-450) from pig liver, *Biochemistry* **15:**4925–4930.

76. Omura, T., Sadano, H., Hasegawa, T., Yoshida, Y., and Kominami, S., 1984, Hemoprotein H-450 identified as a form of cytochrome P-450 having an endogenous cysteinate ligand at the 6th coordination position of the heme, *J. Biochem.* **96:**1491–1500.

77. Svastits, E. W., Alberta, J. A., Kim, I. C., and Dawson, J. H., 1989, Magnetic circular dichroism studies of the active site structure of hemoprotein H450: Comparison to cytochrome P450 and sensitivity to pH effects, *Biochem. Biophys. Res. Commun.* **165:**1170–1176.

78. Ishihara, S., Morohashi, K. I., Sadano, H., Kawabata, S. I., Gotoh, O., and Omura, T., 1990, Molecular cloning and sequence analysis of cDNA coding for rat liver hemoprotein H450, *J. Biochem.* **108:**899–902.

79. Griffin, B. W., 1991, in: *Peroxidases in Chemistry and Biology*, Vol II (J. Everse, K. E. Everse, and M. B. Grisham, eds.), CRC Press, Boca Raton, FL, pp. 85–137.

80. Blanke, S. R., and Hager, L. P., 1988, Identification of the fifth axial heme ligand of chloroperoxidase, *J. Biol. Chem.* **263:**18739–18743.

81. Hahn, J. E., Hodgson, K. O., Andersson, L. A., and Dawson, J. H., 1982, Endogenous cysteine ligation in ferric and ferrous cytochrome P-450. Direct evidence from x-ray absorption spectroscopy, *J. Biol. Chem.* **257:**10934–10941.

82. Bangcharoenpaurpong, O., Champion, P. M., Hall, K. S., and Hager, L. P., 1986, Resonance Raman studies of isotopically labelled chloroperoxidase, *Biochemistry* **25:**2374–2378.

83. Dugad, L. B., Wang, X., Wang, C. C., Lukat, G. S., and Goff, H. M., 1992, Proton nuclear Overhauser effect study of the heme active site structure of chloroperoxidase, *Biochemistry* **31:**1651–1655.

84. Samokyszyn, V. M., and Ortiz de Montellano, P. R., 1991, Topology of the chloroperoxidase active site: Regioselectivity of heme modification by phenylhydrazine and sodium azide, *Biochemistry* **30:**11646–111653.

85. Lambeir, A. M., Dunford, H. B., and Pickard, M. A., 1987, Kinetics of the oxidation of ascorbic acid, ferrocyanide and p-phenolsulfonic acid by chloroperoxidase compounds I and II, *Eur. J. Biochem.* **163:**123–127.

86. Thomas, J. A., Morris, D. R., and Hager, L. P., 1970, Chloroperoxidase. 8. Formation of peroxide and halide complexes and their relation to the mechanism of the halogenation reaction, *J. Biol. Chem.* **245:**3135–3142.

87. Hollenberg, P. F., Rand-Meier, T., and Hager, L. P., 1974, The reaction of chlorite with horseradish peroxidase and chloroperoxidase. Enzymatic chlorination and spectral intermediates, *J. Biol. Chem.* **249:**5816–5825.

88. Libby, R. D., Shedd, A. L., Phipps, A. K., Beachy, T. M., and Gertsberger, S. M., 1992, Defining the involvement of HOCl or Cl₂ as enzyme-generated intermediates in chloroperoxidase-catalyzed reactions, *J. Biol. Chem.* **267:**1769–1775.

89. McCarthy, M. B., and White, R. E., 1983, Functional differences between peroxidase compound I and the cytochrome P-450 reactive oxygen intermediate, *J. Biol. Chem.* **258:**9153–9158.

90. Geigert, J., Lee, T. D., Dalietos, D. J., Hirano, D. S., and Neidleman, S. L., 1986, Epoxidation of alkenes by chloroperoxidase catalysis, *Biochem. Biophys. Res. Commun.* **136:**778–782.

91. Ortiz de Montellano, P. R., Choe, Y. S., De Pillis, G., and Catalano, C. E., 1987, Structure–mechanism relationships in hemoproteins. Oxygenations catalyzed by chloroperoxidase and horseradish-peroxidase, *J. Biol. Chem.* **262:**11641–11646.

92. Fu, H., Kondo, H., Ichikawa, Y., Cook, G. C., and Wong, C. H., 1992, Chloroperoxidase-catalyzed asymmetric synthesis: Enantio-selective reaction of chiral hydroperoxides with sulfides and bromo-hydration of glycals, *J. Org. Chem.* **57:**7265–7270.

93. Colonna, S., Gaggero, N., Casella, L., Carrea, G., and Pasta, P., 1993, Enantioselective epoxidation of styrene derivatives by chloroperoxidase catalysis, *Tetrahedron Asymmetry* **4:**1325–1330.

94. Kobayashi, S., Nakano, M., Goto, T., Kimura, T., and Schaap, A. P., 1986, An evidence of the peroxidase dependent oxygen transfer from hydrogen peroxide to sulfides, *Biochem. Biophys. Res. Commun.* **135:**166–171.

95. Colonna, S., Gaggero, N., Manfredi, A., Casella, L., Gullotti, M., Carrea, G., and Pasta, P., 1990, Enantioselective oxidations of sulfides catalyzed by chloroperoxidase, *Biochemistry* **29:**10465–10468.

96. Doerge, D. R., and Corbett, M. D., 1991, Peroxygenation mechanism for chloroperoxidase-catalyzed N-oxidation of arylamines, *Chem. Res. Toxicol.* **4:**556–560.

97. Boddupalli, S. S., Estabrook, R. W., and Peterson, J. A., 1990, Fatty acid monooxygenation by cytochrome P450 BM-3, *J. Biol. Chem.* **265:**4233–4239.

98. Urban, P., Werck-Reichhart, D., Teutsch, H. G., Durst, F., Regnier, S., Kazmaier, M., and Pompon, D., 1994, Characterization of recombinant plant cinnamate 4-hydroxylase produced in yeast. Kinetic and spectral properties of the major plant P450 of the phenylpropanoid pathway, *Eur. J. Biochem.* **222:**843–850.

99. Richards, M. K., and Marletta, M. A., 1994, Characterization of neuronal nitric oxide synthase and a C415H mutant purified from a baculovirus overexpression system, *Biochemistry* **33:**14723–14732.

100. Jousserandot, A., Boucher, J. L., Desseaux, C., Delaforge, M., and Mansuy, D., 1995, Formation of nitrogen oxides including NO from oxidative cleavage of C=N(OH) bonds: a general cytochrome P450-dependent reaction, *Bioorg. Med. Chem. Lett.* **5:**423–426.

101. Mansuy, D., Boucher, J. L., and Clement, B., 1995, On the mechanism of nitric oxide formation upon oxidative cleavage of C=N(OH) bonds by NO-synthases and cytochromes P450, *Biochimie*, in press.

Cytochrome P450 Nomenclature and Alignment of Selected Sequences

DAVID R. NELSON

1. Doing the Numbers

This chapter is a summary of sequence data available for cytochrome P450 genes and proteins. It serves as a vehicle to update the nomenclature since the last major listing was published in January of 1993.[1] It is now April, 1994, about 16 months since that list appeared, and many new P450 sequences have been determined. Table I consists of 368 entries, covering 281 genes and 15 pseudogenes. Also included are 29 different expressed sequence tags (ESTs) or PCR fragments from *Caenorhabditis elegans, Arabidopsis thaliana,* and *Catharanthus roseus* that have not yet been assigned to P450 families. Some of the *Arabidopsis* sequences are probably from the same gene, so the 29 EST sequences really represent less than 29 genes. Nevertheless, at least 325 different cytochrome P450 sequences are listed in Table I. This is amazing progress since the first P450 sequence was published in 1982.

The sequences belong to 50 families and 82 subfamilies, and they come from 67 species. Fourteen new families have appeared since the 1993 nomenclature update.[1] There are 19 vertebrates (12 mammals, 5 fish, and 2 birds). Invertebrates include six insect species, the nematode worm *C. elegans,* with five ESTs but no complete sequences, a spiny lobster, and a pond snail. Plants have 13 species represented and there are 11 lower eukaryote P450s. The bacteria now have 8 genera and 15 species including gram-positive as well as gram-negative bacteria. A cyanobacterium is present and there is a partial sequence from the heat-tolerant bacterium *Streptomyces thermotolerans,* but no sequence is available from an archaebacterium.

DAVID R. NELSON • Department of Biochemistry, The University of Tennessee, Memphis, Memphis, Tennessee 38163.

Cytochrome P450: Structure, Mechanism, and Biochemistry (Second Edition), edited by Paul R. Ortiz de Montellano. Plenum Press, New York, 1995.

In a single species, rats have 45 genes and 2 pseudogenes. Rats will have CYP1B1, CYP8, and CYP21 genes, but these are not sequenced yet. This gives a total of 50 genes in the rat. Humans have 34 genes and 7 pseudogenes so far. Nineteen of the human sequences are included in Fig. 1. Humans appear to have fewer sequences in the 2B, 2C, and 2D subfamilies when compared to rats. It is interesting to estimate the number of P450s in *C. elegans,* assuming random sampling. About 3–4% of the genome of *C. elegans* has been sequenced, and there are four different P450 ESTs represented. This extrapolates to about 100 P450 sequences in *C. elegans.* Of course, the numbers are very small and the estimate could be off by quite a bit. A more representative value will be available when about 10% of the genome is done, probably within a year.

2. The Scope of Table I

Table I does not reproduce all of the data in the 1993 nomenclature update. That would be a great waste, since there are over 700 references and 1200 accession numbers given there. The table is meant to be used in conjunction with that earlier tabulation to obtain comprehensive coverage through the end of 1993 and the early part of 1994. Table I does include all of the gene names from the 1993 list, plus any new entries made since then. New entries include reference to journal articles and Genbank accession numbers if they are available. References from the 1993 update that were incomplete or lacked accession numbers, have been included again to complete them. The 1993 update should be consulted for trivial names (Tables 2 and 4) and chromosomal mapping data (Table 1). This old information has not changed since 1993.

3. Cytochrome P450 Sequence Alignment

The sequences in this alignment represent 65 of the 82 subfamilies and 41 of the 50 cytochrome P450 families. Those families and subfamilies not included were confidential or unavailable at the time the alignment was made. The sequences are arranged in the same order as the branching pattern of the phylogenetic tree in Fig. 2. This automatically clusters the most similar sequences together for ease of comparison. On the left, the sequences are numbered for identification. On the right, the number of amino acids up to that point is given. To the far right on the first page, the Genbank accession numbers are given for each sequence. Two sequences have an insert that has been removed to save space. In sequence 22 CYP5 human, 27 amino acids have been removed between amino acids 298 and 299—IVRDVFSSTGCKPNPSRQHQPSPMARP. Sequence 24 CYP6B1 had 7 amino acids removed between 280 and 281—EDVKALE. These inserts occur before the I-helix in a highly variable part of the P450 sequences.

4. Phylogenetic Tree of Cytochrome P450 Sequences

This tree was computed using a PAM250 scoring matrix. This tree and more detailed trees of individual families and groups of families are the basis for the nomenclature described in detail in the 1993 nomenclature update.[1] The percent similarity between

```
 1 2C9   mp4       human         ----------------------MDS--------------LVV-LVLCLS  12 M21940
 2 2E1   j         human         ----------------------MFALGV-----------TVA-LLVWAA  15 J02843
 3 2H1   CHP3      chicken       ----------------------MDFLGLP----------TIL-LLVCIS  16 M13454
 4 2F1   IIF1      human         ----------------------MDSI-------------STA-ILLLLL  13 J02906
 5 2B6   LM2       human         ----------------------MELS-------------VLL-FLALLT  13 X16864
 6 2G1   olf1      rat           ----------------------MALGGAF----------SIF-MTLCLS  16 J04715
 7 2A6   IIA3      human         ----------------------MLASGML----------LVA-LLVCLT  16 X13929
 8 2K1             trout         --------------MSLIEDILQTSSTVT----------LLG-TVLFLL  24 L11528
 9 2J1   ib        rabbit        --------------MVAALSSLAAALGAGLH-------PKT-LLLGAV  26 D90405
10 2D6   db1       human         ------------------MGL---EALV-----------PLA-VIVAIF 16 Y00300
11 1A1   P1        human         ------------------MLFPISMSATEFL-------LAS-VIFCLV  22 X04300
12 17    17a       human         ----------------------------------MWEL-VALLLL  10 M14564
13 21    C21       human         ----------------------------------M-LLLGLL   7 M17252
14 71A1            avocado       ------------------MAIL----------VSLLF-LAIALT  15 M32885
15 71B1  Thlaspi   arvense       ------------------MDLL-----------L-YIV-AALVIF  14 L24438
16 75A1            petunia       ------------------MMLLTELG--------AATSI-FLIAHI  19 X71130
17 76A1            eggplant      -----------------------------------------  0 X71658
18 73    Jerusalem artichoke     ------------------MDLLLI----------EKTLV-ALFAAI  27 Z17369
19 78              Zea mays      ----------------------------------ALLAW-ATSPGG  11 L23029
20 77A1            eggplant      ----------------------------------IFTAF-SLLFSL  11 X71655
21 3A3   HLp       human         ------------------MALIPDLAME-------TWLL-LAVSLV  20 M13785
22 5A1   TXS       human         --------------MMEALGFLKLEVNGPMV-------TVAL-SVALLA 27 M74055
23 6A1   Musca     domestica     ------------------MDFGSFLLYA-------LGVL----ASL  17 M25367
24 6B1             Butterfly     -------------------MLYLLAL-------VTVL----AGL  14 M80828
25 4A1             human         -------------------MSVSVLSPSRLLGDVSGILQ-AASLLI 26 L04751
26 4B1   IVB1      human         -------------------MSGTATMVPSFLSLSFSSLG-LWASGL 26 J02871
27 4F3             human         -------------------MPQLSLSSLGLWPMAASPWL-LLLLVG 26 D12620
28 4C1             cockroach     --------------------MEFITILLS-TALFHS  15 M63798
29 4E1             Drosophila    -----------------------------------------  0 K00045
30 4D1             Drosophila    -------------------MFLVIGAILA-SALFVG  16 X67645
31 102   BM3       B. megaterium -----------------------------------------  0 J04832
32 110   Anabaena  cyanobact.    -----------------------------------------  0 M38044
33 72A1  Catharanthus ros.       ----------MEMDMDTIRKAIAATIFALVMAWAWRV-LDWAWF-TPKRIE 39 L10081
34 53    bphA      Asperg. niger ------------------------MLAL------LLS-PYGAYLG-LAL-LV 19 X52521
35 10              pond snail    MAIMKKFIHHSLKQLIKPNLTSTKRVVSTSPRKEQGVAAISLEP-SEMAQC 50 S46130
36 11A1  scc       human         ------------------MLAKGLPPRSVLVKGCQTFL-SAPREG  26 M28253
37 11B1  11B1      human         ------------------MALRAKAEVCM-AVPWLS  17 X55764
38 24    25OH D3   24hyd rat     -------------MSCPIDKRRTLIAFLRRLRDLGQPPRSVTSK-AS-ASR  36 L04618
39 27    27OH      human         -----------MAALGCARLRWAL---RGAGRGLCPHGARAK-AAIPAA  34 M62401
40 51    14DM      S. cerevisiae ----MSATKSIVGEALEYVNIGLSHFL-ALPLAQ-----RISLI-III-PF 39 M15663
41 7     Chol7a    human         -------------------MMTTSLI---WGIAIAACCC-LWLIL-  22 X56088
42 56    DIT2      S. cerevisiae ----------------------MELL-------K-LLCLILF-LTLSYV  18 X55713
43 52A1  alk1      C. tropicalis ----MSSSPSIAQEFLATITPYVEYCQ-ENYTKWYYFIPLVILS-LNLISM 45 M15945
44 52D1  ALK4-A    C. maltosa    ------------MAIF--------------TPELWLICFAVT-VYIFDY  22 D12716
45 52B1  alk6      C. tropicalis ----------MSLTETTATFIY-------NYWYIIFHLYFYTTSKIIKY  32 Z13013
46 52C1  alk7      C. tropicalis ------------------MYQLFCFLAGIIVV-YKAAQY  20 Z13014
47 19    arom      human         ------------------MVLE------------MLN-PIHYNI  13 M28420
48 105A  SU1       Str. griseolus -----------------------------TDTATTP-QTTDAP  13 M32238
49 105D  P450soy   St.griseus    ----------------------------MTESTTDPARQNLDPT-SPAPAT  22 X63601
50 105B  SU2       Str. griseolus --------------------------TTAERTA-PPDALT  13 M32239
51 105C  Streptomyces sp.        ----------------------------MTQA-------AP   6 M31939
52 105E  Rhodococ.fascians       -----------------------------------MAGTA   5 Z29635
53 107A1 eryF      Sacc.eryth.   -----------------------------M-TT--V-PDLESD  10 M54983
54 55    Fusarium  oxysporum     -----------------------------MASGAP-------   6 M63340
55 112   BJ-1      Bradyrhiz.jap. --------------------------------MS-EQQPLP   8 L02323
56 114   BJ-3      Bradyrhiz.jap. -----------------------------------------  0 L12971
57 113             eryK Sacc.eryth. --------------VCADVETTCCA-RRTLT-TIDEVP  22 L05776
58 106A1 BM1       B.megaterium  ----------------------------MNKE-------VIP   7 X16610
59 109   Bacillus  subtilis      ----------------------------MTNQTAR-SSKKER  13 M36988
60 108   terp                    ---------------------------MDARATIPE-HIARTV  13 M91440
61 111   Pseudomonas sp.         ------------------------MER-PDLKNP   9 L23310
62 101   cam       P. putida     ----------------------------TTETIQS-NANLAP  13 M12546
63 103   pinF1     A.tumefaciens ----------------------------MIANSST-DVSVAD  13 M19352
64 104   pinF2     A.tumefaciens ----------------------------M-EERRV-SISSIT  12 M19352
65 74    Allene    ox.syn. flax  -----------------------------------------  0 U00428
```

FIGURE 1. An alignment of 65 cytochrome P450 sequences.

```
 1 CLL--L-LSLWR------------QSSGRGKLPPGPTPL-PVIGNI--LQIGIKDI----SKSLTNLSK--VYG----PVF  65
 2 FLL--L-VSMWR------------QVHSSWNLPPGPFPL-PIIGNL--FQLELKNI----PKSFTRLAQ--RFG----PVF  68
 3 CL---L-IAAWR------------STSQRGKEPPGPTPI-PIIGNV--FQLNPWDL----MGSFKELSK--KYG----PIF  68
 4 ALV--C-LLLTL------------SSRDKGKLPPGPRPL-SILGNL--LLLCSQDM----LTSLTKLSK--EYG----SMY  66
 5 GLL--L-LLVQR------------HPNTHDRLPPGPRPL-PLLGNL--LQMDRRGL----LKSFLRFRE--KYG----DVF  66
 6 CLL--I-LIAWK------------RTSRGGKLPPGPTPI-PFLGNL--LQVRIDAT----FQSFLKLQK--KYG----SVF  69
 7 VMV--L-MSVWQ------------QRKSKGKLPPGPTPL-PFIGNY--LQLNTEQM----YNSLMKISE--RYG----PVF  69
 8 VLY--L-RSSGS------------SSEEQGKEPPGPRPL-PLLGNM--LQLDLKKP----YCTLCELSK--KYG----SIF  77
 9 AFL--F-FAYFL------------KTRRPKNYPPGPWPL-PFLGNL--FTLDMEKS----HLQLQQFVK--KYG----NLF  79
10 LLL--V-DLMHR------------RQRWAARYPPGPLPL-PGLGNL--LHVDFQNT----PYCFDQLRR--RFG----DVF  69
11 FWV--I-RASRP------------QVPKGLKNPPGPWGW-PLIGHM--LTLG-KNP----HLALSRMSQ--QYG----DVL  74
12 TLA--Y-LFWPK------------RRCPGAKYPKSLLSL-PLVGSL-PFLPRHGHM----HNNFFKLQK--KYG----PIY  64
13 LL---P-LLAGA------------RLLWNWWKLRSLHLP-PLAPGFL--HLLQPDL-----PIYLLGLTQ--KFG----PIY  59
14 FFL----LKLNE------------KREKKPNLPPSPPN-LPIIGNL--HQLGN-LP----HRSLRSLAN--ELG----PLI  66
15 ASL----LIAKS------------KRKPKKNLPPGPPR-LPIIGNL--HQLGE-KP----HRAMVELSK--TYG----PLM  65
16 IIS----TLI--------------SKTTGRHLPPGPRGW-PVIGAL--PLLGA-MP----HVSLAKMAK--KYG----AIM  68
17 ----------------------------------------FGNM-FDLAGS-AP----YKKIACLKE--KYG----PIL  27
18 IGA----ILISK------------LRGKKFKLPPGPIPV-PIFGNW--LQVGDDLN----HRNLTDLAK--RFG----EIL  69
19 PAW--T-NGRGA------------SASLLSWDPVVCPC-SAASSRC--PGAAAPRP----RRDGPRRRP--RAK----ELM  64
20 FIF----LLTRK------------PKSKTPNLPPGPPGW-PIVGNL-FQVAGSGKQ----FFEYIRDLKP--KYG----SIF  65
21 LLY--L-YGTHS------------HGLFKKLGIPGPTPL-PFLGNI--LSY-HKGF----CMFDMECHK--KYG----KVW  72
22 LLK--W-YSTSA------------FSRLFKLGLRHPKPS-PFIGNL--TFF-RQGF----WESQMELRK--LYG----PLC  79
23 ALY--F-VRWNF------------GYWKRR-GIPHEEPH-LVMGNV--KGL-RSKY--HIGEIIADYYR--KFK-GSDPLP  73
24 LHY--Y-FTRTF------------NYWKKR-NVAGPKPV-PFFGNL--KDS-VLRR---KPQVMVYKSI--YDEFPNEKVV  70
25 LLLLL----IKAVQLYLHRQ-----WLLKALQQFPCPPSH-WLFGHIQELQQD-QEL----QRIQKWVET--FPS----ACP  87
26 ILVLGF---LKLIHLLLRRR-----TLAKAMDKFPGPPTH-WLFGHALEIQET-GSL----DKVVSWAHQ--FPY----AHP  88
27 ASWLL----ARILAWTYTFY-----DNCCRLRCFPQPPKRNWFLGHLGLIHSSEEGL----LYTQSLACT--FGD----MCC  89
28 HLPVPV-------PSGAKRA-----RFVYLVNKLPGPTAY-PVVGNAIEAIVPRNKL----FQVFDRRAK--LYG----PLY  74
29 ----------------------------------------------------------------------------------  0
30 LLL----------YHLKFK-----RLIDLISYMPGGPVL-PLVGHGHHFIGKPPHE-MVKKIFEFMETI--SKD----QVL  74
31 ----------------------------MTIKEMPQPKTF-GELKNL-PLLNTDK-P----VQALMKIAD--ELG----EIF  41
32 ---------------------------MLTQLPNPISV-PSWWQL-INWIAD--P----IGFQKKYSK--KYG----NIF  39
33 KRLR---QQGFRGNPYRFLV-----GDVKESGKMMQEALS-KPMEF-NNDIVPR--L----MPHINHTIN--TYG----RNS  99
34 -LYYLL-P--------YLKR-----AHLRDIPAP-GLAAF-TNFWLL--LQT-RRGHR---FVVVDNAHK--KYG-----K  70
35 PFRKSIDTFTETTNA-VKAPGMTEVQPFERIPG-PKGLPIVGTLFDY-FKKDGPKFS--KMFEVYRQRAL--EFG----NIY 121
36 LGRLRVPTGEGAGIS-TRSP-----RPFNEIPS-PGDNGW-LNLYHF-WRETGTHKV----HLHHVQNFQ--KYG----PIY  89
37 LQRAQALGTRAARV--PRTV-----LPFEAMPR-RPGNRW-LRLLQI-WREQGYEDL----HLEVHQTFQ--ELG----PIF  79
38 APKEVPLCPLMTD-GETRN------VTSLPGPT-NWPLLG-SLLEIF-WKGGLK-------KQHDTLAEYHKKYG----QIF  97
39 LPSDKATGAPGAGPGVRRRQ-----RSLEEIPR-LGQ----LRFFFQ-LFVQGYAL-----QLHQLQVLYKAKYG----PMW  96
40 I----Y-NIVWQ----LLY-----SLRK--DRPPLVFYWIPWVGSA--VVYGMKP-----YEFFEECQK--KYG----DIF  91
41 --------------GIRRR------QTGEPPLENGLIPYLGCA--LQFGANP-----LEFLRANQR--KHG----HVF  67
42 -AFAII-V-------PPLN-------FPKN-IP-TIPFYVVFLPVI--FPI-DQTEL--YDLYIRESME--KYG----AVK  70
43 LHTKYL-ERKFKA----KPL-----AVYVQDYTFGLITPL------V--LIY-YKSKGYVMQFACDLWDKVSDPKA--KTI 105
44 IYTKYL-MYKLGA----KPI-----THVIDDGFFGFRLPF------L--ITL-ANNQGRLIEFSVKRF---LSSPH--QTF  79
45 HHTTYL-MIKFKA----SPP-----LNYINKGFFGIQATF------T--ELK-HLICHTSIDYAIDQFNNVPFPHV--HTF  92
46 YKRRTL-VTKFHC----KPA-----RISPNKSWLEYLGIA------S--VVH-ADEMIRKGGLYSEIDGRFKSLDV--STF  80
47 TSIVPEAMPAATMSVLLLTG-----LFLLLWNYEGTSSIPGPGYCMGIGPLISHGRFLWM-GSRSACNYYNRVYGE--FMR  85
48 AFPS--------------------NRS-CPYQLPDGYAQLRD--------TPGPLH--------------------R  41
49 SFP---------------------QDRGCPYHPPAGYAPLRE--------GRPLS--------------------R  49
50 VPAS--------------------RAPGCPFDPAPDVTEAAR--------TEPVT--------------------R  41
51 VTFS--------------------TVRENYFGPPAEMQALRH--------KAPVT--------------------R  34
52 DLPL--------------------EMRRNGLNPTEELAQVRD--------RDGVI--------------------P  33
53 ----------------------DSFHV-DWYRTYAELRE--------TAPVT--------------------P  32
54 SFP---------------------FSRASGPEPPAEFAKLRA--------TNPVS--------------------Q  33
55 TLP---------------------MWRVDHIEPSPEMLALRA--------N--------------------R  31
56 ----------------------------------------------------------------------------------  0
57 G----------------------MADET-ALLDWLGTMRE------------K--------------------Q  41
58 VTE--------------------IPKFQSRAEEFFPIQWYK-------EMLNNS--------------------P  35
59 YANLIPMEEL-------------HSEKDRLFPFPIYDKLRR------------ES--------------------P  44
60 ILPQ-------------------GYAD-DEVIYPAFKWLRD--------EQPLA--------------------M  42
61 ----------------------DLYTQQVPHDIFARLRR--------EEPVY--------------------W  32
62 L-PPHVPEHLVFDFDMY-------NPSNLSAGVQEAWAVLQE--------SNVPD--------------------L  53
63 QKFLNVAKSNQIDPDAV------PISRLDSEGHSIFAEWRP--------KR--------------------P  51
64 W---RFPMLFAPVDDVT-------TIDDLTLDPYPIYRRMRV------------QN--------------------P  47
65 ------SQPPP-----SSD-----ETTLPIRQIPGDYGL-PGIGPIQDRLDYFYNQ--GREEFFKSRLQ--KYKS---TVY  57
```

FIGURE 1. (*continued*)

```
 1 TLYFGLKPIVVLHGYEAVKEAL-IDLGEEFSG-------------RGI-FPLAERANR------GFGIVFS--NGKKWKE 122
 2 TLYVGSQRMVVMHGYKAVKEAL-LDYKDEFSG-------------RGD-LPAF-HAHR------DRGIIFN--NGPTWKD 124
 3 TIHLGPKKIVVLYGYDIVKEAL-IDNGEAFSG-------------RGI-LPLIEKLFK------GTGIVTS--NGETWRQ 125
 4 TVHLGPRRVVVLSGYQAVKEAL-VDQGEEFSG-------------RGD-YPAFFNFTK------GNGIAFS--SGDRWKV 123
 5 TVHLGPRPVVMLCGVEAIREAL-VDKAEAFSG-------------RGK-IAMVDPFFR------GYGVIFA--NGNRWKV 123
 6 TVYLA-KAVVILCGHEAVKEAL-VDQADDFSG-------------RGE-MPTLEKNFQ------GYGLALS--NGERWKI 125
 7 TIHLGPRRVVVLCGHDAVREAL-VDQAEEFSG-------------RGE-QATFDWVFK------GYGVVFS--NGERAKQ 126
 8 TVHFGPKKVVVLAGYKTVKQAL-VNQAEDFGD-------------RDI-TPVFYDFNQ------GHGILFA--NGDSWKE 134
 9 CLDLAGKSIVIVTGLPLIKEVL-VHMDQNFIN-------------RPV-PPIRERSFK------KNGLIMS--SGQLWKE 136
10 SLQLAWTPVVVLNGLAAVREAL-VTHGEDTAD-------------RPP-VPITQILGFGPR----SQGVFLAR-YGPAWRE 130
11 QIRIGSTPVVVLSGLDTIRQAL-VRQGDDFKG-------------RPD-LYTFTLISN------GQSMSFSPDSGPVWAA 133
12 SVRMGTKTTVIVGHHQLAKEVL-IKKGKDFSG-------------RPQ-MATLDIASNN-----RKGIAFA-DSGAHWQL 123
13 RLHLGLQDVVVLNSKRTIEEAM-VKKWADFAG-------------RPE-PLTYKLVSKN-----YPDLSLG-DYSLLWKA 116
14 LLHLGHIPTLIVSTAEIAEEIL-KTHDLIFAS-------------RPS-TTAARRIFYD-----CTDVAFS-PYGEYWRQ 125
15 SLKLGSVTTVVATSVETVRDVL-KTYDLECCS-------------RPY-MTYPARITYN-----LKDLVFS-PYDKYWRQ 124
16 YLKVGTCGMAVASTPDAAKAFL-KTLDINFSN-------------RPP-NAGATHLAYN-----AQDMVFA-HYGPRWKL 127
17 WLKIGSMNTMVIQTANSASELF-RNHDVSFSD-------------RPI-VDVNLAHNYY-----KGSMALA-PYGNYWRF  86
18 LLRMGGQRNLVVVSSPELAKEVL-HTQGVEFGS-------------RTR-NVVFDIFTCK-----CQDMVFT-VYGEHWRK 128
19 AFSVGDTPAVVSSCPATAREVL---AHPSFAD-------------RPV-KRSARELMF------ARAIGFA-PNGEYWRR 120
20 TLKMGSRTMIIVASAELAHEAL-IQKGQIFAS-------------RPR-ENPTRTIFSC-----NKFSVNAAVYGPVWRS 125
21 GFYDGGQPVLAITDPDMIKLVL-VK--ECYSV-------------FTN-REPFGPVGFM-----KSAISIA--EDEEWKR 128
22 GYYLGRRMFIVISEPDMIKQVL-V---ENFSV-------------FTN-RMASGLEFKSV-----ADSVLFL--RDKRWEE 125
23 GIFLGHKPAAVVLDKELRKRVL-IKDFSNFAN-------------RGL-YYNEKDDPLT-----GHLVMV---EGEKWRS 130
24 GIYRMTTPSVLLRDLDIIKHVL-IKDFESFAD-------------RGV-EFSL--DGLG-----ANIFHA---DGDRWRS 125
25 HWLWGGKVRVQLYDPDYMKVIL-GR--SDPKS-------------HGS-YRFLAPWI-------GYGLLLL--NGQTWFQ 141
26 LWFGQFIGFLNIYEPDYAKAVY-SR--GDPKA-------------PDV-YDFFLQWI-------GRGLLVL--EGPKWLQ 142
27 WWVGPWHAIVRIFHPTYIKPVL-FA--PAAIVPKD----------KVF-YSFLKPWL-------GDGLLLS--AGEKWSR 146
28 RIWAGPIAQVGLTRPEHVELIL-RD--TKHIDK------------SLV-YSFIRPWL-------GEGLLTG--TGAKWHS 129
29 -----------------------------------------------------------------------------   0
30 KVWLGPELNVLMGNPKDVEVVL-GT--LRFNDK------------AGE-YKALEPWL-------KEGLLVS--RGRKWHK 129
31 KFEAPGRVTRYLSSQRLIKEAC-DESRFDKNL------------SQALKFVRDFA--------GDGLFTSWTHEKNWKK  99
32 SMQLAGIGSFVILEPQALQEIFTQDSRFDVG--------------------------------------DGDRHRR  77
33 FTWMGRIPRIHVMEPELIKEVL-THSSKYQKN-------------FDV-HNPLVKFL-------LTGVGSF--EGAKWSK 155
34 LVRIAPRHT-SIADDGAIQAVY-GH--G--NG-------------FLK-SDFYDAFVSI-----HRGLFNTR-DRAEHTR 124
35 YEKVGHFHCVVISSPGEYSR-L-VHAERQ----------------YPN-RREMVPIAYYRKQKGFDLGVVNS--QGEEWYR 181
36 REKLGNVESVYVIDPEDVAL-L-FKSEGP----------------NPE-RFLIPPWVAYHQYYQRPIGVLK--KSAAWKK 149
37 RYDLGGGAGMVCVMLPEDVEK-L-QQVDSL---------------HPH-RMSLEPWVAYRQHRGHKCGVFLL--NGPEWRF 139
38 RMKLGSFDSVHLGSPSLLEA-L-YRTESA----------------HPQ-RLEIKPWKAYRDHRNEAYGLMIL--RGQEWQR 157
39 MSYLGPQMHVNLASAPLLEQ-V-MRQEGK----------------YPV-RNDMELWKEHRDQHDLTYGPFTT--EGHHWYQ 156
40 SFVLLGRVMTVYLGPKGHEFVF-NAKLADVSA-------------EAA-YAHLTTPVF------GKGVIY------DCPN 144
41 TCKLMGKYVHFITNPLSYHKVL-CHGKYFD---------------WKK-FHFATSAKAFGHR---SIDPM--DGNTTEN 124
42 FFFGSRWNILVSRSEYLAQIFK-DE--DTFAK-------------SGN-QKKIPYSALAAYT---GDNVISA--YGAVWRN 129
43 GLKILGIPLIETKDPENVKAIL-ATQFNDFSL-------------GTR-HDFLYSLL-------GDGIFTL--DGAGWKH 161
44 MNRAFGIPIILTRDPVNIKAML-AVQFDEFSL-------------GLR-YNQFEPLL-------GNGIFTS--DGEPWKH 135
45 VTKVLGNELIMTKDPENIKVLL-RFPVFDKFD-------------YGTR-SSAVQPSL------GMGIFTL--EGENWKA 149
46 KSITLGKTTYVTKDIENIRHILSATEMNSWNL-------------GAR-PIALRPFI-------GDGIFAS--EGQSWKH 137
47 VWISGEETLIISKSSSMFHIMK-HNHYSSRFG-------------SKL-GLQ--CIGMH-----EKGIIFN-NNPELWKT 143
48 VTLYDGRQAWVVTKHEAARKLL-GDPR-LSSNRTD---------DNFFATSP-RFEAVRESPQA-FI-----GAL----DPPEHGT 104
49 VTLFDGRPVWAVTGHALARRLL-ADPR-LSTDRSHP------DFPVPAE-RFAGAQRRRVA-LL-----GV----DDPEHNT 112
50 ATLWDGSSCWVTVRHQDVRAVL-GDPR-FSADAHR------TGFPFLTA-GGREIIGTNPT-FL-----RM----DDPEHAR 104
51 TAFADGRPGWLVTGYADARAVL-SDSR-FTARGER------EHPAVPR-AATLEDERCRR-LI------------AGQFT  92
52 VGELYGAPAFLVCRYEDVRRIL-ADSN-RFSNAHTPMFA--IPSGGDVI-EDELAAMRAGN-LI-----GL----DPPDHTR 100
53 VRFL-GQDAWLVTGYDEAKAAL-SDLR-LSSDPKKKYPGVEVEFPAYLG-FPEDVRNYFAT-NM----GTS----DPPTHTR 101
54 VKLFDGSLAWLVTKHKDVCFVA-TSEK-LSKVRTR------QGFPELSA-SGKQAAKAKPT-FV-----DM----DPPEHMH  96
55 VRFPSGHEGWWVTGDREAKAVL-SDAA-FRPAGMP------PAAFT-PDSVILGSPGW-LV-----SH----EGREHAR  91
56 ----------LSRHADIFWAF-KATG-DAFRGPAPGE----LARYFSR-AATSPSLNLLA-ST-----LAMK--DPPTHTR  56
57 PVWQDRYGVWHVFRHADVQTVL-RDTATFSSDPTR-------FSSDPTR-VIEGASPTPGM-IH-----EI----DPPEHRA 104
58 VYFHEETNTWNVFQYEHVKQVL-SNYDFFSSDGQR-------TTIFVGDN-SKKKSTSPITN-LT-----NL----DPPDHRK  99
59 VRYDPLRDCWDVFKYDDVQFVL-KNPKLFSSKRG------------IQTESI-LT------M---DPPKHTK  93
60 AHIEGYDPMWIATKHADVMQIG-KQPGLFSNAEGSEILYDQNNEAFMRS-ISGGCPHVIDS-LT-----SM----DPPTHTA 112
61 NPESDGSGFWAVLRHKDIIEVS-RQPLLFSSAYENGGHRI-FNENEVGL-TNAGEAAVGVP-FI-----SL----DPPVHTQ 101
62 VWTRCNGGHWIATRGQLIREAY-EDYRHFSS----------ECPFIPR---EAGEAYDFI-PT-----SM----DPPEQRQ 110
63 FLRREDG-IFLVLRADHIFLLG-TDPRTRQI----------ETELMLNR-GVKAGAVFDFI-DH-----SMLF-SNGETHGK 113
64 VVHVASVRRTFLTKAFDTKMVK-DDPSRFSSDDPS------TPMKPAFQ-AHTLMRK-------------DGTEHAR 103
65 RANMPPGPFIASNPRVIVLLDA-KSFPVL---------------FDM-SKVEKKDL-------FTGTYM---PSTELTG 109
```

FIGURE 1. (*continued*)

```
 1 IRRFSLMTLRNFGMGK-------RSIEDRVQEEARCLVEELRKTKASPCD---------PTFILGCAPCNVICSIIF----H   184
 2 IRRFSLTTLRNYGMGK-------QGNESRIQREAHFLLEALRKTQGQPFD---------PTFLIGCAPCNVIADILF----R   186
 3 LRRFALTTLRDFGMGK-------KGIEERIQEEAHFLVERIRKTHEEPFN---------PGKFLIHAVANIICSIVF----G   187
 4 LRQFSIQILRNFGMGK-------RSIEERILEEGSFLLADVRKTEGEPFD---------PTFVLSRSVSNIICSVLF----G   185
 5 LRRFSVTTMRDFGMGK-------RSVEERIQEEAQCLIEELRKSKGALMD---------PTFLFQSITANIICSIVF----G   185
 6 LRRFSLTVLRNFGMGK-------RSIEERIQEEAGYLLEELHKVKGAPID---------PTFYLSRTVSNVICSVVF----G   187
 7 LRRFSIATLRDFGVGK-------RGIEERIQEEAGFLIDALRGTGGANID---------PTFFLSRTVSNVISSIVF----G   188
 8 MRRFALTNLRDFGMGK-------KGSEEKILEEIPYLIEVFEKHEGKAFD---------TTQSVLYAVSNIISAIVY----G   196
 9 QRRFALMTLRNFGLGK-------KSLEERIQEEARHLTEAMEKEGGQPFD---------AHFKINNAVSNIICSITF----G   198
10 QRRFSVSTLRNLGLGK-------KSLEQWVTEEAACLCAAFANHSGRPFR---------PNGLLDKAVSNVIASLTC----G   192
11 RRRLAQNGLKSFSIASDPASSTSCYLEEHVSKEAEVLISTLQELMAGPGHFN------PYRYVVVSVTNVICAICF----G   204
12 HRRLAMATFALFKDGD--------QKLEKIICQEISTLCDMLATHNGQSID--------ISFPVFVAVTNVISLICF----N   185
13 HKKLTRSALLLGIRDS---------MEPVVEQLTQEFCERMRAQPGTPVA---------IEEEFSLLTCSIICYLTF----G   191
14 VRKICVLELLSIKRVN-------SYRSIREEEVGLMMERISQS-CSTGEA-------VNLSELLLLLSSGTITRVAF----G   188
15 VRKLTVVELYTAKRVQ-------SFRHIREEEVASFVRFNKQA-ASSEET-------VNLSQKILKMSGSVICRIGF----G   187
16 LRKLSNLHMLGGKALE-------NWANVRANELGHMLKSMSDM-SREGQR-VV------VAEMLTFAMANMIGQVMLS---K   191
17 SRRICTVEMFVHKRIN-------ETTNIRQESVDKMLRLDEEKASSSGGGGEGIE----VTRYMFLASFNMVGNMIFS---K   154
18 MRRIMTVPFFTNKVVQ-------QYRYGWEAEAAAVVDDVKKNPAAATEG-IV------IRRRLQLMMYNNMFRIMF----D   192
19 LRRVASTHLFSPRRVA-------SHEPGRQGDAEAMLRSIAAE-QSASGA-VA------LRPHLQAAALNNIMGSVF----G   183
20 LRRNMVQNMLSPSRLK-------EFREFREIAMDKLIERIRVD-AKENND-VV------WALKNARFAVFYILVAMCF----G   189
21 LRSLLSPTFTSGKLKE--------MVPIIAQYGDVLVRNLRRE-RETGKPVT------LKDVFGAYSMDVITSSSF----G   190
22 VRGALMSAFSPEKLNE--------MVPLISQACDLLLAHLKRY-AESGDAFD------IQRCYCNYTTDVVASVPF----G   197
23 LRTKLSPTFTAGKMKY--------MYNTVLEVGQRLLEVMYEK-LEVSSELD------MRDILARFNTDVIGSVAF----G   192
24 LRNRFTPLFTSGKLKS--------MLPLMSQVGDRFINSIDEV-SQTQPEQS------IHNLVQKFTMTNIAACVF----G   187
25 HRRMLTPAFHYDILKP---------YVGLMADSVRVMLDKWEELLGQDSP-------LEVFQHVSLMTLDTIMKCAFSHQ-G   206
26 HRKLLTPGFHYDVLKP---------YVAVFTESTRIMLDKWEEKAREGKS-------FDIFCDVGHMALNTLMKCTFGRG-D   207
27 HRRMLTPAFHFNILKP---------YMKIFNESVNIMHAKWQLLASEGSAR------LDMFEHISLMTLDSLQKCVFSFD-S   212
28 HRKMITPTFHFKILDI---------FVDVFVEKSEILVKKLQSKVG-GKD-------FDIYPFITHCALDIICETAMGIQ-M   193
29 -----------------------------------------------------------------------------------  0
30 RRKIITPAFHFKILDQ---------FVEVFEKGSRDLLRNMEQDRLKHGDSG-----FSLYDWINLCTMDTICETAMGVS-I   196
31 AHNILLPSFSQQAMKG---------YHAMMVDIAVQLVQKWERLNADEH------IEVPEDMTRLTLDTIGLCGFNYR-F   163
32 ERKLLMPPFHGERLQA---------YAMEKGIQAYAQQICLITNQIASEWQIGQP--VFARSAMQKLSLEVIIQIVFGLADG  148
33 HRRIISPAFTLEKLKS---------MLPAFAICYHDMLTKWEKIAEKQGSHE-----VDIFPTFDVLTSDVISKVAFG---S  220
34 KRKTVSHTFSMKSIGQ---------FEQYIHGNIELFVKWWNRMADTQRNPKTGFASLDALNWFNYLAFDIIGDLAF---G  193
35 QRTVVSKKMLKLAEVS---NFST--QMGEVSDDFVKRLSHVRDSHGEI-------PALERELFKWAMESIGTFLF----E  245
36 DRVALNQEVMAPEATK---NFLP--LLDAVSRDFVSVLHRRIKKAGSGNYS------GDISDDLFRFAFESITNVIF----G  216
37 NRLRLNPEVLSPNAVQ---RFLP--MVDAVARDFSQALKKKVLQNARGSLT------LDVQPSIFHYTIEASNLALF----G  206
38 VRSAFQKKLMKPVEIM---KLDK--KINEVLADFLERMDELCDERGRI---------PDLYSELNKWSFESICLVLY----E  221
39 LRQALNQRLLKPAEAA---LYTD--AFNEVIDDFMTRLDQLRAESASGNQV------SDMAQLFYYFALEAICYILF----E  223
40 SRLMEQKKFVKGALTK--EAFKS--YVPLIAEEVYKYFRDSKNFRLNERTTGT----IDVMVTQPEMTIFTASRSLL----G  214
41 INDTFIKTLQGHALNS---------LTESMMENLQRIMRPPVSSNSKTA---------AWVTEGMY-SFCYRVMF------  180
42 YRNAVTNGLQHFDDAP---------IFKNAKILCTLIKNRLLEGQTSI--------PMGPLSQ-RMALDNISQVAL----G  188
43 SRTMLRPQFAREQVSH----VKL--LEPHMQVLFKHIRKHHGQT---------F-IQELFFRLTVDSATEFLLG---E  220
44 SRIMLRPQFIKSQVSH----VNR--LEPHFNLLQKNITAQTDNY------------FDIQTLFFRFTLDTATEFLFG---Q  195
45 TRSVLRNMFDRKSIDK----VHD--FEPHFKTLQKRIDG-KVGY------------FDIQQEFLKLGLELSIEFIFG---Q  208
46 SRIMLRPVFAKEHVKQ----ITS--MEPYVQLIIKNHEGQP---------------LEFQTLAHLFTIDYSTDFLLG---E  197
47 TRPFFMKALSGPGLVR---------MVTVCAESLKTHLDRLEEVTNESG--------YVDVLTLL-RRVMLDTS------   199
48 RRRMTISEFTVKRIKG---------MRPEVEEVVHGFLDEMLA-AGPT-----------ADLVSQF-ALPVPSMV-----  157
49 QRRMLIPTFSVKRIGA---------LRPRIQETVDRLLDAMER-GPP------------AELVSAF-ALPVPSMV-----  165
50 LRRMLTADFIVKKVEA---------MRPEVQRLADDLVDRMTT-GRTS-----------ADLVTEF-ALPLPSLV-----  157
51 ARRMRQLTGRTERIVR---------EH------------LDAMEH-MGSP----------ADLVEHF-ALPVPSLV------  135
52 LRHILAAEFSVHRLSR---------LQPRIAEIVDSALDGLEQ-AGQP-----------ADLMDRY-ALPVSLLV------  153
53 LRKLVSQEFTVRRVVA---------MRPRVEQITAELLDEVGD-SGV-----------VDIVDRF-AHPLPIKV------  153
54 QRSMVEPTFTPEAVKN---------LQPYIQRTVDDLLEQMKQ-KGCANGP-------VDLVKEF-ALPVPSYI------  152
55 LRAIVAPAFSDRRVKL---------LVQQVEAIAAHLFETLAA-QPQP----------ADLRRHL-SFPLPAMV------  144
56 LRRLISRDFTMGQIDN---------LRRIVAARLDGITPALER-GEA----------VDLHREF-ALALPMLV------  108
57 LRKVVSSAFTPRTISD---------LEPRIRDVTRSLLAD-AG--ES----------FDLVDVL-AFPLPVTI------  154
58 ARSLLAAAFTPRSLKN---------WEPRIKQIAADLVEAIQK--NST---------INIVDDL-SSPFPSLV------  151
59 LRALVSRAFTPKAVKQ---------LETRIKDVTAFLLQEARQ--KST---------IDIIEDF-AGPLPVII------  145
60 YRGLTLNWFQPASIRK---------LEENIRRIAQASVQRLLD-FDGE---------CDFMTDC-ALYYPLHV------  165
61 YRKVIMPALSPARLQD---------IEQRIRVRAEALIERIPL--GEE---------VDLVPLL-SAFPLTI------  153
62 FRALANQVVGMPVVDK---------LENRIQELACSLIESLRP-QGQ---------CNFTEDY-AEPFPIRI------  162
63 RRSGLSKAFSFRMVEA---------LRPEIAKITECLWDDLQK-VDD---------FNFTEMY-ASQLPALT------  165
64 ERMAMARAFAPKAIAD---------HWAPIYRDIVNEYLDRLP-RGDT---------VDLFAEI-CGPVAARI------  156
65 GYRILSYLDPSEPNHT---------KLKQLLFNLIKNRRDYVIP-----------EFSSSFTDLCEVVEYDLATKG---K  166
```

FIGURE 1. (*continued*)

```
 1 KRF--DYKDQQFLNLMEKLNENIK---------ILSSPWIQI-CNNFSPIIDY----FPGTH--------------NKLL 234
 2 KHF--DYNDEKFLRLMYLFNENFH---------LLSTPWLQL-YNNFPSFLHY----LPGSH--------------RKVI 236
 3 DRF--DYEDKKFLDLIEMLEENNK---------YQNRIQTLL-YNFFPTILDS----LPGPH--------------KTLI 237
 4 SRF--DYDDERLLTIIRLINDNFQ---------IMSSPWGEL-YDIFPSLLDW----VPGPH--------------QRIF 235
 5 KRF--HYQDQEFLKMLNLFYQTFS---------LISSVFGQL-FELFSGFLKY----FPGAH--------------RQVY 235
 6 KRF--DYEDQRFRSLMKMINESFV---------EMSMPWAQL-YDMYWGVIQY----FPGRH--------------NRLY 237
 7 DRF--DYKDKEFLSLLRMMLGIFQ---------FTSTSTGQL-YEMFSSVMKH----LPGPQ--------------QQAF 238
 8 SRF--EYTDPLFTGMADRAKESIH---------LTGSASIQM-YNMFPWLGPW----INNLT--------------RLKK 246
 9 ERF--EYHDGQFQELLKLFDEVMY---------LEASMLCQL-YNIFPWIMKF----LPGAH--------------QTLF 248
10 RRF--EYDDPRFLRLLDLAQEGLK---------EESGFLREV-LNAVPVLLH-----IPALA--------------GKVL 241
11 RRY--DHNHQELLSLVNL-NNNFG---------EVVG-SGNP-ADFIPI-LRY----LPNPS--------------LNAFK 252
12 TSY--KNGDPELNVIQNY-NEGII---------DNLS-KDSL-VDLVPW-LKI----FPNKT--------------LEKLK 233
13 DKI---KDDNLMPAYYKCIQEVLK---------TWSH-WSIQIVDVIPF-LRF----FPNPG--------------LRRLK 227
14 KKYEGEEERKNKFADLATELTTLM---------GAFFVGDYFPSFAWVD-VLT-------GM--------------DARLK 238
15 INLEG-SKLENTYQEIIVQAFEVL---------GSLAAVDYFPVIGTIIDRIT-------GL--------------HAKCE 237
16 RVFVDKGVEVNEFKDMVVELMTIA---------GYFNIGDFIPCLAWMD-LQ--------GI--------------EKRMK 240
17 DLTDPESKQGSEFFNAMIGIMEWA---------GVPNISDIFPCLKMFD-VQ--------GL--------------RKKME 203
18 RRF--ESEDDPLFLKLKALNCERS---------RLAQSFEYNYGDFIPI-LRP-------FL--------------RNYLK 240
19 TRYDVTSGAGAAEAEHLKSMVREGFELL------GAFNWSDHLP-WLAHLYDPSN--------------------VTRRC 236
20 VEMDNEEMIERV--DQMMKDVLIV---------LDPRIDDFLPILRLFV-GY--------KQ--------------RKRVN 236
21 VNV--DSLNNPQDPLVENTKKLLR---------FDFLDPFFLSITVFPF-L-------IPILE--------------VLNIC 239
22 TPV--DSWQAPEDPFVKHCKRFFE---------FCIPRPILVLLLSFPS-I------MVPLA--------------RILPN 246
23 IEC--NSLRNPHDRFLAMGRKSIEV--------PRHNALIMAFIDSFPE-L---------SR--------------KLGMR 239
24 LNL----DEGMLKTLEDLDKHIFTV--------NYSAELDMM----YPG-I---------LK--------------KLNGS 228
25 SIQVD-RNSQSYIQAISDLNNLVFSRV------RNAFHQNDTIYSLTSA-GRW----THRA--------------CQLAH 260
26 TG-LG-HRDSSYYLAVSDLTLLMQQRL------VSFQYHNDFIYWLTPH-GRR----FLRA--------------CQVAH 260
27 HCQ---EKPSEYIAAILELSALVTKRH------QQILLYIDFLYYLTPD-GQR----FRRACRL-----------VHDFT 267
28 NAQ-E-ESESEYVKAVYEISELTMQRS------VRPWLHPKVIFDLTTM-GKR----YAECLRILHGFT-------NKVIQ 254
29 --------------------------------------------------------------------------------- 0
30 NAQ-S-NADSEYVQAVKTISMVLHKRM------FNILYRFDLTYMLTPL-ARA----EKKALNVVHQF--------TEKII 256
31 NSF---YRDQPHPFITSM--------------VRALDEAMNKLQRANP--------DDPAY--------------DENKR 204
32 ERY---QQIKPLFTDWLN--------------MTDSPLRSSMLFLKSL-QKD----WG--------------TWTPW 189
33 TYEEGGKIF-RLLKELMDLTIDCM--------RDVYIPGWSYLPTKRN-KRM----KEIN--------------KEITD 271
34 APF--GMLDKGKD-FAEMRKTPDSPPSYVQAVEVLNRRGEVSATLGCYPA-L------KPFAKYL-----------PDSFF 253
35 ERI---GCLGQETSPMAQTFIANL--------EGFFKTLQPLMYNLPT-YKL----WS--------------TKLWK 292
36 ERQ---GMLEEVVNPEAQRFIDAI--------YQMFHTSVPMLNLPPDLFRL----FR--------------TKTWK 264
37 ERL---GLVGHSPSSASLNFLHAL--------EVMFKSTVQLMFMPRSLSRW----TS--------------PKVWK 254
38 KRF---GLLQKETEEEALTFITAI--------KTMMSTFGKMMVTPVELHKR----LN--------------TKVWQ 269
39 KRI---GCLQRSIPEDTVTFVRSI--------GLMFQNSLYATFLSWTRPV--------------------LPFWK 269
40 KEM--------RAKLDTDFAYLY--------SDLDKGFTPINFVFPN--------LP--------------LEHYR 252
41 -EA--GYLTIFGRDLTRRDTQKAH--------ILNNLDNFKQFDKV---FPA----LVAGLP-----------IHMFR 229
42 FDF--GALTHEKNAFHEHLIRIKK--------QIF-HPFFLT---FPF-L------DVLPIP--------------SRKKA 234
43 SAESLRDESVGLTPTTKDFDGRNEFAD------AFNYSQTNQAYRFLLQ-QMY----WILN--------------GSEFR 275
44 SVHSL----------NDGENSLQFLE------AFTKSQAILATRANLH-ELY----FLAD--------------GIKFR 239
45 VV------------SEDVPHYDDFTQ------AWDRCQDYMMLRLLLG-DFY----WMAN--------------DWRYK 250
46 SCDSLKDFLGEESNSTLDTSLRLAFAS------QFNKTQQQMTIRFMLG-KLA----FLMY--------------PKSFQ 252
47 -NT--LFLRIPLDESAIVVKIQGY---------FDAWQALLIKPDIF---FKI--------------------SWLYK 242
48 -IC--RLLGVP----------------------YADHEF-FQDAS----KRLVQ-----------STDAQ 186
49 -IC--ALLGVP----------------------YADHAF-FEERS----QRLLR-----------GPGAD 194
50 -IC--LLLGVP----------------------YEDHAF-FQERS----RVLLTL----------RSTPE 187
51 -IA--ELLGVP----------------------PADREQ-FQH-----DTLRW-----------RST-E 161
52 -LC--ELLGVP----------------------YADRDE-LRDRT----ARLLDL----------SASAE 182
53 -IC--ELLGVD----------------------EKYRGE-FGRWS----SEILVMD---------PERAE 184
54 -IY--TLLGVP----------------------FNDLEY-LTQ----QNAIRTNG---------SSTAR 182
55 -IS--ALMGVL----------------------YEDHAF-FAGLS----DEVMTHQHES------GPRSA 178
56 -FA--ELFGMP----------------------QDDMFE-LAAGI----GTILEGLGPHAS-----DPQLA 144
57 -VA--ELLGLP----------------------PMDHEQ-FGDWSGALVDIQMDDPTDPALA----ERIAD 195
58 -IA--DLFGVP----------------------VKDRYQ-FKKWV----DILFQ-PYDQERL----EEIEQ 187
59 -IA--EMLGAP----------------------IEDRHL-IKTYS----DVLVAGAKDSSDKAV---ADMVH 184
60 -VM--TALGVP----------------------EDDEPL-MLKLTQDFFGVHEPDEQAVAAPRQSADEAAR 210
61 -LA--ELLGLD----------------------PDCWYE-LYNWT----NAFVG--EDDPEF----RKSPE 188
62 -FM--LLAGLP----------------------EEDIPH-LKYLT----DQMTR-----------PDGSM 191
63 -IA--SVLGLP----------------------SEDTPF-FTRLV----YKVSRCL-SPSWR----DEEFE 201
64 -LA--HILGIC----------------------EASDVE-IIRWS----QRLIDGAGNFGWR----SELFE 193
65 AAFNDPAEQAAFNFLSRAFFGVKPIDT------PLGKDAPSLISKWVLF-NLA----PILS--------------VGLPK 221
```

FIGURE 1. (*continued*)

```
 1 KNVAFMKSYILEKVKEHQESMDMNNPQD---FIDCFLM-KMEKEK--HNQPSE----FTIESLENTAVDLFGAGTETTSTTL  306
 2 KNVAEVKEYVSERVKEHHQSLDPNCPRD---LTDCLLV-EMEKEK--HSAERL----YTMDGITVTVADLFFAGTETTSTTL  308
 3 KNTETVDDFIKEIVIAHQESFDASCPRD---FIDAFIN-KMEQEK--EN--SY----FTVESLTRTTLDLFLAGTGTTSTTL  307
 4 QNFKCLRDLIAHSVHDHQASLDPRSPRD---FIQCFLT-KMAAEK--EDPLSH----FHMDTLLMTTHNLLFGGTKTVSTTL  307
 5 KNLQEINAYIGHSVEKHRETLDPSAPKD---LIDTYLL-HMEKEK--SNAHSE----FSHQNLNLNTLSLFFAGTETTSTTL  307
 6 NLIEELKDFIASRVKINEASFDPSNPRD---FIDCFLI-KMYQDK--SDPHSE----FNLKNLVLTTLNLFFAGTETVSSTL  309
 7 QLLQGLENFIAKKVEHNQRTLDPNSPRD---FIDSFLI-RMQEEE--KNPNTE----FYLKNLVMTTLNLFIGGTETVSTTL  310
 8 NIADMKMEVI-ELVRGLKETLNPHMCRG---FVDSFLV-RKLEES--GHMDSF----YHDDNLVFSVGNLFSAGTDTTGTTL  317
 9 SNWKKLELFVSRMLENHKKDWNPAETRD---FIDAYLK-EMSKYP--GSATSS----FNEENLICSTLDLFLAGTETTS-DM  319
10 RFQKAFLTQLDELLTEHRMTWDPQPPRD---LTEAFLA-EMEKAK--GNPESS----FNDENLRIVVADLFSAGMVTTSTTL  313
11 DLNEKFYSFMQKMVKEHYKTFEKGHIRD---ITDSLIE-HCQEKQLDENANVQ----LSDEKIINIVLDLFGAGFDTVTTAI  326
12 SHVKIRNDLLNKILENYKEKFRSDSITN---MLDTLMQ-AKMNSDNGNAGPDQDSELLSDNHILTTIGDIFGAGVETTTSVV  311
13 QAIEKRDHIVEMQLRQHKESLVAGQWRD---MMDYMLQ-GVAQPSMEE-GSGQ----LLEGHVHMAAVDLLIGGTETTANTL  300
14 RNHGELDAFVDHVIDDHLLSRKANGSDG---VEQKDLVDVLLHLQKDS-SLGVH---LNRNNLKAVILDMFSGGTDTTAVTL  313
15 KVFHGIDSFFDQAIQRHI-------DDP---SIKDDIIDLLLKMERGEGSLGEYE--LTREHTKGILMNILTAGIDTSAQTM  307
16 RLHKKFDALLTKMFDEHKATTYERKGKP-------DFLDVVME-NGDNSEGER----LSTTNIKALLLNLFTAGTDTSSSAI  310
17 RDMGKGKEIT-KKFIEERIEERKKGEKN---RSIKDLLDVLIDFEGSGKDEPDK---LSEDEITVIILEMFLAGTETTSSSV  278
18 LCKEVKDKRIQLFKDYFVDERKKIGSTK---KMDNNQLKCAIDHILEAKEKGE----INEDNVLYIVENINVAAIETTLWSI  315
19 AALPRVQTFVRGVIDEHRRRRQ-NSAAL---NDNADFVDVLLSLEGDEK-------LGDDDMVAILWEMVFRGTDTTALLT   306
20 EVRKRQIETL-VPLIEKRRSVVQN-PGSD---KTAASFSYLDTLFDVKVEGRKSG---PTNAELVTLCSEFLNGGTDTTATAL  311
21 VFPREVTNFLRKAVKRMKESRLEDTQKH---RVD-FLQLMIDSHKNSKETESHKA--LSDLELVAQSIIFIFAGYETTSSVL  315
22 KNRDELNGFFNKLIRNVIALRDQQAAEE---RRD-FLQMVLDAR-HSASPMGVQDFDLTVDEIVGQAFIFLIAGYEIITNTL  323
23 VLPEDVHQFFMSSIKETVDYREKNNIR----RND-FLDLVLDLKNNPESISKLGG--LTFNELAAQVFVFFLGGFETSSSTM  314
24 LFPKVVSKFFDNLTKNVLEMRKGTPSY----QKD-MIDLIQELREKKTLELSRKHENLTDGVISAQMFIFYMAGYETSATTM  305
25 QHTDQVIQLRKAQLQKEGELEKIKRKR----HLD-FLDILLL--AKMENGSI-----LSDKDLRAEVDTFMFEGHDTTASGI  330
26 DHTDQVIRERKAALQDEKVRKKIQNRR----HLD-FLDILLG--ARDEDDIK-----LSDADLRAEVDTFMFEGHDTTTSGI  330
27 DDVIQERRRTLPSQGVDDFLQAKAKSK----TLD-FIDVLLL--SKDEDGKK-----LSDEDIRAEADTFMFEGHDTTASGL  337
28 ERKSLRQMTGMKPTISNEEDELLGKKK----RLA-FLDLLLE---ASENGTK-----MSDTDIREEVDTFMFEGHDTTSAGI  323
29 ----------------------------KMA-FLDTLLS---SKVDGRP-----LTSQELNEEVSTFMFEGHDTTTSGV   42
30 VQRREELIREGSSQESSNDDADVGAKR----KMA-FLDILLQ---STVDERP-----LSNLDIREEVDTFMFKGHDTTSSAL  325
31 QFQEDIKVMNDLVDKIIADRKASG-EQ----SDD-LLTHMLN--GKDETGEP-----LDDENIRYQIITFLIAGHETTSGLL  273
32 GQKHKQRSIYDLLQAEIEEKRTKENEQ----RGD-VLSLMMA--ARDENGQA-----MTDEELKDELLTILFAGHETTATTI  259
33 MLRFIINKRMKALKAGEPGEDDLLGVL----LES-NIQEIQK--QGNKKDGG-----MSINDVIEECKLFYFAGQETTGVLL  341
34 RDGIQAVEDLAGIAVARVNERLRPEVMANNTRVD-LLARLMEGKDSNGEK-------LGRAELTAEALTQLIAGSDTTSNTS  327
35 QFENYSDNVIDIGRSLVEKKWH--PCKMEV-TQN-LHLISYL--VNNGS--------MSTKEVTGLIVDLMLAAVETTSSAT  360
36 DHVAAWDVIFSKADIYTQNFYWELRQKGSVHHDY-RGILYRL--LGDSK--------MSFEDIKANVTEMLAGGVDTTSMTL  323
37 EHFEAWDCIFQYGDNCIQKIYQELAF-SRPQQ-Y-TSIVAEL--LLNAE-------LSPDAIKANSMELTAGSVDTTVFPL  323
38 AHTLAWDTIFKSVKPCIDNRLQ---------RYS-QQPGADF--LCDIYQQDH----LSKKELYAAVTELQLAAVETTANSL  335
39 RYLDGWNAIFSFGKKLIDEKLEDMEAQLQAAGPD-GIQVSGY--LHFLLASGQ----LSPREAMGSLPELLMAGVDTTSNTL  344
40 KRDHAQKAISGTYMSLIKERRKNNDIQD----RD-LIDSLMK--NSTYKDGVK----MTDQEIANLLIGVLMGGQHTSAATS  323
41 TAHNAREKLAESLRHENLQKRESISEL----ISLRMFLNDTLST-----------FDDLEKAKTHLVVLWASQANTIPAT  294
42 FKDVVVSFRELLVKRVQDELVNNYKFEQTTFAASDLIRAHNNE--IIDYKQ-------LTD-----NIVIILVAGHENPQLLF  302
43 KSIAIVHKFADHYVQKALELTDEDLEK----KEG-YVFLFEL--AKQT--------RDPKVLRDQLNILVAGRDTTAGLL  341
44 QYNKMVQDFSQRCVDKVLNMSNSEIDK----LDR-YFFLYEM--VKIT--------RNPQVLRDQCLNILLAGRDTTASLL  305
45 QSNQIVQAFCDYLVQKSLENTC---------NDK-FVFVHQL--AKHT--------TNKTFIRDQALSLIMASRDTTAELM  311
46 YSIQMQKDFVDVYIDRVVGMSEEELNNH---PKS-YVLLYQL--ARQT--------KNRDILQDELMSILLAGRDTTASLL  319
47 KYEKSVKDLKDAIEVLIAEKRCRISTEEKLEECMDFATELILAEKRGD--------LTRENVNQCILEMLIAAPDTMSVSL  315
48 SALTARNDLAGYLDGLITQFQTEPGAG----LVGALVADQL-------ANGE----IDREELISTAMLLLIAGHETTASMT  252
49 DVNRARDELEEYLGALIDRKRAEPGDG----LLDELIHRDH-------PDGP----VDREQLVAFAVILLIAGHETTANMI  260
50 EVRAAQDELLEYLARLARTKRERPDDA----IISRLV----------ARGE----LDDTQIATMGRLLLVAGHETTANMT  249
51 EVTEAFVSLGGQLRLVRLKRPTDDA----LLSGLIA---------ADPA----LTDEELASIAFLLLVAGHTTAHQI  224
52 QRAVAQREDRRYMATLVTRAQEQPGDD----LLGILARK--------IGDN----LSTDELISIISLIMLGGHRTTASMI  246
53 QRGQAAREVVNFILDLVERRRTEPGDD----LLSALIRVQDD------DDGR----LSADELTSIALVLLLAGFEASVSLI  251
54 EASAANQELLDYLAILVEQRLVEPKDD----IISKLCTEQV------KPGN----IDKSDAVQIAFLLLLVAGNATMVNMI  248
55 SRLAWEEELRAYIRGKMRDKRQDPDDN----LLTDLLAAVD-------QGK-----ASEEEAVGLAWGMLVAGVDTTVAQI  243
56 AADAASARVQAYFGDLIQRKRTDPRRD----IVSMLVGAHDD------DADT----LSDAELISMLWGMLLGGFVTTAASI  211
57 VLNPALAPLTAYLKARCAERRADPGDD----LISRLVLAEV-------DGRA----LDDEEAANFSTALLLAGHITTTVLL  261
58 EKQRAGAEYFQYLYPIVIEKRSNLSDD----IISDLIQAEV-------DGET----FTDEEIVHATMLLLGAGVETTSHAI  253
59 NRRDGHAFLSDYFRDILSKRRAEPKED----LMTMLLQAEI-------DGEY-----LTEEQLIGFCILLLVAGNETTNLI  250
60 RFHETIATFYDYFNGFTVDRRSCPKDD----VMSLLANSKL------DGNY-----IDDKYINAYYVAIATAGHDTTSSSS  276
61 DMAKVLGEFMGFCQELFESRRANPGPD----IATLLANAEI-------NGQP-----VALRDFIGNLTLTLVGGNETTRNSI  254
62 TFAEAKEALYDYLIPIIEQRRQRKPGTD---AISIVANGQV-------NGRP-----ITSDEAKRMCGLLLVGGLDTVVNFL  257
63 EIEASAIELQDYVRSVIADSGRRMRDD----FLSRYLKAVR-------EAGT-----LSPIEEIMQLMLILAGSDTTRTAM  267
64 RSDEANAEMNCLFNDLVKKHRSAPNPSA---FATMLN----------APDP-----IPLSQIYANIKIAIGGGVNEPRDAL  256
65 EVEEATLHSVRLPPLLVQNDYHRLYEF----FTS-AAGSVLD--EAEQSGIS---RDEACHNILFAVCFNSWGGFKILLPSL  293
```

FIGURE 1. (continued)

```
 1 RYALLLLLKH-PEVTAKVQEEIERVIG----RNR------SPCMQDRSHMPYTDAVVHEVQRYIDLLPTSL--PHAVTCDIK   375
 2 RYGLLILMKY-PEIEEKLHEEIDRVIG----PSR------IPAIKDRQEMPYMHAVVHEIQRFITLVPSNL--PHEATRDTI   377
 3 RYGLLILLKH-PEIEEKMHKEIDRVVG----RDR------SPCMADRSQLPYTDAVIHEIQRFIDFLPLNV--PHAVIKDTK   376
 4 HHAFLALMKY-PKVQARVQEEIDLVVG----RAR------LPALKDRAAMPYTDAVIHEVQRFADIIPMNL--PHRVTRDTA   376
 5 RYGFLLMLKY-PHVAERVYREIEQVIG----PHR------PPELHDRAKMPYTEAVIYEIQRFSDLLPMGV--PHIVTQHTS   376
 6 RYGFLLLMKY-PEVEAKIHEEINQVIG----THR------TPRVDDRAKMPYTDAVIHEIQRLTDIVPLGV--PHNVIRDTH   378
 7 RYGFLLLMKH-PEVEAKVHEEIDRVIG----KNR------QPKFEDRAKMPYMEAVIHEIQRFGDVIPMSL--ARRVKKDTK   379
 8 RWGLLLLMTKY-PHIQDQVQEEISRVIG-----SR------QTLVEDRKNLPYTDAVIHETQRLANIVPMSV--PHTTSRDVT   385
 9 RWGLLFMALY-PEIQEKVHAEIDSVIG----QWQ------QPSMASRESLPYTNAVIHEVQRMGNILPLNV--PREVTVDTT   388
10 AWGLLLMILH-PDVQRRVQQEIDDVIG----QVR------RPEMGDQAHMPYTTAVIHEVQRFGDIVPLGM--THMTSRDIE   382
11 SWSLMYLVMN-PRVQRKIQEELDTVIG----RSR------RPRLSDRSHLPYMEAFILETFRHSSFVPFTI--PHSTTRDTS   395
12 KWTLAFLLHN-PQVKKKLYEEIDQNVG----FSR------TPTISDRNRLLLLEATIREVLRLRPMELI--PHKANVDSS   380
13 SWAVVFLLHH-PEIQQRLQEELDHELGPGASSSR------VP-YKDRARLPLLNATIAEVLRLRPVVPLAL--PHRTTRPSS   372
14 EWAMAELIKH-PDVMEKAQQEVRRVVG----KKA------KVEEEDLHQLHYLKLIIKETLRLHPVAPLLV--PRESTRDVV   382
15 TWAMTHLLAN-PRVMKKLQAEIREKIK----NID------EITDDDVEQLDYFKLVLKETFRISPIVPVLV--PRVAAKDLK   376
16 EWALAEMMKN-PAILKKAQAEMDQVIG----RNR------RLLESDIPNLPYLRAICKETFRKHPSTPLNL--PRISNEPCI   379
17 EWALTELLRH-PQAMAKVKLEILQVIG----PNK------KFEECDIDSLPYMQAVLKEQLRLHPPLPLLI--PRKAIQDTK   347
18 EWGIAELVNH-PEIQAKLRHELDTKLG----PGV------QITEPDVQNLPYLQAVVKETLRLRMAIPLLV--PHMNLHDAK   384
19 EWCMAELVRH-PAVQARVRAEVDAAVG----AGG------CPTDADVARMPYLQAVVKETLRAHPPGPLLSWARLATADVPL   377
20 EWGIGRLMEN-PTIQNQLYQEIKTIVG-----DK------KVDENDIEKMPYLNAVVKELLRKHPPTYFTL--THSVTEPVK   379
21 SFIMYELATH-PDVQQKLQEEIDAVLP----NKA------PPTYDTVLQMEYLDMVVNETLRLFPIAMRLE---RVCKKDVE   383
22 SFATYLLATN-PDCQEKLLREVDVFKEK---HMA------PEFCSLEEGLPYLDMVIAETLRMYPPAFRFT---REAAQDCE   392
23 GFALYELAQN-QQLQDRLREEVNEVFDQF---KED------NISYDALMNIPYLDQVLNETLRKYPVGSALT---RQTLNDYV   384
24 TYLFYELAKN-PDIQDKLIAEIDEVLSR---HDG------NITYECLSEMTYLSKVFDETLRKYPVADFTQ---RNAKTDYV   374
25 SWILYALATH-PKHQERCREEIHSLLG----DGA------SITWNHLDQMPYTTMCIKEALRLYPPVPGIG--RELSTPVTF   399
26 SWFLYCMALY-PEHQHRVREEVREILG----DQD------FFQWDDLGKMTYLYMCIKESFRLYPPVPQVY--RQLSKPVTF   399
27 SWVLYHLAKH-PEYQERCRQEVQELLKDR--EPK------EIEWDDLAQLPFLTMCIKESLRLHPPVPAVS--RCCTQDIVL   408
28 CWALFLLGSH-PEIQDKVYEELDHIFQG---SDR------STTMRDLADMKYLERVIKESLRLFPSVPFIG--RVLKEDTKI   393
29 GFAVYLLSRH-PDEQEKLFNEQCDVMGA---SGLGR----DATFQEISTMKHLDLFIKEAQRLYPSVPFIG--RFTEKDYVI   114
30 MFFFYNIATH-PEAQKKCFEEIRSVVG----NDKST----PVSYELLNQLHYVDLCVKETLRMYPSVPLLG--RKVLEDCEI   396
31 SFALYFLVKN-PHVLQKAAEEAARVLV----DP-------VPSYKQVKQLKYVFMVLNEALRLWPTAPAFS--LYAKEDTVL   341
32 AWAFYQILKN-VNVQEKLQQELDR-LGA---NP-------NP--MEIAQLPYLTAVSQETLRMYPVLPTLF--PRITKSSIN   325
33 TWTTILLSKH-PEWQERAREEVLQAFG----KN-------KPEFERLNHLKYVSMILYEVLRLYPPVIDLT--KIVHKDTKL   409
34 CAILYWCMRT-PGVIEKLHKALDEAIP----QDVD-----VPTHAMVKDIPYLQWVIWETMRIHSTSAMGL--PREIPAGNP   397
35 VWCLYNLAKN-PQVQEKLFQEITEAQA-----KNNG----TISAEDLCKLPMVKAVVKETLRLYPITYSTS--RNI-AEDME   429
36 QWHLYEMARN-LKVQDMLRAEVLAARH-----QAQG----DMATM-LQLVPLLKASIKETLRLHPISVTLQ--RYL-VNDLV   403
37 LMTLFELARN-PNVQQALRQESLAAAA-----SISE----HPQKA-TTELPLLRAALKETLRLYPVGLFLE--RVA-SSDLV   391
38 MWILYNLSRN-PQAQRRLLQEVQSVLP-----DNQT----PRAED-LRNMPYLKACLKESMRLTPSVPFTT--RTL-DKPTV   403
39 TWALYHLSKN-PEIQEALHEEVVGVVP-----AGV-----PQHKDFAHMPLLKAVLKETLRLYPVVPTNS--RII-EKEIE   412
40 AWILLHLAER-PDVQQELYEEQMRVLD--GGK-K------ELTYDLLQEMPLLNQTIKETLRMHHPLHSLF--RKVMKDMHV   393
41 FWSLFQMIRN-PEAMKAATEEVKRTLENAGQKVSLEGNPICLSQAELNDLPVLNSIIKESLRLSSASLNI---RTA-KEDFT   371
42 NSSLYLLAKYSNEWQEKLRKEVNGITD----PKG------LAD------LPLLNAFLFEVVRMYPPLSTII--NRCTTKTCK   366
43 SFLFFELSRN-PEIFLKREEIENKFGLGQ-DARVE----EISFETLKSCEYLKAVINETLRIYPSVPHNF--RVATRNTTL   415
44 SFAFFELALN-EPIWIKLRTEVLHVFQ----TSL-E----LITFDLLKKCPYLQAILHETLRLYPSVPRNA--RFSKKNTTL   375
45 AFTILELSRK-SHHLGKLREEIDANFGL---ESP-D----LLTFDSLRKFKYVQAILNETLRMYPGVPRNM--KTAKCTTTL   382
46 TFLFFELSHH-PEVFNKLKEEIERHFP----DVE------SVTFGTIQRCDYLQWCINETMRLHSPYPFF--RTAANDTVI   388
47 FFMLFLIAKH-PNVEEAIIKEIQTVIG-----ER------DIKIDDIQKLKVMEIFMIYESMRYQPVVDLVM-RKA-LEDDV   382
48 SLSVITLLDH-PEQYAALRADRS--------------------------LVPGAVEELLRYLAIADIAGG-RVA-TADIE   303
49 SLGTFTLLSH-PEQLAALRAGGT--------------------------STAVVVEELLRFLSIAEGLQ--RLA-TEDME   310
50 ALSTLVLLRN-PDQLARLRAEPA--------------------------LVKGAVEELLRYLTIVHNG-VPRIA-TEDVL   300
51 ALGAFLLLEH-PDQLAALRADPA--------------------------LTGSAVEELLRHLSVVHHGPT-RAA-LQDAD   275
52 GLSVLALLHH-PEQAAMMIEDPN--------------------------CVNSGIEELLRWLSVAHSQ-PPRMA-VTEVQ   297
53 GIGTYLLLTH-PDQLALVRRDPS--------------------------ALPNAVEEILRYIAPPETTT--RFA-AEEVE   301
54 ALGVATLAQH-PDQLAQLKANPS--------------------------LAPQFVEELCRYHTASALAIK-RTA-KEDVM   299
55 EFGLHAMFRH-PQQRERLVGDPS--------------------------LVDKAVQEILRMYPPGWDGIM-RYP-RTDVT   294
56 DHAVLAMLAY-PEQRHWLQADAA--------------------------RVRAFVEEVLRCDAPAMFSSIPRIA-QRDIE   264
57 GNIVRTLDEH-PAHWDAAAEDPA--------------------------RIPAIVEEVLRYRPPFPQMQ--RTT-TKATE   311
58 ANMFYSFLYD-DKSLYELRNNRES--------------------------LAPKAVEEMLRYRFHISRRD--RTV-VQDNE   304
59 ANAVRYLTED-SVVQQQVRQNTD--------------------------NVANVIEETLRYYSPVQAIG--RVA-TEDTE   300
60 GGAIIGLSRN-PEQLALAKSDPA--------------------------LIPRLVDEAVRWTAPVKSFM--RTA-LADTE   326
61 SHTIVTLSQQ-PDQWDILRQRPE--------------------------LLKTATAEMVRHASPVLHMR--RTA-MEDTE   304
62 SFSMEFLAKS-PEHRQELIERPE--------------------------RIPAACEELLRFSLVADG--RIL-TSDYE   304
63 VMVTALALQN-PALWSSLRGNQS--------------------------YVAAAVEEGLRFEPPVGSFP--RLA-LKDID   317
64 GTILTGLLTN-PEQLEEVKRQQ---------------------------CWGQAFEEGLRWVAPIQASS--RLV-REDTE   305
65 MKWIGRAGLE-LHTKLAQEIRSAIQST----GGG------KVTMAAMEQMPLMKSVVYETLRIEPPVALQY--GKAKKDFIL   362
```

FIGURE 1. (*continued*)

```
 1 FRN--------YLIPKGTTILISLTSVLHDNKE-FPNP-EMFDPHHFLDEGG------------------NFKKSKYFMP  427
 2 FRG--------YIIPKGTVIVPTLDSVLYDNQE-FPDP-EKFKPEHFLNENG------------------KFKYSDYFKP  429
 3 LRD--------YFIPKDTMIFPLLSPILQDCKE-FPNP-EKFDPGHFLNANG------------------TFRRSDYFMP  428
 4 FRG--------FLIPKGTDVITLLNTVHYDPSQ-FLTP-QEFNPEHFLDANQ------------------SFKKSPAFMP  428
 5 FRG--------YIIPKDTEVFLILSTALHDPHY-FEKP-DAFNPDHFLDANG------------------ALKKTEAFIP  428
 6 FRG--------YFLPKGTDVYPLIGSVLKDPKY-FRYP-EAFYPQHFLDEQG------------------RFKKNDAFVA  430
 7 FRD--------FFLPKGTEVYPMLGSVLRDPSF-FSNP-QDFNPQHFLNEKG------------------QFKKSDAFVP  431
 8 FQG--------YFIKKGTSVIPLLTSVLQDDSE-WESP-NTFNPSHFLDEQG------------------GFVKRDAFMA  437
 9 LAG--------YHLPKGTVVLTNLTALHKDPEE-WATP-DTFNPEHFL-ENG------------------QFKKKEAFIP  439
10 VQG--------FRIPKGTTLITNLSSVLKDEAV-WEKP-FRFHPEHFLDAQG------------------HFVKPEAFLP  434
11 LKG--------FYIPKGRCVFVNQWQINHDQKL-WVNP-SEFLPERFLTPDG----------------AIDKVLSEKVII  449
12 IGE--------FAVDKGTEVIINLWALHHNEKE-WHQP-DQFMPERFLNPAG-----------------TQLISPSVSYLP  434
13 ISG--------YDIPEGTVIIPNLQGAHLDETV-WERP-HEFWPDRFLEPGK--------------------NSRALA  420
14 IRG--------YHIPAKTRVFINAWAIGRDPKS-WENA-EEFLPERFVNNSVD---------------FKGQDFQLIP  435
15 IAG--------YDVPEKTWIHVNMWAVHMSPSI-WKDP-ETFNPERFIDNQTD---------------FKGLNFELLP  429
16 VDG--------YYIPKNTRLSVNIWAIGRDPQV-WENP-LEFNPERFLSGRNSKID--------------PRGNDFELIP  435
17 FMG--------YDIPKGTQVLVNAWAIGRDPEY-WDNP-FEFKPERFLSKVD------------------VKGQNYELIP  400
18 LGG--------FDIPAESKILVNAWWLANNPDQ-WKKP-EEFRPERFLEEEAKVE---------------ANGNDFRYLP  439
19 CNG--------MVVPAGTTAMVNMWAITHDAAV-WADP-DAFAPERFLPSEGGADVD--------------VRGVDLRLAP  434
20 LAG--------YDIPMDTNVEFFVHGISHDPNV-WSDP-EKFDPDRFLSGREDADI---------------TGVKEVKMMP  435
21 ING--------MFIPKGWVVMIPSYALHRDPKY-WTEP-EKFLPERFSKKNK------------------DNIDPYIYTP  435
22 VLG--------QRIPAGAVLEMAVGALHHDPEH-WPSP-ETFNPERFTAEAR------------------QQHRPFTYLP  444
23 VPHNPK-----YVLPKGTLVFIPVLGIHYDPEL-YPNP-EEFDPERFSPEMV-----------------KQRDSVDWLG  439
24 FPGTD------ITIKKGQTIIVSTWGIQNDPKY-YPNP-EKFDPERFNPENV-----------------KDRHPCAYLP  428
25 PDG--------RSLPKGIMVLLSIYGLHHNPKV-NPNP-EVFDPFRFAPG------------------SAQHSHAFLP  449
26 VDG--------RSLPAGSLISMHIYALHRNSAV-WPDP-EVFDSLRFSTENA----------------SKRHPFAFMP  451
27 PDG--------RVIPKGIICLISVFGTHHNPAV-WPDP-EVYDPFRFDPKNI----------------KERSPLAFIP  460
28 -GD--------YLVPAGCMMNLQIYHVHRNQDQ-YPNP-EAFNPDNFLPERV----------------AKRHPYAYVP  444
29 -DG--------DIVPKGTTLNLGLLMLGYNDRV-FMDP-HKFQPERFDR------------------EKPGPFEYVP  162
30 -NG--------KLIPAGTNIGISPLYLGRREEL-FSEP-NIFKPERFDVVTTA---------------EKLNPYAYIP  448
31 GGE--------YPLEKGDELMVLIPQLHRDKTI-WGDDVEEFRPERFENP-------------------SAIPQHAFKP  392
32 IAG--------YQLEPDTTLMASIYLIHYREDL-YPNP-QQFRPERFIE--------------------RQTSPSEYIP  374
33 -GS--------YTIPAGTQVMLPTVMLHREKSI-WGEDAMEFNPMRFVDGVAN---------------ATKNNVTYLP  462
34 PVTISG-----HTFYPGDVVSVPSYTIHRSKEI-WGPDAEQFVPERWDPARLT------------------PRQKAAFIP  453
35 LGG--------YTIPAGTHVQANLYGMYRDPSL-FPEP-EGILPERWLRMNGSQMDA--------------TIKSTSQLV  485
36 LRD--------YMIPAKTLVQVAIYALGREPTF-FFDP-ENFDPTRWLSKDK------------------NITYFRNLG  454
37 LQN--------YHIPAGTLVRVFLYSLGRNAL-FDP-PERYNPQRWLDIKG------------------SGRNFYHVP  442
38 LGE--------YALPKGTVLTLNTQVLGSSEDN-FEDS-HKFRPERWLQKEK------------------KINPFAHLP  454
39 VDG--------FLFPKNTQFVFCHYVVSRDPTA-FSEP-ESFQPHRWLRNSQPATPR--------------IQHPFGSVP  468
40 PNTS-------YVIPAGYHVLVSPGYTHLRDEY-FPNA-HQFNIHRWN----KDSASSYSVGEEVDYGFGAISKGVSSPYLP  462
41 LHLEDGS----YNIRKDSIIALYPQLMHLDPEI-YPDP-LTFKYDRYLDENGKTKTTFYCNG---------LKLKYYYMP  436
42 LGAE-------IVIPKGVYVGYNNFGTSHDPKT-WGTTADDFKPERWGSDIETIRKNWRM------------AKNRCAVTG  427
43 PRGGGEGGLSPIAIKKGQVVMYTILATHRDKDI-YGEDAYVFRPERWFEPE------------------TRKLGWAYVP  475
44 PHGGGVDGMSPILIKKGQPVAYFICATHVDKEF-YTKDALIFRPERWCEPLI-----------------KKNLAWSYLP  436
45 PKGGGPDGQDPILVKKGQSVGFISIATHLDPVLNFGSDAHVFRPDRWFDSS----------------MKNLGCKYLP  443
46 PRGGGKSCTDPILVHKGEQVLFSFYSVNREEKY-FGTNTDKFAPERWSESL----------------RRTEFIP  445
47 IDG--------YPVKKGTNIILNIGRMHR-LEF-FPKP-NEFTLENFAK------------------NVPYRYFQP  429
48 VEG--------HLIRAGEGVIVVNSIANRDGTV-YEDP-DALDIHRSA---------------------RHHLA  346
49 VDG--------ATIRKGEGVVFSTSLINRDADV-FPRA-ETLDWDRPA---------------------RHHLA  353
50 IGG--------RTIAAGEGVLCMISSANRDAEV-FPGG-DDLDVARDA---------------------RRHVA  343
51 IEG--------TPVKAGEVVVVSLGAANRDPAR-FERP-DAVDVTRED--------------------TGHLA  318
52 IAG--------VTIPAGSFVIPSLLAANRDSNL-TDRP-DDLDITRGV--------------------AGHLA  340
53 IGG--------VAIPQYSTVLVANGAANRDPKQ-FPDP-HRFDVTRDT--------------------RGHLS  344
54 IGD--------KLVRANEGIIASNQSANRDEEV-FENP-DEFNMNRKWPP------------------QDPLG  344
55 IAG--------EHIPAESKVLVGLPATSFDPHH-FDDP-EIFDIERQE--------------------KPHLA  337
56 LGG--------VVIPKNADVRVLIASGNRDPDA-FADP-DRFDPARFYGTSPGMSTDG-----------KIMLS  316
57 VAG--------VPIPADVMVNTWVLSANRDSDA-HDDP-DRFDPSAQVRPA-----------------PRTSS  357
58 LLG--------VKLKKGDVVIAWMSACNMDETM-FENP-FSVDIHRPTN-----------------KKHLT  348
59 LGG--------VFIKKGSSVISWIASANRDEDK-FCKP-DCFKIDRPS------------------YPHLS  343
60 VRG--------QNIKRGDRIMLSYPSANRDEV-FSNP-DEFDITRFP-------------------NRHLG  369
61 IGG--------QAIAKGDKVVLWYASGNRDESV-FSDA-DRFDVTRTG-------------------VQHVG  347
62 FHG--------VQLKKGDQILLPQMLSGLDERE-NACP-MHVDFSRQK--------------------VSHTT  349
63 LDG--------YVLPKGSLLALSVMSGLRDEKH-YEHP-QLFDVGRQQM------------------RWHLG  361
64 IRG--------FIVPKGDIVMTIQASANRDKPI-FDRP-EEFVADRFVGEGVKLMEYV------------SAHQS  348
65 ESHEAA-----YQVKEGEMLFGYQPFATKDPKI-FDRP-EEFVADRFVGEGVKLMEYV------------MWSNGPETET  423
```

FIGURE 1. (continued)

```
 1 FSAGKRICVGEALAGMELFLF-LTSILQNF-NLKSLVDPKNLDTTPVVNGFASVPPFYQLCFI-PV------------ 490
 2 FSTGKRVCAGEGLARMELFLL-LCAILQHF-NLKPLVDPKDIDLSPIHIGFGCIPPRYKLCVI-PRS----------- 493
 3 FSAGKRICAGEGLARMEIFLF-LTSILQNF-SLKPVKDRKDIDISPIITSLANMPRPYEVSFI-PR------------ 491
 4 FSAGRRLCLGELLARMELFLY-LTAILQSF-SLQPLGAPEDIDLTPLSSGLGNLPRPFQLCLR-PR------------ 491
 5 FSLGKRICLGEGIARAELFLF-FTTILQNF-SMASPVAPEDIDLTPQECGVGKIPPTYQIRFL-PR------------ 491
 6 FSSGKRICVGEALARMELFLY-FTSILQRF-SLRSLVPPADIDIAHKISGFGNIPPTYELCFM-AR------------ 493
 7 FSIGKRNCFGEGLARMELFLF-FTTVMQNF-RLKSSQSPKDIDVSPKHVGFATIPRNYTMSFL-PR------------ 494
 8 FSAGRRVCLGEGLARMELFLF-FTSLLQRF-RFSPPGVTEDDLDLTPLLGFTLNPSPHQLCAV-SRV----------- 501
 9 FSIGKRACLGEQLAKSELFIF-FTSLMQKF-TFKPPSDEKLTLNFRM--GITLSPVKHRICAI-PRA----------- 501
10 FSAGRRRACLGEPLARMELFLF-FTSLLQHF-SFSVPTG-QPRPSHHGVFAFLVSPSPYELCAV-PR----------- 496
11 FGMGKRKCIGETIARWEVFLF-LAILLQRV-EFSVPLG-VKVDMTPIYGLTMKHACCEHFQMQ-LRS----------- 512
12 FGAGPRSCIGEILARQELFLI-MAWLLQRF-DLEVPDD-GQLPSLEGIPKVVFLIDSFKVKIK-VRQAWREAQAEGST- 508
13 FGCGARVCLGEPLARLELFVV-LTRLLQAF-TLLPS-GDALPSLQPLPHCSVILKMQPFQVRLQPRGMGAHSPGQNQ-- 494
14 FGAGRRGCPGIAFGISSVEIS-LANLLYWF-NWELPGI-------------------------------------- 471
15 FGSGRRMCPGMGMGLAVVHLT-LINLLYRF-DWKLPNGMKAEE--LSIEENYGLICVKKLPLEAIPVLTQWT------ 497
16 FGAGRRICAGTRMGIVMVEYI-LGTLVHSF-DWKLPSEVIELNMEEAFGLALQKAVPLEAMVTPRLQLDVYVP----- 506
17 FGAGRRMCVGLPLGHRMMHFT-FGSLLHEF-DWELPHNVSPKSINMEESMGITARKKQPLKVIPKKA----------- 465
18 FGVGRRSCPGIILALPILGIT-IGRLVQNF-ELLPPPGQSKIDTDEKGGQFSLHILKHSTIVAKPRSF---------- 505
19 FGAGRRVCPGKNLGLTTVGLW-VARLVHAF-QWALPDGAAAVCLDEVLKLSLEMKTPLVAA--AIPRTA--------- 499
20 FGVGRRICPGLGMATVHVNLM-LARMVQEF-EWFAYPGNNKVDFSEKLEFTVVMKNPLRAKVKLRI------------ 499
21 FGSGPRNCIGMRFALMNMKLA-LIRVLQNF-SFKPCKE-TQIPLKLSLGGLLQ-PEKPVVLKVESRDGTVSGA----- 504
22 FGAGPRSCLGVRLGLLEVKLT-LLHVLHKF-RFQACPE-TQVPLQLESKSALG-PKNGVYIKIVSR----------- 506
23 FGDGPRNCIGMRFGKMQSRLG-LALVIRHF-RFTVCSR-TDIPMQINPESLAWTPKKNNLYLKNVQAIRKKIK----- 507
24 FSAGPRNCLGMRFAKWQSEVC-IMKVLSKY-RVEPSMK-SSGEFKFDPMRLFALPKGGIYVNLVRR----------- 491
25 FSGGSRNCIGKQFAMNELKVA-TALTLLRF-ELLP--DPTRIPI-PIAR-LVLKSKNGIHLRLRRLPNPCEDKDQL--- 519
26 FSAGPRNCIGQQFAMSEMKVV-TAMCLLRF-EFSL--DPSRLPI-KMPQ-LVLRSKNGFHLHLKPLG----------- 512
27 FSAGPRNCIGQAFAMAEMKVV-LGLTLLAF-RVLP--DHTEPRR--KPE-LVLRAEGGLWLRVEPLS----------- 520
28 FSAGPRNCIGQKFATLEEKTV-LSSILRNF-KVRS--IEKREDL-TLMNELILRPESGIKVELIPRLPADAC------ 511
29 FSAGPRNCIGQKFALLEIKTV-VSKIIRNF-EVLPALDELYDPI-LSAS-MTLKSENGLHLRMKQRLVCDST------ 230
30 FSAGPRNCIGQKFAMLEIKPS-WPMCSGTT-RLTL--WATSFGT-TRADRRTYSAYQGPLSSRCGRVY--------- 511
31 FGNGQRACIGQQFALHEATLV-LGMMLKHF-DFE---DHTNYEL-DIKETLTLKPEGFVVKAKSKKIPLGGIP----- 459
32 FGGGSRRCLGIALALLEIKLV-IATVLSNY-QLAL---AEDKPVN-VQRRGFTLAPDGGVRVIMTGKKSLKFEQSSKIFN 448
33 FSWGPRVCLGQNFALLQAKLG-LAMILQRF-KFDV--APSYVHA-PFTI-LTVQPQFGSHVIYKKLES---------- 524
34 FSTGPRACVGRNVAEMELLVI-CGTVFRLF-EFEMQQ-EGPME---TREGFLR---KPLGLQVGMKRRQPGSA------ 517
35 WGHGARMCLGRRIAEQEMHIT-LSKIIQNF-TLSYN-H-DDVE-PILNTMLT--PDRPVRIEFKPRQ---------- 545
36 FGWGVRQCLGRRIAELEMTIF-LINMLENF-RVEIQ-HLSDVG-TTFNLILM--PEKPISFTFWPFNQEATQQ------ 521
37 FGFGMRQCLGRRLAEVEMLLL-LHHVLKNF-LVETL-EQEDIK-MVYRFILM--PSTLPLFTFRAIQ---------- 503
38 FGIGKRMCIGRRLAELQLHLA-LCWIIQKY-DIVAT-DNEPVE-MLHLGILV--PSRELPIAFRPR---------- 514
39 FGYGVRACLGRRIAELEMQLL-LARLIQKY-KVVLAPETGELK-SVARIVLV--PNKKVGLQFLQRQ---------- 530
40 FGGGRHRCIGEHFAYCQLGVL-MSIFIRTL-KWHYPEG-KTVPP-PDFTSMVTLPTGPAKIIWEKRNPEQKI------- 530
41 FGSGATICPGRLFAIHEIKQF-LILMLSYF-ELELIEGQ--AKCPPLDQSRAGLGILPPLNDIEFKYKFKHL------- 504
42 FHGGRRACLGEKLALTEMRIS-LAEMLKQF-RWSL---DPEWEEKLTPAGPLC----PLNLKLKFENIME-------- 489
43 FNGGPRICLGQQFALTEASYV-TVRLLQEF-GNLKQ-DPNTEYPPKLQNTLTLSLFEGAEVQMYLIL---------- 539
44 FNGGPRICLGQQFALTEASYV-LTRLAQCY-TKISLQPNSFEYPPKKQVHLTMSLLDGVHVKISNLSIS--------- 503
45 FNAGPRTCLGQQYTLIEASYL-LVRLAQTY-ETVES-HPDSVYPRRKALINMCAADGVDVKFHRL------------ 506
46 FSAGPRACLGQGLPRVEASYV-TIRLLQTF-HGLHN-ASKQ-YPPNRVVAATMRLTDGCNVCFI------------- 505
47 FGFGPRGCAGKYIAMVMMKAI-LVTLLRRF-HVKTLQGGCVESIQKIHDLSLHPDETKNMLEMIFTPRNSDRCLEH--- 503
48 FGFGVHQCLGQNLARLELEVI-LNALMDRVPTLRLAVP---VEQLVLRPGTTIQGVNELPVTW-------------- 405
49 FGFGVHQCLGQNLARAELDIA-MRTLFERLPGLRLAVP---AHEIRHKPGDTIQGLLDLPVAW-------------- 412
50 FGFGVHQCLGQPLARVELQIA-IETLLRRLPDLRLAVP---HEEIPFRGDMAIYGVHSLPIAW-------------- 402
51 FGHGMHQCLGRQLARIELRVA-LTALLERFPHLRLACP---AAEIPLRHDMQVYGADRLPVAW-------------- 377
52 FGHGVHFCLGHSLARMTLRTA-VPAVLRRFPDLALS-P---SHDVRLRSASIVLGLEELQLTW-------------- 399
53 FGQGIHFCMGRPLAKLEGEVA-LRALFGRFPDLSLGID---ADDVVWRRSLLLRGIDHLPVRLDG------------ 405
54 FGFGDHRCIAEHLAKAELTTV-FSTLYQKFPDLKVAVP---LGKINYTPLNRDVGIVDLPVIF------------- 403
55 FSYGPHACIGVALARLELKVV-FGSIFQRLPALRLAVA---PEQLKLRKEIITGGFEQFPVLW-------------- 396
56 FGHGIHFCLGAQLARVQLAES-LPRIQARFPTLAF------AGQPTREPSAFLRTFRTLPVRLHAQGS--------- 377
57 FGHGVHFCLAAPLARLENRVA-LEEIIARFGRLTVDRD---DERLRHFEQIVL-GTRHLPVLAGSSPRQSA------- 423
58 FGNGPHFCLGAPLARLEMKII-LEAFLEAFSHIEPFED--FELEPHLTASATGQSLTYLPMTVYR---------- 410
59 FGFGIHFCLGAPLARLEANIA-LSSLLSMSACIEKAAH---DEKLEAIPSPFVFGVKRLPVRITFK---------- 405
60 FGWGAHMCLGQHLAKLEMKIF-FEELLPKLKSVELSGP---PRLV---ATNFVGGPKNVPIRFTKA---------- 428
61 FGSGQHVCVGSRLAEMQLRVV-FEILSTRVKRFELCSK---SRRFRSNFLNGLKNLNVVLVPK------------- 406
62 FGHGSHLCLGQHLARREIIVT-LKEWLTRIPDFSIAPG----AQIQ-HKSGIVSGVQALPLVWDPATTKAV------ 414
63 FGAGVHRCLGETLARIELQEG-LRTLLRRAPNLAVVGD---WPRMM-GHGGIRRATDMMVKLSFDL----------- 422
64 FGSGPHHCPGAQISRQTVGAIMLPILFDRFPDMILPHP---ELVQW--RGFGFRGPINLPVTLR------------- 407
65 PSVANKQCAGKDFVVMAARLF-VVELFKRY-DSFD---IEVGTSSLGATITLTSLKRSTF------------- 478
```

FIGURE 1. (*continued*)

TABLE I
Cytochrome P450 Sequences, Including New Entries since the 1993 Nomenclature Update[1]

CYP1	*Opsanus tau* (toadfish)	
	Morrison, H., Sogin, M., and Stegeman, J.	
	submitted to Nomenclature Committee	
CYP1A1	*Oncorhynchus mykiss* (rainbow trout)[2]	see 1993 update
CYP1A1	human[3-5]	D12525 D01198
CYP1A1	human[3,4]	D10855 D01150
CYP1A1	*Macaca fascicularis* (monkey)	see 1993 update
CYP1A1	rabbit	see 1993 update
CYP1A1	dog	see 1993 update
CYP1A1	*Cavia cobaya* (guinea pig)[6]	D11043
CYP1A1	rat	see 1993 update
CYP1A1	hamster	see 1993 update
Cyp1a-1	mouse[7]	K02588
CYP1A2	human	see 1993 update
CYP1A2	*Macaca fascicularis* (monkey)	see 1993 update
CYP1A2	rabbit	see 1993 update
CYP1A2	dog	see 1993 update
CYP1A2	rat	see 1993 update
CYP1A2	hamster	see 1993 update
Cyp1a-2	mouse[7]	K02589
CYP1A2	chicken	see 1993 update
CYP1A2	*Oncorhynchus mykiss* (rainbow trout)[2]	
Cyp1b-1	mouse[8]	U02479
	Note: only 104 amino acids by PCR.	
CYP1B1	human	
	Sutter, T.	
	submitted to Nomenclature Committee	
CYP2A1	rat	see 1993 update
CYP2A2	rat	see 1993 update
CYP2A3	rat	see 1993 update
Cyp2a-4	mouse	see 1993 update
Cyp2a-5	mouse	see 1993 update
CYP2A6	human	see 1993 update
CYP2A7	human	see 1993 update
CYP2A8	hamster	see 1993 update
CYP2A9	hamster	see 1993 update
CYP2A10	rabbit[9]	L10236
CYP2A11	rabbit[9]	L10237
Cyp2a-12	mouse[10]	L06463
	Note: called 7 alpha hydroxylase, but this sequence is very different from CYP7 sequences. It is actually a 2A sequence.	
CYP2A	bovine (fragment)	see 1993 update
CYP2B1	rat	see 1993 update
CYP2B2	rat[11]	S51970
	promoter region, no coding sequence	
CYP2B2	rat	L28169
	Shephard, E.E.A.	
	unpublished (1993)	
	promoter region	
CYP2B3	rat	see 1993 update

TABLE I
(Continued)

CYP2B3	rat	see 1993 update
CYP2B4	rabbit[12]	L10912
CYP2B4P	rabbit pseudogene	see 1993 update
CYP2B5	rabbit	see 1993 update
CYP2B6	human	see 1993 update
CYP2B7P	human pseudogene	see 1993 update
CYP2B8	rat	see 1993 update
Cyp2b-9	mouse	see 1993 update
Cyp2b-10	mouse	see 1993 update
CYP2B11	dog	see 1993 update
CYP2B12	rat[13]	S48369 X63545
Cyp2b-13	mouse	see 1993 update
CYP2B14	rat	see 1993 update
CYP2B14P	rat	see 1993 update
CYP2B15	rat[14]	D17343→D17349
	most similar to 2B12, 89% identical	
CYP2B	rat fragment	see 1993 update
CYP2B	*Cavia cobaya* (guinea pig) fragment	see 1993 update
CYP2B	sheep fragment	see 1993 update
CYP2C1	rabbit	D26152
	Noshiro, M., Ishida, H., and Okuda, K. unpublished (1993)	
CYP2C2	rabbit	see 1993 update
CYP2C3	rabbit	see 1993 update
CYP2C4	rabbit	see 1993 update
CYP2C5	rabbit	see 1993 update
CYP2C6	rat	see 1993 update
CYP2C7	rat	see 1993 update
CYP2C8	human	see 1993 update
CYP2C9	human[15]	S46963
CYP2C9	human[16,17]	L16877→L16883
CYP2C10	human	see 1993 update
CYP2C11	rat	see 1993 update
CYP2C12	rat	see 1993 update
CYP2C13	rat	see 1993 update
CYP2C14	rabbit	see 1993 update
CYP2C15	rabbit	see 1993 update
CYP2C16	rabbit	see 1993 update
CYP2C17	human splice variant of 2C18 and 2C19	see 1993 update
CYP2C18	human[16–18]	L16869→L16876
CYP2C19	human	see 1993 update
CYP2C20	*Macaca fascicularis* (monkey)[19] MKmp13	S53046
CYP2C21	dog	see 1993 update
CYP2C22	rat	see 1993 update
CYP2C23	rat[20]	U04733
CYP2C23	rat[21]	S67064
CYP2C24	rat[22]	S59652

(continued)

TABLE I
(Continued)

CYP2C25	*Mesocricetus auratus* (Syrian hamster)[23]	X63022
CYP2C26	*Mesocricetus auratus* (Syrian hamster)[23]	D11435
CYP2C27	*Mesocricetus auratus* (Syrian hamster)[23]	D11436
CYP2C28	*Mesocricetus auratus* (Syrian hamster)	D11437
Cyp2c-29	mouse	
	Gonzalez, F.J.	
	submitted to Nomenclature Committee	
CYP2C30	rabbit	D26153
	Noshiro, M., Ishida, H., and Okuda, K.	
	unpublished (1993)	
CYP2C31	*Capra hircus* (goat)	X76502
	Zeilmaker, W.M., Van't Klooster, G.A.E., Gremmels-Gehrmann, F.J., Van Miert, A.S.J.P.A., and Horbach, G.J.M.J.	
	unpublished	
CYP2C	human	see 1993 update
CYP2C	rat	see 1993 update
Cyp2c	mouse	see 1993 update
CYP2D1	rat	see 1993 update
CYP2D2	rat	see 1993 update
CYP2D3	rat	see 1993 update
CYP2D4	rat	see 1993 update
CYP2D5	rat	see 1993 update
CYP2D6	human	see 1993 update
CYP2D7P	human pseudogene	see 1993 update
CYP2D7AP	human pseudogene[24]	X58467
	Note: CYP2D7AP is 94.7% identical to CYP2D7P.	
CYP2D7BP	human pseudogene[24]	X58468
	Note: CYP2D7BP is a chimeric gene composed of part of CYP2D7AP and part of CYP2D6. There are only 14 base changes in 13,677 base pairs relative to these parents. This gene is different from CYP2D8P.	
CYP2D8P	human pseudogene	see 1993 update
Cyp2d-9	mouse	see 1993 update
Cyp2d-10	mouse	see 1993 update
Cyp2d-11	mouse	see 1993 update
Cyp2d-12	mouse	see 1993 update
Cyp2d-13	mouse	see 1993 update
CYP2D14	bovine[25]	S45538 X68013
CYP2D	rat fragment	see 1993 update
CYP2D	rat fragment	see 1993 update
CYP2E1	human	see 1993 update
CYP2E1	*Macaca fascicularis (monkey)*[6]	S55205
	MKj1	
CYP2E1	rabbit	see 1993 update
Cyp2e-1	mouse[26]	L11650
CYP2E1	rat[27]	S48325
CYP2E2	rabbit	see 1993 update

TABLE I
(Continued)

CYP2F1	human	see 1993 update
Cyp2f-2	mouse	see 1993 update
CYP2G1	rat	see 1993 update
CYP2G1	rabbit	see 1993 update
CYP2H1	chicken	see 1993 update
CYP2H2	chicken	see 1993 update
CYP2J1	rabbit	see 1993 update
CYP2K1	*Oncorhynchus mykiss*(rainbow trout)[28]	L11528
CYP2L	spiny lobster	
	James, M.O.	
	submitted to Nomenclature Committee	
CYP2	*Caenorhabditis elegans* EST 14F12	see 1993 update
CYP3A1	rat[29]	D13912
	Note: only 11 differences and two missing amino acids with 3A1. All changes are limited to a 113-amino-acid region, probably represents a gene conversion.	
CYP3A2	rat	see 1993 update
CYP3A3	human	see 1993 update
CYP3A4	human[30]	D11131
CYP3A5	human	see 1993 update
CYP3A5P	human pseudogene	L26985
	Schuetz, J.D.	
	unpublished (1994)	
CYP3A6	rabbit	see 1993 update
CYP3A7	human	see 1993 update
CYP3A8	*Macaca fascicularis (monkey)*[6]	S53047
	MKnf2	
CYP3A9	rat	see 1993 update
CYP3A10	hamster	see 1993 update
Cyp3a-11	mouse	see 1993 update
CYP3A12	dog	see 1993 update
Cyp3a-13	mouse	see 1993 update
CYP3A14	*Cavia cobaya* (guinea pig)	D16363
	Mori, T., Itoh, S., and Kamataki, T.	
	unpublished (1993)	
CYP3A15	*Cavia cobaya* (guinea pig)	D26487
	Mori, T.	
	submitted to Nomenclature Committee	
Cyp3a-16	mouse	
	Itoh, S.	
	submitted to Nomenclature Committee	
CYP3A	rat fragment	see 1993 update
CYP3A	sheep fragment	see 1993 update
CYP3A	*Capra hircus* (goat)	X76503
	Zeilmaker, W. M., Van't Klooster, G.A.E., Gremmels-Gehrmann, F.J., Van Miert, A.S.J.P.A., and Horbach, G.J.M.J.	
	unpublished	

(continued)

TABLE I
(Continued)

CYP4A1	rat	see 1993 update
CYP4A2	rat	see 1993 update
CYP4A3	rat	see 1993 update
CYP4A4	rabbit[31]	L04758
	Note: The translation is different from the 4A4 sequence in the alignment after AALLG. A frameshift occurs between nucleotides 1201 and 1203, with an extra "G" present. After 1320 P450 similarity stops. 1320–1378 is an intron sequence.	
CYP4A5	rabbit	see 1993 update
CYP4A6	rabbit	see 1993 update
CYP4A7	rabbit	see 1993 update
CYP4A8	rabbit	see 1993 update
CYP4A9	human	see 1993 update
Cyp4a-10	mouse	X69296
	Henderson, C.J., Bammler, T.K., and Wolf, C.R. unpublished	
Cyp4a-10	mouse[32]	X71478
CYP4A11	human[33]	L04751
CYP4A11	human[32]	X71480
CYP4A11	human[34]	S67580
Cyp4a-12	mouse[32]	X71479
CYP4A13	*Cavia cobaya* (guinea pig)[32]	X71481
CYP4B1	human	see 1993 update
CYP4C1	*Blaberus discoidalis* (cockroach)	see 1993 update
CYP4D1	*Drosophila melanogaster*[35]	X67645
CYP4D2	*Drosophila melanogaster*	Z23005 X75955
	Frolov, M.V., and Alatortsev, V.E. Unpublished (1993)	
CYP4F1	rat[36]	S53039 M94548
CYP4F2	human	U02388
	Chen, Z., and Hardwick, J.P. unpublished	
CYP4F3	human[37]	D12620 D12621
	Note: three overlapping clones F-22, A-4, and M-6 are identical. Clone S-8 is different at four nucleotides and one amino acid.	
CYP4G1	*Drosophila melanogaster*	
	Waters, L.C.	
	submitted to Nomenclature Committee	
	Note: called P450-A1 starts at amino acid 192 and goes to the end.	
CYP4	*Caenorhabditis elegans* EST 8B12	see 1993 update
CYP4	*Caenorhabditis elegans* EST wEST00713	see 1993 update
CYP4	*Caenorhabditis elegans* (nematode worm) EST CEMSH9lR	
	C-terminal fragment with coding region identical to wEST00713 CYP4 fragment region before the translated "AGPRNCIG" may be an intron sequence.	

TABLE I

(Continued)

CYP5A1	human	see 1993 update
CYP5A1	rat	D28773
	Tone, Y.	
	unpublished	
	submitted to Nomenclature Committee	
CYP6A1	*Musca domestica* (housefly)	see 1993 update
CYP6A2	*Drosophila melanogaster*[38]	S51248 M88009
	Note: This is P450-B1, the sister sequence P450-A1 is CYP4G1.	
CYP6A3	*Musca domestica* (housefly)	
	Cohen, M.B., and Feyereisen, R.	
	submitted to Nomenclature Committee	
CYP6A4	*Musca domestica* (housefly)	
	Cohen, M.B., and Feyereisen, R.	
	submitted to Nomenclature Committee	
CYP6A5	*Musca domestica* (housefly)	
	Cohen, M.B., and Feyereisen, R.	
	submitted to Nomenclature Committee	
CYP6A6	*Musca domestica* (housefly)	
	Cohen, M.B., and Feyereisen, R.	
	submitted to Nomenclature Committee	
CYP6B1	*Papilio polyxenes* (black swallowtail butterfly)[39]	M80828 M83117 S48952
CYP6B2	*Helicoverpa armigera* (Australian cotton bollworm)	
	Hobbs, A.	
	submitted to Nomenclature Committee	
CYP6C1	*Musca domestica* (housefly)	
	Cohen, M.B., and Feyereisen, R.	
	submitted to Nomenclature Committee	
CYP6C2	*Musca domestica* (housefly)	
	Cohen, M.B., and Feyereisen, R.	
	submitted to Nomenclature Committee	
CYP6D1	*Musca domestica* (housefly)	
	Scott, J.	
	submitted to Nomenclature Committee	
CYP7	human[40]	L04629→L04634
CYP7	human[41–43]	L07951
CYP7	rat	see 1993 update
CYP7	rabbit	see 1993 update
CYP7	bovine	see 1993 update
CYP8	bovine	
	Tanabe, T.	
	submitted to Nomenclature Committee	
	Note: prostacyclin synthase.	
CYP9	*Heliothis virescens* (tobacco budworm)	
	Rose, R.	
	submitted to Nomenclature Committee	
CYP10	*Lymnaea stagnalis* (pond snail)[44]	S46130
CYP11A1	human[45,46]	see 1993 update

(continued)

TABLE I
(Continued)

CYP11A1	bovine	see 1993 update
CYP11A1	pig	see 1993 update
CYP11A1	rat	see 1993 update
CYP11A1	rabbit[47]	S59219
CYP11A1	*Oncorhynchus mykiss* (rainbow trout)[48]	S57305
CYP11A1	chicken	see 1993 update
CYP11B1	human	see 1993 update
CYP11B1	bovine	see 1993 update
Cyp11b-1	mouse	see 1993 update
CYP11B1	rat[49]	D10107 S58847
	This is the sequence of a mutant salt-sensitive form of 11B1 with five amino acid changes relative to wild type.	
CYP11B1	rat[50]	S58858
	Note: Fig. 3 has six errors. 1,2) aldo-46 amino acids 2 and 3 are incorrectly translated gctctc = AL not HS. 3) 11B3 codon at 559–561 incorrectly translated gac = D not N. 4–6) 11beta-62, 11B1, and 11B3 codon at 964–966 tcc = Ser not Pro.	
CYP11B1	rat[51]	D14086→D14108 S58849
CYP11B1	rat[49]	D11354
	Note: Two versions of 11B1 are seen in rats with only one amino acid difference at amino acid 84 G (normal) changed to E. G is found in 11B2 and 11B3.	
CYP11B2	human	see 1993 update
CYP11B2	bovine	see 1993 update
Cyp11b-2	mouse	see 1993 update
CYP11B2	rat[50]	S58859
CYP11B2	rat[51]	D14086→D14108 S58850
CYP11B3	rat[50]	
CYP11B3	rat[51]	D14086→D14108 S59144
	Note: only one amino acid difference with Nomura's 11B3.	
CYP11B4	bovine	see 1993 update
CYP11B5P	bovine pseudogene	see 1993 update
CYP11B6P	bovine pseudogene	see 1993 update
CYP11B7P	bovine pseudogene	see 1993 update
CYP11B8P	rat pseudogene[51]	D14086→D14108
	Note: authors call this sequence 11B4.	
CYP17	human EST	Z19875
	UK-HGMP (United Kingdom human genome mapping project) covers amino acids 270–348 when translating the complementary strand. The fragment goes through at least six frameshifts. sequence ID AAAAWEO	
CYP17	human EST	Z20209

TABLE I
(Continued)

	UK-HGMP (United Kingdom human genome mapping project) covers amino acids 265–349 when translating the complementary strand. The fragment goes through at least eight frameshifts. sequence ID AAABPSZ	
CYP17	rat[52]	S50146 Z11902
Cyp17	mouse	see 1993 update
CYP17	Cavia cobaya (guinea pig)	see 1993 update
CYP17	pig[53]	Z11855 S40341
CYP17	chicken	see 1993 update
CYP17	Oncorhynchus mykiss (rainbow trout)[54]	S50356
CYP17	dogfish Trant, J. submitted to Nomenclature Committee	
CYP19	human[55]	S52034 S52789 S52793 S52794
CYP19	human[56]	D14473 S59092 S59095 S59171
CYP19	rat[57] Note: promoter.	S59505
Cyp19	mouse	see 1993 update
CYP19	pig Choi, I., Ko, Y., Green, M.L., Simmen, F.A., and Simmen, R.C. Cytochome p450 aromatase gene expression in peri-implantation porcine embryos: Correlation with insulinlike growth factors during development. unpublished (1993) Note: This is only a fragment of 80 amino acids, including helix K and the EXXR conserved sequence.	L15471
CYP19	chicken (three different strains of chicken)[58]	M73277→M73285 M73286→M73294 M73295→M73303
CYP19	Japanese quail[59] Note: only three amino acid differences with chicken.	S46949
CYP19	Oncorhynchus mykiss (rainbow trout)	see 1993 update
CYP19	channel catfish Trant, J. submitted to Nomenclature Committee	
CYP19	goldfish	see 1993 update
CYP21A1	bovine	see 1993 update
CYP21A1P	human with congenital adrenal hyperplasia[60]	M26857 X05445
CYP21A1P	human pseudogene[61]	S60612
CYP21A2	human with congenital adrenal hyperplasia[60] Note: mutant gene, two amino acid differences.	M28548 X05449
CYP21A2	human with congenital adrenal hyperplasia[60] Note: normal gene.	M26856 X05448
Cyp21a-1	mouse	see 1993 update
CYP21	sheep[62]	S42095
CYP21	sheep[62]	S42096

(continued)

TABLE I
(Continued)

CYP21	sheep[62]	S42097
CYP21	pig[63]	S53049 M83939
Cyp21a-2p	mouse pseudogene	see 1993 update
CYP24	rat[64]	L04608→L04619
		S52625→S52636
CYP24	human[65]	L13286
CYP27	human	see 1993 update
CYP27	rabbit	see 1993 update
CYP27	rat	see 1993 update
Lower eukaryotes are numbered 51 through 69		
CYP51	*Saccharomyces cerevisiae*	see 1993 update
CYP51	*Candida tropicalis*	see 1993 update
CYP51	*Candida albicans*	see 1993 update
CYP52A1	*Candida tropicalis*	see 1993 update
CYP52A2	*Candida tropicalis*	see 1993 update
CYP52A3	*Candida maltosa*[66]	S64322
CYP52A3	*Candida maltosa*[67]	D12475 D01168
	Note: This is CYP52A3-b, one of two alleles in	
	C. maltosa.	
CYP52A4	*Candida maltosa*	see 1993 update
CYP52A5	*Candida maltosa*	see 1993 update
CYP52A6	*Candida tropicalis*[68]	Z13010
CYP52A7	*Candida tropicalis*[68]	Z13011
CYP52A8	*Candida tropicalis*[68]	Z13012
CYP52A9	*Candida maltosa*	see 1993 update
CYP52A10	*Candida maltosa*	see 1993 update
CYP52A11	*Candida maltosa*	see 1993 update
CYP52B1	*Candida tropicalis*	see 1993 update
CYP52C1	*Candida tropicalis*	see 1993 update
CYP52C2	*Candida maltosa*	see 1993 update
CYP52D1	*Candida maltosa*	see 1993 update
CYP52E1	*Candida apicola*	X76225
	Lottermoser, K.	
	unpublished (1994)	
CYP53	*Aspergillus niger*	see 1993 update
CYP54	*Neurospora crassa*	see 1993 update
CYP55	*Fusarium oxysporum*	D14517
	Tomura, D., Obika, K., Fukamizu, A., and Shoun, H.	
	probable lateral transfer from *Streptomyces* bacteria.	
	unpublished (1993)	
CYP56	*Saccharomyces cerevisiae*	see 1993 update
CYP57A1v1	*Nectria haematococca*	
	McCkluskey, K., and Vanetten, H.	
	submitted to Nomenclature Committee	
	PDAT9 pisatin demethylase	
CYP57A1v2	*Nectria haematococca*	
	McCkluskey, K., and Vanetten, H.	
	submitted to Nomenclature Committee	
	pda4 pisatin demethylase	

TABLE I
(Continued)

	Note: 98% identical to CYP57A1v1, probable allele.	
CYP57A2	*Nectria haematococca*	
	McCkluskey, K., and Vanetten, H.	
	submitted to Nomenclature Committee	
	pda6 pisatin demethylase	
	high similarity to CYP57A1v1, same subfamily, new member	
CYP58	*Fusarium sporotrichioides* (filamentous fungus)	
	Hohn, T.	
	submitted to Nomenclature Committee	
	Note: called TOX4.	
CYP59	*Emericella nidulans*[69]	L27825

Plant P450 families are numbered CYP71 to CYP99. Many of the sequences included here are PCR fragments or expressed sequence tags that are quite short. These fragments have not been assigned to families, unless a clear family resemblance was present. In the *Arabidopsis* expressed sequence tags, some are from the C-terminal and some are from the N-terminal or middle region of the P450s. Clearly some are from the same sequence, but the relationship cannot be known until the whole sequence is complete, or until PCR primers from the fragments can be used to identify which fragments are in the same sequence.

CYP71A1	*Persea americana* (avocado)	see 1993 update
CYP71A2	*Solanum melongena* cv. Sinsadoharanasu (eggplant)[70]	D14990 X71654
	Note: clone name 154 also called CYPEG4.	
CYP71A3	*Solanum melongena* cv. Sinsadoharanasu (eggplant)[70]	X70982
	Note: clone name F151 also called CYPEG3.	
CYP71A4	*Solanum melongena* cv. Sinsadoharanasu (eggplant)[70]	X70981
	Note: clone name E138 also called CYPEG2.	
	Incorrectly called CYP71A1 in Genbank entry	
CYP71B1	*Thlaspi arvense*	L24438
	Udvardi, M.K., Metzger, J.D., Krishnapillai, V.V., Peacock, J., and Dennis, E.S.	
	unpublished (1994)	
CYP71C1	*Zea mays* (maize)	
	Frey, M.	
	submitted to Nomenclature Committee	
CYP71C2	*Zea mays* (maize)	
	Frey, M.	
	submitted to Nomenclature Committee	
CYP71C3	*Zea mays* (maize)	
	Frey, M.	
	submitted to Nomenclature Committee	
CYP71C4	*Zea mays* (maize)	
	Frey, M.	
	submitted to Nomenclature Committee	
CYP72A1	*Catharanthus roseus* L. (Madagascar periwinkle)[71]	L10081
CYP72A1	*Catharanthus roseus* L. (Madagascar periwinkle)[72]	X69775
	PCR fragment	
	Note: sequence 3 from Fig. 2 (identical to 72A1).	
CYP72A1v1	*Catharanthus roseus*	L19074
	Mangold, U., Eichel, J., Batschauer, A., Lanz, T., Kaiser, T., Spangenberg, G., Werck-Reichhart, D., and Schroeder, J.	

(continued)

TABLE I
(Continued)

	Closely related cytochrome P450 proteins from Madagascar periwinkle (*Catharanthus roseus*): gene, cDNA, and transgenic expression in tobacco and *Arabidopsis thaliana*. unpublished (1993) Note: sequence called CYP72B, 8 amino acid differences with CYP72A1.	
CYP72A1v2	*Catharanthus roseus* Mangold, U., Eichel, J., Batschauer, A., Lanz, T., Kaiser, T., Spangenberg, G., Werck-Reichhart, D., and Schroeder, J. Closely related cytochrome P450 proteins from Madagascar periwinkle (*Catharanthus roseus*): gene, cDNA, and transgenic expression in tobacco and *Arabidopsis thaliana*. unpublished (1993) Note: sequence called CYP72C, 14 amino acid differences with CYP72A1 plus a 3-amino-acid C-terminal extension.	L19075
CYP72	*Catharanthus roseus* (Madagascar periwinkle) PCR fragment[72] Note: sequence 1 from Fig. 2.	X69789
CYP72	*Catharanthus roseus* (Madagascar periwinkle) PCR fragment[72] Note: sequence 2 from Fig. 2.	X69790
CYP72	*Arabidopsis thaliana* EST N-terminal fragment	T13009
CYP73	*Phaseolus aureus* (mung bean)[73] cinnamate 4-hydroxylase	L07634
CYP73	*Medicago sativa* (alfalfa)[74] cinnamate 4-hydroxylase	L11046
CYP73	*Helianthus tuberosus* (Jerusalem artichoke)[75] cinnamate 4-hydroxylase	Z17369
CYP73	*Catharanthus roseus* (Madagascar periwinkle) PCR fragment[72] Note: sequence 16 from Fig. 2.	X69788
CYP73	*Arabidopsis thaliana* EST C-terminal fragment	T04086
CYP74	*Linum usitatissimum* (flaxseed)[76] Note: This sequence has two exceptions to the P450 motif.	U00428
CYP74	*Parthenium argentatum* (guayule, a desert shrub) Backhaus, R. submitted to Nomenclature Committee	
CYP75A1	*Petunia hybrida* cv. Blue Star[77]	X71130
CYP75A1	*Petunia hybrida*[78]	Z22544
CYP75A2	*Solanum melongena* cv. Sinsadoharanasu (eggplant)[79]	X70824
CYP75A3	*Petunia hybrida*[78]	Z22545
CYP76A1	*Solanum melongena* cv. Sinsadoharanasu (eggplant)[80] Note: clone name F94 also called CYPEG8.	X71658
CYP76A2	*Solanum melongena* cv. Sinsadoharanasu (eggplant)[80]	X71657

TABLE I
(Continued)

	Note: clone name G17 also called CYPEG7.	
CYP77A1	*Solanum melongena* cv. Sinsadoharanasu (eggplant)[81]	X71656
	Note: clone name H1.	
CYP77A2	*Solanum melongena* cv. Sinsadoharanasu (eggplant)[81]	X71655
	Note: clone name H2.	
CYP78	*Zea mays* (maize)	L23209
	Larkin, J.	
	unpublished (1993)	
CYP79	*Sorghum bicolor*	
	Lindberg Møller, B.	
	submitted to Nomenclature Committee	
PCR fragments from *Catharanthus roseus* not assigned to families		
	Catharanthus roseus (Madagascar periwinkle) PCR fragment[72]	X69776
	Note: sequence 4 from Fig. 2.	
	Possible 71 family member	
	Catharanthus roseus (Madagascar periwinkle) PCR fragment[72]	X69777
	Note: sequence 5 from Fig. 2.	
	Possible 71 family member	
	Catharanthus roseus (Madagascar periwinkle) PCR fragment[72]	X69778
	Note: sequence 6 from Fig. 2.	
	Possible 71 family member	
	Catharanthus roseus (Madagascar periwinkle) PCR fragment[72]	X69779
	Note: sequence 7 from Fig. 2.	
	Possible 71 family member	
	Catharanthus roseus (Madagascar periwinkle) PCR fragment[72]	X69780
	Note: sequence 8 from Fig. 2.	
	Possible 71 family member	
	Catharanthus roseus (Madagascar periwinkle) PCR fragment[72]	X69781
	Note: sequence 9 from Fig. 2.	
	Possible 71 family member	
	Catharanthus roseus (Madagascar periwinkle) PCR fragment[72]	X69782
	Note: sequence 10 from Fig. 2.	
	Possible 71 family member	
	Catharanthus roseus (Madagascar periwinkle) PCR fragment[72]	X69783
	Note: sequence 11 from Fig. 2	
	Catharanthus roseus (Madagascar periwinkle) PCR fragment[72]	X69784
	Note: sequence 12 from Fig. 2.	
	Catharanthus roseus (Madagascar periwinkle) PCR fragment[72]	X69785
	Note: sequence 13 from Fig. 2.	
	Possible 71 family member	

(continued)

TABLE I
(Continued)

Catharanthus roseus (Madagascar periwinkle) PCR fragment[72]	X69786
Note: sequence 14 from Fig. 2.	
Possible 71 family member	
Catharanthus roseus (Madagascar periwinkle) PCR fragment[72]	X69787
Note: sequence 15 from Fig. 2.	

Arabidopsis thaliana ESTs (expressed sequence tags) not assigned to families

Arabidopsis thaliana EST	Z18072
amino acids from 124 to 255 compared to 71A1	
Possible 71 family member	
Arabidopsis thaliana EST	Z17988 Z26124
includes K helix to just before the thiolate Cys	
Possible 71 family member	
Arabidopsis thaliana EST	T04013 T04016
C-terminal fragment	
Arabidopsis thaliana EST	T04014 T04015
C-helix region	
Arabidopsis thaliana EST	T04172
C-helix region	
Arabidopsis thaliana EST	T04186
C-helix region	
Arabidopsis thaliana EST	T04417
N-terminal fragment	
Arabidopsis thaliana EST	T04541 T13587
C-terminal fragment	
Arabidopsis thaliana EST	T04613 T12851
between C-helix and I-helix	
Arabidopsis thaliana EST	T04714 T04452
C-terminal fragment	T13082
Arabidopsis thaliana EST	T04814 T13586
N-terminal fragment	
Arabidopsis thaliana EST	T14112
N-terminal fragment similar to T04814 and T13586	
Arabidopsis thaliana EST	T14211 T13569
C-helix region probable but very poor sequence, many unreadable bases.	
Arabidopsis thaliana EST	Z24511
C-terminal fragment	
Arabidopsis thaliana EST	Z26103
C-terminal fragment	
Arabidopsis thaliana EST	Z27299
C-terminal fragment with a small deletion after the Cys	

Bacterial sequences are numbered 101 and higher.

CYP101	*Pseudomonas putida* cam	see 1993 update
CYP102	*Bacillus megaterium* BM-3[82]	new X-ray crystal structure

TABLE I
(Continued)

CYP103	*Agrobacterium tumefaciens* pinF1	see 1993 update
CYP104	*Agrobacterium tumefaciens* pinF2	see 1993 update
CYP105A1	*Streptomvces griseolus* SU1	see 1993 update
CYP105B1	*Streptomyces griseolus* SU2	see 1993 update
CYP105C1	*Streptomyces* spp. choP	see 1993 update
CYP105D1	*Streptomvces griseus*[83] $P450_{soy}$	S45823 X63601
CYP105E1	*Rhodococcus fascians*	Z29635
	Crespi, M., Vereecke, D.M., Temmerman, W.G., Van Montagu, M., and Desomer, J.	
	unpublished (1994)	
CYP106A1	*Bacillus megaterium* BM-1	see 1993 update
CYP106A2	*Bacillus megaterium*[84] $P450_{meg}$	Z21972
CYP107A1	*Saccharopolyspora erythraea* eryF	see 1993 update
CYP107B1	*Saccharopolyspora erythraea* orf405	see 1993 update
CYP108	*Pseudomonas* spp. $P450_{terp}$	see 1993 update
CYP109	*Bacillus subtilis* ORF405	see 1993 update
CYP110	*Anabaena* spp. ORF3	see 1993 update
CYP111	*Pseudomonas incognita*[85]	L23310
CYP112	*Bradyrhizobium japonicum*[86]	L02323 L12971
CYP113	*Saccharopolyspora erythraea*[87]	L05776 S51613
	eryK erythromycin C-12 hydroxylase	
	Note: two different database entries have different start codons. Neither is ATG.	
CYP114	*Bradyrhizobium japonicum*[86]	L02323 L12971
CYP115P	*Bradyrhizobium japonicum*	L02323 L12971
	Tully, R.E., and Keister, D.L. A cluster of cytochrome P-450 genes from *Bradyrhizobium japonicum.*	
	unpublished (1993)	
	Note: This is called a pseudogene by the authors, but it looks to be free of frameshifts and stop codons in the segment reported.	
CYP116	*Rhodococcus* sp.	
	De Mot, R.	
	submitted to Nomenclature Committee	
Unidentified fragments		
CEL10E1	*Caenorhabditis elegans* EST	M88882
	2nd frame contains PPGP and KKYG from N-terminal region of many P450s.	
	Streptomyces thermotolerans[88]	M80346
	Note: P450 fragment called carX.	

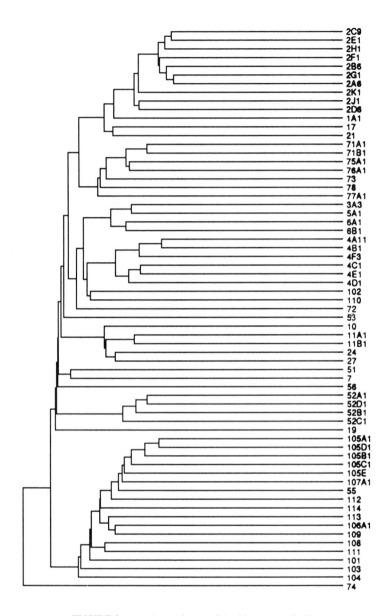

FIGURE 2. A phylogenetic tree of the 65 sequences in Fig. 1.

sequences for family and subfamily relationships has been described. These relationships are seen in the tree, by dropping a vertical line (or a ruler) down the page just to the left of the CYP2 family cluster. This line intersects distinct families and leaves subfamilies in clusters on the right. For example, 71A and 71B cluster to the right of this imaginary line, and so do 11A and 11B. Since this tree is mainly a family tree, not too many subfamily clusters are seen. They are much easier to see when larger numbers of sequences are included. Obviously, some judgment is needed to prevent undesired results. If the line were moved slightly to the left, then the 3 and 5 families would cluster together in a single family. On the other hand, a strict vertical line outside the CYP2 cluster would separate the CYP4 family into mammalian and insect groups, when it is clear that these belong to the same cluster. The 52 family would also split when it should be preserved as a distinct group.

The last sequence in the tree is the allene oxide synthetase CYP74. This enzyme has a substrate that carries its own oxygen and, therefore, it does not need the molecular oxygen binding features in and around the I-helix. This region is usually highly conserved, but it has degenerated in CYP74. As a result, CYP74 becomes the odd sequence in the collection. Just recently, two other non-oxygen-requiring P450 sequences have been determined, and they too have lost sequence conservation in the I-helix region. Because the sequences are confidential, I cannot identify them.

5. P450s on Internet

The data in this chapter, as well as other more extensive sequence alignments and lists of accession numbers, are available on the internet via Gopher (copyright University of Minnesota). If you would like to browse this information, get an internet connection to your computer and install the Gopher software, or find someone who has it. Once you have Gopher, search the following path to get to the P450 sequence data depository. Search gophers in North America, then Gophers in the USA, then Gophers in Maryland, then Computational Biology (Genome database project), then search databases at Hopkins, then sequence databases, and finally Cytochrome P450 alignment and nomenclature. This site currently has a 213-sequence alignment and a table of accession numbers and bibliographic information. For the most current information on cytochromes P450 consult the World Wide Web server at http://drnelson.utmem.edu/homepage.html.

ACKNOWLEDGMENT. Many thanks to those who submitted their new sequences to the Committee on Standardized Cytochrome P450 Nomenclature. There are 31 sequences in the current list that would not be here otherwise.

References

1. Nelson, D. R., Kamataki, T., Waxman, D. J., Guengerich, F. P., Estabrook, R. W., Feyereisen, R., Gonzalez, F. J., Coon, M. J., Gunsalus, I. C., Gotoh, O., Okuda, K., and Nebert, D. W., 1993, The P450 superfamily: Update on new sequences, gene mapping, accession numbers, early trivial names of enzymes and nomenclature, *DNA Cell Biol.* **12**:1–51.
2. Berndtson, A. K., and Chen, T. T., 1994, Two unique CYP1 genes are expressed in responses to 3-methylcholanthrene treatment in rainbow trout, *Arch. Biochem. Biophys.* **310**:187–195.

3. Kawajiri, K., Watanabe, J., Gotoh, O., Tagashira, Y., and Sogawa, K., 1986, Structure and drug inducibility of the human cytochrome P-450c gene, *Eur. J. Biochem.* **159:**219–225.

4. Kubota, M., Sogawa, K., Kaizu, Y., Sawaya, T., Watanabe, J., Kawajiri, K., Gotoh, O., and Fujii-Kuriyama, Y., 1991, Xenobiotic responsive element in the 5′-upstream region of the human P-450c gene, *J. Biochem.* **110:**232–236.

5. Hayashi, S.-i., Watanabe, J., Nakachi, K., and Kawajiri, K., 1991, Genetic linkage of lung cancer-associated MspI polymorphisms with amino acid replacement in the heme binding region of the human cytochrome P450IA1 gene, *J. Biochem.* **110:**407–411.

6. Ohgiya, S., Ishizaki, K., and Shinriki, N., 1993, Molecular cloning of guinea pig CYP1A1: Complete primary structure and fast mobility of expressed protein on electrophoresis, *Biochim. Biophys. Acta* **1216:**237–244.

7. Kimura, S., Gonzalez, F. J., and Nebert, D. W., 1984, The murine Ah locus, *J. Biol. Chem.* **259:**10705–10713.

8. Shen, Z., Wells, R., Liu, J., and Elkind, M. M., 1993, Identification of a cytochrome P450 gene by reverse transcription-PCR using degenerate primers containing inosine, *Proc. Natl. Acad. Sci. USA* **90:**11483–11487.

9. Peng, H.-M., Coon, M. J., and Ding, X., 1993, Isolation and heterologous expression of cloned cDNAs for two rabbit nasal microsomal proteins CYP2A10 and CYP2A11 that are related to nasal microsomal cytochrome P-450 form a, *J. Biol. Chem.* **268:**17253–17260.

10. Iwasaki, M., Juvonen, R., Lindberg, R., and Negishi, M. M., 1993, Site-directed mutagenesis of mouse steroid 7 alpha-hydroxylase cytochrome P-450 (7 alpha): Role of residue 209 in determining steroid–cytochrome P-450 interaction, *Biochem. J.* **291:**569–573.

11. Hoffmann, M., Mager, W. H., Scholte, B. J., Civil, A., and Planta, R. J., 1992, Analysis of the promoter of the cytochrome P-450 2B2 gene in the rat, *Gene Expr.* **2:**353–363.

12. Ryan, R., Grimm, S. W., Kedzie, K. M., Halpert, J. R., and Philpot, R. M., 1993, Cloning, sequencing, and functional studies of phenobarbital-inducible forms of cytochrome P450 2B and 4B expessed in rabbit kidney, *Arch. Biochem. Biophys.* **304:**454–463.

13. Friedberg, T., Grassow, M. A., Bartlomowicz-Oesch, B., Siegert, P., Arand, M., Adesnik, M., and Oesch, F., 1992, Sequence of a novel cytochrome CYP2B cDNA coding for a protein which is expressed in a sebaceous gland, but not in the liver, *Biochem. J.* **287:**775–783.

14. Nakayama, K., Suwa, Y., Mizukami, Y., Sogawa, K., and Fujii-Kuriyama, Y., 1993, Cloning and sequencing of a novel rat cytochrome P450 2B-encoding gene, *Gene* **136:**333–336.

15. Ohgiya, S., Komori, M., Ohi, H., Shiramatsu, K., Shinriki, N., and Kamataki, T., 1992, Six-base deletion occurring in messages of human cytochrome P-450 in the CYP2C subfamily results in reduction of tolbutamide hydroxylase activity, *Biochem. Int.* **27:**1073–1081.

16. Goldstein, J. A., Raucy, J. L., Blaisdell, J. A., Faletto, M. B., and Romkes, M., 1991, Cloning and expression of complementary DNAs for multiple members of the human cytochrome P450IIC subfamily, *Biochemistry* **30:**3247–3255.

17. de Morais, S. M., Schweikl, H., Blaisdell, J. A., and Goldstein, J. A., 1993, Gene structure and upstream regulatory regions of human CYP2C9 and CYP2C18, *Biochem. Biophys. Res. Commun.* **194:**194–201.

18. Romkes, M., Faletto, M. B., Blaisdell, J. A., Raucy, J. L., and Goldstein, J. A., 1993, Correction: Cloning and expression of complementary DNAs for multiple members of the human cytochrome P450IIC subfamily, *Biochemistry* **32:**1390.

19. Komori, M., Kikuchi, O., Sakuma, T., Funaki, J., Kitada, M., and Kamataki, T., 1992, Molecular cloning of monkey liver cytochrome P-450 cDNAs: Similarity of the primary sequences to human cytochromes P-450, *Biochim. Biophys. Acta* **1171:** 141–146.

20. Karara, A., Makita, K., Jacobson, H. R., Falck, J. R., Guengerich, F. P., DuBois, R. N., and Capdevila, J. H., 1993, Molecular cloning, expression, and enzymatic characterization of the rat kidney cytochrome P-450 arachidonic acid epoxygenase, *J. Biol. Chem.* **268:**13565–13570.

21. Imaoka, S., Wedlund, P. J., Ogawa, H., Kimura, S., Gonzalez, F. J., and Kim, H. Y., 1993, Identification of CYP2C23 expressed in rat kidney as an arachidonic acid epoxygenase, *J. Pharmacol. Exp. Ther.* **267:**1012–1016.

22. Zaphiropoulos, P. G., 1993, Differential expression of cytochrome P450 2C24 and transcripts in rat kidney and prostate: Evidence indicative of alternative and possibly trans splicing events, *Biochem. Biophys. Res. Commun.* **192:**778–786.

23. Sakuma, T., Masaki, K., Itoh, S., Yokoi, T., and Kamataki, T., 1994, Sex-related difference in the expression of cytochrome P450 in hamsters: cDNA cloning and examination of the expression of three distinct CYP2C cDNAs, *Mol. Pharmacol.* **45:**228–236.

24. Heim, M. H., and Meyer, U. A., 1992, Evolution of a highly polymorphic human gene locus for a drug metabolizing enzyme, *Genomics* **14:**49–58.

25. Tsuneoka, Y., Matsuo, Y., Higuchi, R., and Ichikawa, Y., 1992, Characterization of the cytochrome P-450IID subfamily in bovine liver. Nucleotide sequences and microheterogeneity, *Eur. J. Biochem.* **208:**739–746.

26. Davis, J. F., and Felder, M. R., 1993, Mouse ethanol-inducible cytochrome P450 (P450IIE1). Characterization of cDNA clones and testosterone induction in kidney tissue, *J. Biol. Chem.* **268:**16584–16589.

27. Richardson, T. H., Schenkman, J. B., Turcan, R., Goldfarb, P. S., and Gibson, G. G., 1992, Molecular cloning of a cDNA for rat diabetes-inducible cytochrome P450RLM6: Hormonal regulation and similarity to the cytochrome P4502E1 gene, *Xenobiotica* **22:**621–631.

28. Buhler, D. R., Yang, Y.-H., Dreher, T. W., Miranda, C. L., and Wang, J.-L., 1994, Cloning and sequencing of the major rainbow trout constitutive cytochrome P450 (P450 2K1). Identification of a new P450 gene subfamily and its expression in mature rainbow trout liver and trunk kidney. *Arch. Biochem. Biophys.* **312:**45–51.

29. Kirita, S., and Matsubara, T., 1993, cDNA cloning and characterization of a novel member of steroid-induced cytochrome P450 3A in rats, *Arch. Biochem. Biophys.* **307:**253–258.

30. Hashimoto, H., Toide, K., Kitamura, R., Fujita, M., Tagawa, S., Kamataki, T., and Itoh, S., 1993, Gene structure of CYP3A4 an adult-specific form of cytochrome P450 in human livers and its transcriptional control, *Eur. J. Biochem.* **218:**585–595.

31. Palmer, C. N., Griffin, K. J., and Johnson, E. F., 1993, Rabbit prostaglandin omega-hydroxylase (CYP4A4): Gene structure and expression, *Arch. Biochem. Biophys.* **300:**670–676.

32. Bell, D. R., Plant, N. J., Rider, C. G., Na, L., Brown, S., Ateitalla, I., Acharya, S. K., Davies, M. H., Elias, E. E., Jenkins, N. A., Gilbert, D. J., Copeland, N. G., and Elcombe, C. R., 1993, Species-specific induction of cytochrome P-450 4A RNAs: PCR cloning of partial guinea-pig, human and mouse cDNAs, *Biochem. J.* **294:**173–180.

33. Palmer, C. N., Richardson, T. H., Griffin, K. J., Hsu, M.-H., Muerhoff, A. S., Clark, J. E., and Johnson, E. F., 1993, Characterization of a cDNA encoding a human kidney, cytochrome P450 4A fatty acid omega-hydroxylase and the cognate enzyme expressed in Escherichia coli, *Biochim. Biophys. Acta* **1172:**161–166.

34. Imaoka, S., Ogawa, H., Kimura, S., and Gonzalez, F. J., 1993, Complete cDNA sequence and cDNA-directed expression of CYP4A11, a fatty acid omega-hydroxylase expressed in human kidney, *DNA Cell Biol.* **12:**893–899.

35. Gandhi, R., Varak, E., and Goldberg, M. L., 1992, Molecular analysis of a cytochrome P450 gene of family 4 on the Drosophila X chromosome, *DNA Cell Biol.* **11:**397–404.

36. Chen, L., and Hardwick, J. P., 1993, Identification of a new P450 subfamily, CYP4F1, expressed in rat hepatic tumors, *Arch. Biochem. Biophys.* **300:**18–23.

37. Kikuta, Y., Kusunose, E., Endo, K., Yamamoto, S., Sogawa, K., Fujii-Kuriyama, Y., and Kusunose, M., 1993, A novel form of cytochrome P-450 family 4 in human polymorphonuclear leukocytes. cDNA cloning and expression of leukotriene B4 omega-hydroxylase, *J. Biol. Chem.* **268:**9376–9380.

38. Waters, L. C., Zelhof, A. C., Shaw, B. J., and Ch'ang, L.-Y., 1992, Possible involvement of the LTR of transposable element 17.6 in regulating expression of an insecticide resistance associated P450

gene in Drosophila. *Proc. Natl. Acad. Sci. USA* **89**:4855–4859; correction: 1992, *Proc. Natl. Acad. Sci. USA* **89**:12209.

39. Cohen, M. B., Schuler, M. A., and Berenbaum, M. R., 1992, A host-inducible cytochrome P450 from a host specific caterpillar: Molecular cloning and evolution, *Proc. Natl. Acad. Sci. USA* **89**:10920–10924.

40. Nishimoto, M., Noshiro, M., and Okuda, K., 1993, Structure of the gene encoding human liver cholesterol 7 alpha-hydroxylase, *Biochim. Biophys. Acta* **1172**:147–150.

41. Thompson, J. F., Lira, M. E., Lloyd, D. B., Hayes, L. S., Williams, S., and Elsenboss, L., 1993, Cholesterol 7 alpha-hydroxylase promoter separated from cyclophilin pseudogene by alu sequence, *Biochim. Biophys. Acta* **1168**:239–242.

42. Cohen, J. C., Cali, J. J., Jelinek, D. F., Mehrabian, M., Sparkes, R. S., Lusis, A. J., Russell, D. W., and Hobbs, H. H., 1992, Cloning of the human cholesterol 7 alpha hydroxylase gene (CYP7) and localization to chromosome 8q11-q12, *Genomics* **14**:153–161.

43. Molowa, D. T., Chen, W. S., Cimis, G. M., and Tan, C. P., 1992, Transcriptional regulation of the human cholesterol 7 alpha hydroxylase gene, *Biochemistry* **31**:2539–2544.

44. Teunissen, Y., Geraerts, W. P., van Heerikhuizen, H., Planta, R. J., and Joosse, J., 1992, Molecular cloning of a cDNA encoding a member of a novel cytochrome P450 family in the mollusc Lymnaea stagnalis, *J. Biochem.* **112**:249–252.

45. Moore, C. C., Hum, D. W., and Miller, W. L., 1992, Identification of positive and negative placenta-specific basal elements and a cyclic adenosine 3',5'-monophosphate response element in the human gene for P450scc, *Mol. Endocrinol.* **6**:2045–2058.

46. Hum, D. W., Staels, B., Black, S. M., and Miller, W. L., 1993, Basal transcriptional activity and cyclic adenosine 3',5'-monophosphate responsiveness of the human cytochrome P450scc promoter transfected into MA-10 Leydig cells, *Endocrinology* **132**:546–552.

47. Yang, X., Iwamoto, K., Wang, M., Artwhol, J., Mason, J. I., and Pang, S., 1993, Inherited congenital adrenal hyperplasia in the rabbit is caused by a deletion in the gene encoding cytochrome P450 cholesterol side chain cleavage enzyme, *Endocrinology* **132**:1977–1982.

48. Takahashi, M., Tanaka, M., Sakai, N., Adachi, S., Miller, W. L., and Nagahama, Y., 1993, Rainbow trout ovarian cholesterol side-chain cleavage cytochrome P450 (P450scc). cDNA cloning and mRNA expression during oogenesis, *FEBS Lett.* **319**:45–48.

49. Matsukawa, N., Nonaka, Y., Higaki, J., Nagano, M., Mikami, H., Ogihara, T., and Okamoto, M., 1993, Dahl's salt-resistant normotensive rat has mutations in cytochrome P450 (11 beta), but the salt-sensitive hypertensive rat does not, *J. Biol. Chem.* **268**:9117–9121.

50. Nomura, M., Morohashi, K.-i., Kirita, S., Nonaka, Y., Okamoto, M., Nawata, H., and Omura, T., 1993, Three forms of rat CYP11B genes: 11 beta-hydroxylase gene, aldosterone synthase gene, and a novel gene, *J. Biochem.* **113**:144–152.

51. Mukai, K., Imai, M., Shimada, H., and Ishimura, Y., 1993, Isolation and characterization of rat CYP11B genes involved in late steps of mineralo- and glucocorticoid syntheses, *J. Biol. Chem.* **268**:9130–9137.

52. Nason, T. F., Han, X.-G., and Hall, P. F., 1992, Cyclic AMP regulates expression of the rat gene for steroid 17 alpha-hydroxylase/C17-20 lyase P-450 (CYP17) in rat Leydig cells, *Biochim. Biophys. Acta* **1171**:73–80.

53. Zhang, P., Nason, T. F., Han, X.-G., and Hall, P. F., 1992, Gene for 17 alpha-hydroxylase/C(17-20) lyase P-450: Complete nucleotide sequence of the porcine gene and 5' upstream sequence of the rat gene, *Biochim. Biophys. Acta* **1131**:345–348.

54. Sakai, N., Tanaka, M., Adachi, S., Miller, W. L., and Nagahama, Y., 1992, Rainbow trout cytochrome P-450c17 (17 alpha-hydroxylase/17-20 lyase) cDNA cloning, enzymatic properties and temporal pattern of ovarian P-450c17 mRNA expression during oogenesis, *FEBS Lett.* **301**:60–64.

55. Harada, N., 1992, A unique aromatase (P-450AROM) mRNA formed by alternative use of tissue-specific exons 1 in human skin fibroblasts, *Biochem. Biophys. Res. Commun.* **189**:1001–1007.

56. Toda, K., and Shizuta, Y., 1993, Molecular cloning of a cDNA showing alternative splicing of the 5'-untranslated sequence of mRNA for human aromatase P-450, *Eur. J. Biochem.* **213**:383–389.

57. Fitzpatrick, S. L., and Richards, J. S., 1993, Cis-acting elements of the rat aromatase promoter required for adenosine 3',5'-monophosphate induction in ovarian granulosa cells and constitutive expression in R2C Leydig cells, *Mol. Endocrinol.* **7**:341–354.

58. Matsumine, H., Herbst, M., Ou, S.-H. I., Wilson, J. D., and McPhaul, M. J., 1991, Aromatase mRNA in the extragonadal tissues of chickens with the henny-feathering trait is derived from a distinctive promoter structure that contains a segment of a retroviral long terminal repeat, *J. Biol. Chem.* **266**: 19900–19907.

59. Harada, N., Yamada, K., Foidart, A., and Balthazart, J., 1992, Regulation of aromatase cytochrome P-450 (estrogen synthetase) transcripts in the quail brain by testosterone, *Brain Res. Mol. Brain Res.* **15**:19–26.

60. Rodrigues, N. R., Dunham, I., Yu, C. Y., Carroll, M. C., Porter, R. R., and Campbell, R. D., 1987, Molecular characterization of the HLA-linked steroid 21-hydroxylase B gene from an individual with congenital adrenal hyperplasia, *EMBO J.* **6**:1653–1661.

61. Collier, S., Tassabehji, M., and Strachan, T., 1993, A de novo pathological point mutation at the 21-hydroxylase locus: Implications for gene conversion in the human genome, *Nature Genet.* **3**:260–265.

62. Crawford, R. J., Hammond, V. E., Connell, J. M., and Coghlan, J. P., 1992, The structure and activity of two cytochrome P450c21 proteins encoded in the ovine adrenal cortex, *J. Biol. Chem.* **267**:16212–16218.

63. Burghelle-Mayeur, C., Geffrotin, C., and Vaiman, M., 1992, Sequences of the swine 21-hydroxylase gene (Cyp21) and a portion of the opposite-strand overlapping gene of unknown function previously described in human, *Biochim. Biophys. Acta* **1171**:153–161.

64. Ohyama, Y., Noshiro, M., Eggertsen, G., Gotoh, O., Kato, Y., Bjorkhem, I., and Okuda, K., 1993, Structural characterization of the gene encoding rat 25-hydroxyvitamin D$_3$ 24-hydroxylase, *Biochemistry* **32**:76–82.

65. Chen, K.-S., Prahl, J. M., and DeLuca, H. F., 1993, Isolation and expression of 1,25-dihydroxyvitamin D$_3$ 24-hydroxylase cDNA, *Proc. Natl. Acad. Sci. USA* **90**:4543–4547.

66. Schunck, W. H., Vogel, F., Gross, B., Kargel, E., Mauersberger, S., Kopke, K., Gengnagel, C., and Muller, H. G., 1991, Comparison of two cytochromes P-450 from Candida maltosa: Primary structures, substrate specificities and effects of their expression in Saccharomyces cerevisiae on the proliferation of the endoplasmic reticulum, *Eur. J. Cell Biol.* **55**:336–345.

67. Ohkuma, M., Hijiki, T., Tanimoto, T., Schunck, W. H., Muller, H. G., Yano, K., and Takagi, M., 1991, Evidence that more than one gene encodes n-alkane-inducible cytochrome P-450s in *Candida maltosa*, found by two-step gene disruption, *Agric. Biol. Chem.* **55**:1757–1764.

68. Seghezzi, W., Meili, C., Ruffiner, R., Kuenzi, R., Sanglard, D., and Fiechter, A., 1992, Identification and characterization of additional members of the cytochrome P450 multigene family CYP52 of Candida tropicalis, *DNA Cell Biol.* **11**:767–780.

69. Keller, N. P., Kantz, N. J., and Adams, T. H., 1994, *Aspergillus nidulans* verA is required for production of the mycotoxin sterigmatocystin, *Appl. Environ. Microbiol.* **60**:1444–1450.

70. Umemoto, N., Kobayashi, O., Ishizaki-Nishizawa, O., and Toguri, T., 1993, cDNA sequences encoding cytochrome P450 (CYP71 family) from eggplant seedlings, *FEBS Lett.* **330**:169–173.

71. Vetter, H.-P., Mangold, U., Schroeder, G., Marner, F.-J., Werck-Reichhart, D., and Schroeder, J., 1992, Molecular analysis and heterologous expression of an inducible cytochrome P-450 protein from periwinkle (*Catharanthus roseus* L.), *Plant Physiol.* **100**:998–1007.

72. Meijer, A. H., Souer, E., Verpoorte, R., and Hoge, J. H. C., 1993, Isolation of cytochrome P450 cDNA clones from the higher plant *Catharanthus roseus* by a PCR strategy, *Plant Mol. Biol.* **22**:379–383.

73. Mizutani, M., Ward, E. R., DiMaio, J., Ryals, J. A., Ohta, D., and Sato, R., 1993, Molecular cloning and sequencing of a cDNA encoding mung bean cytochrome P450 (P450C4H) possessing cinnamate 4-hydroxylase activity, *Biochem. Biophys. Res. Commun.* **190**:875–880.

74. Fahrendorf, T., and Dixon, R. A., 1993, Stress responses in alfalfa (*Medicago sativa* L.). XVIII. Molecular cloning and expression of the elicitor-inducible cinnamic acid 4-hydroxylase cytochrome P450, *Arch. Biochem. Biophys.* **305**:509–515.

75. Teutsch, H. G., Hasenfratz, M. P., Lesot, A., Stoltz, C., Garnier, J.-M., Jeltsch, J.-M., Durst, F., and Werck-Reichhart, D., 1993, Isolation and sequence of a cDNA encoding the Jerusalem artichoke cinnamate 4-hydroxylase, a major plant cytochrome P450 involved in the general phenylpropanoid pathway, *Proc. Natl. Acad. Sci. USA* **90**:4102–4106.

76. Song, W.-C., Funk, C. D., and Brash, A. R., 1993, Molecular cloning of an allene oxide synthase: A cytochrome P450 specialized for the metabolism of fatty acid hydroperoxides, *Proc. Natl. Acad. Sci. USA* **90**:8519–8523.

77. Toguri, T., Azuma, M., and Ohtani, T., 1993, Characterization of a cDNA encoding a cytochrome P450 from the flowers of *Petunia hybrida, Plant Sci.* **94**:119–126.

78. Holton, T. A., Brugliera, F., Lester, D. R., Tanaka, Y., Hyland, C. D., Menting, J. G. T., Lu, C.-Y., Farcy, E., Stevenson, T. W., and Cornish, E. C., 1993, Cloning and expression of cytochrome P450 genes controlling flower colour, *Nature* **366**:276–279.

79. Toguri, T., Umemoto, N., Kobayashi, O., and Ohtani, T., 1993, Activation of anthocyanin synthesis genes by white light in eggplant hypocotyl tissues, and identification of an inducible P-450 cDNA, *Plant Mol. Biol.* **23**:933–946.

80. Toguri, T., Kobayashi, O., and Umemoto, N., 1993, The cloning of eggplant seedling cDNAs encoding proteins from a novel cytochrome P450 family (CYP76), *Biochim. Biophys. Acta* **1216**:165–169.

81. Toguri, T., and Tokugawa, K., 1994, Cloning of eggplant hypocotyl cDNAs encoding cytochrome P450 belonging to a novel family (CYP77), *FEBS Lett.* **338**:290–294.

82. Ravichandran, K. G., Boddupalli, S. S., Hasemann, C. A., Peterson, J. A., and Deisenhofer, J., 1993, Crystal structure of hemoprotein domain of P450BM-3, a prototype for microsomal P450s, *Science* **261**:731–736.

83. Trower, M. K., Lenstra, R., Omer, C., Buchholz, S. E., and Sariaslani, F. S., 1992, Cloning, nucleotide sequence determination and expression of the genes encoding cytochrome P-450soy (soyC) and ferredoxinsoy (soyB) from *Streptomyces griseus, Mol. Microbiol.* **6**:2125–2134.

84. Rauschenbach, R., Isernhagen, M., Noeske-Jungblut, C., Boidol, W., and Siewert, G., 1993, Cloning, sequencing and expression of the gene for cytochrome P450meg, the steroid-15β-monooxygenase from *Bacillus megaterium* ATCC 13368, *Mol. Gen. Genet.* **241**:170–176.

85. Ropp, J. D., Gunsalus, I. C., and Sligar, S. G., 1993, Cloning and expression of a member of a new cytochrome P-450 family: Cytochrome P450lin (CYP111) from *Pseudomonas incognita, J. Bacteriol.* **175**:6028–6037.

86. Tully, R. E., and Keister, D. L., 1993, Cloning and mutagenesis of a cytochrome P-450 locus from *Bradyrhizobium japonicum* that is expressed anaerobically and symbiotically, *Appl. Environ. Microbiol.* **59**:4136–4142.

87. Stassi, D. L., Donadio, S., Staver, M. J., and Katz, L., 1993, Identification of a *Saccharopolyspora erythraea* gene required for the final hydroxylation step in erythromycin biosynthesis, *J. Bacteriol.* **175**:182–189.

88. Schoner, B. E., Geistlich, M., Rosteck, P., Rao, R. N., Seno, E., Reynolds, P., Cox, K., Burgett, S., and Hershberger, C. L., 1992, Sequence similarity between macrolide resistance determinants and ATP binding transport proteins, *Gene* **115**:93–96.

Rat and Human Liver Cytochromes P450

Substrate and Inhibitor Specificities and Functional Markers

MARIA ALMIRA CORREIA

The tables in this appendix summarize the substrate/inhibitor specificities and functional markers that may be used for diagnostic characterization of individual rat and human liver cytochromes P450 (P450s). They have been compiled with the sole overall objective of practical utility rather than comprehensiveness. Wherever feasible, the literature cited has been that which provides methodological details of assay procedures, sometimes at the expense of accurate chronological citation of the literature as initially reported. For greater details and/or further discussion, the reader is referred to Chapter 9 (P450 inhibition), Chapter 12 (biosynthetic steroid-metabolizing P450s), Chapter 14 (human liver P450s), and Appendix A.

MARIA ALMIRA CORREIA • Department of Molecular and Cellular Pharmacology, University of California, San Francisco, San Francisco, California 94143.

Cytochrome P450: Structure, Mechanism, and Biochemistry (Second Edition), edited by Paul R. Ortiz de Montellano. Plenum Press, New York, 1995.

TABLE I
Rat Liver P450s: Current Nomenclatures, Past Designations, and Functional Markers[a]

Current P450 symbol	PIR database accession No.[b]	Past P450 designations				Functional marker
		Levin[c]	Guengerich[d]	Waxman[e]	Other[f]	
CYP1A1	A00185	c	BNF-B	—	P450MC, P448	Ethoxyresorufin O-deethylase[i], Acetanilide hydroxylase[j], R-Warfarin 8-hydroxylase[d]
CYP1A2	A22562, A20963	d	ISF-G	—	ISF-P450	Theophylline 3-demethylase[k], Caffeine 3-demethylase[k], Methoxyresorufin O-demethylase[i]
CYP2A1	A29560, A34272	a	UT-F	3	RLM2b	Testosterone 7α-hydroxylase[l]
CYP2A2	A31887	m	—	—	RLM2	Progesterone 15α-hydroxylase[m]
CYP2B1[g]	A00176, A22363	b	PB-B	PB-4	PB-1, P450PB	Pentoxyresorufin O-deethylase[n], Testosterone 16β-hydroxylase[l]
CYP2B2[g]	A00177, A21872	e	PB-D	PB-5	PBRLM6	Testosterone 16β-hydroxylase[l], Progesterone 16α-hydroxylase[o]
CYP2C6	A25954, A28516	k	PB-C	PB-1	RLM5a, PB1a/PB1b	S-Warfarin 7-hydroxylase[p], Progesterone 21-hydroxylase[o]
CYP2C7	A25585	f	—	—	RLM5b	Testosterone 16α-hydroxylase[l], Progesterone 16α-hydroxylase[o]
CYP2C11	A29421	h	UT-A	PB2c	RLM5, P450 male	Testosterone 2α-hydroxylase[l], Progesterone 2α-hydroxylase[o], Testosterone 16α-hydroxylase[l]
CYP2C12	A32140	i	UT-I	PB2d	P450 female, fRLM4	Steroid disulfate 15β-hydroxylase[q]
CYP2C13	A32470	g	—	—	RLM3, UT-5	Testosterone 6β-hydroxylase[l], Progesterone 16α-hydroxylase[o], Progesterone 6β-hydroxylase[o]

CYP	A (PIR)					Functional markers
CYP2D1	A31579		UT-H	UT-7 / db1	—	*Debrisoquine 4-hydroxylase*[t]
CYP2D2	C31579		—	db2	—	*Debrisoquine 4-hydroxylase*[t]
CYP2E1	A28145	j	—	RLM6 / P450ac	—	**N-nitrosodimethylamine N-demethylase**[s] / **Aniline p-hydroxylase**[t]
CYP3A1[h]	A22631	p	PCN-E	6β-4 / PCNa	—	**Nifedipine oxidase**[u] / *Erythromycin N-demethylase*[v] / *Testosterone 6β-hydroxylase*[l] / *Testosterone 2β-hydroxylase*[l]
CYP3A2[h]	A25222	1	—	6β-1/3	PB2a / PCNb/c	*Erythromycin N-demethylase*[v] / *Testosterone 6β-hydroxylase*[l] / *Testosterone 2β-hydroxylase*[l]
CYP4A1	A26137	w	—	LAω	—	**Lauric acid ω-hydroxylase**[w]

[a] Relatively selective functional markers are indicated in bold. Activities that can serve as functional markers but are not selective are indicated in italic.
[b] The accession numbers given are for the PIR database only. Accession numbers to other databases may be obtained from Nelson et al.[1]
[c] References 2–4a.
[d] Guengerich[5] and references therein.
[e] Waxman[6] and references therein.
[f] Only a few of the trivial names commonly encountered have been included in this category. For a more comprehensive listing refer to Nelson et al.[1]
[g] P450s 2B1 and 2B2 may be structurally distinguished by form-specific monoclonal antibodies, as well as by their relative SDS–PAGE mobility.[7,8]
[h] P450s 3A1 and 3A2 may be functionally distinguished by their relative ratios of testosterone 6β-hydroxylase/testosterone 2β-hydroxylase activities which are on the order of ~5.6 and ≥23, respectively. The testosterone 2β-hydroxylase activities of both 3A1 and 3A2 are comparable.[9] Similarly, differential ratios may also be derived from the data of Sonderfan et al.[10] and Underwood et al.[11] These P450s may also be distinguished structurally by form-specific monoclonal antibodies.[12]
[i] Reference 13.
[j] Reference 14.
[k] Reference 15.
[l] References 16–19.
[m] Reference 20.
[n] Reference 21.
[o] Reference 22.
[p] References 23–25.
[q] References 26, 27.
[r] Reference 28.
[s] References 29–31.
[t] Reference 29.
[u] Reference 32.
[v] Reference 33.
[w] Reference 34.

TABLE II
Rat Liver P450s: Sex Specificity and Inducibility[a]

P450	Sex specificity	Inducibility[b]	Prototype inducers	Dose[c] (mg/kg/day, i.p.)	Refs.
CYP1A1	None	++++	β-NF[d]	80	35
		++++	3-Methylcholanthrene	25	3
		++	Arochlor 1254	300	3
		+++	TCDD	10 μg	3
		+++	Phenothiazine	100	3
		++	Chlorpromazine	40	3
CYP1A2	None	+++	Isosafrole	150	3
		++	β-NF	80	35
		++	Arochlor 1254	300	3
		++	3-Methylcholanthrene	25	3
		++	Phenothiazine	100	3
CYP2A1	F>M	+	β-NF	80	35
		+	3-Methylcholanthrene	25	3
		+	TCDD	10 μg	3
CYP2A2	M	—[e]	—		
CYP2B1	None	++++	Phenobarbital	80	3
		+++	γ-Chlordane	50	3
		++	AIA	200	36
		+++	trans-Stilbene oxide	400	3
		++	Arochlor 1254	300	3
CYP2B2	None	+++	Phenobarbital	80	3
		+++	γ-Chlordane	50	3
		+++	trans-Stilbene oxide	400	3
CYP2C6	None	+	Phenobarbital	80	37, 38
CYP2C7	F>M	++	Phenobarbital[f]	80	37
		+	Kepone	20	37
CYP2C11	M	—[e]	—		
CYP2C12	F	—[e]	—		

TABLE II
(Continued)

P450	Sex specificity	Inducibility[b]	Prototype inducers	Dose[c] (mg/kg/day, i.p.)	Refs.
CYP2C13	M	—[e]	—		
CYP2E1	None	++	Acetone	5% v/v/drinking water/10d	39
		+++	Ethanol	6.4%/isocaloric diet/3wk	31, 40
		+++	Isoniazid	0.1% w/v/drinking water/10d	31, 40
		+++	Pyridine	100	31, 41
		++	Pyrazole	200	31
		++	4-Methylpyrazole	200	31
CYP3A1	M[g]	+++	PCN	25	33, 42
		+++	Dexamethasone	100	9, 33, 42
		+++	TAO	300	33, 42
		++	Phenobarbital	80	9, 33, 42
		+++	Clotrimazole	75	43
CYP3A2	M[h]	++	Phenobarbital	80	9, 33, 42
		++	PCN	25	33, 42
		+++	Dexamethasone	100	9, 33, 42
CYP4A1	None	+++	Clofibrate	320	44
		+++	Clobuzarit	50	44
		+++	DEHP	1250	44
		+++	MEHP	100	44

[a]Abbreviations used: AIA, allylisopropylacetamide; β-NF, β-naphthoflavone; DEHP, di(2-ethylhexyl)phthalate; MEHP, mono-(2-ethylhexyl)phthalate; PCN, pregnenolone 16α-carbonitrile; TAO, troleandomycin; TCDD, 2,3,7,8-tetrachlorodibenzo-*p*-dioxin.

[b]The extent of induction is indicated by ++++ for highest (>20–30 fold), +++ for good (≈ 5–10 fold), ++ for modest (2–4 fold),and + for low (1.5–2 fold).

[c]Unless otherwise indicated, the treatment is for 3–4 days.

[d]Young male rats (<100 g body weight) exhibit maximal induction.

[e]The expression of this constitutive enzyme in untreated rat liver is repressed by treatment with certain xenobiotic inducers.

[f]In immature male rats.

[g]Inducible in male rats, although extremely low constitutive levels may be present. Undetectable in untreated female rats but selectively induced by dexamethasone, TAO, PCN.

[h]Constitutive in male rats and inducible by the above inducers in male, but not female rats.

TABLE III
General Substrates for Rat Liver Cytochromes P450

Trivial name	Chemical structure	Type of oxidation	Reported P450 catalysts	Refs.
Aminopyrine		*N*-Demethylation	Multiple (2B1, 2B2, 2C11, 2C6, 3A, 1A1, 1A2)	46
Benzphetamine		*N*-Demethylation	Multiple (2B1, 2C11, 2C6, 3A)	47
Benzo(*a*)pyrene		Ring hydroxylation	1A1, 1A2	48
7-Ethoxycoumarin		*O*-Deethylation	2B1, 1A1, 1A2	14, 49

| Ethylmorphine | | N-Demethylation | Multiple (3A, 2B1, 2C11, 2C6) | 47 |
| p-Chloro-N-methylaniline | | N-Demethylation | 1A1, 1A2 | 50 |

TABLE IV

General Inhibitors of Rat Liver Cytochromes P450

Trivial name	Chemical structure	Mode of inhibition	P450s inhibited/inactivated	Refs.
Carbon monoxide	$C=O$	Fe^{2+}heme complexation	All	Chapter 9
Cimetidine		Fe^{3+}heme complexation MI complexation	Multiple P450 2C11	51 54
Ketoconazole		Fe^{3+}heme complexation	Multiple	55
Metyrapone		Fe^{3+}heme complexation	Multiple *(known to activate 2C6)*	56 57
Piperonyl butoxide		MI complexation	Multiple	58, 59

Compound	Structure	Mechanism	P450	Ref.
SKF-525A	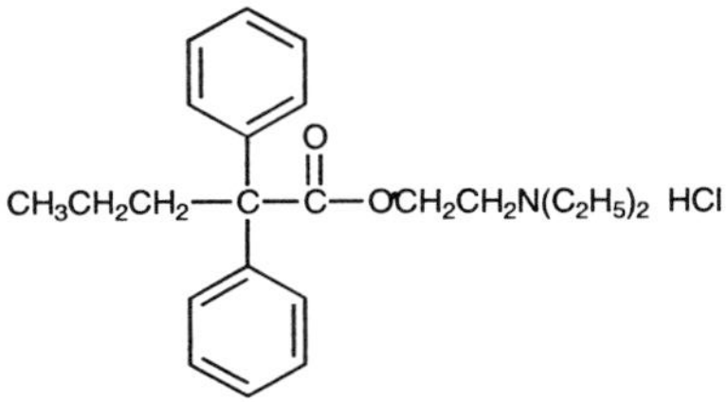 CH$_3$CH$_2$CH$_2$—C—OCH$_2$CH$_2$N(C$_2$H$_5$)$_2$ · HCl	Tight binding (*MI complexation of N-demethylated metabolites*)	Multiple	60
Allylisopropyl-acetamide	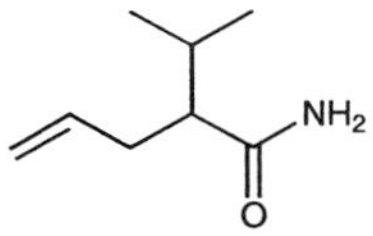	Mechanism-based (Heme *N*-alkylation)	Multiple (2B1, 2C11, 3A, 2C6)	47
1-Aminobenzotriazole	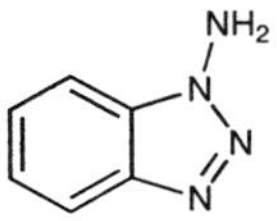	Mechanism-based (Heme *N*-arylation)	Multiple (2C11, 2B1, 3A, 2C6)	62
3,5-Dicarbethoxy-2,6-dimethyl-4-ethyl-1,4-dihydropyridine (DDEP)	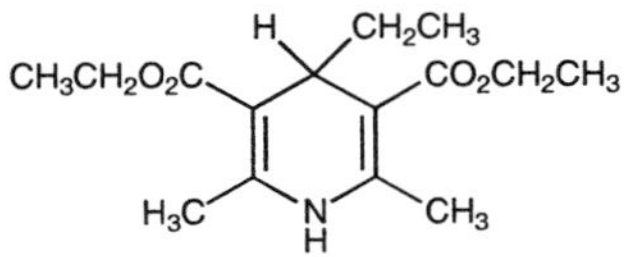	Mechanism-based (Heme *N*-ethylation) (Heme-protein modification)	Multiple (2C11,2C6) (3A)	63

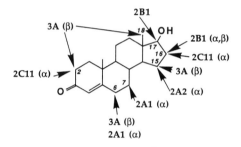

2B1, R= C$_5$H$_{11}$
1A1, R= C$_2$H$_5$
1A2, R= CH$_3$

7-ALKOXYPHENOXAZONES[a]

TESTOSTERONE[b]

WARFARIN[c]

FIGURE 1. Chemical probes useful as diagnostic functional markers for rat liver cytochromes P450. In each case, the sites of oxidation by rat liver cytochromes P450 are indicated by arrows. Their relative selectivity for a given oxidation site is indicated in Table I. For assay conditions, refer to: a, Ref. 13; b, Refs., 11, 45; c, Refs. 23–25.

TABLE V
Rat Liver P450s: Relatively Selective Inhibitors

P450s inhibited/ inactivated	Inhibitor (trivial name)	Mode of inhibition	Refs.
1A1	1-Ethynylpyrene	Mechanism-based	64, 65
1A2	Furafylline[a]	Mechanism-based	66
2B1	Secobarbital	Mechanism-based	8, 67
	2-p-BPCFA[b]	Mechanism-based	68
	2-p-NPCFA	Mechanism-based	68
2C6	Sulfaphenazole?[c]	Competitive	–
	Pregn-4,20-diene-3-one	Mechanism-based	69
2C11	Cannabidiol?[d]	Competitive	70
2D	4-Allyloxymetamphetamine	Mechanism-based	71
2E1	Disulfiram[e]	Mechanism-based	72
	Diallylsulfide	Mechanism-based	73, 74
	Diallylsulfoxide	Mechanism-based	74
	PEITC	Mechanism-based	75
3A1/2	Troleandomycin	Quasi-irreversible/ MI complexation	10, 33
	Gestodene?[f]	Mechanism-based	–
	6β-Thiotestosterone	Mechanism-based	11
	2β-Thiotestosterone	Mechanism-based	11
4A1	10-Undecynoic acid	Mechanism-based	76

[a] Although furafylline apparently is a considerably less selective inhibitor of human 1A2 (Guengerich, Chapter 14), at 500 μM, it inhibits phenacetin O-deethylase, a rat 1A2 functional marker, but not aryl hydrocarbon hydroxylase, a rat 1A1 functional probe.[66]

[b] The abbreviations used are: 2-p-BPCFA, N-(2-p-bromophenethyl)chlorofluoroacetamide; 2-p-NPCFA, N-(2-p-nitrophenethyl)chlorofluoroacetamide; PEITC, phenethyl isothiocyanate.

[c] A relatively selective inhibitor of human P450 2C9[25] and possibly of its rat liver ortholog 2C6.

[d] Although cannabidiol has been reported to be a selective chemical inhibitor of 2C11, apparently it also appreciably inhibits P450s 3A at comparable concentrations [30 μM (L.M. Bornheim, personal communication)]. 2α-Thiotestosterone was synthesized as a rationally designed inhibitor but was found to inhibit P450s 3A but not 2C11 when tested.[11] The only inhibitors to date that are selective are anti-2C11 polyclonal IgGs cross immunoreacted against common epitopes of other P450s 2C and monoclonal IgGs to specific 2C11 epitopes.[77,78]

[e] Selectivity may be dose- or concentration-dependent.

[f] Given the high similarities in primary structure and substrate preference of the rat 3A enzymes and their human 3A4 ortholog, although no reported evidence exists, it is likely that gestodene could be used as a potent inhibitor of rat P450s 3A.

TABLE VI
Rat Biosynthetic P450s: Current and Conventional Nomenclatures and Human Orthologs[a]

Current P450 symbol	PIR database accession No.	Other P450 designations	Substrate[b]	Function[b]	Human CYP ortholog
CYP7	S06632 A36450	7α	Cholesterol	Hydroxylase	7
CYP11A1	A27321	scc	Cholesterol	Side chain cleavage enzyme	11A1
CYP11B1	S05666	11β	11-Deoxycorticosterone	11β-Hydroxylase	11B1
CYP11B2	A35342	Aldo-2	Corticosterone	18-Dehydrogenase	11B2/11β-2
CYP17	A27659 A33980	17α	Pregnenolone Progesterone	17α-Hydroxylase	17
CYP19	A36121	aromatase	Androstenedione Testosterone	Ring A aromatization	19
CYP21			Progesterone 17α-OH-Progesterone	21-Hydroxylation	21A2/c21B
CYP24	S13918	cc24	Vitamin D_3	25-Hydroxylation	
CYP27	A36239 A34558	25-hydroxylase	Vitamin D	25/27-Hydroxylation	27-hydroxylase

[a] Nomenclature, PIR database accession numbers, and corresponding human CYP ortholog designations have been derived from Nelson *et al.*[1]
[b] Except for CYP7,[79,80] relevant information pertaining to the function and regulation of individual P450s is discussed by Kagawa and Waterman (Chapter 12).

TABLE VII

Substrates and Inhibitors Frequently Used as Diagnostic Probes of Select Human Liver Cytochromes P450: Chemical Structures and Mechanistic Features[a]

P450	Substrate	Type of oxidation	Inhibitor	Mode of inhibition	Selectivity[b]
1A1	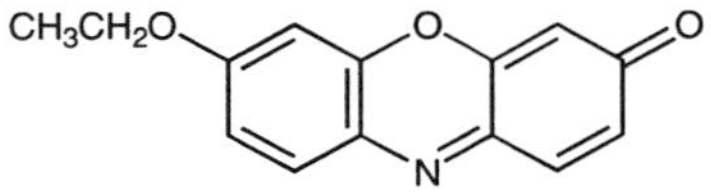 7-Ethoxyresorufin	O-Deethylation[c]	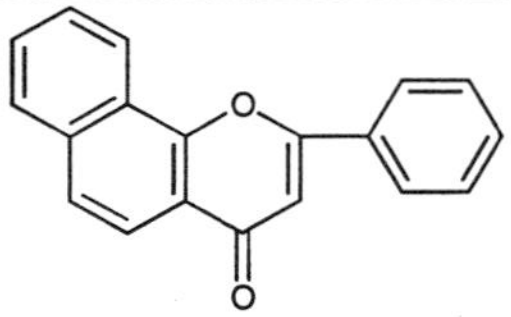 7,8-Benzoflavone	Competitive	NS[d]
			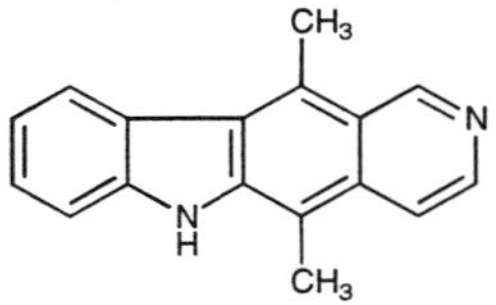 Ellipticine	Competitive	RS[e]?
1A2	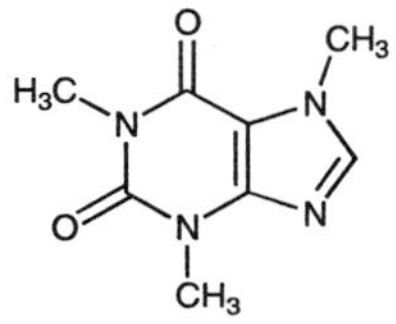 Caffeine	3-N-Demethylation[f]	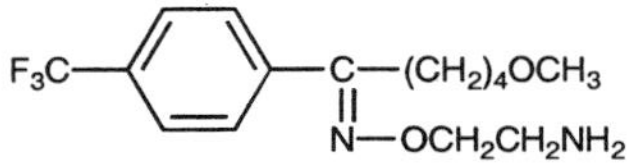 Fluvoxamine	Competitive	RS[g]?
			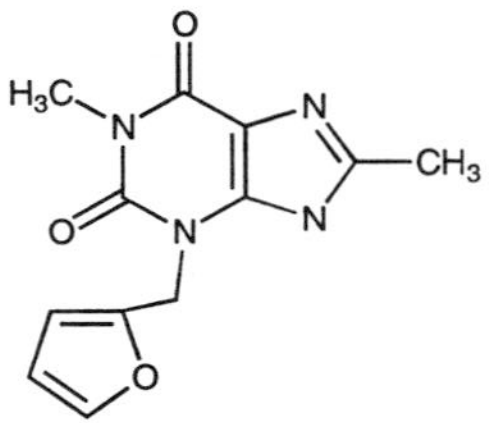 Furafylline	Irreversible/ mechanism-based	RS[h]

(continued)

TABLE VII
(*Continued*)

P450	Substrate	Type of oxidation	Inhibitor	Mode of inhibition	Selectivity[b]
2A6	Coumarin	7-Hydroxylation[i]	(C$_2$H$_5$)$_2$N—C(=S)—SH Diethyldithiocarbamate	Irreversible/ mechanism-based? (*a proven mechanism-based inhibitor of 2E1*)	NS[j]
2C9	Tolbutamide	Benzylic oxidation[k]	Sulfaphenazole	Competitive (heme complexation)	RS[l]
			Tienilic acid	Irreversible/ mechanism-based (protein modification)	RS[m]

2C19	4′-Hydroxylation[n]	(S)-Mephenytoin		
2D6	4-Hydroxylation[o]	Debrisoquine	Quinidine	Reversible (heme complexation)	RS[p]
	1′-Hydroxylation[q]	Bufuralol		
2E1	6-Hydroxylation[r]	Chlorzoxazone	4-Methylpyrazole	Reversible (heme complexation)	RS[s]
	Demethylation[t]	N-Nitrosodimethylamine (NDMA)	Diethyldithiocarbamate	Irreversible/mechanism-based	NS[u]

(continued)

TABLE VII
(Continued)

P450	Substrate	Type of oxidation	Inhibitor	Mode of inhibition	Selectivity[b]
3A4					

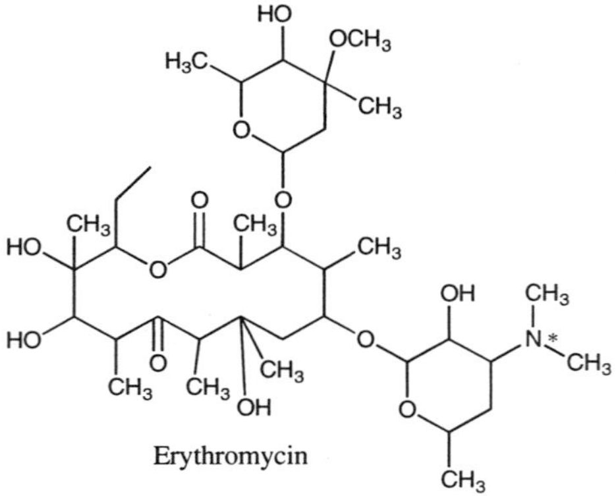

Erythromycin

N-Demethylation[v]

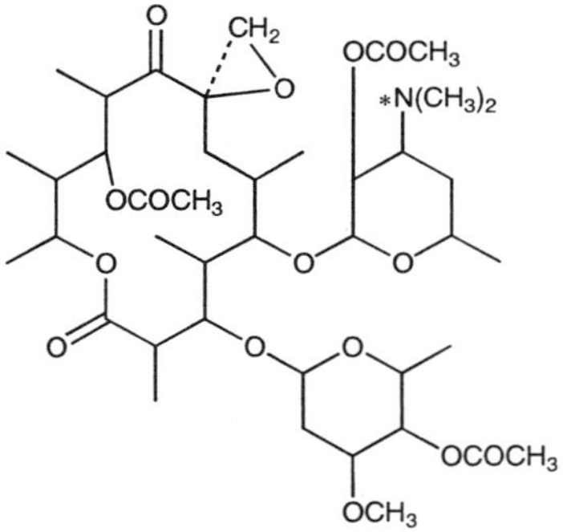

Troleandomycin

Quasi-irreversible/
mechanism-based
(MI complexation)

RS[w]

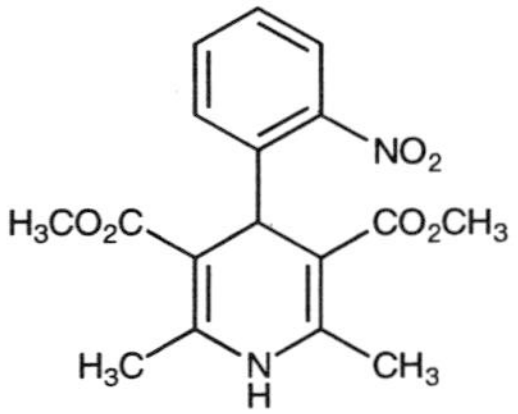

Nifedipine

N-Oxidation[x]

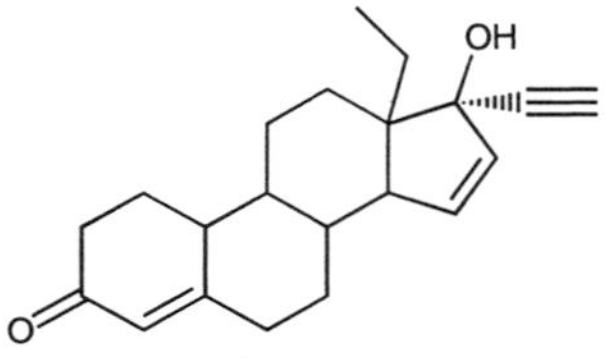

Gestodene

Irreversible/mechanism-
based
(Heme modification?)

RS[y]

4A11

H_3C ⌇⌇⌇ COOH ω-Oxidation[z]

Lauric acid

7

Cholesterol — 7α-Hydroxylation[aa]

HO

[a]Unless otherwise indicated, the substrates listed are relatively selective for a given subfamily of human liver P450s. However, not all of the inhibitors listed are as selective and thus useful as definitive diagnostic probes. For other details and references, see Chapters 9 and 14.

[b]The relative selectivity of a given inhibitor is indicated by RS, relatively selective, or NS, nonselective.

[c]Reference 13.

[d]Inhibits 1A1 and 1A2 as well as 2C8 and 2C9.[81]

[e]Is a potent inhibitor of 1A1 and 1A2.[82,83]

[f]References 84–87.

[g]Reference 88.

[h]References 66, 89, 90. Slight inhibition (<20%) of 2E1 has been reported.[90a]

[i]References 91–93.

[j]Inhibits several P450s, but may be rendered more selective by lowering the concentrations to 10 μM.[81]

[k]References 94–96.

[l]Relatively selective competitive inhibitor.[25,90a]

[m]References 97, 98.

[n]Reference 99.

[o]References 100–102.

[p]References 103, 104. Very slight inhibition of 3A4 at higher concentrations.[90a]

[q]Reference 105.

[r]A good functional probe in vitro,[106] but apparently not in vivo, since 1A2 also catalyzes the reaction, albeit to a lesser extent.[107,108]

[s]References 109, 110. Is also found to inhibit 2D6 strongly and at higher concentrations, 1A2, 2C9, and 3A4 less strongly.[90a]

[t]References 30, 31.

[u]References 90a, 111. Recently shown to inhibit several other P450s at concentrations required to inhibit 2E1.[81]

[v]References 112, 113.

[w]References 81, 114. Ketoconazole is another potent inhibitor, albeit competitive and somewhat nonselective.[90a]

[x]References 32,115.

[y]References 116, 117.

[z]Reference 118.

[aa]Reference 80.

ACKNOWLEDGMENT. The compilation of this appendix was supported by NIH grants DK-26506 and GM-44037.

References

1. Nelson, D. R., Kamataki, T., Waxman, D. J., Guengerich, F. P., Estabrook, R. W., Feyereisen, R., Gonzalez, F. J., Coon, M. J., Gunsalus, I. C., Gotoh, O., Okuda, K., and Nebert, D. W., 1993, The P450 superfamily: Update on new sequences, gene mapping, accession numbers, early trivial names of enzymes, and nomenclature, *DNA Cell Biol.* **12:**1–51.
2. Levin, W., 1990, The 1988 Bernard B. Brodie Award lecture: Functional diversity of hepatic cytochromes P450, *Drug Metab. Dispos.* **18:**824–852.
3. Thomas, P. E., Reik, L. M., Ryan, D. E., and Levin, W., 1983, Induction of two immunochemically related rat liver cytochrome P-450 isozymes, cytochromes P-450c and P-450d, by structurally diverse xenobiotics, *J. Biol. Chem.* **258:**4590–4598.
4. Thomas, P. E., Reik, L. M., Ryan, D. E., Bandiera, S., and Levin, W., 1985, Polyclonal and monoclonal antibody probes of the structure and function of hepatic cytochrome P-450 isozymes, in: *Drug Metabolism* (G. Siest, ed.), Pergamon Press, Elmsford, NY, pp. 131–139.
4a. Thomas, P. E., Bandiera, S., Reik, L. M., Maines, S. L., Ryan, D. E., and Levin, W., 1987, Polyclonal and monoclonal antibodies as probes of rat hepatic cytochrome P-450 isozymes, *Fed. Proc.* **46:**2563–2566.
5. Guengerich, F. P., 1987, Enzymology of rat liver cytochromes P450, in: *Mammalian Cytochromes P450* (F. P. Guengerich, ed.), CRC Press, Boca Raton, FL, Vol. I, pp. 1–54.
6. Waxman, D. J., 1986, Appendix: Rat hepatic cytochrome P-450: Comparative study of multiple isozymic forms, in: *Cytochrome P-450: Structure, Mechanism, and Biochemistry* (P. R. Ortiz de Montellano, ed.), Plenum Press, New York, pp. 525–539.
7. Reik, L. M., Levin, W., Ryan, D. E., Maines, S. L., and Thomas, P. E., 1985, Monoclonal antibodies distinguish among isozymes of the cytochrome P-450b subfamily, *Arch. Biochem. Biophys.* **242:**365–382.
8. Waxman, D. J., and Walsh, C., 1982, Phenobarbital-induced rat liver cytochrome P-450, *J. Biol. Chem.* **257:**10446–10457.
9. Nagata, K., Gonzalez, F. J., Yamazoe, Y., and Kato, R., 1990, Purification and characterization of four catalytically active testosterone 6β-hydroxylase P-450s from rat liver microsomes: Comparison of a novel form with three structurally and functionally related forms, *J. Biochem.* **107:**718–725.
10. Sonderfan, A. J., Arlotto, M. P., Dutton, D. R., McMillen, S. K., and Parkinson, A., 1987, Regulation of testosterone hydroxylation by rat liver microsomal cytochrome P-450, *Arch. Biochem. Biophys.* **255:**27–41.
11. Underwood, M. C., Cashman, J. R., and Correia, M. A., 1992, Specifically designed thiosteroids as active site-directed probes for functional dissection of cytochrome P-450 3A isozymes, *Chem. Res. Toxicol.* **5:**42–53.
12. Cooper, K. O., Reik, L. M., Jayyosi, Z., Bandiera, S., Kelley, M., Ryan, D. E., Daniel, R., McCluskey, S. A., Levin, W., and Thomas, P. E., 1993, Regulation of two members of the steroid-inducible cytochrome P450 subfamily (3A) in rats, *Arch. Biochem. Biophys.* **301:**345–354.
13. Rodrigues, A. D., and Prough, R. A., 1992, Induction of cytochromes P450IA1 and P450IA2 and measurement of catalytic activities, *Methods Enzymol.* **206:**423–431.
14. Guengerich, F. P., Dannan, G. A., Wright, S. T., Martin, M. V., and Kaminsky, L. S., 1982, Purification and characterization of liver microsomal cytochromes P-450: Electrophoretic, spectral, catalytic, and immunochemical properties and inducibility of eight isozymes isolated from rats treated with phenobarbital or β-naphthoflavone, *Biochemistry* **21:**6019–6030.

15. Fuhr, U., Doehmer, J., Battula, N., Wolfel, C., Kudla, C., Keita, Y., and Staib, A. H., 1992, Biotransformation of caffeine and theophylline in mammalian cell lines genetically engineered for expression of single cytochrome P450 isoforms, *Biochem. Pharmacol.* **43**:225–235.

16. Levin, W., Thomas, P. E., Ryan, D. E., and Wood, A. W., 1987, Isozyme specificity of testosterone 7α-hydroxylation in rat hepatic microsomes: Is cytochrome P-450a the sole catalyst? *Arch. Biochem. Biophys.* **258**:630–635.

17. Waxman, D. J., Ko, A., and Walsh, C., 1983, Regioselectivity and stereoselectivity of androgen hydroxylations catalyzed by cytochrome P-450 isozymes purified from phenobarbital-induced rat liver, *J. Biol. Chem.* **258**:11937–11947.

18. Waxman, D. J., 1988, Interactions of hepatic cytochromes P-450 with steroid hormones. Regioselectivity and stereoselectivity of steroid metabolism and hormonal regulation of rat P-450 enzyme expression, *Biochem. Pharmacol.* **37**:71–84.

19. Wood, A. W., Ryan, D. R., Thomas, P. E., and Levin, W., 1983, Regio- and stereoselective metabolism of two C19 steroids by five highly purified and reconstituted rat hepatic cytochrome P450 isozymes, *J. Biol. Chem.* **258**:8839–8847.

19a. Ryan, D. E., Iida, S., Wood, A. W., Thomas, P. E., Lieber, C. S., and Levin, W., 1984, Characterization of three highly purified cytochromes P-450 from hepatic microsomes of adult male rats, *J. Biol. Chem.* **259**:1239–1250.

20. Jansson, I., and Schenkman, J. B., 1985, Characterization of a new form (RLM2) of liver microsomal cytochrome P-450 from untreated rat, *Fed. Proc.* **44**:1206.

21. Lubet, R. A., Mayer, R. T., Cameron, J. W., Nims, R. W., Burke, M. D., Wolff, T., and Guengerich, F. P., 1985, Dealkylation of pentoxyresorufin: A rapid and sensitive assay for measuring induction of cytochrome(s) P-450 by phenobarbital and other xenobiotics in the rat, *Arch. Biochem. Biophys.* **238**:43–48.

22. Swinney, D. C., Ryan, D. E., Thomas, P. E., and Levin, W., 1987, Regioselective progesterone hydroxylation catalyzed by eleven rat hepatic cytochrome P-450 isozymes, *Biochemistry* **26**:7073–7083.

23. Kaminsky L. S., 1989, Warfarin as a probe of cytochromes P450 function, *Drug Metab. Rev.* **20**:479–487.

24. Rettie, A. E., Heimark, L., Mayer, R. T., Burke, M. D., Trager, W. F., and Juchau, M. R., 1985, Stereoselective and regioselective hydroxylation of warfarin and selective O- dealkylation of phenoxazone ethers in human placenta, *Biochem. Biophys. Res. Commun.* **126**:1013–1021.

25. Rettie, A. E., Korzekwa, K. R., Kunze, K. L., Lawrence, R. F., Eddy, A. C., Aoyama, T., Gelboin, H. V., Gonzalez, F. J., and Trager, W. F., 1992, Hydroxylation of warfarin by human cDNA-expressed cytochrome P-450: A role for P-4502C9 in the etiology of (*S*)-warfarin-drug interactions, *Chem. Res. Toxicol.* **5**:54–59.

26. Ryan, D. E., Dixon, R., Evans, R. H., Ramanathan, L., Thomas, P. E., Wood, A. W., and Levin, W., 1984, Rat hepatic cytochrome P-450 isozyme specificity for the metabolism of the steroid sulfate, 5α-androstane-3α,17β-diol-3,17-disulfate, *Arch. Biochem. Biophys.* **233**:636–642.

27. MacGeoch, C., Morgan, E. T., Halpert, J., and Gustafsson, J.-A., 1984, Purification, characterization, and pituitary regulation of the sex-specific cytochrome P-450 15β-hydroxylase from liver microsomes of untreated female rat, *J. Biol. Chem.* **259**:15433–15439.

28. Larrey, D., Distlerath, L. M., Dannan, G. A., Wilkinson, G. R., and Guengerich, F. P., 1984, Purification and characterization of the rat liver microsomal cytochrome P-450 involved in the 4-hydroxylation of debrisoquine, a prototype for genetic variation in oxidative drug metabolism, *Biochemistry* **23**:2787–2795.

29. Ryan, D. E., Ramanathan, L., Iida, S., Thomas, P. E., Haniu, M., Shively, J. E., Lieber, C. S., and Levin, W., 1985, Characterization of a major form of rat hepatic microsomal cytochrome P-450 induced by isoniazid, *J. Biol. Chem.* **260**:6385–6393.

30. Yang, C. S., Yoo, J.-S. H., Ishizaki, H., and Hong, J., 1990, Cytochrome P450IIE1: Roles in nitrosamine metabolism and mechanisms of regulation, *Drug Metab. Rev.* **22**:147–159.

31. Yang, C. S., Patten, C. J., Ishizaki, H., and Yoo, J.-S. H., 1992, Induction, purification, and characterization of cytochrome P450IIE1, *Methods Enzymol.* **206**:595–603.

32. Guengerich, F. P., Martin, M. V., Beaune, P. H., Kremers, P., Wolff, T., and Waxman, D. J., 1986, Characterization of rat and human liver microsomal cytochrome P-450 forms involved in nifedipine oxidation, a prototype for genetic polymorphism in oxidative drug metabolism, *J. Biol. Chem.* **261**:5051–5060.

33. Wrighton, S. A., Maurel, P., Schuetz, E. G., Watkins, P. B., Young, B., and Guzelian, P. S., 1985, Identification of the cytochrome P-450 induced by macrolide antibiotics in rat liver as the glucocorticoid responsive cytochrome P-450p, *Biochemistry* **24**:2171–2178.

34. Gibson, G. G., Orton, T. C., and Tamburini, P. P., 1982, Cytochrome P-450 induction by clofibrate: Purification and properties of a hepatic cytochrome P-450 relatively specific for the 12- and 11-hydroxylation of dodecanoic acid (lauric acid), *Biochem. J.* **203**:161–168.

35. Lau, P. P., and Strobel, H. W., 1982, Multiple forms of cytochrome P-450 in liver microsomes from β-naphthoflavone-pretreated rats, *J. Biol. Chem.* **257**:5257–5262.

36. De Matteis, F., 1971, Loss of haem in rat liver caused by the porphyrogenic agent 2-allyl-2-isopropylacetamide, *Biochem. J.* **124**:767–777.

37. Bandiera, S., Ryan, D. E., Levin, W., and Thomas, P. E., 1986, Age- and sex-related expression of cytochromes P450f and P450g in rat liver, *Arch. Biochem. Biophys.* **248**:658–676.

38. Wolf, C. R., Seilman, S., Oesch, F., Mayer, R. T., and Burke, M. D., 1986, Multiple forms of cytochrome P-450 related to forms induced marginally by phenobarbital, *Biochem. J.* **240**:27–33.

39. Song, B.-J., Veech, R. L., Park, S. S., Gelboin, H. V., and Gonzalez, F. J., 1989, Induction of rat hepatic N-nitrosodimethylamine demethylase by acetone is due to protein stabilization, *J. Biol. Chem.* **264**:3568–3572.

40. Ryan, D. E., Koop, D. R., Thomas, P. E., Coon, M. J., and Levin, W., 1986, Evidence that isoniazid and ethanol induce the same microsomal cytochrome P-450 in rat liver, an isozyme homologous to rabbit liver cytochrome P-450 isozyme 3a, *Arch. Biochem. Biophys.* **246**:633–644.

41. Kim, S. G., and Novak, R. F., 1990, Induction of rat hepatic P450IIE1 (CYP 2E1) by pyridine: Evidence for a role of protein synthesis in the absence of transcriptional activation, *Biochem. Biophys. Res. Commun.* **166**:1072–1079.

42. Halpert, J. R., 1988, Multiplicity of steroid-inducible cytochromes P-450 in rat liver microsomes, *Arch. Biochem. Biophys.* **263**:59–68.

43. Ritter, J. K., and Franklin, M. R., 1987, High magnitude hepatic cytochrome P-450 induction by an N-substituted imidazole antimycotic, clotrimazole, *Biochem. Pharmacol.* **36**:2783–2787.

44. Sharma, R., Lake, B. G., Foster, J., and Gibson, G. G., 1988, Microsomal cytochrome P-452 induction and peroxisome proliferation by hypolipidaemic agents in rat liver, *Biochem. Pharmacol.* **37**:1193–1201.

45. Arlotto, M. P., Trant, J. M., and Estabrook, R. W., 1992, Measurement of steroid hydroxylation reactions by high-performance liquid chromatography as indicator of P450 identity and function, *Methods Enzymol.* **206**:454–462.

46. Imaoka, S., Inoue, K., and Funae, Y., 1988, Aminopyrine metabolism by multiple forms of cytochrome P-450 from rat liver microsomes: Simultaneous quantitation of four aminopyrine metabolites by high-performance liquid chromatography, *Arch. Biochem. Biophys.* **265**:159–170.

47. Bornheim, L. M., Underwood, M. C., Caldera, P., Rettie, A. E., Trager, W. F., Wrighton, S. A., and Correia, M. A., 1987, Inactivation of multiple hepatic cytochrome P-450 isozymes in rats by allylisopropylacetamide: Mechanistic implications, *Mol. Pharmacol.* **32**:299–308.

48. Nebert, D. W., 1982, Genetic differences in the induction of monooxygenase activities by polycyclic aromatic compounds, in: *Hepatic Cytochrome P-450 Monooxygenase System* (J.B. Schenkman and D. Kupfer, eds.), Pergamon Press, Elmsford, NY, pp. 269–291.

49. Prough, R. A., Burke, M. D., and Mayer, R. T., 1978, Direct fluorometric methods for measuring mixed-function oxidase activity, *Methods Enzymol.* **52**:372–377.

50. Correia, M. A., and Mannering, G. J., 1973, Reduced diphosphopyridine nucleotide synergism of the reduced triphosphopyridine nucleotide dependent mixed function oxidase of hepatic microsomes. Role of the type I drug binding site of cytochrome P-450, *Mol. Pharmacol.* **9:**470–485.

51. Winzor, D. J., Ioannoni, B., and Reilly, P. E., 1986, The nature of microsomal monooxygenase inhibition by cimetidine, *Biochem. Pharmacol.* **35:**2157–2161.

52. Wang, R. W., Miwa, G. T., Argenbright, L. S., and Lu, A. Y. H., 1988, In vitro studies on the interaction of famotidine with liver microsomal cytochrome P-450, *Biochem. Pharmacol.* **37:**3049–3053.

53. Ritter, J. K., and Franklin, M. R., 1987, Induction and inhibition of rat hepatic drug metabolism by *N*-substituted imidazole drugs, *Drug Metab. Dispos.* **15:**335–343.

54. Bellward, G. D., and Levine, M., 1994, Inhibition of hepatic cytochrome P-450 by cimetidine: Evidence for formation of a metabolic-intermediate complex, in: *Abstracts, 10th International Symposium on Microsomes Drug Oxidations,* Toronto, Canada, July 18–21.

55. Pasanen, M., Taskinen, T., Iscan, M., Sotaniemi, E. A., Kairaluoma, M., and Pelkonen, O., 1988, Inhibition of human hepatic and placental xenobiotic monooxygenases by imidazole antimycotics, *Biochem. Pharmacol.* **37:**3861–3866.

56. Testa, B., and Jenner, P., 1981, Inhibitors of cytochrome P-450s and their mechanism of action, *Drug Metab. Rev.* **12:**1–117.

57. Waxman, D. J., and Walsh, C., 1983, Cytochrome P-450 isozyme 1 from phenobarbital-induced rat liver: Purification, characterization, and interactions with metyrapone and cytochrome b_5, Biochemistry **22:**4846–4855.

58. Franklin, M. R., 1972, Inhibition of hepatic oxidative xenobiotic metabolism by piperonyl butoxide, *Biochem. Pharmacol.* **21:**3287–3299.

59. Hodgson, E., and Philpot, R. M., 1974, Interaction of methylenedioxyphenyl (1,3-benzodioxole) compounds with enzymes and their effects on mammals, *Drug. Metab. Rev.* **3:**231–301.

60. Buening, M. K., and Franklin, M. R., 1974, The formation of complexes absorbing at 455 nm from cytochrome P-450 and metabolites of compounds related to SKF 525-A, *Drug Metab. Dispos.* **2:**386–390.

61. Waxman, D. J., Lapenson, D. P., Park, S. S., Attisano, C., and Gelboin, H. V., 1987, Monoclonal antibodies inhibitory to rat hepatic cytochromes P-450: P-450 form specificities and use as probes for cytochrome P-450-dependent steroid hydroxylations, *Mol. Pharmacol.* **32:**615–624.

62. Ortiz de Montellano, P. R., and Costa, A. K., 1986, Dissociation of cytochrome P-450 inactivation and induction, *Arch. Biochem. Biophys.* **251:**514–524.

63. Correia, M. A., Decker, C., Sugiyama, K., Caldera, P., Bornheim, L., Wrighton, S. A., Rettie, A. E., and Trager, W. F., 1987, Degradation of rat hepatic cytochrome P-450 heme by 3,5-dicarbethoxy-2,6-dimethyl-4-ethyl-1,4-dihydropyridine to irreversibly bound protein adducts, *Arch. Biochem. Biophys.* **258:**436–451.

64. Hopkins, N. E., Foroozesh, M. K., and Alworth, W. L., 1992, Suicide inhibitors of cytochrome P450 1A1 and P450 2B1, *Biochem. Pharmacol.* **44:**787–796.

65. Chan, W. K., Sui, Z., and Ortiz de Montellano, P. R., 1993, Determinants of protein modification versus heme alkylation: Inactivation of cytochrome P450 1A1 by 1-ethynylpyrene and phenylacetylene, *Chem. Res. Toxicol.* **6:**38–45.

66. Sesardic, D., Boobis, A., Murray, B., Murray, S., Segura, J., De La Torre, R., and Davies, D., 1990, Furafylline is a potent and selective inhibitor of cytochrome P450 1A2 in man, *Br. J. Clin. Pharmacol.* **29:**651–663.

67. Lunetta, J. M., Sugiyama, K., and Correia, M. A., 1989, Secobarbital-mediated inactivation of rat liver cytochrome P-450b: A mechanistic reappraisal, *Mol. Pharmacol.* **35:**10–17.

68. Halpert, J., Jaw, J.-Y., Balfour, C., and Kaminsky, L. S., 1990, Selective inactivation by chlorofluoroacetamides of the major phenobarbital-inducible form(s) of rat liver cytochrome P-450, *Drug Metab. Dispos.* **18:**168–174.

69. Halpert, J., Jaw, J.-Y., and Balfour, C., 1989, Speciffc inactivation by 17β-substituted steroids of rabbit and rat liver cytochromes P-450 responsible for progesterone 21-hydroxylation, *Mol. Pharmacol.* **34:**148–156.

70. Narimatsu, S., Watanabe, K., Yamamoto, I., and Yoshimura, H., 1988, Mechanism for inhibitory effect of cannabidiol on microsomal testosterone oxidation in male rat liver, *Drug Metab. Dispos.* **16:**880–889.

71. Lin, L. Y., Fujimoto, M., DiStefano, E. W., Schmitz, D. A., Jayasinghe, A., and Cho, A. K., 1994, Selective mechanism-based inactivation of rat CYP2D by 4-allyloxymethamphetamine, *FASEB J.* **8:**A100.

72. Brady, J. F., Xiao, F., Wang, M.-H., Li, Y., Ning, S. M., Gapac, J. M., and Yang, C. S., 1991, Effects of disulfiram on hepatic P450IIE1, other microsomal enzymes, and hepatotoxicity in rats, *Toxicol Appl. Pharmacol.* **108:**366–373.

73. Brady, J. F., Ishizaki, H., Fukuto, J. M., Lin, M. C., Fadel, A., Gapac, J. M., and Yang, C. S., 1991, Inhibition of cytochrome P-450 2E1 by diallyl sulfide and its metabolites, *Chem. Res. Toxicol.* **4:**642–647.

74. Brady, J. F., Wang, M.-H., Hong, J.-Y., Xiao, F., Li, Y., Yoo, J.-S.H., Ning, S.M., Lee, M.-J., Fukuto, J. M., Gapac, J. M., and Yang, C. S., 1991, Modulation of rat hepatic microsomal monooxygenase enzymes and cytotoxicity by diallyl sulfide, *Toxicol Appl. Pharmacol.* **10:**342–354.

75. Ishizaki, H., Brady, J. F., Ning, S. M., and Yang, C. S., 1990, Effect of phenethyl isothiocyanate on microsomal *N*-nitrosodimethylamine metabolism and other monooxygenase activities, *Xenobiotica* **20(3):**255–264.

76. CaJacob, C. A., Chan, W. K., Shepard, E., and Ortiz de Montellano, P. R., 1988, The catalytic site of rat hepatic lauric acid omega-hydroxylase. Protein versus prosthetic heme alkylation in the omega-hydroxylation of acetylenic fatty acids, *J. Biol. Chem.* **263:**18640–18649.

77. Morgan, E. T., Ronnholm, M., and Gustafsson, J.-A., 1987, Preparation and characterization of monoclonal antibodies recognizing unique epitopes on sexually differentiated rat liver cytochrome P450 isozymes, *Biochemistry* **26:**4194–4200.

78. Ryan, D. E., Thomas, P. E., Levin, W., Maines, S. L., Bandiera, S., and Reik, L. M., 1993, Monoclonal antibodies of differentiating specificities as probes of cytochrome P450h (2C11), *Arch. Biochem. Biophys.* **301:**282–293.

79. Li, Y. C., Wang, D. P., and Chiang, J. Y. L., 1990, Regulation of cholesterol 7α-hydroxylase in the liver: Cloning, sequencing, and regulation of cholesterol 7α-hydroxylase mRNA, *J. Biol. Chem.* **265:**12012–12019.

80. Nishimoto, M., Noshiro, M., and Okuda, K., 1993, Structure of the gene encoding human liver cholesterol 7α-hydroxylase, *Biochim. Biophys. Acta* **1172:**147–150.

81. Chang, T. K. H., Gonzalez, F. J., and Waxman, D. J., 1994, Evaluation of triacetyloleandomycin, α-naphthoflavone and diethyldithiocarbamate as selective chemical probes for inhibition of human cytochromes P450, *Arch. Biochem. Biophys.* **311:**437–442.

82. Lesca, P., Rafidinarivo, E., LeCointe, P., and Mansuy, D., 1979, A class of strong inhibitors of microsomal monooxygenases: The ellipticines, *Chem. Biol. Interact.* **24:**189–198.

83. Tassaneeyakul, W., Birkett, D. J., Veronese, M. E., McManus, M. E., Tukey, R. H., Quattrochi, L. C., Gelboin, H. V., and Miners, J. O., 1993, Specificity of substrate and inhibitor probes for human cytochromes P450 1A1 and 1A2, *J. Pharmacol. Exp. Ther.* **265:**401–407.

84. Butler, M. A., Iwasaki, M., Guengerich, F. P., and Kadlubar, F. F., 1989, Human cytochrome P-450$_{PA}$ (P-450IA2), the phenacetin O-deethylase, is primarily responsible for the hepatic 3-demethylation of caffeine and N-oxidation of carcinogenic arylamines, *Proc. Natl. Acad. Sci. USA* **86:**7696–7700.

85. Butler, M. A., Lang, N. P., Young, J. F., Caporaso, N. E., Vineis, P., Hayes, R. B., Teitel, C. H., Massengill, J. P., Lawsen, M. F., and Kadlubar, F. F., 1992, Determination of CYP1A2 and NAT2 phenotypes in human populations by analysis of caffeine urinary metabolites, *Pharmacogenetics* **2:**116–127.

86. Tassaneeyakul, W., Mohamed, Z., Birkett, D. J., McManus, M. E., Veronese, M. E., Tukey, R. H., Quattrochi, L. C., Gonzalez, F. J., and Miners, J. O., 1992, Caffeine as a probe for human cytochromes P450: Validation using cDNA expression, immunoinhibition and microsomal kinetic and inhibitor techniques, *Pharmacogenetics* **2:**173–183.

87. Kalow, W., and Tang, B. K., 1993, The use of caffeine for enzyme assays: A critical appraisal, *Clin. Pharmacol. Ther.* **53:**503–514.

88. Brosen, K., Skjelbo, E., Rasmussen, B. B., Poulsen, H. E., and Loft, S., 1993, Fluvoxamine is a potent inhibitor of cytochrome P4501A2, *Biochem. Pharmacol.* **45:**1211–1214.

89. Kunze, K. L., and Trager, W. F., 1993, Isoform-selective mechanism-based inhibition of human cytochrome-P450 1A2 by furafylline, *Chem. Res. Toxicol.* **6:**649–656.

90. Kunze, K. L., Wienkers, L. C., Thummel, K. E., and Trager, W. F., 1993, The use of furafylline together with liver screening techniques and enzyme kinetics to evaluate the role of P450 1A2 in the metabolism of theophylline at therapeutic concentrations in human liver, in: *Abstracts, 5th North American ISSX Meeting,* Tucson, Vol. 4, p. 138.

90a. Newton, D. J., Wang, R. W., and Lu, A. Y. H., 1995, Cytochrome P450 inhibitors, *Drug Metab. Dispos.* **23:**154–158.

91. Yamano, S., Tatsuno, J., and Gonzalez, F. J., 1990, The CYP2A3 gene product catalyzes coumarin 7-hydroxylation in human liver microsomes, *Biochemistry* **29:**1322–1329.

92. Yun, C.-H., Shimada, T., and Guengerich, F. P., 1991, Purification and characterization of human liver microsomal cytochrome P-450 2A6, *Mol. Pharmacol.* **40:**679–685.

93. Waxman, D. J., Lapenson, D. P., Aoyama, T., Gelboin, H. V., Gonzalez, F. J., and Korzekwa, K., 1991, Steroid hormone hydroxylase specificities of eleven cDNA-expressed human cytochrome P450s, *Arch. Biochem. Biophys.* **290:**160–166.

94. Shimada, T., Misono, K. S., and Guengerich, F. P., 1986, Human liver microsomal cytochrome P-450 mephenytoin 4-hydroxylase, a prototype of genetic polymorphism in oxidative drug metabolism. Purification and characterization of two similar forms involved in the reaction, *J. Biol. Chem.* **261:**909–921.

95. Relling, M. V., Aoyama, T., Gonzalez, F. J., and Meyer, U. A., 1989, Tolbutamide and mephenytoin hydroxylation by human cytochrome P450s in the CYP2C subfamily, *J. Pharmacol. Exp. Ther.* **252:**442–447.

96. Srivastava, P. K., Yun, C.-H., Beaune, P. H., Ged, C., and Guengerich, F. P., 1991, Separation of human liver tolbutamine hydroxylase and (*S*)-mephenytoin 4′-hydroxylase cytochrome P-450 enzymes, *Mol. Pharmacol.* **40:**69–79.

97. Lopez-Garcia, M. P., Dansette, P. M., Valadon, P., Amar, C., Beaune, P. H., Guengerich, F. P., and Mansuy, D., 1993, Human liver P450s expressed in yeast as tools for reactive metabolite formation studies: Oxidative activation of tienilic acid by P450 2C9 and P450 2C10, *Eur. J. Biochem.* **213:**223–232.

98. Lopez-Garcia, M. P., Dansette, P. M., and Mansuy, D., 1994, Thiophene derivatives as new mechanism-based inhibitors of cytochromes P-450: Inactivation of yeast-expressed human liver cytochrome P-450 2C9 by tienilic acid, *Biochemistry* **33:**166–175.

99. Romkes, M., Faletto, M. B., Blaisdell, J. A., Raucy, J. L., and Goldstein, J. A., 1991, Cloning and expression of complementary DNAs for multiple members of the human cytochrome P450IIC subfamily, *Biochemistry* **30:**3247–3255.

100. Distlerath, L. M., Reilly, P. E. B., Martin, M. V., Davis, G. G., Wilkinson, G. R., and Guengerich, F. P., 1985, Purification and characterization of the human liver cytochromes P-450 involved in debrisoquine 4-hydroxylation and phenacetin O-deethylation, two prototypes for genetic polymorphism in oxidative drug metabolism, *J. Biol. Chem.* **260:**9057–9067.

101. Gonzalez, F. J., Matsunaga, T., Nagata, K., Meyer, U. A., Nebert, D. W., Pastewka, J., Kozak, C. A., Gillette, J., Gelboin, H. V., and Hardwick, J. P., 1987, Debrisoquine 4-hydroxylase: Characterization of a new P450 gene subfamily, regulation, chromosomal mapping, and molecular analysis of the DA rat polymorphism, *DNA* **6:**149–161.

102. Gonzalez, F. J., Skoda, R. C., Kimura, S., Umeno, M., Zanger, U. M., Nebert, D. W., Gelboin, H. V., Hardwick, J. P., and Meyer, U. A., 1988, Characterization of the common genetic defect in humans deficient in debrisoquine metabolism, *Nature* **331**:442–446.

103. Inaba, T., Tyndale, R. E., and Mahon, W. A., 1986, Quinidine: Potent inhibition of sparteine and debrisoquine oxidation in vivo, *Br. J. Clin. Pharmacol.* **22**:199–200.

104. Leemann, T., Dayer, P., and Meyer, U. A., 1986, Single-dose quinidine treatment inhibits metoprolol oxidation in extensive metabolizers, *Eur. J. Clin. Pharmacol.* **29**:739–741.

105. Gut, J., Gasser, R., Dayer, P., Kronbach, T., Catin, T., and Meyer, U. A., 1984, Debrisoquine-type polymorphism of drug oxidation: Purification from human liver of a cytochrome P450 isozyme with high activity for bufuralol hydroxylation, *FEBS Lett.* **173**:287–290.

106. Peter, R., Böcker, R. G., Beaune, P. H., Iwasaki, M., Guengerich, F. P., and Yang, C.-S., 1990, Hydroxylation of chlorzoxazone as a specific probe for human liver cytochrome P-450 IIE1, *Chem. Res. Toxicol.* **3**:566–573.

107. Berthou, F., Carriere, V., Ratanasavanh, D., Goasduff, T., Morel, F., Gautier, J. C., Guillouzo, A., and Beaune, P., 1993, On the specificity of chlorzoxazone as drug probe of cytochrome P4502E1, in: *Abstracts, 5th European ISSX Meeting,* Tours, Vol. 3, p. 116.

108. Hotta, H., Hatanaka, T., Tsutsui, M., Gonzalez, F. J., Ono, S., and Satoh, T., 1994, Chlorzoxazone is metabolized by human CYP1A2 as well as by human CYP2E1, in: *Abstracts, 10th International Symposium on Microsomes Drug Oxidations,* Toronto, Canada, p. 485.

109. Feierman, D. E., and Cederbaum, A. I., 1986, Inhibition of microsomal oxidation of ethanol by pyrazole and 4-methylpyrazole *in vitro.* Increased effectiveness after induction by pyrazole and 4-methylpyrazole, *Biochem. J.* **239**:671–677.

110. Yamazaki, H., Inui, Y., Yun, C.-H., Mimura, M., Guengerich, F. P., and Shimada, T., 1992, Cytochrome P450 2E1 and 2A6 enzymes as major catalysts for metabolic activation of *N*-nitrosodialkylamines and tobacco-related nitrosamines in human liver microsomes, *Carcinogenesis* **13**:1789–1794.

111. Guengerich, F. P., Kim, D.-H., and Iwasaki, M., 1991, Role of human cytochrome P-450; IIE1 in the oxidation of several low molecular weight cancer suspects, *Chem. Res. Toxicol.* **4**:168–179.

112. Watkins, P. B., Wrighton, S. A., Maurel, P., Schuetz, E. G., Mendez-Picon, G., Parker, G. A., and Guzelian, P. S., 1985, Identification of an inducible form of cytochrome P-450 in human liver, *Proc. Natl. Acad. Sci. USA* **82**:6310–6314.

113. Watkins, P. B., Murray, S. A., Winkelman, L. G., Heuman, D. M., Wrighton, S. A., and Guzelian, P. S., 1989, Erythromycin breath test as an assay of glucocorticoid-inducible liver cytochromes P-450, *J. Clin. Invest.* **83**:688–697.

114. Pessayre, D., Konstantinova-Mitcheva, M., Descatoire, V., Cobert, B., Wandscheer, J.-C., Level, R., Feldmann, G., Mansuy, D., and Benhamou, J.-P., 1981, Hypoactivity of cytochrome P-450 after triacetyloleandomycin administration, *Biochem. Pharmacol.* **30**:559–564.

115. Guengerich, F. P., Brian, W. R., Iwasaki, M., Sari, M.-A., Bäärnhielm, C., and Berntsson, P., 1991, Oxidation of dihydropyridine calcium channel blockers and analogues by human liver cytochrome P-450 IIIA4, *J. Med. Chem.* **34**:1838–1844.

116. Guengerich, F. P., 1990, Mechanism-based inactivation of human liver cytochrome P-450 IIIA4 by gestodene, *Chem. Res. Toxicol.* **3**:363–371.

117. Wrighton, S. A., Brian, W. R., Sari, M. A., Iwasaki, M., Guengerich, F. P., Raucy, J. L., Molowa, D. T., and VandenBranden, M., 1990, Studies on the expression and metabolic capabilities of human liver cytochrome P450IIIA5 (HLp3), *Mol. Pharmacol.* **38**:207–213.

118. Kawashima H., Kusunose, E., Kikuta, Y., Kinoshita, H., Tanaka, S., Yamamoto, S., Kishimoto, T., and Kusunose, M., 1994, Purification and cDNA cloning of human liver CYP4A fatty acid ω-hydroxylase, *J. Biochem.* **116**:74–80.

Index